ROBUST CONTROL, THEORY AND APPLICATIONS

Edited by **Andrzej Bartoszewicz**

Robust Control, Theory and Applications
Edited by Andrzej Bartoszewicz

Published by InTech

Janeza Trdine 9, 51000 Rijeka, Croatia

Publishing Process Manager Katarina Lovrecic

Technical Editor Teodora Smiljanic

Cover Designer InTech Design Team

Additional hard copies can be obtained from orders@intechopen.com

Robust Control, Theory and Applications, Edited by Andrzej Bartoszewicz
p. cm.
ISBN 978-953-307-229-6

For EU product safety concerns:
IN TECH d.o.o., Prolaz Marije Krucifikse Kozulić 3, 51000 Rijeka, Croatia,
info@intechopen.com or visit our website at intechopen.com.

Meet the editor

Andrzej Bartoszewicz received both his M.Sc. degree in 1987 and the Ph.D. degree in 1993 from the Technical University of Łódź, Poland. He obtained the degree of doktor habilitowany in control engineering and robotics from Academy of Mining and Metallurgy in Cracow, Poland. He was a visiting scholar at the Purdue University, West Lafayette, USA and at the Strathclyde University, Glasgow, UK. For one year he was at the University of Leicester, UK. Currently he is a professor at the Technical University of Łódź, vice dean of department of electrical, electronic, computer and control engineering, the head of Electric Drive and Industrial Automation Group and vice-director of the Institute of Automatic Control. He has published three monographs and over 250 papers, primarily in the fields of sliding mode control and congestion control in data transmission networks.

Contents

Preface

The main purpose of control engineering is to steer the regulated plant in such a way that it operates in a required manner. The desirable performance of the plant should be obtained despite the unpredictable influence of the environment on all parts of the control system, including the plant itself, and no matter if the system designer knows precisely all the parameters of the plant. Even though the parameters may change with time, load and external circumstances, still the system should preserve its nominal properties and ensure the required behaviour of the plant. In other words, the principal objective of control engineering is to design control (or regulation) systems which are robust with respect to external disturbances and modelling uncertainty. This objective may very well be obtained in a number of ways which are discussed in this monograph.

The monograph is divided into five sections. In section 1 some principal issues of the field are presented. That section begins with a general introduction presenting well developed robust control techniques, then discusses the problem of robust hybrid control and concludes with some new insights into stability and control of linear interval parameter plants. These insights are made both from an engineering (quantitative) perspective and from the population (community) ecology point of view. The next two sections, i.e. section 2 and section 3 are devoted to new results in the framework of two important robust control techniques, namely: H-infinity and sliding mode control. The two control concepts are quite different from each other, however both are nowadays very well grounded theoretically, verified experimentally, and both are regarded as fundamental design techniques in modern control theory. Section 4 presents various other significant developments in the theory of robust control. It begins with three contributions related to the design of continuous and discrete time robust proportional integral derivative controllers. Next, the section discusses selected problems in passive and active fault tolerant control, and presents some important issues of robust model predictive and fuzzy control. Recent developments in quantitative feedback theory, stabilizability and detectability of variational control systems, control of multi agent systems and control of flat systems are also the topics considered in the same section. The monograph is concerned not only with a wide spectrum of theoretical issues in robust control domain, but it also demonstrates a number of successful, recent engineering and non-engineering applications of the theory. These are described in section 5 and include internet based switching control, and applications of robust control techniques in electric drives, power electronics, bilateral teleoperation systems, automotive industry, wastewater treatment, thermostatic baths, multi-channel sound reproduction systems, inventory management and biological processes.

In conclusion, the main objective of this monograph is to present a broad range of well worked out, recent theoretical and application studies in the field of robust control system analysis and design. We believe, that thanks to the authors and to the Intech Open Access Publisher, this ambitious objective has been successfully accomplished. The editor and authors truly hope that the result of this joint effort will be of significant interest to the control community and that the contributions presented here will advance the progress in the field, and motivate and encourage new ideas and solutions in the robust control area.

Andrzej Bartoszewicz
Institute of Automatic Control,
Technical University of Łódź
Poland

Part 1

Fundamental Issues in Robust Control

1

Introduction to Robust Control Techniques

Khaled Halbaoui[1,2], Djamel Boukhetala[2] and Fares Boudjema[2]
[1]*Power Electronics Laboratory, Nuclear Research Centre of Brine CRNB, BP 180 Ainoussera 17200, Djelfa*
[2]*Laboratoire de Commande des Processus, ENSP, 10 avenue Pasteur, Hassan Badi, BP 182 El-Harrach Algeria*

1. Introduction

The theory of "Robust" Linear Control Systems has grown remarkably over the past ten years. Its popularity is now spreading over the industrial environment where it is an invaluable tool for analysis and design of servo systems. This rapid penetration is due to two major advantages: its applied nature and its relevance to practical problems of automation engineer.

To appreciate the originality and interest of robust control tools, let us recall that a control has two essential functions:

- shaping the response of the servo system to give it the desired behaviour,
- maintaining this behaviour from the fluctuations that affect the system during operation (wind gusts for aircraft, wear for a mechanical system, configuration change to a robot.).

This second requirement is termed "robustness to uncertainty". It is critical to the reliability of the servo system. Indeed, control is typically designed from an idealized and simplified model of the real system.

To function properly, it must be robust to the imperfections of the model, i.e. the discrepancies between the model and the real system, the excesses of physical parameters and the external disturbances.

The main advantage of robust control techniques is to generate control laws that satisfy the two requirements mentioned above. More specifically, given a specification of desired behaviour and frequency estimates of the magnitude of uncertainty, the theory evaluates the feasibility, produces a suitable control law, and provides a guaranty on the range of validity of this control law (strength). This combined approach is systematic and very general. In particular, it is directly applicable to Multiple-Input Multiple Output systems.

To some extent, the theory of Robust Automatic Control reconciles dominant frequency (Bode, Nyquist, PID) and the Automatic Modern dominated state variables (Linear Quadratic Control, Kalman).

It indeed combines the best of both. From Automatic Classic, it borrows the richness of the frequency analysis systems. This framework is particularly conducive to the specification of performance objectives (quality of monitoring or regulation), of band-width and of robustness. From Automatic Modern, it inherits the simplicity and power of synthesis

methods by the state variables of enslavement. Through these systematic synthesis tools, the engineer can now impose complex frequency specifications and direct access to a diagnostic feasibility and appropriate control law. He can concentrate on finding the best compromise and analyze the limitations of his system.
This chapter is an introduction to the techniques of Robust Control. Since this area is still evolving, we will mainly seek to provide a state of the art with emphasis on methods already proven and the underlying philosophy. For simplicity, we restrict to linear time invariant systems (linear time-invariant, LTI) continuous time. Finally, to remain true to the practice of this theory, we will focus on implementation rather than on mathematical and historical aspects of the theory.

2. Basic concepts

The control theory is concerned with influencing systems to realize that certain output quantities take a desired course. These can be technical systems, like heating a room with output *temperature*, a boat with the output quantities *heading* and *speed*, or a power plant with the output *electrical power*. These systems may well be social, chemical or biological, as, for example, the system of *national economy* with the output *rate of inflation*. The nature of the system does not matter. Only the dynamic behaviour is of great importance to the control engineer. We can describe this behaviour by differential equations, difference equations or other functional equations. In classical control theory, which focuses on technical systems, the system that will be influenced is called the (*controlled*) plant.
In which kinds in manners can we influence the system? Each system is composed not only of output quantities, but as well of input quantities. For the heating of a room, this, for example, will be the position of the valve, for the boat the power of the engine and angle of the rudder. These input variables have to be adjusted in a manner that the output variables take the desired course, and they are called *actuating variables*. In addition to the actuating variables, the *disturbance variables* affect the system, too. For instance, a heating system, where the temperature will be influenced by the number of people in the room or an open window, or a boat, whose course will be affected by water currents.
The desired course of output variables is defined by the *reference variables*. They can be defined by operator, but they can also be defined by another system. For example, the autopilot of an aircraft calculates the reference values for altitude, the course, and the speed of the plane. But we do not discuss the generation of reference variables here. In the following, we take for them for granted. Just take into account that the reference variables do not necessarily have to be constant; they can also be time-varying.
Of which information do have we need to calculate the actuating variables to make the output variables of the system follow the variables of reference? Clearly the reference values for the output quantities, the behavior of the plant and the time-dependent behavior of the disturbance variables must be known. With this information, one can theoretically calculate the values of the actuating variables, which will then affect the system in a way that the output quantities will follow the desired course. This is the principle of a *steering mechanism* (Fig. 1). The input variable of the steering mechanism is the reference variable ω, its output quantity actuating variable u , which again - with disturbance variable w forms the input value of the plant. y represents the output value of the system.

The disadvantage of this method is obvious. If the behavior of the plant is not in accordance with the assumptions which we made about it, or if unforeseen disruptions, then the quantities of output will not continue to follow the desired course. A steering mechanism cannot react to this deviation, because it does not know the output quantity of the plant.

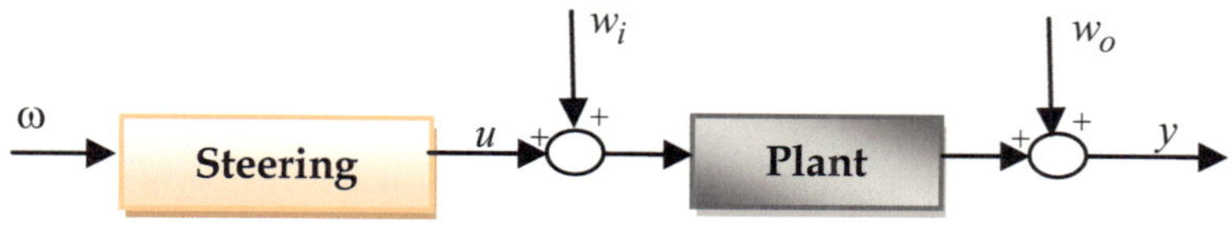

Fig. 1. Principle of a steering mechanism

A improvement which can immediately be made is the principle of an (*automatic*) *control* (Fig. 2). Inside the automatic check, the reference variable ω is compared with the measured output variable of the plant y (*control variable*), and a suitable output quantity of the controller u (*actuating variable*) are calculated inside the control unit of the difference Δy (*control error*).

During old time the control unit itself was called the controller, but the modern controllers, including, between others, the adaptive controllers (Boukhetala et al., 2006), show a structure where the calculation of the difference between the actual and wished output value and the calculations of the control algorithm cannot be distinguished in the way just described. For this reason, the tendency today is towards giving the name *controller* to the section in which the variable of release is obtained starting from the reference variable and the measured control variable.

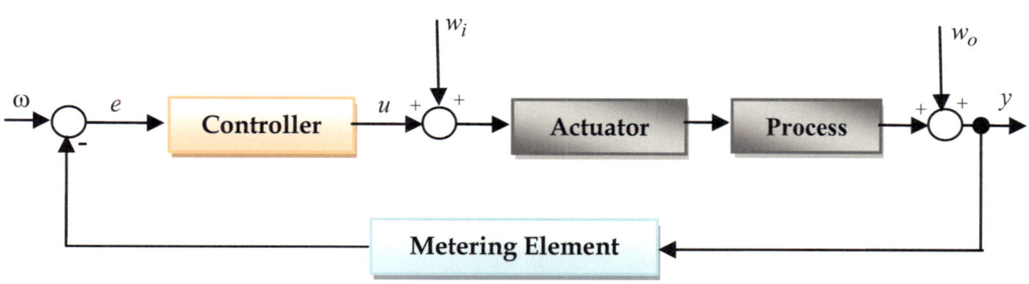

Fig. 2. Elements of a control loop

The quantity u is usually given as low-power signal, for example as a digital signal. But with low power, it is not possible to tack against a physical process. How, for example, could be a boat to change its course by a rudder angle calculated numerically, which means a sequence of zeroes and ones at a voltage of 5 V? Because it's not possible directly, a static inverter and an electric rudder drive are necessary, which may affect the rudder angle and the boat's route. If the position of the rudder is seen as actuating variable of the system, the static inverter, the electric rudder drive and the rudder itself from the *actuator* of the system. The actuator converts the controller output, a signal of low power, into the actuating variable, a signal of high power that can directly affect the plant.

Alternatively, the output of the static inverter, that means the armature voltage of the rudder drive, could be seen as actuating variable. In this case, the actuator would consist only of static converter, whereas the rudder drive and the rudder should be added to the plant. These various views already show that a strict separation between the actuator and the process is not possible. But it is not necessary either, as for the design of the controller;

we will have to take every transfer characteristic from the controller output to the control variable into account anyway. Thus, we will treat the actuator as an element of the plant, and henceforth we will employ the actuating variable to refer to the output quantity of the controller.

For the feedback of the control variable to the controller the same problem is held, this time only in the opposite direction: a signal of high power must be transformed into a signal of low power. This happens in the measuring element, which again shows dynamic properties that should not be overlooked.

Caused by this feedback, a crucial problem emerges, that we will illustrate by the following example represented in (Fig. 3). We could formulate strategy of a boat's automatic control like this: the larger the deviation from the course is, the more the rudder should be steered in the opposite direction. At a glance, this strategy seems to be reasonable. If for some reason a deviation occurs, the rudder is adjusted. By steering into the opposite direction, the boat receives a rotatory acceleration in the direction of the desired course.

The deviation is reduced until it disappears finally, but the rotating speed does not disappear with the deviation, it could only be reduced to zero by steering in the other direction. In this example, because of the rotating speed of the boat will receive a deviation in the other direction after getting back to the desired course. This is what happened after the rotating speed will be reduced by counter-steering caused by the new deviation. But as we already have a new deviation, the whole procedure starts again, only the other way round. The new deviation could be even greater than the first.

The boat will begin zigzagging its way, if worst comes to worst, with always increasing deviations. This last case is called *instability*. If the amplitude of vibration remains the same, it is called *borderline of stability*.

Only if the amplitudes decrease the system is *stable*. To receive an acceptable control algorithm for the example given, we should have taken the dynamics of the plant into account when designing the control strategy.

A suitable controller would produce a counter-steering with the rudder right in time to reduce the rotating speed to zero at the same time the boat gets back on course.

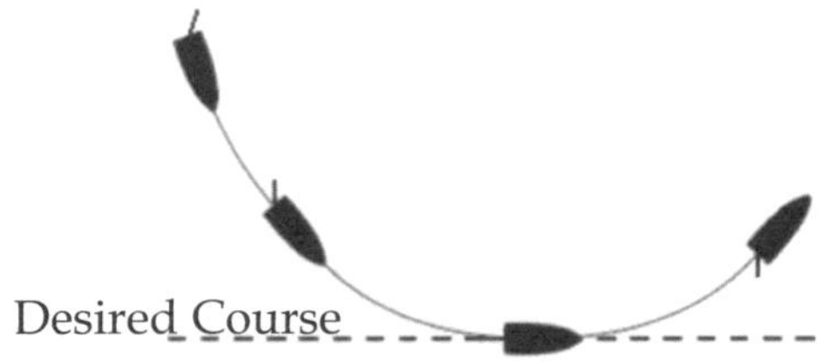

Fig. 3. Automatic cruise control of a boat

This example illustrates the requirements with respect to the controlling devices. A requirement is accuracy, i.e. the control error should be also small as possible once all the initial transients are finished and a stationary state is reached. Another requirement is the speed, i.e. in the case of a changing reference value or a disturbance; the control error should be eliminated as soon as possible. This is called the *response behavior*. The requirement of the third and most important is the stability of the whole system. We will see that these conditions are contradicted, of this fact of forcing each kind of controller (and therefore fuzzy controllers, too) to be a compromise between the three.

3. Frequency response

If we know a plant's transfer function, it is easy to construct a suitable controller using this information. If we cannot develop the transfer function by theoretical considerations, we could as well employ statistical methods on the basis of a sufficient quantity of values measured to determine it. This method requires the use of a computer, a plea which was not available during old time. Consequently, in these days a different method frequently employed in order to describe a plant's dynamic behavior, *frequency response* (Franklin et al., 2002). As we shall see later, the frequency response can easily be measured. Its good graphical representation leads to a clear method in the design process for simple PID controllers. Not to mention only several criteria for the stability, which as well are employed in connection with fuzzy controllers, root in frequency response based characterization of a plant's behavior.

The easiest way would be to define the frequency response to be the transfer function of a linear transfer element with purely imaginary values for s.

Consequently, we only have to replace the complex variable s of the transfer function by a variable purely imaginary. $j\omega$: $G(j\omega) = G(s)|_{s=j\omega}$. The frequency response is thus a complex function of the parameter ω. Due to the restriction of s to purely imaginary values; the frequency response is only part of the transfer function, but a part with the special properties, as the following theorem shows:

Theorem 1 *If a linear transfer element has the frequency response* $G(j\omega)$, *then its response to the input signal* $x(t) = a\sin\omega t$ *will be-after all initial transients have settled down-the output signal*

$$y(t) = a|G(j\omega)|\sin(\omega t + \varphi(G(j\omega))) \tag{1}$$

If the following equation holds:

$$\int_0^\infty |g(t)|\,dt \prec \infty \tag{2}$$

$|G(j\omega)|$ *is obviously the ratio of the output sine amplitude to the input sine amplitude ((transmission) gain or amplification).* $\phi(G(j\omega)$ *is the phase of the complex quantity* $G(j\omega)$ *and shows the delay of the output sine in relation to the input sine (phase lag).* $g(t)$ *is the impulse response of the plant. In case the integral given in (2) does not converge, we have to add the term* $r(t)$ *to the right hand side of (1), which will, even for* $t \to \infty$, *not vanish.*

The examination of this theorem shows clearly what kind of information about the plant the frequency response gives: Frequency response characterizes the system's behavior for any frequency of the input signal. Due to the linearity of the transfer element, the effects caused by single frequencies of the input signal do not interfere with each other. In this way, we are now able to predict the resulting effects at the system output for each single signal component separately, and we can finally superimpose these effects to predict the overall system output.

Unlike the coefficients of a transfer function, we can measure the amplitude and phase shift of the frequency response directly: The plant is excited by a sinusoidal input signal of a certain frequency and amplitude. After all initial transients are installed we obtain a sinusoidal signal at the output plant, whose phase position and amplitude differ from the input signal. The quantities can be measured, and depending to (1), this will also instantly provide the amplitude and phase lag of the frequency response $G(j\omega)$. In this way, we can construct a table for different input frequencies that give the principle curve of the frequency response. Take of measurements for negative values of ω, i.e. for negative frequencies, which is obviously not possible, but it is not necessary either, delay elements for the transfer functions rational with real coefficients and for $G(j\omega)$ will be conjugate complex to $G(-j\omega)$. Now, knowing that the function $G(j\omega)$ for $\omega \geq 0$ already contains all the information needed, we can omit an examination of negative values of ω.

4. Tools for analysis of controls

4.1 Nyquist plot

A Nyquist plot is used in automatic control and signal processing for assessing the stability of a system with feedback. It is represented by a graph in polar coordinates in which the gain and phase of a frequency response are plotted. The plot of these phasor quantities shows the phase as the angle and the magnitude as the distance from the origin (see. Fig.4). The Nyquist plot is named after Harry Nyquist, a former engineer at Bell Laboratories.

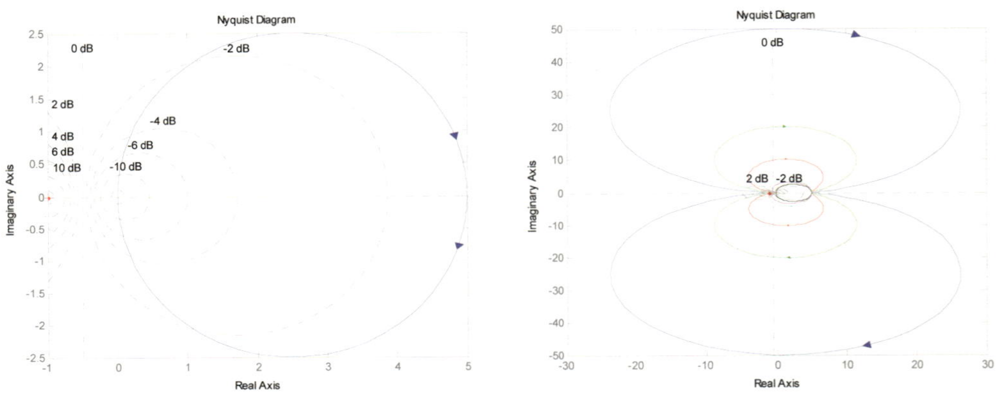

First-order system Second-order systems

Fig. 4. Nyquist plots of linear transfer elements

Assessment of the stability of a closed-loop negative feedback system is done by applying the Nyquist stability criterion to the Nyquist plot of the open-loop system (i.e. the same system without its feedback loop). This method is easily applicable even for systems with delays which may appear difficult to analyze by means of other methods.

Nyquist Criterion: We consider a system whose open loop transfer function (OLTF) is $G(s)$; when placed in a closed loop with feedback $H(s)$, the closed loop transfer function (CLTF) then becomes $\dfrac{G}{1+G.H}$. The case where $H=1$ is usually taken, when investigating stability, and then the *characteristic equation*, used to predict stability, becomes $G+1=0$.

We first construct *The Nyquist Contour*, a contour that encompasses the right-half of the complex plane:

- a path traveling up the $j\omega$ axis, from $0\ -j\infty$ to $0\ +j\infty$.
- a semicircular arc, with radius $r \to \infty$, that starts at $0\ +j\infty$ and travels clock-wise to $0\ -j\infty$

The Nyquist Contour mapped through the function $1+G(s)$ yields a plot of $1+G(s)$ in the complex plane. By the Argument Principle, the number of clock-wise encirclements of the origin must be the number of zeros of $1+G(s)$ in the right-half complex plane minus the poles of $1+G(s)$ in the right-half complex plane. If instead, the contour is mapped through the open-loop transfer function $G(s)$, the result is the Nyquist plot of $G(s)$. By counting the resulting contour's encirclements of -1, we find the difference between the number of poles and zeros in the right-half complex plane of $1+G(s)$. Recalling that the zeros of $1+G(s)$ are the poles of the closed-loop system, and noting that the poles of $1+G(s)$ are same as the poles of $G(s)$, we now state *The Nyquist Criterion*:

Given a Nyquist contour Γ_s, let P be the number of poles of $G(s)$ encircled by Γ_s and Z be the number of zeros of $1+G(s)$ encircled by Γ_s. Alternatively, and more importantly, Z is the number of poles of the closed loop system in the right half plane. The resultant contour in the $G(s)$-plane, $\Gamma_{G(s)}$ shall encircle (clock-wise) the point $(-1+j0)$ N times such that $N = Z - P$. For stability of a system, we must have $Z = 0$, i.e. the number of closed loop poles in the right half of the s-plane must be zero. Hence, the number of counterclockwise encirclements about $(-1+j0)$ must be equal to P, the number of open loop poles in the right half plane (Faulkner, 1969), (Franklin, 2002).

4.2 Bode diagram

A Bode plot is a plot of either the magnitude or the phase of a transfer function $T(j\omega)$ as a function of ω. The magnitude plot is the more common plot because it represents the gain of the system. Therefore, the term "Bode plot" usually refers to the magnitude plot (Thomas, 2004),(William, 1996),(Willy, 2006). The rules for making Bode plots can be derived from the following transfer function:

$$T(s) = K\left(\frac{s}{\omega_0}\right)^{\pm n}$$

where n is a positive integer. For $+n$ as the exponent, the function has n zeros at $s=0$. For $-n$, it has n poles at $s=0$. With $s = j\omega$, it follows that $T(j\omega) = Kj^{\pm n}(\omega/\omega_0)^{\pm n}$, $|T(j\omega)| = Kj(\omega/\omega_0)^{\pm n}$ and $\angle T(j\omega) = \pm n \times 90^\circ$. If ω is increased by a factor of 10, $|T(j\omega)|$ changes by a factor of $10^{\pm n}$. Thus a plot of $|T(j\omega)|$ versus ω on $\log-\log$ scales has a slope of $\log\left(10^{\pm n}\right) = \pm n\ decades/decade$. There are $20dBs$ in a $decade$, so the slope can also be expressed as $\pm 20n\ dB/decade$.

In order to give an example, (Fig. 5) shows the Bode diagrams of the first order and second order lag. Initial and final values of the phase lag courses can be seen clearly. The same holds for the initial values of the gain courses. Zero, the final value of these courses, lies at negative infinity, because of the logarithmic representation. Furthermore, for the second order lag the resonance magnification for smaller dampings can be see at the resonance frequency ω_0.

Even with a transfer function being given, a graphical analysis using these two diagrams might be clearer, and of course it can be tested more easily than, for example, a numerical analysis done by a computer. It will almost always be easier to estimate the effects of changes in the values of the parameters of the system, if we use a graphical approach instead of a numerical one. For this reason, today every control design software tool provides the possibility of computing the Nyquist plot or the Bode diagram for a given transfer function by merely clicking on a button.

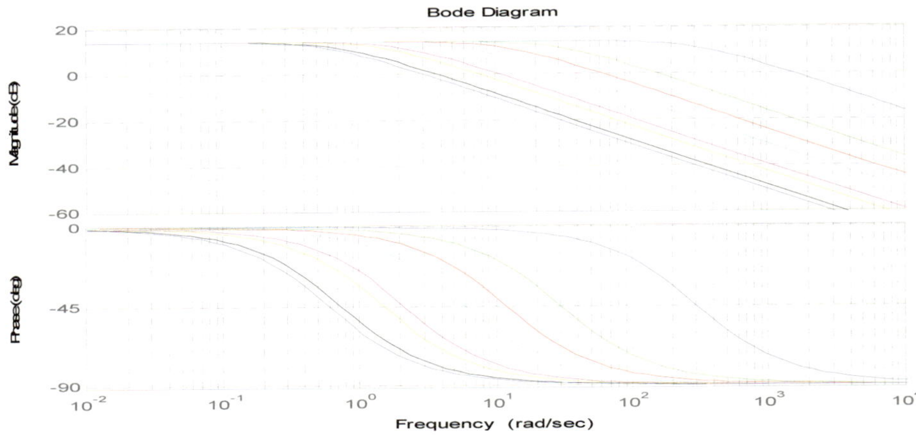

First-order systems

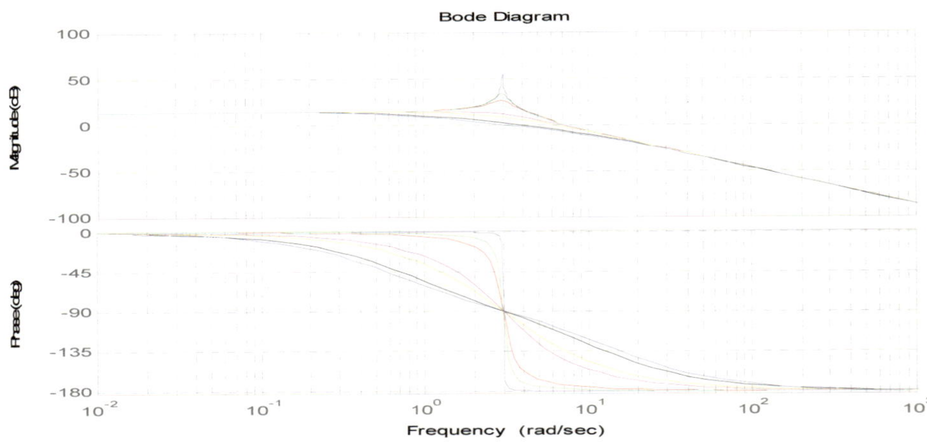

Second-order systems

Fig. 5. Bode diagram of first and second-order systems

4.3 Evans root locus

In addition to determining the stability of the system, the root locus can be used to design for the damping ratio and natural frequency of a feedback system (Franklin et al., 2002). Lines of constant damping ratio can be drawn radially from the origin and lines of constant natural frequency can be drawn as arcs whose center points coincide with the origin (see. Fig. 6). By selecting a point along the root locus that coincides with a desired damping ratio and natural frequency a gain, *K*, can be calculated and implemented in the controller. More elaborate techniques of controller design using the root locus are available in most control textbooks: for instance, lag, lead, PI, PD and PID controllers can be designed approximately with this technique.

The definition of the damping ratio and natural frequency presumes that the overall feedback system is well approximated by a second order system, that is, the system has a dominant pair of poles. This often doesn't happen and so it's good practice to simulate the final design to check if the project goals are satisfied.

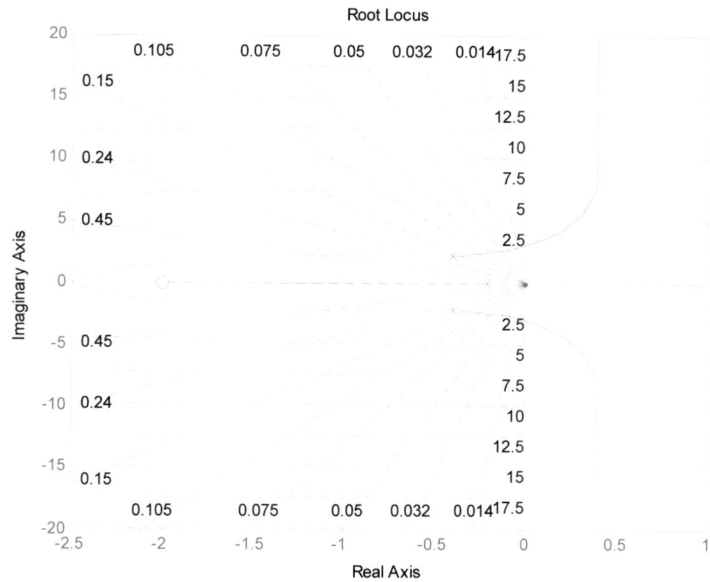

Fig. 6. Evans root locus of a second-order system

Suppose there is a plant (process) with a transfer function expression $P(s)$, and a forward controller with both an adjustable gain K and a transfer function expression $C(s)$. A unity feedback loop is constructed to complete this feedback system. For this system, the overall transfer function is given by:

$$T(s) = \frac{K.C(s).P(s)}{1 + K.C(s).P(s)} \tag{3}$$

Thus the closed-loop poles of the transfer function are the solutions to the equation $1 + K.C(s).P(s) = 0$. The principal feature of this equation is that roots may be found wherever $K.C.P = -1$. The variability of K, the gain for the controller, removes amplitude from the equation, meaning the complex valued evaluation of the polynomial in s $C(s).P(s)$ needs to have net phase of 180 deg, wherever there is a closed loop pole. The geometrical construction adds angle contributions from the vectors extending from each of the poles of KC to a prospective closed loop root (pole) and subtracts the angle contributions from similar vectors extending from the zeros, requiring the sum be 180. The vector formulation arises from the fact that each polynomial term in the factored CP, $(s-a)$ for example, represents the vector from a which is one of the roots, to s which is the prospective closed loop pole we are seeking. Thus the entire polynomial is the product of these terms, and according to vector mathematics the angles add (or subtract, for terms in the denominator) and lengths multiply (or divide). So to test a point for inclusion on the root locus, all you do is add the angles to all the open loop poles and zeros. Indeed a form of protractor, the "spirule" was once used to draw exact root loci.

From the function $T(s)$, we can also see that the zeros of the open loop system (CP) are also the zeros of the closed loop system. It is important to note that the root locus only gives the location of closed loop poles as the gain K is varied, given the open loop transfer function. The zeros of a system cannot be moved.

Using a few basic rules, the root locus method can plot the overall shape of the path (locus) traversed by the roots as the value of K varies. The plot of the root locus then gives an idea of the stability and dynamics of this feedback system for different values of K.

5. Ingredients for a robust control

The design of a control consists in adjusting the transfer function of the compensator so as to obtain the properties and the behavior wished in closed loop. In addition to the constraint of stability, we look typically the best possible performance. This task is complicated by two principal difficulties. On the one hand, the design is carried out on a idealized model of the system. We must therefore ensure the robustness to imperfections in the model, i.e. to ensure that the desired properties for a family of systems around the reference model. On the other hand, it faces inherent limitations like the compromise between performances and robustness.

This section shows how these objectives and constraints can be formulated and quantified in a consistent framework favorable to their taking into systematic account.

5.1 Robustness to uncertainty

The design of a control is carried out starting from a model of the real system often called **nominal model** or **reference model**. This model may come from the equations of physics or a process identification. In any case, this model is only one approximation of reality. Its deficiencies can be multiple: dynamic nonlinearities neglected, uncertainty on certain physical parameters, assumptions simplifying, errors of measurement to the identification, etc.. In addition, some system parameters can vary significantly with time or operating conditions. Finally, from the unforeseeable external factors can come to disturb the operation of the control system.

It is thus insufficient to optimize control compared to the nominal model: it is also necessary to be guarded against the uncertainty of modeling and external risks. Although these factors are poorly known, one has information in general on their maximum amplitude or their statistical nature. For example, the frequency of the oscillation, maximum intensity of the wind, or the terminals min and max on the parameter value. It is from this basic knowledge that one will try to carry out a robust control.

There are two classes of uncertain factors. A first class includes the uncertainty and external disturbances. These are signals or actions randomness that disrupt the controlled system. They are identified according to their point of entry into the loop. Referring again to (Fig. 2) there are basically:

- the disruption of the control w_i which can come from errors of discretization or quantification of the control or parasitic actions on the actuators.
- Disturbances at exit w_o corresponding to external effects on the output or unpredictable on the system, e.g. the wind for a airplane, an air pressure change for a chemical reactor, etc..

It should be noted that these external actions do not modify the dynamic behavior interns system, but only the "trajectory" of its outputs.

A second class of uncertain factors joins together imperfections and variations of the dynamic model of the system. Recall that the robust control techniques applied to finite dimensional linear models, while real systems are generally non-linear and infinite dimensional. Typically, the model used thus neglects non-linear ties and is valid only in one limited frequency band. It depends of more than physical parameters whose value can fluctuate and is often known only roughly. For practical reasons, one will distinguish:

- **the dynamic uncertainty** which gathers the dynamic ones neglected in the model. There is usually only an upper bound on the amplitude of these dynamics. One must thus assume and guard oneself against worst case in the limit of this marker.
- **the parametric uncertainty** or **structured** which is related to the variations or errors in estimation on certain physical parameters of the system, or with uncertainties of dynamic nature, but entering the loop at different points. Parametric uncertainty intervenes mainly when the model is obtained starting from the equations of physics. The way in which the parameters influential on the behavior of the system determines the "structure" of the uncertainty.

5.2 Representation of the modeling uncertainty

The dynamic uncertainty (unstructured) can encompass of physical phenomena very diverse (linear or nonlinear, static or time-variant, frictions, hysteresis, etc.). The techniques discussed in this chapter are particularly relevant when one does not have any specific information if not an estimate of the maximum amplitude of dynamic uncertainty, In other words, when uncertainty is reasonably modeled by a ball in the space of bounded operators of ℓ_2 in ℓ_2.

Such a model is of course very rough and tends to include configurations with physical sense. If the real system does not comprise important nonlinearities, it is often preferable to be restricted with a stationary purely linear model of dynamic uncertainty. We can then balance the degree of uncertainty according to the frequency and translate the fact that the system is better known into low than in high frequency. Uncertainty is then represented as a disturbing system **LTI** $\Delta G(s)$ which is added to the nominal model $G(s)$ of the real system:

$$G_{true}(s) = G(s) + \Delta G(s) \tag{4}$$

This system must be BIBO-stable (bounded ℓ_2 in ℓ_2), and it usually has an estimate of the maximum amplitude of $\Delta G(j\omega)$ in each frequency band. Typically, this amplitude is small at lower frequencies and grows rapidly in the high frequencies where the dynamics neglected become important. This profile is illustrated in (Fig. 7). It defines a family of systems whose envelope on the Nyquist diagram is shown in (Fig. 8) (case SISO). The radius of the disk of the frequency uncertainty ω is $|\Delta G(j\omega)|$.

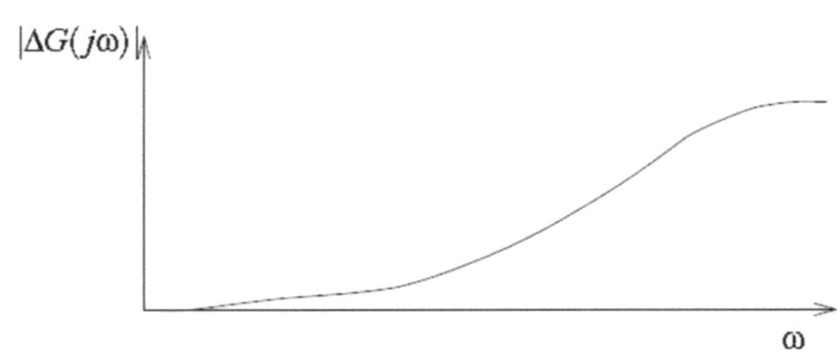

Fig. 7. Standard profile for $|\Delta G(j\omega)|$.

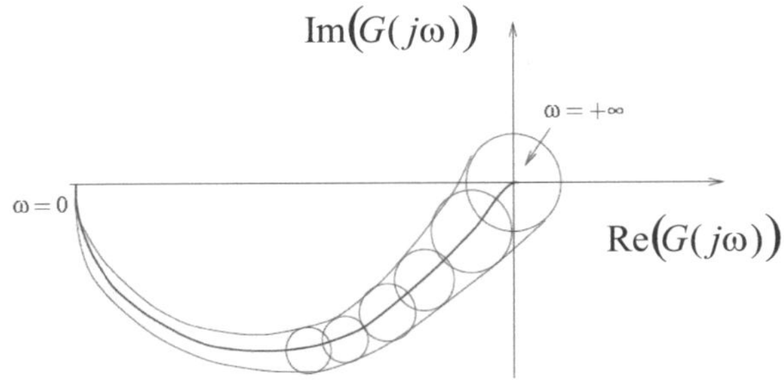

Fig. 8. Family of systems

The information on the amplitude $|\Delta G(j\omega)|$ of the uncertainty can be quantified in several ways:

- **additive uncertainty**: the real system is of the form:

$$G_{true}(s) = G(s) + \Delta(s) \tag{5}$$

Where $\Delta(s)$ is a stable transfer function satisfying:

$$\|W_l(\omega)\Delta(j\omega)W_r(\omega)\|_\infty \prec 1 \tag{6}$$

for certain models $W_l(s)$ and $W_r(s)$. These weighting matrices make it possible to incorporate information on the frequential dependence and directional of the maximum amplitude of $\Delta(s)$ (see singular values).

- **multiplicative uncertainty at the input**: the real system is of the form:

$$G_{true}(s) = G(s).(I + \Delta(s)) \tag{7}$$

where $\Delta(s)$ is like above. This representation models errors or fluctuations on the behavior in input.

- **multiplicative uncertainty at output**: the real system is of the form:

$$G_{true}(s) = (I + \Delta(s)).G(s) \tag{8}$$

This representation is adapted to modeling of the errors or fluctuations in the output behavior. According to the data on the imperfections of the model, one will choose one or the other of these representations. Let us note that multiplicative uncertainty has a relative character.

5.3 Robust stability

Let the linear system be given by the transfer function

$$G(s) = \frac{b_m s^m + b_{m-1}s^{m-1} + \ldots + b_1 s + b_0}{a_n s^n + a_{n-1}s^{n-1} + \ldots + a_1 s + a_0} e^{T_L s}$$

$$= \frac{b_m}{a_n} \frac{\prod_{\mu=1}^{m}(s - n_\mu)}{\prod_{\nu=1}^{n}(s - p_\nu)} e^{T_L s} = V \frac{\prod_{\mu=1}^{m}(\frac{-s}{n_\mu} + 1)}{\prod_{\nu=1}^{n}(\frac{-s}{p_\nu} + 1)} e^{-T_L s} \quad \text{where } m \le n \tag{9}$$

with the gain

$$V = \frac{b_0}{a_0} \tag{10}$$

First we must explain what we mean by stability of a system. Several possibilities exist to define the term, two of which we will discuss now. A third definition by the Russian mathematician Lyapunov will be presented later. The first definition is based on the step response of the system:

Definition 1 *A system is said to be stable if, for* $t \to \infty$, *its step response converges to a finite value. Otherwise, it is said to be instable.*

This unit step function has been chosen to stimulate the system does not cause any restrictions, because if the height of the step is modified by the factor *k*, the values to the system output will change by the same factor *k*, too, according to the linearity of the system. Convergence towards a finite value is therefore preserved.

A motivation for this definition can be the idea of following illustration: If a system converges towards a finished value after strong stimulation that a step in the input signal represents, it can suppose that it will not be wedged in permanent oscillations for other kinds of stimulations.

It is obvious to note that according to this definition the first order and second order lag is stable, and that the integrator is instable.

Another definition is attentive to the possibility that the input quantity may be subject to permanent changes:

Definition 2 *A linear system is called stable if for an input signal with limited amplitude, its output signal will also show a limited amplitude. This is the BIBO-Stability (bounded input - bounded output).*

Immediately, the question on the connection between the two definitions arises, that we will now examine briefly. The starting point of discussion is the convolution integral, which gives the relationship between the system's input and the output quantity (the impulse response):

$$y(t) = \int_{\tau=0}^{t} g(t-\tau)x(\tau)d\tau = \int_{\tau=0}^{t} g(\tau)x(t-\tau)d\tau \tag{11}$$

$x(t)$ is bounded if and only if $|x(t)| \leq k$ holds (with $k > 0$) for all t. This implies:

$$|y(t)| \leq \int_{\tau=0}^{t} |g(\tau)||x(t-\tau)|d\tau \leq k \int_{\tau=0}^{t} g(\tau)d\tau \tag{12}$$

Now, with absolute convergence of the integral of the impulse response,

$$\int_{\tau=0}^{\infty} |g(\tau)|d\tau = c < \infty \tag{13}$$

$y(s)$ will be limited by kc, also, and thus the whole system will be BIBO-stable. Similarly it can be shown that the integral (13) converges absolutely for all BIBO-stable systems. BIBO

stability and the absolute convergence of the impulse response integral are the equivalent properties of system.
Now we must find the conditions under which the system will be stable in the sense of a finite step response *(Definition 2)*: Regarding the step response of a system in the frequency domain,

$$y(s) = G(s)\frac{1}{s} \tag{14}$$

If we interpret the factor $1/s$ as an integration (instead of the Laplace transform of the step signal), we obtain

$$y(s) = \int_{\tau=0}^{t} g(\tau)d\tau \tag{15}$$

in the time domain for $y(0) = 0.y(t)$ converge to a finite value only if the integral converges:

$$\int_{\tau=0}^{t} g(\tau)d\tau = c < \infty \tag{16}$$

Convergence is obviously a weaker criterion than absolute convergence. Therefore, each BIBO-stable system will have a finite step response. To treat the stability always in the sense of the BIBO-stability is tempting because this stronger definition makes other differentiations useless. On the other hand, we can simplify the following considerations much if we use the finite-step-response-based definition of stability (Christopher, 2005),(Arnold, 2006). In addition to this, the two definitions are equivalent as regards the transfer functions anyway. Consequently, henceforth we will think of stability as characterized in *(Definition 2)*.
Sometimes stability is also defined while requiring that the impulse response to converge towards zero for $t \to \infty$. A glance at the integral (16) shows that this criterion is necessary but not sufficient condition for stability as defined by *(Definition 2)*, while *(Definition 2)* is the stronger definition. If we can prove a finite step response, then the impulse response will certainly converge to zero.

5.3.1 Stability of a transfer function

If we want to avoid having to explicitly calculate the step response of a system in order to prove its stability, then a direct examination of the transfer function of the system's, trying to determine criteria for the stability, seems to suggest itself (Levine, 1996). This is relatively easy concerning all ideas that we developed up to now about the step response of a rational transfer function. The following theorem is valid:
Theorem 2 *A transfer element with a rational transfer function is stable in the sense of (Definition 2) if and only if all poles of the transfer function have a negative real part.*
According to equation (17), the step response of a rational transfer element is given by:

$$y(t) = \sum_{\lambda=1}^{i} h_\lambda(t)e^{s_\lambda t} \tag{17}$$

For each pole s_λ of multiplicity n_λ, we obtain a corresponding operand $h_\lambda(t)e^{s_\lambda t}$, which $h_\lambda(t)$ is a polynomial of degree $n_\lambda - 1$. For a pole with a negative real part, this summand disappears to increase t, as the exponential function converges more quickly towards zero than the polynomial $h_\lambda(t)$ can increase. If all the poles of the transfer function have a negative real part, then all corresponding terms disappear. Only the summand $h_i(t)e^{s_i t}$ for the simple pole $s_i = 0$ remains, due to the step function. The polynomial $h_i(t)$ is of degree $n_i - 1 = 0$, i.e. a constant, and the exponential function is also reduced to a constant. In this way, this summand form the finite final value of the step function, and the system is stable.

We omit the proof in the opposite direction, i.e. a system is instable if at least one pole has a positive real part because it would not lead to further insights. It is interesting that *(Theorem 2)* holds as well for systems with delay according to (9). The proof of this last statement will be also omitted.

Generally, the form of the initial transients as reaction to the excitations of outside will also be of interest besides that the fact of stability. If a plant has, among others, a complex conjugate pair of poles s_λ, $\overline{s_\lambda}$, the ratio $|\mathrm{Re}(s_\lambda)|\sqrt{\mathrm{Re}(s_\lambda)^2 + \mathrm{Im}(s_\lambda)^2}$ is equal to the damping ratio D and therefore responsible for the form of the initial transient corresponding to this pair of poles. In practical applications one will therefore pay attention not only to that the system's poles have a negative real part, but also to the damping ratio D having a sufficiently high value, i.e. that a complex conjugate pair of poles lies at a reasonable distance to the axis of imaginaries.

5.3.2 Stability of a control loop

The system whose stability must be determined will in the majority of the cases be a closed control loop (Goodwin, 2001), as shown in (Fig. 2). A simplified structure is given in (Fig. 9). Let the transfer function of the control unit is $K(s)$, the plant will be given by $G(s)$ and the metering element by $M(s)$. To keep further derivations simple, we set $M(s)$ to 1, i.e. we neglect the dynamic behavior of the metering element, for simple cases, but it should normally be no problem to take the metering element also into consideration.

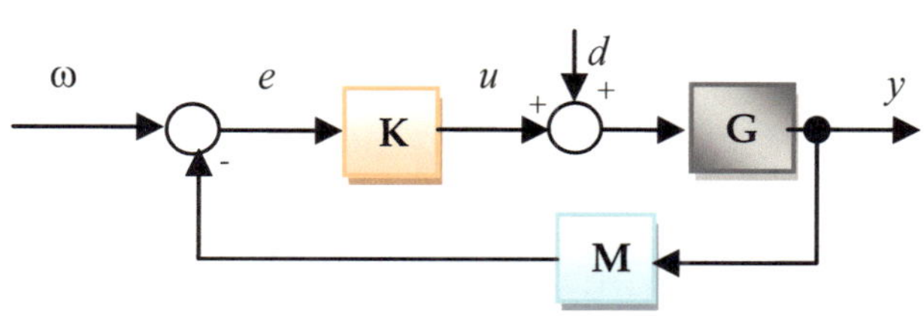

Fig. 9. Closed-loop system

We summarize the disturbances that could affect the closed loop system to virtually any point, into a single disturbance load that we impressed at the plant input. This step simplifies the theory without the situation for the controller easier than it would be in practical applications. Choose the plant input as the point where the disturbance affects the plant is most unfavorable: The disturbance can affect plants and no countermeasure can be applied, as the controller can only counteract after the changes at the system output.

To be able to apply the criteria of stability to this system we must first calculate the transfer function that describes the transfer characteristic of the entire system between the input quantity ω and the output quantity y. This is the transfer function of the closed loop, which is sometimes called the *reference (signal) transfer function*. To calculate it, we first set d to zero. In the frequency domain we get

$$y(s) = G(s)u(s) = G(s)K(s)(\omega(s) - y(s)) \tag{18}$$

$$T(s) = \frac{y(s)}{\omega(s)} = \frac{G(s)K(s)}{G(s)K(s)+1} \tag{19}$$

In a similar way, we can calculate a *disturbance transfer function,* which describes the transfer characteristic between the disturbance d and the output quantity y:

$$S(s) = \frac{y(s)}{d(s)} = \frac{G(s)K(s)}{G(s)K(s)+1} \tag{20}$$

The term $G(s)K(s)$ has a special meaning: if we remove the feedback loop, so this term represents the transfer function of the resulting open circuit. Consequently, $G(s)K(s)$ is sometimes called the *open-loop transfer function*. The gain of this function (see (9)) is called *open-loop gain*.

We can see that the reference transfer function and the disturbance transfer function have the same denominator $G(s)K(s)+1$. On the other hand, by *(Theorem 2)*, it is the denominator of the transfer function that determines the stability. It follows that only the open-loop transfer function affects the stability of a system, but not the point of application of an input quantity. We can therefore restrict an analysis of the stability to a consideration of the term $G(s)K(s)+1$.

However, since both the numerator and denominator of the two transfer functions $T(s)$ and $S(s)$ are obviously relatively prime to each other, the zeros of $G(s)K(s)+1$ are the poles of these functions, and as a direct consequence of *(Theorem 2)* we can state:

***Theorem* 3** *A closed-loop system with the open-loop transfer function* $G(s)K(s)$ *is stable if and only if all solutions of the characteristic equation have a negative real part.*

$$G(s)K(s)+1=0 \tag{21}$$

Computing these zeros in an analytic way will no longer be possible if the degree of the plant is greater than two, or if an exponential function forms a part of the open-loop transfer function. Exact positions of the zeros, though, are not necessary in the analysis of stability. Only the fact whether the solutions have a positive or negative real part is of importance. For this reason, in the history of the control theory criteria of stability have been developed that could be used to determine precisely without having to make complicated calculations (Christopher, 2005), (Franklin, 2002).

5.3.3 Lyapunov's stability theorem

We state below a variant of Lyapunov's direct method that establishes global asymptotic stability.

Theorem 4 Consider the dynamical system $\dot{x}(t) = f(x(t))$ and let $x = 0$ be its unique equilibrium point. If there exists a continuously differentiable function $V : \Re^n \to \Re$ such that

$$V(0) = 0 \tag{22}$$

$$V(x) \succ 0 \quad \forall x \neq 0 \tag{23}$$

$$\|x\| \to \infty \Rightarrow V(x) \to \infty \tag{24}$$

$$\dot{V}(x) \prec 0 \quad \forall x \neq 0, \tag{25}$$

then $x = 0$ is globally asymptotically stable.
Condition (25) is what we refer to as the *monotonicity requirement* of Lyapunov's theorem. In the condition, $\dot{V}(x)$ denotes the derivation of $V(x)$ along the trajectories of $\dot{x}(t)$ and is given by

$$\dot{V}(x) = < \frac{\partial V(x)}{\partial x}, f(x) >,$$

where <.,.> denotes the standard inner product in $\Re^n$ and $\frac{\partial V(x)}{\partial x} \in \Re^n$ is the gradient of $V(x)$. As far as the first two conditions are concerned, it is only needed to assume that $V(x)$ is lower bounded and achieves its global minimum at $x = 0$ .There is no conservatism, however, in requiring (22) and (23). A function satisfying condition (24) is called *radially unbounded*. We refer the reader to (Khalil, 1992) for a formal proof of this theorem and for an example that shows condition (24) cannot be removed. Here, we give the geometric intuition of Lyapunov's theorem, which essentially carries all of the ideas behind the proof.

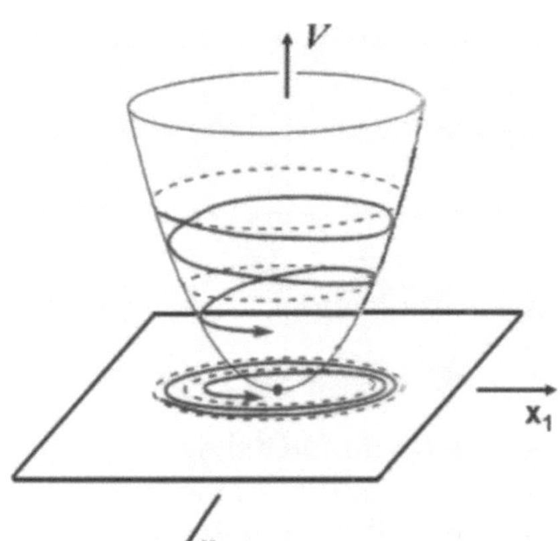

Fig. 10. Geometric interpretation of Lyapunov's theorem.

(Fig. 10) shows a hypothetical dynamical system in $\Re^2$. The trajectory is moving in the (x_1, x_2) plane but we have no knowledge of where the trajectory is as a function of time. On the other hand, we have a scalar valued function $V(x)$, plotted on the z-axis, which has the guaranteed property that as the trajectory moves the value of this function along the trajectories strictly decreases. Since $V(x(t))$ is lower bounded by zero and is strictly decreasing, it must converge to a nonnegative limit as time goes to infinity. It takes a relatively straightforward argument appealing to continuity of $V(x)$ and $\dot{V}(x)$) to show that the limit of $V(x(t))$ cannot be strictly positive and indeed conditions (22)-(25) imply

$$V(x(t)) \rightarrow 0 \text{ as } t \rightarrow \infty$$

Since $x = 0$ is the only point in space where $V(x)$ vanishes, we can conclude that $x(t)$ goes to the origin as time goes to infinity.
It is also insightful to think about the geometry in the (x_1, x_2) plane. The level sets of $V(x)$ are plotted in (Fig. 10) with dashed lines. Since $V(x(t))$ decreases monotonically along trajectories, we can conclude that once a trajectory enters one of the level sets, say given by $V(x) = c$, it can never leave the set $\Omega_c := \{x \in \Re^n \,|\, V_x \leq c\}$ .This property is known as *invariance of sub-level sets*.
Once again we emphasize that the significance of Lyapunov's theorem is that it allows stability of the system to be verified without explicitly solving the differential equation. Lyapunov's theorem, in effect, turns the question of determining stability into a search for a so-called Lyapunov function, a positive definite function of the state that decreases monotonically along trajectories. There are two natural questions that immediately arise. First, do we even know that Lyapunov functions always exist?
Second, if they do in fact exist, how would one go about finding one? In many situations, the answer to the first question is positive. The type of theorems that prove existence of Lyapunov functions for every stable system are called *converse theorems*. One of the well known converse theorems is a theorem due to Kurzweil that states if *f* in *(Theorem 4)* is continuous and the origin is globally asymptotically stable, then there exists an infinitely differentiable Lyapunov function satisfying conditions of *(Theorem 4)*. We refer the reader to (Khalil, 1992) and (Bacciotti & Rosier,2005) for more details on converse theorems. Unfortunately, converse theorems are often proven by assuming knowledge of the solutions of *(Theorem 4)* and are therefore useless in practice. By this we mean that they offer no systematic way of finding the Lyapunov function. Moreover, little is known about the connection of the dynamics *f* to the Lyapunov function *V*. Among the few results in this direction, the case of linear systems is well settled since a stable linear system always admits a quadratic Lyapunov function. It is also known that stable and smooth homogeneous systems always have a homogeneous Lyapunov function (Rosier, 1992).

5.3.4 Criterion of Cremer, Leonhard and Michailow

Initially let us discuss a criterion which was developed independently by Cremer , Leonhard and Michailov during the years 1938-1947. The focus of interest is the phase shift of the Nyquist plot of a polynomial with respect to the zeros of the polynomial (Mansour, 1992). Consider a polynomial of the form

$$P(s) = s^n + a_{n-1}s^{n-1} + \ldots + a_1 s + a_0 = \prod_{\nu=1}^{n}(s - s_\nu) \tag{26}$$

be given. Setting $s = j\omega$ and substituting we obtain

$$\begin{aligned} P(j\omega) &= \prod_{\nu=1}^{n}(j\omega - s_\nu) = \prod_{\nu=1}^{n}(|j\omega - s_\nu|)e^{j\varphi_\nu(\omega)}) \\ &= \prod_{\nu=1}^{n}|j\omega - s_\nu| e^{j\sum_{\nu=1}^{n}\varphi_\nu(\omega)} = |P(j\omega)| e^{j\varphi(\omega)} \end{aligned} \tag{27}$$

We can see, that the frequency response $P(j\omega)$ is the product of the vectors $(j\omega - s_\nu)$, where the phase $\varphi(\omega)$ is given by the sum of the angles $\varphi_\nu(\omega)$ of those vectors. (Fig.11) shows the situation corresponding to a pair of complex conjugated zeros with negative real part and one zero with a positive real part.

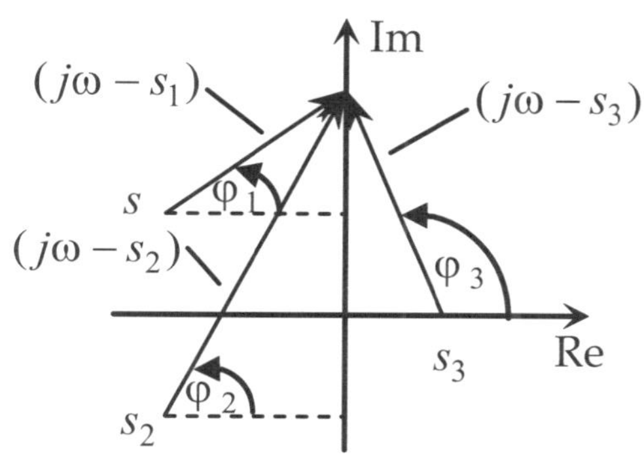

Fig. 11. Illustration to the Cremer-Leonhard-Michailow criterion

If the parameter ω traverses the interval $(-\infty,\infty)$, it causes the end point of the vectors $(j\omega - s_\nu)$ to move along the axis of imaginaries in positive direction. For zeros with negative real part, the corresponding angle φ_ν traverses the interval from $-\frac{\pi}{2}$ to $+\frac{\pi}{2}$, for zeros with positive real part the interval from $+\frac{3\pi}{2}$ to $+\frac{\pi}{2}$. For zeros lying on the axis of imaginaries the corresponding angle φ_ν initially has the value $-\frac{\pi}{2}$ and switches to the value $+\frac{\pi}{2}$ at $j\omega = s_\nu$.

We will now analyze the phase of frequency response, i.e. the entire course which the angle $\varphi(\omega)$ takes. This angle is just the sum of the angles $\varphi u_\nu(\omega)$. Consequently, each zero with a negative real part contributes an angle of $+\pi$ to the phase shift of the frequency response, and each zero with a positive real part of the angle $-\pi$. Nothing can be said about zeros located on the imaginary axis because of the discontinuous course where the values of the phase to take. But we can immediately decide zeros or not there watching the Nyquist plot of the polynomial $P(s)$. If she got a zero purely imaginary $s = s_\nu$, the corresponding Nyquist plot should pass through the origin to the frequency $\omega = |s_\nu|$. This leads to the following theorem:

***Theorem* 5** *A polynomial P(s) of degree n with real coefficients will have only zeros with negative real part if and only if the corresponding Nyquist plot does not pass through the origin of the complex plane and the phase shift* $\Delta\varphi$ *of the frequency response is equal to* $n\pi$ *for* $-\infty < \omega < +\infty$. *If* ω *traverses the interval* $0 \le \omega < +\infty$ *only, then the phase shift needed will be equal to* $\frac{n}{2}\pi$.

We can easily prove the fact that for $0 \le \omega < +\infty$ the phase shift needed is only $\frac{n}{2}\pi$ —only half the value:

For zeros lying on the axis of reals, it is obvious that their contribution to the phase shift will be only half as much if ω traverses only half of the axis of imaginaries (from 0 to ∞). The zeros with an imaginary part different from zero are more interesting. Because of the polynomial's real-valued coefficients, they can only appear as a pair of complex conjugated zeros. (Fig. 12) shows such a pair with $s_1 = s_2$ and $\alpha_1 = -\alpha_2$. For $-\infty < \omega < +\infty$ the contribution to the phase shift by this pair is 2π. For $0 \le \omega < +\infty$, the contribution of s_1 is $\frac{\pi}{2} + |\alpha_1|$ and the one for s_2 is $\frac{\pi}{2} - |\alpha_1|$. Therefore, the overall contribution of this pair of poles is π, so also for this case the phase shift is reduced by one half if only the half axis of imaginaries is taken into consideration.

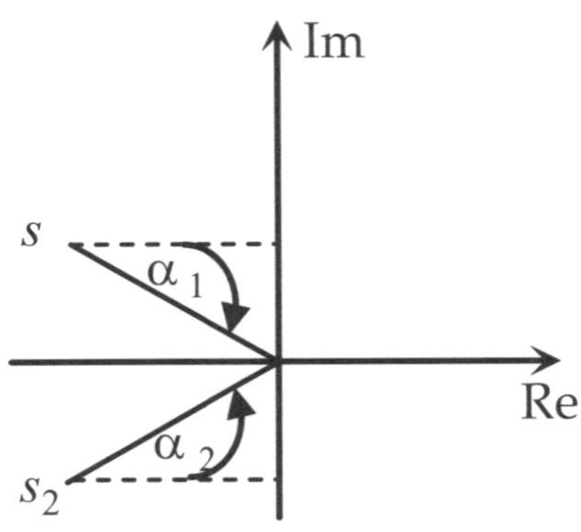

Fig. 12. Illustration to the phase shift for a complex conjugated pair of poles

6. Beyond this introduction

There are many good textbooks on Classical Robust Control. Two popular examples are (Dorf & Bishop, 2004) and (Franklin et al., 2002). A less typical and interesting alternative is the recent textbook (Goodwin et al., 2000). All three of these books have at least one chapter devoted to the Fundamentals of Control Theory. Textbooks devoted to Robust and Optimal Control are less common, but there are some available. The best known is probably (Zhou et al.1995). Other possibilities are (Aström & Wittenmark, 1996),(Robert, 1994)(Joseph et al, 2004). An excellent book about the Theory and Design of Classical Control is the one by Aström and Hägglund (Aström & Hägglund, 1995). Good references on the limitations of control are (Looze & Freudenberg, 1988). Bode's book (Bode, 1975) is still interesting, although the emphasis is on vacuum tube circuits.

7. References

Aström, K. J. & Hägglund, T. (1995). *PID Controllers: Theory, Design and Tuning,* International Society for Measurement and Control, Seattle, WA, , 343p, 2nd edition. ISBN: 1556175167.

Aström, K. J. & Wittenmark, B. (1996). *Computer Controlled Systems,* Prentice-Hall, Englewood Cliffs, NJ, 555p, 3rd edition. ISBN-10: 0133148998.

Arnold Zankl (2006). *Milestones in Automation: From the Transistor to the Digital Factory,* Wiley-VCH, ISBN 3-89578-259-9.

Bacciotti, A. & Rosier, L. (2005). *Liapunov functions and stability in control theory*, Springer, 238 p, ISBN:3540213325.

Bode, H. W. (1975). *Network Analysis and Feedback Amplifier Design,* R. E. Krieger Pub. Co., Huntington, NY. Publisher: R. E. Krieger Pub. Co; 577p, 14th print. ISBN 0882752421.

Christopher Kilian (2005). *Modern Control Technology. Thompson Delmar Learning,* ISBN 1-4018-5806-6.

Dorf, R. C. & Bishop, R. H. (2005). *Modern Control Systems,* Prentice-Hall, Upper Saddle River, NJ, 10th edition. ISBN 0131277650.

Faulkner, E.A. (1969): *Introduction to the Theory of Linear Systems,* Chapman & Hall; ISBN 0-412-09400-2.

Franklin, G. F.; Powell, J. D. & Emami-Naeini, A. (2002). *Feedback Control of Dynamical Systems,* Prentice-Hall, Upper Saddle River, NJ, 912p, 4th edition. ISBN: 0-13-032393-4.

Joseph L. Hellerstein; Dawn M. Tilbury, & Sujay Parekh (2004). *Feedback Control of Computing Systems,* John Wiley and Sons. ISBN 978-0-471-26637-2.

Boukhetala, D.; Halbaoui, K. and Boudjema, F.(2006). *Design and Implementation of a Self tuning Adaptive Controller for Induction Motor Drives.*International Review of Electrical Engineering, 260-269, ISSN: 1827- 6660.

Goodwin, G. C, Graebe, S. F. & Salgado, M. E. (2000). *Control System Design,* Prentice-Hall, Upper Saddle River, NJ.,908p, ISBN: 0139586539.

Goodwin, Graham .(2001). *Control System Design,* Prentice Hall. ISBN 0-13-958653-9.

Khalil, H. (2002). *Nonlinear systems,* Prentice Hall, New Jersey, 3rd edition. ISBN 0130673897.

Looze, D. P & Freudenberg, J. S. (1988). *Frequency Domain Properties of Scalar and Multivariable Feedback Systems,* Springer-Verlag, Berlin. *, 281p, ISBN:038718869.*

Levine, William S., ed (1996). *The Control Handbook,* New York: CRC Press. ISBN 978-0-849-38570-4.

Mansour, M. (1992). *The principle of the argument and its application to the stability and robust stability problems,* Springer Berlin – Heidelberg, Vo 183, 16-23, ISSN 0170-8643, ISBN 978-3-540-55961-0.

Robert F. Stengel (1994). *Optimal Control and Estimation,* Dover Publications. ISBN 0-486-68200-5.

Rosier, L. (1992). *Homogeneous Lyapunov function for homogeneous continuous vector field,* Systems Control Lett, 19(6):467-473. ISSN 0167-6911

Thomas H. Lee (2004).*The design of CMOS radio-frequency integrated circuits,* (Second Edition ed.). Cambridge UK: Cambridge University Press. p. §14.6 pp. 451–453.ISBN 0-521-83539-9.

Zhou, K., Doyle J. C., & Glover, K.. (1995). *Robust and Optimal Control*, Prentice-Hall, Upper Saddle River, NJ., 596p, ISBN: 0134565673.

William S Levine (1996).*The control handbook: the electrical engineering handbook series*, (Second Edition ed.). Boca Raton FL: CRC Press/IEEE Press. p. §10.1 p. 163. ISBN 0849385709.

Willy M C Sansen (2006).*Analog design essentials*, Dordrecht, The Netherlands: Springer. p. §0517-§0527 pp. 157–163.ISBN 0-387-25746-2.

2

Robust Control of Hybrid Systems

Khaled Halbaoui[1,2], Djamel Boukhetala[2] and Fares Boudjema[2]
[1]*Power Electronics Laboratory, Nuclear Research Centre of Brine CRNB, BP 180 Ain oussera 17200, Djelfa,*
[2]*Laboratoire de Commande des Processus, ENSP, 10 avenue Pasteur, Hassan Badi, BP 182 El-Harrach,*
Algeria

1. Introduction

The term "hybrid systems" was first used in 1966 Witsenhausen introduced a hybrid model consisting of continuous dynamics with a few sets of transition. These systems provide both continuous and discrete dynamics have proven to be a useful mathematical model for various physical phenomena and engineering systems. A typical example is a chemical batch plant where a computer is used to monitor complex sequences of chemical reactions, each of which is modeled as a continuous process. In addition to the discontinuities introduced by the computer, most physical processes admit components (eg switches) and phenomena (eg collision), the most useful models are discrete. The hybrid system models arise in many applications, such as chemical process control, avionics, robotics, automobiles, manufacturing, and more recently molecular biology.
The control design for hybrid systems is generally complex and difficult. In literature, different design approaches are presented for different classes of hybrid systems, and different control objectives. For example, when the control objective is concerned with issues such as safety specification, verification and access, the ideas in discrete event control and automaton framework are used for the synthesis of control.
One of the most important control objectives is the problem of stabilization. Stability in the continuous systems or not-hybrid can be concluded starting from the characteristics from their fields from vectors. However, in the hybrid systems the properties of stability also depend on the rules of commutation. For example, in a hybrid system by commutation between two dynamic stable it is possible to obtain instabilities while the change between two unstable subsystems could have like consequence stability. The majority of the results of stability for the hybrid systems are extensions of the theories of Lyapunov developed for the continuous systems. They require the Lyapunov function at consecutive switching times to be a decreasing sequence. Such a requirement in general is difficult to check without calculating the solution of the hybrid dynamics, and thus losing the advantage of the approach of Lyapunov.
In this chapter, we develop tools for the systematic analysis and robust design of hybrid systems, with emphasis on systems that require control algorithms, that is, hybrid control systems. To this end, we identify mild conditions that hybrid equations need to satisfy so that their behavior captures the effect of arbitrarily small perturbations. This leads to new concepts of global solutions that provide a deep understanding not only on the robustness

properties of hybrid systems, but also on the structural properties of their solutions. Alternatively, these conditions allow us to produce various tools for hybrid systems that resemble those in the stability theory of classical dynamical systems. These include general versions of theorems of Lyapunov stability and the principles of invariance of LaSalle.

2. Hybrid systems: Definition and examples

Different models of hybrid systems have been proposed in the literature. They mainly differ in the way either the continuous part or the discrete part of the dynamics is emphasized, which depends on the type of systems and problems we consider. A general and commonly used model of hybrid systems is the hybrid automaton (see e.g. (Dang, 2000) and (Girard, 2006)). It is basically a finite state machine where each state is associated to a continuous system. In this model, the continuous evolutions and the discrete behaviors can be considered of equal complexity and importance. By combining the definition of the continuous system, and discrete event systems hybrid dynamical systems can be defined:

Definition 1 *A hybrid system* H *is a collection* $H := (Q, X, \Sigma, U, F, R)$, *where*

- Q *is a finite set, called the set of discrete states;*
- $X \subseteq \Re^n$ *is the set of continuous states;*
- Σ *is a set of discrete input events or symbols;*
- $X \subseteq \Re^m$ *is the set of continuous inputs;*
- $F : Q \times X \times U \rightarrow \Re^n$ *is a vector field describing the continuous dynamics;*
- $R : Q \times X \times \Sigma \times U \rightarrow Q \times X$ *describes the discrete dynamics.*

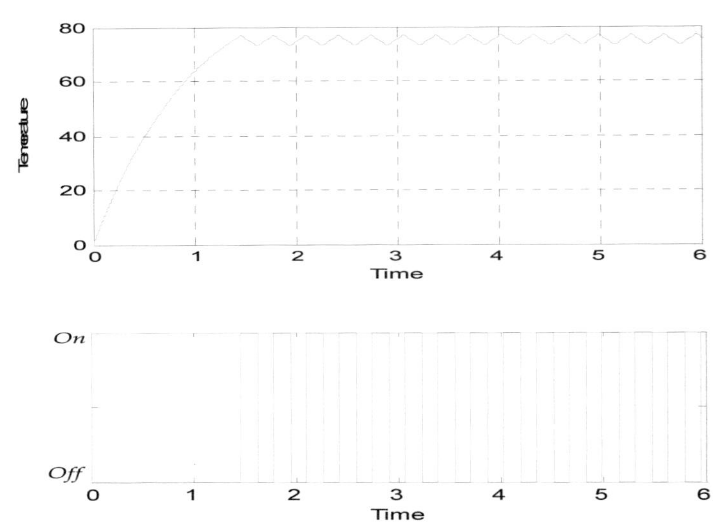

Fig. 1. A trajectory of the room temperature.

Example 1 **(Thermostat).** The thermostat consists of a heater and a thermometer which maintain the temperature of the room in some desired temperature range (Rajeev, 1993). The lower and upper thresholds of the thermostat system are set at x_m and x_M such that $x_m \prec x_M$. The heater is maintained on as long as the room temperature is below x_M, and it is turned off whenever the thermometer detects that the temperature reaches x_M. Similarly, the heater remains off if the temperature is above x_m and is switched on whenever the temperature falls to x_m (Fig. 1). In practical situations, exact threshold detection is impossible due to sensor imprecision. Also, the reaction time of the on/off switch is usually non-zero. The effect of these inaccuracies is that we cannot guarantee switching exactly at the nominal values x_m and x_M. As we will see, this causes non-determinism in the discrete evolution of the temperature.

Formally we can model the thermostat as a hybrid automaton shown in (Fig. 2). The two operation modes of the thermostat are represented by two locations '*on*' and '*off*'. The *on/off* switch is modeled by two discrete transitions between the locations. The continuous variable x models the temperature, which evolves according to the following equations.

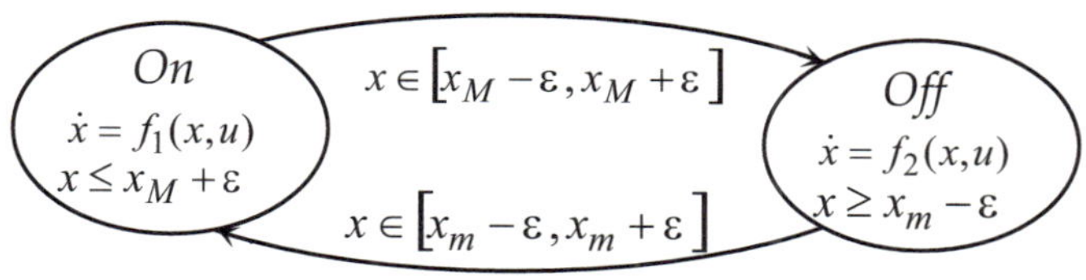

Fig. 2. Model of the thermostat.

- If the thermostat is on, the evolution of the temperature is described by:

$$\dot{x} = f_1(x,u) = -x + 4 + u \tag{1}$$

- When the thermostat is off, the temperature evolves according to the following differential equation:

$$\dot{x} = f_2(x,u) = -x + u$$

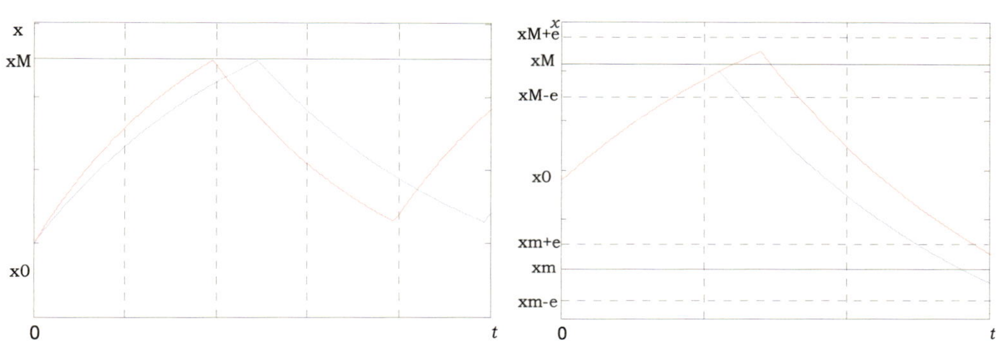

Fig. 3. Two different behaviors of the temperature starting at x_0.

The second source of non-determinism comes from the continuous dynamics. The input signal u of the thermostat models the fluctuations in the outside temperature which we cannot control. (Fig. 3 left) shows this continuous non-determinism. Starting from the initial temperature x_0, the system can generate a "*tube*" of infinite number of possible trajectories, each of which corresponds to a different input signal u. To capture uncertainty of sensors, we define the first guard condition of the transition from '*on*' to '*off*' as an interval $[x_M - \varepsilon, x_M + \varepsilon]$ with $\varepsilon \succ 0$. This means that when the temperature enters this interval, the thermostat can either turn the heater off immediately or keep it on for some time provided that $x \le x_M + \varepsilon$. (Fig. 3 right) illustrates this kind of non-determinism. Likewise, we define the second guard condition of the transition from 'off' to 'on' as the interval $[x_m - \varepsilon, x_m + \varepsilon]$. Notice that in the thermostat model, the temperature does not change at the switching points, and the reset maps are thus the identity functions.

Finally we define the two staying conditions of the 'on' and 'off' locations as $x \le x_M + \varepsilon$ and $x \ge x_M - \varepsilon$ respectively, meaning that the system can stay at a location while the corresponding staying conditions are satisfied.

Example 2 (Bouncing Ball). Here, the ball (thought of as a point-mass) is dropped from an initial height and bounces off the ground, dissipating its energy with each bounce. The ball exhibits continuous dynamics between each bounce; however, as the ball impacts the ground, its velocity undergoes a discrete change modeled after an inelastic collision. A mathematical description of the bouncing ball follows. Let $x_1 := h$ be the height of the ball and $x_2 := \dot{h}$ (Fig. 4). A hybrid system describing the ball is as follows:

$$g(x) := \begin{bmatrix} 0 \\ -\gamma . x_2 \end{bmatrix}, D := \{x : x_1 = 0,\ x_2 \prec 0\}\ f(x) := \begin{bmatrix} x_2 \\ -g \end{bmatrix}, C := \{x : x_1 \ge 0\} \backslash D . \tag{2}$$

This model generates the sequence of hybrid arcs shown in (Fig. 5). However, it does not generate the hybrid arc to which this sequence of solutions converges since the origin does not belong to the jump set D. This situation can be remedied by including the origin in the jump set D. This amounts to replacing the jump set D by its closure. One can also replace the flow set C by its closure, although this has no effect on the solutions.

It turns out that *whenever the flow set and jump set are closed, the solutions of the corresponding hybrid system enjoy a useful compactness property: every locally eventually bounded sequence of solutions has a subsequence converging to a solution.*

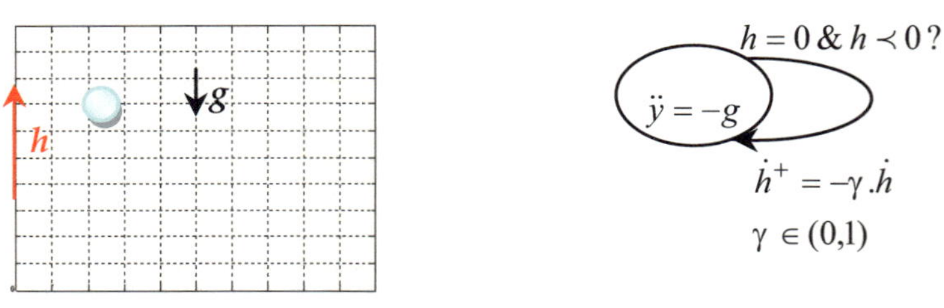

Fig. 4. Diagram for the bouncing ball system

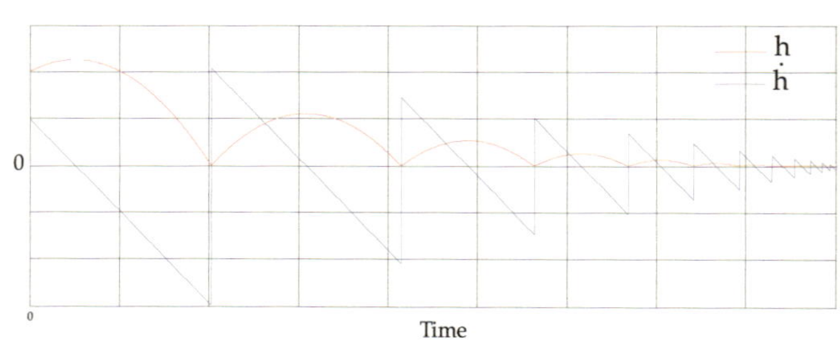

Fig. 5. Solutions to the bouncing ball system

Consider the sequence of hybrid arcs depicted in (Fig. 5). They are solutions of a hybrid "bouncing ball" model showing the position of the ball when dropped for successively lower heights, each time with zero velocity. The sequence of graphs created by these hybrid arcs converges to a graph of a hybrid arc with hybrid time domain given by $\{0\} \times$ {nonnegative integers} where the value of the arc is zero everywhere on its domain. If this hybrid arc is a solution then the hybrid system is said to have a "compactness" property. This attribute for the solutions of hybrid systems is critical for robustness properties. It is the hybrid generalization of a property that automatically holds for continuous differential equations and difference equations, where nominal robustness of asymptotic stability is guaranteed.

Solutions of hybrid systems are hybrid arcs that are generated in the following way: Let C and D be subsets of $\Re^n$ and let f , respectively g , be mappings from C , respectively D , to $\Re^n$. The hybrid system $H := (f,g,C,D)$ can be written in the form

$$\begin{aligned} \dot{x} &= f(x) \quad && x \in C \\ x^+ &= g(x) \quad && x \in D \end{aligned} \tag{3}$$

The map f is called the "flow map", the map g is called the "jump map", the set C is called the "flow set", and the set D is called the "jump set". The state x may contain variables taking values in a discrete set (logic variables), timers, etc. Consistent with such a situation is the possibility that $C \cup D$ is a strict subset of $\Re^n$. For simplicity, assume that f and g are continuous functions. At times it is useful to allow these functions to be set-valued mappings, which will denote by F and G, in which case F and G should have a closed graph and be locally bounded, and F should have convex values.

In this case, we will write

$$\begin{aligned} \dot{x} &\in F \quad && x \in C \\ x^+ &\in G \quad && x \in D \end{aligned} \tag{4}$$

A solution to the hybrid system (4) starting at a point $x_0 \in C \cup D$ is a hybrid arc x with the following properties:

1. $x(0,0) = x_0$;
2. given $(s,j) \in dom\ x$, if there exists $\tau \succ s$ such that $(\tau,j) \in dom\ x$, then, for all $t \in [s,\tau]$, $x(t,j) \in C$ and, for almost all $t \in [s,\tau]$, $\dot{x}(t,j) \in F(x(t,j))$;
3. given $(t,j) \in dom\ x$, if $(t,j+1) \in dom\ x$ then $x(t,j) \in D$ and $x(t,j+1) \in G(x(t,j))$.

Solutions from a given initial condition are not necessarily unique, even if the flow map is a smooth function.

3. Approaches to analysis and design of hybrid control systems

The analysis and design tools for hybrid systems in this section are in the form of Lyapunov stability theorems and LaSalle-like invariance principles. Systematic tools of this type are the base of the theory of systems for purely of the continuous-time and discrete-time systems. Some similar tools available for hybrid systems in (Michel, 1999) and (DeCarlo, 2000), the tools presented in this section generalize their conventional versions of continuous-time and discrete-time hybrid systems development by defining an equivalent concept of stability and provide extensions intuitive sufficient conditions of stability asymptotically.

3.1 LaSalle-like invariance principles

Certain principles of invariance for the hybrid systems have been published in (Lygeros et al., 2003) and (Chellaboina et al., 2002). Both results require, among other things, unique solutions which is not generic for hybrid control systems. In (Sanfelice et al., 2005), the general invariance principles were established that do not require uniqueness. The work in (Sanfelice et al., 2005) contains several invariance results, some involving integrals of functions, as for systems of continuous-time in (Byrnes & Martin, 1995) or (Ryan, 1998), and some involving nonincreasing energy functions, as in work of LaSalle (LaSalle, 1967) or (LaSalle, 1976). Such a result will be described here.

Suppose we can find a continuously differentiable function $V:\Re^n \to \Re$ such that

$$\begin{aligned} u_c(x) &:= \langle \nabla V(x), f(x) \rangle \leq 0 \qquad && \forall x \in \overline{C} \\ u_d(x) &:= V(g(x)) - V(x) \leq 0 && \forall x \in \overline{D} \end{aligned} \tag{5}$$

Consider $x(\cdot,\cdot)$ a bounded solution with an unbounded hybrid time. Then there exists a value r in the range V so that x tends to the largest weakly invariant set inside the set

$$M_r := V^{-1}(r) \cap \left(u_c^{-1}(0) \cup \left(u_d^{-1}(0) g(u_d^{-1}(0)) \right) \right) \tag{6}$$

where $u_d^{-1}(0)$: the set of points x satisfying $u_d(x) = 0$ and $g(u_d^{-1}(0))$ corresponds to the set of points $g(y)$ where $y \in u_d^{-1}(0)$.

The naive combination of continuous-time and discrete-time results would omit the intersection with $g(u_d^{-1}(0))$. This term, however, can be very useful for zeroing in set to which trajectories converge.

3.2 Lyapunov stability theorems

Some preliminary results on the existence of the non-smooth Lyapunov function for the hybrid systems published in (DeCarlo, 2000). The first results on the existence of smooth Lyapunov functions, which are closely related to the robustness, published in (Cai et al., 2005). These results required open basins of attraction, but this requirement has since been relaxed in (Cai et al. 2007). The simplified discussion here is borrowed from this posterior work.

Let $\mathcal{O}$ be an open subset of the state space containing a given compact set A and let $\omega:\mathcal{O} \to \Re_{\geq 0}$ be a continuous function which is zero for all $x \in A$, is positive otherwise, which grows without limit as its argument grows without limit or near the limit $\mathcal{O}$. Such a function is called a suitable indicator for the compact set A in the open set $\mathcal{O}$. An example of such a function is the standard function on $\Re^n$ which is an appropriate indicator of origin. More generally, the distance to a compact set A is an appropriate indicator for all A on $\Re^n$.

Given an open set $\mathcal{O}$, an appropriate indicator ω and hybrid data (f,g,C,D), a function $V:\mathcal{O} \to \Re_{\geq 0}$ is called a smooth Lyapunov function for $(f,g,C,D,\omega,\mathcal{O})$ if it is smooth and there exist functions α_1, α_2 belonging to the class-$\mathcal{K}_\infty$, such as

$$\begin{aligned} &\alpha_1(\omega(x)) \leq V(x) \leq \alpha_2(\omega(x)) \qquad && \forall x \in \mathcal{O} \\ &\langle \nabla V(x), f(x) \rangle \leq -V(x) && \forall x \in \overline{C} \cap \mathcal{O} \\ &V(g(x)) \leq e^{-1} V(x) && \forall x \in \overline{D} \cap \mathcal{O} \end{aligned} \tag{7}$$

Suppose that such a function exists, it is easy to verify that all solutions for the hybrid system (f,g,C,D) from $\mathcal{O} \cap \left(\overline{C} \cup \overline{D} \right)$ satisfied

$$\omega(x(t,j)) \leq \alpha_1^{-1}\left(e^{-t-j}\alpha_2(\omega(x(0,0)))\right) \qquad \forall (t,j) \in dom\ x \tag{8}$$

In particular,

- (pre-stability of A) for each $\varepsilon \succ 0$ there exists $\delta \succ 0$ such that $x(0,0) \in A + \delta B$ implies, for each generalized solution, that $x(t,j) \in A + \varepsilon B$ for all $(t,j) \in dom\ x$, and
- (before attractive A on $\mathcal{O}$) any generalized solution from $\mathcal{O} \cap \left(\overline{C} \cup \overline{D}\right)$ is bounded and if its time domain is unbounded, so it converges to A.

According to one of the principal results in (Cai et al., 2006) *there exists a smooth Lyapunov function for* $(f, g, C, D, \omega, \mathcal{O})$ *if and only if the set* A *is pre-stable and pre-attractive on* $\mathcal{O}$ *and* $\mathcal{O}$ *is forward invariant* (i.e., $x(0,0) \in \mathcal{O} \cap \left(\overline{C} \cup \overline{D}\right)$ implies $x(t,j) \in \mathcal{O}$ for all $(t,j) \in dom\ x$).

One of the primary interests in inverse Lyapunov theorems is that they can be employed to establish the robustness of the asymptotic stability of various types of perturbations.

4. Hybrid control application

In system theory in the 60s researchers were discussing mathematical frameworks so to study systems with continuous and discrete dynamics. Current approaches to hybrid systems differ with respect to the emphasis on or the complexity of the continuous and discrete dynamics, and on whether they emphasize analysis and synthesis results or analysis only or simulation only. On one end of the spectrum there are approaches to hybrid systems that represent extensions of system theoretic ideas for systems (with continuous-valued variables and continuous time) that are described by ordinary differential equations to include discrete time and variables that exhibit jumps, or extend results to switching systems. Typically these approaches are able to deal with complex continuous dynamics. Their main emphasis has been on the stability of systems with discontinuities. On the other end of the spectrum there are approaches to hybrid systems embedded in computer science models and methods that represent extensions of verification methodologies from discrete systems to hybrid systems. Several approaches to robustness of asymptotic stability and synthesis of hybrid control systems are represented in this section.

4.1 Hybrid stabilization implies input-to-state stabilization

In the paper (Sontag, 1989) it has been shown, for continuous-time control systems, that smooth stabilization involves smooth input-to-stat stabilization with respect to input additive disturbances. The proof was based on converse Lyapunov theorems for continuous-time systems. According to the indications of (Cai et al., 2006), and (Cai et al. 2007), the result generalizes to hybrid control systems via the converse Lyapunov theorem.

In particular, if we can find a hybrid controller, with the type of regularity used in sections 4.2 and 4.3, to achieve asymptotic stability, then the input-to-state stability with respect to input additive disturbance can also be achieved.

Here, consider the special case where the hybrid controller is a logic-based controller where the variable takes values in the logic of a finite set. Consider the hybrid control system

$$\mathcal{H} := \begin{cases} \dot{\xi} = f_q(\xi) + \eta_q(\xi)(u_q + \upsilon_q d) & \xi \in C_q,\ q \in Q \\ \begin{bmatrix} \xi \\ q \end{bmatrix}^{+} \in G_q(\xi) & \xi \in D_q,\ q \in Q \end{cases} \tag{9}$$

where Q is a finite index set, for each $q \in Q$, f_q, $\eta_q : C_q \to \Re^n$ are continuous functions, C_q and D_q are closed and G_q has a closed graph and is locally bounded. The signal u_q is the control, and d is the disturbance, while υ_q is vector that is independent of the state, input, and disturbance. Suppose $\mathcal{H}$ is stabilizable by logic-based continuous feedback; that is, for the case where $d = 0$, there exist continuous functions k_q defined on C_q such that, with $u_q := k_q(\xi)$, the nonempty and compact set $A = \cup_{q\in Q} A_q \times \{q\}$ is pre-stable and globally pre-attractive. Converse Lyapunov theorems can then be used to establish the existence of a logic-based continuous feedback that renders the closed-loop system input-to-state stable with respect to d. The feedback has the form

$$u_q := k_q(\xi) - \varepsilon.\eta_q^T(\xi)\nabla V_q(\xi) \tag{10}$$

where $\varepsilon \succ 0$ and $V_q(\xi)$ is a smooth Lyapunov function that follows from the assumed asymptotic stability when $d \equiv 0$. There exist class-$\mathcal{K}_\infty$ functions α_1 and α_2 such that, with this feedback control, the following estimate holds:

$$|\xi(t,j)|_{A(t,j)} \le \max\left\{ \alpha_1^{-1}\left(2.\exp(-t-j).\alpha_2\left(|\xi(0,0)|_{Aq(0,0)}\right)\right), \alpha_1^{-1}\left(\frac{\max_{q\in Q}|\upsilon_q|^2}{2.\varepsilon}\|d\|_\infty^2\right)\right\} \tag{11}$$

where $\|d\|_\infty := \sup_{(s,i)\in \text{dom}\, d} |d(s,i)|$.

4.2 Control Lyapunov functions

Although the control design using a continuously differentiable control-Lyapunov function is well established for input-affine nonlinear control systems, it is well known that not all controllable input-affine nonlinear control system function admits a continuously differentiable control-Lyapunov function. A well known example in the absence of this control-Lyapunov function is the so-called "Brockett", or "nonholonomic integrator". Although this system does not allow continuously differentiable control Lyapunov function, it has been established recently that admits a good "patchy" control-Lyapunov function.

The concept of control-Lyapunov function, which was presented in (Goebel et al., 2009), is inspired not only by the classical control-Lyapunov function idea, but also by the approach to feedback stabilization based on patchy vector fields proposed in (Ancona & Bressan, 1999). The idea of control-Lyapunov function was designed to overcome a limitation of discontinuous feedbacks, such as those from patchy feedback, which is a lack of robustness to measurement noise. In (Goebel et al., 2009) it has been demonstrated that any asymptotically controllable nonlinear system admits a smooth patchy control-Lyapunov function if we admit the possibility that the number of patches may need to be infinite. In addition, it was shown how to construct a robust stabilizing hybrid feedback from a patchy control-Lyapunov function. Here the idea when the number of patches is finite is outlined and then specialized to the nonholonomic integrator.

Generally , a global patchy smooth control-Lyapunov function for the origin for the control system $\dot{x} = f(x,u)$ in the case of a finite number of patches is a collection of functions V_q and sets Ω_q and Ω'_q where $q \in Q := \{1,\ldots,m\}$, such as

a. for each $q \in Q$, Ω_q and Ω'_q are open and

- $\mathcal{O} := \Re^n \setminus \{0\} = \cup_{q\in Q}\Omega_q = \cup_{q\in Q}\Omega'_q$
- for each $q \in Q$, the outward unit normal to $\partial\Omega_q$ is continuous on $\left(\partial\Omega_q \setminus \cup_{r\succ q}\Omega'_r\right) \cap \mathcal{O}$,

- for each $q \in Q$, $\bar{\Omega}'_q \cap \mathcal{O} \subset \Omega_q$;
  b. for each $q \in Q$, V_q is a smooth function defined on a neighborhood (relative to $\mathcal{O}$) of Ω_q.
  c. there exist a continuous positive definite function α and class-$\mathcal{K}_\infty$ functions $\bar{\gamma}$ and $\underline{\gamma}$ such that
- $\underline{\gamma}(|x|) \le V_q(x) \le \bar{\gamma}(|x|) \qquad V_q \forall q \in Q,\ x \in \overline{\left(\Omega_q \backslash \cup_{r \succ q} \Omega'_r\right)} \cap \mathcal{O}$;
- for each $q \in Q$ and $x \in \left(\Omega_q \backslash \cup_{r \succ q} \Omega'_r\right)$ there exists $u_{x,q}$ such that

$$\left\langle \nabla V_q(x), f(x,u_x,q) \right\rangle \le -\alpha(x)$$

- for each $q \in Q$ and $x \in \left(\Omega_q \backslash \cup_{r \succ q} \Omega'_r\right) \cap \mathcal{O}$ there exists $u_{x,q}$ such that

$$\left\langle \nabla V_q(x), f(x,u_x,q) \right\rangle \le -\alpha(x)$$
$$\left\langle n_q(x), f(x,u_x,q) \right\rangle \le -\alpha(x)$$

where $x \mapsto n_q(x)$ denotes the outward unit normal to $\partial\Omega_q$.
From this patchy control-Lyapunov function one can construct a robust hybrid feedback stabilizer, at least when the set $\{\ u, \upsilon.f(x,u) \le c\ \}$ is convex for each real number c and every real vector υ, with the following data

$$u_q := k_q(x)\,, C_q = \overline{\left(\Omega_q \backslash \cup_{r \succ q} \Omega'_r\right)} \cap \mathcal{O} \tag{12}$$

where k_q is defined on C_q, continuous and such that

$$\begin{aligned} &\left\langle \nabla V_q(x), f(x,k_q(x)) \right\rangle \le -0.5\alpha(x) && \forall x \in C_q \\ &\left\langle n_q(x), f(x,k_x(x)) \right\rangle \le -0.5\alpha(x) && \forall x \in \left(\partial\Omega_q \backslash \cup_{r \succ k} \Omega'_r\right) \cap \mathcal{O} \end{aligned} \tag{13}$$

The jump set is given by

$$D_q = \left(\mathcal{O} \backslash \Omega_q\right) \cup \left(\cup_{r \succ q} \overline{\Omega'_r} \cap \mathcal{O}\right) \tag{14}$$

and the jump map is

$$G_q(x) = \begin{cases} \left\{\ r \in Q : x \in \overline{\Omega'_r} \cap \mathcal{O},\ r \succ q \right\} & x \in \left(\cup_{r \succ q} \overline{\Omega'_r} \cap \mathcal{O}\right) \cap \Omega_q \\ \left\{\ r \in Q : x \in \overline{\Omega'_r} \cap \mathcal{O} \right\} & x \in \mathcal{O} \backslash \Omega_q \end{cases} \tag{15}$$

With this control, the index increases with each jump except probably the first one. Thus, the number of jumps is finite, and the state converges to the origin, which is also stable.

4.3 Throw-and-catch control

In (Prieur, 2001), it was shown how to combine local and global state feedback to achieve global stabilization and local performance. The idea, which exploits hysteresis switching (Halbaoui et al., 2009b), is completely simple. Two continuous functions, k_{global} and k_{local} are shown when the feedback $u = k_{global}(x)$ render the origin of the control system $\dot{x} = f(x,u)$ globally asymptotically stable whereas the feedback $u = k_{local}(x)$ makes the origin of the control system locally asymptotically stable with basin of attraction containing the open set $\mathcal{O}$, which contains the origin. Then we took C_{local} a compact subset of the $\mathcal{O}$ that contains the origin in its interior and one takes D_{global} to be a compact subset of C_{local}, again containing the origin in its interior and such that, when using the controller k_{local}, trajectories starting in D_{global} never reach the boundary of C_{local} (Fig. 6). Finally, the hybrid control which achieves global asymptotic stabilization while using the controller k_q for small signals is as follows

$$\begin{array}{ll} u := k_q(x) & C := \left\{ (x,q) : x \in C_q \right\} \\ g(q,x) := \text{toggle } (q) & D := \left\{ (x,q) : x \in D_q \right\} \end{array} \tag{16}$$

In the problem of uniting of local and global controllers, one can view the global controller as a type of "bootstrap" controller that is guaranteed to bring the system to a region where another controller can control the system adequately.

A prolongation of the idea of combine local and global controllers is to assume the existence of continuous bootstrap controller that is guaranteed to introduce the system, in finite time, in a vicinity of a set of points, not simply a vicinity of the desired final destination (the controller doesn't need to be able to maintain the state in this vicinity); moreover, these sets of points form chains that terminate at the desired final destination and along which controls are known to steer (or "throw") form one point in the chain at the next point in the chain. Moreover, in order to minimize error propagation along a chain, a local stabilizer is known for each point, except perhaps those points at the start of a chain. Those can be employed "to catch" each jet.

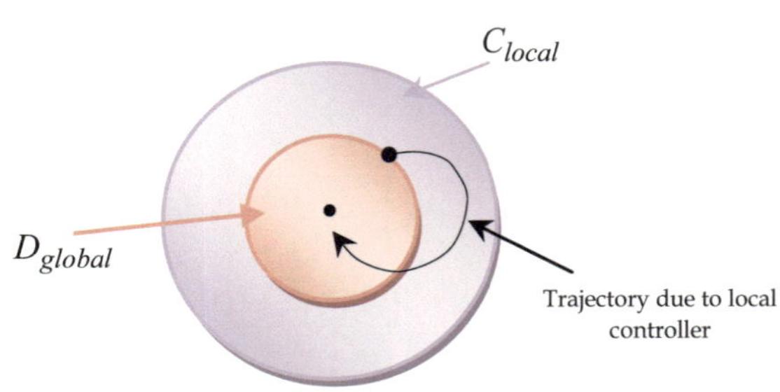

Fig. 6. Combining local and global controllers

4.4 Supervisory control

In this section, we review the supervisory control framework for hybrid systems. One of the main characteristics of this approach is that the plant is approximated by a discrete-event system and the design is carried out in the discrete domain. The hybrid control systems in the supervisory control framework consist of a continuous (state, variable) system to be controlled, also called the plant, and a discrete event controller connected to the plant via an interface in a feedback configuration as shown in (Fig. 7). It is generally assumed that the dynamic behavior of the plant is governed by a set of known nonlinear ordinary differential equations

$$\dot{x}(t) = f(x(t), r(t)) \tag{17}$$

where $x \in \Re^n$ is the continuous state of the system and $r \in \Re^m$ is the continuous control input. In the model shown in (Fig. 7), the plant contains all continuous components of the hybrid control system, such as any conventional continuous controllers that may have been developed, a clock if time and synchronous operations are to be modeled, and so on. The controller is an event driven, asynchronous discrete event system (DES), described by a finite state automaton. The hybrid control system also contains an interface that provides the means for communication between the continuous plant and the DES controller.

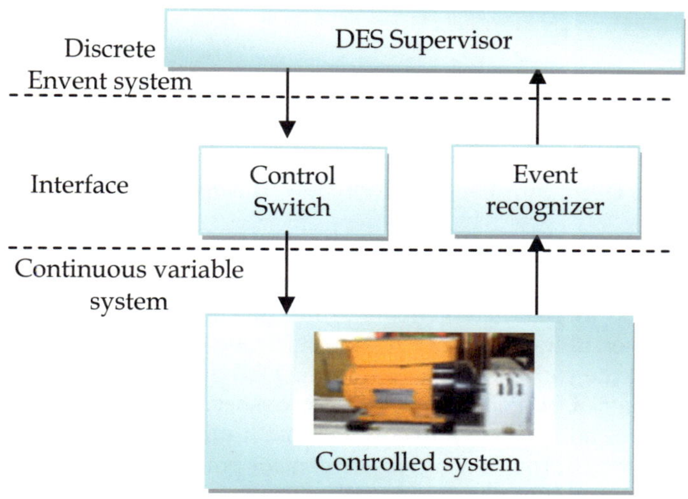

Fig. 7. Hybrid system model in the supervisory control framework.

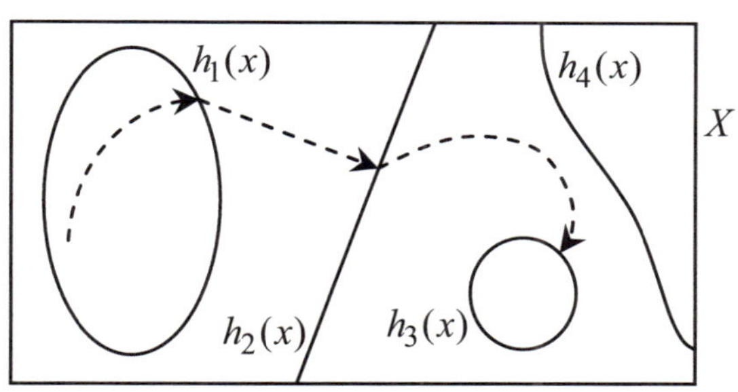

Fig. 8. Partition of the continuous state space.

The interface consists of the generator and the actuator as shown in (Fig. 7). The generator has been chosen to be a partitioning of the state space (see Fig. 8). The piecewise continuous command signal issued by the actuator is a staircase signal as shown in (Fig. 9), not unlike the output of a zero-order hold in a digital control system. The interface plays a key role in determining the dynamic behavior of the hybrid control system. Many times the partition of the state space is determined by physical constraints and it is fixed and given. Methodologies for the computation of the partition based on the specifications have also been developed.

In such a hybrid control system, the plant taken together with the actuator and generator, behaves like a discrete event system; it accepts symbolic inputs via the actuator and produces symbolic outputs via the generator. This situation is somewhat analogous to the

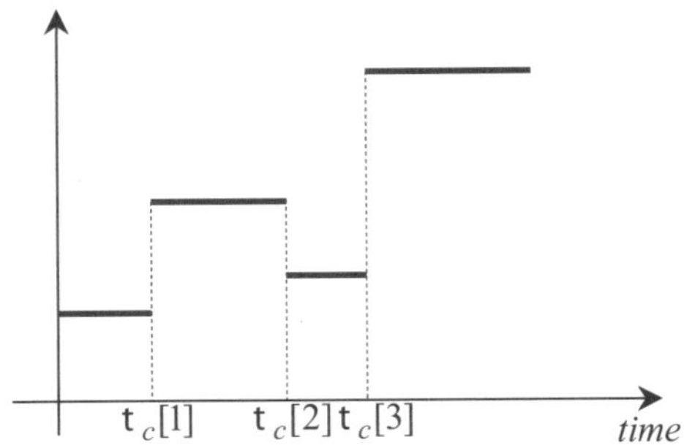

Fig. 9. Command signal issued by the interface.

way a continuous time plant, equipped with a zero-order hold and a sampler, "looks" like a discrete-time plant. The DES which models the plant, actuator, and generator is called the DES plant model. From the DES controller's point of view, it is the DES plant model which is controlled.

The DES plant model is an approximation of the actual system and its behavior is an abstraction of the system's behavior. As a result, the future behavior of the actual continuous system cannot be determined uniquely, in general, from knowledge of the DES plant state and input. The approach taken in the supervisory control framework is to incorporate all the possible future behaviors of the continuous plant into the DES plant model. A conservative approximation of the behavior of the continuous plant is constructed and realized by a finite state machine. From a control point of view this means that if undesirable behaviors can be eliminated from the DES plant (through appropriate control policies) then these behaviors will be eliminated from the actual system. On the other hand, just because a control policy permits a given behavior in the DES plant, is no guarantee that that behavior will occur in the actual system.

We briefly discuss the issues related to the approximation of the plant by a DES plant model.

A *dynamical system* Σ can be described as a triple $T;W;B$ with $T \subseteq \Re$ the *time axis*, W the *signal space*, and $B \subset W^T$ (the set of all functions $f: T \to W$) the *behavior*. The behavior of the DES plant model consists of all the pairs of plant and control symbols that it can generate. The time axis T represents here the occurrences of events. A necessary condition for the DES plant model to be a valid approximation of the continuous plant is that the behavior of the continuous plant model B_c is contained in the behavior of the DES plant model, i.e. $B_c \subseteq B_d$.

The main objective of the controller is to restrict the behavior of the DES plant model in order to specify the control specifications. The specifications can be described by a behavior B_{spec} . *Supervisory control of hybrid systems is based on the fact that if undesirable behaviors can be eliminated from the DES plant then these behaviors can likewise be eliminated from the actual system.* This is described formally by the relation

$$B_d \cap B_s \subseteq B_{spec} \Rightarrow B_c \cap B_s \subseteq B_{spec} \quad (18)$$

and is depicted in (Fig. 10). The challenge is to find a discrete abstraction with behavior B_d which is a approximation of the behavior B_c of the continuous plant and for which is possible to design a supervisor in order to guarantee that the behavior of the closed loop system satisfies the specifications B_{spec}. A more accurate approximation of the plant's behavior can be obtained by considering a finer partitioning of the state space for the extraction of the DES plant.

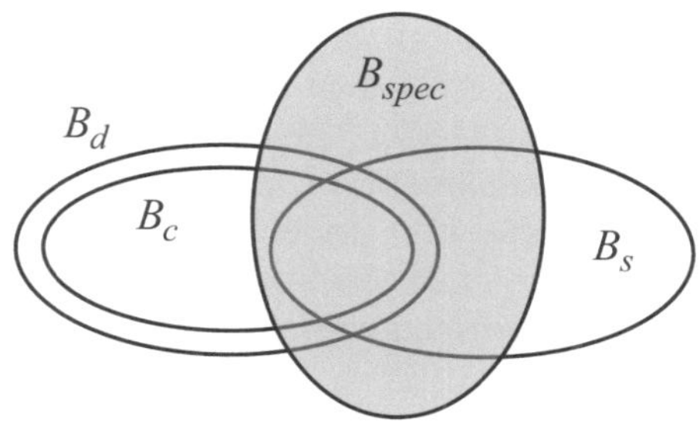

Fig. 10. The DES plant model as an approximation.

An interesting aspect of the DES plant's behavior is that it is distinctly nondeterministic. This fact is illustrated in (Fig.11). The figure shows two different trajectories generated by the same control symbol. Both trajectories originate in the same DES plant state $\tilde{p}_1$. (Fig.11) shows that for a given control symbol, there are at least two possible DES plant states that can be reached from $\tilde{p}_1$. Transitions within a DES plant will usually be nondeterministic unless the boundaries of the partition sets are invariant manifolds with respect to the vector fields that describe the continuous plant.

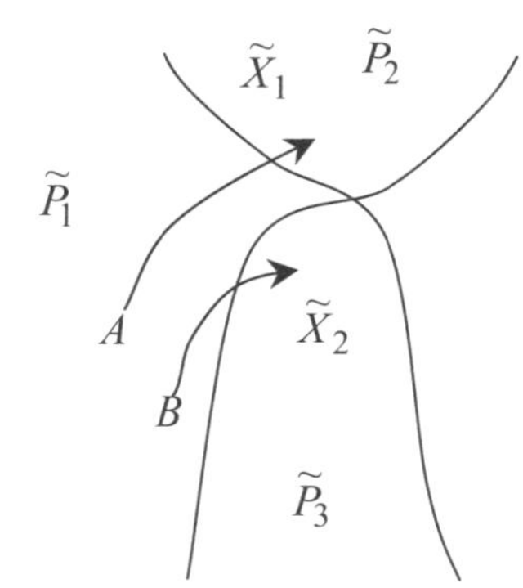

Fig. 11. Nondeterminism of the DES plant model.

There is an advantage to having a hybrid control system in which the DES plant model is deterministic. It allows the controller to drive the plant state through any desired sequence of regions provided, of course, that the corresponding state transitions exist in the DES plant model. If the DES plant model is not deterministic, this will not always be possible. This is because even if the desired sequence of state transitions exists, the sequence of inputs which achieves it may also permit other sequences of state transitions. Unfortunately, given a continuous-time plant, it may be difficult or even impossible to design an interface that leads to a DES plant model which is deterministic. Fortunately, it is not generally necessary to have a deterministic DES plant model in order to control it. The supervisory control problem for hybrid systems can be formulated and solved when the DES plant model is nondeterministic. This work builds upon the frame work of supervisory control theory used in (Halbaoui et al., 2008) and (Halbaoui et al., 2009a).

5. Robustness to perturbations

In control systems, several perturbations can occur and potentially destroy the good behavior for which the controller was designed for. For example, noise in the measurements of the state taken by controller arises in all implemented systems. It is also common that when a controller is designed, only a simplified model of the system to control exhibiting the most important dynamics is considered. This simplifies the control design in general. However, sensors/actuators that are dynamics unmodelled can substantially affect the behavior of the system when in the loop. In this section, it is desired that the hybrid controller provides a certain degree of robustness to such disturbances. In the following sections, general statements are made in this regard.

5.1 Robustness via filtered measurements

In this section, the case of noise in the measurements of the state of the nonlinear system is considered. Measurement noise in hybrid systems can lead to nonexistence of solutions. This situation can be corrected, at least for the small measurement noise, if under global existence of solutions, C_c and D_c always "overlap" while ensuring that the stability properties still hold. The "overlap" means that for every $\xi \in O$, either $\xi + e \in C_c$ or $\xi + e \in D_c$ all or small e. There exist generally always inflations of C and D that preserve the semiglobal practices asymptotic stability, but they do not guarantee the existence of solutions for small measurement noise.
Moreover, the solutions are guaranteed to exist for any locally bounded measurement noise if the measurement noise does not appear in the flow and jump sets. This can be carried out by filtering measures. (Fig. 12) illustrates this scenario. The state x is corrupted by the noise e and the hybrid controller H_c measures a filtered version of $x + e$.

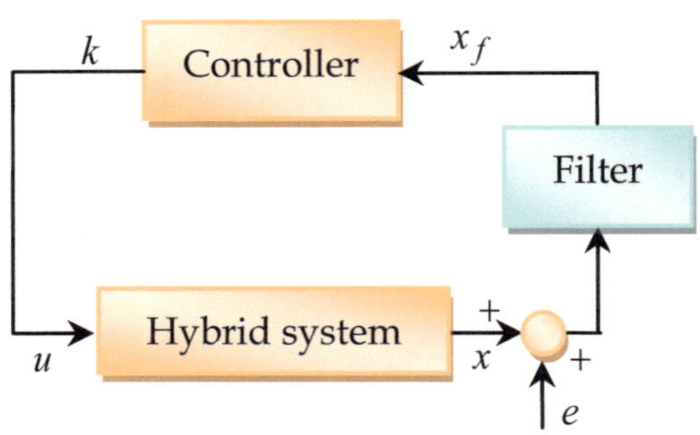

Fig. 12. Closed-loop system with noise and filtered measurements.

The filter used for the noisy output $y = x + e$ is considered to be linear and defined by the matrices A_f, B_f, and L_f, and an additional parameter $\varepsilon_f > 0$. It is designed to be asymptotically stable. Its state is denoted by x_f which takes value in R^{n_f}. At the jumps, x_f is given to the current value of y. Then, the filter has flows given by

$$\varepsilon_f \dot{x}_f = A_f x_f + B_f y, \tag{19}$$

and jumps given by

$$x_f^+ = A_f^{-1} B_f x_f + B_f y. \tag{20}$$

The output of the filter replaces the state x in the feedback law. The resulting closed-loop system can be interpreted as family of hybrid systems which depends on the parameter ε_f. It is denoted by $H_{cl}^{\varepsilon_f}$ and is given by

$$H_{cl}^{\varepsilon_f}:\begin{cases}\left.\begin{aligned}&\dot{x}=f_p(x+\kappa(L_f x_f,x_c))\\&\dot{x}_c=f_c(L_f x_f,x_c)\\&\varepsilon_f\dot{x}_f=A_f x_f+B_f(x+e)\end{aligned}\right\} & (L_f x_f,x_c)\in C_c\\ \left.\begin{aligned}&x^+=x\\&x_c^+\in G_c(L_f x_f,x_c)\\&x_f^+=-A_f^{-1}B_f(x+e)\end{aligned}\right\} & (L_f x_f,x_c)\in D_c\end{cases} \tag{21}$$

5.2 Robustness to sensor and actuator dynamics

This section reviews the robustness of the closed-loop H_{cl} when additional dynamics, coming from sensors and actuators, are incorporated. (Fig. 13) shows the closed loop H_{cl} with two additional blocks: a model for the sensor and a model for the actuator. Generally, to simplify the controller design procedure, these dynamics are not included in the model of the system $\dot{x}=f_p(x,u)$ when the hybrid controller H_c is conceived. Consequently, it is important to know whether the stability properties of the closed-loop system are preserved, at least semiglobally and practically, when those dynamics are incorporated in the closed loop.

The sensor and actuator dynamics are modeled as stable filters. The state of the filter which models the sensor dynamics is given by $x_s\in R^{n_s}$ with matrices (A_s,B_s,L_s) , the state of the filter that models the actuator dynamics is given by $x_a\in R^{n_a}$ with matrices (A_a,B_a,L_a), and $\varepsilon_d>0$ is common to both filters.

Augmenting H_{cl} by adding filters and temporal regularization leads to a family $H_{cl}^{\varepsilon_d}$ given as follows

$$H_{cl}^{\varepsilon_d}:\begin{cases}\left.\begin{aligned}&\dot{x}=f_p(x,L_a x_a)\\&\dot{x}_c=f_c(L_s x_s,x_c)\\&\dot{\tau}=-\tau+\tau^*\\&\varepsilon_d\dot{x}_s=A_s x_s+B_s(x+e)\\&\varepsilon_d\dot{x}_a=A_a x_a+B_a\kappa(L_s x_s,x_c)\end{aligned}\right\} & (L_s x_s,x_c)\in C_c \text{ or } \tau\le\bar{\tau}\\ \left.\begin{aligned}&x^+=x\\&x_c^+\in G_c(L_s x_s,x_c)\\&x_s^+=x_s\\&x_a^+=x_a\\&\tau^+=0\end{aligned}\right\} & (L_s x_s,x_c)\in D_c \text{ and } \tau\ge\bar{\tau}\end{cases} \tag{22}$$

where τ^* is a constant satisfying $\tau^*>\bar{\tau}$.

The following result states that for fast enough sensors and actuators, and small enough temporal regularization parameter, the compact set A is semiglobally practically asymptotically stable.

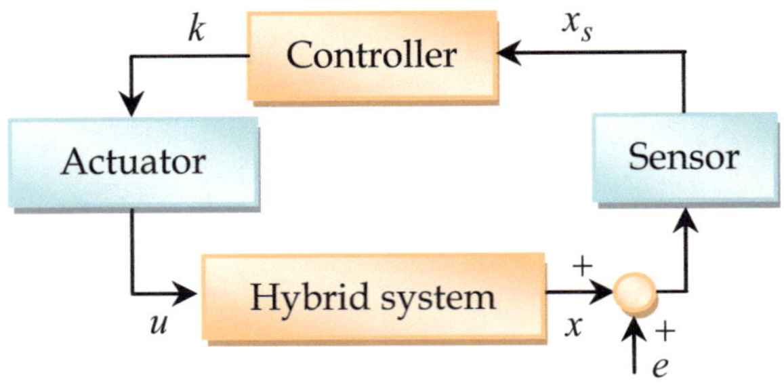

Fig. 13. Closed-loop system with sensor and actuator dynamics.

5.3 Robustness to sensor dynamics and smoothing

In many hybrid control applications, the state of the controller is explicitly given as a continuous state ξ and a discrete state $q \in Q := \{1,...,n\}$, that is, $x_c := [\xi \; q]^T$. Where this is the case and the discrete state q chooses a different control law to be applied to the system for for various values of q, then the control law generated by the hybrid controller H_c can have jumps when q changes. In many scenarios, it is not possible for the actuator to switch between control laws instantly. In addition, particularly when the control law $\kappa(\cdot,\cdot,q)$ is continuous for each $q \in Q$, it is desired to have a smooth transition between them when q changes.

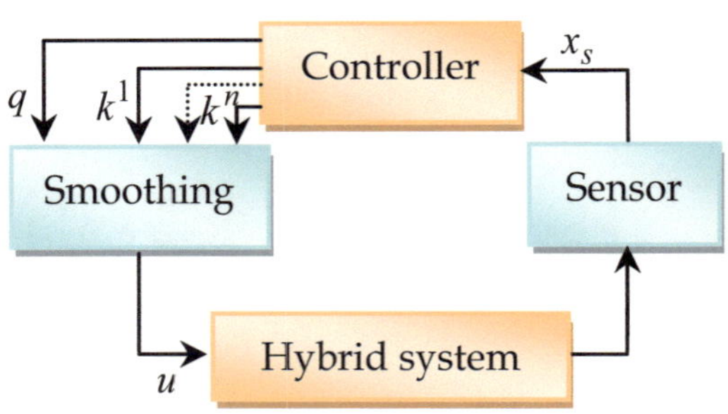

Fig. 14. Closed-loop system with sensor dynamics and control smoothing.

(Fig. 14) shows the closed-loop system, noted that $H_{cl}^{\varepsilon_d}$, resulting from adding a block that makes the smooth transition between control laws indexed by q and indicated by κ^q. The smoothing control block is modeled as a linear filter for the variable q. It is defined by the parameter ε_u and the matrices (A_u, B_u, L_u).
The output of the control smoothing block is given by

$$\alpha(x, x_c, L_u x_u) = \sum_{q \in Q} \lambda_q(L_u x_u)\kappa(x, x_c, q) \tag{23}$$

where for each $q \in Q, \lambda_q : R \to [0,1]$, is continuous and $\lambda_q(q) = 1$. Note that the output is such that the control laws are smoothly "blended" by the function λ_q.
In addition to this block, a filter modeling the sensor dynamics is also incorporated as in section 5.2. The closed loop $H_{cl}^{\varepsilon_f}$ can be written as

$$H_{cl}^{\varepsilon_f}: \begin{cases} \left.\begin{array}{l} \dot{x} = f_p(x + \alpha(x, x_c, L_u x_u)) \\ \dot{x}_c = f_c(L_s x_s, x_c) \\ \dot{q} = 0 \\ \dot{\tau} = -\tau + \bar{\tau}^* \\ \varepsilon_u \dot{x}_s = A_s x_s + B_s(x) \\ \varepsilon_u \dot{x}_u = A_u x_u + B_u q \end{array}\right\} (L_s x_s, x_c) \in C_c \text{ or } \tau \le \bar{\tau} \\ \left.\begin{array}{l} x^+ = x \\ \begin{bmatrix} \xi^+ \\ q^+ \end{bmatrix}^+ \in G_c(L_s x_s, x_c) \\ x_s^+ = x_s \\ x_u^+ = x_u \\ \tau^+ = 0 \end{array}\right\} (L_s x_s, x_c) \in D_c \text{ and } \tau \ge \bar{\tau} \end{cases} \quad (24)$$

6. Conclusion

In this chapter, a dynamic systems approach to analysis and design of hybrid systems has been continued from a robust control point of view. Stability and convergence tools for hybrid systems presented include hybrid versions of the traditional Lyapunov stability theorem and of LaSalle's invariance principle.

The robustness of asymptotic stability for classes of closed-loop systems resulting from hybrid control was presented. Results for perturbations arising from the presence of measurement noise, unmodeled sensor and actuator dynamics, control smoothing.

It is very important to have good software tools for the simulation, analysis and design of hybrid systems, which by their nature are complex systems. Researchers have recognized this need and several software packages have been developed.

7. References

Rajeev, A.; Thomas, A. & Pei-Hsin, H.(1993). *Automatic symbolic verification of embedded systems*, In IEEE Real-Time Systems Symposium, 2-11, DOI: 10.1109/REAL.1993.393520 .

Dang, T. (2000). *Vérification et Synthèse des Systèmes Hybrides*. PhD thesis, Institut National Polytechnique de Grenoble.

Girard, A. (2006). *Analyse algorithmique des systèmes hybrides*. PhD thesis, Universitè Joseph Fourier (Grenoble-I).

Ancona, F. & Bressan, A. (1999). *Patchy vector fields and asymptotic stabilization*, ESAIM: Control, Optimisation and Calculus of Variations, 4:445–471, DOI: 10.1051/cocv:2004003.

Byrnes, C. I. & Martin, C. F. (1995). *An integral-invariance principle for nonlinear systems*, IEEE Transactions on Automatic Control, 983–994, ISSN: 0018-9286.

Cai, C.; Teel, A. R. & Goebel, R. (2007). *Results on existence of smooth Lyapunov functions for asymptotically stable hybrid systems with nonopen basin of attraction*, submitted to the 2007 American Control Conference, 3456 - 3461, ISSN: 0743-1619.

Cai, C.; Teel, A. R. & Goebel, R. (2006). *Smooth Lyapunov functions for hybrid systems* Part I: *Existence is equivalent to robustness* & Part II: *(Pre-)asymptotically stable compact sets*, 1264-1277, ISSN 0018-9286.

Cai, C.; Teel, A. R. & Goebel, R. (2005). *Converse Lyapunov theorems and robust asymptotic stability for hybrid systems*, Proceedings of 24th American Control Conference, 12–17, ISSN: 0743-1619.

Chellaboina, V.; Bhat, S. P. & HaddadWH. (2002). *An invariance principle for nonlinear hybrid and impulsive dynamical systems.* Nonlinear Analysis, Chicago, IL, USA, 3116 - 3122,ISBN: 0-7803-5519-9.

Goebel, R.; Prieur, C. & Teel, A. R. (2009). *smooth patchy control Lyapunov functions.* Automatica (Oxford) Y, 675-683 ISSN : 0005-1098.

Goebel, R. & Teel, A. R. (2006). *Solutions to hybrid inclusions via set and graphical convergence with stability theory applications.* Automatica, 573–587, DOI: 10.1016/j.automatica.2005.12.019.

LaSalle, J. P. (1967). *An invariance principle in the theory of stability, in Differential equations and dynamical systems.* Academic Press, New York.

LaSalle, J. P. (1976) *The stability of dynamical systems.* Regional Conference Series in Applied Mathematics, SIAM ISBN-13: 978-0-898710-22-9.

Lygeros, J.; Johansson, K. H., Simi´c, S. N.; Zhang, J. & Sastry, S. S. (2003). *Dynamical properties of hybrid automata.* IEEE Transactions on Automatic Control, 2–17 ,ISSN: 0018-9286.

Prieur, C. (2001). *Uniting local and global controllers with robustness to vanishing noise,* Mathematics Control, Signals, and Systems, 143–172, DOI: 10.1007/PL00009880

Ryan, E. P. (1998). *An integral invariance principle for differential inclusions with applications in adaptive control.* SIAM Journal on Control and Optimization, 960–980, ISSN 0363-0129.

Sanfelice, R. G.; Goebel, R. & Teel, A. R. (2005). *Results on convergence in hybrid systems via detectability and an invariance principle.* Proceedings of 2005 American Control Conference, 551–556, ISSN: 0743-1619.

Sontag, E. (1989). *Smooth stabilization implies coprime factorization.* IEEE Transactions on Automatic Control, 435–443, ISSN: 0018-9286.

DeCarlo, R.A.; Branicky, M.S.; Pettersson, S. & Lennartson, B.(2000). *Perspectives and results on the stability and stabilizability of hybrid systems.* Proc. of IEEE, 1069–1082, ISSN: 0018-9219.

Michel, A.N.(1999). *Recent trends in the stability analysis of hybrid dynamical systems.* IEEE Trans. Circuits Syst. - I. Fund. Theory Appl., 120–134,ISSN: 1057-7122.

Halbaoui, K.; Boukhetala, D. and Boudjema, F.(2008). *New robust model reference adaptive control for induction motor drives using a hybrid controller.*International Symposium on Power Electronics, Electrical Drives, Automation and Motion, Italy, 1109 - 1113 ISBN: 978-1-4244-1663-9.

Halbaoui, K.; Boukhetala, D. and Boudjema, F.(2009a). *Speed Control of Induction Motor Drives Using a New Robust Hybrid Model Reference Adaptive Controller.* Journal of Applied Sciences, 2753-2761, ISSN:18125654.

Halbaoui, K.; Boukhetala, D. and Boudjema, F.(2009b). *Hybrid adaptive control for speed regulation of an induction motor drive,* Archives of Control Sciences,V2.

3

Robust Stability and Control of Linear Interval Parameter Systems Using Quantitative (State Space) and Qualitative (Ecological) Perspectives

Rama K. Yedavalli and Nagini Devarakonda
The Ohio State University
United States of America

1. Introduction

The problem of maintaining the stability of a nominally stable linear time invariant system subject to linear perturbation has been an active topic of research for quite some time. The recent published literature on this `robust stability' problem can be viewed mainly from two perspectives, namely i) transfer function (input/output) viewpoint and ii) state space viewpoint. In the transfer function approach, the analysis and synthesis is essentially carried out in frequency domain, whereas in the state space approach it is basically carried out in time domain. Another perspective that is especially germane to this viewpoint is that the frequency domain treatment involves the extensive use of `polynomial' theory while that of time domain involves the use of 'matrix' theory. Recent advances in this field are surveyed in [1]-[2].

Even though in typical control problems, these two theories are intimately related and qualitatively similar, it is also important to keep in mind that there are noteworthy differences between these two approaches ('polynomial' vs 'matrix') and this chapter (both in parts I and II) highlights the use of the direct matrix approach in the solution to the robust stability and control design problems.

2. Uncertainty characterization and robustness

It was shown in [3] that modeling errors can be broadly categorized as i) parameter variations, ii) unmodeled dynamics iii) neglected nonlinearities and finally iv) external disturbances. Characterization of these modeling errors in turn depends on the representation of dynamic system, namely whether it is a frequency domain, transfer function framework or time domain state space framework. In fact, some of these can be better captured in one framework than in another. For example, it can be argued convincingly that real parameter variations are better captured in time domain state space framework than in frequency domain transfer function framework. Similarly, it is intuitively clear that unmodeled dynamics errors can be better captured in the transfer function framework. By similar lines of thought, it can be safely agreed that while neglected nonlinearities can be better captured in state space framework, neglected disturbances can be captured with equal ease in both frameworks. Thus it is not surprising that most of the robustness studies of uncertain dynamical systems with real parameter variations are being carried out in time domain state space framework and hence in this chapter, we emphasize the aspect of robust stabilization and control of linear dynamical systems with real parameter uncertainty.

Stability and performance are two fundamental characteristics of any feedback control system. Accordingly, stability robustness and performance robustness are two desirable (sometimes necessary) features of a robust control system. Since stability robustness is a prerequisite for performance robustness, it is natural to address the issue of stability robustness first and then the issue of performance robustness.

Since stability tests are different for time varying systems and time invariant systems, it is important to pay special attention to the nature of perturbations, namely time varying perturbations versus time invariant perturbations, where it is assumed that the nominal system is a linear time invariant system. Typically, stability of linear time varying systems is assessed using Lyapunov stability theory using the concept of quadratic stability whereas that of a linear time invariant system is determined by the Hurwitz stability, i.e. by the negative real part eigenvalue criterion. This distinction about the nature of perturbation profoundly affects the methodologies used for stability robustness analysis.

Let us consider the following linear, homogeneous, time invariant asymptotically stable system in state space form subject to a linear perturbation E:

$$\dot{x} = (A_0 + E)x \qquad x(0) = x_0 \tag{1}$$

where A_0 is an n×n asymptotically stable matrix and E is the error (or perturbation) matrix. The two aspects of characterization of the perturbation matrix E which have significant influence on the scope and methodology of any proposed analysis and design scheme are i) the temporal nature and ii) the boundedness nature of E. Specifically, we can have the following scenario:

i. Temporal Nature:

Time invariant error E = constant	vs	Time varying error $E = E(t)$

ii. Boundedness Nature:

Unstructured (Norm bounded)	vs	Structured (Elemental bounds)

The stability robustness problem for linear time invariant systems in the presence of linear time invariant perturbations (i.e. robust Hurwitz invariance problem) is basically addressed by testing for the negativity of the real parts of the eigenvalues (either in frequency domain or in time domain treatments), whereas the time varying perturbation case is known to be best handled by the time domain Lyapunov stability analysis. The robust Hurwitz invariance problem has been widely discussed in the literature essentially using the polynomial approach [4]-[5]. In this section, we address the time varying perturbation case, mainly motivated by the fact that any methodology which treats the time varying case can always be specialized to the time invariant case but not vice versa. However, we pay a price for the same, namely conservatism associated with the results when applied to the time invariant perturbation case. A methodology specifically tailored to time invariant perturbations is discussed and included by the author in a separate publication [6].

It is also appropriate to discuss, at this point, the characterization with regard to the boundedness of the perturbation. In the so called 'unstructured' perturbation, it is assumed that one cannot clearly identify the location of the perturbation within the nominal matrix and thus one has simply a bound on the norm of the perturbation matrix. In the 'structured' perturbation, one has information about the location(s) of the perturbation and thus one can think of having bounds on the individual elements of the perturbation matrix. This approach can be labeled as 'Elemental Perturbation Bound Analysis (EPBA)'. Whether 'unstructured' norm bounded perturbation or 'structured' elemental perturbation is appropriate to consider depends very much on the application at hand. However, it can be safely argued that 'structured' real parameter perturbation situation has extensive applications in many engineering disciplines as the elements of the matrices of a linear state space description contain parameters of interest in the evolution of the state variables and it is natural to look for bounds on these real parameters that can maintain the stability of the state space system.

3. Robust stability and control of linear interval parameter systems under state space framework

In this section, we first give a brief account of the robust stability analysis techniques in 3.1 and then in subsection 3.2 we discuss the robust control design aspect.

3.1 Robust stability analysis

The starting point for the problem at hand is to consider a linear state space system described by

$$\dot{x}(t) = [A_0 + E]x(t)$$

where x is an n dimensional state vector, asymptotically stable matrix and E is the 'perturbation' matrix. The issue of 'stability robustness measures' involves the determination of bounds on E which guarantee the preservation of stability of (1). Evidently, the characterization of the perturbation matrix E has considerable influence on the derived result. In what follows, we summarize a few of the available results, based on the characterization of E.

1. **Time varying (real) unstructured perturbation with spectral norm: Sufficient bound**

For this case, the perturbation matrix E is allowed to be time varying, i.e. $E(t)$ and a bound on the spectral norm ($\sigma_{\max}(E(t))$ where $\sigma(\cdot)$ is the singular value of $(\cdot)$) is derived. When a bound on the norm of E is given, we refer to it as 'unstructured' perturbation. This norm produces a spherical region in parameter space. The following result is available for this case [7]-[8]:

$$\sigma_{\max}(E(t)) < \frac{1}{\sigma_{\max}(P)} \tag{2}$$

where P is the solution to the Lyapunov matrix

$$PA_0 + A_0^T P + 2I = 0 \tag{3}$$

See Refs [9],[10],[11] for results related to this case.

2. Time varying (real) structured variation

Case 1: Independent variations (sufficient bound) [12]-[13]

$$E_{ij}(t) \leq \forall_t \left|E_{ij}(t)\right|_{\max} = \varepsilon_{ij} \tag{4}$$

$$\varepsilon = Max_{ij}\varepsilon_{ij}$$

$$\varepsilon_{ij} < \frac{1}{\sigma_{\max}\left(P_m U_e\right)_s} U_{eij} \tag{5}$$

where P satisfies equation (3) and $U_{oij} = \varepsilon_{ij} / \varepsilon$. For cases when ε_{ij} are not known, one can take $U_{eij} = |A_{oij}| / |A_{oij}|_{max}$. $(\cdot)_m$ denotes the matrix with all modulus elements and $(\cdot)_s$ denotes the symmetric part of $(\cdot)$.

3. Time invariant, (real) structured perturbation E_{ij} = Constant

Case i: Independent Variations [13]-[15]: (Sufficient Bounds). For this case, E can be characterized as

$$E = S_1 D S_2 \tag{6}$$

where S_1 and S_2 are constant, known matrices and $|D_{ij}| \leq d_{ij}d$ with $d_{ij} \geq 0$ are given and $d > 0$ is the unknown. Let U be the matrix elements $U_{ij} = d_{ij}$. Then the bound on d is given by [13]

$$d < \frac{1}{\underset{\omega>0}{Sup}\, \rho\left(\left[S_2\left(j\omega I - A_0\right)^{-1} S_1\right]_m U\right)} = \mu_J = \mu_Q \tag{7}$$

Notice that the characterization of E (with time invariant) in (4) is accommodated by the characterization in [15]. $\rho(\cdot)$ is the spectral radius of $(\cdot)$.

Case ii: Linear Dependent Variation: For this case, E is characterized (as in (6) before), by

$$E = \sum_{i=1}^{r} \beta_i E_i \tag{8}$$

and bounds on $|\beta_i|$ are sought. Improved bounds on $|\beta_i|$ are presented in [6].

This type of representation represents a 'polytope of matrices' as discussed in [4]. In this notation, the interval matrix case (i.e. the independent variation case) is a special case of the above representation where Ei contains a single nonzero element, at a different place in the matrix for different i.

For the time invariant, real structured perturbation case, there are no computationally tractable necessary and sufficient bounds either for polytope of matrices or for interval matrices (even for a 2 x 2 case). Even though some derivable necessary and sufficient conditions are presented in [16] for any general variation in E (not necessarily linear dependent and independent case), there are no easily computable methods available to determine the necessary and sufficient bounds at this stage of research. So most of the research, at this point of time, seems to aim at getting better (less conservative) sufficient bounds. The following example compares the sufficient bounds given in [13]-[15] for the linear dependent variation case.

Let us consider the example given in [15] in which the perturbed system matrix is given by

$$(A_0 + BKC) = \begin{bmatrix} -2+k_1 & 0 & -1+k_1 \\ 0 & -3+k_2 & 0 \\ -1+k_1 & -1+k_2 & -4+k_1 \end{bmatrix}$$

Taking the nominally stable matrix to be

$$A_0 = \begin{bmatrix} -2 & 0 & -1 \\ 0 & -3 & 0 \\ -1 & -1 & -4 \end{bmatrix}$$

the error matrix with k_1 and k_2 as the uncertain parameters is given by

$$E = k_1 E_1 + k_2 E_2$$

where

$$E_1 = \begin{bmatrix} 1 & 0 & 1 \\ 0 & 0 & 0 \\ 1 & 0 & 1 \end{bmatrix} \text{ and } E_2 = \begin{bmatrix} 0 & 0 & 0 \\ 0 & 1 & 0 \\ 0 & 1 & 0 \end{bmatrix}$$

The following are the bounds on $|k_1|$ and $|k_2|$ obtained by [15] and the proposed method.

μ_y	μ_Q	ZK [14]	μ_d [6]
0.815	0.875	1.55	1.75

3.2 Robust control design for linear systems with structured uncertainty

Having discussed the robustness analysis issue above, we now switch our attention to the robust control design issue. Towards this direction, we now present a linear robust control design algorithm for linear deterministic uncertain systems whose parameters vary within given bounded sets. The algorithm explicitly incorporates the structure of the uncertainty into the design procedure and utilizes the elemental perturbation bounds developed above. A linear state feedback controller is designed by parameter optimization techniques to maximize (in a given sense) the elemental perturbation bounds for robust stabilization.

There is a considerable amount of literature on the aspect of designing linear controllers for linear tine invariant systems with small parameter uncertainty. However, for uncertain systems whose dynamics are described by interval matrices (i.e., matrices whose elements are known to vary within a given bounded interval), linear control design schemes that guarantee stability have been relatively scarce. Reference [17] compares several techniques for designing linear controllers for robust stability for a class of uncertain linear systems. Among the methods considered are the standard linear quadratic regulator (LQR) design, Guaranteed Cost Control (GCC) method of [18], Multistep Guaranteed Cost Control (MGCC) of [17]. In these methods, the weighting on state in a quadratic cost function and the Riccati equation are modified in the search for an appropriate controller. Also, the parameter uncertainty is assumed to enter linearly and restrictive conditions are imposed on the bounding sets. In [18], norm inequalities on the bounding sets are given for stability but they are conservative since they do not take advantage of the system structure. There is no guarantee that a linear state feedback controller exists. Reference [19] utilizes the concept of 'Matching conditions (MC)' which in essence constrain the manner in which the uncertainty is permitted to enter into the dynamics and show that a linear state feedback control that guarantees stability exists provided the uncertainty satisfies matching conditions. By this method large bounding sets produce large feedback gains but the existence of a linear controller is guaranteed. But no such guarantee can be given for general 'mismatched' uncertain systems. References [20] and [21] present methods which need the testing of definiteness of a Lyapunov matrix obtained as a function of the uncertain parameters. In the multimodel theory approach, [22] considers a discrete set of points in the parameter uncertainty range to establish the stability. This paper addresses the stabilization problem for a continuous range of parameters in the uncertain parameter set (i.e. in the context of interval matrices). The proposed approach attacks the stability of interval matrix problem directly in the matrix domain rather than converting the interval matrix to interval polynomials and then testing the Kharitonov polynomials.

Robust control design using perturbation bound analysis [23],[24]

Consider a linear, time invariant system described by

$$\dot{x} = Ax + Bu \qquad x(0) = x_0$$

Where x is $n \times 1$ state vector, the control u is $m \times 1$. The matrix pair (A, B) is assumed to be completely controllable.

U=Gx

For this case, the nominal closed loop system matrix is given by

$$\bar{A} = A + BG\,,\; G = {-R_0^{-1}B^T K}\big/{\rho_c}$$

and

$$KA + A^T K - KB\frac{R_0^{-1}}{\rho_c}B^T K + Q = 0$$

and $\bar{A}$ is asymptotically stable.

Here *G* is the Riccati based control gain where *Q*,and R_0 are any given weighting matrices which are symmetric, positive definite and ρ_c is the design variable.

The main interest in determining *G* is to keep the nominal closed loop system stable. The reason Riccati approach is used to determine *G* is that it readily renders (*A*+*BG*) asymptotically stable with the above assumption on *Q* and R_0.

Now consider the perturbed system with linear time varying perturbations $E_A(t)$ and $E_B(t)$ respectively in matrices *A* and *B*

i.e., $\dot{x} = [A + E_A(t)]x(t) + [B + E_B(t)]u(t)$

Let ΔA and ΔB be the perturbation matrices formed by the maximum modulus deviations expected in the individual elements of matrices *A* and *B* respectively. Then one can write

$$\begin{aligned} \Delta A &= \varepsilon_a U_{ea} \\ \Delta B &= \varepsilon_b U_{eb} \end{aligned} \quad \text{(Absolute variation)}$$

where ε_a is the maximum of all the elements in ΔA and ε_b is the maximum of all elements in ΔB. Then the total perturbation in the linear closed loop system matrix of (10) with nominal control $u = Gx$ is given by

$$\Delta = \Delta A + \Delta BG_m = \varepsilon_a U_{ea} + \varepsilon_b U_{eb} G_m$$

Assuming the ratio is $\varepsilon_b / \varepsilon_a = \bar{\varepsilon}$ known, we can extend the main result of equation (3) to the linear state feedback control system of (9) and (10) and obtain the following design observation.

Design observation 1:

The perturbed linear system is stable for all perturbations bounded by ε_a and ε_b if

$$\varepsilon_a < \frac{1}{\sigma_{\max}\left[P_m\left(U_{ea} + \bar{\varepsilon} U_{eb} G_m\right)\right]_s} \equiv \mu \tag{9}$$

and $\varepsilon_b < \bar{\varepsilon}\,\mu$ where

$$P(A+BG) + (A+BG)^T P + 2I_n = 0$$

Remark: If we suppose ΔA = 0, ΔB = 0 and expect some control gain perturbations ΔG, where we can write

$$\Delta G = \varepsilon_g U e_g \tag{10}$$

then stability is assured if

$$\varepsilon_g < \frac{1}{\sigma_{\max}\left(P_m B_m U_{eg}\right)_s} \equiv \mu_g \tag{11}$$

In this context μg can be regarded as a "gain margin".

For a given ε_{aij} and ε_{bij}, one method of designing the linear controller would be to determine G of (3.10) by varying ρ_c of (3.10) such that μ is maximum. For an aircraft control example which utilizes this method, see Reference [9].

4. Robust stability and control of linear interval parameter systems using ecological perspective

It is well recognized that natural systems such as ecological and biological systems are highly robust under various perturbations. On the other hand, engineered systems can be made highly optimal for good performance but they tend to be non-robust under perturbations. Thus, it is natural and essential for engineers to delve into the question of as to what the underlying features of natural systems are, which make them so robust and then try to apply these principles to make the engineered systems more robust. Towards this objective, the interesting aspect of qualitative stability in ecological systems is considered in particular. The fields of population biology and ecology deal with the analysis of growth and decline of populations in nature and the struggle of species to predominate over one another. The existence or extinction of a species, apart from its own effect, depends on its interactions with various other species in the ecosystem it belongs to. Hence the type of interaction is very critical to the sustenance of species. In the following sections these interactions and their nature are thoroughly investigated and the effect of these qualitative interactions on the quantitative properties of matrices, specifically on three matrix properties, namely, eigenvalue distribution, normality/condition number and robust stability are presented. This type of study is important for researchers in both fields since qualitative properties do have significant impact on the quantitative aspects. In the following sections, this interrelationship is established in a sound mathematical framework. In addition, these properties are exploited in the design of controllers for engineering systems to make them more robust to uncertainties such as described in the previous sections.

4.1 Robust stability analysis using principles of ecology

4.1.1 Brief review of ecological principles

In this section a few ecological system principles that are of relevance to this chapter are briefly reviewed. Thorough understanding of these principles is essential to appreciate their influence on various mathematical results presented in the rest of the chapter.

In a complex community composed of many species, numerous interactions take place. These interactions in ecosystems can be broadly classified as i) Mutualism, ii) Competition, iii) Commensalism/Ammensalism and iv) Predation (Parasitism). Mutualism occurs when both species benefit from the interaction. When one species benefits/suffers and the other one remains unaffected, the interaction is classified as Commensalism/Ammensalism. When species compete with each other, that interaction is known as Competition. Finally, if one species is benefited and the other suffers, the interaction is known as Predation (Parasitism). In ecology, the magnitudes of the mutual effects of species on each other are seldom precisely known, but one can establish with certainty, the types of interactions that are present. Many mathematical population models were proposed over the last few decades to study the dynamics of eco/bio systems, which are discussed in textbooks [25]-[26]. The most significant contributions in this area come from the works of Lotka and Volterra. The following is a model of a predator-prey interaction where x is the prey and y is the predator.

$$\begin{aligned} \dot{x} &= xf(x,y) \\ \dot{y} &= yg(x,y) \end{aligned} \tag{12}$$

where it is assumed that $\partial f(x,y) / \partial y < 0$ and $\partial g(x,y) / \partial x > 0$

This means that the effect of y on the rate of change of x ($\dot{x}$) is negative while the effect of x on the rate of change of y ($\dot{y}$) is positive.

The stability of the equilibrium solutions of these models has been a subject of intense study in life sciences [27]. These models and the stability of such systems give deep insight into the balance in nature. If a state of equilibrium can be determined for an ecosystem, it becomes inevitable to study the effect of perturbation of any kind in the population of the species on the equilibrium. These small perturbations from equilibrium can be modeled as linear state space systems where the state space plant matrix is the 'Jacobian'. This means that technically in the Jacobian matrix, one does not know the actual magnitudes of the partial derivatives but their signs are known with certainty. That is, the *nature* of the interaction is known but not the strengths of those interactions. As mentioned previously, there are four classes of interactions and after linearization they can be represented in the following manner.

Interaction type	Digraph representation	Matrix representation
Mutualism	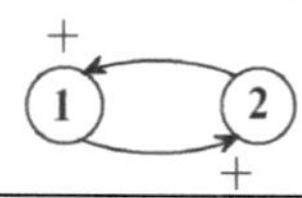	$\begin{bmatrix} * & + \\ + & * \end{bmatrix}$
Competition	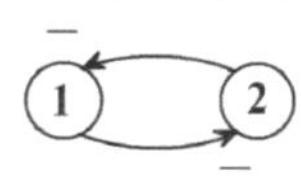	$\begin{bmatrix} * & - \\ - & * \end{bmatrix}$
Commensalism	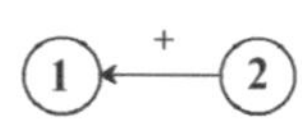	$\begin{bmatrix} * & + \\ 0 & * \end{bmatrix}$
Ammensalism	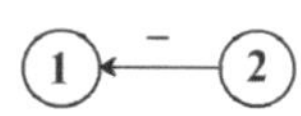	$\begin{bmatrix} * & - \\ 0 & * \end{bmatrix}$
Predation (Parasitism)	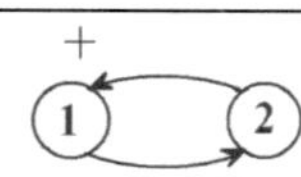	$\begin{bmatrix} * & + \\ - & * \end{bmatrix}$

Table 1. Types of interactions between two species in an ecosystem

In Table 1, column 2 is a visual representation of such interactions and is known as a directed graph or 'digraph' [28] while column 3 is the matrix representation of the interaction between two species. '*' represents the effect of a species on itself.
In other words, in the Jacobian matrix, the 'qualitative' information about the species is represented by the signs +, - or 0. Thus, the $(i,j)^{th}$ entry of the state space (Jacobian) matrix simply consists of signs +, –, or 0, with the + sign indicating species j having a positive influence on species i, - sign indicating negative influence and 0 indicating no influence. The diagonal elements give information regarding the effect of a species on itself. Negative sign means the species is 'self-regulatory', positive means it aids the growth of its own population and zero means that it has no effect on itself. For example, in the Figure 1 below, sign pattern matrices A_1 and A_2 are the Jacobian form while D_1 and D_2 are their corresponding digraphs.

a) $$A_1 = \begin{bmatrix} 0 & + & + \\ - & 0 & 0 \\ - & 0 & - \end{bmatrix} \longrightarrow D_1$$

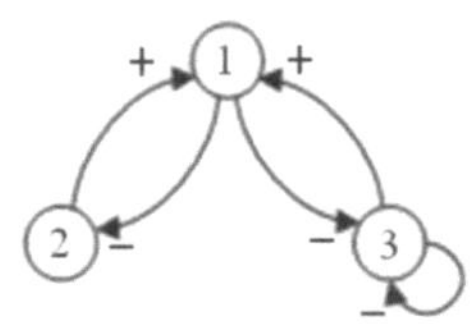

b) $$A_2 = \begin{bmatrix} 0 & - & 0 & 0 & 0 \\ + & 0 & - & 0 & 0 \\ 0 & + & - & - & 0 \\ 0 & 0 & + & 0 & - \\ 0 & 0 & 0 & + & 0 \end{bmatrix} \longrightarrow D_2$$

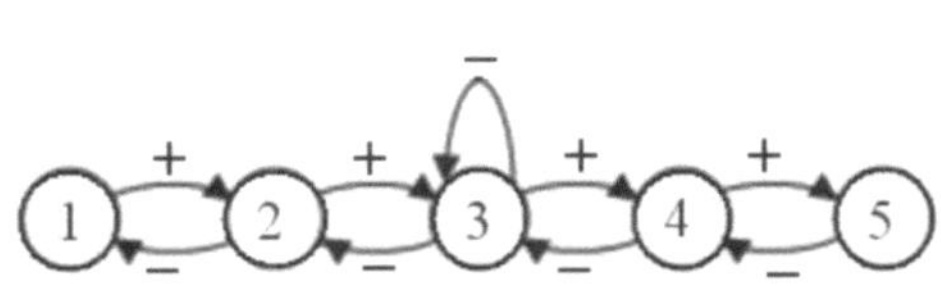

Fig. 1. Various sign patterns and their corresponding digraphs representing ecological systems; a) three species system b) five species system

4.1.2 Qualitative or sign stability

Since traditional mathematical tests for stability fail to analyze the stability of such ecological models, an extremely important question then, is whether it can be concluded, just from this sign pattern, whether the system is stable or not. If so, the system is said to be 'qualitatively stable' [29-31]. In some literature, this concept is also labeled as 'sign stability'. In what follows, these two terms are used interchangeably. It is important to keep in mind that the systems (matrices) that are qualitatively (sign stable) stable are also stable in the ordinary sense. That is, qualitative stability implies Hurwitz stability (eigenvalues with negative real part) in the ordinary sense of engineering sciences. *In other words, once a particular sign matrix is shown to be qualitatively (sign) stable, any magnitude can be inserted in those entries and for all those magnitudes the matrix is automatically Hurwitz stable.* This is the most attractive feature of a sign stable matrix. However, the converse is not true. Systems that are not qualitatively stable can still be stable in the ordinary sense for certain appropriate magnitudes in the entries. From now on, to distinguish from the concept of 'qualitative stability' of life sciences literature, the label of 'quantitative stability' for the standard Hurwitz stability in engineering sciences is used.

These conditions in matrix theory notation are given below

i. $a_{ii} \le 0 \ \forall\, i$

ii. and $a_{ii} < 0$ for at least one i

iii. $a_{ij}a_{ji} \le 0 \ \forall\, i,j \quad i \ne j$

iv. $a_{ij}a_{jk}a_{kl}...a_{mi} = 0$ for any sequence of three or more distinct indices i,*j*,*k*,…*m*.

v. $\mathrm{Det}(A) \ne 0$

vi. Color test (Elaborated in [32],[33])

Note: In graph theory $a_{ij}a_{ji}$ are referred to as *l*-cycles and $a_{ij}a_{jk}a_{kl}...a_{mi}$ are referred to as *k*-cycles. In [34], [35], *l*-cycles are termed 'interactions' while *k*-cycles are termed 'interconnections' (which essentially are all zero in the case of sign stable matrices).

With this algorithm, all matrices that are sign stable can be stored apriori as discussed in [36]. If a sign pattern in a given matrix satisfies the conditions given in the above papers (thus in the algorithm), it is an ecological stable sign pattern and hence that matrix is Hurwitz stable for any magnitudes in its entries. A subtle distinction between 'sign stable' matrices and 'ecological sign stable' matrices is now made, emphasizing the role of nature of interactions. Though the property of Hurwitz stability is held in both cases, ecosystems sustain solely because of interactions between various species. In matrix notation this means that the nature of off-diagonal elements is essential for an ecosystem. Consider a strictly upper triangular 3×3 matrix

$$A = \begin{bmatrix} - & 0 & + \\ 0 & - & 0 \\ 0 & 0 & - \end{bmatrix}$$

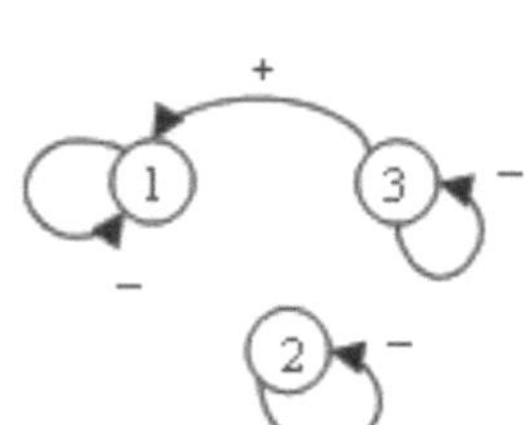

From quantitative viewpoint, it is seen that the matrix is Hurwitz stable for any magnitudes in the entries of the matrix. This means that it is indeed (qualitatively) sign stable. But since there is no predator-prey link and in fact no link at all between species 1&2 and 3&2, *such a*

digraph cannot represent an ecosystem. Therefore, though a matrix is sign stable, it need not belong to the class of ecological sign stable matrices. In Figure 2 below, these various classes of sign patterns and the corresponding relationship between these classes is depicted. So, every ecological sign stable sign pattern is sign stable but the converse is not true.

With this brief review of ecological system principles, the implications of these ecological qualitative principles on three quantitative matrix theory properties, namely eigenvalues, normality/condition number and robust stability are investigated. In particular, in the next section, new results that clearly establish these implications are presented. As mentioned in the previous section, the motivation for this study and analysis is to exploit some of these desirable features of ecological system principles to design controllers for engineering systems to make them more robust.

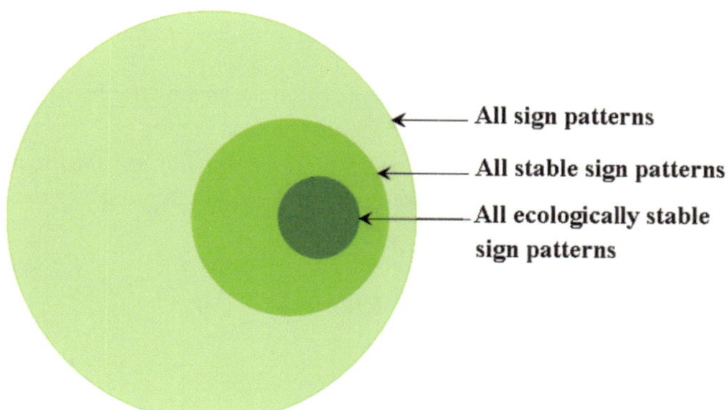

Fig. 2. Classification of sign patterns

4.2 Ecological sign stability and its implications in quantitative matrix theory

In this major section of this chapter, focusing on the ecological sign stability aspect discussed above, its implications in the quantitative matrix theory are established. In particular, the section offers three explicit contributions to expand the current knowledge base, namely i) Eigenvalue distribution of ecological sign stable matrices ii) Normality/Condition number properties of sign stable matrices and iii) Robustness properties of sign stable matrices. These three contributions in turn help in determining the role of magnitudes in quantitative ecological sign stable matrices. This type of information is clearly helpful in designing robust controllers as shown in later sections. With this motivation, a 3-species ecosystem is thoroughly analyzed and the ecological principles in terms of matrix properties that are of interest in engineering systems are interpreted. This section is organized as follows: First, new results on the eigenvalue distribution of ecological sign stable matrices are presented. Then considering ecological systems with only predation-prey type interactions, it is shown how selection of appropriate magnitudes in these interactions imparts the property of normality (and thus highly desirable condition numbers) in matrices. In what follows, for each of these cases, concepts are first discussed from an ecological perspective and then later the resulting matrix theory implications from a quantitative perspective are presented

Stability and eigenvalue distribution

Stability is the most fundamental property of interest to all dynamic systems. Clearly, in time invariant matrix theory, stability of matrices is governed by the negative real part

nature of its eigenvalues. It is always useful to get bounds on the eigenvalue distribution of a matrix with as little computation as possible, hopefully as directly as possible from the elements of that matrix. It turns out that sign stable matrices have interesting eigenvalue distribution bounds. A few new results are now presented in this aspect.
In what follows, the quantitative matrix theory properties for an n-species ecological system is established, i.e., an n×n sign stable matrix with predator-prey and commensal/ammensal interactions is considered and its eigenvalue distribution is analyzed. In particular, various cases of diagonal elements' nature, which are shown to possess some interesting eigenvalue distribution properties, are considered.

Bounds on real part of eigenvalues

Based on several observations the following theorem for eigenvalue distribution along the real axis is stated.

Theorem 1 [37]

(Case of all negative diagonal elements):

For all n×n sign stable matrices, with all negative diagonal elements, the bounds on the real parts of the eigenvalues are given as follows:

The lower bound on the magnitude of the real part is given by the minimum magnitude diagonal element and the upper bound is given by the maximum magnitude diagonal element in the matrix.

That is, for an n×n ecological sign stable matrix $A=[a_{ij}]$,

$$\left|a_{ii}\right|_{\min} \le \left|\mathrm{Re}(\lambda)\right|_{\min} \le \left|\mathrm{Re}(\lambda)\right|_{\max} \le \left|a_{ii}\right|_{\max} \tag{13}$$

Corollary

(Case of some diagonal elements being zero):

If the ecological sign stable matrix has zeros on the diagonal, the bounds are given by

$$\left|a_{ii}\right|_{\min}(=0) < \left|\mathrm{Re}(\lambda)\right|_{\min} \le \left|\mathrm{Re}(\lambda)\right|_{\max} \le \left|a_{ii}\right|_{\max} \tag{14}$$

The sign pattern in Example 1 has all negative diagonal elements. In this example, the case discussed in the corollary where one of the diagonal elements is zero, is considered. This sign pattern is as shown in the matrix below.

$$A=\begin{bmatrix} - & - & - \\ 0 & - & 0 \\ + & + & 0 \end{bmatrix}$$

Bounds on imaginary part of eigenvalues [38]

Similarly, the following theorem can be stated for bounds on the imaginary parts of the eigenvalues of an n×n matrix. Before stating the theorem, we present the following lemma.
Theorem 2
For all n×n ecologically sign stable matrices, bound on the imaginary part of the eigenvalues is given by

$$\mu\big|_{imagss} = \left|\mathrm{Imag}(\lambda_i)\right|_{\max} = \sqrt{\sum_{i,j=1}^{n} -a_{ij}a_{ji}} \quad \forall\, i \ne j \tag{15}$$

Above results are illustrated in figure 3.

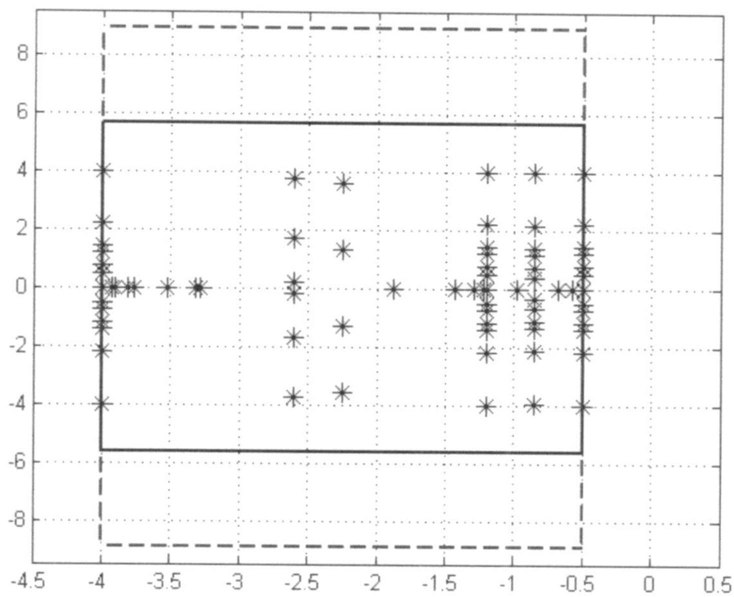

Fig. 3. Eigenvalue distribution for sign stable matrices

Theorem 3

For all n×n matrices, with all k-cycles being zero and with only commensal or ammensal interactions, the eigenvalues are simply the diagonal elements.

It is clear that these theorems offer significant insight into the eigenvalue distribution of n×n ecological sign stable matrices. Note that the bounds can be simply read off from the magnitudes of the elements of the matrices. This is quite in contrast to the general quantitative Hurwitz stable matrices where the lower and upper bounds on the eigenvalues of a matrix are given in terms of the singular values of the matrix and/or the eigenvalues of the symmetric part and skew-symmetric parts of the matrices (using the concept of field of values), which obviously require much computation, and are complicated functions of the elements of the matrices.

Now label the ecological sign stable matrices with magnitudes inserted in the elements as 'quantitative ecological sign stable matrices'. Note that these magnitudes can be arbitrary in each non zero entry of the matrix! It is interesting and important to realize that these bounds, based solely on sign stability, do not reflect diagonal dominance, which is the typical case with general Hurwitz stable matrices. Taking theorems 4, 5, 6 and their respective corollaries into consideration, we can say that it is the 'diagonal connectance' that is important in these quantitative ecological sign stable matrices and not the 'diagonal dominance' which is typical in the case of general Hurwitz stable matrices. This means that *interactions* are critical to system stability even in the case of general n×n matrices.

Now the effect on the quantitative property of normality is presented.

Normality and condition number

Based on this new insight on the eigenvalue distribution of sign stable matrices, other matrix theory properties of sign stable matrices are investigated. The first quantitative matrix theory property is that of *normality/condition number*. But this time, the focus is only on ecological sign stable matrices with pure predator-prey links with no other types of interactions.

A zero diagonal element implies that a species has no control over its growth/decay rate. So in order to regulate the population of such a species, it is essential that, in a sign stable ecosystem model, this species be connected to at least one predator-prey link. In the case where all diagonal elements are negative, the matrix represents an ecosystem with all self-regulating species. If every species has control over its regulation, a limiting case for stability is a system with no interspeciel interactions. This means that there need not be any predator-prey interactions. This is a trivial ecosystem and such matrices actually belong to the only 'sign-stable' set, not to ecological sign stable set.

Apart from the self-regulatory characteristics of species, the phenomena that contribute to the stability of a system are the type of interactions. Since a predator-prey interaction has a regulating effect on both the species, predator-prey interactions are of interest in this stability analysis. In order to study the role played by these interactions, henceforth focus is on systems with n-1 pure predator-prey links in specific places. This number of links and the specific location of the links are critical as they connect all species at the same time preserving the property of ecological sign stability. For a matrix A, pure predator-prey link structure implies that

1. $A_{ij}A_{ji} \leq 0 \;\; \forall \, i,j$
2. $A_{ij}A_{ji} = 0 \text{ iff } A_{ij} = A_{ji} = 0$

Hence, in what follows, matrices with all negative diagonal elements and with pure predator-prey links are considered.

Consider sign stable matrices with identical diagonal elements (negative) and pure predator-prey links of equal strengths.

Normality in turn implies that the modal matrix of the matrix is orthogonal resulting in it having a condition number of one, which is an extremely desirable property for all matrices occurring in engineering applications.

The property of normality is observed in higher order systems too. An ecologically sign stable matrix with purely predator-prey link interactions is represented by the following digraph for a 5-species system. The sign pattern matrix A represents this digraph.

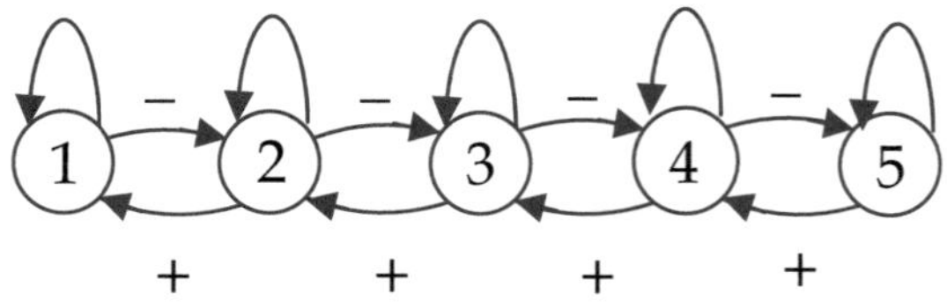

$$A = \begin{bmatrix} - & + & 0 & 0 & 0 \\ - & - & + & 0 & 0 \\ 0 & - & - & + & 0 \\ 0 & 0 & - & - & + \\ 0 & 0 & 0 & - & - \end{bmatrix}$$

Theorem 4

An nxn matrix A with equal diagonal elements and equal predation prey interaction strengths for each predation-prey link is a normal matrix.

The property of $\kappa \equiv 1$ is of great significance in the study of robustness of stable matrices. This significance will be explained in the next section eventually leading to a robust control design algorithm

Robustness

The third contribution of this section is related to the connection between ecological sign stability and robust stability in engineering systems.

As mentioned earlier, the most interesting feature of ecological sign stable matrices is that the stability property is independent of the magnitude information in the entries of the matrix. Thus the nature of interactions, which in turn decide the signs of the matrix entries and their locations in the matrix, are sufficient to establish the stability of the given sign matrix. Clearly, it is this independence (or non-dependence) from magnitude information that imparts the property of robust stability to engineering systems. This aspect of robust stability in engineering systems is elaborated next from quantitative matrix theory point of view.

Robustness as a result of independence from magnitude information

In mathematical sciences, the aspect of 'robust stability' of families of matrices has been an active topic of research for many decades. This aspect essentially arises in many applications of system and control theory. When the system is described by linear state space representation, the plant matrix elements typically depend on some uncertain parameters which vary within a given bounded interval.

Robust stability analysis of a class of interval matrices [39]:

Consider the 'interval matrix family' in which each individual element varies independently within a given interval. Thus the interval matrix family is denoted by

$A \in [A^L, A^U]$ as the set of all matrices A that satisfy

$$\left(A^L\right)_{ij} \leq A_{ij} \leq \left(A^U\right)_{ij} \quad for \ \ every \ \ i,j$$

Now, consider a special 'class of interval matrix family' in which *for each element that is varying, the lower bound i.e.* $(A^L)_{ij}$ *and the upper bound i.e.* $(A^U)_{ij}$ *are of the same sign.*

For example, consider the interval matrix given by

$$A = \begin{bmatrix} 0 & a_{12} & a_{13} \\ a_{21} & 0 & 0 \\ a_{31} & 0 & a_{33} \end{bmatrix} \qquad \begin{matrix} 2 \leq a_{12} \leq 5 \\ 1 \leq a_{13} \leq 4 \\ -3 \leq a_{21} \leq -1 \\ -4 \leq a_{31} \leq -2 \\ -5 \leq a_{33} \leq -0.5 \end{matrix}$$

with the elements a_{12}, a_{13}, a_{21}, a_{31} and a_{33} being uncertain varying in some given intervals as follows:

Qualitative stability as a 'sufficient condition' for robust stability of a class of interval matrices: A link between life sciences and engineering sciences

It is clear that ecological sign stable matrices have the interesting feature that once the sign pattern is a sign stable pattern, the stability of the matrix is independent of the magnitudes of the elements of the matrix. That this property has direct link to stability robustness of matrices with structured uncertainty was recognized in earlier papers on this topic [32] and [33]. In these papers, a viewpoint was put forth that advocates using the 'qualitative stability' concept as a means of achieving 'robust stability' in the standard uncertain matrix theory and offer it as a 'sufficient condition' for checking the robust stability of a class of interval matrices. This argument is illustrated with the following examples.

Consider the above given 'interval matrix'.

Once it is recognized that the signs of the interval entries in the matrix are not changing (within the given intervals), the sign matrix can be formed. The `sign' matrix for this interval matrix is given by

$$A = \begin{bmatrix} 0 & + & + \\ - & 0 & 0 \\ - & 0 & - \end{bmatrix}$$

The above 'sign' matrix is known to be 'qualitative (sign) stable'. Since sign stability is independent of magnitudes of the entries of the matrix, it can be concluded that the above interval matrix is robustly stable in the given interval ranges. Incidentally, if the 'vertex algorithm' of [40] is applied for this problem, it can be also concluded that this 'interval matrix family' is indeed Hurwitz stable in the given interval ranges.

In fact, more can be said about the 'robust stability' of this matrix family using the 'sign stability' application. This matrix family is indeed robustly stable, not only for those given interval ranges above, but it is also robustly stable for *any large 'interval ranges' in those elements as long as those interval ranges are such that the elements do not change signs in those interval ranges.*

In the above discussion, the emphasis was on exploiting the sign pattern of a matrix in robust stability analysis of matrices. Thus, the tolerable perturbations are direction sensitive. Also, no perturbation is allowed in the structural zeroes of the ecological sign stable matrices. In what follows, it is shown that ecological sign stable matrices can still possess superior robustness properties even under norm bounded perturbations, in which perturbations in structural zeroes are also allowed in ecological sign stable matrices.

Towards this objective, the stability robustness measures of linear state space systems as discussed in [39] and [2] are considered. In other words, a linear state space plant matrix A, which is assumed to be Hurwitz stable, is considered. Then assuming a perturbation matrix E in the A matrix, the question as to how much of norm of the perturbation matrix E can be tolerated to maintain stability is asked. Note that in this norm bounded perturbation discussion, the elements of the perturbation matrix can vary in various directions without any restrictions on the signs of the elements of that matrix. When bounds on the norm of E are given to maintain stability, it is labeled as robust stability for unstructured, norm bounded uncertainty. We now briefly recall two measures of robustness available in the literature [2] for robust stability of time varying real parameter perturbations.

Norm bounded robustness measures

Consider a given Hurwitz stable matrix A_0 with perturbation E such that

$$A = A_0 + E \tag{16}$$

where A is any one of the perturbed matrices.

A sufficient bound μ for the stability of the perturbed system is given on the spectral norm of the perturbation matrix as

$$\|E\| < \frac{\alpha_s}{\kappa} = \frac{\left|\mathrm{Re}(\lambda(A_0))\right|_{\min}}{\kappa} = \mu_d \tag{17}$$

where α_s is the real part of the dominant eigenvalue, also known as stability degree and κ is the condition number of the modal matrix of A_0.

Theorem 5[38]

$$\begin{aligned} &\left|\mathrm{Re}(\lambda(A_{NN}))\right|_{\min} > \left|\mathrm{Re}(\lambda(B_{NN}))\right|_{\min} \\ &\qquad\qquad i.e., \mu(A_{NN}) > \mu(B_{NN}) \end{aligned} \tag{18}$$

In other words, a unit norm, normal ecological sign stable matrix is more robust that a unit norm, normal non-ecological sign stable Hurwitz stable matrix.

The second norm bound based on the solution of the Lyapunov matrix equation [7] is given as

$$\|E\| < \mu_p = \frac{1}{\sigma_{\max}(P)} \tag{19}$$

where

P is the solution of the Lyapunov equation of the nominal stable matrix A_0 given by

$$A_0^T P + PA_0 + 2I = 0$$

Based on this bound, the following Lemma is proposed:

Theorem 6

The norm bound μ_p *on a target SS matrix S is d, where d is the magnitude of diagonal element of S i.e.,*

$$\mu_p = \frac{1}{\sigma_{\max}(P)} = d \tag{20}$$

This means that for *any given value of* μ_p*, we can, by mere observation, determine a corresponding stable matrix A!*

This gives impetus to design controllers that drive the closed loop system to a target matrix. Towards this objective, an algorithm for the design of a controller based on concepts from ecological sign stability is now presented.

4.3 Robust control design based on ecological sign stability

Extensive research in the field of robust control design has lead to popular control design methods in frequency domain such as H_∞ and μ-synthesis., Though these methods perform well in frequency domain, they become very conservative when applied to the problem of accommodating real parameter uncertainty. On the other hand, there are very limited robust control design methods in time domain methods that explicitly address real parameter uncertainty [41-47]. Even these very few methods tend to be complex and demand some specific structure to the real parameter uncertainty (such as matching conditions). Therefore, as an alternative to existing methods, the distinct feature of this control design method inspired by ecological principles is its problem formulation in which the robustness measure appears explicitly in the design methodology.

4.3.1 Problem formulation

The problem formulation for this novel control design method is as follows:

For a given linear system

$$\dot{x}(t) = Ax(t) + Bu(t) \tag{21}$$

design a full-state feedback controller

$$u = Gx \tag{22}$$

where the closed loop system

$$A_{n\times n} + B_{n\times m}G_{m\times n} = A_{cl\,n\times n} \tag{23}$$

possesses a desired robustness bound μ (there is no restriction on the value this bound can assume).

Since eigenvalue distribution, condition number (normality) and robust stability properties have established the superiority of target matrices, they become an obvious choice for the closed loop system matrix A_{cl} . Note that the desired bound $\mu = \mu_d = \mu_p$. Therefore, the robust control design method proposed in the next section addresses the three viewpoints of robust stability *simultaneously*!

4.3.2 Robust control design algorithm

Consider the LTI system

$$\dot{x} = Ax + Bu$$

Then, for a full-state feedback controller, the closed loop system matrix is given by

$$\begin{gathered} A_{nxn} + B_{nxm}G_{mxn} = A_{cl\,nxn}(= A_t) \\ \text{Let} \quad A_{cl} - A = A_a \end{gathered} \tag{24}$$

The control design method is classified as follows:

1. Determination of Existence of the Controller[38]
2. Determination of Appropriate Closed loop System[38]
3. Determination of Control Gain Matrix[48]

Following example illustrates this simple and straightforward control design method.

Application: Satellite formation flying control problem

The above control algorithm is now illustrated for the application discussed in [32],[33] and [49].

$$\begin{bmatrix} \dot{x} \\ \ddot{x} \\ \dot{y} \\ \ddot{y} \end{bmatrix} = \begin{bmatrix} 0 & 0 & 1 & 0 \\ 0 & 0 & 0 & 1 \\ 0 & 0 & 0 & 2\omega \\ 0 & 3\omega^2 & -2\omega & 0 \end{bmatrix} \begin{bmatrix} x \\ \dot{x} \\ y \\ \dot{y} \end{bmatrix} + \begin{bmatrix} 0 & 0 \\ 0 & 0 \\ 1 & 0 \\ 0 & 1 \end{bmatrix} \begin{bmatrix} T_x \\ T_y \end{bmatrix} \tag{25}$$

where $x, \dot{x}, y$ and $\dot{y}$ are the state variables, T_x and T_y are the control variables.

For example, when $\omega = 1$, the system becomes

$$A = \begin{bmatrix} 0 & 0 & 1 & 0 \\ 0 & 0 & 0 & 1 \\ 0 & 0 & 0 & 2 \\ 0 & 3 & -2 & 0 \end{bmatrix} \text{ and } B = \begin{bmatrix} 0 & 0 \\ 0 & 0 \\ 1 & 0 \\ 0 & 1 \end{bmatrix}$$

Clearly, the first two rows of A_{cl} cannot be altered and hence a target matrix with all non-zero elements cannot be achieved. Therefore, a controller such that the closed loop system has as many features of a target SS matrix as possible is designed as given below.

Accordingly, an ecological sign stable closed loop system is chosen such that

i The closed loop matrix has as many pure predator-prey links as possible.

ii It also has as many negative diagonal elements as possible.

Taking the above points into consideration, the following sign pattern is chosen which is appropriate for the given A and B matrices:

$$A_{cl_{ss}} = \begin{bmatrix} 0 & 0 & + & 0 \\ 0 & 0 & 0 & + \\ - & 0 & - & + \\ 0 & - & - & - \end{bmatrix} \qquad A_{cl} = \begin{bmatrix} 0 & 0 & 1 & 0 \\ 0 & 0 & 0 & 1 \\ -1 & 0 & -1 & 2 \\ 0 & -1 & -2 & -1 \end{bmatrix}$$

The magnitudes of the entries of the above sign matrix are decided by the stability robustness analysis theorem discussed previously i.e.,

i All non-zero a_{ii} are identical.

ii $a_{ij} = -a_{ji}$ for all non-zero a_{ij} else $a_{ij} = a_{ji} = 0$

Hence, all the pure predator-prey links are of equal interaction strengths and the non-zero diagonal elements have identical self-regulatory intensities. Using the algorithm given above, the gain matrix is computed as shown below.

From the algorithm,

$$G_{es} = \begin{bmatrix} -1.0 & 0 & -1.0 & 0 \\ 0 & -4.0 & 0 & -1.0 \end{bmatrix}$$

The closed loop matrix A_{cl} (= $A+BG_{es}$) is sign-stable and hence can tolerate any amount of variation in the magnitudes of the elements with the sign pattern kept constant.

In this application, it is clear that all non-zero elements in the open loop matrix (excluding elements A_{13} and A_{24} since they are dummy states used to transform the system into a set of first order differential equations) are functions of the angular velocity ω. Hence, real life perturbations in this system occur only due to variation in angular velocity ω. Therefore, a perturbed satellite system is simply an A matrix generated by a different ω. This means that not every randomly chosen matrix represents a physically perturbed system and that for practical purposes, stability of the matrices generated as mentioned above (by varying ω) is sufficient to establish the robustness of the closed loop system. It is only because of the ecological perspective that these structural features of the system are brought to light. Also, it is the application of these ecological principles that makes the control design for satellite formation flying this simple and insightful.

Ideally, we would like A_t to be the eventual closed loop system matrix. However, it may be difficult to achieve this objective for any given controllable pair (A,B). Therefore, we propose to achieve a closed loop system matrix that is close to A_t. Thus the closed loop system is expressed as

$$A_{cl} = A + BG = A_t + \Delta A \tag{26}$$

Noting that ideally we like to aim for $\Delta A = 0$, we impose this condition. Then, $A_{cl} = A_t = A+BG$.

i. When B is square and invertible: As given previously,

$$A_{cl} = A_t \text{ and } G = B^{-1}(A - A_t)$$

ii. When B is not square, but has full rank:
Consider $B\dagger$, the pseudo inverse of B

where, for $B_{n\times m}$, if n > m, $B\dagger = \left(B^T B\right)^{-1} B^T$

Then $G = B\dagger\left(A - A_t\right)$

Because of errors associated with pseudo inverse operation, the expression for the closed loop system is as follows [34]:

$$A_t + \Delta E = A + BG$$

$$A_t + \Delta E = A + B\left(B^T B\right)^{-1} B^T \left(A_t - A\right) \tag{27}$$

Let $B\left(B^T B\right)^{-1} B^T = B_{aug}$

Then $\Delta E = \left(A - A_t\right) + B_{aug}\left(A_t - A\right) = -\left(A_t - A\right) + B_{aug}\left(A_t - A\right) = \left(B_{aug} - I\right)\left(A_t - A\right)$

$$\therefore \Delta E = \left(B_{aug} - I\right)\left(A - A_t\right) \tag{28}$$

which should be as small as possible. Therefore, the aim is to minimize the norm of ΔE. Thus, for a given controllable pair (A,B), we use the elements of the desired closed loop matrix A_t as design variables to minimize the norm of ΔE.
We now apply this control design method to aircraft longitudinal dynamics problem.
Application: Aircraft flight control
Consider the following short period mode of the longitudinal dynamics of an aircraft [50].

$$A = \begin{bmatrix} -0.334 & 1 \\ -2.52 & -0.387 \end{bmatrix} \quad B = \begin{bmatrix} -0.027 \\ -2.6 \end{bmatrix} \tag{29}$$

	Open loop A	Target matrix A_t	Close loop A_{cl}
Matrix	$\begin{bmatrix} -0.334 & 1 \\ -2.52 & -0.387 \end{bmatrix}$	$\begin{bmatrix} -0.3181 & 1.00073 \\ -1.00073 & -0.3181 \end{bmatrix}$	$\begin{bmatrix} -0.3182 & 1.00073 \\ -1.00073 & -0.319 \end{bmatrix}$
Eigenvalues	$\begin{bmatrix} -0.3605 + j1.5872 \\ -0.3605 - j1.5872 \end{bmatrix}$	$\begin{bmatrix} -0.3181 + j1.00073 \\ -0.3181 - j1.00073 \end{bmatrix}$	$\begin{bmatrix} -0.31816 + j1.000722 \\ -0.31816 - j1.000722 \end{bmatrix}$
Norm bound	0.2079	0.3181	0.3181426

The open loop matrix properties are as follows:
Note that the open loop system matrix is stable and has a Lyapunov based robustness bound $\mu_{op} = 0.2079$.
Now for the above controllable pair (A,B), we proceed with the proposed control design procedure discussed before, with the target PS matrix A_t elements as design variables, which very quickly yields the following results:
A_t is calculated by minimizing the norm of $\sigma_{max}(\Delta E)$.

Here $\sigma_{\max}(\Delta E) = 1.2381 \times 10^{-4}$
For this value, following are the properties of the target matrix.
From the expression for G, we get

$$G = \begin{bmatrix} -0.5843 & -0.0265 \end{bmatrix}$$

With this controller, the closed loop matrix A_{cl} is determined.
It is easy to observe that the eventual closed loop system matrix is extremely close to the target *PS* matrix (since $\sigma_{\max}$ (ΔE) ≈ 0) and hence the resulting robustness bounds can be simply read off from the diagonal elements of the target *SS* matrix, which in this example is also equal to the eventual closed loop system matrix. As expected, this robustness measure of the closed loop system is appreciably greater than the robustness measure of the open loop system.
This robust controller methodology thus promises to be a desirable alternative to the other robustness based controllers encompassing many fields of application.

5. Conclusions and future directions

In this book chapter, robust control theory is presented essentially from a state space perspective. We presented the material in two distinct parts. In the first part of the chapter, robust control theory is presented from a quantitative (engineering) perspective, making extensive use of state space models of dynamic systems. Both robust stability analysis as well as control design were addressed and elaborated. Robust stability analysis involved studying and quantifying the tolerable bounds for maintaining the stability of a nominally stable dynamic system. Robust control design dealt with the issue of synthesizing a controller to keep the closed loop systems stable under the presence of a given set of perturbations. This chapter focused on characterizing the perturbations essentially as \`real parameter' perturbations and all the techniques presented accommodate this particular modeling error. In the second part of the chapter, robustness is treated from a completely new perspective, namely from concepts of Population (Community) Ecology, thereby emphasizing the \`qualitative' nature of the stability robustness problem. In this connection, the analysis and design aspects were directed towards studying the role of \`signs' of the elements of the state space matrices in maintaining the stability of the dynamic system. Thus the concept of 'sign stability' from the field of ecology was brought out to the engineering community. This concept is relatively new to the engineering community. The analysis and control design for engineering systems using ecological principles as presented in this chapter is deemed to spur exciting new research in this area and provide new directions for future research. In particular, the role of \`interactions and interconnections' in engineering dynamic systems is shown to be of paramount importance in imparting robustness to the system and more research is clearly needed to take full advantage of these promising ideas. This research is deemed to pave the way for fruitful collaboration between population (community) ecologists and control systems engineers.

6. References

[1] Dorato, P., "Robust Control", IEEE Press, New York, N.Y., 1987
[2] Dorato, P., and Yedavalli, R. K., (Eds) Recent Advances in Robust Control, IEEE Press, 1991, pp. 109-111.

[3] Skelton, R. "Dynamic Systems Control," John Wiley and Sons, New York, 1988
[4] Barmish, B. R., "New Tools for Robustness of Linear Systems", Macmillan Publishing Company, New York, 1994
[5] Bhattacharya, S. P., Chapellat, H., and Keel, L. H., "Robust Control: The Parameteric Approach", Prentice Hall, 1995
[6] Yedavalli, R. K., "Robust Control of Uncertain Dynamic Systems: A Linear State Space Approach", Springer, 2011 (to be published).
[7] Patel, R.V. and Toda, M., "Quantitative Measures of Robustness for Multivariable Systems," Proceedings of Joint Automatic Control Conference, TP8-A, 1980.
[8] Yedavalli, R.K., Banda, S.S., and Ridgely, D.B., "Time Domain Stability Robustness Measures for Linear Regulators," AIAA Journal of Guidance, Control and Dynamics, pp. 520-525, July-August 1985.
[9] Yedavalli, R.K. and Liang, Z., "Reduced Conservation in Stability Robustness Bounds by State Transformation," IEEE Transactions on Automatic Control, Vol. AC-31, pp. 863-866, September 1986.
[10] Hyland, D.C., and Bernstein, D.S., "The Majorant Lyapunov Equations: A Nonnegative Matrix Equation for Robust Stability and Performance of Large Scale Systems," IEEE Transactions on Automatic Control, Vol. AC-32, pp. 1005-1013, November 1987.
[11] Yedavalli, R.K., "Improved Measures of Stability-Robustness for Linear State Space Models," IEEE Transactions on Automatic Control, Vol. AC-30, pp. 577-579, June 1985.
[12] Yedavalli, R.K., "Perturbation Bounds for Robust Stability in Linear State Space Models," International Journal of Control, Vol. 42, No. 6, pp. 1507-1517, 1985.
[13] Qiu, L. and Davison, E.J., "New Perturbation Bounds for the Robust Stability of Linear State Space Models," Proceedings of the 25th Conference on Decision and Control, Athens, Greece, pp. 751-755, 1986.
[14] Hinrichsen, D. and Pritchard, A. J., "Stability Radius for Structured Perturbations and the Algebraic Riccati Equation", Systems and Control Letters, Vol. 8, pp: 105-113, 198.
[15] Zhou, K. and Khargonekar, P., "Stability Robustness Bounds for Linear State Space models with Structured Uncertainty," IEEE Transactions on Automatic Control, Vol. AC-32, pp. 621-623, July 1987.
[16] Tesi, A. and Vicino, A., "Robustness Analysis for Uncertain Dynamical Systems with Structured Perturbations," Proceedings of the International Workshop on 'Robustness in Identification and Control,' Torino, Italy, June 1988.
[17] Vinkler, A. and Wood, L. J., "A comparison of General techniques for designing Controllers of Uncertain Dynamics Systems", Proceedings of the Conference on Decision and Control, pp: 31-39, San Diego, 1979
[18] Chang, S. S. L. and Peng, T. K. C., "Adaptive Guaranteed Cost Control of Systems with Uncertain Parameters," IEEE Transactions on Automatic Control, Vol. AC-17, Aug. 1972
[19] Thorp, J. S., and Barmish B. R., "On Guaranteed Stability of Uncertain Linear Systems via Linear Control", Journal of Optimization Theory and Applications, pp: 559-579, December 1981.

[20] Hollot, C. V. and Barmish, B. R., "Optimal Quadratic Stabilizability of Uncertain Linear Systems" Proceedings of 18th Allerton Conference on Communication, Control and Computing, University of Illinois, 1980.

[21] Schmitendorf, W. E., "A design methodology for Robust Stabilizing Controllers", AIAA Guidance and Control Conference, West Virginia, 1986

[22] Ackermann, J., "Multimodel Approaches to Robust Control Systems Design", IFAC Workshop on Model Error Concepts and Compensations, Boston, 1985.

[23] Yedavalli, R. K., "Time Domain Control Design for Robust Stability of Linear Regulators: Applications to Aircraft Control", Proceedings of the American control Conference, 914-919, Boston 1985.

[24] Yedavalli, R. K., "Linear Control design for Guaranteed Stability of Uncertain Linear Systems", Proceedings of the American control Conference, 990-992, Seattle, 1986.

[25] Leah Edelstein-Keshet., *Mathematical models in Biology*, McGraw Hill, 1988. pp. 234-236.

[26] Hofbauer, J., and Sigmund, K., "Growth Rates and Ecological Models: ABC on ODE," *The Theory of Evolutions and Dynamical Systems*, Cambridge University Press, London, 1988, pp. 29-59.

[27] Dambacher, J. M., Luh, H-K., Li, H. W., and Rossignol, P. A., "Qualitative Stability and Ambiguity in Model Ecosystems," *The American Naturalist*, Vol. 161, No. 6, June 2003, pp. 876-888.

[28] Logofet, D., *Matrices and Graphs: Stability Problems in Mathematical Ecology*, CRC Press, Boca Raton, 1992R, pp. 1-35.

[29] Quirk, J., and Ruppert. R., "Qualitative Economics and the Stability of Equilibrium," *Reviews in Economic Studies*, 32, 1965, pp. 311-326.

[30] May. R., *Stability and Complexity in Model Ecosystems*, Princeton University Press, Princeton, N.J., 1973, pp. 16-34.

[31] Jeffries, C., "Qualitative Stability and Digraphs in Model Ecosystems," *Ecology*, Vol. 55, 1974, pp. 1415-1419.

[32] Yedavalli, R. K., "Qualitative Stability Concept from Ecology and its Use in the Robust Control of Engineering Systems," *Proceedings of the American Control Conference, Minneapolis*, June 2006, pp. 5097-5102.

[33] Yedavalli, R. K., "New Robust and Non-Fragile Control Design Method for Linear Uncertain Systems Using Ecological Sign Stability Approach," *Proceedings of the International Conference on Advances in Control and Optimisation of Dynamical Systems. (ACODS07)*, Bangalore, India, 2007, pp. 267-273.

[34] Devarakonda, N. and Yedavalli, R. K., "A New Robust Control design for linear systems with norm bounded time varying real parameter uncertainty", *Proceedings of ASME Dynamic Systems and Controls Conference*, Boston, MA, 2010.

[35] Devarakonda, N. and Yedavalli, R. K., "A New Robust Control design for linear systems with norm bounded time varying real parameter uncertainty", *IEEE Conference on Decision and Control*, Atlanta, GA, 2010.

[36] Yedavalli, R. K., "Robust Control Design for Linear Systems Using an Ecological Sign Stability Approach," *AIAA Journal of Guidance, Control and Dynamics*, Vol. 32, No. 1, Jan-Feb, 2009, pp 348-352.

[37] Yedavalli, R. K. and Devarakonda, N. "Sign Stability Concept of Ecology for Control Design with Aerospace Applications", AIAA Journal of Guidance Control and Dynamics, Vol. 33, No. 2, pp. 333-346, 2010.

[38] Devarakonda, N. and Yedavalli, R. K., "Engineering Perspective of Ecological Sign Stability and its Application in Control Design", *Proceedings of American Control Conference*, Baltimore, MD, July, 2010.

[39] Yedavalli, R. K., "Flight Control Application of New Stability Robustness Bounds for Linear Uncertain Systems," *Journal of Guidance, Control and Dynamics*, Vol. 16, No. 6, Nov.-Dec. 1993.

[40] Yedavalli, R. K., "Robust Stability of Linear Interval Parameter Matrix Family Problem Revisited with Accurate Representation and Solution," *Proceedings of American Automatic Control Conference*, June, 2009, pp 3710-3717

[41] Yedavalli, R. K., 1989. "Robust Control Design for Aerospace Applications". IEEE Transactions on Aerospace and Electronic Systems, 25(3), pp. 314

[42] Keel, L. H., Bhattacharya, S. P., and Howze, J. W., 1988. "Robust Control with Structured Perturbations", IEEE Transactions on Automatic Control, AC-33(1), pp. 68

[43] Zhou, K., and P. Khargonekar, 1988. "Robust Stabilization of Linear Systems with Norm Bounded Time Varying Uncertainty", Systems and Control Letters, 8, pp. 17.

[44] Petersen, I, R., and Hollot, C. V., 1986. "A Riccati Equation Based Approach to the Stabilization of Uncertain Linear Systems", Automatica, 22(4), pp. 397.

[45] Petersen, I, R., 1985. "A Riccati Equation Approach to the Design of Stabilizing Controllers and Observers for a Class of Uncertain Linear Systems", IEEE Transactions AC-30(9)

[46] Barmish, B. R., Corless, M., and Leitmann, G., 1983. "A New Class of Stabilizing Controllers for Uncertain Dynamical Systems", SIAM J. Optimal Control, 21, pp.246-255.

[47] Barmish, B. R., Petersen, I. R., and Feuer, A., 1983. "Linear Ultimate Boundedness Control for Uncertain Dynamical Systems", Automatica, 19, pp. 523-532.

[48] Yedavalli, R. K., and Devarakonda, N., "Ecological Sign Stability and its Use in Robust Control Design for Aerospace Applications," *Proceedings of IEEE Conference on Control Applications*, Sept. 2008.

[49] Yedavalli, R. K., and Sparks. A., "Satellite Formation Flying and Control Design Based on Hybrid Control System Stability Analysis," Proceedings of the American Control Conference, June 2000, pp. 2210.

[50] Nelson, R., Flight Stability and Automatic Control. McGraw Hill. Chap. 1998.

Part 2

H-infinity Control

4

Robust H_∞ PID Controller Design Via LMI Solution of Dissipative Integral Backstepping with State Feedback Synthesis

Endra Joelianto
Bandung Institute of Technology
Indonesia

1. Introduction

PID controller has been extensively used in industries since 1940s and still the most often implemented controller today. The PID controller can be found in many application areas: petroleum processing, steam generation, polymer processing, chemical industries, robotics, unmanned aerial vehicles (UAVs) and many more. The algorithm of PID controller is a simple, single equation relating proportional, integral and derivative parameters. Nonetheless, these provide good control performance for many different processes. This flexibility is achieved through three adjustable parameters of which values can be selected to modify the behaviour of the closed loop system. A convenient feature of the PID controller is its compatibility with enhancement that provides higher capabilities with the same basic algorithm. Therefore the performance of a basic PID controller can be improved through judicious selection of these three values.

Many tuning methods are available in the literature, among with the most popular method the Ziegler-Nichols (Z-N) method proposed in 1942 (Ziegler & Nichols, 1942). A drawback of many of those tuning rules is that such rules do not consider load disturbance, model uncertainty, measurement noise, and set-point response simultaneously. In general, a tuning for high performance control is always accompanied by low robustness (Shinskey, 1996). Difficulties arise when the plant dynamics are complex and poorly modeled or, specifications are particularly stringent. Even if a solution is eventually found, the process is likely to be expensive in terms of design time. Varieties of new methods have been proposed to improve the PID controller design, such as analytical tuning (Boyd & Barrat, 1991; Hwang & Chang, 1987), optimization based (Wong & Seborg, 1988; Boyd & Barrat, 1991; Astrom & Hagglund, 1995), gain and phase margin (Astrom & Hagglund, 1995; Fung et al., 1998). Further improvement of the PID controller is sought by applying advanced control designs (Ge et al., 2002; Hara et al., 2006; Wang et al., 2007; Goncalves et al., 2008).

In order to design with robust control theory, the PID controller is expressed as a state feedback control law problem that can then be solved by using any full state feedback robust control synthesis, such as Guaranteed Cost Design using Quadratic Bound (Petersen et al., 2000), H_∞ synthesis (Green & Limebeer, 1995; Zhou & Doyle, 1998), Quadratic Dissipative Linear Systems (Yuliar et al., 1997) and so forth. The PID parameters selection by transforming into state feedback using linear quadratic method was first proposed by Williamson and Moore in (Williamson & Moore, 1971). Preliminary applications were investigated in (Joelianto & Tomy, 2003) followed the work in (Joelianto et al., 2008) by extending the method in (Williamson & Moore, 1971) to H_∞ synthesis with dissipative integral backstepping. Results showed that the robust H_∞ PID controllers produce good tracking responses without overshoot, good load disturbance responses, and minimize the effect of plant uncertainties caused by non-linearity of the controlled systems.

Although any robust control designs can be implemented, in this paper, the investigation is focused on the theory of parameter selection of the PID controller based on the solution of robust H_∞ which is extended with full state dissipative control synthesis and integral backstepping method using an algebraic Riccati inequality (ARI). This paper also provides detailed derivations and improved conditions stated in the previous paper (Joelianto & Tomy, 2003) and (Joelianto et al., 2008). In this case, the selection is made via control system optimization in robust control design by using linear matrix inequality (LMI) (Boyd et al., 1994; Gahinet & Apkarian, 1994). LMI is a convex optimization problem which offers a numerically tractable solution to deal with control problems that may have no analytical solution. Hence, reducing a control design problem to an LMI can be considered as a practical solution to this problem (Boyd et al., 1994). Solving robust control problems by reducing to LMI problems has become a widely accepted technique (Balakrishnan & Wang, 2000). General multi objectives control problems, such as H_2 and H_∞ performance, peak to peak gain, passivity, regional pole placement and robust regulation are notoriously difficult, but these can be solved by formulating the problems into linear matrix inequalities (LMIs) (Boyd et al., 1994; Scherer et al., 1997)).

The objective of this paper is to propose a parameter selection technique of PID controller within the framework of robust control theory with linear matrix inequalities. This is an alternative method to optimize the adjustment of a PID controller to achieve the performance limits and to determine the existence of satisfactory controllers by only using two design parameters instead of three well known parameters in the PID controller. By using optimization method, an absolute scale of merits subject to any designs can be measured. The advantage of the proposed technique is implementing an output feedback control (PID controller) by taking the simplicity in the full state feedback design. The proposed technique can be applied either to a single-input-single-output (SISO) or to a multi-inputs-multi-outputs (MIMO) PID controller.

The paper is organised as follows. Section 2 describes the formulation of the PID controller in the full state feedback representation. In section 3, the synthesis of H_∞ dissipative integral backstepping is applied to the PID controller using two design parameters. This section also provides a derivation of the algebraic Riccati inequality (ARI) formulation for the robust control from the dissipative integral backstepping synthesis. Section 4 illustrates an application of the robust PID controller for time delay uncertainties compensation in a network control system problem. Section 5 provides some conclusions.

2. State feedback representation of PID controller

In order to design with robust control theory, the PID controller is expressed as a full state feedback control law. Consider a single input single output linear time invariant plant described by the linear differential equation

$$\dot{x}(t) = Ax(t) + B_2u(t)$$
$$y(t) = C_2x(t) \tag{1}$$

with some uncertainties in the plant which will be explained later. Here, the states $x \in R^n$ are the solution of (1), the control signal $u \in R^1$ is assumed to be the output of a PID controller with input $y \in R^1$. The PID controller for regulator problem is of the form

$$u(t) = K_1 \int_0^t y(t)d(t) + K_2y(t) + K_3\frac{d}{dt}y(t) \tag{2}$$

which is an output feedback control system and $K_1 = K_p / T_i$, $K_2 = K_p$, $K_3 = K_pT_d$ of which K_p, T_i and T_d denote proportional gain, time integral and time derivative of the well known PID controller respectively. The structure in equation (2) is known as the standard PID controller (Astrom & Hagglund, 1995).

The control law (2) is expressed as a state feedback law using (1) by differentiating the plant output y as follows

$$y = C_2x$$
$$\dot{y} = C_2Ax + C_2B_2u$$
$$\ddot{y} = C_2A^2x + C_2AB_2u + C_2B_2\dot{u}$$

This means that the derivative of the control signal (2) may be written as

$$(1 - K_3C_2B_2)\dot{u} - (K_3C_2A^2 + K_2C_2A + K_1C_2)x - (K_3C_2AB_2 + K_2C_2B_2)u = 0 \tag{3}$$

Using the notation $\hat{K}$ as a normalization of K, this can be written in more compact form

$$\hat{K} = [\hat{K}_1 \quad \hat{K}_2 \quad \hat{K}_3] = (1 - K_3C_2B_2)^{-1}[K_1 \quad K_2 \quad K_3] \tag{4}$$

or $\hat{K} = cK$ where c is a scalar. This control law is then given by

$$\dot{u} = \hat{K}[C_2^T \quad A^TC_2^T \quad (A^2)^TC_2^T]^T x + \hat{K}[0 \quad B_2^TC_2^T \quad B_2^TA^TC_2^T]^T u \tag{5}$$

Denote $K_x = \hat{K}[C_2^T \quad A^TC_2^T \quad (A^2)^TC_2^T]^T$ and $K_u = \hat{K}[0 \quad B_2^TC_2^T \quad B_2^TA^TC_2^T]^T$, the block diagram of the control law (5) is shown in Fig. 1. In the state feedback representation, it can be seen that the PID controller has interesting features. It has state feedback in the upper loop and pure integrator backstepping in the lower loop. By means of the internal model principle (IMP) (Francis & Wonham, 1976; Joelianto & Williamson, 2009), the integrator also guarantees that the PID controller will give zero tracking error for a step reference signal.

Equation (5) represents an output feedback law with constrained state feedback. That is, the control signal (2) may be written as

$$u_a = K_ax_a \tag{6}$$

where

$$u_a = \dot{u}, \quad x_a = \begin{bmatrix} x \\ u \end{bmatrix}$$

$$K_a = \hat{K}\left[[C_2^T \quad A^T C_2^T \quad (A^2)^T C_2^T]^T \qquad [0 \quad B_2^T C_2^T \quad B_2^T A^T C_2^T]^T\right]$$

Arranging the equation and eliminating the transpose lead to

$$K_a = \hat{K}\begin{bmatrix} C_2 & 0 \\ C_2 A & C_2 B \\ C_2 A^2 & C_2 A B_2 \end{bmatrix} = \hat{K}\Gamma \tag{7}$$

The augmented system equation is obtained from (1) and (7) as follows

$$\dot{x}_a = A_a x_a + B_a u_a \tag{8}$$

where

$$A_a = \begin{bmatrix} A & B_2 \\ 0 & 0 \end{bmatrix}; \; B_a = \begin{bmatrix} 0 \\ 1 \end{bmatrix}$$

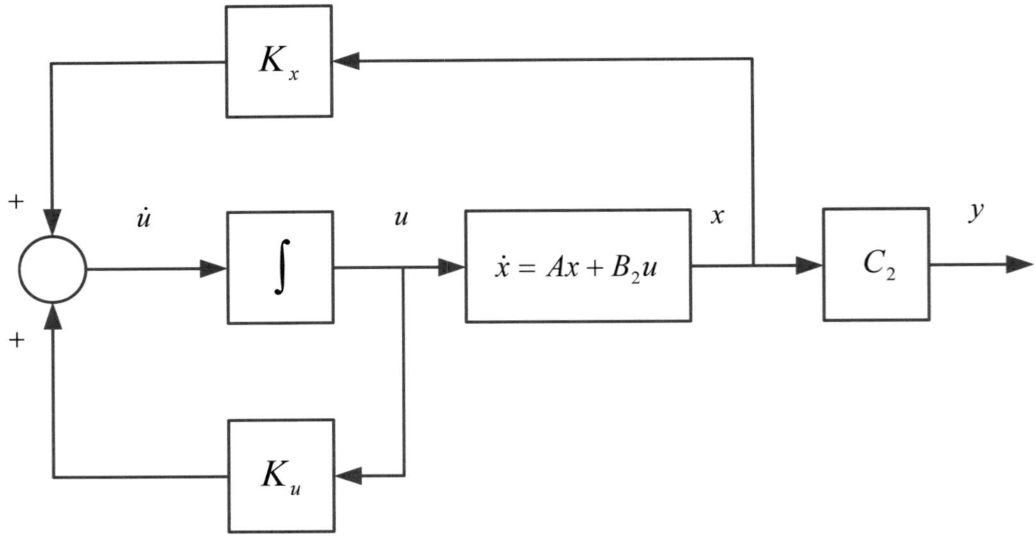

Fig. 1. Block diagram of state space representation of PID controller

Equation (6), (7) and (8) show that the PID controller can be viewed as a state variable feedback law for the original system augmented with an integrator at its input. The augmented formulation also shows the same structure known as the integral backstepping method (Krstic et al., 1995) with one pure integrator. Hence, the selection of the parameters of the PID controller (6) via full state feedback gain is inherently an integral backstepping control problems. The problem of the parameters selection of the PID controller becomes an optimal problem once a performance index of the augmented system (8) is defined. The parameters of the PID controller are then obtained by solving equation (7) that requires the inversion of the matrix Γ. Since Γ is, in general, not a square matrix, a numerical method should be used to obtain the inverse.

For the sake of simplicity, the problem has been set-up in a single-input-single-output (SISO) case. The extension of the method to a multi-inputs-multi-outputs (MIMO) case is straighforward. In MIMO PID controller, the control signal has dimension m, $u \in R^m$ is assumed to be the output of a PID controller with input has dimension p, $y \in R^p$. The parameters of the PID controller K_1, K_2, and K_3 will be square matrices with appropriate dimension.

3. H∞ dissipative integral backstepping synthesis

The backstepping method developed by (Krstic et al., 1995) has received considerable attention and has become a well known method for control system designs in the last decade. The backstepping design is a recursive algorithm that steps back toward the control input by means of integrations. In nonlinear control system designs, backstepping can be used to force a nonlinear system to behave like a linear system in a new set of coordinates with flexibility to avoid cancellation of useful nonlinearities and to focus on the objectives of stabilization and tracking. Here, the paper combines the advantage of the backstepping method, dissipative control and H∞ optimal control for the case of parameters selection of the PID controller to develop a new robust PID controller design.

Consider the single input single output linear time invariant plant in standard form used in H∞ performance by the state space equation

$$\begin{aligned} \dot{x}(t) &= Ax(t) + B_1 w(t) + B_2 u(t), \;\; x(0) = x_0 \\ z(t) &= C_1 x(t) + D_{11} w(t) + D_{12} u(t) \\ y(t) &= C_2 x(t) + D_{21} w(t) + D_{22} u(t) \end{aligned} \tag{9}$$

where $x \in R^n$ denotes the state vector, $u \in R^1$ is the control input, $w \in R^p$ is an external input and represents driving signals that generate reference signals, disturbances, and measurement noise, $y \in R^1$ is the plant output, and $z \in R^m$ is a vector of output signals related to the performance of the control system.

Definition 1.

A system is dissipative (Yuliar et al., 1998) with respect to supply rate $r(z(t), w(t))$ for each initial condition x_0 if there exists a storage function V, $V : R^n \to R^+$ satisfies the inequality

$$V(x(t_0)) + \int_{t_0}^{t_1} r(z(t), w(t))dt \geq V(x(t_1)) \,,\, \forall (t_1, t_0) \in R^+ \,,\, x_0 \in R^n \tag{10}$$

and $t_0 \leq t_1$ and all trajectories (x, y, z) which satisfies (9).

The supply rate function $r(z(t), w(t))$ should be interpreted as the supply delivered to the system. If in the interval $[t_0, t_1]$ the integral $\int_{t_0}^{t_1} r(z(t), w(t))dt$ is positive then work has been done to the system. Otherwise work is done by the system. The supply rate determines not only the dissipativity of the system but also the required performance index of the control system. The storage function V generalizes the notion of an energy function for a dissipative system. The function characterizes the change of internal storage $V(x(t_1)) - V(x(t_0))$ in any interval $[t_0, t_1]$, and ensures that the change will never exceed the amount of the supply into

the system. The dissipative method provides a unifying tool as index performances of control systems can be expressed in a general supply rate by selecting values of the supply rate parameters.
The general quadratic supply rate function (Hill & Moylan, 1977) is given by the following equation

$$r(z,w) = \frac{1}{2}(w^T Qw + 2w^T Sz + z^T Rz) \tag{11}$$

where Q and R are symmetric matrices and

$$\bar{Q}(x) = Q + SD_{11}(x) + D_{11}^T(x)S^T + D_{11}^T(x)RD_{11}(x)$$

such that $\bar{Q}(x) > 0$ for $x \in R^n$ and $R \le 0$ and $\inf_{x\in R^n}\{\sigma_{\min}(\bar{Q}(x))\} = k > 0$. This general supply rate represents general problems in control system designs by proper selection of matrices Q, R and S (Hill & Moylan, 1977; Yuliar et al., 1997): finite gain (H_∞) performance ($Q = \gamma^2 I$, $S = 0$ and $R = -I$); passivity ($Q = R = 0$ and $S = I$); and mixed H_∞-positive real performance ($Q = \theta\gamma^2 I$, $R = -\theta I$ and $S = (1-\theta)I$ for $\theta \in [0,1]$).
For the PID control problem in the robust control framework, the plant (Σ) is given by the state space equation

$$\Sigma = \begin{cases} \dot{x}(t) = Ax(t) + B_1 w(t) + B_2 u(t), x(0) = x_0 \\ z(t) = \begin{pmatrix} C_1 x(t) \\ D_{12} u(t) \end{pmatrix} \end{cases} \tag{12}$$

with $D_{11} = 0$ and $\gamma > 0$ with the quadratic supply rate function for H_∞ performance

$$r(z,w) = \frac{1}{2}(\gamma^2 w^T w - z^T z) \tag{13}$$

Next the plant (Σ) is added with integral backstepping on the control input as follows

$$\begin{aligned} \dot{x}(t) &= Ax(t) + B_1 w(t) + B_2 u(t) \\ u_a(t) &= \dot{u}(t) \\ z(t) &= \begin{pmatrix} C_1 x(t) \\ D_{12} u(t) \\ \rho u_a(t) \end{pmatrix} \end{aligned} \tag{14}$$

where ρ is the parameter of the integral backstepping which act on the derivative of the control signal $\dot{u}(t)$. In equation (14), the parameter $\rho > 0$ is a tuning parameter of the PID controller in the state space representation to determine the rate of the control signal. Note that the standard PID controller in the state feedback representation in the equations (6), (7) and (8) corresponds to the integral backstepping parameter $\rho = 1$. Note that, in this design the gains of the PID controller are replaced by two new design parameters namely γ and ρ which correspond to the robustness of the closed loop system and the control effort.
The state space representation of the plant with an integrator backstepping in equation (14) can then be written in the augmented form as follows

$$\begin{bmatrix} \dot{x}(t) \\ \dot{u}(t) \end{bmatrix} = \begin{bmatrix} A & B_2 \\ 0 & 0 \end{bmatrix} \begin{bmatrix} x(t) \\ u(t) \end{bmatrix} + \begin{bmatrix} B_1 \\ 0 \end{bmatrix} w(t) + \begin{bmatrix} 0 \\ 1 \end{bmatrix} u_a(t)$$
$$z(t) = \begin{bmatrix} C_1 & 0 \\ 0 & D_{12} \\ 0 & 0 \end{bmatrix} \begin{bmatrix} x(t) \\ u(t) \end{bmatrix} + \begin{bmatrix} 0 \\ 0 \\ \rho \end{bmatrix} u_a(t) \tag{15}$$

The compact form of the augmented plant (Σ_a) is given by

$$\begin{aligned} \dot{x}_a(t) &= A_a x_a(t) + B_w w(t) + B_a u_a(t); x_a(0) = x_{a0} \\ z(t) &= C_a x_a(t) + D_{1a} w(t) + D_{2a} u_a(t) \end{aligned} \tag{16}$$

where

$$x_a = \begin{bmatrix} x \\ u \end{bmatrix},\ A_a = \begin{bmatrix} A & B_2 \\ 0 & 0 \end{bmatrix},\ B_w = \begin{bmatrix} B_1 \\ 0 \end{bmatrix},\ B_a = \begin{bmatrix} 0 \\ 1 \end{bmatrix},\ C_a = \begin{bmatrix} C_1 & 0 \\ 0 & D_{12} \\ 0 & 0 \end{bmatrix},\ D_{2a} = \begin{bmatrix} 0 \\ 0 \\ \rho \end{bmatrix}$$

Now consider the full state gain feedback of the form

$$u_a(t) = K_a x_a(t) \tag{17}$$

The objective is then to find the gain feedback K_a which stabilizes the augmented plant (Σ_a) with respect to the dissipative function V in (10) by a parameter selection of the quadratic supply rate (11) for a particular control performance. Fig. 2. shows the system description of the augmented system of the plant and the integral backstepping with the state feedback control law.

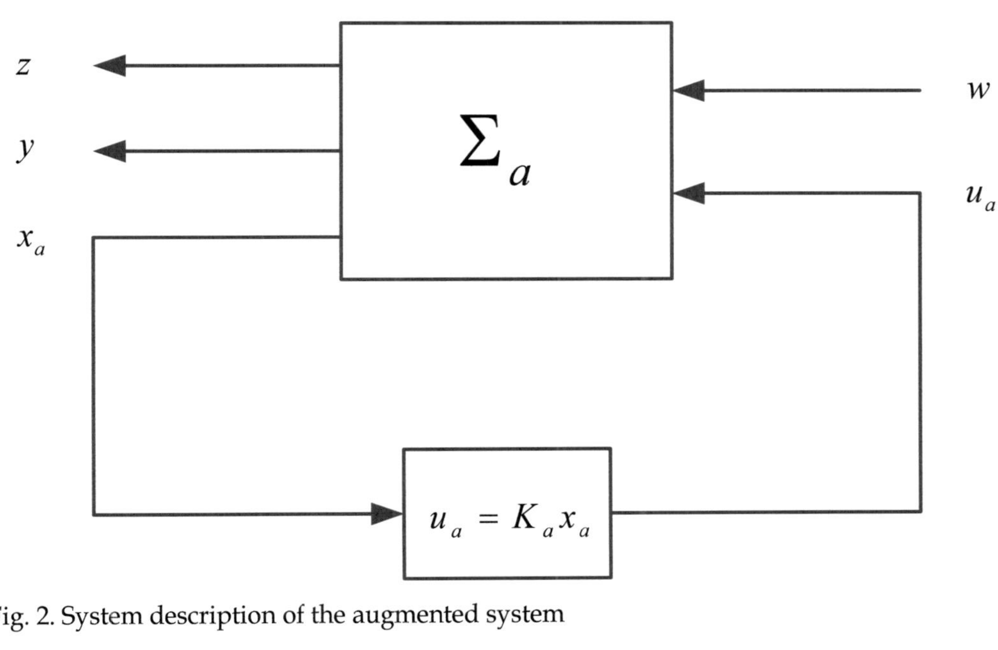

Fig. 2. System description of the augmented system

The following theorem gives the existence condition and the formula of the stabilizing gain feedback K_a.

Theorem 2.

Given $\gamma > 0$ and $\rho > 0$. If there exists $X = X^T > 0$ of the following Algebraic Riccati Inequality

$$X\begin{bmatrix} A & B_2 \\ 0 & 0 \end{bmatrix} + \begin{bmatrix} A^T & 0 \\ B_2^T & 0 \end{bmatrix} X - X\left(\rho^{-2}\begin{bmatrix} 0 & 0 \\ 0 & 1 \end{bmatrix} - \gamma^{-2}\begin{bmatrix} B_1B_1^T & 0 \\ 0 & 0 \end{bmatrix}\right)X + \begin{bmatrix} C_1^TC_1 & 0 \\ 0 & D_{12}^TD_{12} \end{bmatrix} < 0 \tag{18}$$

Then the full state feedback gain

$$K_a = -\rho^{-2}B_a^TX = -\rho^{-2}\begin{bmatrix} 0 & 1 \end{bmatrix}X \tag{19}$$

leads to $\|\Sigma\|_\infty < \gamma$

Proof.

Consider the standard system (9) with the full state feedback gain

$$u(t) = Kx(t)$$

and the closed loop system

$$\dot{x}(t) = A^u x(t) + B_1 w(t),\ \ x(0) = x_0$$
$$z(t) = C^u x(t) + D_{11} w(t)$$

where $D_{11} = 0$, $A^u = A + B_2K$, $C^u = C_1 + D_{12}K$ is strictly dissipative with respect to the quadratic supply rate (11) such that the matrix A^u is asymptotically stable. This implies that the related system

$$\dot{x}(t) = \tilde{A}x(t) + \tilde{B}_1 w(t),\ \ x(0) = x_0$$
$$z(t) = \tilde{C}_1 x(t)$$

where $\tilde{A} = A^u - B_1\bar{Q}^{-1}\bar{S}C^u$, $\tilde{B}_1 = B_1\bar{Q}^{-1/2}$ and $\tilde{C}_1 = (\bar{S}^T\bar{Q}^{-1}\bar{S} - R)^{1/2}C^u$ has H_∞ norm strictly less than 1, which implies there exits a matrix $X > 0$ solving the following Algebraic Riccati Inequality (ARI) (Petersen et al. 1991)

$$\tilde{A}^TX + X\tilde{A} + X\tilde{B}_1\tilde{B}_1^TX + \tilde{C}_1^T\tilde{C}_1 < 0 \tag{20}$$

In terms of the parameter of the original system, this can be written as

$$(A^u)^TX + XA^u + [XB_1 - (C^u)^T\bar{S}^T]\bar{Q}^{-1}[B_1^TX - \bar{S}C^u] - (C^u)^TRC^u < 0 \tag{21}$$

Define the full state feedback gain

$$K = -\bar{E}^{-1}\left((B_2 - B_1\bar{Q}^{-1}\bar{S}D_{12})^T X + D_{12}\bar{R}C_1\right) \tag{22}$$

By inserting

$$
\begin{aligned}
&A^u = A + B_2K, \;\; C^u = C_1 + D_{12}K \\
&\bar{S} = S + D_{11}^T R, \;\; \bar{Q} = Q + SD_{11} + D_{11}^T S^T + D_{11}^T R D_{11} \\
&\bar{R} = \bar{S}^T \bar{Q}^{-1} \bar{S} - R, \;\; \bar{E} = D_{12}^T \bar{R} D_{12} \\
&\bar{B} = B_2 - B_1 \bar{Q}^{-1} \bar{S} D_{12}, \;\; \bar{D} = I - D_{12} \bar{E}^{-1} D_{12}^T \bar{R}
\end{aligned}
$$

into (21) , completing the squares and removing the gain K give the following ARI

$$
\begin{aligned}
&X(A - \bar{B}\bar{E}^{-1}D_{12}^T \bar{R} C_1 - B_1 \bar{Q}\bar{S}C_1) + (A - \bar{B}\bar{E}^{-1}D_{12}^T \bar{R} C_1 - B_1 \bar{Q}\bar{S}C_1)X - \\
&-X(\bar{B}\bar{E}^{-1}\bar{B}^T - B_1 \bar{Q}^{-1} B_1^T)X + C_1^T \bar{D}^T \bar{R} \bar{D} C_1 < 0
\end{aligned} \tag{23}
$$

Using the results (Scherer, 1990), if there exists $X > 0$ satisfies (23) then K given by (22) is stabilizing such that the closed loop system $A^u = A + B_2K$ is asymptotically stable.
Now consider the augmented plant with integral backstepping in (16). In this case, $D_{1a} = \begin{bmatrix} 0 & 0 & 0 \end{bmatrix}^T$. Note that $D_{2a}^T C_a = 0$ and $D_{1a} = 0$. The H_∞ performance is satisfied by setting the quadratic supply rate (11) as follow:

$$
\begin{aligned}
&\bar{S} = 0, \;\; \bar{R} = -R = I, \;\; \bar{E} = D_{2a}^T \bar{R} D_{2a} = D_{2a}^T D_{2a}, \;\; \bar{B} = B_a, \\
&\bar{D} = I - D_{2a}(D_{2a}^T D_{2a})^{-1} D_{2a}^T
\end{aligned}
$$

Inserting the setting to the ARI (23) yields

$$
\begin{aligned}
&X(A_a - B_a(D_{2a}^T D_{2a})^{-1} D_{2a}^T I C_a - B_w Q^{-1} 0 C_a) + \\
&+(A_a - B_a(D_{2a}^T D_{2a})^{-1} D_{2a}^T I C_a - B_w Q^{-1} 0 C_a)X - \\
&-X(B_a(D_{2a}^T D_{2a})^{-1} B_a^T - B_w Q^{-1} B_w^T)X + \\
&+(C_a^T (I - D_{2a}(D_{2a}^T D_{2a})^{-1} D_{2a}^T)^T x \;\; (I - D_{2a}(D_{2a}^T D_{2a})^{-1} D_{2a}^T) C_a) < 0
\end{aligned}
$$

The equation can then be written in compact form

$$
XA_a + A_a^T X - X(\rho^{-2} B_a B_a^T - \gamma^{-2} B_w B_w^T)X + C_a^T C_a < 0 \tag{24}
$$

this gives (18). The full state feedback gain is then found by inserting the setting into (22)

$$
K_a = -\bar{E}^{-1}\left((B_a - B_w \bar{Q}^{-1} \bar{S} D_{2a})^T X - D_{2a}^T \bar{R} C_a \right)
$$

that gives $\|\Sigma\|_\infty < \gamma$ (Yuliar et al., 1998; Scherer & Weiland, 1999). This completes the proof.
The relation of the ARI solution (8) to the ARE solution is shown in the following. Let the transfer function of the plant (9) is represented by

$$
\begin{bmatrix} z(s) \\ y(s) \end{bmatrix} = \begin{bmatrix} P_{11}(s) & P_{12}(s) \\ P_{21}(s) & P_{22}(s) \end{bmatrix} \begin{bmatrix} w(s) \\ u(s) \end{bmatrix}
$$

and assume the following conditions hold:
(A1). (A, B_2, C_2) is stabilizable and detectable
(A2). $D_{22} = 0$

(A3). D_{12} has full column rank, D_{21} has full row rank
(A4). $P_{12}(s)$ and $P_{21}(s)$ have no invariant zero on the imaginary axis
From (Gahinet & Apkarian, 1994), equivalently the Algebraic Riccati Equation (ARE) given by the formula

$$X(A-\overline{B}\overline{E}^{-1}D_{12}^{T}\overline{R}C_1-B_1\overline{Q}\overline{S}C_1)+(A-\overline{B}\overline{E}^{-1}D_{12}^{T}\overline{R}C_1-B_1\overline{Q}\overline{S}C_1)X-$$
$$X(\overline{B}\overline{E}^{-1}\overline{B}^{T}-B_1\overline{Q}^{-1}B_1^{T})X+C_1^{T}\overline{D}^{T}\overline{R}\overline{D}C_1=0 \tag{25}$$

has a (unique) solution $X_{\infty}\geq 0$, such that

$$A+B_2K+B_1\overline{Q}^{-1}[B_1^{T}X-\overline{S}(C_1+D_{12}K)]$$

is asymptotically stable. The characterization of feasible γ using the Algebraic Riccati Inequality (ARI) in (24) and ARE in (25) is immediately where the solution of ARE (X_{∞}) and ARI (X_0) satisfy $0\leq X_{\infty}<X_0$, $X_0=X_0^{T}>0$ (Gahinet & Apkarian, 1994).
The Algebraic Riccati Inequality (24) by Schur complement implies

$$\begin{bmatrix} A_a^{T}X+XA_a+C_a^{T}C_a & XB_a & XB_w \\ B_a^{T}X & \rho^2 I & 0 \\ B_w^{T}X & 0 & -\gamma^2 I \end{bmatrix}<0 \tag{26}$$

Ther problem is then to find $X>0$ such that the LMI given in (26) holds. The LMI problem can be solved using the method (Gahinet & Apkarian, 1994) which implies the solution of the ARI (18) (Liu & He, 2006). The parameters of the PID controller which are designed by using H_{∞} dissipative integral backstepping can then be found by using the following algorithm:

1. Select $\rho>0$
2. Select $\gamma>0$
3. Find $X_0>0$ by solving LMI in (26)
4. Find K_a using (19)
5. Find $\hat{K}$ using (7)
6. Compute K_1, K_2 and K_3 using (4)
7. Apply in the PID controller (2)
8. If it is needed to achieve γ minimum, repeat step 2 and 3 until $\gamma=\gamma_{\min}$ then follows the next step

4. Delay time uncertainties compensation

Consider the plant given by a first order system with delay time which is common assumption in industrial process control and further assume that the delay time uncertainties belongs to an a priori known interval

$$Y(s)=\frac{1}{\tau s+1}e^{-Ls}U(s),\ L\in[a,b] \tag{27}$$

The example is taken from (Joelianto et al., 2008) which represents a problem in industrial process control due to the implementation of industrial digital data communication via ethernet networks with fieldbus topology from the controller to the sensor and the actuator (Hops et al., 2004; Jones, 2006, Joelianto & Hosana, 2009). In order to write in the state space representation, the delay time is approximated by using the first order Pade approximation

$$Y(s) = \frac{1}{\tau s+1}\frac{-ds+1}{ds+1}U(s)\,,\; d = L/2 \tag{28}$$

In this case, the values of τ and d are assumed as follows: $\tau = 1$ s and $d_{nom} = 3$ s. These pose a difficult problem as the ratio between the delay time and the time constant is greater than one ($(d/\tau) > 1$). The delay time uncertainties are assumed in the interval $d \in [2,4]$.
The delay time uncertainty is separated from its nominal value by using linear fractional transformation (LFT) that shows a feedback connection between the nominal and the uncertainty block.

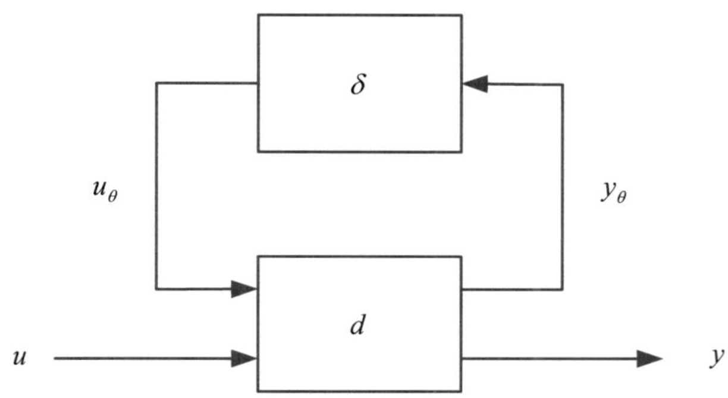

Fig. 3. Separation of nominal and uncertainty using LFT

The delay time uncertainties can then be represented as

$$d = \alpha d_{nom} + \beta\delta\,,\; -1 < \delta < 1$$

$$d = F_u\left(\begin{bmatrix} 0 & 1 \\ \beta & \alpha \end{bmatrix}, \delta\right)$$

After simplification, the delay time uncertainties of any known ranges have a linear fractional transformation (LFT) representation as shown in the following figure.

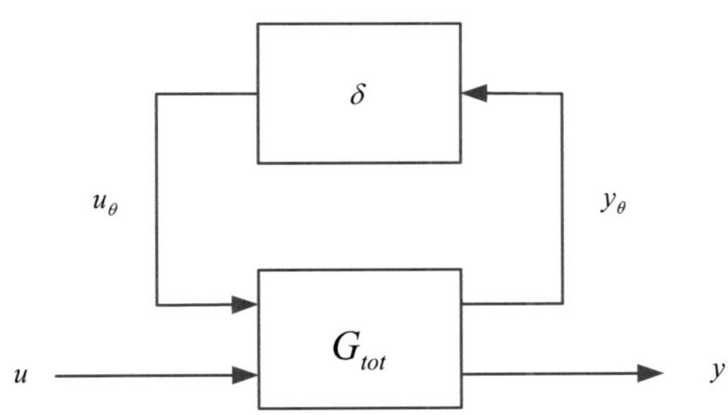

Fig. 4. First order system with delay time uncertainty

The representation of the plant augmented with the uncertainty is

$$G_{tot}(s) = \begin{bmatrix} A_x & B_x \\ C_x & D_x \end{bmatrix} = \begin{bmatrix} A & B_1 & B_2 \\ C_1 & D_{11} & D_{12} \\ C_2 & D_{21} & D_{22} \end{bmatrix} \tag{29}$$

The corresponding matrices in (29) are

$$A_x = \begin{bmatrix} A_{x11} & 0 \\ 1 & -1 \end{bmatrix}, B_x = \begin{bmatrix} B_{x11} & B_{x2} \\ 0 & 1 \end{bmatrix}, C_x = \begin{bmatrix} C_{x11} & 0 \\ 0 & 1 \end{bmatrix}, \ D_x = \begin{bmatrix} D_{x11} & D_{x12} \\ 0 & 0 \end{bmatrix}$$

In order to incorporate the integral backstepping design, the plant is then augmented with an integrator as follows

$$A_a = \begin{bmatrix} A & B_2 \\ 0 & 0 \end{bmatrix} = \begin{bmatrix} A_{x11} & 0 & B_{x11} \\ 1 & -1 & 0 \\ 0 & 0 & 0 \end{bmatrix}$$

$$B_w = \begin{bmatrix} B_1 \\ 0 \end{bmatrix} = \begin{bmatrix} B_{x11} \\ 0 \\ 0 \end{bmatrix},$$

$$B_a = \begin{bmatrix} 0 \\ I \end{bmatrix} = \begin{bmatrix} 0 \\ 0 \\ 1 \end{bmatrix},$$

$$C_a = \begin{bmatrix} C_{x11} & 0 \\ 0 & D_{x12} \\ 0 & 0 \end{bmatrix},$$

$$D_{2a} = \begin{bmatrix} 0 \\ 0 \\ \rho \end{bmatrix}$$

The problem is then to find the solution $X > 0$ and $\gamma > 0$ of ARI (18) and to compute the full state feedback gain given by

$$u_a(t) = K_a x_a(t) = -\rho^{-2}\left(\begin{bmatrix} 0 & 1 \end{bmatrix} X\right)\begin{bmatrix} x(t) \\ u(t) \end{bmatrix}$$

which is stabilizing and leads to the infinity norm $\| \Sigma \|_\infty < \gamma$.
The state space representation for the nominal system is given by

$$A_{nom} = \begin{bmatrix} -1.6667 & -0.6667 \\ 1 & 0 \end{bmatrix}, \ B_{nom} = \begin{bmatrix} 1 \\ 0 \end{bmatrix}, \ C_{nom} = \begin{bmatrix} -1 & 0.6667 \end{bmatrix}$$

In this representation, the performance of the closed loop system will be guaranteed for the specified delay time range with fast transient response (z). The full state feedback gain of the PID controller is given by the following equation

$$\begin{bmatrix} K_1 \\ K_2 \\ K_3 \end{bmatrix} = \left(1 - K_3\begin{bmatrix}-1 & 0.6667\end{bmatrix}\right)\begin{bmatrix}1 \\ 0\end{bmatrix}\begin{bmatrix}\hat{K}_1 \\ \hat{K}_2 \\ \hat{K}_3\end{bmatrix}$$

For different γ, the PID parameters and transient performances, such as: settling time (T_s) and rise time (T_r) are calculated by using LMI (26) and presented in Table 1. For different ρ but fixed γ, the parameters are shown in Table 2. As comparison, the PID parameters are also computed by using the standard H_∞ performance obtained by solving ARE in (25). The results are shown Table 3 and Table 4 for different γ and different ρ respectively. It can be seen from Table 1 and 2 that there is no clear pattern either in the settling time or the rise time. Only Table 1 shows that decreasing γ decreases the value of the three parameters. On the other hand, the calculation using ARE shows that the settling time and the rise time are decreased by reducing γ or ρ. Table 3 shows the same result with the Table 1 when the value of γ is decreased.

γ	**ρ**	**K_p**	**K_i**	**K_d**	**T_r (s)**	**T_s 5% (s)**
0.1	1	0.2111	0.1768	0.0695	10.8	12.7
0.248	1	0.3023	0.2226	0.1102	8.63	13.2
0.997	1	0.7744	0.3136	0.2944	4.44	18.8
1.27	1	10.471	0.5434	0.4090	2.59	9.27
1.7	1	13.132	0.746	0.5191	1.93	13.1

Table 1. Parameters and transient response of PID for different γ with LMI

γ	**ρ**	**K_p**	**K_i**	**K_d**	**T_r (s)**	**T_s 5% (s)**
0.997	0.66	11.019	0.1064	0.3127	39.8	122
0.997	0.77	0.9469	0.2407	0.3113	13.5	39.7
0.997	1	0.7744	0.3136	0.2944	4.44	18.8
0.997	1.24	0.4855	0.1369	0.1886	21.6	56.8
0.997	1.5	0.2923	0.0350	0.1151	94.4	250

Table 2. Parameters and transient response of PID for different ρ with LMI

γ	ρ	K_p	K_i	K_d	T_r (s)	T_s 5% (s)
0.1	1	0.2317	0.055	0.1228	55.1	143
0.248	1	0.2319	0.0551	0.123	55.0	141
0.997	1	0.2373	0.0566	0.126	53.8	138
1.27	1	0.2411	0.0577	0.128	52.6	135
1.7	1	0.2495	0.0601	0.1327	52.2	130

Table 3. Parameters and transient response of PID for different γ with ARE

γ	ρ	K_p	K_i	K_d	T_r (s)	T_s 5% (s)
0.997	0.66	0.5322	0.1396	0.2879	21.9	57.6
0.997	0.77	0.4024	0.1023	0.2164	29.7	77.5
0.997	1	0.2373	0.0566	0.126	39.1	138
0.997	1.24	0.1480	0.0332	0.0777	91.0	234
0.997	1.5	0.0959	0.0202	0.0498	150.0	383

Table 4. Parameters and transient response of PID for different ρ with ARE

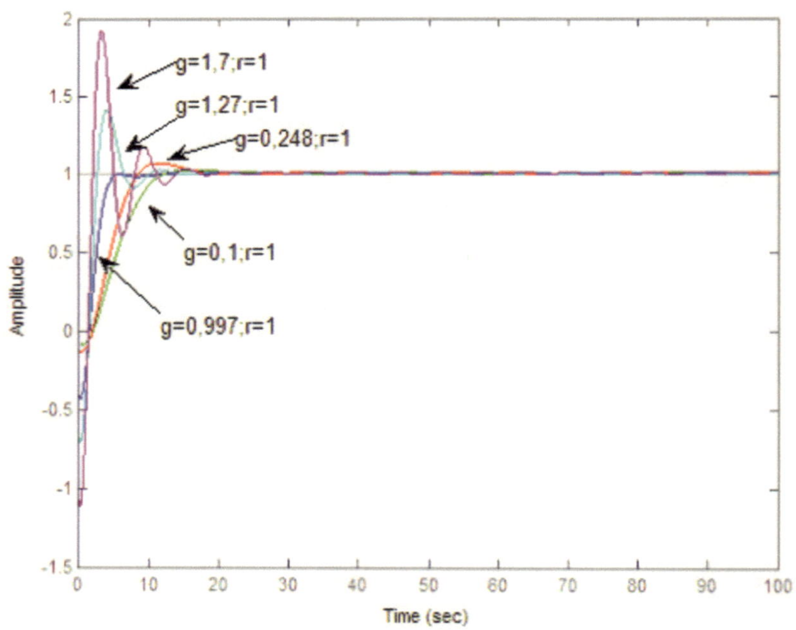

Fig. 5. Transient response for different γ using LMI

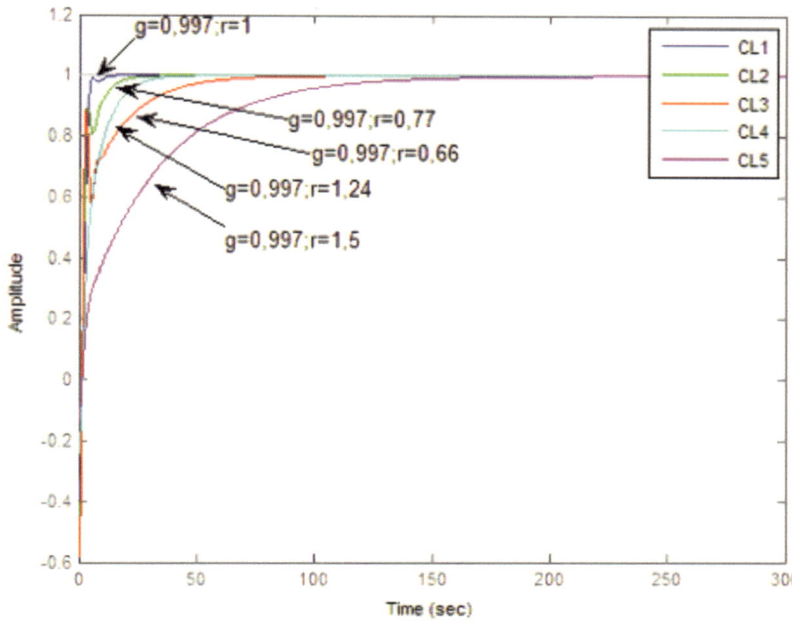

Fig. 6. Transient response for different ρ using LMI

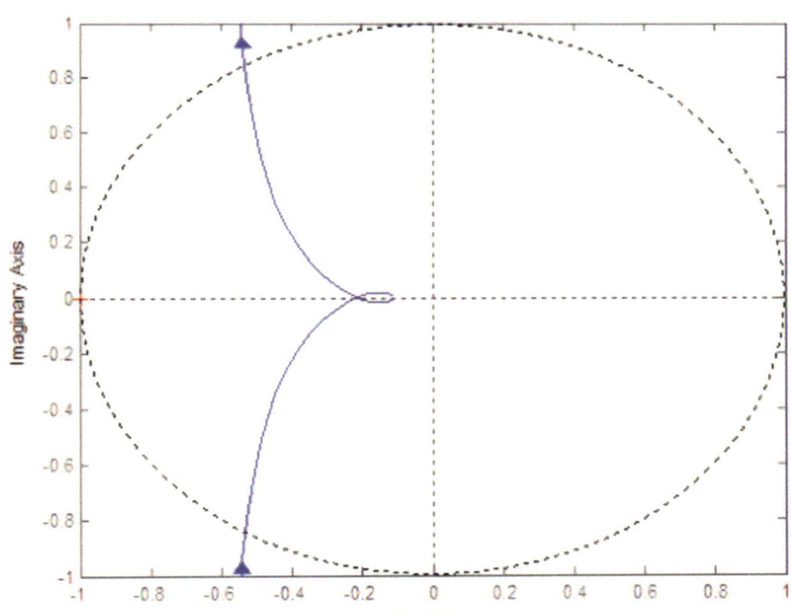

Fig. 7. Nyquist plot $\gamma = 0.248$ and $\rho = 1$ using LMI

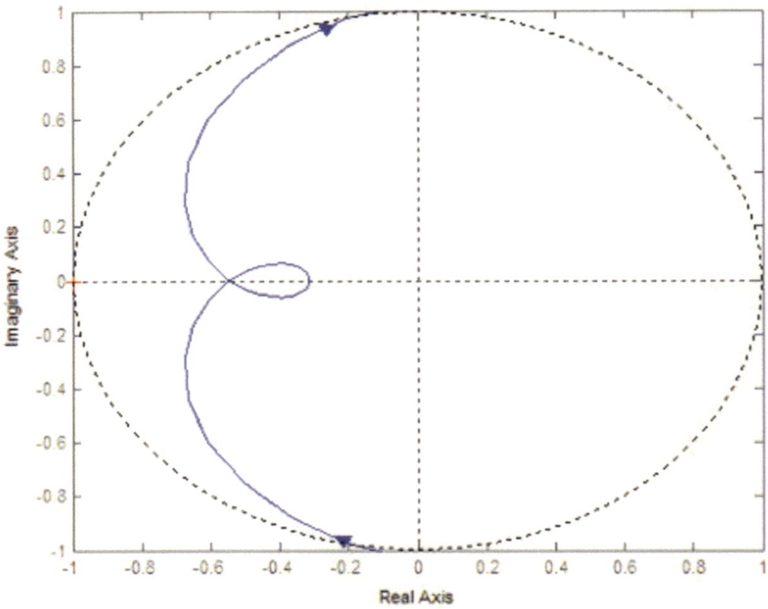

Fig. 8. Nyquist plot $\gamma = 0.997$ and $\rho = 0.66$ using LMI

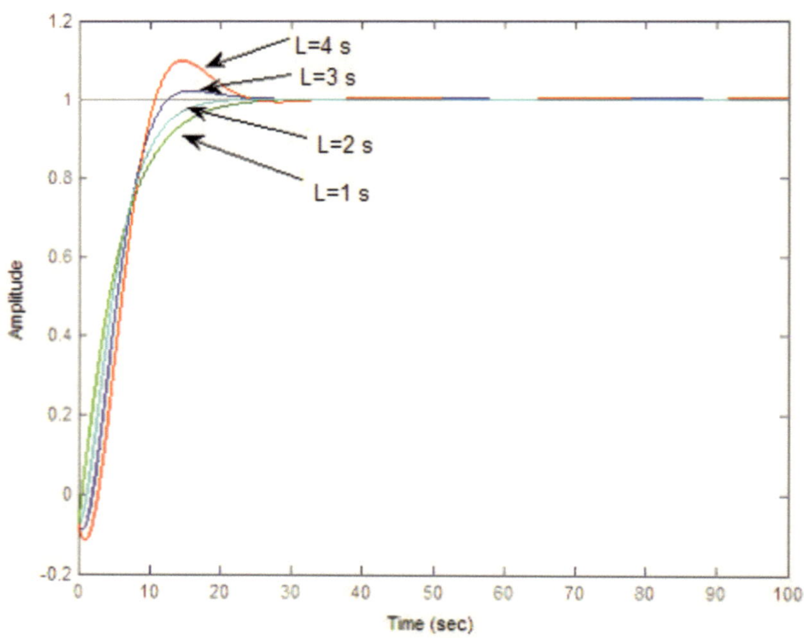

Fig. 9. Transient response for different d using LMI

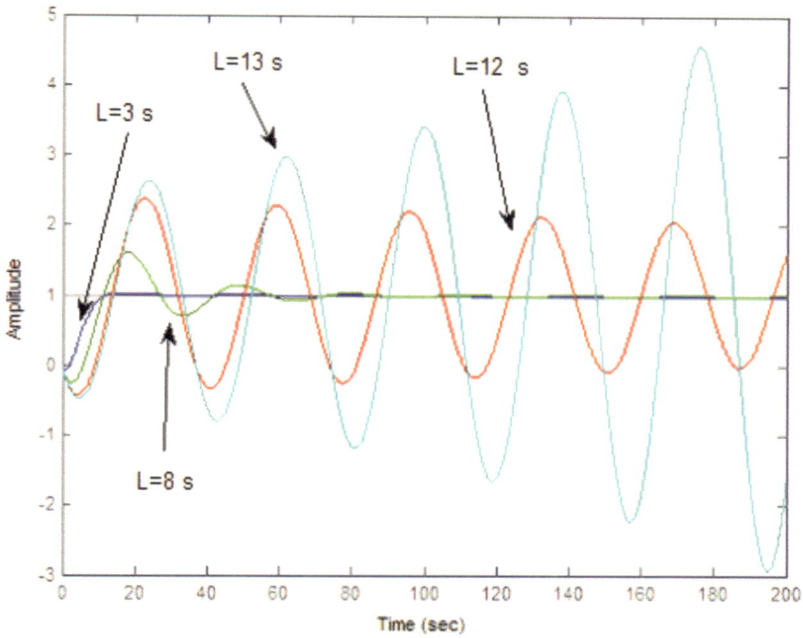

Fig. 10. Transient response for different bigger d using LMI

The simulation results are shown in Figure 5 and 6 for LMI, with γ and ρ are denoted by g and r respectively in the figure. The LMI method leads to faster transient response compared to the ARE method for all values of γ and ρ. Nyquist plots in Figure 7 and 8 show that the LMI method has small gain margin. In general, it also holds for phase margin except at $\gamma = 0.997$ and $\rho = 1.5$ where LMI has bigger phase margin.
In order to test the robustness to the specified delay time uncertainties, the obtained robust PID controller with parameter γ =0.1 and $\rho = 1$ is tested by perturbing the delay time in the range value of $d \in [1,4]$. The results of using LMI are shown in Figure 9 and 10 respectively. The LMI method yields faster transient responses where it tends to oscillate at bigger time delay. With the same parameters γ and ρ, the PID controller is subjected to bigger delay time than the design specification. The LMI method can handle the ratio of delay time and time constant $L/\tau \leq 12$ s while the ARE method has bigger ratio $L/\tau \leq 43$ s. In summary, simulation results showed that LMI method produced fast transient response of the closed loop system with no overshoot and the capability in handling uncertainties. If the range of the uncertainties is known, the stability and the performance of the closed loop system will be guaranteed.

5. Conclusion

The paper has presented a model based method to select the optimum setting of the PID controller using robust H_∞ dissipative integral backstepping method with state feedback synthesis. The state feedback gain is found by using LMI solution of Algebraic Riccati Inequality (ARI). The paper also derives the synthesis of the state feedback gain of robust H_∞ dissipative integral backstepping method. The parameters of the PID controller are

calculated by using two new parameters which correspond to the infinity norm and the weighting of the control signal of the closed loop system.
The LMI method will guarantee the stability and the performance of the closed loop system if the range of the uncertainties is included in the LFT representation of the model. The LFT representation in the design can also be extended to include plant uncertainties, multiplicative perturbation, pole clustering, etc. Hence, the problem will be considered as multi objectives LMI based robust H_∞ PID controller problem. The proposed approach can be directly extended for MIMO control problem with MIMO PID controller.

6. References

Astrom, K.J. & Hagglund, T. (1995). *PID Controllers: Theory, Design, and Tuning*, second ed., Instrument Society of America, ISBN 1-55617-516-7, Research Triangle Park, North Carolina - USA

Boyd, S.P. & Barrat, C.H. (1991). *Linear Controller Design: Limits of Performance*, Prentice Hall Inc., ISBN 0-1353-8687-X, New Jersey

Boyd, S.; El Ghaoui, L., Feron, E. & Balakrishnan, V. (1994). *Linear Matrix Inequalities in System and Control Theory*, SIAM Studies 15, ISBN 0-89871-485-0, Philadelphia

Balakrishnan, V. & Wang, F. (2000). Semidefinite programming in systems and control, In: *Handbook on Semidefinite Programming*, Wolkowics, H; Saigal, R. & Vandenberghe, L. (Ed.), pp. 421-441, Kluwer Academic Pub., ISBN 0-7923-7771-0, Boston

Francis, B.A. & Wonham, W.M. (1976). The internal model principle of control theory, *Automatica*, Vol. 12, pp. 457-465, ISSN 0005-1098

Fung, H.W.; Wang, Q.G. & Lee, T.H. (1998). PI tuning in terms of gain and phase margins, *Automatica*, Vol. 34, pp. 1145-1149, ISSN 0005-1098

Gahinet, P. & Apkarian, P. (1994). A linear matrix inequality approach to $H\infty$control, *Inter. Journal of Robust Nonlinear Control*, Vol. 4, pp. 421-448, ISSN 1099-1239

Ge, M.; Chiu, M.S. & Wang, Q.G. (2002). Robust PID controller design via LMI approach, *Journal of Process Control*, Vol. 12, pp. 3-13, ISSN 0959-1524

Green, M. & Limebeer, D.J. (1995). *Linear Robust Control*, Englewood Cliffs, Prentice Hall Inc., ISBN 0-13-102278-4, New Jersey

Goncalves, E.N.; Palhares, R.M. & Takahashi, R.H.C. (2008). A novel approach for $H_2/H\infty$ robust PID synthesis for uncertain systems, *Journal of Process Control*, Vol. 18, pp. 19-26, ISSN 0959-1524

Hara, S.; Iwasaki, T. & Shiokata, D. (2006). Robust PID control using generalized KYP synthesis, *IEEE Control Systems Magazine*, Feb., pp. 80-91, ISSN 0272-1708

Hill, D.J. & Moylan, P.J. (1977). Stability results for nonlinear feedback systems, *Automatica*, Vol. 13, pp. 377-382, ISSN 0005-1098

Hops, J.; Swing, B., Phelps, B., Sudweeks, B., Pane, J. & Kinslow, J. (2004). Non-deterministic DUT behavior during functional testing of high speed serial busses: challenges and solutions, *Proceedings International Test Conference*, ISBN 0-7803-8581-0, 26-28 Oct. 2004, IEEE, Charlotte, NC, USA

Hwang, S.H. & Chang, H.C. (1987). A theoritical examination of closed-loop properties and tuning methods of single-loop PI controllers, *Chemical Engineering Science*, Vol. 42, pp. 2395-2415, ISSN 0009-2509

Joelianto, E. & Tommy. (2003). A robust DC to DC buckboost converter using PID H∞-backstepping controller, *Proceedings of Int. Confer. on Power Electronics and Drive Systems (PEDS)*, pp. 591-594, ISBN 0-7803-7885-7, 17-20 Nov. 2003, Singapore

Joelianto, E.; Sutarto, H.Y. & Wicaksono, A. (2008). Compensation of delay time uncertainties industrial control ethernet networks using LMI based robust H∞ PID controller, *Proceedings of 5th Int. Confer. Wireless Optical Comm. Networks (WOCN)*, pp. 1-5, ISBN 978-1-4244-1979-1, 5-7 May 2008, Surabaya - Indonesia

Joelianto, E. & Hosana (2009). Loop-back action latency performance of an industrial data communication protocol on a PLC ethernet network, *Internetworking Indonesia Journal*, Vol. I, No.1, pp. 11-18, ISSN 1942-9703

Joelianto, E. & Williamson, D. (2009). Transient response improvement of feedback control systems using hybrid reference control, *International Journal of Control*, Vol. 81, No. 10, pp. 1955-1970, ISSN 0020-7179 print/ISSN 1366-5820 online

Jones, M. (2006). Designning for real time embedded ethernet, *The Industrial Ethernet Book* , Vol. 35, pp. 38-41

Krstic, M.; Kanellakopoulos, I. & Kokotovic, P. (1995). *Nonlinear and Adaptive Control Design*, John Wiley and Sons Inc., ISBN 0-4711-2732-9, USA

Liu, K.Z. & He, R. (2006). A simple derivation of ARE solutions to the standard H∞ control problem based on LMI solution, *Systems & Control Letters*, Vol. 5 , pp. 487-493, ISSN 0167-6911

Petersen, I.R.; Anderson, B.D.O. & Jonkheere, E. (1991). A first principles solution to the non-singular H∞ control problem, *Inter. Journal of Robust and Nonlinear Control*, Vol. 1, pp. 171-185, ISSN 1099-1239

Petersen, I.R.; Ugrinovskii, V.A. & Savkin, A.V. (2000). *Robust Control Design using H∞ Methods*, Springer, ISBN 1-8523-3171-2, London

Scherer, C. (1990). *The Riccati Inequality and State-Space H∞-Optimal Control*, PhD Dissertation, Bayerischen Julius Maximilans, Universitat Wurzburg, Wurzburg

Scherer, C.; Gahinet, P. & Chilali, M. (1997). Multiobjective output-feedback control via LMI optimization, *IEEE Trans. on Automatic Control*, Vol. 42, pp. 896-911, ISSN 0018-9286

Scherer, C. & Weiland, S. (1999). *Linear Matrix Inequalities in Control*, Lecture Notes DISC Course, version 2.0. http://www.er.ele.tue.nl/SWeiland/lmi99.htm

Shinskey, F.G. (1996). *Process Control Systems: Application, Design and Tuning*, fourth ed., McGraw-Hill, ISBN 0-0705-7101-5, Boston

Wang, Q.G.; Lin, C., Ye, Z., Wen, G., He, Y. & Hang, C.C. (2007). A quasi-LMI approach to computing stabilizing parameter ranges of multi-loop PID controllers, *Journal of Process Control*, Vol. 17, pp. 59-72, ISSN 0959-1524

Williamson, D. & Moore, J.B. (1971). Three term controller parameter selection using suboptimal regulator theory, *IEEE Trans. on Automatic Control*, Vol. 16, pp. 82-83, ISSN 0018-9286

Wong, S.K.P. & Seborg, D.E. (1988). Control strategy for single-input-single-output nonlinear systems with time delays, *International Journal of Control*, Vol. 48, pp. 2303-2327, ISSN 0020-7179 print/ISSN 1366-5820 online

Yuliar, S.; James, M.R. & Helton, J.W. (1998). Dissipative control systems synthesis with full state feedback, *Mathematics of Control, Signals, and Systems*, Vol. 11, pp. 335-356, ISSN 0932-4194 print/ISSN 1435-568X online

Yuliar, S.; Samyudia, Y. & Kadiman, K. (1997). General linear quadratic dissipative output feedback control system synthesis, *Proceedings of 2nd Asian Control Confererence (ASCC)*, pp. 659-662, ISBN 89-950038-0-4-94550, Seoul-Korea, ACPA

Zhou, K. & Doyle, J.C. (1998). *Essential of Robust Control*, Prentice Hall Inc., ISBN 0-13-790874-1, USA

Ziegler, J.G. & Nichols, N.B. (1942). Optimum setting for automatic controllers, *Transactions of The ASME*, Vol. 64, pp. 759-769

5

Robust H_∞ Tracking Control of Stochastic Innate Immune System Under Noises

Bor-Sen Chen, Chia-Hung Chang and Yung-Jen Chuang
National Tsing Hua University
Taiwan

1. Introduction

The innate immune system provides a tactical response, signaling the presence of 'non-self' organisms and activating B cells to produce antibodies to bind to the intruders' epitopic sites. The antibodies identify targets for scavenging cells that engulf and consume the microbes, reducing them to non-functioning units (Stengel et al., 2002b). The antibodies also stimulate the production of cytokines, complement factors and acute-phase response proteins that either damage an intruder's plasma membrane directly or trigger the second phase of immune response. The innate immune system protects against many extracellular bacteria or free viruses found in blood plasma, lymph, tissue fluid, or interstitial space between cells, but it cannot clean out microbes that burrow into cells, such as viruses, intracellular bacteria, and protozoa (Janeway, 2005; Lydyard et al., 2000; Stengel et al., 2002b). The innate immune system is a complex system and the obscure relationships between the immune system and the environment in which several modulatory stimuli are embedded (e.g. antigens, molecules of various origin, physical stimuli, stress stimuli).This environment is noisy because of the great amount of such signals. The immune noise has therefore at least two components: (a) the internal noise, due to the exchange of a network of molecular and cellular signals belonging to the immune system during an immune response or in the homeostasis of the immune system. The concept of the internal noise might be viewed in biological terms as a status of sub-inflammation required by the immune response to occur; (b) the external noise, the set of external signals that target the immune system (and hence that add noise to the internal one) during the whole life of an organism.

For clinical treatment of infection, several available methods focus on killing the invading microbes, neutralizing their response, and providing palliative or healing care to other organs of the body. Few biological or chemical agents have just one single effect; for example, an agent that kills a virus may also damage healthy 'self' cells. A critical function of drug discovery and development is to identify new compounds that have maximum intended efficacy with minimal side effects on the general population. These examples include antibiotics as microbe killers; interferons as microbe neutralizers; interleukins, antigens from killed (i.e. non-toxic) pathogens, and pre-formed and monoclonal antibodies as immunity enhancers (each of very different nature); and anti-inflammatory and anti-histamine compounds as palliative drugs (Stengel et al., 2002b).

Recently, several models of immune response to infection (Asachenkov, 1994; Nowak & May, 2000; Perelson & Weisbuch, 1997; Rundell et al., 1995) with emphasis on the human-immunodeficiency virus have been reported (Nowak et al., 1995; Perelson et al., 1993; Perelson et al., 1996; Stafford et al., 2000). Norbert Wiener (Wiener, 1948) and Richard Bellman (Bellman, 1983) appreciated and anticipated the application of mathematical analysis for treatment in a broad sense, and Swan made surveys on early optimal control applications to biomedical problems (Swan, 1981). Kirschner (Kirschner et al., 1997) offers an optimal control approach to HIV treatment, and intuitive control approaches are presented in (Bonhoeffer et al., 1997; De Boer & Boucher, 1996; Wein et al., 1998; Wodarz & Nowak, 1999, 2000).

The dynamics of drug response (pharmacokinetics) are modeled in several works (Robinson, 1986; van Rossum et al., 1986) and control theory is applied to drug delivery in other studies (Bell & Katusiime, 1980; Carson et al., 1985; Chizeck & Katona, 1985; Gentilini et al., 2001; Jelliffe, 1986; Kwong et al., 1995; Parker et al., 1996; Polycarpou & Conway, 1995; Schumitzky, 1986). Recently, Stengel (Stengel et al., 2002a) presented a simple model for the response of the innate immune system to infection and therapy, reviewed the prior method and results of optimization, and introduced a significant extension to the optimal control of enhancing the immune response by solving a two-point boundary-value problem via an iterative method. Their results show that not only the progression from an initially life-threatening state to a controlled or cured condition but also the optimal history of therapeutic agents that produces that condition. In their study, the therapeutic method is extended by adding linear-optimal feedback control to the nominal optimal solution. However, the performance of quadratic optimal control for immune systems may be decayed by the continuous exogenous pathogen input, which is considered as an environmental disturbance of the immune system. Further, some overshoots may occur in the optimal control process and may lead to organ failure because the quadratic optimal control only minimizes a quadratic cost function that is only the integration of squares of states and allows the existence of overshoot (Zhou et al., 1996).

Recently, a minimax control scheme of innate immune system is proposed by the dynamic game theory approach to treat the robust control with unknown disturbance and initial condition (Chen et al., 2008). They consider unknown disturbance and initial condition as a player who wants to destroy the immune system and a control scheme as another player to protect the innate immune system against the disturbance and uncertain initial condition. However, they assume that all state variables are available. It is not the case in practical application.

In this study, a robust H_∞ tracking control of immune response is proposed for therapeutic enhancement to track a desired immune response under stochastic exogenous pathogen input, environmental disturbances and uncertain initial states. Furthermore, the state variables may not be all available and the measurement is corrupted by noises too. Therefore, a state observer is employed for state estimation before state feedback control of stochastic immune systems. Since the statistics of these stochastic factors may be unknown or unavailable, the H_∞ observer-based control methodology is employed for robust H_∞ tracking design of stochastic immune systems. In order to attenuate the stochastic effects of stochastic factors on the tracking error, their effects should be considered in the stochastic H_∞ tracking control procedure from the robust design perspective. The effect of all possible stochastic factors on the tracking error to a desired immune response, which is generated by a desired model, should be controlled below a prescribed level for the enhanced immune systems, i.e. the proposed robust H_∞ tracking control need to be designed from the stochastic H_∞ tracking perspective. Since the stochastic innate immune system is highly nonlinear, it is not easy to solve the robust observer-based tracking control problem by the stochastic nonlinear H∞ tracking method directly.

Recently, fuzzy systems have been employed to efficiently approximate nonlinear dynamic systems to efficiently treat the nonlinear control problem (Chen et al., 1999,2000; Li et al., 2004; Lian et al., 2001). A fuzzy model is proposed to interpolate several linearized stochastic immune systems at different operating points to approximate the nonlinear stochastic innate immune system via smooth fuzzy membership functions. Then, with the help of fuzzy approximation method, a fuzzy H∞ tracking scheme is developed so that the H∞ tracking control of stochastic nonlinear immune systems could be easily solved by interpolating a set of linear H∞ tracking systems, which can be solved by a constrained optimization scheme via the linear matrix inequality (LMI) technique (Boyd, 1994) with the help of Robust Control Toolbox in Matlab (Balas et al., 2007). Since the fuzzy dynamic model can approximate any nonlinear stochastic dynamic system, the proposed H∞ tracking method via fuzzy approximation can be applied to the robust control design of any model of nonlinear stochastic immune system that can be T-S fuzzy interpolated. Finally, a computational simulation example is given to illustrate the design procedure and to confirm the efficiency and efficacy of the proposed H∞ tracking control method for stochastic immune systems under external disturbances and measurement noises.

2. Model of innate immune response

A simple four-nonlinear, ordinary differential equation for the dynamic model of infectious disease is introduced here to describe the rates of change of pathogen, immune cell and antibody concentrations and as an indicator of organic health (Asachenkov, 1994; Stengel et al., 2002a). In general, the innate immune system is corrupted by environmental noises. Further, some state variable cannot be measured directly and the state measurement may be corrupted by measurement noises. A more general dynamic model will be given in the sequel.

$$\begin{aligned}
\dot{x}_1 &= (a_{11} - a_{12}x_3)x_1 + b_1u_1 + w_1 \\
\dot{x}_2 &= a_{21}(x_4)a_{22}x_1x_3 - a_{23}(x_2 - x_2^*) + b_2u_2 + w_2 \\
\dot{x}_3 &= a_{31}x_2 - (a_{32} + a_{33}x_1)x_3 + b_3u_3 + w_3 \\
\dot{x}_4 &= a_{41}x_1 - a_{42}x_4 + b_4u_4 + w_4 \\
y_1 &= c_1x_2 + n_1,\ y_2 = c_2x_3 + n_2,\ y_3 = c_3x_4 + n_3 \\
a_{21}(x_4) &= \begin{cases} \cos(\pi x_4), & 0 \le x_4 \le 0.5 \\ 0, & 0.5 \le x_4 \end{cases}
\end{aligned} \tag{1}$$

where x_1 denotes the concentration of a pathogen that expresses a specific foreign antigen; x_2 denotes the concentration of immune cells that are specific to the foreign antigen; x_3 denotes the concentration of antibodies that bind to the foreign antigen; x_4 denotes the characteristic of a damaged organ [x_4=0: healthy, $x_4 \ge 1$: dead]. The combined therapeutic control agents and the exogenous inputs are described as follows: u_1 denotes the pathogen killer's agent; u_2 denotes the immune cell enhancer; u_3 denotes the antibody enhancer; u_4 denotes the organ healing factor (or health enhancer); w_1 denotes the rate of continuing introduction of exogenous pathogens; $w_2 \sim w_4$ denote the environmental disturbances or unmodeled errors and residues; $w_1 \sim w_4$ are zero mean white noises, whose covariances are uncertain or unavailable; and $a_{21}(x_4)$ is a nonlinear function that describes the mediation of immune cell generation by the damaged cell organ. And if there is no antigen, then the immune cell maintains the steady equilibrium value of x_2^*. The parameters have been chosen to produce a system that recovers naturally from the pathogen infections (without treatment) as a function of initial conditions during a period of times. Here, y_1, y_2, y_3 are the measurements of the corresponding states; c_1, c_2, c_3 are the measurement scales; and n_1, n_2, n_3 are the measurement noises. In this study, we assume the measurement of pathogen x_1 is unavailable. For the benchmark example in (1), both parameters and time units are abstractions, as no specific disease is addressed. The state and control are always positive because concentrations cannot go below zero, and organ death is indicated when $x_4 \geq 1$. The structural relationship of system variables in (1) is illustrated in Fig. 1. Organ health mediates immune cell production, inferring a relationship between immune response and fitness of the individual. Antibodies bind to the attacking antigens, thereby killing pathogenic microbes directly, activating complement proteins, or triggering an attack by phagocytic cells, e.g. macrophages and neutrophils. Each element of the state is subject to an independent control, and new microbes may continue to enter the system. In reality, however, the concentration of invaded pathogens is hardly to be measured. We assume that only the rest of three elements can be measured with measurement noises by medical devices or other biological techniques such as an immunofluorescence microscope, which is a technique based on the ability of antibodies to recognize and bind to specific molecules. It is then possible to detect the number of molecules easily by using a fluorescence microscope (Piston, 1999).

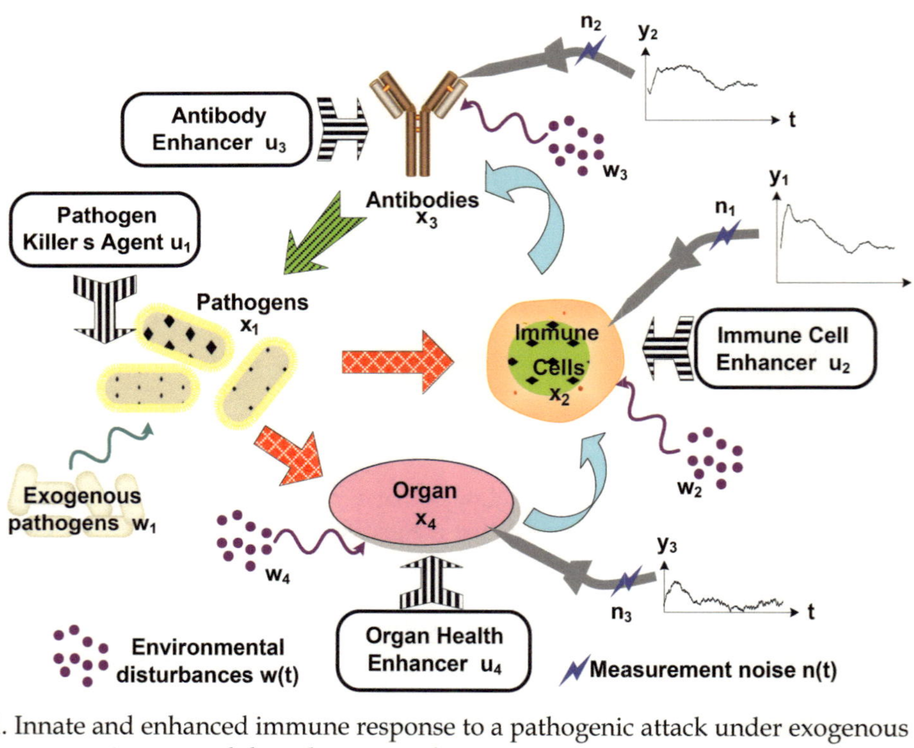

Fig. 1. Innate and enhanced immune response to a pathogenic attack under exogenous pathogens, environmental disturbances, and measurement noises.

Several typical uncontrolled responses to increasing levels of initial pathogen concentration under sub-clinical, clinical, chronic, and lethal conditions have been discussed and shown in Fig. 2 (Stengel et al., 2002a). In general, the sub-clinical response would not require medical examination, while the clinical case warrants medical consultation but is self-healing without intervention. Pathogen concentration stabilizes at non-zero values in the chronic case, which is characterized by permanently degraded organ health, and pathogen concentration diverges without treatment in the lethal case and kills the organ (Stengel et al., 2002b). Finally, a more general disease dynamic model for immune response could be represented as

$$\begin{aligned}\dot{x}(t) &= f(x(t)) + g(x(t))u(t) + Dw(t), \quad x(0) = x_0 \\ y(t) &= c(x(t)) + n(t)\end{aligned} \tag{2}$$

where $x(t) \in \mathbf{R}^{n\times 1}$ is the state vector; $u(t) \in \mathbf{R}^{m\times 1}$ is the control agent; $w(t) \in \mathbf{R}^{n\times 1}$ includes exogenous pathogens, environmental disturbances or model uncertainty. $y(t) \in \mathbf{R}^{l\times 1}$ is the measurement output; and $n(t) \in \mathbf{R}^{l\times 1}$ is the measurement noises. We assume that $w(t)$ and $n(t)$ are independent stochastic noises, whose covariances may be uncertain or unavailable. All possible nonlinear interactions in the immune system are represented by $f(x(t))$.

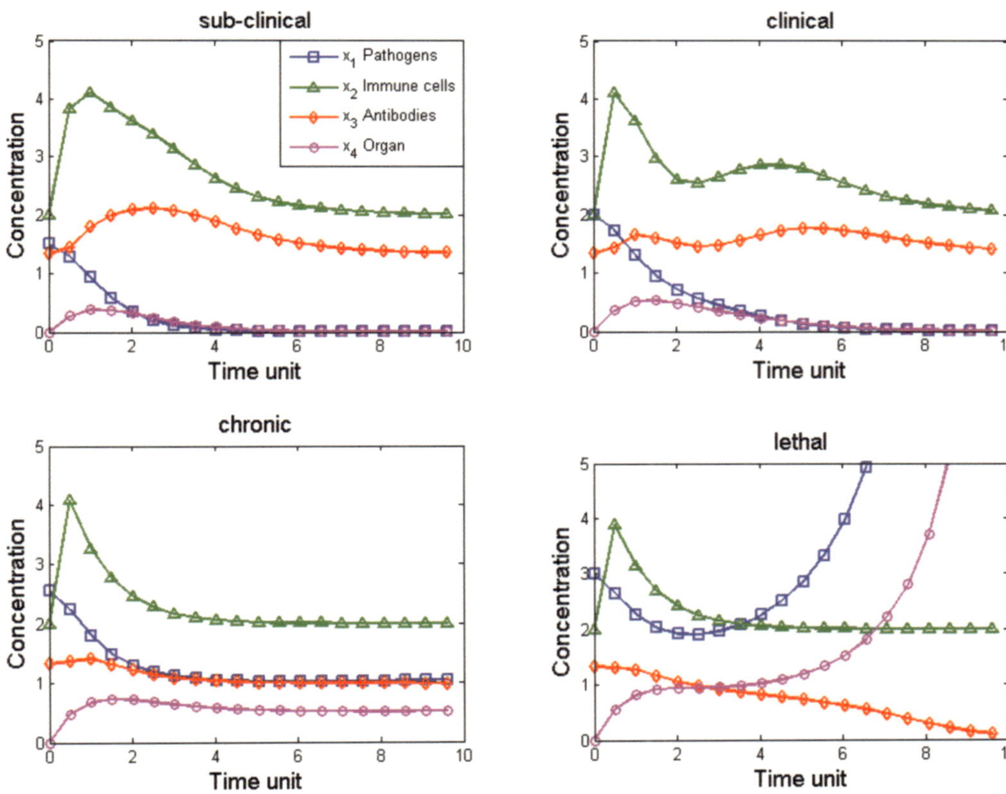

Fig. 2. Native immune responses to attack by different pathogens which are sub-clinical, clinical, chronic, and lethal conditions (Stengel et al., 2002a).

3. Robust H$_\infty$ Therapeutic Control of Stochastic Innate Immune Response

Our control design purpose for nonlinear stochastic innate immune system in (2) is to specify a state feedback control $u(t)=k(x(t)-x_d(t))$ so that the immune system can track the desired response $x_d(t)$. Since the state variables are unavailable for feedback tracking control, the state variables have to be estimated for feedback tracking control $u(t)=k(\hat{x}(t)-x_d(t))$. Suppose the following observer-based control with $y(t)$ as input and $u(t)$ as output is proposed for robust H$_\infty$ tracking control.

$$\begin{aligned}\dot{\hat{x}}(t)&=f(\hat{x}(t))+g(\hat{x}(t))u(t)+l(\hat{x}(t))(y(t)-c(\hat{x}(t)))\\ u(t)&=k(\hat{x}(t)-x_d(t))\end{aligned} \tag{3}$$

where the observer-gain $l(\hat{x}(t))$ is to be specified so that the estimation error $e(t)=x(t)-\hat{x}(t)$ can be as small as possible and control gain $k(\hat{x}(t)-x_d(t))$ is to be specified so that the system states $x(t)$ can come close to the desired state responses $x_d(t)$ from the stochastic point of view.

Consider a reference model of immune system with a desired time response described as

$$\dot{x}_d(t)=A_d x_d(t)+r(t) \tag{4}$$

where $x_d(t)\in\mathbf{R}^{n\times 1}$ is the reference state vector; $A_d\in\mathbf{R}^{n\times n}$ is a specific asymptotically stable matrix and $r(t)$ is a desired reference signal. It is assumed that $x_d(t)$, $\forall t>0$ represents a desired immune response for nonlinear stochastic immune system in (2) to follow, i.e. the therapeutic control is to specify the observer-based control in (3) such that the tracking error $\tilde{x}(t)=x(t)-x_d(t)$ must be as small as possible under the influence of uncertain exogenous pathogens and environmental disturbances $w(t)$ and measurement noises $n(t)$. Since the measurement noises $n(t)$, the exogenous pathogens and environmental disturbances $w(t)$ are uncertain and the reference signal $r(t)$ could be arbitrarily assigned, the robust H$_\infty$ tracking control design in (3) should be specified so that the stochastic effect of three uncertainties $w(t)$, $n(t)$ and $r(t)$ on the tracking error could be set below a prescribed value ρ^2, i.e. both the stochastic H$_\infty$ reference tracking and H$_\infty$ state estimation should be achieved simultaneously under uncertain $w(t)$, $n(t)$ and $r(t)$.

$$\frac{\mathbf{E}\left[\int_0^{t_f}(\tilde{x}^T(t)Q_1\tilde{x}(t)+e^T(t)Q_2e(t))dt\right]}{\mathbf{E}\left[\int_0^{t_f}(w^T(t)w(t)+n^T(t)n(t)+r^T(t)r(t))dt\right]}\le\rho^2 \tag{5}$$

where the weighting matrices Q_i are assumed to be diagonal as follows

$$Q_i=\begin{bmatrix}q_{11}^i & 0 & 0 & 0\\ 0 & q_{22}^i & 0 & 0\\ 0 & 0 & q_{33}^i & 0\\ 0 & 0 & 0 & q_{44}^i\end{bmatrix},\quad i=1,2.$$

The diagonal element q_{ii}^j of Q_i denotes the punishment on the corresponding tracking error and estimation error. Since the stochastic effect of $w(t)$, $r(t)$ and $n(t)$ on tracking error $\tilde{x}(t)$

and estimation error $e(t)$ is prescribed below a desired attenuation level ρ^2 from the energy point of view, the robust H∞ stochastic tracking problem of equation (5) is suitable for the robust H∞ stochastic tracking problem under environmental disturbances $w(t)$, measurement noises $n(t)$ and changeable reference $r(t)$, which are always met in practical design cases.

Remark 1:

If the environmental disturbances $w(t)$and measurement noises $n(t)$ are deterministic signals, the expectative symbol $\mathbf{E}[\bullet]$ in (5) can be omitted.

Let us denote the augmented vector $\bar{x} = \begin{bmatrix} e(t) \\ x(t) \\ x_d(t) \end{bmatrix}$, then we get the dynamic equation of the augmented stochastic system as

$$\dot{\bar{x}}(t) = \begin{bmatrix} \dot{e}(t) \\ \dot{x}(t) \\ \dot{x}_d(t) \end{bmatrix} = \begin{bmatrix} f(x) - f(\hat{x}) + k(\hat{x} - x_d)(g(x) - g(\hat{x})) + l(\hat{x})(c(x) - c(\hat{x})) \\ f(x) + k(\hat{x} - x_d)g(x) \\ A_d x_d \end{bmatrix} + \begin{bmatrix} I & 0 & 0 \\ 0 & D & 0 \\ 0 & 0 & I \end{bmatrix} \begin{bmatrix} n(t) \\ w(t) \\ r(t) \end{bmatrix} \tag{6}$$

The augmented stochastic system above can be represented in a general form by

$$\dot{\bar{x}}(t) = F(\bar{x}(t)) + \bar{D}\bar{w}(t) \tag{7}$$

where $F(\bar{x}(t)) = \begin{bmatrix} f(x) - f(\hat{x}) + k(\hat{x} - x_d)(g(x) - g(\hat{x})) + l(\hat{x})(c(x) - c(\hat{x})) \\ f(x) + k(\hat{x} - x_d)g(x) \\ A_d x_d \end{bmatrix}$, $\bar{w}(t) = \begin{bmatrix} n(t) \\ w(t) \\ r(t) \end{bmatrix}$, and $\bar{D} = \begin{bmatrix} I & 0 & 0 \\ 0 & D & 0 \\ 0 & 0 & I \end{bmatrix}$.

The robust H∞ stochastic tracking performance in (5) can be represented by

$$\frac{\mathbf{E}\left[\int_0^{t_f} \bar{x}^T(t)\bar{Q}\bar{x}(t)dt\right]}{\mathbf{E}\left[\int_0^{t_f} \bar{w}^T(t)\bar{w}(t)dt\right]} \le \rho^2 \text{ if } \bar{x}(0) = 0 \tag{8}$$

or $\mathbf{E}\left[\int_0^{t_f} \bar{x}^T(t)\bar{Q}\bar{x}(t)dt\right] \le \rho^2 \mathbf{E}\left[\int_0^{t_f} \bar{w}^T(t)\bar{w}(t)dt\right]$ where $\bar{Q} = \begin{bmatrix} Q_2 & 0 & 0 \\ 0 & Q_1 & -Q_1 \\ 0 & -Q_1 & Q_1 \end{bmatrix}$

If the stochastic initial condition $\bar{x}(0) \ne 0$ and is also considered in the H∞ tracking performance, then the above stochastic H∞ inequality should be modified as

$$\mathbf{E}\left[\int_0^{t_f} \bar{x}^T(t)\bar{Q}\bar{x}(t)dt\right] \le \mathbf{E}[V(\bar{x}(0))] + \rho^2 \mathbf{E}\left[\int_0^{t_f} \bar{w}^T(t)\bar{w}(t)dt\right] \tag{9}$$

for some positive function $V(\bar{x}(0))$. Then we get the following result.
Theorem 1: If we can specify the control gain $k(\hat{x}-x_d)$ and observer gain $l(\hat{x})$ in the observer-based control law in (3) for stochastic immune system (2) such that the following HJI has a positive solution $V(\bar{x}(t))>0$

$$\bar{x}(t)^T\bar{Q}\bar{x}(t)+\left(\frac{\partial V(\bar{x}(t))}{\partial \bar{x}(t)}\right)^T F(\bar{x}(t))+\frac{1}{4\rho^2}\left(\frac{\partial V(\bar{x}(t))}{\partial \bar{x}(t)}\right)^T \bar{D}\bar{D}^T\left(\frac{\partial V(\bar{x})}{\partial \bar{x}(t)}\right)<0 \tag{10}$$

Then the robust stochastic H_∞ tracking performance in (5) is achieved for a prescribed tracking performance ρ^2.
Proof: see *Appendix A*.
Since ρ^2 is a prescribed noise attenuation level of H_∞ tracking performance in (5), based on the analysis above, the optimal stochastic H_∞ tracking performance still need to minimize ρ^2 as follows

$$\rho_0^2=\min_{V(\bar{x})>0}\rho^2 \tag{11}$$

subject to $V(\bar{x}(t))>0$ and equation (10).
At present, there does not exist any analytic or numerical solution for (10) or (11) except in very simple cases.

4. Robust fuzzy observer-based tracking control design for stochastic innate immune system

Because it is very difficult to solve the nonlinear HJI in (10), no simple approach is available to solve the constrained optimization problem in (11) for robust model tracking control of stochastic innate immune system. Recently, the fuzzy T-S model has been widely applied to approximate the nonlinear system via interpolating several linearized systems at different operating points (Chen et al., 1999,2000; Takagi & Sugeno, 1985). Using fuzzy interpolation approach, the HJI in (10) can be replaced by a set of linear matrix inequalities (LMIs). In this situation, the nonlinear stochastic H_∞ tracking problem in (5) could be easily solved by fuzzy method for the design of robust H_∞ tracking control for stochastic innate immune response systems.
Suppose the nonlinear stochastic immune system in (1) can be represented by the Takagi-Sugeno (T-S) fuzzy model (Takagi & Sugeno, 1985). The T-S fuzzy model is a piecewise interpolation of several linearized models through membership functions. The fuzzy model is described by fuzzy If-Then rules and will be employed to deal with the nonlinear H_∞ tracking problem by fuzzy observer-based control to achieve a desired immune response under stochastic noises. The i-th rule of fuzzy model for nonlinear stochastic immune system in (1) is in the following form (Chen et al., 1999,2000).
Plant Rule *i*:

If $z_1(t)$ is F_{i1} and … and $z_g(t)$ is F_{ig},

$$\text{then } \begin{aligned}\dot{x}(t)&=\mathbf{A}_i x(t)+\mathbf{B}_i u(t)+Dw(t),\ i=1,2,3,\cdots,L\\ y(t)&=C_i x(t)+n(t)\end{aligned} \tag{12}$$

in which F_{ij} is the fuzzy set; $\mathbf{A}_i$, $\mathbf{B}_i$, and C_i are known constant matrices; L is the number of If-Then rules; g is the number of premise variables; and $z_1(t), z_2(t), \ldots, z_g(t)$ are the premise variables. The fuzzy system is inferred as follows (Chen et al., 1999,2000; Takagi & Sugeno, 1985)

$$\begin{aligned}\dot{x}(t) &= \frac{\sum_{i=1}^{L} \mu_i(z(t))[\mathbf{A}_i x(t) + \mathbf{B}_i u(t) + Dw(t)]}{\sum_{i=1}^{L} \mu_i(z(t))} \\ &= \sum_{i=1}^{L} h_i(z(t))[\mathbf{A}_i x(t) + \mathbf{B}_i u(t) + Dw(t)] \\ y &= \sum_{i=1}^{L} h_i(z(t)) C_i x(t) + n(t)\end{aligned} \tag{13}$$

where $\mu_i(z(t)) = \prod_{j=1}^{g} F_{ij}(z_j(t))$, $h_i(z(t)) = \frac{\mu_i(z(t))}{\sum_{i=1}^{L} \mu_i(z(t))}$, $z(t) = [z_1(t), z_2(t), \ldots, z_g(t)]$, and $F_{ij}(z_j(t))$ is the grade of membership of $z_j(t)$ in F_{ij}.

We assume

$$\mu_i(z(t)) \geq 0 \text{ and } \sum_{i=1}^{L} \mu_i(z(t)) > 0 \tag{14}$$

Therefore, we get

$$h_i(z(t)) \geq 0 \text{ and } \sum_{i=1}^{L} h_i(z(t)) = 1 \tag{15}$$

The T-S fuzzy model in (13) is to interpolate L stochastic linear systems to approximate the nonlinear system in (1) via the fuzzy basis functions $h_i(z(t))$. We could specify the parameter $\mathbf{A}_i$ and $\mathbf{B}_i$ easily so that $\sum_{i=1}^{L} h_i(z(t))\mathbf{A}_i x(t)$ and $\sum_{i=1}^{L} h_i(z(t))\mathbf{B}_i$ in (13) can approximate $F(x(t))$ and $g(x(t))$ in (2) by the fuzzy identification method (Takagi & Sugeno, 1985).

By using fuzzy *If-Then* rules interpolation, the fuzzy observer is proposed to deal with the state estimation of nonlinear stochastic immune system (1).

Observer Rule *i*:

If $z_1(t)$ is F_{i1} and … and $z_g(t)$ is F_{ig},

$$\text{then } \dot{\hat{x}}(t) = \mathbf{A}_i \hat{x}(t) + \mathbf{B}_i u(t) + L_i(y(t) - \hat{y}(t)),\ i = 1, 2, 3, \cdots, L \tag{16}$$

where L_i is the observer gain for the ith observer rule and $\hat{y}(t) = \sum_{i=1}^{L} h_i(z(t)) C_i \hat{x}(t)$.

The overall fuzzy observer in (16) can be represented as (Chen et al., 1999,2000)

$$\dot{\hat{x}}(t) = \sum_{i=1}^{L} h_i(z(t))\left[\mathbf{A}_i \hat{x}(t) + \mathbf{B}_i u(t) + L_i(y(t) - \hat{y}(t))\right] \tag{17}$$

Suppose the following fuzzy observer-based controller is employed to deal with the above robust H_∞ tracking control design

Control Rule *j*:

If $z_1(t)$ is F_{j1} and ... and $z_g(t)$ is F_{jg},

$$\text{then } u = \sum_{j=1}^{L} h_j(z(t))K_j(\hat{x}(t) - x_d(t)) \tag{18}$$

Remark 2:

1. The premise variables $z(t)$ can be measurable stable variables, outputs or combination of measurable state variables (Ma et al., 1998; Tanaka et al., 1998; Wang et al., 1996). The limitation of this approach is that some state variables must be measurable to construct the fuzzy observer and fuzzy controller. This is a common limitation for control system design of T-S fuzzy approach (Ma et al., 1998; Tanaka et al., 1998). If the premise variables of the fuzzy observer depend on the estimated state variables, i.e., $\hat{z}(t)$ instead of $z(t)$ in the fuzzy observer, the situation becomes more complicated. In this case, it is difficult to directly find control gains K_j and observer gains L_i. The problem has been discussed in (Tanaka et al., 1998).
2. The problem of constructing T-S fuzzy model for nonlinear systems can be found in (Kim et al., 1997; Sugeno & Kang, 1988).

Let us denote the estimation errors as $e(t) = x(t) - \hat{x}(t)$. The estimation errors dynamic is represented as

$$\begin{aligned}\dot{e}(t) &= \dot{x}(t) - \dot{\hat{x}}(t)\\ &= \sum_{i=1}^{L}\sum_{j=1}^{L} h_i(z(t))h_j(z(t))[\mathbf{A}_i x(t) + \mathbf{B}_i u(t) + Dw(t)] - \left[\mathbf{A}_i\hat{x}(t) + \mathbf{B}_i u(t) + L_i(y(t) - C_j\hat{x}(t))\right]\\ &= \sum_{i=1}^{L}\sum_{j=1}^{L} h_i(z(t))h_j(z(t))[(\mathbf{A}_i - L_iC_j)e(t) - L_in(t)] + Dw(t)\end{aligned}$$

After manipulation, the augmented system in (6) can be expressed as the following fuzzy approximation form

$$\dot{\bar{x}}(t) = \sum_{i=1}^{L} h_i(z(t))\sum_{j=1}^{L} h_j(z(t))\left[\bar{\mathbf{A}}_{ij}\bar{x}(t) + \bar{E}_i\bar{w}(t)\right] \tag{19}$$

where $\bar{\mathbf{A}}_{ij} = \begin{bmatrix} \mathbf{A}_i - L_iC_j & 0 & 0 \\ -\mathbf{B}_iK_j & \mathbf{A}_i + \mathbf{B}_iK_j & -\mathbf{B}_iK_j \\ 0 & 0 & A_d \end{bmatrix}$, $\bar{x}(t) = \begin{bmatrix} e(t) \\ x(t) \\ x_d(t) \end{bmatrix}$, $\bar{w}(t) = \begin{bmatrix} n(t) \\ w(t) \\ r(t) \end{bmatrix}$, $\bar{E}_i = \begin{bmatrix} -L_i & D & 0 \\ 0 & D & 0 \\ 0 & 0 & I \end{bmatrix}$.

Theorem 2: In the nonlinear stochastic immune system of (2), if $\bar{P} = \bar{P}^T > 0$ is the common solution of the following matrix inequalities:

$$\bar{\mathbf{A}}_{ij}^T\bar{P} + \bar{P}\bar{\mathbf{A}}_{ij} + \frac{1}{\rho^2}\bar{P}\bar{E}_i\bar{E}_i^T\bar{P} + \bar{Q} < 0,\; i, j = 1,2,\cdots,\mathrm{L} \tag{20}$$

then the robust H_∞ tracking control performance in (8) or (9) is guaranteed for a prescribed ρ^2.

In the above robust H∞ tracking control design, we don't need the statistics of disturbances, measurement noises and initial condition. We only need to eliminate their effect on the tracking error and state estimation error below a prescribed level ρ^2. To obtain the best H∞ tracking performance, the optimal H∞ tracking control problem can be formulated as the following minimization problem.

$$\rho_0^2 = \min_{\overline{P}>0} \rho^2 \tag{21}$$

subject to $\overline{P} > 0$ and equation (20).

Proof: see *Appendix B*.

In general, it is not easy to analytically determine the constrained optimization problem in (21). Fortunately, the optimal H∞ tracking control problem in (21) can be transferred into a minimization problem subject to some linear matrix inequalities (LMIs). The LMIP can be solved by a computationally efficient method using a convex optimization technique (Boyd, 1994) as described in the following.

By the Schur complements (Boyd, 1994), equation (20) is equivalent to

$$\begin{bmatrix} \overline{\mathbf{A}}_{ij}^T\overline{P} + \overline{P}\overline{\mathbf{A}}_{ij} + \overline{Q} & \overline{P}\overline{L} \\ \overline{L}^T\overline{P} & -\rho^2(HH^T)^{-1} \end{bmatrix} < 0 \tag{22}$$

where $\overline{L} = \begin{bmatrix} L_i & I & 0 \\ 0 & I & 0 \\ 0 & 0 & I \end{bmatrix}$ and $H = \begin{bmatrix} -I & 0 & 0 \\ 0 & D & 0 \\ 0 & 0 & I \end{bmatrix}$.

For the convenience of design, we assume $\overline{P} = \begin{bmatrix} P_{11} & 0 & 0 \\ 0 & P_{22} & 0 \\ 0 & 0 & P_{33} \end{bmatrix}$ and substitute it into (22) to obtain

$$\begin{bmatrix} S_{11} & S_{12} & 0 & Z_i & P_{11} & 0 \\ S_{21} & S_{22} & S_{23} & 0 & P_{22} & 0 \\ 0 & S_{32} & S_{33} & 0 & 0 & P_{33} \\ Z_i^T & 0 & 0 & -\rho^2 I & 0 & 0 \\ P_{11} & P_{22} & 0 & 0 & -\rho^2(DD^T)^{-1} & 0 \\ 0 & 0 & P_{33} & 0 & 0 & -\rho^2 I \end{bmatrix} < 0 \tag{23}$$

where $S_{11} = \mathbf{A}_i^T P_{11} + P_{11}\mathbf{A}_i - C_j^T Z_i^T - Z_i C_j + Q_2$

$S_{12} = S_{21}^T = -P_{22}\mathbf{B}_i K_j$

$S_{22} = (\mathbf{A}_i + \mathbf{B}_i K_j)^T P_{22} + P_{22}(\mathbf{A}_i + \mathbf{B}_i K_j) + Q_1$

$S_{23} = S_{32}^T = -P_{22}\mathbf{B}_i K_j - Q_1$

$S_{33} = A_d^T P_{33} + P_{33} A_d + Q_1$ and $Z_i = P_{11} L_i$.

Since five parameters P_{11}, P_{22}, P_{33}, K_j, and L_i should be determined from (23) and they are highly coupled, there are no effective algorithms for solving them simultaneously till now. In the following, a decoupled method (Tseng, 2008) is provided to solve these parameters simultaneously.

Note that (23) can be decoupled as

$$\begin{bmatrix} S_{11} & S_{12} & 0 & Z_i & P_{11} & 0 \\ S_{21} & S_{22} & S_{23} & 0 & P_{22} & 0 \\ 0 & S_{32} & S_{33} & 0 & 0 & P_{33} \\ Z_i^T & 0 & 0 & -\rho^2 I & 0 & 0 \\ P_{11} & P_{22} & 0 & 0 & -\rho^2(DD^T)^{-1} & 0 \\ 0 & 0 & P_{33} & 0 & 0 & -\rho^2 I \end{bmatrix} =$$

$$\begin{bmatrix} \begin{matrix}\mathbf{A}_i^T P_{11} + P_{11}\mathbf{A}_i - C^T Z_i^T \\ -Z_i C + Q_2 + \gamma P_{22}\end{matrix} & 0 & 0 & Z_i & P_{11} & 0 \\ 0 & -\gamma_1 I + Q_1 & -Q_1 & 0 & 0 & 0 \\ 0 & -Q_1 & \begin{matrix} A_d^T P_{33} + P_{33} A_d \\ +Q_1 + \gamma P_{22}\end{matrix} & 0 & 0 & P_{33} \\ Z^T & 0 & 0 & -\rho^2 I & 0 & 0 \\ P_{11} & 0 & 0 & 0 & \begin{matrix} -\rho^2(DD^T)^{-1} \\ +\gamma P_{22}\end{matrix} & 0 \\ 0 & 0 & P_{33} & 0 & 0 & -\rho^2 I \end{bmatrix} + \tag{24}$$

$$+\begin{bmatrix} -\gamma P_{22} & -P_{22}\mathbf{B}_i K_j & 0 & 0 & 0 & 0 \\ (-P_{22}\mathbf{B}_i K_j)^T & \begin{matrix} (\mathbf{A}_i + \mathbf{B}_i K_j)^T P_{22} \\ +P_{22}(\mathbf{A}_i + \mathbf{B}_i K_j) + \gamma_1 I \end{matrix} & -P_{22}\mathbf{B}_i K_j & 0 & P_{22} & 0 \\ 0 & (-P_{22}\mathbf{B}_i K_j)^T & -\gamma P_{22} & 0 & 0 & 0 \\ 0 & 0 & 0 & 0 & 0 & 0 \\ 0 & P_{22} & 0 & 0 & -\gamma P_{22} & 0 \\ 0 & 0 & 0 & 0 & 0 & 0 \end{bmatrix}$$

where γ and γ_1 are some positive scalars.

Lemma 1:

If

$$\begin{bmatrix} a_{11} & 0 & 0 & a_{14} & a_{15} & 0 \\ 0 & a_{22} & a_{23} & 0 & 0 & 0 \\ 0 & a_{32} & a_{33} & 0 & 0 & a_{36} \\ a_{41} & 0 & 0 & a_{44} & 0 & 0 \\ a_{51} & 0 & 0 & 0 & a_{55} & 0 \\ 0 & 0 & a_{63} & 0 & 0 & a_{66} \end{bmatrix} < 0 \tag{25}$$

and

$$\begin{bmatrix} b_{11} & b_{12} & 0 & 0 \\ b_{21} & b_{22} & b_{23} & b_{24} \\ 0 & b_{32} & b_{33} & 0 \\ 0 & b_{42} & 0 & b_{44} \end{bmatrix} < 0 \tag{26}$$

then

$$\begin{bmatrix} a_{11} & 0 & 0 & a_{14} & a_{15} & 0 \\ 0 & a_{22} & a_{23} & 0 & 0 & 0 \\ 0 & a_{32} & a_{33} & 0 & 0 & a_{36} \\ a_{41} & 0 & 0 & a_{44} & 0 & 0 \\ a_{51} & 0 & 0 & 0 & a_{55} & 0 \\ 0 & 0 & a_{63} & 0 & 0 & a_{66} \end{bmatrix} + \begin{bmatrix} b_{11} & b_{12} & 0 & 0 & 0 & 0 \\ b_{21} & b_{22} & b_{23} & 0 & b_{24} & 0 \\ 0 & b_{32} & b_{33} & 0 & 0 & 0 \\ 0 & 0 & 0 & 0 & 0 & 0 \\ 0 & b_{42} & 0 & 0 & b_{44} & 0 \\ 0 & 0 & 0 & 0 & 0 & 0 \end{bmatrix} < 0 \, . \tag{27}$$

Proof: see *Appendix C.*
From the above lemma, it is obvious that if

$$\begin{bmatrix} \begin{matrix} \mathbf{A}_i^T P_{11} + P_{11}\mathbf{A}_i - C_j^T Z_i^T \\ -Z_i C_j + Q_2 + \gamma P_{22} \end{matrix} & 0 & 0 & Z_i & P_{11} & 0 \\ 0 & -\gamma_1 I + Q_1 & -Q_1 & 0 & 0 & 0 \\ 0 & -Q_1 & \begin{matrix} A_d^T P_{33} + P_{33} A_d \\ +Q_1 + \gamma P_{22} \end{matrix} & 0 & 0 & P_{33} \\ Z^T & 0 & 0 & -\rho^2 I & 0 & 0 \\ P_{11} & 0 & 0 & 0 & \begin{matrix} -\rho^2 (DD^T)^{-1} \\ +\gamma P_{22} \end{matrix} & 0 \\ 0 & 0 & P_{33} & 0 & 0 & -\rho^2 I \end{bmatrix} < 0 \tag{28}$$

and

$$\begin{bmatrix} -\gamma P_{22} & -P_{22}\mathbf{B}_i K_j & 0 & 0 \\ (-P_{22}\mathbf{B}_i K_j)^T & \begin{matrix} (\mathbf{A}_i + \mathbf{B}_i K_j)^T P_{22} \\ +P_{22}(\mathbf{A}_i + \mathbf{B}_i K_j) + \gamma_1 I \end{matrix} & -P_{22}\mathbf{B}_i K_j & P_{22} \\ 0 & (-P_{22}\mathbf{B}_i K_j)^T & -\gamma P_{22} & 0 \\ 0 & P_{22} & 0 & -\gamma P_{22} \end{bmatrix} < 0 \tag{29}$$

then (23) holds.
Remark 3:
Note that (28) is related to the observer part (i.e., the parameters are P_{11}, P_{22}, P_{33}, and L_i) and (29) is related to the controller part (i.e., the parameters are P_{22} and K_j), respectively. Although the parameters P_{22}, K_j and γ are coupled nonlinearly, seven parameters P_{11}, P_{22}, P_{33}, K_j, L_i, γ and γ_1 can be determined by the following arrangement.

Note that, by the Schur complements (Boyd, 1994) equation (28) is equivalent to

$$\begin{bmatrix} \mathbf{A}_i^T P_{11} + P_{11}\mathbf{A}_i - C^T Z_i^T - Z_i C + Q_2 & 0 & 0 & Z_i & P_{11} & 0 & I & 0 & 0 \\ 0 & -\gamma_1 I + Q_1 & -Q_1 & 0 & 0 & 0 & 0 & 0 & 0 \\ 0 & -Q_1 & A_d^T P_{33} + P_{33} A_d + Q_1 & 0 & 0 & P_{33} & 0 & I & 0 \\ Z_i^T & 0 & 0 & -\rho^2 I & 0 & 0 & 0 & 0 & 0 \\ P_{11} & 0 & 0 & 0 & -\rho^2 (DD^T)^{-1} & 0 & 0 & 0 & I \\ 0 & 0 & P_{33} & 0 & 0 & -\rho^2 I & 0 & 0 & 0 \\ I & 0 & 0 & 0 & 0 & 0 & -\gamma^{-1} W_{22} & 0 & 0 \\ 0 & 0 & I & 0 & 0 & 0 & 0 & -\gamma^{-1} W_{22} & 0 \\ 0 & 0 & 0 & 0 & I & 0 & 0 & 0 & -\gamma^{-1} W_{22} \end{bmatrix} < 0 \tag{30}$$

where $W_{22} = P_{22}^{-1}$, and equation (29) is equivalent to

$$\begin{bmatrix} -\gamma W_{22} & -B_i Y_j & 0 & 0 & 0 \\ (-B_i Y_j)^T & W_{22}\mathbf{A}_i^T + \mathbf{A}_i W_{22} + Y_j^T \mathbf{B}_i^T + \mathbf{B}_i Y_j & -B_i Y_j & W_{22} & W_{22} \\ 0 & (-B_i Y_j)^T & -\gamma W_{22} & 0 & 0 \\ 0 & W_{22} & 0 & -\gamma W_{22} & 0 \\ 0 & W_{22} & 0 & 0 & -\gamma_1^{-1} I \end{bmatrix} < 0 \tag{31}$$

where $Y_j = K_j W_{22}$.

Therefore, if (30) and (31) are all held then (23) holds. Recall that the attenuation ρ^2 can be minimized so that the optimal H_∞ tracking performance in (21) is reduced to the following constrained optimization problem.

$$\rho_0^2 = \min_{P_{11}, P_{22}, P_{33}} \rho^2 \tag{32}$$

subject to $P_{11} > 0$, $P_{22} > 0$, $P_{33} > 0$, $\gamma > 0$, $\gamma_1 > 0$ and (30)-(31).
which can be solved by decreasing ρ^2 as small as possible until the parameters $P_{11} > 0$, $P_{22} > 0$, $P_{33} > 0$, $\gamma > 0$ and $\gamma_1 > 0$ do not exist.

Remark 4:

Note that the optimal H∞ tracking control problem in (32) is not a strict LMI problem since it is still a bilinear form in (30)-(31) of two scalars γ and γ_1 and becomes a standard linear matrix inequality problem (LMIP) (Boyd, 1994) if γ and γ_1 are given in advance. The decoupled method (Tseng, 2008) bring some conservatism in controller design. However, the parameters P_{11}, $P_{22} = W_{22}^{-1}$, P_{33}, $K_j = Y_j W_{22}^{-1}$ and $L_i = P_{11}^{-1} Z_i$ can be determined simultaneously from (32) by the decoupled method if scalars γ and γ_1 are given in advance. The useful software packages such as Robust Control Toolbox in Matlab (Balas et al., 2007) can be employed to solve the LMIP in (32) easily.

In general, it is quite easy to determine scalars γ and γ_1 beforehand to solve the LMIP with a smaller attenuation level ρ^2. In this study, the genetic algorithm (GA) is proposed to deal with the optimal H∞ tracking control problem in (32) since GA, which can simultaneously evaluate many points in the parameters space, is a very powerful searching algorithm based on the mechanics of natural selection and natural genetics. More details about GA can be found in (Jang et al., 1997).

According to the analysis above, the H∞ tracking control of stochastic innate immune system via fuzzy observer-based state feedback is summarized as follows and the structural diagram of robust fuzzy observer-based tracking control design has shown in Fig. 3.

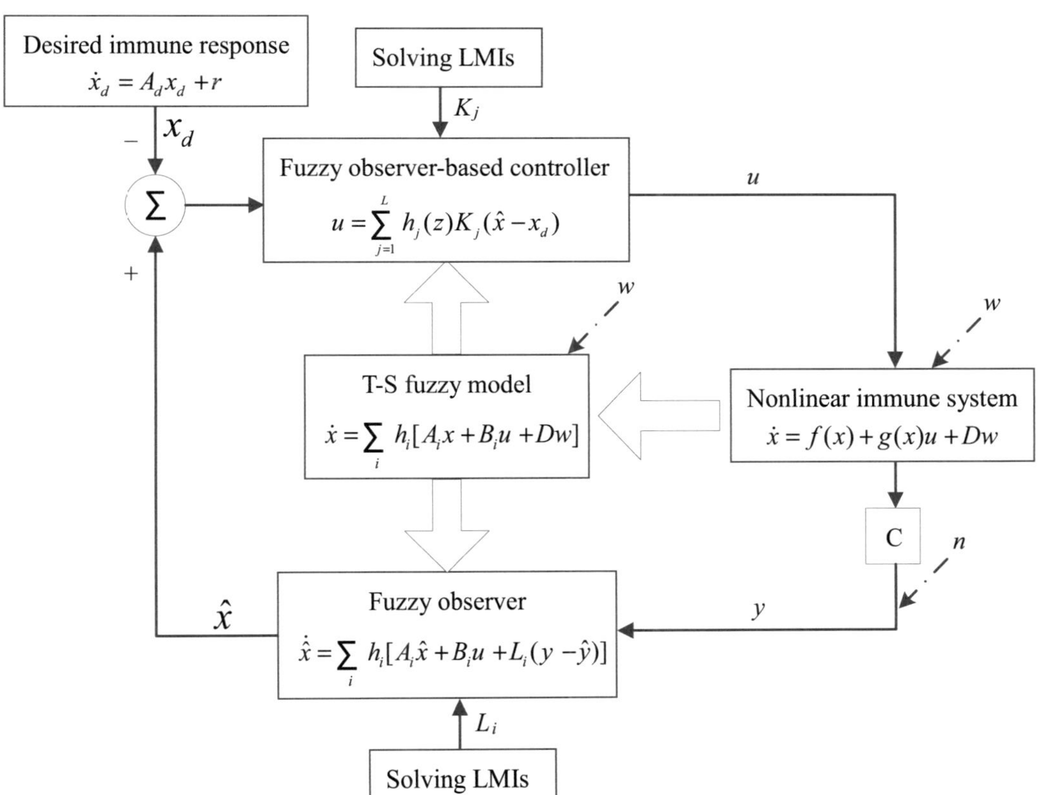

Fig. 3. Structural diagram of robust fuzzy observer-based tracking control design.

Design Procedure:

1. Provide a desired reference model in (4) of the immune system.
2. Select membership functions and construct fuzzy plant rules in (12).
3. Generate randomly a population of binary strings: With the binary coding method, the scalars γ and γ_1 would be coded as binary strings. Then solve the LMIP in (32) with scalars γ and γ_1 corresponding to binary string using Robust Control Toolbox in Matlab by searching the minimal value of ρ^2. If the LMIP is infeasible for the corresponding string, this string is escaped from the current generation.
4. Calculate the fitness value for each passed string: In this step, the fitness value is calculated based on the attenuation level ρ^2.
5. Create offspring strings to form a new generation by some simple GA operators like reproduction, crossover, and mutation: In this step, (i) strings are selected in a mating pool from the passed strings with probabilities proportional to their fitness values, (ii) and then crossover process is applied with a probability equal to a prescribed crossover rate, (iii) and finally mutation process is applied with a probability equal to a prescribed mutation rate. Repeating (i) to (iii) until enough strings are generated to form the next generation.
6. Repeat Step 3 to Step 5 for several generations until a stop criterion is met.
7. Based on the scalars γ and γ_1 obtained from above steps, one can obtain the attenuation level ρ^2 and the corresponding P_{11}, $P_{22} = W_{22}^{-1}$, P_{33}, $K_j = Y_j W_{22}^{-1}$ and $L_i = P_{11}^{-1} Z_i$, simultaneously.
8. Construct the fuzzy observer in (17) and fuzzy controller in (18).

5. Computational simulation example

Parameter	Value	Description
a_{11}	1	Pathogens reproduction rate coefficient
a_{12}	1	The suppression by pathogens coefficient
a_{22}	3	Immune reactivity coefficient
a_{23}	1	The mean immune cell production rate coefficient
x_2^*	2	The steady-state concentration of immune cells
a_{31}	1	Antibodies production rate coefficient
a_{32}	1.5	The antibody mortality coefficient
a_{33}	0.5	The rate of antibodies suppress pathogens
a_{41}	0.5	The organ damage depends on the pathogens damage possibilities coefficient
a_{42}	1	Organ recovery rate
b_1	-1	Pathogen killer's agent coefficient
b_2	1	Immune cell enhancer coefficient
b_3	1	Antibody enhancer coefficient
b_4	-1	Organ health enhancer coefficient
c_1	1	Immune cell measurement coefficient
c_2	1	Antibody measurement coefficient
c_3	1	Organ health measurement coefficient

Table 1. Model parameters of innate immune system (Marchuk, 1983; Stengel et al., 2002b).

We consider the nonlinear stochastic innate immune system in (1), which is shown in Fig. 1. The values of the parameters are shown in Table 1. The stochastic noises of immune systems are mainly due to measurement errors, modeling errors and process noises (Milutinovic & De Boer, 2007). The rate of continuing introduction of exogenous pathogen and environmental disturbances $w_1 \sim w_4$ are unknown but bounded signals. Under infectious situation, the microbes infect the organ not only by an initial concentration of pathogen at the beginning but also by the continuous exogenous pathogens invasion w_1 and other environmental disturbances $w_2 \sim w_4$. In reality, however, the concentration of invaded pathogens is hardly to be measured. So, we assume that only immune cell, antibody, and organ health can be measured with measurement noises by medical devices or other biological techniques (e.g. immunofluorescence microscope). And then we can detect the numbers of molecules easily by using a fluorescence microscope (Piston, 1999).

The dynamic model of stochastic innate immune system under uncertain initial states, environmental disturbances and measurement noises is controlled by a combined therapeutic control as

$$\begin{aligned}
\dot{x}_1 &= (1 - x_3)x_1 - u_1 + w_1 \\
\dot{x}_2 &= 3a_{21}(x_4)x_1x_3 - (x_2 - 2) - u_2 + w_2 \\
\dot{x}_3 &= x_2 - (1.5 + 0.5x_1)x_3 + u_3 + w_3 \\
\dot{x}_4 &= 0.5x_1 - x_4 + u_4 + w_4 \\
y_1 &= x_2 + n_1,\ y_2 = x_3 + n_2,\ y_3 = x_4 + n_3 \\
a_{21}(x_4) &= \begin{cases} \cos(\pi x_4), & 0 \le x_4 \le 0.5 \\ 0, & 0.5 \le x_4 \end{cases}
\end{aligned} \tag{33}$$

A set of initial condition is assumed $x(0) = \begin{bmatrix} 3.5 & 2 & 1.33 & 0 \end{bmatrix}^T$. For the convenience of simulation, we assume that $w_1 \sim w_4$ are zero mean white noises with standard deviations being all equal to 2. The measurement noises $n_1 \sim n_3$ are zero mean white noises with standard deviations being equal to 0.1. In this example, therapeutic controls $u_1 \sim u_4$ are combined to enhance the immune system. The measurable state variables $y_1 \sim y_3$ with measurement noises by medical devices or biological techniques are shown in Fig. 4.

Our reference model design objective is that the system matrix A_d and $r(t)$ should be specified beforehand so that its transient responses and steady state of reference system for stochastic innate immune response system are desired. If the real parts of eigenvalues of A_d are more negative (i.e. more robust stable), the tracking system will be more robust to the environmental disturbances. After some numerical simulations for clinical treatment, the desired reference signals are obtained by the following reference model, which is shown in Fig. 5.

$$\dot{x}_d(t) = \begin{bmatrix} -1.1 & 0 & 0 & 0 \\ 0 & -2 & 0 & 0 \\ 0 & 0 & -4 & 0 \\ 0 & 0 & 0 & -1.5 \end{bmatrix} x_d(t) + B_d u_{step}(t) \tag{34}$$

where $B_d = \begin{bmatrix} 0 & 4 & 16/3 & 0 \end{bmatrix}^T$ and $u_{step}(t)$ is the unit step function. The initial condition is given by $x_d(0) = \begin{bmatrix} 2.5 & 3 & 1.1 & 0.8 \end{bmatrix}^T$.

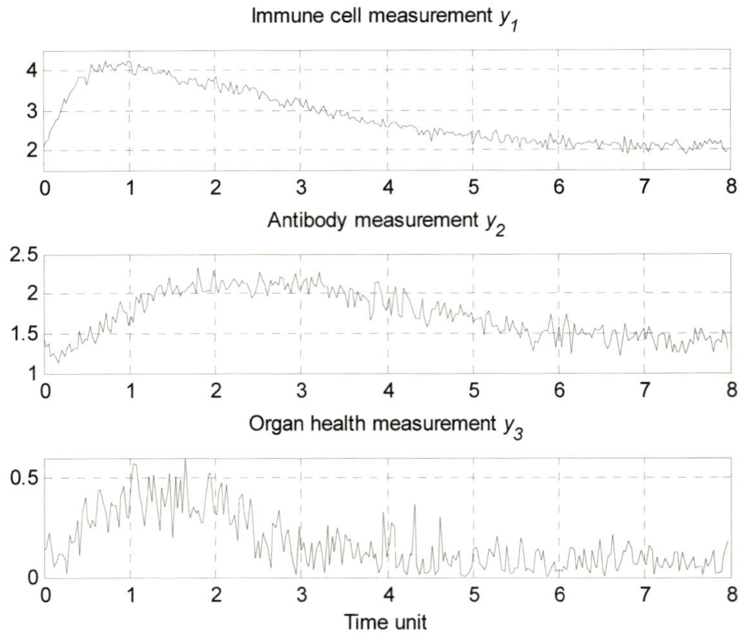

Fig. 4. The measurable state variables $y_1 \sim y_3$ with measurement noises $n_1 \sim n_3$ by medical devices or biological technique.

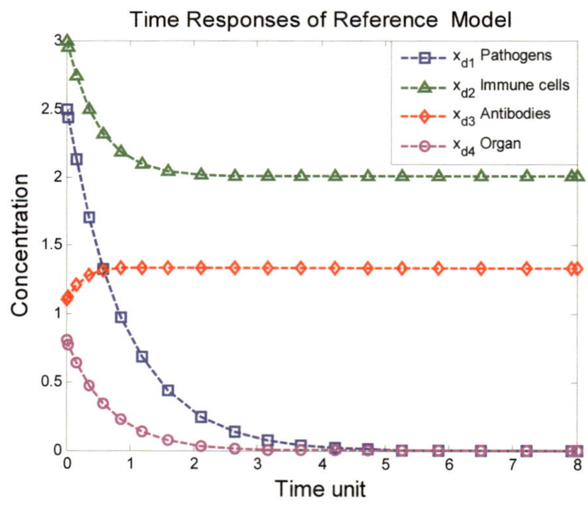

Fig. 5. The desired reference model with four desired states in (34): pathogens (x_{d1}, blue, dashed square line), immune cells (x_{d2}, green, dashed triangle line), antibodies (x_{d3}, red, dashed diamond line) and organ (x_{d4}, magenta, dashed, circle line)

We consider the lethal case of uncontrolled stochastic immune system in Fig. 6. The pathogen concentration increases rapidly causing organ failure. We aim at curing the organ before the organ health index excesses one after a period of pathogens infection. As shown in Fig. 6, the black dashed line is a proper time to administrate drugs.

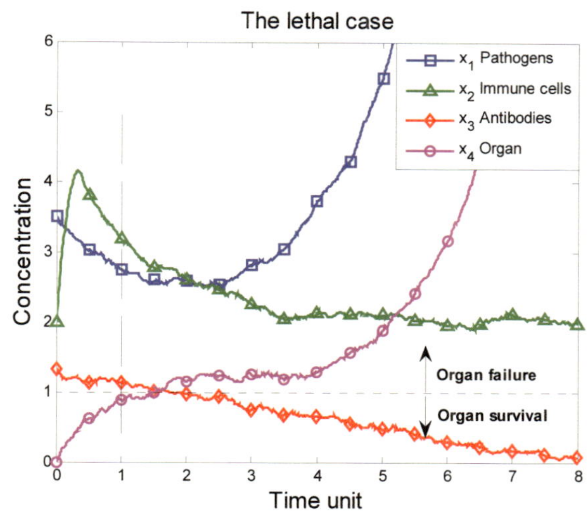

Fig. 6. The uncontrolled stochastic immune responses (lethal case) in (33) are shown to increase the level of pathogen concentration at the beginning of the time period. In this case, we try to administrate a treatment after a short period of pathogens infection. The cutting line (black dashed line) is an optimal time point to give drugs. The organ will survive or fail based on the organ health threshold (horizontal dotted line) [x_4<1: survival, x_4>1: failure].

To minimize the design effort and complexity for this nonlinear innate immune system in (33), we employ the T-S fuzzy model to construct fuzzy rules to approximate nonlinear immune system with the measurement output y_3 and y_4 as premise variables.

Plant Rule *i*:

If y_3 is F_{i1} and y_4 is F_{i2}, then

$$\dot{x}(t) = \mathbf{A}_i x(t) + \mathbf{B}u(t) + Dw(t), \; i = 1,2,3,\cdots,L$$

$$y(t) = Cx(t) + n(t)$$

To construct the fuzzy model, we must find the operating points of innate immune response. Suppose the operating points for y_3 are at $\overline{y}_{31} = -0.333$, $\overline{y}_{32} = 1.667$, and $\overline{y}_{33} = 3.667$. Similarly, the operating points for y_4 are at $\overline{y}_{41} = 0$, $\overline{y}_{42} = 1$, and $\overline{y}_{43} = 2$. For the convenience, we can create three triangle-type membership functions for the two premise variables as in Fig. 7 at the operating points and the number of fuzzy rules is $L = 9$. Then, we can find the fuzzy linear model parameters $\mathbf{A}_i$ in the Appendix D as well as other parameters $\mathbf{B}$, C and D. In order to accomplish the robust H∞ tracking performance, we should adjust a set of weighting matrices Q_1 and Q_2 in (8) or (9) as

$$Q_1 = \begin{bmatrix} 0.01 & 0 & 0 & 0 \\ 0 & 0.01 & 0 & 0 \\ 0 & 0 & 0.01 & 0 \\ 0 & 0 & 0 & 0.01 \end{bmatrix}, \; Q_2 = \begin{bmatrix} 0.01 & 0 & 0 & 0 \\ 0 & 0.01 & 0 & 0 \\ 0 & 0 & 0.01 & 0 \\ 0 & 0 & 0 & 0.01 \end{bmatrix}.$$

After specifying the desired reference model, we need to solve the constrained optimization problem in (32) by employing Matlab Robust Control Toolbox. Finally, we obtain the feasible parameters $\gamma = 40$ and $\gamma_1 = 0.02$, and a minimum attenuation level $\rho_0^2 = 0.93$ and a common positive-definite symmetric matrix $\overline{P}$ with diagonal matrices P_{11}, P_{22} and P_{33} as follows

$$P_{11} = \begin{bmatrix} 0.23193 & -1.5549e-4 & 0.083357 & -0.2704 \\ -1.5549e-4 & 0.010373 & -1.4534e-3 & -7.0637e-3 \\ 0.083357 & -1.4534e-3 & 0.33365 & 0.24439 \\ -0.2704 & -7.0637e-3 & 0.24439 & 0.76177 \end{bmatrix},$$

$$P_{22} = \begin{bmatrix} 0.0023082 & 9.4449e-6 & -5.7416e-5 & -5.0375e-6 \\ 9.4449e-6 & 0.0016734 & 2.4164e-5 & -1.8316e-6 \\ -5.7416e-5 & 2.4164e-5 & 0.0015303 & 5.8989e-6 \\ -5.0375e-6 & -1.8316e-6 & 5.8989e-6 & 0.0015453 \end{bmatrix}$$

$$P_{33} = \begin{bmatrix} 1.0671 & -1.0849e-5 & 3.4209e-5 & 5.9619e-6 \\ -1.0849e-5 & 1.9466 & -1.4584e-5 & 1.9167e-6 \\ 3.4209e-5 & -1.4584e-5 & 3.8941 & -3.2938e-6 \\ 5.9619e-6 & 1.9167e-6 & -3.2938e-6 & 1.4591 \end{bmatrix}$$

The control gain K_j and the observer gain L_i can also be solved in the Appendix D.

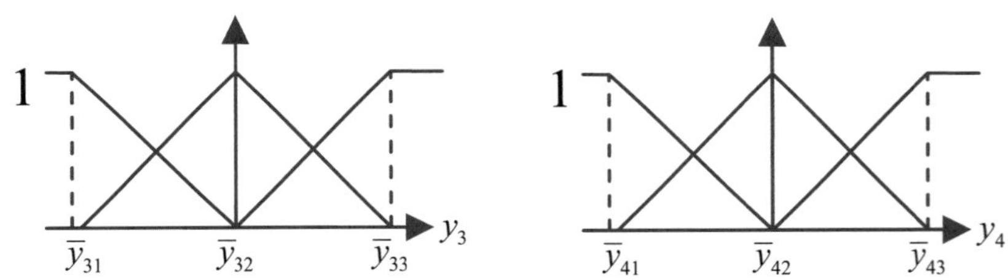

Fig. 7. Membership functions for two premise variables y_3 and y_4.

Figures 8-9 present the robust H_∞ tracking control of stochastic immune system under the continuous exogenous pathogens, environmental disturbances and measurement noises. Figure 8 shows the responses of the uncontrolled stochastic immune system under the initial concentrations of the pathogens infection. After the one time unit (the black dashed line), we try to provide a treatment by the robust H_∞ tracking control of pathogens infection. It is seen that the stochastic immune system approaches to the desired reference model quickly. From the simulation results, the tracking performance of the robust model tracking control via T-S fuzzy interpolation is quite satisfactory except for pathogens state x_1 because the pathogens concentration cannot be measured. But, after treatment for a specific period, the pathogens are still under control. Figure 9 shows the four combined therapeutic control agents. The performance of robust H_∞ tracking control is estimated as

$$\frac{\mathbf{E}\left[\int_0^{t_f}(\tilde{x}^T(t)Q_1\tilde{x}(t)+e^T(t)Q_2e(t))dt\right]}{\mathbf{E}\left[\int_0^{t_f}(w^T(t)w(t)+n^T(t)n(t)+r^T(t)r(t))dt\right]} \approx 0.033 \le \rho^2 = 0.93$$

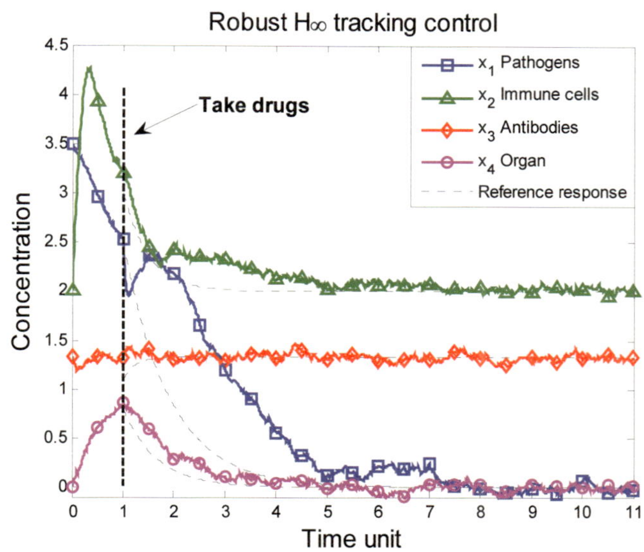

Fig. 8. The robust H∞ tracking control of stochastic immune system under the continuous exogenous pathogens, environmental disturbances and measurement noises. We try to administrate a treatment after a short period (one time unit) of pathogens infection then the stochastic immune system approach to the desired reference model quickly except for pathogens state x_1.

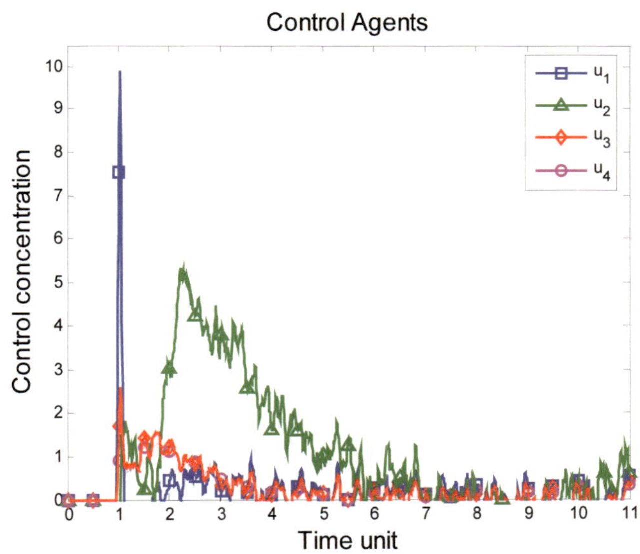

Fig. 9. The robust H∞ tracking control in the simulation example. The drug control agents u_1 (blue, solid square line) for pathogens, u_2 for immune cells (green, solid triangle line), u_3 for antibodies (red, solid diamond line) and u_4 for organ (magenta, solid circle line).

Obviously, the robust H∞ tracking performance is satisfied. The conservative results are due to the inherent conservation of solving LMI in (30)-(32).

6. Discussion and conclusion

In this study, we have developed a robust H_∞ tracking control design of stochastic immune response for therapeutic enhancement to track a prescribed immune response under uncertain initial states, environmental disturbances and measurement noises. Although the mathematical model of stochastic innate immune system is taken from the literature, it still needs to compare quantitatively with empirical evidence in practical application. For practical implementation, accurate biodynamic models are required for treatment application. However, model identification is not the topic of this paper. Furthermore, we assume that not all state variables can be measured. In the measurement process, the measured states are corrupted by noises. In this study, the statistic of disturbances, measurement noises and initial condition are assumed unavailable and cannot be used for the optimal stochastic tracking design. Therefore, the proposed H_∞ observer design is employed to attenuate these measurement noises to robustly estimate the state variables for therapeutic control and H_∞ control design is employed to attenuate disturbances to robustly track the desired time response of stochastic immune system simultaneity. Since the proposed H_∞ observer-based tracking control design can provide an efficient way to create a real time therapeutic regime despite disturbances, measurement noises and initial condition to protect suspected patients from the pathogens infection, in the future, we will focus on applications of robust H_∞ observer-based control design to therapy and drug design incorporating nanotechnology and metabolic engineering scheme.

Robustness is a significant property that allows for the stochastic innate immune system to maintain its function despite exogenous pathogens, environmental disturbances, system uncertainties and measurement noises. In general, the robust H_∞observer-based tracking control design for stochastic innate immune system needs to solve a complex nonlinear Hamilton-Jacobi inequality (HJI), which is generally difficult to solve for this control design. Based on the proposed fuzzy interpolation approach, the design of nonlinear robust H_∞ observer-based tracking control problem for stochastic innate immune system is transformed to solve a set of equivalent linear H_∞ observer-based tracking problem. Such transformation can then provide an easier approach by solving an LMI-constrained optimization problem for robust H_∞ observer-based tracking control design. With the help of the Robust Control Toolbox in Matlab instead of the HJI, we could solve these problems for robust H_∞ observer-based tracking control of stochastic innate immune system more efficiently. From the *in silico* simulation examples, the proposed robust H_∞ observer-based tracking control of stochastic immune system could track the prescribed reference time response robustly, which may lead to potential application in therapeutic drug design for a desired immune response during an infection episode.

7. Appendix

7.1 Appendix A: Proof of Theorem 1

Before the proof of *Theorem 1*, the following lemma is necessary.

Lemma 2:

For all vectors $\alpha, \beta \in \mathbf{R}^{n\times 1}$, the following inequality always holds

$\alpha^T\beta + \beta^T\alpha \le \frac{1}{\rho^2}\alpha^T\alpha + \rho^2\beta^T\beta$ for any scale value $\rho > 0$.

Let us denote a Lyapunov energy function $V(\bar{x}(t)) > 0$. Consider the following equivalent equation:

$$\mathbf{E}\left[\int_0^{t_f}\bar{x}^T(t)\bar{Q}\bar{x}(t)dt\right]=\mathbf{E}\left[V(\bar{x}(0))\right]-\mathbf{E}\left[V(\bar{x}(\infty))\right]+\mathbf{E}\left[\int_0^{t_f}\left(\bar{x}^T(t)\bar{Q}\bar{x}(t)+\frac{dV(\bar{x}(t))}{dt}\right)dt\right] \quad \text{(A1)}$$

By the chain rule, we get

$$\frac{dV(\bar{x}(t))}{dt}=\left(\frac{\partial V(\bar{x}(t))}{\partial \bar{x}(t)}\right)^T\frac{d\bar{x}(t)}{dt}=\left(\frac{\partial V(\bar{x}(t))}{\partial \bar{x}(t)}\right)^T\left(F(\bar{x}(t))+\bar{D}\bar{w}(t)\right) \quad \text{(A2)}$$

Substituting the above equation into (A1), by the fact that $V(\bar{x}(\infty))\geq 0$, we get

$$\mathbf{E}\left[\int_0^{t_f}\bar{x}^T(t)\bar{Q}\bar{x}(t)dt\right]\leq\mathbf{E}\left[V(\bar{x}(0))\right]+\mathbf{E}\left[\int_0^{t_f}\left(\bar{x}^T(t)\bar{Q}\bar{x}(t)+\left(\frac{\partial V(\bar{x}(t))}{\partial \bar{x}(t)}\right)^T\left(F(\bar{x}(t))+\bar{D}\bar{w}(t)\right)\right)dt\right] \quad \text{(A3)}$$

By *Lemma 2*, we have

$$\begin{aligned}\left(\frac{\partial V(\bar{x}(t))}{\partial \bar{x}(t)}\right)^T\bar{D}\bar{w}(t)&=\frac{1}{2}\left(\frac{\partial V(\bar{x}(t))}{\partial \bar{x}(t)}\right)^T\bar{D}\bar{w}(t)+\frac{1}{2}\bar{w}^T(t)\bar{D}^T\frac{\partial V(\bar{x}(t))}{\partial \bar{x}(t)}\\&\leq\frac{1}{4\rho^2}\left(\frac{\partial V(\bar{x}(t))}{\partial \bar{x}(t)}\right)^T\bar{D}\bar{D}^T\frac{\partial V(\bar{x}(t))}{\partial \bar{x}(t)}+\rho^2\bar{w}^T(t)\bar{w}(t)\end{aligned} \quad \text{(A4)}$$

Therefore, we can obtain

$$\begin{aligned}\mathbf{E}\left[\int_0^{t_f}\bar{x}^T(t)\bar{Q}\bar{x}(t)dt\right]\leq\mathbf{E}\left[V(\bar{x}(0))\right]+\mathbf{E}\Bigg[\int_0^{t_f}\Bigg(&\bar{x}^T(t)\bar{Q}\bar{x}(t)+\left(\frac{\partial V(\bar{x}(t))}{\partial \bar{x}(t)}\right)^T F(\bar{x}(t))\\&+\frac{1}{4\rho^2}\left(\frac{\partial V(\bar{x}(t))}{\partial \bar{x}(t)}\right)^T\bar{D}\bar{D}^T\frac{\partial V(\bar{x}(t))}{\partial \bar{x}(t)}+\rho^2\bar{w}^T(t)\bar{w}(t)\Bigg)dt\Bigg]\end{aligned} \quad \text{(A5)}$$

By the inequality in (10), then we get

$$\mathbf{E}\left[\int_0^{t_f}\bar{x}^T(t)\bar{Q}\bar{x}(t)dt\right]\leq\mathbf{E}\left[V(\bar{x}(0))\right]+\rho^2\mathbf{E}\left[\int_0^{t_f}\bar{w}^T(t)\bar{w}(t)dt\right] \quad \text{(A6)}$$

If $\bar{x}(0)=0$, then we get the inequality in (8).

7.2 Appendix B: Proof of Theorem 2

Let us choose a Lyapunov energy function $V(\bar{x}(t))=\bar{x}^T(t)\bar{P}\bar{x}(t)>0$ where $\bar{P}=\bar{P}^T>0$. Then equation (A1) is equivalent to the following:

$$\begin{aligned}&\mathbf{E}\left[\int_0^{t_f}\bar{x}^T(t)\bar{Q}\bar{x}(t)dt\right]=\mathbf{E}\left[V(\bar{x}(0))\right]-\mathbf{E}\left[V(\bar{x}(\infty))\right]+\mathbf{E}\left[\int_0^{t_f}\left(\bar{x}^T(t)\bar{Q}\bar{x}(t)+2\bar{x}^T(t)\bar{P}\dot{\bar{x}}(t)\right)dt\right]\\&\leq\mathbf{E}\left[V(\bar{x}(0))\right]+\mathbf{E}\left[\int_0^{t_f}\left(\bar{x}^T(t)\bar{Q}\bar{x}(t)+2\bar{x}^T(t)\bar{P}\left(\sum_{i=1}^{L}h_i(z(t))\sum_{j=1}^{L}h_j(z(t))\left[\bar{\mathbf{A}}_{ij}\bar{x}(t)+\bar{E}_i\bar{w}(t)\right]\right)\right)dt\right]\\&=\mathbf{E}\left[V(\bar{x}(0))\right]+\mathbf{E}\left[\int_0^{t_f}\left(\bar{x}^T(t)\bar{Q}\bar{x}(t)+\sum_{i=1}^{L}h_i(z(t))\sum_{j=1}^{L}h_j(z(t))\left[2\bar{x}^T(t)\bar{P}\bar{\mathbf{A}}_{ij}\bar{x}(t)+2\bar{x}^T(t)\bar{P}\bar{E}_i\bar{w}(t)\right]\right)dt\right]\end{aligned} \quad \text{(A7)}$$

By *Lemma 2*, we have

$$2\overline{x}^T(t)\overline{P}\overline{E}_i\overline{w}(t)=\overline{x}^T(t)\overline{P}\overline{E}_i\overline{w}(t)+\overline{w}^T(t)\overline{E}_i^T\overline{P}\overline{x}(t)\leq\frac{1}{\rho^2}\overline{x}^T(t)\overline{P}\overline{E}_i\overline{E}_i^T\overline{P}\overline{x}(t)+\rho^2\overline{w}^T(t)\overline{w}(t) \quad \text{(A8)}$$

Therefore, we can obtain

$$\begin{aligned}\mathbf{E}\left[\int_0^{t_f}\overline{x}^T(t)\overline{Q}\overline{x}(t)dt\right]&\leq\mathbf{E}\left[V(\overline{x}(0))\right]+\mathbf{E}\left[\int_0^{t_f}\left(\overline{x}^T(t)\overline{Q}\overline{x}(t)+\sum_{i=1}^L h_i(z(t))\sum_{j=1}^L h_j(z(t))\left[\overline{x}^T(\overline{P}\overline{\mathbf{A}}_{ij}+\overline{\mathbf{A}}_{ij}^T\overline{P})\overline{x}+\right.\right.\right.\\&\left.\left.\left.+\frac{1}{\rho^2}\overline{x}^T(t)\overline{P}\overline{E}_i\overline{E}_i^T\overline{P}\overline{x}(t)\right]+\rho^2\overline{w}^T(t)\overline{w}(t)\right)dt\right]=\\&=\mathbf{E}\left[V(\overline{x}(0))\right]+\mathbf{E}\left[\int_0^{t_f}\left(\overline{x}^T(t)\overline{Q}\overline{x}(t)+\sum_{i=1}^L h_i(z(t))\sum_{j=1}^L h_j(z(t))\overline{x}^T(t)\left[\overline{P}\overline{\mathbf{A}}_{ij}+\overline{\mathbf{A}}_{ij}^T\overline{P}+\right.\right.\right.\\&\left.\left.\left.+\frac{1}{\rho^2}\overline{P}\overline{E}_i\overline{E}_i^T\overline{P}\right]\overline{x}(t)+\rho^2\overline{w}^T(t)\overline{w}(t)\right)dt\right]\end{aligned} \quad \text{(A9)}$$

By the inequality in (20), then we get

$$\mathbf{E}\left[\int_0^{t_f}\overline{x}^T(t)\overline{Q}\overline{x}(t)dt\right]\leq\mathbf{E}\left[V(\overline{x}(0))\right]+\rho^2\mathbf{E}\left[\int_0^{t_f}\overline{w}^T(t)\overline{w}(t)dt\right] \quad \text{(A10)}$$

This is the inequality in (9). If $\overline{x}(0)=0$, then we get the inequality in (8).

7.3 Appendix C: Proof of Lemma 1

For $\begin{bmatrix}e_1 & e_2 & e_3 & e_4 & e_5 & e_6\end{bmatrix}\neq 0$, if (25)-(26) hold, then

$$\begin{bmatrix}e_1\\e_2\\e_3\\e_4\\e_5\\e_6\end{bmatrix}^T\left\{\begin{bmatrix}a_{11}&0&0&a_{14}&a_{15}&0\\0&a_{22}&a_{23}&0&0&0\\0&a_{32}&a_{33}&0&0&a_{36}\\a_{41}&0&0&a_{44}&0&0\\a_{51}&0&0&0&a_{55}&0\\0&0&a_{63}&0&0&a_{66}\end{bmatrix}+\begin{bmatrix}b_{11}&b_{12}&0&0&0&0\\b_{21}&b_{22}&b_{23}&0&b_{24}&0\\0&b_{32}&b_{33}&0&0&0\\0&0&0&0&0&0\\0&b_{42}&0&0&b_{44}&0\\0&0&0&0&0&0\end{bmatrix}\right\}\begin{bmatrix}e_1\\e_2\\e_3\\e_4\\e_5\\e_6\end{bmatrix}$$

$$=\begin{bmatrix}e_1\\e_2\\e_3\\e_4\\e_5\\e_6\end{bmatrix}^T\begin{bmatrix}a_{11}&0&0&a_{14}&a_{15}&0\\0&a_{22}&a_{23}&0&0&0\\0&a_{32}&a_{33}&0&0&a_{36}\\a_{41}&0&0&a_{44}&0&0\\a_{51}&0&0&0&a_{55}&0\\0&0&a_{63}&0&0&a_{66}\end{bmatrix}\begin{bmatrix}e_1\\e_2\\e_3\\e_4\\e_5\\e_6\end{bmatrix}+\begin{bmatrix}e_1\\e_2\\e_3\\e_5\end{bmatrix}^T\begin{bmatrix}b_{11}&b_{12}&0&0\\b_{21}&b_{22}&b_{23}&b_{24}\\0&b_{32}&b_{33}&0\\0&b_{42}&0&b_{44}\end{bmatrix}\begin{bmatrix}e_1\\e_2\\e_3\\e_5\end{bmatrix}<0$$

This implies that (27) holds. Therefore, the proof is completed.

7.4 Appendix D: Parameters of the Fuzzy System, control gains and observer gains

The nonlinear innate immune system in (33) could be approximated by a Takagi-Sugeno Fuzzy system. By the fuzzy modeling method (Takagi & Sugeno, 1985), the matrices of the local linear system $\mathbf{A}_i$, the parameters $\mathbf{B}$, C, D, K_j and L_i are calculated as follows:

$$\mathbf{A}_1=\begin{bmatrix}0&0&0&0\\3&-1&0&0\\-0.5&1&-1.5&0\\0.5&0&0&-1\end{bmatrix},\ \mathbf{A}_2=\begin{bmatrix}0&0&0&0\\3&-1&0&0\\-0.5&1&-1.5&0\\0.5&0&0&-1\end{bmatrix},\ \mathbf{A}_3=\begin{bmatrix}0&0&0&0\\3&-1&0&0\\-0.5&1&-1.5&0\\0.5&0&0&-1\end{bmatrix},$$

$$\mathbf{A}_4=\begin{bmatrix}-2&0&0&0\\9&-1&0&0\\-1.5&1&-1.5&0\\0.5&0&0&-1\end{bmatrix},\ \mathbf{A}_5=\begin{bmatrix}-2&0&0&0\\-9&-1&0&0\\-1.5&1&-1.5&0\\0.5&0&0&-1\end{bmatrix},\ \mathbf{A}_6=\begin{bmatrix}-2&0&0&0\\9&-1&0&0\\-1.5&1&-1.5&0\\0.5&0&0&-1\end{bmatrix},$$

$$\mathbf{A}_7=\begin{bmatrix}-4&0&0&0\\15&-1&0&0\\-2.5&1&-1.5&0\\0.5&0&0&-1\end{bmatrix},\ \mathbf{A}_8=\begin{bmatrix}-4&0&0&0\\-15&-1&0&0\\-2.5&1&-1.5&0\\0.5&0&0&-1\end{bmatrix},\ \mathbf{A}_9=\begin{bmatrix}-4&0&0&0\\15&-1&0&0\\-2.5&1&-1.5&0\\0.5&0&0&-1\end{bmatrix}$$

$$\mathbf{B}=\begin{bmatrix}-1&0&0&0\\0&-1&0&0\\0&0&1&0\\0&0&0&-1\end{bmatrix},\ C=\begin{bmatrix}0&1&0&0\\0&0&1&0\\0&0&0&1\end{bmatrix},\ D=\begin{bmatrix}1&0&0&0\\0&1&0&0\\0&0&1&0\\0&0&0&1\end{bmatrix}$$

$$K_j=\begin{bmatrix}17.712&0.14477&-0.43397&0.18604\\0.20163&18.201&0.37171&-0.00052926\\0.51947&-0.31484&-13.967&-0.052906\\0.28847&0.0085838&0.046538&14.392\end{bmatrix},\ j=1,\cdots,9$$

$$L_i=\begin{bmatrix}12.207&-26.065&22.367\\93.156&-8.3701&7.8721\\-8.3713&20.912&-16.006\\7.8708&-16.005&14.335\end{bmatrix},\ i=1,\cdots,9.$$

8. References

Asachenkov, A.L. (1994) *Disease dynamics*. Birkhäuser Boston.

Balas, G., Chiang, R., Packard, A. & Safonov, M. (2007) *MATLAB: Robust Control Toolbox 3 User's Guide*. The MathWorks, Inc.

Bell, D.J. & Katusiime, F. (1980) A Time-Optimal Drug Displacement Problem, *Optimal Control Applications & Methods*, 1, 217-225.

Bellman, R. (1983) *Mathematical methods in medicine*. World Scientific, Singapore.

Bonhoeffer, S., May, R.M., Shaw, G.M. & Nowak, M.A. (1997) Virus dynamics and drug therapy, *Proc. Natl. Acad. Sci. USA*, 94, 6971-6976.

Boyd, S.P. (1994) *Linear matrix inequalities in system and control theory*. Society for Industrial and Applied Mathematics, Philadelphia.

Carson, E.R., Cramp, D.G., Finkelstein, F. & Ingram, D. (1985) Control system concepts and approaches in clinical medicine. In Carson, E.R. & Cramp, D.G. (eds), *Computers and Control in Clinical Medicine*. Plenum Press, New York, 1-26.

Chen, B.S., Chang, C.H. & Chuang, Y.J. (2008) Robust model matching control of immune systems under environmental disturbances: dynamic game approach, *J Theor Biol*, 253, 824-837.

Chen, B.S., Tseng, C.S. & Uang, H.J. (1999) Robustness design of nonlinear dynamic systems via fuzzy linear control, *IEEE Transactions on Fuzzy Systems*, 7, 571-585.

Chen, B.S., Tseng, C.S. & Uang, H.J. (2000) Mixed H-2/H-infinity fuzzy output feedback control design for nonlinear dynamic systems: An LMI approach, *IEEE Transactions on Fuzzy Systems*, 8, 249-265.

Chizeck, H.J. & Katona, P.G. (1985) Closed-loop control. In Carson, E.R. & Cramp, D.G. (eds), *Computers and Control in Clinical Medicine*. Plenum Press, New York, 95-151.

De Boer, R.J. & Boucher, C.A. (1996) Anti-CD4 therapy for AIDS suggested by mathematical models, *Proc. Biol. Sci.* , 263, 899-905.

Gentilini, A., Morari, M., Bieniok, C., Wymann, R. & Schnider, T.W. (2001) Closed-loop control of analgesia in humans. *Proc. IEEE Conf. Decision and Control*. Orlando, 861-866.

Janeway, C. (2005) *Immunobiology : the immune system in health and disease*. Garland, New York.

Jang, J.-S.R., Sun, C.-T. & Mizutani, E. (1997) *Neuro-fuzzy and soft computing : a computational approach to learning and machine intelligence*. Prentice Hall, Upper Saddle River, NJ.

Jelliffe, R.W. (1986) Clinical applications of pharmacokinetics and control theory: planning, monitoring, and adjusting regimens of aminoglycosides, lidocaine, digitoxin, and digoxin. In Maronde, R.F. (ed), *Topics in clinical pharmacology and therapeutics*. Springer, New York, 26-82.

Kim, E., Park, M. & Ji, S.W. (1997) A new approach to fuzzy modeling, *IEEE Transactions on Fuzzy Systems*, 5, 328-337.

Kirschner, D., Lenhart, S. & Serbin, S. (1997) Optimal control of the chemotherapy of HIV, *J. Math. Biol.*, 35, 775-792.

Kwong, G.K., Kwok, K.E., Finegan, B.A. & Shah, S.L. (1995) Clinical evaluation of long range adaptive control for meanarterial blood pressure regulation. *Proc. Am. Control Conf.*, Seattle, 786-790.

Li, T.H.S., Chang, S.J. & Tong, W. (2004) Fuzzy target tracking control of autonomous mobile robots by using infrared sensors, *IEEE Transactions on Fuzzy Systems*, 12, 491-501.

Lian, K.Y., Chiu, C.S., Chiang, T.S. & Liu, P. (2001) LMI-based fuzzy chaotic synchronization and communications, *IEEE Transactions on Fuzzy Systems*, 9, 539-553.

Lydyard, P.M., Whelan, A. & Fanger, M.W. (2000) *Instant notes in immunology*. Springer, New York.

Ma, X.J., Sun, Z.Q. & He, Y.Y. (1998) Analysis and design of fuzzy controller and fuzzy observer, *IEEE Transactions on Fuzzy Systems*, 6, 41-51.

Marchuk, G.I. (1983) *Mathematical models in immunology*. Optimization Software, Inc. Worldwide distribution rights by Springer, New York.

Milutinovic, D. & De Boer, R.J. (2007) Process noise: an explanation for the fluctuations in the immune response during acute viral infection, *Biophys J*, 92, 3358-3367.

Nowak, M.A. & May, R.M. (2000) *Virus dynamics : mathematical principles of immunology and virology*. Oxford University Press, Oxford.

Nowak, M.A., May, R.M., Phillips, R.E., Rowland-Jones, S., Lalloo, D.G., McAdam, S., Klenerman, P., Koppe, B., Sigmund, K., Bangham, C.R. & et al. (1995) Antigenic oscillations and shifting immunodominance in HIV-1 infections, *Nature*, 375, 606-611.

Parker, R.S., Doyle, J.F., III., Harting, J.E. & Peppas, N.A. (1996) Model predictive control for infusion pump insulin delivery. *Proceedings of the 18th Annual International Conference of the IEEE Engineering in Medicine and Biology Society*. Amsterdam, 1822-1823.

Perelson, A.S., Kirschner, D.E. & De Boer, R. (1993) Dynamics of HIV infection of CD4+ T cells, *Math. Biosci.*, 114, 81-125.

Perelson, A.S., Neumann, A.U., Markowitz, M., Leonard, J.M. & Ho, D.D. (1996) HIV-1 dynamics in vivo: virion clearance rate, infected cell life-span, and viral generation time, *Science*, 271, 1582-1586.

Perelson, A.S. & Weisbuch, G. (1997) Immunology for physicists, *Reviews of Modern Physics*, 69, 1219-1267.

Piston, D.W. (1999) Imaging living cells and tissues by two-photon excitation microscopy, *Trends Cell Biol*, 9, 66-69.

Polycarpou, M.M. & Conway, J.Y. (1995) Modeling and control of drug delivery systems using adaptive neuralcontrol methods. *Proc. Am. Control Conf.*, Seattle, 781-785.

Robinson, D.C. (1986) Topics in clinical pharmacology and therapeutics. In Maronde, R.F. (ed), *Principles of Pharmacokinetics*. Springer, New York, 1-12.

Rundell, A., HogenEsch, H. & DeCarlo, R. (1995) Enhanced modeling of the immune system to incorporate naturalkiller cells and memory. *Proc. Am. Control Conf.*, Seattle, 255-259.

Schumitzky, A. (1986) Stochastic control of pharmacokinetic systems. In Maronde, R.F. (ed), *Topics in clinical pharmacology and therapeutics*. Springer, New York, 13-25.

Stafford, M.A., Corey, L., Cao, Y., Daar, E.S., Ho, D.D. & Perelson, A.S. (2000) Modeling plasma virus concentration during primary HIV infection, *J. Theor. Biol.*, 203, 285-301.

Stengel, R.F., Ghigliazza, R., Kulkarni, N. & Laplace, O. (2002a) Optimal control of innate immune response, *Optimal Control Applications & Methods*, 23, 91-104.

Stengel, R.F., Ghigliazza, R.M. & Kulkarni, N.V. (2002b) Optimal enhancement of immune response, *Bioinformatics*, 18, 1227-1235.

Sugeno, M. & Kang, G.T. (1988) Structure identification of fuzzy model, *Fuzzy Sets and Systems*, 28, 15-33.

Swan, G.W. (1981) Optimal-Control Applications in Biomedical-Engineering - a Survey, *Optimal Control Applications & Methods*, 2, 311-334.

Takagi, T. & Sugeno, M. (1985) Fuzzy Identification of Systems and Its Applications to Modeling and Control, *IEEE Transactions on Systems Man and Cybernetics*, 15, 116-132.

Tanaka, K., Ikeda, T. & Wang, H.O. (1998) Fuzzy regulators and fuzzy observers: Relaxed stability conditions and LMI-based designs, *IEEE Transactions on Fuzzy Systems*, 6, 250-265.

Tseng, C.S. (2008) A novel approach to H-infinity decentralized fuzzy-observer-based fuzzy control design for nonlinear interconnected systems, *IEEE Transactions on Fuzzy Systems*, 16, 1337-1350.

van Rossum, J.M., Steyger, O., van Uem, T., Binkhorst, G.J. & Maes, R.A.A. (1986) Pharmacokinetics by using mathematical systems dynamics. In Eisenfeld, J. & Witten, M. (eds), *Modelling of biomedical systems*. Elsevier, Amsterdam, 121-126.

Wang, H.O., Tanaka, K. & Griffin, M.F. (1996) An approach to fuzzy control of nonlinear systems: Stability and design issues, *IEEE Transactions on Fuzzy Systems*, 4, 14-23.

Wein, L.M., D'Amato, R.M. & Perelson, A.S. (1998) Mathematical analysis of antiretroviral therapy aimed at HIV-1 eradication or maintenance of low viral loads, *J. Theor. Biol.*, 192, 81-98.

Wiener, N. (1948) *Cybernetics; or, Control and communication in the animal and the machine*. Technology Press, Cambridge.

Wodarz, D. & Nowak, M.A. (1999) Specific therapy regimes could lead to long-term immunological control of HIV, *Proc. Natl. Acad. Sci. USA*, 96, 14464-14469.

Wodarz, D. & Nowak, M.A. (2000) CD8 memory, immunodominance, and antigenic escape, *Eur. J. Immunol.*, 30, 2704-2712.

Zhou, K., Doyle, J.C. & Glover, K. (1996) *Robust and optimal control*. Prentice Hall, Upper Saddle River, N.J.

6

Robust H_∞ Reliable Control of Uncertain Switched Nonlinear Systems with Time-varying Delay

Ronghao Wang[1], Jianchun Xing[1], Ping Wang[1], Qiliang Yang[1] and Zhengrong Xiang[2]
[1]*PLA University of Science and Technology*
[2]*Nanjing University of Science and Technology*
China

1. Introduction

Switched systems are a class of hybrid system consisting of subsystems and a switching law, which define a specific subsystem being activated during a certain interval of time. Many real-world processes and systems can be modeled as switched systems, such as the automobile direction-reverse systems, computer disk systems, multiple work points control systems of airplane and so on. Therefore, the switched systems have the wide project background and can be widely applied in many domains (Wang, W. & Brockett, R. W., 1997; Tomlin, C. et al., 1998; Varaiya, P., 1993). Besides switching properties, when modeling a engineering system, system uncertainties that occur as a result of using approximate system model for simplicity, data errors for evaluation, changes in environment conditions, etc, also exit naturally in control systems. Therefore, both of switching and uncertainties should be integrated into system model. Recently, study of switched systems mainly focuses on stability and stabilization (Sun, Z. D. & Ge, S. S., 2005; Song, Y. et al., 2008; Zhang, Y. et al., 2007). Based on linear matrix inequality technology, the problem of robust control for the system is investigated in the literature (Pettersson, S. & Lennartson, B., 2002). In order to guarantee H_∞ performance of the system, the robust H_∞ control is studied using linear matrix inequality method in the literature (Sun, W. A. & Zhao, J., 2005).

In many engineering systems, the actuators may be subjected to faults in special environment due to the decline in the component quality or the breakage of working condition which always leads to undesirable performance, even makes system out of control. Therefore, it is of interest to design a control system which can tolerate faults of actuators. In addition, many engineering systems always involve time delay phenomenon, for instance, long-distance transportation systems, hydraulic pressure systems, network control systems and so on. Time delay is frequently a source of instability of feedback systems. Owing to all of these, we shouldn't neglect the influence of time delay and probable actuators faults when designing a practical control system. Up to now, research activities of this field for switched system have been of great interest. Stability analysis of a class of linear switching systems with time delay is presented in the literature (Kim, S. et al., 2006). Robust H_∞ control for discrete switched systems with time-varying delay is discussed in the literature (Song, Z. Y. et al., 2007). Reliable guaranteed-cost control for a class of uncertain switched linear systems with time delay is investigated in the literature (Wang, R. et al., 2006). Considering that the nonlinear disturbance could not be avoided in several applications, robust reliable control for uncertain switched nonlinear systems with time delay is studied in the literature (Xiang, Z. R. & Wang, R. H., 2008). Furthermore, Xiang and Wang (Xiang, Z. R. & Wang, R. H., 2009) investigated robust L_∞ reliable control for uncertain nonlinear switched systems with time delay.

Above the problems of robust reliable control for uncertain nonlinear switched time delay systems, the time delay is treated as a constant. However, in actual operation, the time delay is usually variable as time. Obviously, the system model couldn't be described appropriately using constant time delay in this case. So the paper focuses on the system with time-varying delay. Besides, it is considered that H_∞ performance is always an important index in control system. Therefore, in order to overcome the passive effect of time-varying delay for switched systems and make systems be anti-jamming and fault-tolerant, this paper addresses the robust H_∞ reliable control for nonlinear switched time-varying delay systems subjected to uncertainties. The multiple Lyapunov-Krasovskii functional method is used to design the control law. Compared with the multiple Lyapunov function adopted in the literature (Xiang, Z. R. & Wang, R. H., 2008; Xiang, Z. R. & Wang, R. H., 2009), the multiple Lyapunov-Krasovskii functional method has less conservation because the more system state information is contained in the functional. Moreover, the controller parameters can be easily obtained using the constructed functional.

The organization of this paper is as follows. At first, the concept of robust reliable controller, γ -suboptimal robust H_∞ reliable controller and γ -optimal robust H_∞ reliable controller are presented. Secondly, fault model of actuator for system is put forward. Multiple Lyapunov-Krasovskii functional method and linear matrix inequality technique are adopted to design robust H_∞ reliable controller. Meanwhile, the corresponding switching law is proposed to guarantee the stability of the system. By using the key technical lemma, the design problems of γ -suboptimal robust H_∞ reliable controller and γ -optimal robust H_∞ reliable controller can be transformed to the problem of solving a set of the matrix inequalities. It is worth to point that the matrix inequalities in the γ -optimal problem are not linear, then we make use of variable substitute method to acquire the controller gain matrices and γ -optimal problem can be transferred to solve the minimal upper bound of the scalar γ . Furthermore, the iteration solving process of optimal disturbance attenuation performance γ is presented. Finally, a numerical example shows the effectiveness of the proposed method. The result illustrates that the designed controller can stabilize the original system and make it be of H_∞ disturbance attenuation performance when the system has uncertain parameters and actuator faults.

Notation Throughout this paper, A^T denotes transpose of matrix A , $L_2[0,\infty)$ denotes space of square integrable functions on $[0,\infty)$. $\|x(t)\|$ denotes the Euclidean norm. I is an identity matrix with appropriate dimension. $diag\{a_i\}$ denotes diagonal matrix with the diagonal elements a_i , $i=1,2,\cdots,q$. $S<0$ (or $S>0$) denotes S is a negative (or positive) definite symmetric matrix. The set of positive integers is represented by Z^+ . $A\le B$ (or $A\ge B$) denotes $A-B$ is a negative (or positive) semi-definite symmetric matrix. $*$ in $\begin{bmatrix} A & B \\ * & C \end{bmatrix}$ represents the symmetric form of matrix, i.e. $*=B^T$.

2. Problem formulation and preliminaries

Consider the following uncertain switched nonlinear system with time-varying delay

$$\dot{x}(t) = \hat{A}_{\sigma(t)}x(t) + \hat{A}_{d\sigma(t)}x(t-d(t)) + \hat{B}_{\sigma(t)}u^f(t) + D_{\sigma(t)}w(t) + f_{\sigma(t)}(x(t),t) \quad (1)$$

$$z(t) = C_{\sigma(t)}x(t) + G_{\sigma(t)}u^f(t) + N_{\sigma(t)}w(t) \quad (2)$$

$$x(t) = \phi(t), t \in [-\rho, 0] \quad (3)$$

where $x(t) \in R^m$ is the state vector, $w(t) \in R^q$ is the measurement noise, which belongs to $L_2[0,\infty)$, $z(t) \in R^p$ is the output to be regulated, $u^f(t) \in R^l$ is the control input of actuator fault. The function $\sigma(t):[0,\infty) \to \underline{N} = \{1,2,\cdots,N\}$ is switching signal which is deterministic, piecewise constant, and right continuous, i.e. $\sigma(t):\{(0,\sigma(0)),(t_1,\sigma(t_1)),\cdots,(t_k,\sigma(t_k))\}, k \in Z^+$, where t_k denotes the kth switching instant. Moreover, $\sigma(t) = i$ means that the ith subsystem is activated, N is the number of subsystems. $\phi(t)$ is a continuous vector-valued initial function. The function $d(t)$ denotes the time-varying state delay satisfying $0 \le d(t) \le \rho < \infty, \dot{d}(t) \le \mu < 1$ for constants ρ, μ, and $f_i(\bullet,\bullet)$: $R^m \times R \to R^m$ for $i \in \underline{N}$ are unknown nonlinear functions satisfying

$$\|f_i(x(t),t)\| \le \|U_i x(t)\| \quad (4)$$

where U_i are known real constant matrices.
The matrices $\hat{A}_i$, $\hat{A}_{di}$ and $\hat{B}_i$ for $i \in \underline{N}$ are uncertain real-valued matrices of appropriate dimensions. The matrices $\hat{A}_i$, $\hat{A}_{di}$ and $\hat{B}_i$ can be assumed to have the form

$$[\hat{A}_i, \hat{A}_{di}, \hat{B}_i] = [A_i, A_{di}, B_i] + H_i F_i(t)[E_{1i}, E_{di}, E_{2i}] \quad (5)$$

where $A_i, A_{di}, B_i, H_i, E_{1i}, E_{di}$ and E_{2i} for $i \in \underline{N}$ are known real constant matrices with proper dimensions, H_i, E_{1i}, E_{di} and E_{2i} denote the structure of the uncertainties, and $F_i(t)$ are unknown time-varying matrices that satisfy

$$F_i^T(t)F_i(t) \le I \quad (6)$$

The parameter uncertainty structure in equation (5) has been widely used and can represent parameter uncertainty in many physical cases (Xiang, Z. R. & Wang, R. H., 2009; Cao, Y. et al., 1998).
In actual control system, there inevitably occurs fault in the operation process of actuators. Therefore, the input control signal of actuator fault is abnormal. We use $u(t)$ and $u^f(t)$ to represent the normal control input and the abnormal control input, respectively. Thus, the control input of actuator fault can be described as

$$u^f(t) = M_i u(t) \quad (7)$$

where M_i is the actuator fault matrix of the form

$$M_i = diag\{m_{i1}, m_{i2}, \cdots, m_{il}\},\ 0 \le \underline{m}_{ik} \le m_{ik} \le \overline{m}_{ik},\ \overline{m}_{ik} \ge 1,\ k = 1,2,\cdots,l \quad (8)$$

For simplicity, we introduce the following notation

$$M_{i0} = diag\{\tilde{m}_{i1}, \tilde{m}_{i2}, \cdots, \tilde{m}_{il}\} \tag{9}$$

$$J_i = diag\{j_{i1}, j_{i2}, \cdots, j_{il}\} \tag{10}$$

$$L_i = diag\{l_{i1}, l_{i2}, \cdots, l_{il}\} \tag{11}$$

where $\tilde{m}_{ik} = \frac{1}{2}(\overline{m}_{ik} + \underline{m}_{ik})$, $j_{ik} = \frac{\overline{m}_{ik} - \underline{m}_{ik}}{\overline{m}_{ik} + \underline{m}_{ik}}$, $l_{ik} = \frac{m_{ik} - \tilde{m}_{ik}}{\tilde{m}_{ik}}$

By equation (9)-(11), we have

$$M_i = M_{i0}(I + L_i), \ |L_i| \le J_i \le I \tag{12}$$

where $|L_i|$ represents the absolute value of diagonal elements in matrix L_i, i.e.

$$|L_i| = diag\{|l_{i1}|, |l_{i2}|, \cdots, |l_{il}|\}$$

Remark 1 $m_{ik} = 1$ means normal operation of the k th actuator control signal of the i th subsystem. When $m_{ik} = 0$, it covers the case of the complete fault of the k th actuator control signal of the i th subsystem. When $\underline{m}_{ik} > 0$ and $m_{ik} \ne 1$, it corresponds to the case of partial fault of the k th actuator control signal of the i th subsystem.

Now, we give the definition of robust H_∞ reliable controller for the uncertain switched nonlinear systems with time-varying delay.

Definition 1 Consider system (1) with $w(t) \equiv 0$. If there exists the state feedback controller $u(t) = K_{\sigma(t)}x(t)$ such that the closed loop system is asymptotically stable for admissible parameter uncertainties and actuator fault under the switching law $\sigma(t)$, $u(t) = K_{\sigma(t)}x(t)$ is said to be a robust reliable controller.

Definition 2 Consider system (1)-(3). Let $\gamma > 0$ be a positive constant, if there exists the state feedback controller $u(t) = K_{\sigma(t)}x(t)$ and the switching law $\sigma(t)$ such that

i. With $w(t) \equiv 0$, the closed system is asymptotically stable.
ii. Under zero initial conditions, i.e. $x(t) = 0$ $(t \in [-\rho, 0])$, the following inequality holds

$$\|z(t)\|_2 \le \gamma \|w(t)\|_2, \ \forall w(t) \in L_2[0, \infty), \ w(t) \ne 0 \tag{13}$$

$u(t) = K_{\sigma(t)}x(t)$ is said to be γ-suboptimal robust H_∞ reliable controller with disturbance attenuation performance γ. If there exists a minimal value of disturbance attenuation performance γ, $u(t) = K_{\sigma(t)}x(t)$ is said to be γ-optimal robust H_∞ reliable controller.

The following lemmas will be used to design robust H_∞ reliable controller for the uncertain switched nonlinear system with time-varying delay.

Lemma 1 (Boyd, S. P. et al., 1994; Schur complement) For a given matrix $S = \begin{bmatrix} S_{11} & S_{12} \\ S_{21} & S_{22} \end{bmatrix}$ with $S_{11} = S_{11}^T$, $S_{22} = S_{22}^T$, $S_{12} = S_{21}^T$, then the following conditions are equivalent

i $S < 0$

ii $S_{11} < 0$, $S_{22} - S_{21}S_{11}^{-1}S_{12} < 0$

iii $S_{22} < 0$, $S_{11} - S_{12}S_{22}^{-1}S_{21} < 0$

Lemma 2 (Cong, S. et al., 2007) For matrices X and Y of appropriate dimension and $Q>0$, we have

$$X^T Y + Y^T X \le X^T Q X + Y^T Q^{-1} Y$$

Lemma 3 (Lien, C.H., 2007) Let Y, D, E and F be real matrices of appropriate dimensions with F satisfying $F^T = F$, then for all $F^T F \le I$

$$Y + DFE + E^T F^T D^T < 0$$

if and only if there exists a scalar $\varepsilon > 0$ such that

$$Y + \varepsilon DD^T + \varepsilon^{-1} E^T E < 0$$

Lemma 4 (Xiang, Z. R. & Wang, R. H., 2008) For matrices R_1, R_2, the following inequality holds

$$R_1 \Sigma(t) R_2 + R_2^T \Sigma^T(t) R_1^T \le \beta R_1 U R_1^T + \beta^{-1} R_2^T U R_2$$

where $\beta > 0$, $\Sigma(t)$ is time-varying diagonal matrix, U is known real constant matrix satisfying $|\Sigma(t)| \le U$, $|\Sigma(t)|$ represents the absolute value of diagonal elements in matrix $\Sigma(t)$.

Lemma 5 (Peleties, P. & DeCarlo, R. A., 1991) Consider the following system

$$\dot{x}(t) = f_{\sigma(t)}(x(t)) \tag{14}$$

where $\sigma(t):[0,\infty) \to \underline{N} = \{1,2,\cdots,N\}$. If there exist a set of functions $V_i : R^m \to R,\ i \in \underline{N}$ such that

(i) V_i is a positive definite function, decreasing and radially unbounded;

(ii) $dV_i(x(t))/dt = (\partial V_i / \partial x) f_i(x) \le 0$ is negative definite along the solution of (14);

(iii) $V_j(x(t_k)) \le V_i(x(t_k))$ when the i th subsystem is switched to the j th subsystem $i, j \in N$, $i \ne j$ at the switching instant t_k, $k = Z^+$, then system (14) is asymptotically stable.

3. Main results

3.1 Condition of stability

Consider the following unperturbed switched nonlinear system with time-varying delay

$$\dot{x}(t) = A_{\sigma(t)} x(t) + A_{d\sigma(t)} x(t - d(t)) + f_{\sigma(t)}(x(t), t) \tag{15}$$

$$x(t) = \phi(t), t \in [-\rho, 0] \tag{16}$$

The following theorem presents a sufficient condition of stability for system (15)-(16).

Theorem 1 For system (15)-(16), if there exists symmetric positive definite matrices P_i, Q, and the positive scalar δ such that

$$P_i < \delta I \tag{17}$$

$$\begin{bmatrix} A_i^T P_i + P_i A_i + P_j + Q + \delta U_i^T U_i & P_i A_{di} \\ * & -(1-\mu)Q \end{bmatrix} < 0 \tag{18}$$

where $i \neq j,\ i,j \in \underline{N}$, then systems (15)-(16) is asymptotically stable under the switching law

$$\sigma(t) = \arg\min_{i \in \underline{N}}\{x^T(t)P_i x(t)\}\cdot$$

Proof For the i th subsystem, we define Lyapunov-Krasovskii functional

$$V_i(x(t)) = x^T(t)P_i x(t) + \int_{t-d(t)}^{t} x^T(\tau)Qx(\tau)d\tau$$

where P_i, Q are symmetric positive definite matrices. Along the trajectories of system (15), the time derivative of $V_i(t)$ is given by

$$\begin{aligned} \dot{V}_i(x(t)) &= \dot{x}^T(t)P_i x(t) + x^T(t)P_i\dot{x}(t) + x^T(t)Qx(t) - (1-\dot{d}(t))x^T(t-d(t))Qx(t-d(t)) \\ &\le \dot{x}^T(t)P_i x(t) + x^T(t)P_i\dot{x}(t) + x^T(t)Qx(t) - (1-\mu)x^T(t-d(t))Qx(t-d(t)) \\ &= 2x^T(t)P_i[A_i x(t) + A_{di}x(t-d(t)) + f_i(x(t),t)] + x^T(t)Qx(t) \\ &\quad - (1-\mu)x^T(t-d(t))Qx(t-d(t)) \\ &= x^T(t)(A_i^T P_i + P_i A_i + Q)x(t) + 2x^T(t)P_i A_{di}x(t-d(t)) + 2x^T(t)P_i f_i(x(t),t) \\ &\quad - (1-\mu)x^T(t-d(t))Qx(t-d(t)) \end{aligned}$$

From Lemma 2, it is established that

$$2x^T(t)P_i f_i(x(t),t) \le x^T(t)P_i x(t) + f_i^T(x(t),t)P_i f_i(x(t),t)$$

From expressions (4) and (17), it follows that

$$2x^T(t)P_i f_i(x(t),t) \le x^T(t)P_i x(t) + \delta f_i^T(x(t),t) f_i(x(t),t) \le x^T(t)(P_i + \delta U_i^T U_i)x(t)$$

Therefore, we can obtain that

$$\begin{aligned} \dot{V}_i(x(t)) &\le x^T(t)(A_i^T P_i + P_i A_i + Q + P_i + \delta U_i^T U_i)x(t) + 2x^T(t)P_i A_{di}x(t-d(t)) \\ &\quad - (1-\mu)x^T(t-d(t))Qx(t-d(t)) \\ &= \eta^T \Theta_i \eta \end{aligned}$$

where $\eta = \begin{bmatrix} x(t) \\ x(t-d(t)) \end{bmatrix}$, $\Theta_i = \begin{bmatrix} A_i^T P_i + P_i A_i + Q + P_i + \delta U_i^T U_i & P_i A_{di} \\ * & -(1-\mu)Q \end{bmatrix}$

From (18), we have

$$\Theta_i + diag\{P_j - P_i, 0\} < 0 \tag{19}$$

Using η^T and η to pre- and post- multiply the left-hand term of expression (19) yields

$$\dot{V}_i(x(t)) < x^T(t)(P_i - P_j)x(t) \tag{20}$$

The switching law $\sigma(t) = \arg\min_{i\in \underline{N}}\{x^T(t)P_i x(t)\}$ expresses that for $i, j \in \underline{N}, i \neq j$, there holds the inequality

$$x^T(t)P_i x(t) \le x^T(t)P_j x(t) \tag{21}$$

(20) and (21) lead to

$$\dot{V}_i(x(t)) < 0 \tag{22}$$

Obviously, the switching law $\sigma(t) = \arg\min_{i\in \underline{N}}\{x^T(t)P_i x(t)\}$ also guarantees that Lyapunov-Krasovskii functional value of the activated subsystem is minimum at the switching instant. From Lemma 5, we can obtain that system (15)-(16) is asymptotically stable. The proof is completed.

■

Remark 2 It is worth to point that the condition (21) doesn't imply $P_i \le P_j$, for the state $x(t)$ doesn't represent all the state in domain R^m but only the state of the i th activated subsystem.

3.2 Design of robust reliable controller

Consider system (1) with $w(t) \equiv 0$

$$\dot{x}(t) = \hat{A}_{\sigma(t)}x(t) + \hat{A}_{d\sigma(t)}x(t-d(t)) + \hat{B}_{\sigma(t)}u^f(t) + f_{\sigma(t)}(x(t),t) \tag{23}$$

$$x(t) = \phi(t), t \in [-\rho, 0] \tag{24}$$

By (7), for the i th subsystem the feedback control law can be designed as

$$u^f(t) = M_i K_i x(t) \tag{25}$$

Substituting (25) to (23), the corresponding closed-loop system can be written as

$$\dot{x}(t) = \tilde{A}_i x(t) + \hat{A}_{di}x(t-d(t)) + f_i(x(t),t) \tag{26}$$

where $\tilde{A}_i = \hat{A}_i + \hat{B}_i M_i K_i$, $i \in \underline{N}$.

The following theorem presents a sufficient existing condition of the robust reliable controller for system (23)-(24).

Theorem 2 For system (23)-(24), if there exists symmetric positive definite matrices X_i, S, matrix Y_i and the positive scalar λ such that

$$X_i > \lambda I \tag{27}$$

$$\begin{bmatrix} \Psi_i & A_{di}S & H_i & \Phi_i^T & X_i & X_i & X_i U_i^T \\ * & -(1-\mu)S & 0 & SE_{di}^T & 0 & 0 & 0 \\ * & * & -I & 0 & 0 & 0 & 0 \\ * & * & * & -I & 0 & 0 & 0 \\ * & * & * & * & -S & 0 & 0 \\ * & * & * & * & * & -X_j & 0 \\ * & * & * & * & * & * & -\lambda I \end{bmatrix} < 0 \tag{28}$$

where $i \neq j, i, j \in \underline{N}$, $\Psi_i = A_i X_i + B_i M_i Y_i + (A_i X_i + B_i M_i Y_i)^T$, $\Phi_i = E_{1i} X_i + E_{2i} M_i Y_i$, then there exists the robust reliable state feedback controller

$$u(t) = K_{\sigma(t)} x(t),\ K_i = Y_i X_i^{-1} \tag{29}$$

and the switching law is designed as $\sigma(t) = \arg\min_{i \in \underline{N}} \{x^T(t) X_i^{-1} x(t)\}$, the closed-loop system is asymptotically stable.

Proof From (5) and Theorem 1, we can obtain the sufficient condition of asymptotically stability for system (26)

$$P_i < \delta I \tag{30}$$

$$\begin{bmatrix} \Lambda_{ij} & P_i(A_{di} + H_i F_i(t) E_{di}) \\ * & -(1-\mu)Q \end{bmatrix} < 0 \tag{31}$$

and the switching law is designed as $\sigma(t) = \arg\min_{i \in \underline{N}} \{x^T(t) P_i x(t)\}$, where

$$\Lambda_{ij} = P_i[A_i + B_i M_i K_i + H_i F_i(t)(E_{1i} + E_{2i} M_i K_i)] + [A_i + B_i M_i K_i + H_i F_i(t)(E_{1i} + E_{2i} M_i K_i)]^T P_i \\ + P_j + Q + \delta U_i^T U_i$$

Denote

$$Y_{ij} = \begin{bmatrix} P_i(A_i + B_i M_i K_i) + (A_i + B_i M_i K_i)^T P_i + P_j + Q + \delta U_i^T U_i & P_i A_{di} \\ * & -(1-\mu)Q \end{bmatrix} \tag{32}$$

Then (31) can be written as

$$Y_{ij} + \begin{bmatrix} P_i H_i \\ 0 \end{bmatrix} F_i(t) \begin{bmatrix} E_{1i} + E_{2i} M_i K_i & E_{di} \end{bmatrix} + \begin{bmatrix} E_{1i} + E_{2i} M_i K_i & E_{di} \end{bmatrix}^T F_i^T(t) \begin{bmatrix} P_i H_i \\ 0 \end{bmatrix}^T < 0 \tag{33}$$

By Lemma 3, if there exists a scalar $\varepsilon > 0$ such that

$$Y_{ij} + \varepsilon \begin{bmatrix} P_i H_i \\ 0 \end{bmatrix} \begin{bmatrix} P_i H_i \\ 0 \end{bmatrix}^T + \varepsilon^{-1} \begin{bmatrix} E_{1i} + E_{2i} M_i K_i & E_{di} \end{bmatrix}^T \begin{bmatrix} E_{1i} + E_{2i} M_i K_i & E_{di} \end{bmatrix} < 0 \tag{34}$$

then (31) holds.
(34) can also be written as

$$\begin{bmatrix} \Pi_{ij} & P_i A_{di} + \varepsilon^{-1}(E_{1i} + E_{2i} M_i K_i)^T E_{di} \\ * & -(1-\mu)Q + \varepsilon^{-1} E_{di}{}^T E_{di} \end{bmatrix} < 0 \tag{35}$$

where

$$\Pi_{ij} = (A_i + B_i M_i K_i)^T P_i + P_i(A_i + B_i M_i K_i) + \varepsilon P_i H_i H_i^T P_i + \varepsilon^{-1}(E_{1i} + E_{2i} M_i K_i)^T (E_{1i} + E_{2i} M_i K_i) \\ + P_j + Q + \delta U_i^T U_i$$

Using $diag\{\varepsilon^{1/2}, \varepsilon^{1/2}\}$ to pre- and post- multiply the left-hand term of expression (35) and denoting $\bar{P}_i = \varepsilon P_i, \bar{Q} = \varepsilon Q$, we have

$$\begin{bmatrix} \bar{\Pi}_{ij} & \bar{P}_i A_{di} + (E_{1i} + E_{2i} M_i K_i)^T E_{di} \\ * & -(1-\mu)\bar{Q} + E_{di}{}^T E_{di} \end{bmatrix} < 0 \tag{36}$$

where

$$\begin{aligned} \bar{\Pi}_{ij} &= (A_i + B_i M_i K_i)^T \bar{P}_i + \bar{P}_i (A_i + B_i M_i K_i) + \bar{P}_i H_i H_i^T \bar{P}_i + (E_{1i} + E_{2i} M_i K_i)^T (E_{1i} + E_{2i} M_i K_i) \\ &+ \bar{P}_j + \bar{Q} + \varepsilon \delta U_i^T U_i \end{aligned}$$

By Lemma 1, (36) is equivalent to

$$\begin{bmatrix} \bar{\Pi}_{ij}' & \bar{P}_i A_{di} & \bar{P}_i H_i & (E_{1i} + E_{2i} M_i K_i)^T \\ * & -(1-\mu)\bar{Q} & 0 & E_{di}^T \\ * & * & -I & 0 \\ * & * & * & -I \end{bmatrix} < 0 \tag{37}$$

where $\bar{\Pi}_{ij}' = (A_i + B_i M_i K_i)^T \bar{P}_i + \bar{P}_i (A_i + B_i M_i K_i) + \bar{P}_j + \bar{Q} + \varepsilon \delta U_i^T U_i$

Using $diag\{\bar{P}_i^{-1}, \bar{Q}^{-1}, I, I\}$ to pre- and post- multiply the left-hand term of expression (37) and denoting $X_i = \bar{P}_i^{-1}, Y_i = K_i \bar{P}_i^{-1}, S = \bar{Q}^{-1}, \lambda = (\varepsilon\delta)^{-1}$, (37) can be written as

$$\begin{bmatrix} \bar{\Pi}_{ij}'' & A_{di} S & H_i & (E_{1i} X_i + E_{2i} M_i Y_i)^T \\ * & -(1-\mu)S & 0 & S E_{di}^T \\ * & * & -I & 0 \\ * & * & * & -I \end{bmatrix} < 0 \tag{38}$$

where $\bar{\Pi}_{ij}'' = (A_i X_i + B_i M_i Y_i)^T + (A_i + B_i M_i Y_i) + X_i (X_j^{-1} + S^{-1} + \lambda^{-1} U_i^T U_i) X_i$

Using Lemma 1 again, (38) is equivalent to (28). Meanwhile, substituting $X_i = \bar{P}_i^{-1}, \bar{P}_i = \varepsilon P_i$ and $\lambda = (\varepsilon\delta)^{-1}$ to (30) yields (27). Then the switching law becomes

$$\sigma(t) = \arg\min_{i \in \underline{N}} \{x^T(t) X_i^{-1} x(t)\} \tag{39}$$

Based on the above proof line, we know that if (27) and (28) holds, and the switching law is designed as (39), the state feedback controller $u(t) = K_{\sigma(t)} x(t)$, $K_i = Y_i X_i^{-1}$ can guarantee system (23)-(24) is asymptotically stable. The proof is completed. ■

3.3 Design of robust H_∞ reliable controller

Consider system (1)-(3). By (7), for the i th subsystem the feedback control law can be designed as

$$u^f(t) = M_i K_i x(t) \tag{40}$$

Substituting (40) to (1) and (2), the corresponding closed-loop system can be written as

$$\dot{x}(t) = \tilde{A}_i x(t) + \hat{A}_{di} x(t - d(t)) + D_i w(t) + f_i(x(t), t) \tag{41}$$

$$z(t) = \tilde{C}_i x(t) + N_i w(t) \tag{42}$$

where $\tilde{A}_i = \hat{A}_i + \hat{B}_i M_i K_i, \tilde{C}_i = C_i + G_i M_i K_i$, $i \in \underline{N}$.

The following theorem presents a sufficient existing condition of the robust H_∞ reliable controller for system (1)-(3).

Theorem 3 For system (1)-(3), if there exists symmetric positive definite matrices X_i, S, matrix Y_i and the positive scalar λ, ε such that

$$X_i > \lambda I \tag{43}$$

$$\begin{bmatrix} \Psi_i & D_i & (C_i X_i + G_i M_i Y_i)^T & A_{di} S & H_i & \Phi_i^T & X_i & X_i & X_i U_i^T \\ * & -\gamma\varepsilon I & N_i^T & 0 & 0 & 0 & 0 & 0 & 0 \\ * & * & -\frac{\gamma}{\varepsilon} I & 0 & 0 & 0 & 0 & 0 & 0 \\ * & * & * & -(1-\mu)S & 0 & SE_{di}^T & 0 & 0 & 0 \\ * & * & * & * & -I & 0 & 0 & 0 & 0 \\ * & * & * & * & * & -I & 0 & 0 & 0 \\ * & * & * & * & * & * & -S & 0 & 0 \\ * & * & * & * & * & * & * & -X_j & 0 \\ * & * & * & * & * & * & * & * & -\lambda I \end{bmatrix} < 0 \tag{44}$$

where $i \neq j, i, j \in \underline{N}$, $\Psi_i = A_i X_i + B_i M_i Y_i + (A_i X_i + B_i M_i Y_i)^T$, $\Phi_i = E_{1i} X_i + E_{2i} M_i Y_i$, then there exists the robust H_∞ reliable state feedback controller

$$u(t) = K_{\sigma(t)} x(t), \quad K_i = Y_i X_i^{-1} \tag{45}$$

and the switching law is designed as $\sigma(t) = \arg\min_{i \in \underline{N}} \{x^T(t) X_i^{-1} x(t)\}$, the closed-loop system is asymptotically stable with disturbance attenuation performance γ for all admissible uncertainties as well as all actuator faults.

Proof By (44), we can obtain that

$$\begin{bmatrix} \Psi_i & A_{di} S & H_i & \Phi_i^T & X_i & X_i & X_i U_i^T \\ * & -(1-\mu)S & 0 & SE_{di}^T & 0 & 0 & 0 \\ * & * & -I & 0 & 0 & 0 & 0 \\ * & * & * & -I & 0 & 0 & 0 \\ * & * & * & * & -S & 0 & 0 \\ * & * & * & * & * & -X_j & 0 \\ * & * & * & * & * & * & -\lambda I \end{bmatrix} < 0 \tag{46}$$

From Theorem 2, we know that closed-loop system (41) is asymptotically stable.
Define the following piecewise Lyapunov-Krasovskii functional candidate

$$V(x(t)) = V_i(x(t)) = x^T(t)P_i x(t) + \int_{t-d(t)}^{t} x^T(\tau)Qx(\tau)d\tau,\ t \in [t_n,\ t_{n+1}),\ n = 0,1,\cdots \tag{47}$$

where P_i, Q are symmetric positive definite matrices, and $t_0 = 0$. Along the trajectories of system (41), the time derivative of $V_i(x(t))$ is given by

$$\dot{V}_i(x(t)) \le \xi^T \begin{bmatrix} \Lambda_i & P_i D_i & P_i(A_{di} + H_i F_i(t) E_{di}) \\ * & 0 & 0 \\ * & * & -(1-\mu)Q \end{bmatrix} \xi \tag{48}$$

where

$$\xi = \begin{bmatrix} x^T(t) & w^T(t) & x^T(t-d(t)) \end{bmatrix}^T,$$

$$\Lambda_i = P_i[A_i + B_i M_i K_i + H_i F_i(t)(E_{1i} + E_{2i} M_i K_i)] + [A_i + B_i M_i K_i + H_i F_i(t)(E_{1i} + E_{2i} M_i K_i)]^T P_i + P_i + Q + \delta U_i^T U_i$$

By simple computing, we can obtain that

$$\begin{aligned} &\gamma^{-1} z^T(t) z(t) - \gamma w^T(t) w(t) \\ &= \xi^T \begin{bmatrix} \gamma^{-1}(C_i + G_i M_i K_i)^T (C_i + G_i M_i K_i) & \gamma^{-1}(C_i + G_i M_i K_i)^T N_i & 0 \\ * & \gamma^{-1} N_i^T N_i - \gamma I & 0 \\ * & * & 0 \end{bmatrix} \xi \end{aligned} \tag{49}$$

Denote $X_i = \bar{P}_i^{-1}, Y_i = K_i \bar{P}_i^{-1}, S = \bar{Q}^{-1}, \bar{P}_i = \varepsilon P_i, \bar{Q} = \varepsilon Q$. Substituting them to (44), and using Lemma 1 and Lemma 3, through equivalent transform we have

$$\begin{bmatrix} \gamma^{-1}(C_i + G_i M_i K_i)^T (C_i + G_i M_i K_i) + \Lambda_{ij} & \gamma^{-1}(C_i + G_i M_i K_i)^T N_i + P_i D_i & P_i(A_{di} + H_i F_i(t) E_{di}) \\ * & \gamma^{-1} N_i^T N_i - \gamma I & 0 \\ * & * & -(1-\mu)Q \end{bmatrix} < 0 \tag{50}$$

where

$$\Lambda_{ij} = P_i[A_i + B_i M_i K_i + H_i F_i(t)(E_{1i} + E_{2i} M_i K_i)] + [A_i + B_i M_i K_i + H_i F_i(t)(E_{1i} + E_{2i} M_i K_i)]^T P_i + P_j + Q + \delta U_i^T U_i$$

Obviously, under the switching law $\sigma(t) = \arg\min_{i \in \underline{N}}\{x^T(t) P_i x(t)\}$ there is

$$\gamma^{-1} z^T(t) z(t) - \gamma w^T(t) w(t) + \dot{V}_i(x(t)) < 0 \tag{51}$$

Define

$$J = \int_0^{\infty} (\gamma^{-1} z^T(t) z(t) - \gamma w^T(t) w(t)) dt \tag{52}$$

Consider switching signal

$$\sigma(t):\{(0,i^{(0)}),(t_1,i^{(1)}),(t_2,i^{(2)}),\cdots,(t_k,i^{(k)})\}$$

which means the $i^{(k)}$ th subsystem is activated at t_k.

Combining (47), (51) and (52), for zero initial conditions, we have

$$J \le \int_0^{t_1}(\gamma^{-1}z^T(t)z(t)-\gamma w^T(t)w(t)+\dot{V}_{i^{(0)}}(t))dt+\int_{t_1}^{t_2}(\gamma^{-1}z^T(t)z(t)-\gamma w^T(t)w(t)+\dot{V}_{i^{(1)}}(t))dt+\cdots$$
$$<0$$

Therefore, we can obtain $\|z(t)\|_2 < \gamma\|w(t)\|_2$. The proof is completed. ■

When the actuator fault is taken into account in the controller design, we have the following theorem.

Theorem 4 For system (1)-(3), γ is a given positive scalar, if there exists symmetric positive definite matrices X_i, S, matrix Y_i and the positive scalar $\alpha,\varepsilon,\lambda$ such that

$$X_i > \lambda I \tag{53}$$

$$\begin{bmatrix} \Sigma_i & D_i & \Sigma_{1i}^T & A_{di}S & H_i & \Sigma_{2i}^T & X_i & X_i & X_iU_i^T & Y_i^T M_{i0} J_i^{1/2} \\ * & -\gamma\varepsilon I & N_i^T & 0 & 0 & 0 & 0 & 0 & 0 & 0 \\ * & * & \Sigma_{3i} & 0 & 0 & \alpha G_i J_i E_{2i}^T & 0 & 0 & 0 & 0 \\ * & * & * & -(1-\mu)S & 0 & SE_{di}^T & 0 & 0 & 0 & 0 \\ * & * & * & * & -I & 0 & 0 & 0 & 0 & 0 \\ * & * & * & * & * & \Sigma_{4i} & 0 & 0 & 0 & 0 \\ * & * & * & * & * & * & -S & 0 & 0 & 0 \\ * & * & * & * & * & * & * & -X_j & 0 & 0 \\ * & * & * & * & * & * & * & * & -\lambda I & 0 \\ * & * & * & * & * & * & * & * & * & -\alpha I \end{bmatrix} < 0 \tag{54}$$

where $i \ne j, i, j \in \underline{N}$

$$\Sigma_i = A_iX_i + B_iM_{i0}Y_i + (A_iX_i + B_iM_{i0}Y_i)^T + \alpha B_iJ_iB_i^T$$
$$\Sigma_{1i} = C_iX_i + G_iM_{i0}Y_i + \alpha G_iJ_iB_i^T$$
$$\Sigma_{2i} = E_{1i}X_i + E_{2i}M_{i0}Y_i + \alpha E_{2i}J_iB_i^T$$
$$\Sigma_{3i} = -\frac{\gamma}{\varepsilon}I + \alpha G_iJ_iG_i^T,\ \Sigma_{4i} = -I + \alpha E_{2i}J_iE_{2i}^T$$

then there exists the γ-suboptimal robust H_∞ reliable controller

$$u(t) = K_{\sigma(t)}x(t),\ K_i = Y_iX_i^{-1} \tag{55}$$

and the switching law is designed as $\sigma(t) = \arg\min_{i \in \underline{N}}\{x^T(t)X_i^{-1}x(t)\}$, the closed-loop system is asymptotically stable.

Proof By Theorem 3, substituting (12) to (44) yields

$$T_{ij} + \begin{bmatrix} B_iM_{i0}L_iY_i + (B_iM_{i0}L_iY_i)^T & 0 & (G_iM_{i0}L_iY_i)^T & 0 & 0 & (E_{2i}M_{i0}L_iY_i)^T & 0 & 0 & 0 & 0 \\ * & 0 & 0 & 0 & 0 & 0 & 0 & 0 & 0 & 0 \\ * & * & 0 & 0 & 0 & 0 & 0 & 0 & 0 & 0 \\ * & * & * & 0 & 0 & 0 & 0 & 0 & 0 & 0 \\ * & * & * & * & 0 & 0 & 0 & 0 & 0 & 0 \\ * & * & * & * & * & 0 & 0 & 0 & 0 & 0 \\ * & * & * & * & * & * & 0 & 0 & 0 & 0 \\ * & * & * & * & * & * & * & 0 & 0 & 0 \\ * & * & * & * & * & * & * & * & 0 & 0 \\ * & * & * & * & * & * & * & * & * & 0 \end{bmatrix} < 0 \quad (56)$$

where

$$T_{ij} = \begin{bmatrix} \Psi_{i0} & D_i & (C_iX_i + G_iM_{i0}Y_i)^T & A_{di}S & H_i & \Phi_{i0}^T & X_i & X_i & X_iU_i^T \\ * & -\gamma\varepsilon I & N_i^T & 0 & 0 & 0 & 0 & 0 & 0 \\ * & * & -\frac{\gamma}{\varepsilon}I & 0 & 0 & 0 & 0 & 0 & 0 \\ * & * & * & -(1-\mu)S & 0 & SE_{di}^T & 0 & 0 & 0 \\ * & * & * & * & -I & 0 & 0 & 0 & 0 \\ * & * & * & * & * & -I & 0 & 0 & 0 \\ * & * & * & * & * & * & -S & 0 & 0 \\ * & * & * & * & * & * & * & -X_j & 0 \\ * & * & * & * & * & * & * & * & -\lambda I \end{bmatrix}$$

$\Psi_{i0} = A_iX_i + B_iM_{i0}Y_i + (A_iX_i + B_iM_{i0}Y_i)^T$

$\Phi_{i0} = E_{1i}X_i + E_{2i}M_{i0}Y_i$

Denote

$$\Xi_i = \begin{bmatrix} B_iM_{i0}L_iY_i + (B_iM_{i0}L_iY_i)^T & 0 & (G_iM_{i0}L_iY_i)^T & 0 & 0 & (E_{2i}M_{i0}L_iY_i)^T & 0 & 0 & 0 & 0 \\ * & 0 & 0 & 0 & 0 & 0 & 0 & 0 & 0 & 0 \\ * & * & 0 & 0 & 0 & 0 & 0 & 0 & 0 & 0 \\ * & * & * & 0 & 0 & 0 & 0 & 0 & 0 & 0 \\ * & * & * & * & 0 & 0 & 0 & 0 & 0 & 0 \\ * & * & * & * & * & 0 & 0 & 0 & 0 & 0 \\ * & * & * & * & * & * & 0 & 0 & 0 & 0 \\ * & * & * & * & * & * & * & 0 & 0 & 0 \\ * & * & * & * & * & * & * & * & 0 & 0 \\ * & * & * & * & * & * & * & * & * & 0 \end{bmatrix} \quad (57)$$

Notice that M_{i0} and L_i are both diagonal matrices, then we have

$$\Xi_i = \begin{bmatrix} B_i \\ 0 \\ G_i \\ 0 \\ 0 \\ E_{2i} \\ 0 \\ 0 \\ 0 \end{bmatrix} L_i \begin{bmatrix} M_{i0}Y_i & 0 & 0 & 0 & 0 & 0 & 0 & 0 & 0 \end{bmatrix} + \left(\begin{bmatrix} B_i \\ 0 \\ G_i \\ 0 \\ 0 \\ E_{2i} \\ 0 \\ 0 \\ 0 \end{bmatrix} L_i \begin{bmatrix} M_{i0}Y_i & 0 & 0 & 0 & 0 & 0 & 0 & 0 & 0 \end{bmatrix} \right)^T$$

From Lemma 4 and (12), we can obtain that

$$\Xi_i \le \alpha \begin{bmatrix} B_i \\ 0 \\ G_i \\ 0 \\ 0 \\ E_{2i} \\ 0 \\ 0 \\ 0 \end{bmatrix} J_i \begin{bmatrix} B_i \\ 0 \\ G_i \\ 0 \\ 0 \\ E_{2i} \\ 0 \\ 0 \\ 0 \end{bmatrix}^T + \alpha^{-1} \begin{bmatrix} Y_i^T M_{i0} \\ 0 \\ 0 \\ 0 \\ 0 \\ 0 \\ 0 \\ 0 \\ 0 \end{bmatrix} J_i \begin{bmatrix} Y_i^T M_{i0} \\ 0 \\ 0 \\ 0 \\ 0 \\ 0 \\ 0 \\ 0 \\ 0 \end{bmatrix}^T \tag{58}$$

Then the following inequality

$$T_{ij} + \alpha \begin{bmatrix} B_i \\ 0 \\ G_i \\ 0 \\ 0 \\ E_{2i} \\ 0 \\ 0 \\ 0 \end{bmatrix} J_i \begin{bmatrix} B_i \\ 0 \\ G_i \\ 0 \\ 0 \\ E_{2i} \\ 0 \\ 0 \\ 0 \end{bmatrix}^T + \alpha^{-1} \begin{bmatrix} Y_i^T M_{i0} \\ 0 \\ 0 \\ 0 \\ 0 \\ 0 \\ 0 \\ 0 \\ 0 \end{bmatrix} J_i \begin{bmatrix} Y_i^T M_{i0} \\ 0 \\ 0 \\ 0 \\ 0 \\ 0 \\ 0 \\ 0 \\ 0 \end{bmatrix}^T < 0 \tag{59}$$

can guarantee (56) holds.

By Lemma 1, we know that (59) is equivalent to (54). The proof is completed. ■

Remark 3 (54) is not linear, because there exist unknown variables ε, ε^{-1}. Therefore, we consider utilizing variable substitute method to solve matrix inequality (54). Using $diag\{I,\varepsilon^{-1},I,I,I,I,I,I,I\}$ to pre- and post- multiply the left-hand term of expression (54), and denoting $\eta = \varepsilon^{-1}$, (54) can be transformed as the following linear matrix inequality

$$\begin{bmatrix} \Sigma_i & \eta D_i & \Sigma_{1i}^T & A_{di}S & H_i & \Sigma_{2i}^T & X_i & X_i & X_iU_i^T & Y_i^T M_{i0} J_i^{1/2} \\ * & -\eta\gamma I & \eta N_i^T & 0 & 0 & 0 & 0 & 0 & 0 & 0 \\ * & * & \Sigma_{3i} & 0 & 0 & \alpha G_i J_i E_{2i}^T & 0 & 0 & 0 & 0 \\ * & * & * & -(1-\mu)S & 0 & SE_{di}^T & 0 & 0 & 0 & 0 \\ * & * & * & * & -I & 0 & 0 & 0 & 0 & 0 \\ * & * & * & * & * & \Sigma_{4i} & 0 & 0 & 0 & 0 \\ * & * & * & * & * & * & -S & 0 & 0 & 0 \\ * & * & * & * & * & * & * & -X_j & 0 & 0 \\ * & * & * & * & * & * & * & * & -\lambda I & 0 \\ * & * & * & * & * & * & * & * & * & -\alpha I \end{bmatrix} < 0 \tag{60}$$

where $\Sigma_{3i} = -\eta\gamma I + \alpha G_i J_i G_i^T$

Corollary 1 For system (1)-(3), if the following optimal problem

$$\min_{X_i>0,S>0,\alpha>0,\varepsilon>0,\lambda>0,Y_i} \gamma \tag{61}$$

s.t. (53) and (54)

has feasible solution $X_i > 0, S > 0, \alpha > 0, \varepsilon > 0, \lambda > 0, Y_i, i \in \underline{N}$, then there exists the γ-optimal robust H_∞ reliable controller

$$u(t) = K_{\sigma(t)}x(t),\ K_i = Y_i X_i^{-1} \tag{62}$$

and the switching law is designed as $\sigma(t) = \arg\min_{i\in\underline{N}}\{x^T(t)X_i^{-1}x(t)\}$, the closed-loop system is asymptotically stable.

Remark 4 There exist unknown variables $\gamma\varepsilon$, $\gamma\varepsilon^{-1}$ in (54), so it is difficult to solve the above optimal problem. We denote $\theta = \gamma\varepsilon$, $\chi = \gamma\varepsilon^{-1}$, and substitute them to (54), then (54) becomes a set of linear matrix inequalities. Notice that $\gamma \le \frac{\theta+\chi}{2}$, we can solve the following the optimal problem to obtain the minimal upper bound of γ

$$\min_{X_i>0,S>0,\alpha>0,\theta>0,\chi>0,\lambda>0,Y_i} \frac{\theta+\chi}{2} \tag{63}$$

$$\text{s.t.} \begin{bmatrix} \Sigma_i & D_i & \Sigma_{1i}^T & A_{di}S & H_i & \Sigma_{2i}^T & X_i & X_i & X_iU_i^T & Y_i^T M_{i0} J_i^{1/2} \\ * & -\theta I & N_i^T & 0 & 0 & 0 & 0 & 0 & 0 & 0 \\ * & * & \Sigma_{3i} & 0 & 0 & \alpha G_i J_i E_{2i}^T & 0 & 0 & 0 & 0 \\ * & * & * & -(1-\mu)S & 0 & SE_{di}^T & 0 & 0 & 0 & 0 \\ * & * & * & * & -I & 0 & 0 & 0 & 0 & 0 \\ * & * & * & * & * & \Sigma_{4i} & 0 & 0 & 0 & 0 \\ * & * & * & * & * & * & -S & 0 & 0 & 0 \\ * & * & * & * & * & * & * & -X_j & 0 & 0 \\ * & * & * & * & * & * & * & * & -\lambda I & 0 \\ * & * & * & * & * & * & * & * & * & -\alpha I \end{bmatrix} < 0 \tag{64}$$

$$X_i > \lambda I \tag{65}$$

where $\Sigma_{3i} = -\chi I + \alpha G_i J_i G_i^T$, then the minimal value of γ can be acquired based on the following steps

Step 1. From (63)-(65), we solve the minimal value $\gamma^{(0)}$ of $\dfrac{\theta+\chi}{2}$, where $\gamma^{(0)}$ is the first iterative value;

Step 2. Choosing an appropriate step size $\delta = \delta_0$, and let $\gamma^{(1)} = \gamma^{(0)} - \delta_0$, then we substitute

Step 3. $\gamma^{(1)}$ to (60) to solve LMIs. If there is not feasible solution, stop iterating and $\gamma^{(0)}$ is just the optimal performance index; Otherwise, continue iterating until $\gamma^{(k)}$ is feasible solution but $\gamma^{(k+1)}$ is not, then $\gamma = \gamma^{(0)} - k\delta_0$ is the optimal performance index.

4. Numerical example

In this section, an example is given to illustrate the effectiveness of the proposed method. Consider system (1)-(3) with parameter as follows (the number of subsystems $N = 2$)

$$A_1 = \begin{bmatrix} -2 & 0 \\ 0 & 2 \end{bmatrix},\ A_2 = \begin{bmatrix} 1 & -3 \\ 0 & -2 \end{bmatrix},\ A_{d1} = \begin{bmatrix} -5 & 0 \\ 0 & -4 \end{bmatrix},\ A_{d2} = \begin{bmatrix} -3 & -1 \\ 0 & -6 \end{bmatrix},\ B_1 = \begin{bmatrix} -5 & 7 \\ -9 & 0 \end{bmatrix},\ B_2 = \begin{bmatrix} -8 & 2 \\ 2 & 6 \end{bmatrix}$$

$$E_{11} = \begin{bmatrix} 2 & 5 \\ 0 & 0 \end{bmatrix},\ E_{12} = \begin{bmatrix} 1 & 2 \\ 4 & 0 \end{bmatrix},\ E_{21} = \begin{bmatrix} 2 & 0 \\ 3 & 1 \end{bmatrix},\ E_{22} = \begin{bmatrix} 2 & 0 \\ 0.2 & 0 \end{bmatrix},\ E_{d1} = \begin{bmatrix} -1 & 0 \\ 1 & 0.1 \end{bmatrix},\ E_{d2} = \begin{bmatrix} 2 & 0 \\ 1 & 0 \end{bmatrix}$$

$$C_1 = \begin{bmatrix} -0.8 & 0.5 \end{bmatrix},\ C_2 = \begin{bmatrix} 0.3 & -0.8 \end{bmatrix},\ G_1 = \begin{bmatrix} 0.1 & 0 \end{bmatrix},\ G_2 = \begin{bmatrix} -0.1 & 0 \end{bmatrix},\ D_1 = \begin{bmatrix} 2 & -1 \\ -4 & 0 \end{bmatrix}$$

$$D_2 = \begin{bmatrix} 3 & -6 \\ -5 & 12 \end{bmatrix},\ H_1 = H_2 = \begin{bmatrix} 0.1 & 0.2 \\ 0.1 & 0 \end{bmatrix},\ N_1 = N_2 = \begin{bmatrix} 0.01 & 0 \end{bmatrix}$$

The time-varying delay $d(t) = 0.5e^{-t}$, the initial condition $x(t) = \begin{bmatrix} 2 & -1 \end{bmatrix}^T$, $t \in [-0.5, 0]$, uncertain parameter matrices $F_1(t) = F_2(t) = \begin{bmatrix} \sin t & 0 \\ 0 & \sin t \end{bmatrix}$, $\gamma = 0.8$ and nonlinear functions is selected as

$$f_1(x(t),t) = \begin{bmatrix} x_1(t)\cos(t) \\ 0 \end{bmatrix},\ f_2(x(t),t) = \begin{bmatrix} 0 \\ x_2(t)\cos(t) \end{bmatrix}$$

then $U_1 = \begin{bmatrix} 1 & 0 \\ 0 & 0 \end{bmatrix}$, $U_2 = \begin{bmatrix} 0 & 0 \\ 0 & 1 \end{bmatrix}$.

When $M_1 = M_2 = I$, from Theorem 3 and using LMI toolbox in Matlab, we have

$$X_1 = \begin{bmatrix} 0.6208 & 0.0909 \\ 0.0909 & 0.1061 \end{bmatrix},\ X_2 = \begin{bmatrix} 0.2504 & 0.1142 \\ 0.1142 & 0.9561 \end{bmatrix}$$

$$Y_1 = \begin{bmatrix} 0.6863 & 0.5839 \\ -3.2062 & -0.3088 \end{bmatrix},\ Y_2 = \begin{bmatrix} 0.8584 & -1.3442 \\ -0.5699 & -5.5768 \end{bmatrix}$$

$$S = \begin{bmatrix} 0.1123 & 0.0370 \\ 0.0370 & 0.1072 \end{bmatrix}, \ \varepsilon = 19.5408, \ \lambda = 0.0719$$

Then robust H_∞ controller can be designed as

$$K_1 = \begin{bmatrix} 0.3426 & 5.2108 \\ -5.4176 & 1.7297 \end{bmatrix}, \ K_2 = \begin{bmatrix} 4.3032 & -1.9197 \\ 0.4056 & -5.8812 \end{bmatrix}$$

Choosing the switching law $\sigma(t) = \arg\min_{i \in \underline{N}}\{x^T(t)X_i^{-1}x(t)\}$, the switching domain is

$$\Omega_1 = \{x(t) \in R^2 \left| x^T(t)X_1^{-1}x(t) \le x^T(t)X_2^{-1}x(t)\right.\}$$
$$\Omega_2 = \{x(t) \in R^2 \left| x^T(t)X_1^{-1}x(t) > x^T(t)X_2^{-1}x(t)\right.\}$$

i.e.

$$\Omega_1 = \{x(t) \in R^2 \left| x^T(t)\begin{bmatrix} -2.3815 & -1.0734 \\ -1.0734 & 9.6717 \end{bmatrix} x(t) \le 0\right.\}$$
$$\Omega_2 = \{x(t) \in R^2 \left| x^T(t)\begin{bmatrix} -2.3815 & -1.0734 \\ -1.0734 & 9.6717 \end{bmatrix} x(t) > 0\right.\}$$

The switching law is

$$\sigma(t) = \begin{cases} 1 & x(t) \in \Omega_1 \\ 2 & x(t) \in \Omega_2 \end{cases}$$

The state responses of the closed-loop system are shown in Fig. 1.

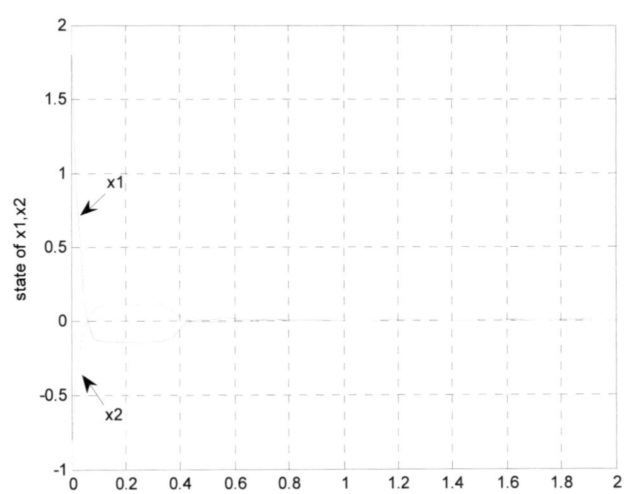

Fig. 1. State responses of the closed-loop system with the normal switched controller when the actuator is normal

The Fig. 1 illustrates that the designed normal switched controller can guarantee system is asymptotically stable when the actuator is normal.
However, in fact, the actuator fault can not be avoided. Here, we assume that the actuator fault model with parameters as follows
For subsystem 1

$$0.04 \le m_{11} \le 1\,,\ 0.1 \le m_{12} \le 1.2$$

For subsystem 2

$$0.1 \le m_{21} \le 1\,,\ 0.04 \le m_{22} \le 1$$

Then we have

$$M_{10} = \begin{bmatrix} 0.52 & 0 \\ 0 & 0.65 \end{bmatrix},\ J_1 = \begin{bmatrix} 0.92 & 0 \\ 0 & 0.85 \end{bmatrix}$$

$$M_{20} = \begin{bmatrix} 0.55 & 0 \\ 0 & 0.52 \end{bmatrix},\ J_2 = \begin{bmatrix} 0.82 & 0 \\ 0 & 0.92 \end{bmatrix}$$

Choosing the fault matrices of subsystem 1 and subsystem 2 are

$$M_1 = \begin{bmatrix} 0.04 & 0 \\ 0 & 0.1 \end{bmatrix},\ M_2 = \begin{bmatrix} 0.1 & 0 \\ 0 & 0.04 \end{bmatrix}$$

Then the above switched controller still be used to stabilize the system, the simulation result of the state responses of closed-loop switched system is shown in Fig. 2.

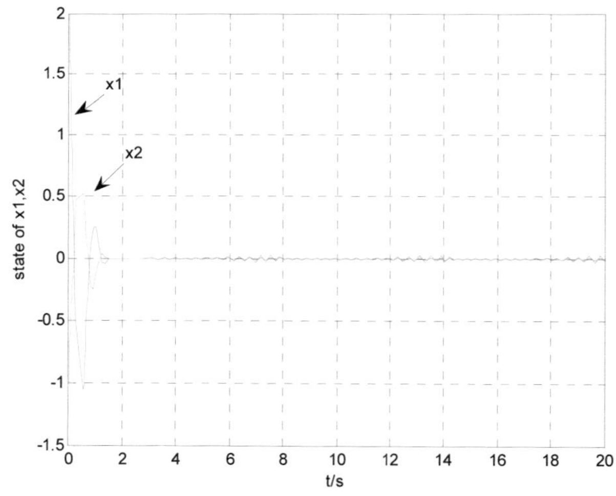

Fig. 2. State responses of the closed-loop system with the normal switched controller when the actuator is failed

Obviously, it can be seen that system state occurs vibration and the system can not be stabilized effectively.
The simulation comparisons of Fig. 1 and Fig. 2 shows that the design method for normal switched controller may lose efficacy when the actuator is failed.

Then for the above fault model, by Theorem 3 and using LMI toolbox in Matlab, we have

$$X_1 = \begin{bmatrix} 0.0180 & 0.0085 \\ 0.0085 & 0.0123 \end{bmatrix},\ X_2 = \begin{bmatrix} 0.0436 & -0.0007 \\ -0.0007 & 0.0045 \end{bmatrix}$$

$$Y_1 = \begin{bmatrix} 0.4784 & 0.6606 \\ -0.5231 & -0.0119 \end{bmatrix},\ Y_2 = \begin{bmatrix} 0.7036 & -0.1808 \\ -0.1737 & -0.5212 \end{bmatrix}$$

$$S = \begin{bmatrix} 0.0130 & 0.0000 \\ 0.0000 & 0.0012 \end{bmatrix},\ \alpha = 0.0416,\ \varepsilon = 946.1561,\ \lambda = 0.0036$$

Then robust H_∞ reliable controller can be designed as

$$K_1 = \begin{bmatrix} 1.8533 & 52.3540 \\ -42.5190 & 28.3767 \end{bmatrix},\ K_2 = \begin{bmatrix} 15.5242 & -37.5339 \\ -5.8387 & -116.0295 \end{bmatrix}$$

Choosing the switching law as

$$\sigma(t) = \begin{cases} 1 & x(t) \in \Omega_1 \\ 2 & x(t) \in \Omega_2 \end{cases}$$

where

$$\Omega_1 = \{x(t) \in R^2 \,|\, x^T(t) \begin{bmatrix} 59.6095 & -60.5390 \\ -60.5390 & -100.9210 \end{bmatrix} x(t) \le 0\}$$

$$\Omega_2 = \{x(t) \in R^2 \,|\, x^T(t) \begin{bmatrix} 59.6095 & -60.5390 \\ -60.5390 & -100.9210 \end{bmatrix} x(t) > 0\}$$

The state responses of the closed-loop system are shown in Fig. 3.

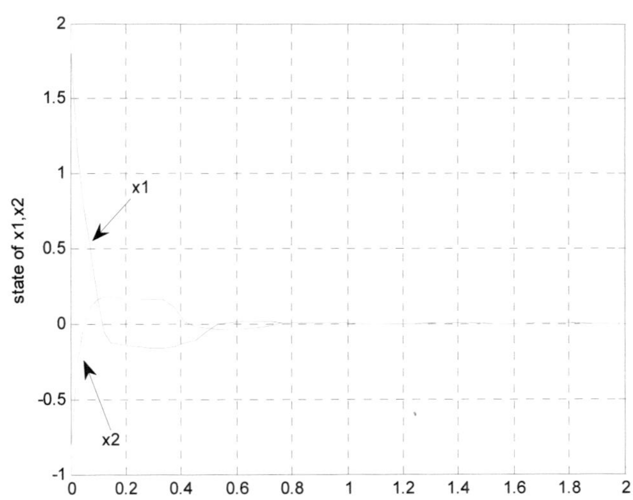

Fig. 3. State responses of the closed-loop system with the reliable switched controller when the actuator is failed

It can be seen that the designed robust H_∞ reliable controller makes the closed-loop switched system is asymptotically stable for admissible uncertain parameter and actuator fault.
The simulation of Fig. 3 also shows that the design method of robust H_∞ reliable controller can overcome the effect of time-varying delay for switched system.
Moreover, by Corollary 1, based on the solving process of Remark 4 we can obtain the optimal H_∞ disturbance attenuation performance $\gamma = 0.54$, the optimal robust H_∞ reliable controller can be designed as

$$K_1 = \begin{bmatrix} 9.7714 & 115.4893 \\ -69.8769 & 41.1641 \end{bmatrix}, \quad K_2 = \begin{bmatrix} 9.9212 & -106.5624 \\ -62.1507 & -608.0198 \end{bmatrix}$$

The parameter matrices X_1, X_2 of the switching law are

$$X_1 = \begin{bmatrix} 0.0031 & 0.0011 \\ 0.0011 & 0.0018 \end{bmatrix}, \quad X_2 = \begin{bmatrix} 0.0119 & -0.0011 \\ -0.0011 & 0.0004 \end{bmatrix}$$

5. Conclusion

In order to overcome the passive effect of time-varying delay for switched systems and make systems be anti-jamming and fault-tolerant, robust H_∞ reliable control for a class of uncertain switched systems with actuator faults and time-varying delays is investigated. At first, the concept of robust reliable controller, γ-suboptimal robust H_∞ reliable controller and γ-optimal robust H_∞ reliable controller are presented. Secondly, fault model of actuator for switched systems is put forward. Multiple Lyapunov-Krasovskii functional method and linear matrix inequality technique are adopted to design robust H_∞ reliable controller. The matrix inequalities in the γ-optimal problem are not linear, then we make use of variable substitute method to acquire the controller gain matrices. Furthermore, the iteration solving process of optimal disturbance attenuation performance γ is presented. Finally, a numerical example shows the effectiveness of the proposed method. The result illustrates that the designed controller can stabilize the original system and makes it be of H_∞ disturbance attenuation performance when the system has uncertain parameters and actuator faults. Our future work will focus on constructing the appropriate multiply Lyapunov-Krasovskii functional to obtain the designed method of time delay dependent robust H_∞ reliable controller.

6. Acknowledgment

The authors are very grateful to the reviewers and to the editors for their helpful comments and suggestions on this paper. This work was supported by the Natural Science Foundation of China under Grant No. 60974027.

7. References

Boyd, S. P.; Ghaoui, L. E. & Feron, et al. (1994). *Linear matrix inequalities in system and control theory*. SIAM.

Cao, Y.; Sun, Y. & Cheng, C. (1998). Delay dependent robust stabilization of uncertain systems with multiple state delays. *IEEE Transactions on Automatic Control*, Vol. 43, No. 11, 1608–1612.

Cong, S.; Fei, S. M. & Li, T. (2007). Exponential stability analysis for the switched system with time delay: multiple Lyapunov functions approach, *Acta Automatica Sinica*, 2007, Vol. 33, No. 9, 985-988 (in Chinese)

Kim, S.; Campbell, S. A. & Liu, X. (2006). Stability of a class of linear switching systems with time delay. *IEEE Transactions on Circuits & Systems-I*, Vol. 53, No. 2, 384-393.

Lien, C. H. (2007). Non-fragile guaranteed cost control for uncertain neutral dynamic systems with time-varying delays in state and control input. *Chaos, Solitons & Fractals*, Vol. 31, No. 4, 889–899.

Peleties, P. & DeCarlo, R. A. (1991). Asymptotic stability of m-switched systems using Lypaunov-like functions. *Proceedings of American Control Conference*, 1991, 1679-1684.

Pettersson, S. & Lennartson, B. (2002). Hybrid system stability and robustness verification using linear matrix inequalities. *International Journal of Control*, Vol. 75, No. 16-17, 1335-1355.

Sun, Z. D. & Ge, S. S. (2005). Analysis and synthesis of switched linear control systems. *Automatica*, Vol. 41, No. 2, 181-195.

Song, Y.; Fan, J. & Fei, M. R., et al. (2008). Robust H_∞ control of discrete switched system with time delay. *Applied Mathematics and Computation*, Vol. 205, No. 1, 159-169.

Sun, W. A. & Zhao, J. (2005). H_∞ robust control of uncertain switched linear systems based on LMIs. *Control and Decision*, Vol. 20, No. 6, 650-655. (in Chinese)

Song, Z. Y.; Nie, H. & Zhao, J. (2007). Robust H_∞ control for discrete switched systems with time-varying delays. *Journal of Northeastern University(Natural Science)*, Vol. 28, No. 4, 469-472. (in Chinese)

Tomlin, C.; Pappas, G. J. & Sastry, S. (1998). Conflict resolution for air traffic management: a study in multiagent hybrid systems. *IEEE Transactions on Automatic Control*, Vol. 43, No. 4, 509–521.

Varaiya, P. (1993). Smart cars on smart roads: problems of control. *IEEE Transactions on Automatic Control*, Vol. 38, No. 2, 195–207.

Wang, W. & Brockett, R. W. (1997). Systems with finite communication bandwidth constraints–part I: State estimation problems. *IEEE Transactions on Automatic Control*, Vol. 42, No. 9, 1294–1299.

Wang, R.; Liu, J. C. & Zhao, J. (2006). Reliable guaranteed-cost control for a class of uncertain switched linear systems with time-delay. *Control Theory and Applications*, Vol. 23, No. 6, 1001-1004. (in Chinese)

Xiang, Z. R. & Wang, R. H. (2008). Robust reliable control for uncertain switched nonlinear systems with time delay. *Proceedings of the 7th World Congress on Intelligent Control and Automation*, pp. 5487-5491.

Xiang, Z. R. & Wang, R. H. (2009). Robust L_∞ reliable control for uncertain nonlinear switched systems with time delay. *Applied Mathematics and Computation*, Vol. 210, No. 1, 202-210.

Zhang, Y.; Liu, X. Z. & Shen, X. M. (2007). Stability of switched systems with time delay. *Nonlinear Analysis: Hybrid Systems*, Vol. 1, No. 1, 44-58.

Part 3

Sliding Mode Control

7

Optimal Sliding Mode Control for a Class of Uncertain Nonlinear Systems Based on Feedback Linearization

Hai-Ping Pang and Qing Yang
Qingdao University of Science and Technology
China

1. Introduction

Optimal control is one of the most important branches in modern control theory, and linear quadratic regulator (LQR) has been well used and developed in linear control systems. However, there would be several problems in employing LQR to uncertain nonlinear systems. The optimal LQR problem for nonlinear systems often leads to solving a nonlinear two-point boundary-value (TPBV) problem (Tang et al. 2008; Pang et al. 2009) and an analytical solution generally does not exist except some simplest cases (Tang & Gao, 2005). Additionally, the optimal controller design is usually based on the precise mathematical models. While if the controlled system is subject to some uncertainties, such as parameter variations, unmodeled dynamics and external disturbances, the performance criterion which is optimized based on the nominal system would deviate from the optimal value, even the system becomes unstable (Gao & Hung, 1993 ; Pang & Wang, 2009).

The main control strategies to deal with the optimal control problems of nonlinear systems are as follows. First, obtain approximate solution of optimal control problems by iteration or recursion, such as successive approximate approach (Tang, 2005), SDRE (Shamma & Cloutier, 2001), ASRE (Cimen & Banks, 2004). These methods could have direct results but usually complex and difficult to be realized. Second, transform the nonlinear system into a linear one by the approximate linearization (i.e. Jacobian linearization), then optimal control can be realized easily based on the transformed system. But the main problem of this method is that the transformation is only applicable to those systems with less nonlinearity and operating in a very small neighborhood of equilibrium points. Third, transform the nonlinear system into a linear one by the exact linearization technique (Mokhtari et al. 2006; Pang & Chen, 2009). This differs entirely from approximate linearization in that the approximate linearization is often done simply by neglecting any term of order higher than 1 in the dynamics while exact linearization is achieved by exact state transformations and feedback.

As a precise and robust algorithm, the sliding mode control (SMC) (Yang & Özgüner, 1997; Choi et al. 1993; Choi et al. 1994) has attracted a great deal of attention to the uncertain nonlinear system control problems. Its outstanding advantage is that the sliding motion exhibits complete robustness to system uncertainties. In this chapter, combining LQR and SMC, the design of global robust optimal sliding mode controller (GROSMC) is concerned. Firstly, the GROSMC is designed for a class of uncertain linear systems. And then, a class of affine nonlinear systems is considered. The exact linearization technique is adopted to transform the nonlinear system into an equivalent linear one and a GROSMC is designed based on the transformed system. Lastly, the global robust optimal sliding mode tracking controller is studied for a class of uncertain affine nonlinear systems. Simulation results illustrate the effectiveness of the proposed methods.

2. Optimal sliding mode control for uncertain linear system

In this section, the problem of robustify LQR for a class of uncertain linear systems is considered. An optimal controller is designed for the nominal system and an integral sliding surface (Lee, 2006; Laghrouche et al. 2007) is constructed. The ideal sliding motion can minimize a given quadratic performance index, and the reaching phase, which is inherent in conventional sliding mode control, is completely eliminated (Basin et al. 2007). Then the sliding mode control law is synthesized to guarantee the reachability of the specified sliding surface. The system dynamics is global robust to uncertainties which satisfy matching conditions. A GROSMC is realized. To verify the effectiveness of the proposed scheme, a robust optimal sliding mode controller is developed for rotor position control of an electrical servo drive system.

2.1 System description and problem formulation

Consider an uncertain linear system described by

$$\dot{x}(t) = (A + \Delta A)x(t) + (B + \Delta B)u(t) + \delta(x,t) \tag{1}$$

where $x(t) \in R^n$ and $u(t) \in R^m$ are the state and the control vectors, respectively. ΔA and ΔB are unknown time-varying matrices representing system parameter uncertainties. $\delta(x,t)$ is an uncertain extraneous disturbance and/or unknown nonlinearity of the system.

Assumption 1. The pair (A,B) is controllable and $\text{rank}(B) = m$.

Assumption 2. ΔA , ΔB and $\delta(x,t)$ are continuously differentiable in x , and piecewise continuous in t .

Assumption 3. There exist unknown continuous functions of appropriate dimension $\Delta\tilde{A}$, $\Delta\tilde{B}$ and $\tilde{\delta}(x,t)$, such that

$$\Delta A = B\Delta\tilde{A}, \ \Delta B = B\Delta\tilde{B}, \ \delta(x,t) = B\tilde{\delta}(x,t).$$

These conditions are the so-called matching conditions.

From these assumptions, the state equation of the uncertain dynamic system (1) can be rewritten as

$$\dot{x}(t) = Ax(t) + Bu(t) + B\tilde{\delta}(x,t), \tag{2}$$

where

Assumption 4. There exist unknown positive constants γ_0 and γ_1 such that

$$\left\|\tilde{\delta}(x,t)\right\| \le \gamma_0 + \gamma_1 \left\|x(t)\right\|.$$

where $\|\bullet\|$ denotes the Euclidean norm.

By setting the uncertainty to zero, we can obtain the dynamic equation of the original system of (1), as

$$\dot{x}(t) = Ax(t) + Bu(t). \tag{3}$$

For the nominal system (3), let's define a quadratic performance index as follows:

$$J_0 = \frac{1}{2}\int_0^{\infty}[x^{\mathrm{T}}(t)Qx(t) + u^{\mathrm{T}}(t)Ru(t)]dt\,, \tag{4}$$

where $Q \in R^{n \times n}$ is a semi-positive definite matrix, the weighting function of states; $R \in R^{m \times m}$ is a positive definite matrix, the weighting function of control variables.
According to optimal control theory and considering Assumption1, there exists an optimal feedback control law that minimizes the index (4). The optimal control law can be written as

$$u^*(t) = -R^{-1}B^T Px(t)\,, \tag{5}$$

where $P \in R^{n \times n}$ is a positive definite matrix solution of Riccati matrix equation:

$$-PA - A^T P + PBR^{-1}B^T P - Q = 0. \tag{6}$$

So the dynamic equation of the closed-loop system is

$$\dot{x}(t) = (A - BR^{-1}B^T P)x(t). \tag{7}$$

Obviously, according to optimal control theory, the closed-loop system is asymptotically stable. However, when the system is subjected to uncertainties such as external disturbances and parameter variations, the optimal system behavior could be deteriorated, even unstable. In the next part, we will utilize sliding mode control strategy to robustify the optimal control law.

2.2 Design of optimal sliding mode controller

2.2.1 Design of optimal sliding mode surface

Considering the uncertain system (2), we chose the integral sliding surface as follows:

$$s(x,t) = G[x(t) - x(0)] - G\int_0^t (A - BR^{-1}B^{\mathrm{T}}P)x(\tau)d\tau = 0. \tag{8}$$

where $G \in R^{m \times n}$, which satisfies that GB is nonsingular, $x(0)$ is the initial state vector. In sliding mode, we have $s(x,t) = 0$ and $\dot{s}(x,t) = 0$. Differentiating (8) with respect to t and considering (1), we obtain

$$\begin{aligned}\dot{s} &= G[(A + \Delta A)x + (B + \Delta B)u + \delta] - G(A - BR^{-1}B^{\mathrm{T}}P)x \\ &= G\Delta Ax + G(B + \Delta B)u + G\delta + GBR^{-1}B^{\mathrm{T}}Px \\ &= G(\Delta Ax + BR^{-1}B^{\mathrm{T}}Px) + G\delta + G(B + \Delta B)u\end{aligned} \tag{9}$$

the equivalent control becomes

$$u_{\mathrm{eq}} = -[G(B + \Delta B)]^{-1}[G(\Delta A + BR^{-1}B^{\mathrm{T}}P)x + G\delta]. \tag{10}$$

Substituting (10) into (1) and considering Assumption3, the ideal sliding mode dynamics becomes

$$\begin{aligned}\dot{x} &= (A+\Delta A)x-(B+\Delta B)[G(B+\Delta B)]^{-1}[G(\Delta A+BR^{-1}B^{\mathrm{T}}P)x+G\delta]+\delta \\ &= (A+B\Delta\tilde{A})x-(B+B\Delta\tilde{B})[G(B+B\Delta\tilde{B})]^{-1}[GB(\Delta\tilde{A}x+\tilde{\delta})+GBR^{-1}B^{\mathrm{T}}Px]+B\tilde{\delta} \\ &= (A-BR^{-1}B^{\mathrm{T}}P)x\end{aligned} \tag{11}$$

Comparing equation (11) with equation (7), we can see that they have the same form. So the sliding mode is asymptotically stable. Furthermore, it can be seen from (11) that the sliding mode is robust to uncertainties which satisfying matching conditions. So we call (8) a robust optimal sliding surface.

2.2.2 Design of sliding mode control law

To ensure the reachability of sliding mode in finite time, we chose the sliding mode control law as follows:

$$\begin{aligned}&u(t)=u_{\mathrm{c}}(t)+u_{\mathrm{d}}(t),\\ &u_{\mathrm{c}}(t)=-R^{-1}B^{\mathrm{T}}Px(t),\\ &u_{\mathrm{d}}(t)=-(GB)^{-1}(\eta+\gamma_0\|GB\|+\gamma_1\|GB\|\|x(t)\|)\mathrm{sgn}(s).\end{aligned} \tag{12}$$

Where $\eta>0$, $u_{\mathrm{c}}(t)$ is the continuous part, used to stabilize and optimize the nominal system; $u_{\mathrm{d}}(t)$ is the discontinuous part, which provides complete compensation for uncertainties of system (1). Let's select a quadratic performance index as follows:

$$J(t)=\frac{1}{2}\int_0^{\infty}[x^{\mathrm{T}}(t)Qx(t)+u_{\mathrm{c}}^{\mathrm{T}}(t)Ru_{\mathrm{c}}(t)]dt. \tag{13}$$

where the meanings of Q and R are as the same as that in (4).

Theorem 1. Consider uncertain linear system (1) with Assumptions 1-4. Let u and sliding surface be given by (12) and (8), respectively. The control law (12) can force the system trajectories with arbitrarily given initial conditions to reach the sliding surface in finite time and maintain on it thereafter.

Proof. Choosing $V=(1/2)s^{\mathrm{T}}s$ as a lyapunov function, and differentiating this function with respect to t and considering Assumptions 1-4, we have

$$\begin{aligned}\dot{V} &= s^{\mathrm{T}}\dot{s}\\ &= s^{\mathrm{T}}\{G[(A+\Delta A)x+(B+\Delta B)u+\delta]-G(A-BR^{-1}B^{\mathrm{T}}P)x\}\\ &= s^{\mathrm{T}}\{GBu+GB\tilde{\delta}+GBR^{-1}B^{\mathrm{T}}Px\}\\ &= s^{\mathrm{T}}\{-GBR^{-1}B^{\mathrm{T}}Px-(\eta+\gamma_0\|GB\|+\gamma_1\|GB\|\|x\|)\mathrm{sgn}(s)+GB\tilde{\delta}+GBR^{-1}B^{\mathrm{T}}Px\}\\ &= s^{\mathrm{T}}\{-(\eta+\gamma_0\|GB\|+\gamma_1\|GB\|\|x\|)\mathrm{sgn}(s)+GB\tilde{\delta}\}\\ &= -\eta\|s\|_1-(\gamma_0\|GB\|+\gamma_1\|GB\|\|x\|)\|s\|_1+s^{\mathrm{T}}GB\tilde{\delta}\\ &\le -\eta\|s\|_1-(\gamma_0\|GB\|+\gamma_1\|GB\|\|x\|)\|s\|_1+(\gamma_0\|GB\|+\gamma_1\|GB\|\|x\|)\|s\|\end{aligned}$$

where $\|\bullet\|_1$ denotes 1-norm. Noting the fact that $\|s\|_1\ge\|s\|$, we get

$$\dot{V}=s^{\mathrm{T}}\dot{s}\le-\eta\|s\| \tag{14}$$

This implies that the sliding mode control law we have chosen according to (12) could ensure the trajectories which start from arbitrarily given points be driven onto the sliding surface (8) in finite time and would not leave it thereafter despite uncertainties. The proof is complete.

Conclusion 1. The uncertain system (1) with the integral sliding surface (8) and the control law (12) achieves global sliding mode, and the performace index (13) is minimized. So the system designed is global robust and optimal.

2.3 Application to electrical servo drive

The speed and position electrical servo drive systems are widely used in engineering systems, such as CNC machines, industrial robots, winding machines and etc. The main properties required for servo systems include high tracking behavior, no overshoot, no oscillation, quick response and good robustness.

In general, with the implementation of field-oriented control, the mechanical equation of an induction motor drive or a permanent synchronous motor drive can be described as

$$J_m\ddot{\theta}(t)+B_m\dot{\theta}(t)+T_{\mathrm{d}}=T_{\mathrm{e}} \tag{15}$$

where θ is the rotor position; J_m is the moment of inertia; B_m is the damping coefficient; T_{d} denotes the external load disturbance, nonlinear friction and unpredicted uncertainties; T_{e} represents the electric torque which defined as

$$T_{\mathrm{e}}=K_{\mathrm{t}}i \tag{16}$$

where K_{t} is the torque constant and i is the torque current command.

Define the position tracking error $e(t)=\theta_{\mathrm{d}}(t)-\theta(t)$, where $\theta_{\mathrm{d}}(t)$ denotes the desired position, and let $x_1(t)=e(t)$, $x_2(t)=\dot{x}_1(t)$, $u=i$, then the error state equation of an electrical servo drive can be described as

$$\begin{bmatrix}\dot{x}_1\\ \dot{x}_2\end{bmatrix}=\begin{bmatrix}0 & 1\\ 0 & -\dfrac{B_m}{J_m}\end{bmatrix}\begin{bmatrix}x_1\\ x_2\end{bmatrix}+\begin{bmatrix}0\\ \dfrac{B_m}{J_m}\end{bmatrix}\dot{\theta}_{\mathrm{d}}+\begin{bmatrix}0\\ 1\end{bmatrix}\ddot{\theta}_{\mathrm{d}}+\begin{bmatrix}0\\ -\dfrac{K_{\mathrm{t}}}{J_m}\end{bmatrix}u+\begin{bmatrix}0\\ \dfrac{1}{J_m}\end{bmatrix}T_{\mathrm{d}}. \tag{17}$$

Supposing the desired position is a step signal, the error state equation can be simplified as

$$\begin{bmatrix}\dot{x}_1\\ \dot{x}_2\end{bmatrix}=\begin{bmatrix}0 & 1\\ 0 & -\dfrac{B_m}{J_m}\end{bmatrix}\begin{bmatrix}x_1\\ x_2\end{bmatrix}+\begin{bmatrix}0\\ -\dfrac{K_{\mathrm{t}}}{J_m}\end{bmatrix}u+\begin{bmatrix}0\\ \dfrac{1}{J_m}\end{bmatrix}T_{\mathrm{d}}. \tag{18}$$

The parameters of the servo drive model in the nominal condition with $T_{\mathrm{d}}=0\mathrm{Nm}$ are (Lin & Chou, 2003):

$$\hat{J}=5.77\times10^{-2}\,\mathrm{Nms^2},$$
$$\hat{B}=8.8\times10^{-3}\,\mathrm{Nms/rad},$$
$$\hat{K}_{\mathrm{t}}=0.667\ \mathrm{Nm/A}.$$

The initial condition is $x(0)=\begin{bmatrix}1 & 0\end{bmatrix}^T$. To investigate the effectiveness of the proposed controller, two cases with parameter variations in the electrical servo drive and load torque disturbance are considered here.

Case 1: $J_m=\hat{J}_m$, $B_m=\hat{B}_m$, $T_{\mathrm{d}}=1(t-10)\mathrm{Nm}-1(t-13)\mathrm{Nm}$.

Case 2: $J_m = 3\hat{J}_m$, $B_m = \hat{B}_m$, $T_d = 0$.
The optimal controller and the optimal robust SMC are designed, respectively, for both cases. The optimal controller is based on the nominal system with a quadratic performance index (4). Here

$$Q = \begin{bmatrix} 1 & 0 \\ 0 & 1 \end{bmatrix}, \quad R = 1$$

In Case 1, the simulation results by different controllers are shown in Fig. 1. It is seen that when there is no disturbance ($t < 10\text{s}$), both systems have almost the same performance.

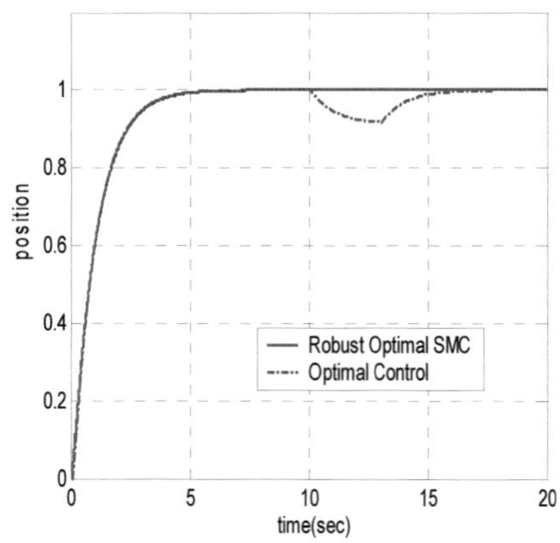

(a) Position responses

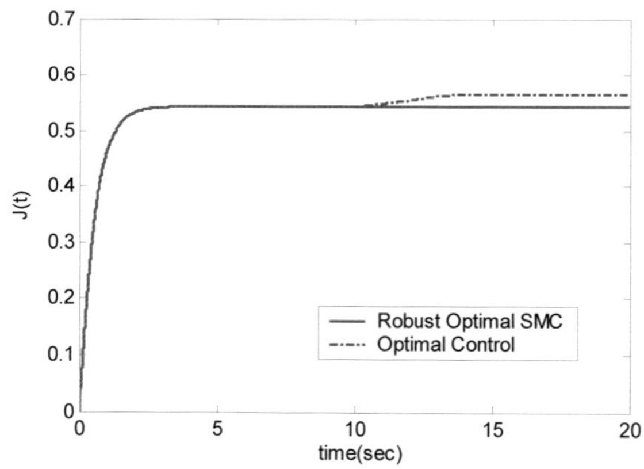

(b) Performance indexes

Fig. 1. Simulation results in Case 1

But when a load torque disturbance occurs at $t = (10 \sim 13)\text{s}$, the position trajectory of optimal control system deviates from the desired value, nevertheless the position trajectory of the robust optimal SMC system is almost not affected.
In Case 2, the simulation results by different controllers are given in Fig.2. It is seen that the robust optimal SMC system is insensitive to the parameter uncertainty, the position trajectory is almost as the same as that of the nominal system. However, the optimal control

system is affected by the parameter variation. Compared with the nominal system, the position trajectory is different, bigger overshoot and the relative stability degrades.

In summery, the robust optimal SMC system owns the optimal performance and global robustness to uncertainties.

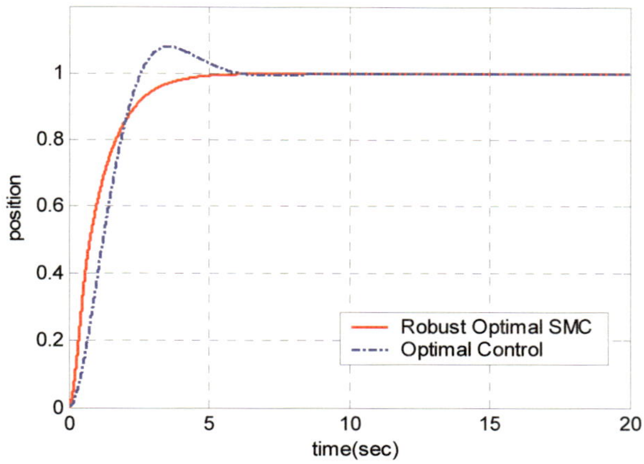

(a) Position responses

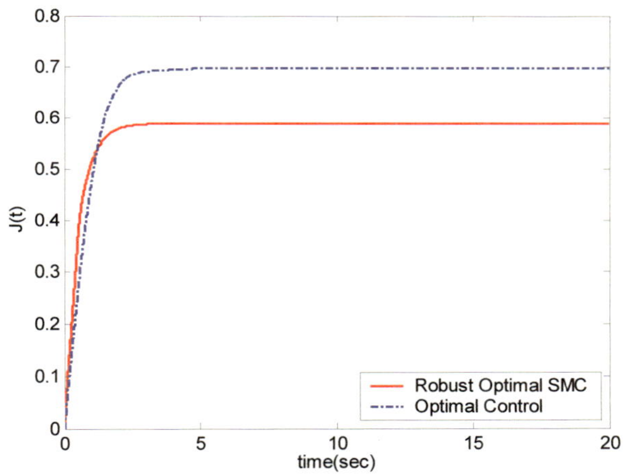

(b) Performance indexes

Fig. 2. Simulation results in Case 2

2.4 Conclusion

In this section, the integral sliding mode control strategy is applied to robustifying the optimal controller. An optimal robust sliding surface is designed so that the initial condition is on the surface and reaching phase is eliminated. The system is global robust to uncertainties which satisfy matching conditions and the sliding motion minimizes the given quadratic performance index. This method has been adopted to control the rotor position of an electrical servo drive. Simulation results show that the robust optimal SMCs are superior to optimal LQR controllers in the robustness to parameter variations and external disturbances.

3. Optimal sliding mode control for uncertain nonlinear system

In the section above, the robust optimal SMC design problem for a class of uncertain linear systems is studied. However, nearly all practical systems contain nonlinearities, there would exist some difficulties if optimal control is applied to handling nonlinear problems (Chiou & Huang, 2005; Ho, 2007, Cimen & Banks, 2004; Tang et al., 2007).In this section, the global robust optimal sliding mode controller (GROSMC) is designed based on feedback linearization for a class of MIMO uncertain nonlinear system.

3.1 Problem formulation

Consider an uncertain affine nonlinear system in the form of

$$\begin{aligned} \dot{x} &= f(x) + g(x)u + d(t,x)\,, \\ y &= H(x)\,, \end{aligned} \tag{19}$$

where $x \in R^n$ is the state, $u \in R^m$ is the control input, and $f(x)$ and $g(x)$ are sufficiently smooth vector fields on a domain $D \subset R^n$.Moreover, state vector x is assumed available, $H(x)$ is a measured sufficiently smooth output function and $H(x) = (h_1, \cdots, h_m)^{\mathrm{T}}$. $d(t,x)$ is an unknown function vector, which represents the system uncertainties, including system parameter variations, unmodeled dynamics and external disturbances.

Assumption 5. There exists an unknown continuous function vector $\delta(t,x)$ such that $d(t,x)$ can be written as

$$d(t,x) = g(x)\delta(t,x)\,.$$

This is called matching condition.

Assumption 6. There exist positive constants γ_0 and γ_1, such that

$$\|\delta(t,x)\| \le \gamma_0 + \gamma_1 \|x\|$$

where the notation $\|\cdot\|$ denotes the usual Euclidean norm.

By setting all the uncertainties to zero, the nominal system of the uncertain system (19) can be described as

$$\begin{aligned} \dot{x} &= f(x) + g(x)u\,, \\ y &= H(x)\,. \end{aligned} \tag{20}$$

The objective of this paper is to synthesize a robust sliding mode optimal controller so that the uncertain affine nonlinear system has not only the optimal performance of the nominal system but also robustness to the system uncertainties. However, the nominal system is nonlinear. To avoid the nonlinear TPBV problem and approximate linearization problem, we adopt the feedback linearization to transform the uncertain nonlinear system (19) into an equivalent linear one and an optimal controller is designed on it, then a GROSMC is proposed.

3.2 Feedback linearization

Feedback linearization is an important approach to nonlinear control design. The central idea of this approach is to find a state transformation $z = T(x)$ and an input transformation

$u = u(x,v)$ so that the nonlinear system dynamics is transformed into an equivalent linear time-variant dynamics, in the familiar form $\dot{z} = Az + Bv$, then linear control techniques can be applied.

Assume that system (20) has the vector relative degree $\{r_1,\cdots,r_m\}$ and $r_1 + \cdots + r_m = n$. According to relative degree definition, we have

$$\begin{aligned} y_i^{(k)} &= L_f^k h_i, \quad 0 \le k \le r_i - 1 \\ y_i^{(r_i)} &= L_f^{r_i} h_i + \sum_{j=1}^{m} g_j (L_f^{r_i-1} h_i) u_j , \end{aligned} \tag{21}$$

and the decoupled matrix

$$E(x) = (e_{ij})_{m\times m} = \begin{bmatrix} L_{g_1}(L_f^{r_1-1}h_1) & \cdots & L_{g_m}(L_f^{r_1-1}h_1) \\ \vdots & \cdots & \vdots \\ L_{g_1}(L_f^{r_m-1}h_m) & \cdots & L_{g_m}(L_f^{r_m-1}h_m) \end{bmatrix}$$

is nonsingular in some domain $\forall x \in X_0$.

Choose state and input transformations as follows:

$$z_i^j = T_i^j(x) = L_f^j h_i, i = 1,\cdots,m; j = 0,1,\cdots,r_i - 1 \tag{22}$$

$$u = E^{-1}(x)[v - K(x)], \tag{23}$$

where $K(x) = (L_f^{r_1} h_1, \cdots, L_f^{r_m} h_m)^{\mathrm{T}}$, v is an equivalent input to be designed later. The uncertain nonlinear system (19) can be transformed into m subsystems; each one is in the form of

$$\begin{bmatrix} \dot{z}_i^0 \\ \dot{z}_i^1 \\ \vdots \\ \dot{z}_i^{r_i-1} \end{bmatrix} = \begin{bmatrix} 0 & 1 & 0 & \cdots & 0 \\ 0 & 0 & 1 & \cdots & 0 \\ \vdots & \vdots & \vdots & \vdots & \vdots \\ 0 & 0 & 0 & \cdots & 1 \\ 0 & 0 & 0 & \cdots & 0 \end{bmatrix} \begin{bmatrix} z_i^0 \\ z_i^1 \\ \vdots \\ z_i^{r_i-1} \end{bmatrix} + \begin{bmatrix} 0 \\ 0 \\ \vdots \\ 0 \\ 1 \end{bmatrix} v_i + \begin{bmatrix} 0 \\ 0 \\ \vdots \\ 0 \\ L_d L_f^{r_i-1} h_i \end{bmatrix}. \tag{24}$$

So system (19) can be transformed into the following equivalent model of a simple linear form:

$$\dot{z}(t) = Az(t) + Bv(t) + \omega(t,z), \tag{25}$$

where $z \in R^n$, $v \in R^m$ are new state vector and input, respectively. $A \in R^{n\times n}$ and $B \in R^{n\times m}$ are constant matrixes, and (A,B) are controllable. $\omega(t,z) \in R^n$ is the uncertainties of the equivalent linear system. As we can see, $\omega(t,z)$ also satisfies the matching condition, in other words there exists an unknown continuous vector function $\tilde{\omega}(t,z)$ such that $\omega(t,z) = B\tilde{\omega}(t,z)$.

3.3 Design of GROSMC

3.3.1 Optimal control for nominal system

The nominal system of (25) is

$$\dot{z}(t) = Az(t) + Bv(t). \tag{26}$$

For (26), let $v = v_0$ and v_0 can minimize a quadratic performance index as follows:

$$J = \frac{1}{2}\int_0^{\infty}[z^{\mathrm{T}}(t)Qz(t) + v_0^{\mathrm{T}}(t)Rv_0(t)]dt \tag{27}$$

where $Q \in R^{n\times n}$ is a symmetric positive definite matrix, $R \in R^{m\times m}$ is a positive definite matrix. According to optimal control theory, the optimal feedback control law can be described as

$$v_0(t) = -R^{-1}B^{\mathrm{T}}Pz(t) \tag{28}$$

with P the solution of the matrix Riccati equation

$$PA + A^{\mathrm{T}}P - PBR^{-1}B^{\mathrm{T}}P + Q = 0. \tag{29}$$

So the closed-loop dynamics is

$$\dot{z}(t) = (A - BR^{-1}B^{\mathrm{T}}P)z(t). \tag{30}$$

The closed-loop system is asymptotically stable.

The solution to equation (30) is the optimal trajectory $z^*(t)$ of the nominal system with optimal control law (28). However, if the control law (28) is applied to uncertain system (25), the system state trajectory will deviate from the optimal trajectory and even the system becomes unstable. Next we will introduce integral sliding mode control technique to robustify the optimal control law, to achieve the goal that the state trajectory of uncertain system (25) is the same as that of the optimal trajectory of the nominal system (26).

3.3.2 The optimal sliding surface

Considering the uncertain system (25) and the optimal control law (28), we define an integral sliding surface in the form of

$$s(t) = G[z(t) - z(0)] - G\int_0^t (A - BR^{-1}B^{\mathrm{T}}P)z(\tau)d\tau \tag{31}$$

where $G \in R^{m\times n}$, which satisfies that GB is nonsingular, $z(0)$ is the initial state vector. Differentiating (31) with respect to t and considering (25), we obtain

$$\begin{aligned}\dot{s}(t) &= G\dot{z}(t) - G(A - BR^{-1}B^{\mathrm{T}}P)z(t)\\ &= G[Az(t) + Bv(t) + \omega(t,z)] - G(A - BR^{-1}B^{\mathrm{T}}P)z(t)\\ &= GBv(t) + GBR^{-1}B^{\mathrm{T}}Pz(t) + G\omega(t,z)\end{aligned} \tag{32}$$

Let $\dot{s}(t) = 0$, the equivalent control becomes

$$v_{\mathrm{eq}}(t) = -(GB)^{-1}\left[GBR^{-1}B^{\mathrm{T}}Pz(t) + G\omega(t,z)\right] \tag{33}$$

Substituting (33) into (25), the sliding mode dynamics becomes

$$\begin{aligned}\dot{z} &= Az - B(GB)^{-1}(GBR^{-1}B^{\mathrm{T}}Pz + G\omega) + \omega \\ &= Az - BR^{-1}B^{\mathrm{T}}Pz - B(GB)^{-1}G\omega + \omega \\ &= Az - BR^{-1}B^{\mathrm{T}}Pz - B(GB)^{-1}GB\tilde{\omega} + B\tilde{\omega} \\ &= (A - BR^{-1}B^{\mathrm{T}}P)z\end{aligned} \tag{34}$$

Comparing (34) with (30), we can see that the sliding mode of uncertain linear system (25) is the same as optimal dynamics of (26), thus the sliding mode is also asymptotically stable, and the sliding motion guarantees the controlled system global robustness to the uncertainties which satisfy the matching condition. We call (31) a global robust optimal sliding surface.
Substituting state transformation $z = T(x)$ into (31), we can get the optimal switching function $s(x,t)$ in the x-coordinates.

3.3.3 The control law

After designing the optimal sliding surface, the next step is to select a control law to ensure the reachability of sliding mode in finite time.
Differentiating $s(x,t)$ with respect to t and considering system (20), we have

$$\dot{s} = \frac{\partial s}{\partial x}\dot{x} + \frac{\partial s}{\partial t} = \frac{\partial s}{\partial x}(f(x) + g(x)u) + \frac{\partial s}{\partial t}. \tag{35}$$

Let $\dot{s} = 0$, the equivalent control of nonlinear nominal system (20) is obtained

$$u_{eq}(t) == -\left[\frac{\partial s}{\partial x}g(x)\right]^{-1}\left[\frac{\partial s}{\partial x}f(x) + \frac{\partial s}{\partial t}\right]. \tag{36}$$

Considering equation (23), we have $u_{eq} = E^{-1}(x)[v_0 - K(x)]$.

Now, we select the control law in the form of

$$\begin{aligned}u(t) &= u_{\mathrm{con}}(t) + u_{\mathrm{dis}}(t), \\ u_{\mathrm{con}}(t) &== -\left[\frac{\partial s}{\partial x}g(x)\right]^{-1}\left[\frac{\partial s}{\partial x}f(x) + \frac{\partial s}{\partial t}\right], \\ u_{\mathrm{dis}}(t) &= -\left[\frac{\partial s}{\partial x}g(x)\right]^{-1}(\eta + (\gamma_0 + \gamma_1\|x\|)\left\|\frac{\partial s}{\partial x}g(x)\right\|)\mathrm{sgn}(s),\end{aligned} \tag{37}$$

where $\mathrm{sgn}(s) = [\mathrm{sgn}(s_1)\ \ \mathrm{sgn}(s_2)\ \ \cdots\ \ \mathrm{sgn}(s_m)]^{\mathrm{T}}$ and $\eta > 0$. $u_{\mathrm{con}}(t)$ and $u_{\mathrm{dis}}(t)$ denote continuous part and discontinuous part of $u(t)$, respectively.
The continuous part $u_{\mathrm{con}}(t)$, which is equal to the equivalent control of nominal system (20), is used to stabilize and optimize the nominal system. The discontinuous part $u_{\mathrm{dis}}(t)$ provides the complete compensation of uncertainties for the uncertain system (19).
Theorem 2. Consider uncertain affine nonlinear system (19) with Assumputions 5-6. Let u and sliding surface be given by (37) and (31), respectively. The control law can force the system trajectories to reach the sliding surface in finite time and maintain on it thereafter.

Proof. Utilizing $V=(1/2)s^{\mathrm{T}}s$ as a Lyapunov function candidate, and taking the Assumption 5 and Assumption 6, we have

$$\begin{aligned}
\dot{V} &= s^{\mathrm{T}}\dot{s} = s^{\mathrm{T}}\left(\frac{\partial s}{\partial x}(f+gu+d)+\frac{\partial s}{\partial t}\right) = \\
&= s^{\mathrm{T}}\left\{\frac{\partial s}{\partial x}f-\left[\frac{\partial s}{\partial x}f+\frac{\partial s}{\partial t}+\left(\eta+(\gamma_0+\gamma_1\|x\|)\left\|\frac{\partial s}{\partial x}g\right\|\right)\mathrm{sgn}(s)\right]+\frac{\partial s}{\partial x}d+\frac{\partial s}{\partial t}\right\} = \\
&= s^{\mathrm{T}}\left\{-\left[\eta+(\gamma_0+\gamma_1\|x\|)\left\|\frac{\partial s}{\partial x}g\right\|\right]\mathrm{sgn}(s)\right\}+s^{\mathrm{T}}\frac{\partial s}{\partial x}d = -\eta\|s\|_1-(\gamma_0+\gamma_1\|x\|)\left\|\frac{\partial s}{\partial x}g\right\|\|s\|_1+s^{\mathrm{T}}\frac{\partial s}{\partial x}g\delta \\
&\le -\eta\|s\|_1-(\gamma_0+\gamma_1\|x\|)\left\|\frac{\partial s}{\partial x}g\right\|\|s\|_1+\left\|\frac{\partial s}{\partial x}g\right\|\|s\|\,\|\delta\| \le \\
&\le -\eta\|s\|_1-(\gamma_0+\gamma_1\|x\|)\left\|\frac{\partial s}{\partial x}g\right\|\|s\|_1+(\gamma_0+\gamma_1\|x\|)\left\|\frac{\partial s}{\partial x}g\right\|\|s\|
\end{aligned} \tag{38}$$

where $\|\bullet\|_1$ denotes the 1-norm. Noting the fact that $\|s\|_1\ge\|s\|$, we get

$$\dot{V}=s^{\mathrm{T}}\dot{s}\le-\eta\|s\|<0 \quad ,\text{for } \|s\|\ne 0. \tag{39}$$

This implies that the trajectories of the uncertain nonlinear system (19) will be globally driven onto the specified sliding surface $s=0$ despite the uncertainties in finite time. The proof is complete.

From (31), we have $s(0)=0$, that is the initial condition is on the sliding surface. According to Theorem2, we know that the uncertain system (19) with the integral sliding surface (31) and the control law (37) can achieve global sliding mode. So the system designed is global robust and optimal.

3.4 A simulation example

Inverted pendulum is widely used for testing control algorithms. In many existing literatures, the inverted pendulum is customarily modeled by nonlinear system, and the approximate linearization is adopted to transform the nonlinear systems into a linear one, then a LQR is designed for the linear system.

To verify the effectiveness and superiority of the proposed GROSMC, we apply it to a single inverted pendulum in comparison with conventional LQR.

The nonlinear differential equation of the single inverted pendulum is

$$\begin{aligned}
\dot{x}_1 &= x_2, \\
\dot{x}_2 &= \frac{g\sin x_1 - amLx_2^2\sin x_1\cos x_1 + au\cos x_1}{L(4/3-am\cos^2 x_1)}+d(t),
\end{aligned} \tag{40}$$

where x_1 is the angular position of the pendulum (rad), x_2 is the angular speed $(\mathrm{rad/s})$, M is the mass of the cart, m and L are the mass and half length of the pendulum, respectively. u denotes the control input, g is the gravity acceleration, $d(t)$ represents the external disturbances, and the coefficient $a=m/(M+m)$. The simulation parameters are as follows: $M=1\,\mathrm{kg}$, $m=0.2\,\mathrm{kg}$, $L=0.5\,\mathrm{m}$, $g=9.8\,\mathrm{m/s^2}$, and the initial state vector is $x(0)=[-\pi/18 \quad 0]^{\mathrm{T}}$.

Two cases with parameter variations in the inverted pendulum and external disturbance are considered here.

Case 1: The m and L are 4 times the parameters given above, respectively. Fig. 3 shows the robustness to parameter variations by the suggested GROSMC and conventional LQR.

Case 2: Apply an external disturbance $d(t) = 0.01\sin 2t$ to the inverted pendulum system at $t = 9\,s$. Fig. 4 depicts the different responses of these two controllers to external disturbance.

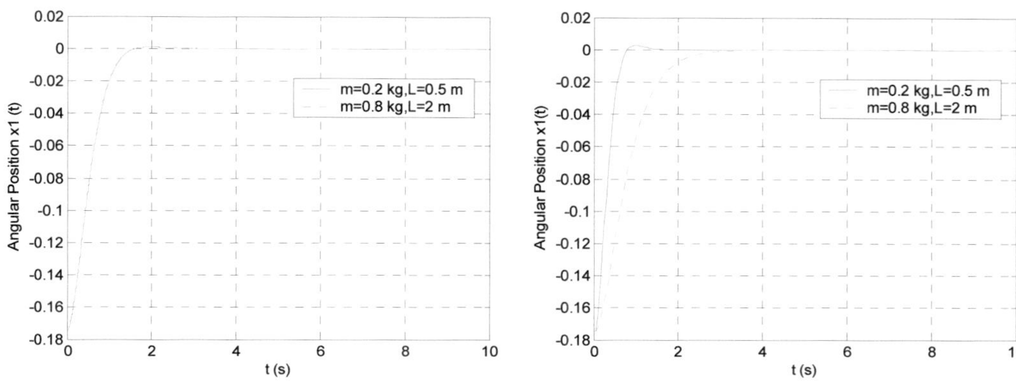

(a) By GROSMC (b) By Conventional LQR.

Fig. 3. Angular position responses of the inverted pendulum with parameter variation

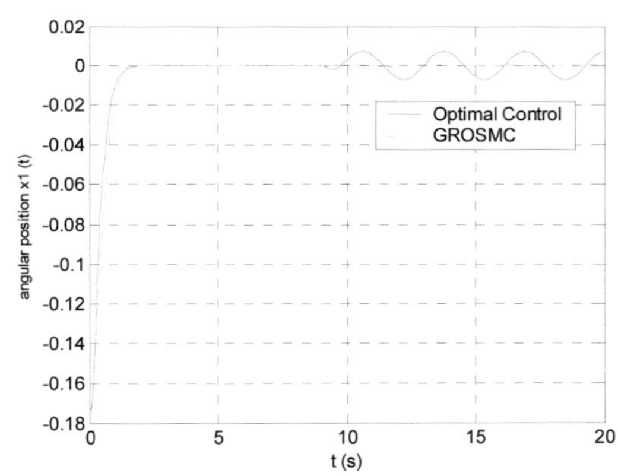

Fig. 4. Angular position responses of the inverted pendulum with external disturbance.

From Fig. 3 we can see that the angular position responses of inverted pendulum with and without parameter variations are exactly same by the proposed GROSMC, but the responses are obviously sensitive to parameter variations by the conventional LQR. It shows that the proposed GROSMC guarantees the controlled system complete robustness to parameter variation. As depicted in Fig. 4, without external disturbance, the controlled system could be driven to the equilibrium point by both of the controllers at about $t = 2\,s$. However, when the external disturbance is applied to the controlled system at $t = 9\,s$, the inverted pendulum system could still maintain the equilibrium state by GROSMC while the LQR not.

The switching function curve is shown in Fig. 5. The sliding motion occurs from the beginning without any reaching phase as can be seen. Thus, the GROSMC provides better features than conventional LQR in terms of robustness to system uncertainties.

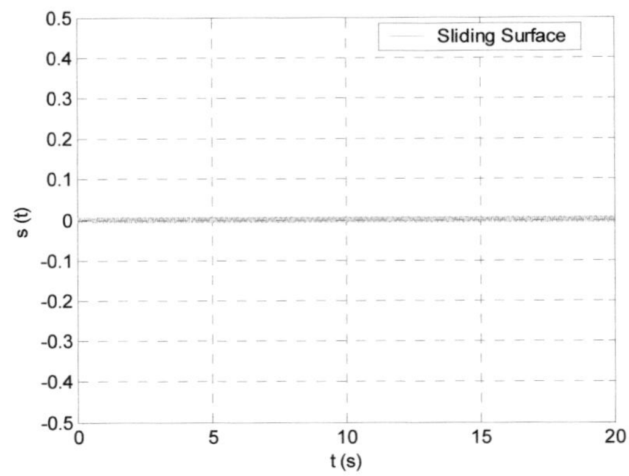

Fig. 5. The switching function s(t)

3.5 Conclusion

In this section, the exact linearization technique is firstly adopted to transform an uncertain affine nonlinear system into a linear one. An optimal controller is designed to the linear nominal system, which not only simplifies the optimal controller design, but also makes the optimal control applicable to the entire transformation region. The sliding mode control is employed to robustfy the optimal regulator. The uncertain system with the proposed integral sliding surface and the control law achieves global sliding mode, and the ideal sliding dynamics can minimized the given quadratic performance index. In summary, the system designed is global robust and optimal.

4. Optimal sliding mode tracking control for uncertain nonlinear system

With the industrial development, there are more and more control objectives about the system tracking problem (Ouyang et al., 2006; Mauder, 2008; Smolders et al., 2008), which is very important in control theory synthesis. Taking the robot as an example, it is often required to follow some special trajectories quickly as well as provide robustness to system uncertainties, including unmodeled dynamics, internal parameter variations and external disturbances. So the main tracking control problem becomes how to design the controller, which can not only get good tracking performance but also reject the uncertainties effectively to ensure the system better dynamic performance. In this section, a robust LQR tracking control based on intergral sliding mode is proposed for a class of nonlinear uncertain systems.

4.1 Problem formulation and assumption

Consider a class of uncertain affine nonlinear systems as follows:

$$\begin{cases} \dot{x} = f(x) + \Delta f(x) + g(x)[u + \delta(x,t,u)] \\ y = h(x) \end{cases} \tag{41}$$

where $x \in R^n$ is the state vector, $u \in R^m$ is the control input with $m=1$, and $y \in R$ is the system output. $f(x)$, $g(x)$, $\Delta f(x)$ and $h(x)$ are sufficiently smooth in domain $D \subset R^n$. $\delta(x,t,u)$ is continuous with respect to t and smooth in (x,u). $\Delta f(x)$ and $\delta(x,t,u)$ represent the system uncertainties, including unmodelled dynamics, parameter variations and external disturbances.

Our goal is to design an optimal LQR such that the output y can track a reference trajectory $y_r(t)$ asymptotically, some given performance criterion can be minimized, and the system can exhibit robustness to uncertainties.

Assumption 7. The nominal system of uncertain affine nonlinear system (41), that is

$$\begin{cases} \dot{x} = f(x) + g(x)u \\ y = h(x) \end{cases} \tag{42}$$

has the relative degree ρ in domain D and $\rho = n$.

Assumption 8. The reference trajectory $y_r(t)$ and its derivations $y_r^{(i)}(t)$ $(i=1,\cdots,n)$ can be obtained online, and they are limited to all $t \geq 0$.

While as we know, if the optimal LQR is applied to nonlinear systems, it often leads to nonlinear TPBV problem and an analytical solution generally does not exist. In order to simplify the design of this tracking problem, the input-output linearization technique is adopted firstly.

Considering system (41) and differentiating y, we have

$$y^{(k)} = L_f^k h(x), \quad 0 \leq k \leq n-1$$

$$y^{(n)} = L_f^n h(x) + L_{\Delta f} L_f^{n-1} h(x) + L_g L_f^{n-1} h(x)[u + \delta(x,t,u)].$$

According to the input-out linearization, choose the following nonlinear state transformation

$$z = T(x) = \begin{bmatrix} h(x) & \cdots & L_f^{n-1} h(x) \end{bmatrix}^{\mathrm{T}}. \tag{43}$$

So the uncertain affine nonlinear system (40) can be written as

$$\begin{aligned} \dot{z}_i &= z_{i+1}, \qquad i = 1,\cdots,n-1 \\ \dot{z}_n &= L_f^n h(x) + L_{\Delta f} L_f^{n-1} h(x) + L_g L_f^{n-1} h(x)[u + \delta(x,t,u)]. \end{aligned}$$

Define an error state vector in the form of

$$e = \begin{bmatrix} z_1 - y_r \\ \vdots \\ z_n - y_r^{(n-1)} \end{bmatrix} = z - \Re,$$

where $\Re = \begin{bmatrix} y_r & \cdots & y_r^{(n-1)} \end{bmatrix}^{\mathrm{T}}$. By this variable substitution $e = z - \Re$, the error state equation can be described as follows:

$$\begin{aligned} \dot{e}_i &= e_{i+1}, \qquad i = 1,\cdots,n-1 \\ \dot{e}_n &= L_f^n h(x) + L_{\Delta f} L_f^{n-1} h(x) + L_g L_f^{n-1} h(x) u(t) + L_g L_f^{n-1} h(x) \delta(x,t,u) - y_r^{(n)}(t). \end{aligned}$$

Let the feedback control law be selected as

$$u(t) = \frac{-L_f^n h(x) + v(t) + y_r^{(n)}(t)}{L_g L_f^{n-1} h(x)} \tag{44}$$

The error equation of system (40) can be given in the following forms:

$$\dot{e}(t) = \begin{bmatrix} 0 & 1 & 0 & \cdots & 0 \\ 0 & 0 & 1 & \cdots & 0 \\ \vdots & \vdots & \vdots & & \vdots \\ 0 & 0 & 0 & \cdots & 1 \\ 0 & 0 & 0 & 0 & 0 \end{bmatrix} e(t) + \begin{bmatrix} 0 \\ 0 \\ 0 \\ \vdots \\ L_{\Delta f} L_f^{n-1} h(x) \end{bmatrix} + \begin{bmatrix} 0 \\ 0 \\ 0 \\ \vdots \\ 1 \end{bmatrix} v(t) + \begin{bmatrix} 0 \\ 0 \\ 0 \\ \vdots \\ L_g L_f^{n-1} h(x) \delta(x,t,u) \end{bmatrix}. \tag{45}$$

Therefore, equation (45) can be rewritten as

$$\dot{e}(t) = Ae(t) + \Delta A + Bv(t) + \Delta\delta. \tag{46}$$

where

$$A = \begin{bmatrix} 0 & 1 & 0 & \cdots & 0 \\ 0 & 0 & 1 & \cdots & 0 \\ \vdots & \vdots & \vdots & & \vdots \\ 0 & 0 & 0 & \cdots & 1 \\ 0 & 0 & 0 & 0 & 0 \end{bmatrix}, \quad B = \begin{bmatrix} 0 \\ 0 \\ 0 \\ \vdots \\ 1 \end{bmatrix},$$

$$\Delta A = \begin{bmatrix} 0 \\ 0 \\ 0 \\ \vdots \\ L_{\Delta f} L_f^{n-1} h(x) \end{bmatrix}, \quad \Delta\delta = \begin{bmatrix} 0 \\ 0 \\ 0 \\ \vdots \\ L_g L_f^{n-1} h(x) \delta(x,t,u) \end{bmatrix}.$$

As can be seen, $e \in R^n$ is the system error vector, $v \in R$ is a new control input of the transformed system. $A \in R^{n\times n}$ and $B \in R^{n\times m}$ are corresponding constant matrixes. ΔA and $\Delta\delta$ represent uncertainties of the transformed system.

Assumption 9. There exist unknown continuous function vectors of appropriate dimensions $\Delta\tilde{A}$ and $\Delta\tilde{\delta}$, such that

$$\Delta A = B\Delta\tilde{A}, \ \Delta\delta = B\Delta\tilde{\delta}$$

Assumption 10. There exist known constants a_m, b_m such that

$$\left\|\Delta\tilde{A}\right\| \le a_m, \ \left\|\Delta\tilde{\delta}\right\| \le b_m$$

Now, the tracking problem becomes to design a state feedback control law v such that $e \to 0$ asymptotically. If there is no uncertainty, i.e. $\delta(t,e) = 0$, we can select the new input as $v = -Ke$ to achieve the control objective and obtain the closed loop dynamics $\dot{e} = (A - BK)e$. Good tracking performance can be achieved by choosing K using optimal control theory so that the closed loop dynamics is asymptotically stable. However, in presence of the uncertainties, the closed loop performance may be deteriorated. In the next section, the integral sliding mode control is adopted to robustify the optimal control law.

4.2 Design of optimal sliding mode tracking controller

4.2.1 Optimal tracking control of nominal system.

Ignoring the uncertainties of system (46), the corresponding nominal system is

$$\dot{e}(t) = Ae(t) + Bv(t). \tag{47}$$

For the nominal system (47), let $v = v_0$ and v_0 can minimize the quadratic performance index as follows:

$$I = \frac{1}{2}\int_0^{\infty}[e^{\mathrm{T}}(t)Qe(t) + v_0{}^{\mathrm{T}}(t)Rv_0(t)]dt \tag{48}$$

where $Q \in R^{n\times n}$ is a symmetric positive definite matrix, $R \in R^{m\times m}$ (here $m = 1$) is a positive definite matrix.

According to optimal control theory, an optimal feedback control law can be obtained as:

$$v_0(t) = -R^{-1}B^{\mathrm{T}}Pe(t) \tag{49}$$

with P the solution of matrix Riccati equation

$$PA + A^{\mathrm{T}}P - PBR^{-1}B^{\mathrm{T}}P + Q = 0.$$

So the closed-loop system dynamics is

$$\dot{e}(t) = (A - BR^{-1}B^{\mathrm{T}}P)e(t). \tag{50}$$

The designed optimal controller for system (47) is sensitive to system uncertainties including parameter variations and external disturbances. The performance index (48) may deviate from the optimal value. In the next part, we will use integral sliding mode control technique to robustify the optimal control law so that the uncertain system trajectory could be same as nominal system.

4.2.2 The robust optimal sliding surface.

To get better tracking performance, an integral sliding surface is defined as

$$s(e,t) = Ge(t) - G\int_0^t (A - BR^{-1}B^{\mathrm{T}}P)e(\tau)d\tau - Ge(0), \tag{51}$$

where $G \in R^{m\times n}$ is a constant matrix which is designed so that GB is nonsingular. And $e(0)$ is the initial error state vector.

Differentiating (51) with respect to t and considering system (46), we obtain

$$\begin{aligned}\dot{s}(e,t) &= G\dot{e}(t) - G(A - BR^{-1}B^{\mathrm{T}}P)e(t)\\ &= G[Ae(t) + \Delta A + Bv(t) + \Delta\delta] - G(A - BR^{-1}B^{\mathrm{T}}P)e(t)\\ &= GBv(t) + GBR^{-1}B^{\mathrm{T}}Pe(t) + G(\Delta A + \Delta\delta).\end{aligned} \tag{52}$$

Let $\dot{s}(e,t)=0$, the equivalent control can be obtained by

$$v_{\mathrm{eq}}(t)=-(GB)^{-1}[GBR^{-1}B^{\mathrm{T}}Pe(t)+G(\Delta A+\Delta\delta)]. \tag{53}$$

Substituting (53) into (46), and considering Assumption 10, the ideal sliding mode dynamics becomes

$$\begin{aligned}\dot{e}(t)&=Ae(t)+\Delta A+Bv_{\mathrm{eq}}(t)+\Delta\delta\\&=Ae(t)+\Delta A-B(GB)^{-1}[GBR^{-1}B^{\mathrm{T}}Pe(t)+G(\Delta A+\Delta\delta)]+\Delta\delta\\&=(A-BR^{-1}B^{\mathrm{T}}P)e(t)-B(GB)^{-1}G[\Delta A+\Delta\delta]+\Delta A+\Delta\delta\\&=(A-BR^{-1}B^{\mathrm{T}}P)e(t)-B(GB)^{-1}GB(\Delta\tilde{A}+\Delta\tilde{\delta})+B(\Delta\tilde{A}+\Delta\tilde{\delta})\\&=(A-BR^{-1}B^{\mathrm{T}}P)e(t).\end{aligned} \tag{54}$$

It can be seen from equation (50) and (54) that the ideal sliding motion of uncertain system and the optimal dynamics of the nominal system are uniform, thus the sliding mode is also asymptotically stable, and the sliding mode guarantees system (46) complete robustness to uncertainties. Therefore, (51) is called a robust optimal sliding surface.

4.2.3 The control law.

For uncertain system (46), we propose a control law in the form of

$$\begin{aligned}v(t)&=v_{\mathrm{c}}(t)+v_{\mathrm{d}}(t),\\v_{\mathrm{c}}(t)&=-R^{-1}B^{\mathrm{T}}Pe(t),\\v_{\mathrm{d}}(t)&=-(GB)^{-1}[ks+\varepsilon\,\mathrm{sgn}(s)].\end{aligned} \tag{55}$$

where v_{c} is the continuous part, which is used to stabilize and optimize the nominal system. And v_{d} is the discontinuous part, which provides complete compensation for system uncertainties. $\mathrm{sgn}(s)=\begin{bmatrix}\mathrm{sgn}(s_1) & \cdots & \mathrm{sgn}(s_m)\end{bmatrix}^{\mathrm{T}}$. k and ε are appropriate positive constants, respectively.

Theorem 3. Consider uncertain system (46) with Assumption9-10. Let the input v and the sliding surface be given as (55) and (51), respectively. The control law can force system trajectories to reach the sliding surface in finite time and maintain on it thereafter if $\varepsilon\geq(a_{\mathrm{m}}+d_{\mathrm{m}})\|GB\|$.

Proof: Utilizing $V=(1/2)s^{\mathrm{T}}s$ as a Lyapunov function candidate, and considering Assumption 9-10, we obtain

$$\begin{aligned}\dot{V}&=s^{\mathrm{T}}\dot{s}=s^{\mathrm{T}}[G\dot{e}(t)-G(A-BR^{-1}B^{\mathrm{T}}P)e(t)]\\&=s^{\mathrm{T}}\left\{G[Ae(t)+\Delta A+Bv(t)+\Delta\delta]-G(A-BR^{-1}B^{\mathrm{T}}P)e(t)\right\}\\&=s^{\mathrm{T}}\left[G\Delta A-GBR^{-1}B^{\mathrm{T}}Pe(t)-\left(ks+\varepsilon\,\mathrm{sgn}(s)\right)+G\Delta\delta+GBR^{-1}B^{\mathrm{T}}Pe(t)\right]\\&=s^{\mathrm{T}}\left\{-[ks+\varepsilon\,\mathrm{sgn}(s)]+G\Delta A+G\Delta\delta\right\}=-k\|s\|_1-\varepsilon\|s\|+s^{\mathrm{T}}(G\Delta A+G\Delta\delta)\\&\leq-k\|s\|_1-\varepsilon\|s\|+(a_{\mathrm{m}}+d_{\mathrm{m}})\|GB\|\|s\|\leq-k\|s\|-\left[\varepsilon-(a_{\mathrm{m}}+d_{\mathrm{m}})\|GB\|\right]\|s\|\end{aligned}$$

where $\|\bullet\|_1$ denotes the 1-norm. Note the fact that for any $\|s\|\neq0$, we have $\|s\|_1\geq\|s\|$. If $\varepsilon\geq(a_m+d_m)\|GB\|$, then

$$\dot{V} = s^{\mathrm{T}}\dot{s} \le -\varepsilon\|s\|_1 + \|G\|\|\delta\|\|s\| \le -(\varepsilon - \|G\|\|\delta\|)\|s\| < 0. \tag{56}$$

This implies that the trajectories of uncertain system (46) will be globally driven onto the specified sliding surface $s(e,t) = 0$ in finite time and maintain on it thereafter. The proof is completed.

From (51), we have $s(0) = 0$, that is to say, the initial condition is on the sliding surface. According to Theorem3, uncertain system (46) achieves global sliding mode with the integral sliding surface (51) and the control law (55). So the system designed is global robust and optimal, good tracking performance can be obtained with this proposed algorithm.

4.3 Application to robots.

In the recent decades, the tracking control of robot manipulators has received a great of attention. To obtain high-precision control performance, the controller is designed which can make each joint track a desired trajectory as close as possible. It is rather difficult to control robots due to their highly nonlinear, time-varying dynamic behavior and uncertainties such as parameter variations, external disturbances and unmodeled dynamics. In this section, the robot model is investigated to verify the effectiveness of the proposed method.

A 1-DOF robot mathematical model is described by the following nonlinear dynamics:

$$\begin{bmatrix} \dot{q} \\ \ddot{q} \end{bmatrix} = \begin{bmatrix} 0 & 1 \\ 0 & -\dfrac{C(q,\dot{q})}{M(q)} \end{bmatrix} \begin{bmatrix} q \\ \dot{q} \end{bmatrix} - \begin{bmatrix} 0 \\ \dfrac{G(q)}{M(q)} \end{bmatrix} + \begin{bmatrix} 0 \\ \dfrac{1}{M(q)} \end{bmatrix} \tau - \begin{bmatrix} 0 \\ \dfrac{1}{M(q)} \end{bmatrix} d(t), \tag{57}$$

where q, $\dot{q}$ denote the robot joint position and velocity, respectively. τ is the control vector of torque by the joint actuators. m and l are the mass and length of the manipulator arm, respectively. $d(t)$ is the system uncertainties. $C(q,\dot{q}) = 0.03\cos(q)$, $G(q) = mgl\cos(q)$, $M(q) = 0.1 + 0.06\sin(q)$. The reference trajectory is $y_{\mathrm{r}}(t) = \sin \pi t$.

According to input-output linearization technique, choose a state vector as follows:

$$z = \begin{bmatrix} z_1 \\ z_2 \end{bmatrix} = \begin{bmatrix} q \\ \dot{q} \end{bmatrix}.$$

Define an error state vector of system (57) as $e = \begin{bmatrix} e_1 & e_2 \end{bmatrix}^{\mathrm{T}} = \begin{bmatrix} q - y_{\mathrm{r}} & \dot{q} - \dot{y}_{\mathrm{r}} \end{bmatrix}^{\mathrm{T}}$, and let the control law $\tau = (v + \ddot{y}_{\mathrm{r}})M(q) + C(q,\dot{q})\dot{q} + G(q)$.

So the error state dynamic of the robot can be written as:

$$\begin{bmatrix} \dot{e}_1 \\ \dot{e}_2 \end{bmatrix} = \begin{bmatrix} 0 & 1 \\ 0 & 0 \end{bmatrix} \begin{bmatrix} e_1 \\ e_2 \end{bmatrix} + \begin{bmatrix} 0 \\ 1 \end{bmatrix} v - \begin{bmatrix} 0 \\ 1/M(q) \end{bmatrix} d(t) \tag{58}$$

Choose the sliding mode surface and the control law in the form of (51) and (55), respectively, and the quadratic performance index in the form of (48). The simulation parameters are as follows: $m = 0.02$, $g = 9.8$, $l = 0.5$, $d(t) = 0.5\sin 2\pi t$, $k = 18$, $\varepsilon = 6$, $G = \begin{bmatrix} 0 & 1 \end{bmatrix}$, $Q = \begin{bmatrix} 10 & 2 \\ 2 & 1 \end{bmatrix}$, $R = 1$. The initial error state vector is $e = \begin{bmatrix} 0.5 & 0 \end{bmatrix}^{\mathrm{T}}$.

The tracking responses of the joint position $q$and its velocity are shown in Fig. 6 and Fig. 7, respectively. The control input τ is displayed in Fig. 8. From Fig. 6 and Fig. 7 it can be seen that the position error can reach the equilibrium point quickly and the position track the reference sine signal y_r well. Simulation results show that the proposed scheme manifest good tracking performance and the robustness to parameter variations and the load disturbance.

4.4 Conclusions

In order to achieve good tracking performance for a class of nonlinear uncertain systems, a sliding mode LQR tracking control is developed. The input-output linearization is used to transform the nonlinear system into an equivalent linear one so that the system can be handled easily. With the proposed control law and the robust optimal sliding surface, the system output is forced to follow the given trajectory and the tracking error can minimize the given performance index even if there are uncertainties. The proposed algorithm is applied to a robot described by a nonlinear model with uncertainties. Simulation results illustrate the feasibility of the proposed controller for trajectory tracking and its capability of rejecting system uncertainties.

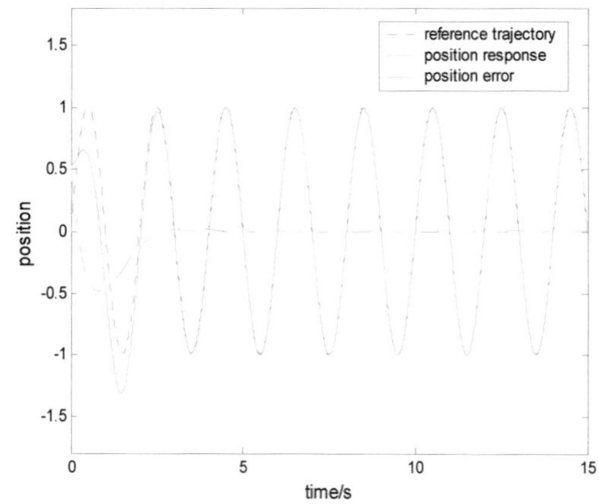

Fig. 6. The tracking response of q

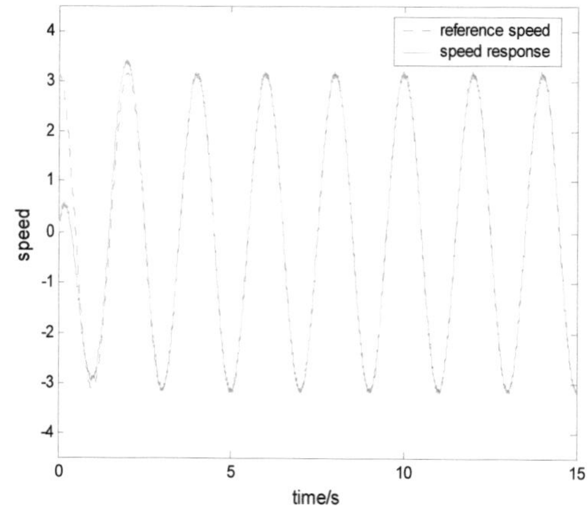

Fig. 7. The tracking response of $\dot{q}$

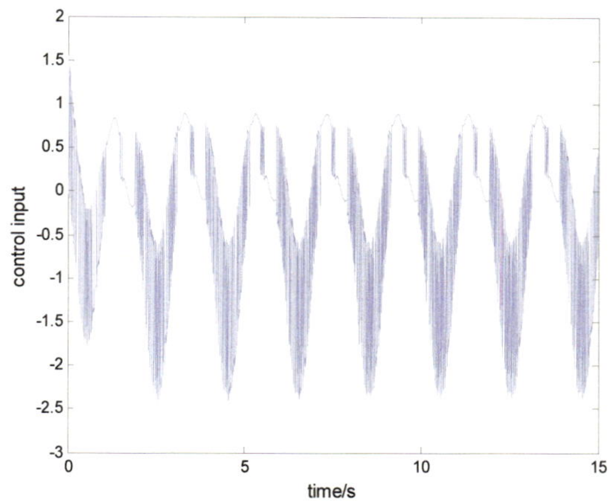

Fig. 8. The control input τ

5. Acknowledgements

This work is supported by National Nature Science Foundation under Grant No. 60940018.

6. References

Basin, M.; Rodriguez-Gonzaleza, J.; Fridman, L. (2007). Optimal and robust control for linear state-delay systems. *Journal of the Franklin Institute*. Vol.344, pp.830–845.

Chen, W. D.; Tang, D. Z.; Wang, H. T. (2004). Robust Tracking Control Of Robot M anipulators Using Backstepping. *Joranal of System Simulation*. Vol. 16, No. 4, pp. 837-837, 841.

Choi, S. B.; Cheong, C. C.; Park, D. W. (1993). Moving switch surfaces for robust control of second-order variable structure systems. *International Journal of Control*. Vol.58, No.1, pp. 229-245.

Choi, S. B.; Park, D. W.; Jayasuriya, S. (1994). A time-varying sliding surface for fast and robust tracking control of second-order uncertain systems. *Automatica*. Vol.30, No.5, pp. 899-904.

Chiou, K. C.; Huang, S. J. (2005). An adaptive fuzzy controller for robot manipulators. *Mechatronics*. Vol.15, pp.151-177.

Cimen, T.; Banks, S. P. (2004). Global optimal feedback control for general nonlinear systems with nonquadratic performance criteria. *System & Control Letters*. Vol.53, pp.327-346.

Cimen, T.; Banks, S. P. (2004). Nonlinear optimal tracking control with application to super-tankers for autopilot design. *Automatica*. Vol.40, No.11, pp.1845 - 1863.

Gao, W. B.; Hung, J C. (1993). Variable structure control of nonlinear systems: A new approach. *IEEE Transactions on Industrial Electronics*. Vol.40, No.1, pp. 45-55.

Ho, H. F.; Wong, Y. K.; Rad, A. B. (2007). Robust fuzzy tracking control for robotic manipulators. *Simulation Modelling Practice and Theory*. Vol.15, pp.801-816.

Laghrouche, S.; Plestan, F.; Glumineau, A. (2007). Higher order sliding mode control based on integral sliding mode. *Automatica*. Vol. 43, pp.531-537.

Lee, J. H. (2006). Highly robust position control of BLDDSM using an improved integral variable structure system. *Automatica*. Vol.42, pp.929-935.

Lin, F. J.; Chou, W. D. (2003). An induction motor servo drive using sliding mode controller with genetic algorithm. *Electric Power Systems Research*. Vol.64, pp.93-108.

Mokhtari, A.; Benallegue, A.; Orlov, Y. (2006). Exact linearization and sliding mode observer for a quadrotor unmanned aerial vehicle. *International Journal of Robotics and Automation*. Vol. 21, No.1, pp.39-49.

Mauder, M. Robust tracking control of nonholonomic dynamic systems with application to the bi-steerable mobile robot. (2008). *Automatica*. Vol.44, No.10, pp.2588-2592.

Ouyang, P. R.; Zhang, W. J.; Madan M. Gupta. (2006). An adaptive switching learning control method for trajectory tracking of robot manipulators. *Mechatronics*. Vol.16, No.1, pp.51-61.

Tang, G. Y.; Sun, H. Y.; Pang, H. P. (2008). Approximately optimal tracking control for discrete time-delay systems with disturbances. *Progress in Natural Science*. Vol.18, pp. 225-231.

Pang, H. P.; Chen, X. (2009). Global robust optimal sliding mode control for uncertain affine nonlinear systems. *Journal of Systems Engineering and Electronics*. Vol.20, No.4, pp. 838-843.

Pang H. P.; Wang L.P. (2009). Global Robust Optimal Sliding Mode Control for a class of Affine Nonlinear Systems with Uncertainties Based on SDRE. *Proceeding of 2009 Second International Workshop on Computer Science and Engineering*. Vol. 2, pp. 276-280.

Pang, H. P.; Tang, G. Y.; Sun, H.Y. (2009). Optimal Sliding Mode Design for a Class of Uncertain Systems with Time-delay. *Information and Control*. Vol.38, No.1, pp.87-92.

Tang, G. Y.; Gao, D. X. (2005). Feedforward and feedback optimal control for nonlinear systems with persistent disturbances. *Control and Decision*. Vol.20, No.4, pp. 366-371.

Tang, G. Y. (2005). Suboptimal control for nonlinear systems: a successive approximation approach. *Systems and Control Letters*. Vol. 54, No.5, pp.429-434.

Tang, G. Y.; Zhao, Y. D.; Zhang, B. L. (2007). Optimal output tracking control for nonlinear systems via successive approximation approach. *Nonlinear Analysis*. Vol.66, No.6, pp.1365–1377.

Shamma, J. S.; Cloutier, J. R. (2001). Existence of SDRE stabilizing feedback. *Proceedings of the American Control Conference*. pp.4253-4257, Arlington VA.

Smolders, K.; Volckaert, M. Swevers, J. (2008). Tracking control of nonlinear lumped mechanical continuous-time systems: A model-based iterative learning approach. *Mechanical Systems and Signal Processing*. Vol.22, No.8, pp.1896–1916.

Yang K. D.; Özgüner. Ü. (1997). Sliding-mode design for robust linear optimal control. *Automatica*. Vol. 33, No. 7, pp. 1313-1323.

8

Robust Delay-Independent/Dependent Stabilization of Uncertain Time-Delay Systems by Variable Structure Control

Elbrous M. Jafarov
Faculty of Aeronautics and Astronautics, Istanbul Technical University
Turkey

1. Introduction

It is well known that many engineering control systems such as conventional oil-chemical industrial processes, nuclear reactors, long transmission lines in pneumatic, hydraulic and rolling mill systems, flexible joint robotic manipulators and machine-tool systems, jet engine and automobile control, human-autopilot systems, ground controlled satellite and networked control and communication systems, space autopilot and missile-guidance systems, etc. contain some time-delay effects, model uncertainties and external disturbances. These processes and plants can be modeled by some uncertain dynamical systems with state and input delays. The existence of time-delay effects is frequently a source of instability and it degrades the control performances. The stabilization of systems with time-delay is not easier than that of systems without time-delay. Therefore, the stability analysis and controller design for uncertain systems with delay are important both in theory and in practice. The problem of robust stabilization of uncertain time-delay systems by various types of controllers such as PID controller, Smith predictor, and time-delay controller, recently, sliding mode controllers have received considerable attention of researchers. However, in contrast to variable structure systems without time-delay, there is relatively no large number of papers concerning the sliding mode control of time-delay systems. Generally, stability analysis can be divided into two categories: delay-independent and delay-dependent. It is worth to mention that delay-dependent conditions are less conservative than delay-independent ones because of using the information on the size of delays, especially when time-delays are small. As known from (Utkin, 1977)-(Jafarov, 2009) etc. sliding mode control has several useful advantages, e.g. fast response, good transient performance, and robustness to the plant parameter variations and external disturbances. For this reason, now, sliding mode control is considered as an efficient tool to design of robust controllers for stabilization of complex systems with parameter perturbations and external disturbances. Some new problems of the sliding mode control of time-delay systems have been addressed in papers (Shyu & Yan, 1993)-(Jafarov, 2005). Shyu and Yan (Shyu & Yan, 1993) have established a new sufficient condition to guarantee the robust stability and β-stability for uncertain systems with single time-delay. By these conditions a variable structure controller is designed to stabilize the time-delay systems with uncertainties. Koshkoei and Zinober (Koshkouei & Zinober, 1996) have designed a new sliding mode controller for MIMO canonical controllable time-delay systems with matched external disturbances by using Lyapunov-Krasovskii functional. Robust stabilization of time-delay systems with uncertainties by using sliding mode control has been considered by Luo, De La Sen and Rodellar (Luo et al., 1997). However, disadvantage of this design approach is that, a variable structure controller is not simple. Moreover, equivalent control term depends on unavailable external disturbances. Li and DeCarlo (Li & De Carlo, 2003) have proposed a new robust four terms sliding mode controller design method for a class of multivariable time-delay systems with unmatched parameter uncertainties and matched external disturbances by using the Lyapunov-Krasovskii functional combined by LMI's techniques. The behavior and design of sliding mode control systems with state and input delays are considered by Perruquetti and Barbot (Perruquetti & Barbot, 2002) by using Lyapunov-Krasovskii functional.

Four-term robust sliding mode controllers for matched uncertain systems with single or multiple, constant or time varying state delays are designed by Gouaisbaut, Dambrine and Richard (Gouisbaut et al., 2002) by using Lyapunov-Krasovskii functionals and Lyapunov-Razumikhin function combined with LMI's techniques. The five terms sliding mode controllers for time-varying delay systems with structured parameter uncertainties have been designed by Fridman, Gouisbaut, Dambrine and Richard (Fridman et al., 2003) via descriptor approach combined by Lyapunov-Krasovskii functional method. In (Cao et al., 2007) some new delay-dependent stability criteria for multivariable uncertain networked control systems with several constant delays based on Lyapunov-Krasovskii functional combined with descriptor approach and LMI techniques are developed by Cao, Zhong and Hu. A robust sliding mode control of single state delayed uncertain systems with parameter perturbations and external disturbances is designed by Jafarov (Jafarov, 2005). In survey paper (Hung et al., 1993) the various type of reaching conditions, variable structure control laws, switching schemes and its application in industrial systems is reported by J. Y.Hung, Gao and J.C.Hung. The implementation of a tracking variable structure controller with boundary layer and feed-forward term for robotic arms is developed by Xu, Hashimoto, Slotine, Arai and Harashima(Xu et al., 1989).A new fast-response sliding mode current controller for boost-type converters is designed by Tan, Lai, Tse, Martinez-Salamero and Wu (Tan et al., 2007). By constructing new types of Lyapunov functionals and additional free-weighting matrices, some new less conservative delay-dependent stability conditions for uncertain systems with constant but unknown time-delay have been presented in (Li et al., 2010) and its references.

Motivated by these investigations, the problem of sliding mode controller design for uncertain multi-input systems with several fixed state delays for delay-independent and delay-dependent cases is addressed in this chapter. A new combined sliding mode controller is considered and it is designed for the stabilization of perturbed multi-input time-delay systems with matched parameter uncertainties and external disturbances. Delay-independent/dependent stability and sliding mode existence conditions are derived by using Lyapunov-Krasovskii functional and Lyapunov function method and formulated in terms of LMI. Delay bounds are determined from the improved stability conditions. In practical implementation chattering problem can be avoided by using saturation function (Hung et al., 1993), (Xu et al., 1989).

Five numerical examples with simulation results are given to illustrate the usefulness of the proposed design method.

2. System description and assumptions

Let us consider a multi-input state time-delay systems with matched parameter uncertainties and external disturbances described by the following state-space equation:

$$\dot{x}(t) = (A_0 + \Delta A_0)x(t) + (A_1 + \Delta A_1)x(t-h_1) + ... + (A_N + \Delta A_N)x(t-h_N) + Bu(t) + Df(t),\ \mathrm{t} > 0$$

$$x(t) = \phi(t)\,,\quad -h \le t \le 0 \tag{1}$$

where $x(t) \in R^n$ is the measurable state vector, $u(t) \in R^m$ is the control input, $A_0, A_1, .., A_N$ and B are known constant matrices of appropriate dimensions, with B of full rank, $h = \max[h_1, h_2, ..., h_N], h_i > 0$, $h_1, h_2, ..., h_N$ are known constant time-delays, $\phi(t)$ is a continuous vector-valued initial function in $-h \le t \le 0$; $\Delta A_0, \Delta A_1, \ldots, \Delta A_N$ and D are the parameter uncertainties, $\phi(t)$ is unknown but norm-bounded external disturbances.
Taking known advantages of sliding mode, we want to design a simple suitable sliding mode controller for stabilization of uncertain time-delay system (1).
We need to make the following conventional assumptions for our design problem.

Assumption 1:

a. (A_0, B) is stabilizable;
b. The parameter uncertainties and external disturbances are matched with the control input, i.e. there exist matrices $E_0(t), E(t), E_1(t), \ldots, E_N(t)$, such that:

$$\Delta A_0(t) = BE_0(t)\ ;\ A_1(t) = BE_1(t)\ ;\ ..., \Delta A_N(t) = BE_N(t)\ ;\ D(t) = BE(t) \tag{2}$$

with norm-bounded matrices:

$$\max_t \left\| \Delta E_0(t) \right\| \le \alpha_0\ ;\ \max_t \left\| \Delta E_1(t) \right\| \le \alpha_1\ ; ..., \max_t \left\| \Delta E_N(t) \right\| \le \alpha_N$$

$$\left\| E(t) \right\| = \alpha$$

$$\left\| \mathrm{G} \right\| = g$$

$$\left\| f(t) \right\| \le f_0 \tag{3}$$

where $\alpha_0, \alpha_1, \alpha_1, ...\alpha_n, g$ and f_0 are known positive scalars.
The control goal is to design a combined variable structure controller for robust stabilization of time-delay system (1) with matched parameter uncertainties and external disturbances.

3. Control law and sliding surface

To achieve this goal, we form the following type of combined variable structure controller:

$$u(t) = u_{lin}(t) + u_{eq}(t) + u_{vs}(t) + u_r(t) \tag{4}$$

where

$$u_{lin}(t) = -Gx(t) \tag{5}$$

$$u_{eq}(t) = -(CB)^{-1}\left[CA_0x(t) + CA_1x(t-h_1) + \ldots + CA_Nx(t-h_N)\right] \tag{6}$$

$$u_{vs}(t) = -\left[k_0\left\|x(t)\right\| + k_1\left\|x(t-h_1)\right\| + \ldots + k_N\left\|x(t-h_N)\right\|\right]\frac{s(t)}{\left\|s(t)\right\|} \tag{7}$$

$$u_r = -\delta\frac{s(t)}{\left\|s(t)\right\|} \tag{8}$$

where $k_0, k_1, \ldots, k_N$ and δ are the scalar gain parameters to be selected; G is a design matrix; $(CB)^{-1}$ is a non-singular $m \times m$ matrix. The sliding surface on which the perturbed time-delay system states must be stable is defined as a linear function of the undelayed system states as follows:

$$s(t) = \Gamma C x(t) \tag{9}$$

where C is a $m \times n$ gain matrix of full rank to be selected; Γ is chosen as identity $m \times m$ matrix that is used to diagonalize the control.
Equivalent control term (6) for non-perturbed time-delay system is determined from the following equations:

$$\dot{s}(t) = C\dot{x}(t) = CA_0x(t) + CA_1x(t-h_1) + \ldots + CA_Nx(t-h_N) + CBu(t) = 0 \tag{10}$$

Substituting (6) into (1) we have a non-perturbed or ideal sliding time-delay motion of the nominal system as follows:

$$\dot{x}(t) = \overline{A}_0x(t) + \overline{A}_1x(t-h_1) + \ldots + \overline{A}_Nx(t-h_N) \tag{11}$$

where

$$(CB)^{-1}C = G_{eq},\ A_0 - BG_{eq}A_0 = \overline{A}_0,\ A_1 - BG_{eq}A_1 = \overline{A}_1, \ldots, A_N - BG_{eq}A_N = \overline{A}_N \tag{12}$$

Note that, constructed sliding mode controller consists of four terms:

1. The linear control term is needed to guarantee that the system states can be stabilized on the sliding surface;
2. The equivalent control term for the compensation of the nominal part of the perturbed time-delay system;
3. The variable structure control term for the compensation of parameter uncertainties of the system matrices;
4. The min-max or relay term for the rejection of the external disturbances.

Structure of these control terms is typical and very simple in their practical implementation.
The design parameters $G, C, k_0, k_1, \ldots, k_N, \delta$ of the combined controller (4) for delay-independent case can be selected from the sliding conditions and stability analysis of the perturbed sliding time-delay system.
However, in order to make the delay-dependent stability analysis and choosing an appropriate Lyapunov-Krasovskii functional first let us transform the nominal sliding time-delay system (11) by using the Leibniz-Newton formula. Since x(t) is continuously differentiable for $t \geq 0$, using the Leibniz-Newton formula, the time-delay terms can be presented as:

$$x(t-h_1) = x(t) - \int_{t-h_1}^{t}\dot{x}(\theta)d\theta, \ldots, x(t-h_N) = x(t) - \int_{t-h_N}^{t}\dot{x}(\theta)d\theta \tag{13}$$

Then, the system (11) can be rewritten as

$$\dot{x}(t)=(\bar{A}_0+\bar{A}_1+...+\bar{A}_N)x(t)-\bar{A}_1\int_{t-h_1}^{t}\dot{x}(\theta)d\theta-...-\bar{A}_N\int_{t-h_N}^{t}\dot{x}(\theta)d\theta \tag{14}$$

Substituting again (11) into (14) yields:

$$\begin{aligned}\dot{x}(t)&=(\bar{A}_0+\bar{A}_1+...+\bar{A}_N)x(t)-\bar{A}_1\int_{t-h_1}^{t}\left[\bar{A}_0x(\theta)+\bar{A}_1x(\theta-h_1)+...+\bar{A}_Nx(\theta-h_N)\right]d\theta\\&\quad-...-\bar{A}_N\int_{t-h_N}^{t}\left[\bar{A}_0x(\theta)+\bar{A}_1x(\theta-h_1)+...+\bar{A}_Nx(\theta-h_N)\right]d\theta\\&=(\bar{A}_0+\bar{A}_1+...+\bar{A}_N)x(t)-\bar{A}_1\bar{A}_0\int_{t-h_1}^{t}x(\theta)d\theta-\bar{A}_1^2\int_{t-h_1}^{t}x(\theta-h_1)d\theta-...-\bar{A}_1\bar{A}_N\int_{t-h_1}^{t}x(\theta-h_N)d\theta\\&\quad-...-\bar{A}_N\bar{A}_0\int_{t-h_N}^{t}x(\theta)d\theta-\bar{A}_N\bar{A}_1\int_{t-h_N}^{t}x(\theta-h_1)d\theta-...-\bar{A}_N^2\int_{t-h_N}^{t}x(\theta-h_N)d\theta\end{aligned} \tag{15}$$

Then in adding to (15) the perturbed sliding time-delay system with control action (4) or overall closed loop system can be formulated as:

$$\begin{aligned}\dot{x}(t)&=(\tilde{A}_0+\bar{A}_1+...+\bar{A}_N)x(t)-\bar{A}_1\bar{A}_0\int_{t-h_1}^{t}x(\theta)d\theta-\bar{A}_1^2\int_{t-h_1}^{t}x(\theta-h_1)d\theta\\&\quad-...-\bar{A}_1\bar{A}_N\int_{t-h_1}^{t}x(\theta-h_N)d\theta-...-\bar{A}_N\bar{A}_0\int_{t-h_N}^{t}x(\theta)d\theta-\bar{A}_N\bar{A}_1\int_{t-h_N}^{t}x(\theta-h_1)d\theta\\&\quad-...-\bar{A}_N^2\int_{t-h_N}^{t}x(\theta-h_N)d\theta+\Delta A_0x(t)\\&\quad+\Delta A_1x(t-h_1)+...+\Delta A_Nx(t-h_N)\\&\quad-B[k_0\|x(t)\|+k_1\|x(t-h_1)\|+...+k_N\|x(t-h_N)\|]\frac{s(t)}{\|s(t)\|}-B\delta\frac{s(t)}{\|s(t)\|}+Df(t)\end{aligned} \tag{16}$$

where $\tilde{A}_0=\bar{A}_0-BG$, the gain matrix G can be selected such that $\tilde{A}_0$ has the desirable eigenvalues.

The design parameters $G,C,k_0,k_1,...,k_N,\delta$ of the combined controller (4) for delay-dependent case can be selected from the sliding conditions and stability analysis of the perturbed sliding time-delay system (16).

4. Robust delay-independent stabilization

In this section, the existence condition of the sliding manifold and delay-independent stability analysis of perturbed sliding time-delay systems are presented.

4.1 Robust delay-independent stabilization on the sliding surface

In this section, the sliding manifold is designed so that on it or in its neighborhood in different from existing methods the perturbed sliding time-delay system (1),(4) is globally asymptotically stable with respect to state coordinates. The perturbed stability results are formulated in the following theorem.

Theorem 1: Suppose that Assumption 1 holds. Then the multivariable time-delay system (1) with matched parameter perturbations and external disturbances driven by combined controller (4) and restricted to the sliding surface s(t)=0 is robustly globally asymptotically delay-independent stable with respect to the state variables, if the following LMI conditions and parameter requirements are satisfied:

$$H = \begin{bmatrix} \widetilde{A_0}^T P + P\widetilde{A_0} + R_1 + \ldots + R_N & P\overline{A}_1 & \ldots & P\overline{A}_N \\ (P\overline{A}_1)^T & -R_1 & \ldots & 0 \\ \vdots & \vdots & \vdots & \vdots \\ (P\overline{A}_N)^T & 0 & \ldots & -R_N \end{bmatrix} < 0 \tag{17}$$

$$CB = B^T PB > 0 \tag{18}$$

$$k_0 = \alpha_0; k_1 = \alpha_1; \ldots; k_N = \alpha_N \tag{19}$$

$$\delta \geq f_o \tag{20}$$

where $P, R_1, \ldots R_N$ are some symmetric positive definite matrices which are a feasible solution of LMI (17) with (18); $\widetilde{A_0} = \overline{A_0} - BG$ in which a gain matrix G can be assigned by pole placement such that $\tilde{A}_0$ has some desirable eigenvalues.

Proof: Choose a Lyapunov-Krasovskii functional candidate as follows:

$$V = x^T(t)Px(t) + \sum_{i=1}^{N} \int_{t-h_i}^{t} x^T(\theta)R_i x(\theta)d\theta \tag{21}$$

The time-derivative of (21) along the state trajectories of time-delay system (1), (4) can be calculated as follows:

$$\begin{aligned}
\dot{V} &= 2x^T(t)P\left[A_0x(t) + A_1x(t-h_1) + \ldots + A_Nx(t-h_N) + \Delta A_0x(t) + \Delta A_1x(t-h_1)\right. \\
&\qquad \left. + \ldots + \Delta A_N x(t-h_N) + Bu(t) + Df(t)\right] \\
&\quad + x^T(t)R_1x(t) - x^T(t-h_1)R_1x(t-h_1) + \ldots + x^T(t)R_Nx(t) - x^T(t-h_N)R_Nx(t-h_N) \\
&= 2x^T(t)P\overline{A}_0x(t) + 2x^T(t)P\overline{A}_1x(t-h_1) + \ldots + 2x^T(t)P\overline{A}_Nx(t-h_N) + 2x^T(t)PBE_0x(t) \\
&\quad + 2x^T(t)PBE_1x(t-h_1) + \ldots + 2x^T(t)PBE_Nx(t-h_N) \\
&\quad - 2x^T(t)PB[k_0\|x(t)\| + k_1\|x(t-h_1)\| + \ldots + k_N\|x(t-h_N)\|]\frac{s(t)}{\|s(t)\|} \\
&\quad - 2x^T(t)PBGx(t) - 2\delta x^T(t)PB\frac{s(t)}{\|s(t)\|} + 2x^T(t)PBEf(t) \\
&\quad + x^T(t)(R_1 + \ldots + R_N)x(t) - x^T(t-h_1)R_1x(t-h_1) - \ldots - x^T(t-h_N)R_Nx(t-h_N)
\end{aligned}$$

Since $x^T(t)PB = s^T(t)$, then we obtain:

$$\begin{aligned}
\dot{V} \le\, & x^T(t)\left[\widetilde{A_0}^T P + P\widetilde{A_0} + R_1 + \ldots + R_N\right]x(t) \\
& + 2x^T(t)P\overline{A}_1 x(t-h_1) + \ldots + 2x^T(t)P\overline{A}_N x(t-h_N) \\
& - x^T(t-h_1)R_1 x(t-h_1) - \ldots - x^T(t-h_1)R_N x(t-h_N) \\
& + 2s^T(t)E_0 x(t) + 2s^T(t)E_1 x(t-h_1) + \ldots + 2s^T(t)E_N x(t-h_N) \\
& + 2s^T(t)Ef(t) - 2s^T(t)[k_0\|x(t)\| + k_1\|x(t-h_1)\| + \ldots + k_N\|x(t-h_N)\|]\frac{s(t)}{\|s(t)\|} - 2\delta s^T(t)\frac{s(t)}{\|s(t)\|} \\
\le & \begin{bmatrix} x(t) \\ x(t-h_1) \\ \vdots \\ x(t-h_N) \end{bmatrix}^T \begin{bmatrix} \left[\widetilde{A_0}^T P + P\widetilde{A_0} + R_1 + \ldots + R_N\right] & P\overline{A}_1 & \ldots & P\overline{A}_N \\ (P\overline{A}_1)^T & -R_1 & \ldots & 0 \\ \vdots & \vdots & \vdots & \vdots \\ (P\overline{A}_N)^T & 0 & \ldots & -R_N \end{bmatrix} \begin{bmatrix} x(t) \\ x(t-h_1) \\ \vdots \\ x(t-h_N) \end{bmatrix} \\
& - [(k_0-\alpha_0)\|x(t)\|\|s(t)\| + (k_1-\alpha_1)\|x(t-h_1)\|\|s(t)\| + \ldots + (k_N-\alpha_N)\|x(t-h_N)\|\|s(t)\|] \\
& - (\delta - f_0)\|s(t)\|
\end{aligned} \tag{22}$$

Since (17)-(20) hold, then (22) reduces to:

$$\dot{V} \le z^T(t)Hz(t) < 0 \tag{23}$$

where $z^T(t) = [x(t)x(t-h_1)\ldots x(t-h_N)]$.

Therefore, we can conclude that the perturbed time-delay system (1), (4) is robustly globally asymptotically delay-independent stable with respect to the state coordinates. Theorem 1 is proved.

4.2 Existence conditions

The final step of the control design is the derivation of the sliding mode existence conditions or the reaching conditions for the perturbed time-delay system (1),(4) states to the sliding manifold in finite time. These results are summarized in the following theorem.

Theorem 2: Suppose that Assumption 1 holds. Then the perturbed multivariable time-delay system (1) states with matched parameter uncertainties and external disturbances driven by controller (4) converge to the siding surface s(t)=0 in finite time, if the following conditions are satisfied:

$$k_0 = \alpha_0 + g; k_1 = \alpha_1; \ldots; k_N = \alpha_N \tag{24}$$

$$\delta \ge f_o \tag{25}$$

Proof: Let us choose a modified Lyapunov function candidate as:

$$V = \frac{1}{2}s^T(t)(CB)^{-1}s(t) \tag{26}$$

The time-derivative of (26) along the state trajectories of time-delay system (1), (4) can be calculated as follows:

$$
\begin{aligned}
\dot{V} &= s^T(t)(CB)^{-1}\dot{s}(t) = s^T(t)(CB)^{-1}C\dot{x}(t) \\
&= s^T(t)(CB)^{-1}C\big[A_0x(t) + A_1x(t-h_1) + \ldots + A_Nx(t-h_N) \\
&\quad + \Delta A_0x(t) + \Delta A_1x(t-h_1) + \ldots + \Delta A_Nx(t-h_N) + Bu(t) + Df(t)\big] \\
&= s^T(t)(CB)^{-1}\big[CA_0x(t) + CA_1x(t-h_1) + \ldots + CA_Nx(t-h_N) \\
&\quad + CBE_0x(t) + CBE_1x(t-h_1) + \ldots + CBE_Nx(t-h_N) \\
&\quad - CB\Big((CB)^{-1}\big[CA_0x(t) + CA_1x(t-h) + \ldots + CA_Nx(t-h_N)\big] \\
&\quad - \big[k_0\|x(t)\| + k_1\|x(t-h_1)\| + \ldots + k_N\|x(t-h_N)\|\big]\frac{s(t)}{\|s(t)\|} \\
&\quad - Gx(t) - \delta\frac{s(t)}{\|s(t)\|}\Big) + CBEf(t)\big] \\
&= s^T(t)[E_0x(t) + E_1x(t-h_1) + \ldots + E_Nx(t-h_N)] \\
&\quad - \big[k_0\|x(t)\| + k_1\|x(t-h_1)\| + \ldots + k_N\|x(t-h_N)\|\big]\frac{s(t)}{\|s(t)\|} \\
&\quad - Gx(t) - \delta\frac{s(t)}{\|s(t)\|} + Ef(t)] \\
&\leq -[(k_0 - \alpha_0 - g)\|x(t)\|\|s(t)\| + (k_1 - \alpha_1)\|x(t-h_1)\|\|s(t)\| \\
&\quad + \ldots + (k_N - \alpha_N)\|x(t-h_N)\|\|s(t)\|] - (\delta - f_0)\|s(t)\|
\end{aligned}
\tag{27}
$$

Since (24), (25) hold, then (27) reduces to:

$$\dot{V} = s^T(t)(CB)^{-1}\dot{s}(t) \leq -(\delta - f_0)\|s(t)\| \leq -\eta\|s(t)\| \tag{28}$$

where

$$\eta = \delta - f_0 \geq 0 \tag{29}$$

Hence we can evaluate that

$$\dot{V}(t) \leq -\eta\sqrt{\frac{2}{\lambda_{\min}(CB)^{-1}}}\sqrt{V(t)} \tag{30}$$

The last inequality (30) is known to prove the finite-time convergence of system (1), (4) towards the sliding surface s(t)=0 (Utkin, 1977), (Perruquetti & Barbot, 2002). Therefore, Theorem 2 is proved.

4.3 Numerical examples and simulation

In order to demonstrate the usefulness of the proposed control design techniques let us consider the following examples.

Example 1: Consider a networked control time-delay system (1), (4) with parameters taking from (Cao et al., 2007):

$$A_0 = \begin{bmatrix} -4 & 0 \\ -1 & -3 \end{bmatrix}, A_1 = \begin{bmatrix} -1.5 & 0 \\ -1 & -0.5 \end{bmatrix}, B = \begin{bmatrix} 2 \\ 2 \end{bmatrix} \tag{31}$$

$$\Delta A_0 = 0.5\sin(t)A_0\ ,\Delta A_1 = 0.5\cos(t)A_1\,, f\ = 0.3\sin(t)$$

The LMI stability and sliding mode existence conditions are computed by MATLAB programming (see Appendix 1) where LMI Control Toolbox is used. The computational results are following:

$$\text{A0hat} = \begin{bmatrix} -1.0866 & 1.0866 \\ 1.9134 & -1.9134 \end{bmatrix};\ \text{A1hat} = \begin{bmatrix} -0.1811 & 0.1811 \\ 0.3189 & -0.3189 \end{bmatrix}$$

$$\text{G1} = \begin{bmatrix} 0.9567 & 1.2933 \end{bmatrix};\ \text{A0til} = \begin{bmatrix} -3.0000 & -1.5000 \\ 0.0000 & -4.5000 \end{bmatrix};\ \text{eigA0til} = \begin{bmatrix} -3.0000 \\ -4.5000 \end{bmatrix}$$

$$\text{eigA0hat} = \begin{bmatrix} 0.0000 \\ -3.0000 \end{bmatrix};\ \text{eigA1hat} = \begin{bmatrix} 0.0000 \\ -0.5000 \end{bmatrix}$$

$$\text{lhs} = \begin{bmatrix} -1.8137 & 0.0020 & -0.1392 & 0.1392 \\ 0.0020 & -1.7813 & 0.1382 & -0.1382 \\ -0.1392 & 0.1382 & -1.7364 & 0.0010 \\ 0.1392 & -0.1382 & 0.0010 & -1.7202 \end{bmatrix};\ \text{eigsLHS} = \begin{bmatrix} -2.0448 \\ -1.7952 \\ -1.7274 \\ -1.4843 \end{bmatrix}$$

$$\text{P} = \begin{bmatrix} 0.6308 & -0.0782 \\ -0.0782 & 0.3891 \end{bmatrix};\ \text{eigP} = \begin{bmatrix} 0.3660 \\ 0.6539 \end{bmatrix};$$

$$\text{R1} = \begin{bmatrix} 1.7364 & -0.0010 \\ -0.0010 & 1.7202 \end{bmatrix};\ \text{eigR1} = \begin{bmatrix} 1.7202 \\ 1.7365 \end{bmatrix}$$

$$\text{BTP} = \begin{bmatrix} 1.1052 & 0.6217 \end{bmatrix};\ \text{BTPB} = 3.4538$$

invBTPB = 0.2895; normG1 = 1.6087

k0= 2.1087; k1=0.5; $\delta \geq 0.3$; H< 0;

The networked control time-delay system is robustly asymptotically delay-independent stable.

Example 2: Consider a time-delay system (1), (4) with parameters:

$$A_0 = \begin{bmatrix} -1 & 0.7 \\ 0.3 & 1 \end{bmatrix}, A_1 = \begin{bmatrix} 0.1 & 0.1 \\ 0 & 0.2 \end{bmatrix},\ A_2 = \begin{bmatrix} 0.2 & 0 \\ 0 & 0.1 \end{bmatrix}, B = \begin{bmatrix} 1 \\ 1 \end{bmatrix}$$

$$h_1 = 0.1, \quad h_2 = 0.2 \tag{32}$$

$$\Delta A_0 = \begin{bmatrix} 0.2\sin(t) & 0 \\ 0 & 0.1\sin(t) \end{bmatrix}, \Delta A_1 = \begin{bmatrix} 0.1\cos(t) & 0 \\ 0 & 0.2\cos(t) \end{bmatrix}, \Delta A_2 = \begin{bmatrix} 0.2\cos(t) & 0 \\ 0 & 0.1\cos(t) \end{bmatrix}.$$

Matching condition for external disturbances is given by:

$$D = \text{B}\ E = \begin{bmatrix} 1 \\ 1 \end{bmatrix} 0.2\cos\ \text{t}\ ;\ f(t) = 0.2\cos t$$

The LMI stability and sliding mode existence conditions are computed by MATLAB programming (see Appendix 2) where LMI Control Toolbox is used. The computational results are following:

$$\text{A0hat} = \begin{bmatrix} -0.3947 & -0.0911 \\ 0.9053 & 0.2089 \end{bmatrix}; \text{A1hat} = \begin{bmatrix} -0.0304 & -0.0304 \\ 0.0696 & 0.0696 \end{bmatrix}; \text{A2hat} = \begin{bmatrix} 0.0607 & -0.0304 \\ -0.1393 & 0.0696 \end{bmatrix}$$

$$\text{Geq} = [\,0.6964 \quad 0.3036]; \text{G} = [\,-4.5759 \quad 12.7902]$$

$$\text{A0til} = \begin{bmatrix} 4.1812 & -12.8812 \\ 5.4812 & -12.5812 \end{bmatrix}; \text{eigA0til} = \begin{bmatrix} -4.2000 + 0.6000i \\ -4.2000 - 0.6000i \end{bmatrix}$$

$$\text{eigA0hat} = \begin{bmatrix} -0.1858 \\ 0.0000 \end{bmatrix}; \text{eigA1hat} = \begin{bmatrix} 0 \\ 0.0393 \end{bmatrix}; \text{eigA2hat} = \begin{bmatrix} 0.0000 \\ 0.1304 \end{bmatrix}$$

$$\text{lhs} = \begin{bmatrix} -0.7085 & -0.5711 & -0.0085 & -0.0085 & 0.0169 & -0.0085 \\ -0.5711 & -0.8257 & 0.0084 & 0.0084 & -0.0167 & 0.0084 \\ -0.0085 & 0.0084 & -1.0414 & -0.2855 & 0 & 0 \\ -0.0085 & 0.0084 & -0.2855 & -1.1000 & 0 & 0 \\ 0.0169 & -0.0167 & 0 & 0 & -1.0414 & -0.2855 \\ -0.0085 & 0.0084 & 0 & 0 & -0.2855 & -1.1000 \end{bmatrix}; \text{eigsLHS} = \begin{bmatrix} -1.3581 \\ -1.3578 \\ -1.3412 \\ -0.7848 \\ -0.7837 \\ -0.1916 \end{bmatrix}$$

$$\text{P} = \begin{bmatrix} 2.0633 & 0.7781 \\ 0.7781 & 0.4592 \end{bmatrix}; \text{eigP} = \begin{bmatrix} 0.1438 \\ 2.3787 \end{bmatrix}; \text{R1} = \begin{bmatrix} 1.0414 & 0.2855 \\ 0.2855 & 1.1000 \end{bmatrix}; \text{R2} = \begin{bmatrix} 1.0414 & 0.2855 \\ 0.2855 & 1.1000 \end{bmatrix}$$

$$\text{eigR1} = \begin{bmatrix} 0.7837 \\ 1.3578 \end{bmatrix}; \text{eigR2} = \begin{bmatrix} 0.7837 \\ 1.3578 \end{bmatrix}$$

$$\text{BTP} = [\,2.8414 \;\; 1.2373]; \text{BTPB} = 4.0788$$

$$\text{invBTPB} = 0.2452; \text{normG} = 13.5841$$

$$\alpha_0 = 0.2;\;\; \alpha_1 = 0.2;\;\; \alpha_2 = 0.2;\; d = \max\|D\| = 0.2;\;\; f_0 = \max\|f(t)\| = 0.2828;$$

$$\text{k0}=13.7841; \text{k1}=0.2; \text{k2}=0.2; \delta \geq 0.2828; \text{H}< 0;$$

Thus, we have designed all the parameters of the combined sliding mode controller.

Aircraft control design example 3: Consider the lateral-directional control design of the DC-8 aircraft in a cruise-flight configuration for M = 0.84, h = 33.000ft, and V = 825ft/s with nominal parameters taken from (Schmidt, 1998):

$$\begin{bmatrix} \dot{r} \\ \dot{\beta} \\ \dot{p} \\ \dot{\phi} \end{bmatrix} = \begin{bmatrix} -0.228 & 2.148 & -0.021 & 0.0 \\ -1.0 & -0.0869 & 0.0 & 0.0390 \\ 0.335 & -4.424 & -1.184 & 0.0 \\ 0.0 & 0.0 & 1.0 & 0.0 \end{bmatrix} \begin{bmatrix} r \\ \beta \\ p \\ \phi \end{bmatrix} + \begin{bmatrix} -1.169 & 0.065 \\ 0.0223 & 0.0 \\ 0.0547 & 2.120 \\ 0.0 & 0.0 \end{bmatrix} \begin{bmatrix} \delta_r \\ \delta_a \end{bmatrix} \tag{33}$$

where β is the sideslip angle, deg., p is the roll rate, deg/s, ϕ is the bank angle, deg., r is the yaw rate, deg/s, δ_r is the rudder control, δ_a is the aileron control. However, some small transient time-delay effect in this equation may occur because of influence of sideslip on aerodynamics flow and flexibility effects of aerodynamic airframe and surfaces in lateral-directional couplings and directional-lateral couplings. The gain constants of gyro, rate gyro and actuators are included in to lateral directional equation of motion. Therefore, it is assumed that lateral direction motion of equation contains some delay effect and perturbed parameters as follows:

$$A_1 = \begin{bmatrix} 0 & 0 & -0.002 & 0.0 \\ 0 & 0 & 0.0 & 0.004 \\ 0.034 & -0.442 & 0 & 0 \\ 0.0 & 0.0 & 0 & 0 \end{bmatrix} \tag{34}$$

$$\Delta A_0 = 0.1 A_0 \sin(t)\,,\ \Delta A_1 = 0.1 A_1 \cos(t)\,,\ D = I_4\,;\ f = 0.2\sin(t)\,;\ h_1 = 0.01 - 0.04s$$

The LMI stability and sliding mode existence conditions are computed by MATLAB programming (see Appendix 3) where LMI Control Toolbox is used. The computational results are following:

$$\text{A0hat} = \begin{bmatrix} -0.0191 & -0.0008 & 0.0000 & 0.0007 \\ -1.0042 & -0.0434 & 0.0003 & 0.0390 \\ 0.0006 & 0.0000 & -0.0000 & -0.0000 \\ 0 & 0 & 1.0000 & 0 \end{bmatrix};\ \text{A1hat} = \begin{bmatrix} -0.0000 & 0.0000 & -0.0000 & 0.0001 \\ -0.0000 & 0.0003 & -0.0000 & 0.0040 \\ 0.0000 & -0.0000 & 0.0000 & -0.0000 \\ 0 & 0 & 0 & 0 \end{bmatrix}$$

$$\text{G} = \begin{bmatrix} -0.8539 & 0.0163 & 0.0262 & 0 \\ 0.0220 & -0.0001 & 0.4710 & 0 \end{bmatrix};\ \text{G1} = \begin{bmatrix} -0.5925 & 0.0890 & 0.1207 & 0.0501 \\ 0.0689 & -0.0086 & 0.3452 & 0.0485 \end{bmatrix}$$

$$\text{A0til} = \begin{bmatrix} -0.7162 & 0.1038 & 0.1187 & 0.0561 \\ -0.9910 & -0.0454 & -0.0024 & 0.0379 \\ -0.1130 & 0.0134 & -0.7384 & -0.1056 \\ 0 & 0 & 1.0000 & 0 \end{bmatrix}$$

$$\text{eigA0til} = \begin{bmatrix} -0.5+0.082i & -0.5-0.082i & -0.3 & -0.2 \end{bmatrix}$$

$$\text{eigA0hat} = \begin{bmatrix} -0.0621 & -0.0004 & -0.0000 & -0.0000 \end{bmatrix}$$

$$\text{eigA1hat} = 1.0\text{e-}003 * \begin{bmatrix} 0.2577 \\ -0.0000 + 0.0000i \\ -0.0000 - 0.0000i \\ 0 \end{bmatrix}$$

$$\text{P} = \begin{bmatrix} 72.9293 & 39.4515 & -2.3218 & 24.7039 \\ 39.4515 & 392.5968 & 10.8368 & -1.4649 \\ -2.3218 & 10.8368 & 67.2609 & -56.4314 \\ 24.7039 & -1.4649 & -56.4314 & 390.7773 \end{bmatrix};\ \text{eigP} = \begin{bmatrix} 57.3353 & 66.3033 & 397.7102 & 402.2156 \end{bmatrix}$$

$$\text{R1} = \begin{bmatrix} 52.5926 & 29.5452 & 0.3864 & 2.5670 \\ 29.5452 & 62.3324 & 3.6228 & -0.4852 \\ 0.3864 & 3.6228 & 48.3292 & -32.7030 \\ 2.5670 & -0.4852 & -32.7030 & 61.2548 \end{bmatrix};\ \text{eigR1} = \begin{bmatrix} 21.3032 & 27.3683 & 86.9363 & 88.9010 \end{bmatrix}$$

$$\text{BTP} = \begin{bmatrix} -84.5015 & -36.7711 & 6.6350 & -31.9983 \\ -0.1819 & 25.5383 & 142.4423 & -118.0289 \end{bmatrix}$$

$$\text{BTPB} = \begin{bmatrix} 98.3252 & 8.5737 \\ 8.5737 & 301.9658 \end{bmatrix};\ \text{invBTPB} = \begin{bmatrix} 0.0102 & -0.0003 \\ -0.0003 & 0.0033 \end{bmatrix}$$

$$\text{gnorm} = 0.8545\text{lhs} = \begin{bmatrix} -41.4566 & -29.8705 & -0.6169 & -2.3564 & -0.0008 & 0.0105 & -0.0016 & 0.1633 \\ -29.8705 & -51.6438 & -3.8939 & 0.8712 & -0.0078 & 0.1015 & -0.015 & 1.5728 \\ -0.6169 & -3.8939 & -38.2778 & 32.1696 & -0.0002 & 0.0028 & -0.0004 & 0.043 \\ -2.3564 & 0.8712 & 32.1696 & -51.6081 & 0 & -0.0002 & 0 & -0.0038 \\ -0.0008 & -0.0078 & -0.0002 & 0 & -52.593 & -29.545 & -0.3864 & -2.567 \\ 0.0105 & 0.1015 & 0.0028 & -0.0002 & -29.545 & -62.333 & -3.6228 & 0.4852 \\ -0.0016 & -0.015 & -0.0004 & 0 & -0.3864 & -3.6228 & -48.33 & 32.703 \\ 0.1633 & 1.5728 & 0.043 & -0.0038 & -2.567 & 0.4852 & 32.703 & -61.255 \end{bmatrix};$$

$$\text{eigsLHS} = \begin{bmatrix} -88.9592 \\ -86.9820 \\ -78.9778 \\ -75.8961 \\ -27.3686 \\ -21.3494 \\ -16.0275 \\ -11.9344 \end{bmatrix}$$

$$k0 = 1.0545; \quad k1 = 0.5; \quad \delta \geq 0.2; \quad H < 0;$$

Thus, we have designed all the parameters of the aircraft control system and the uncertain time-delay system (1), (4) with given nominal (33) and perturbed (34) parameters are simulated by using MATLAB-SIMULINK. The SIMULINK block diagram of the uncertain time-delay system with variable structure contoller (VSC) is given in Fig. 1. Simulation results are given in Fig. 2, 3, 4 and 5. As seen from the last four figures, system time responses to the rudder and aileron pulse functions (0.3 within 3-6 sec) are stabilized very well for example the settling time is about 15-20 seconds while the state time responses of aircraft control action as shown in Fig. 5 are unstable or have poor dynamic characteristics. Notice that, as shown in Fig. 4, control action contains some switching, however it has no high chattering effects because the continuous terms of controller are dominant.

Numerical examples and simulation results show the usefulness and effectiveness of the proposed design approach.

5. Robust delay-dependent stabilization

In this section, the existence condition of the sliding manifold and delay-dependent stability analysis of perturbed sliding time-delay systems are presented.

5.1 Robust delay-dependent stabilization on the sliding surface

In this section the sliding manifold is designed so that on it or in its neighborhood in different from existing methods the perturbed sliding time-delay system (16) is globally asymptotically stable with respect to state coordinates. The stability results are formulated in the following theorem.

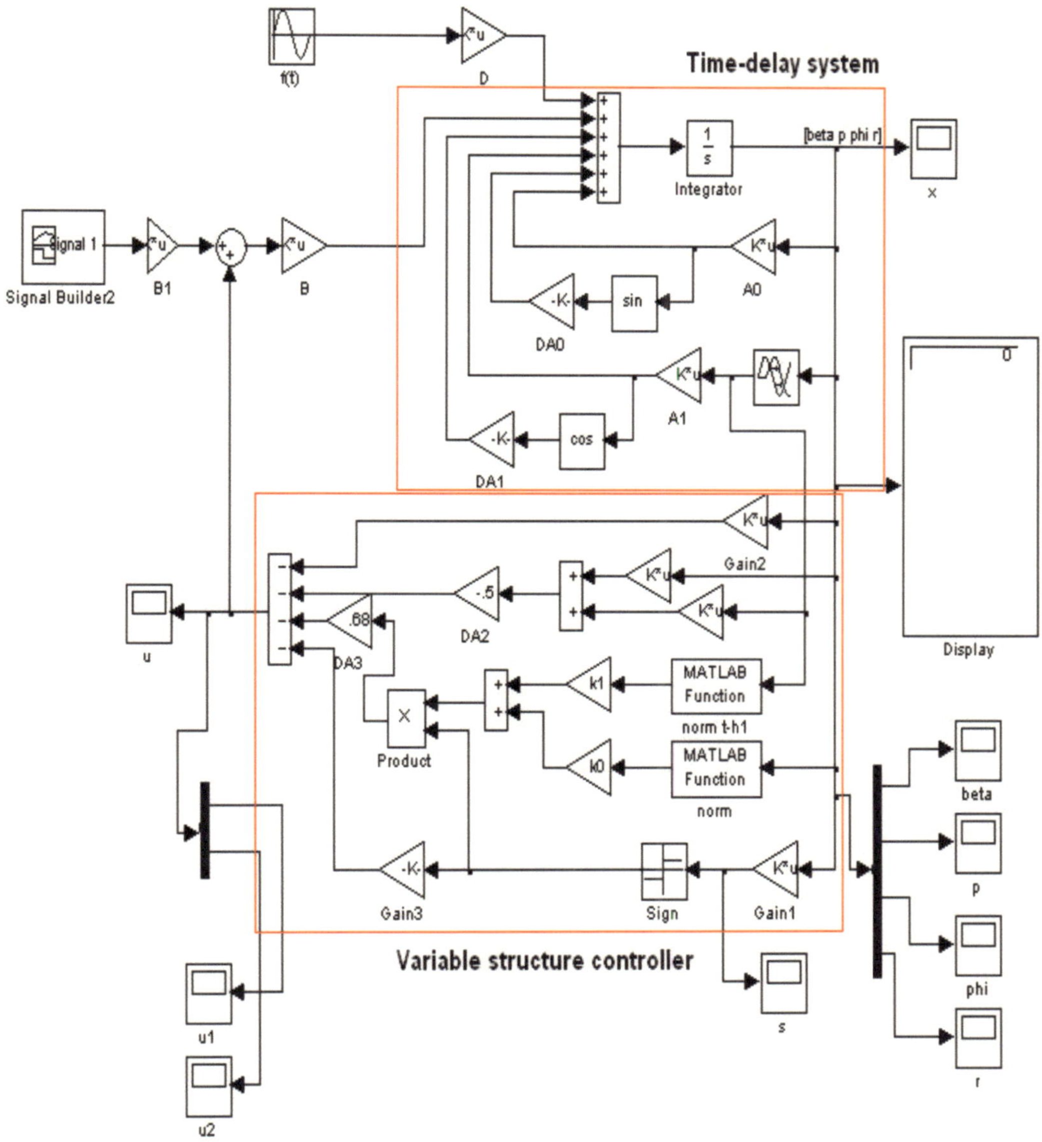

Fig. 1. SIMULINK block diagram of uncertain time-delay system with VSC

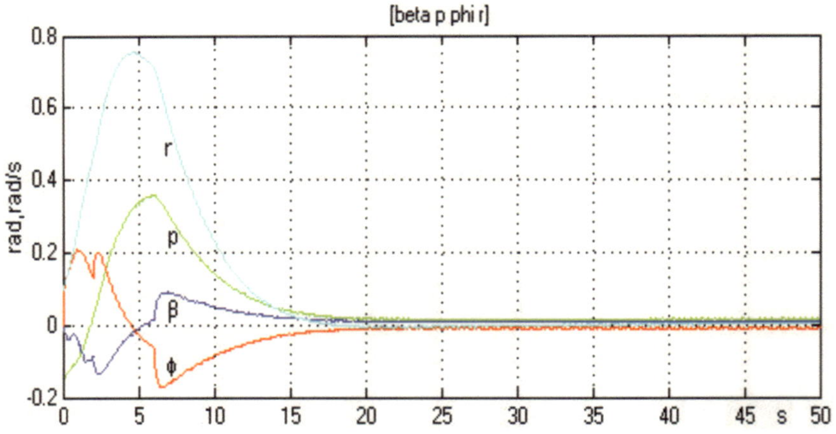

Fig. 2. States' time responses with control

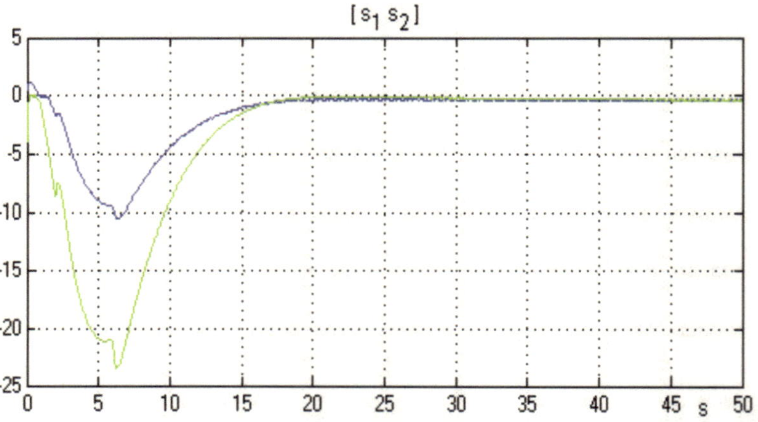

Fig. 3. Sliding functions

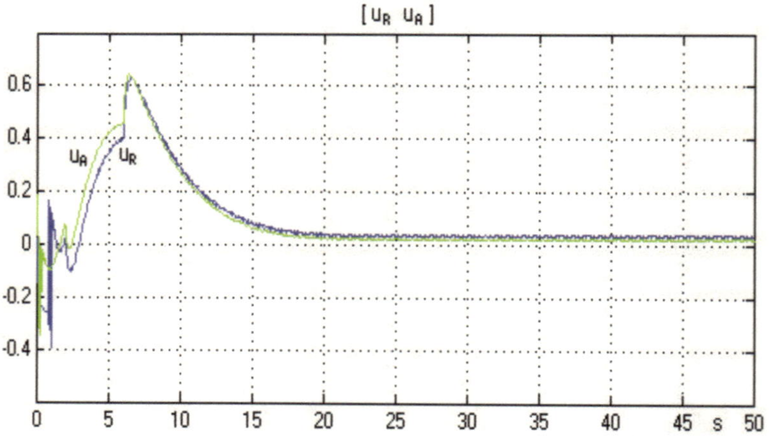

Fig. 4. Control functions

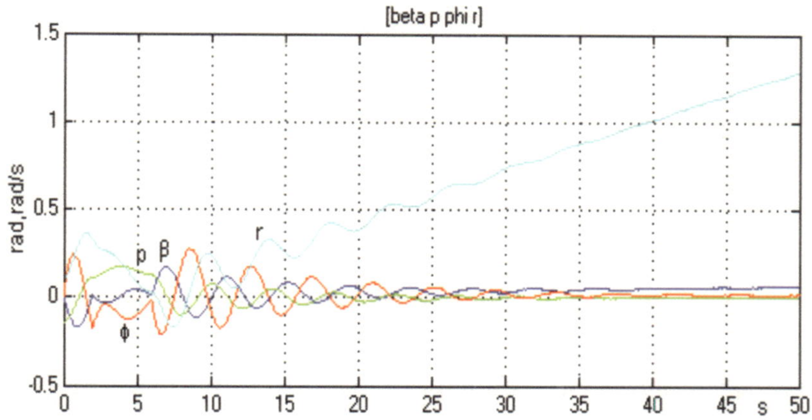

Fig. 5. States' time responses without control

Theorem 3: Suppose that Assumption 1 holds. Then the transformed multivariable sliding time-delay system (16) with matched parameter perturbations and external disturbances driven by combined controller (4) and restricted to the sliding surface s(t)=0 is robustly globally asymptotically delay-dependent stable with respect to the state variables, if the following modified LMI conditions and parameter requirements are satisfied:

$$H=\begin{bmatrix} H_{11} & -P\bar{A}_1\bar{A}_0 & -P\bar{A}_1^2 & \cdots & -P\bar{A}_1\bar{A}_N & -P\bar{A}_N\bar{A}_0 & -P\bar{A}_N\bar{A}_1 & \ldots & -P\bar{A}_N^2 & 0 & 0 & 0 & 0 \\ * & -\frac{1}{h_1}R_1 & 0 & & 0 & 0 & 0 & & 0 & & & & 0 \\ * & 0 & -\frac{1}{h_1}S_1 & & 0 & & & & & & & & 0 \\ \vdots & & & \ddots & 0 & & & & & & & & 0 \\ * & 0 & 0 & 0 & -\frac{1}{h_N}R_N & & & & & & & & 0 \\ * & 0 & 0 & 0 & 0 & & & & & & & & 0 \\ * & & & & & & & & & & & & 0 \\ \vdots & & & & & & & \ddots & & & & & 0 \\ 0 & & & & & & & & -\frac{1}{h_1}S_N & 0 & 0 & 0 & 0 \\ 0 & & & & & & & & 0 & 0 & -T_1 & 0 & 0 \\ 0 & & & & & & & & 0 & 0 & 0 & \ddots & 0 \\ 0 & 0 & 0 & 0 & 0 & 0 & 0 & & 0 & 0 & 0 & 0 & -T_N \end{bmatrix}<0 \quad (35)$$

where

$$H_{11}=(\tilde{A}_0+\bar{A}_1+\ldots+\bar{A}_N)^{T}P+P(\tilde{A}_0+\bar{A}_1+\ldots+\bar{A}_N)+h_1(S_1+R_1)+\ldots+h_N(S_N+R_N)+T_1+\ldots+T_N$$

$$CB=B^{T}PB>0 \quad (36)$$

$$k_0=\alpha_0;k_1=\alpha_1;\ldots k_N=\alpha_N; \quad (37)$$

$$\delta \geq f_o \tag{38}$$

where $P, R_1, \ldots R_N$ are some symmetric positive definite matrices which are a feasible solution of modified LMI (35) with (36); $\widetilde{A_0} = \overline{A_0} - BG$ is a stable matrix.

Proof: Let us choose a special augmented Lyapunov-Krasovskii functional as follows:

$$\begin{aligned} V &= x^T(t)Px(t) + \sum_{i=1}^{N} \int_{-h_i}^{0} \int_{t+\theta}^{t} x^T(\rho)R_i x(\rho)d\rho\, d\theta \\ &+ \sum_{i=1}^{N} \int_{-h_i}^{0} \int_{t+\theta-h_i}^{t} x^T(\rho)S_i x(\rho)d\rho\, d\theta + \sum_{i=1}^{N} \int_{t-h_i}^{t} x^T(\theta)T_i x(\theta)d\theta \end{aligned} \tag{39}$$

The introduced special augmented functional (39) involves three particular terms: first term V_1 is standard Lyapunov function, second and third are non-standard terms, namely V_2 and V_3 are similar, except for the length integration horizon [t-h, t] for V_2 and [t+θ-h, t] for V_3, respectively. This functional is different from existing ones.
The time-derivative of (39) along the perturbed time-delay system (16) can be calculated as:

$$\begin{aligned} \dot{V} &= x^T(t)\Big[(\tilde{A}_0 + \overline{A}_1 + \ldots + \overline{A}_N)^T P + P(\tilde{A}_0 + \overline{A}_1 + \ldots + \overline{A}_N) \\ &\quad + h_1(S_1 + R_1) + \ldots + h_N(S_N + R_N) + T_1 + \ldots + T_N x(t)\Big] \\ &- 2x^T(t)P\overline{A}_1\overline{A}_0 \int_{t-h_1}^{t} x(\theta)d\theta - 2x^T(t)P\overline{A}_1^2 \int_{t-h_1}^{t} x(\theta - h_1)d\theta - \ldots - 2x^T(t)P\overline{A}_1\overline{A}_N \int_{t-h_1}^{t} x(\theta - h_N)d\theta \\ &- \ldots - 2x^T(t)P\overline{A}_N\overline{A}_0 \int_{t-h_N}^{t} x(\theta)d\theta - 2x^T(t)P\overline{A}_N\overline{A}_1 \int_{t-h_N}^{t} x(\theta - h_1)d\theta - \ldots - 2x^T(t)P\overline{A}_N^2 \int_{t-h_N}^{t} x(\theta - h_N)d\theta \\ &- h_1 \int_{t-h_1}^{t} x^T(\theta)R_1 x(\theta)d\theta - \ldots - h_N \int_{t-h_N}^{t} x^T(\theta)R_N x(\theta)d\theta \\ &- h_1 \int_{t-h_1}^{t} x^T(\theta - h_1)S_1 x(\theta - h_1)d\theta - \ldots - h_N \int_{t-h_N}^{t} x^T(\theta - h_N)S_N x(\theta - h_N)d\theta \\ &- x^T(t-h_1)T_1 x(t-h_1) - x^T(t-h_N)T_N x(t-h_N) \\ &+ 2x^T(t)P\Delta A_1 x(t-h_1) + \ldots + 2x^T(t)P\Delta A_N x(t-h_N) - 2x^T(t)PB[k_0\|x(t)\| + k_1\|x(t-h_1)\| \\ &+ \ldots + k_N\|x(t-h_N)\|]\frac{s(t)}{\|s(t)\|} - 2x^T(t)PB\delta\frac{s(t)}{\|s(t)\|} + x^T(t)PDf(t) \end{aligned} \tag{40}$$

Since for some h>0 Noldus inequality holds:

$$\begin{aligned} h_1 \int_{t-h_1}^{t} x^T(\theta)R_1 x(\theta)d\theta &\geq \left[\int_{t-h_1}^{t} x(\theta)d\theta\right]^T R_1 \left[\int_{t-h_1}^{t} x(\theta)d\theta\right] \ldots \\ h_N \int_{t-h_N}^{t} x^T(\theta - h_N)S_N x(\theta - h_N)d\theta &\geq \left[\int_{t-h_N}^{t} x(\theta - h_N)d\theta\right]^T S_N \left[\int_{t-h_N}^{t} x(\theta - h_N)d\theta\right] \end{aligned} \tag{41}$$

and $x^T(t)PB = s^T(t)$ then (40) becomes as:

$$\dot{V} \le x^T(t)\Big[(\tilde{A}_0 + \bar{A}_1 + ... + \bar{A}_N)^T P + P(\tilde{A}_0 + \bar{A}_1 + ... + \bar{A}_N) + h_1(S_1 + R_1) + ... + h_N(S_N + R_N) + T_1 + ... + T_N\Big]x(t)$$

$$-2x^T(t)P\bar{A}_1\bar{A}_0 \int_{t-h_1}^{t} x(\theta)d\theta - 2x^T(t)P\bar{A}_1^2 \int_{t-h_1}^{t} x(\theta - h_1)d\theta$$

$$-... - 2x^T(t)P\bar{A}_1\bar{A}_N \int_{t-h_1}^{t} x(\theta - h_N)d\theta - ... - 2x^T(t)P\bar{A}_N\bar{A}_0 \int_{t-h_N}^{t} x(\theta)d\theta$$

$$-2x^T(t)P\bar{A}_N\bar{A}_1 \int_{t-h_N}^{t} x(\theta - h_1)d\theta - ... - 2x^T(t)P\bar{A}_N^2 \int_{t-h_N}^{t} x(\theta - h_N)d\theta - \frac{1}{h_1}\left[\int_{t-h_1}^{t} x(\theta)d\theta\right]^T R_1 \left[\int_{t-h_1}^{t} x(\theta)d\theta\right]$$

$$-... - \frac{1}{h_N}\left[\int_{t-h_N}^{t} x(\theta)d\theta\right]^T R_N \left[\int_{t-h_N}^{t} x(\theta)d\theta\right] - \frac{1}{h_1}\left[\int_{t-h_1}^{t} x(\theta - h_1)d\theta\right]^T S_1 \left[\int_{t-h_1}^{t} x(\theta - h_1)d\theta\right]$$

$$-... - \frac{1}{h_N}\left[\int_{t-h_N}^{t} x(\theta - h_N)d\theta\right]^T S_N \left[\int_{t-h_N}^{t} x(\theta - h_N)d\theta\right] - x^T(t-h_1)T_1x(t-h_1) - x^T(t-h_N)T_Nx(t-h_N)$$

$$+2x^T(t)PBE_1x(t-h_1) + ... + 2x^T(t)PBE_Nx(t-h_N) - 2x^T(t)PB\big[k_0\|x(t)\| + k_1\|x(t-h_1)\|$$

$$+... + k_N\|x(t-h_N)\|\big]\frac{s(t)}{\|s(t)\|} - 2x^T(t)PB\delta\frac{s(t)}{\|s(t)\|} + x^T(t)PBEf(t)$$

$$\le x^T(t)\big[(\tilde{A}_0 + \bar{A}_1 + ... + \bar{A}_N)^T P + P(\tilde{A}_0 + \bar{A}_1 + ... + \bar{A}_N) + h_1(S_1 + R_1) + ... + h_N(S_N + R_N) + T_1 + ... + T_N\big]x(t)$$

$$-2x^T(t)P\bar{A}_1\bar{A}_0 \int_{t-h_1}^{t} x(\theta)d\theta - 2x^T(t)P\bar{A}_1^2 \int_{t-h_1}^{t} x(\theta - h_1)d\theta - ... - 2x^T(t)P\bar{A}_1\bar{A}_N \int_{t-h_1}^{t} x(\theta - h_N)d\theta$$

$$-... - 2x^T(t)P\bar{A}_N\bar{A}_0 \int_{t-h_N}^{t} x(\theta)d\theta - 2x^T(t)P\bar{A}_N\bar{A}_1 \int_{t-h_N}^{t} x(\theta - h_1)d\theta - ... - 2x^T(t)P\bar{A}_N^2 \int_{t-h_N}^{t} x(\theta - h_N)d\theta$$

$$-\frac{1}{h_1}\left[\int_{t-h_1}^{t} x(\theta)d\theta\right]^T R_1 \left[\int_{t-h_1}^{t} x(\theta)d\theta\right] - ... - \frac{1}{h_N}\left[\int_{t-h_N}^{t} x(\theta)d\theta\right]^T R_N \left[\int_{t-h_N}^{t} x(\theta)d\theta\right]$$

$$-\frac{1}{h_1}\left[\int_{t-h_1}^{t} x(\theta - h_1)d\theta\right]^T S_1 \left[\int_{t-h_1}^{t} x(\theta - h_1)d\theta\right] - ... - \frac{1}{h_N}\left[\int_{t-h_N}^{t} x(\theta - h_N)d\theta\right]^T S_N \left[\int_{t-h_N}^{t} x(\theta - h_N)d\theta\right]$$

$$-x^T(t-h_1)T_1x(t-h_1) - x^T(t-h_N)T_Nx(t-h_N) + 2s^T(t)E_1x(t-h_1) + ... + 2s(t)E_Nx(t-h_N)$$

$$-2s^T(t)\big[k_0\|x(t)\| + k_1\|x(t-h_1)\| + ... + k_N\|x(t-h_N)\|\big]\frac{s(t)}{\|s(t)\|} - 2\delta s^T(t)\frac{s(t)}{\|s(t)\|} + s^T(t)Ef(t)$$

$$= \left[x(t) \quad \int_{t-h_1}^{t} x(\theta)d\theta \quad \int_{t-h_1}^{t} x(\theta - h_1)d\theta \quad \int_{t-h_1}^{t} x(\theta - h_N)d\theta \quad \int_{t-h_N}^{t} x(\theta)d\theta \quad \int_{t-h_N}^{t} x(\theta - h_1)d\theta \quad \int_{t-h_N}^{t} x(\theta - h_N)d\theta \quad x(t-h_1) \quad x(t-h_N)\right]$$

$$
\begin{bmatrix}
H_{11} & -P\bar{A}_1\bar{A}_0 & -P\bar{A}_1^2 & \cdots & -P\bar{A}_1\bar{A}_N & -P\bar{A}_N\bar{A}_0 & -P\bar{A}_N\bar{A}_1 & \ldots & -P\bar{A}_N^2 & 0 & 0 & 0 & 0 \\
* & -\frac{1}{h_1}R_1 & 0 & & 0 & 0 & 0 & & 0 & 0 & 0 & 0 & 0 \\
* & 0 & -\frac{1}{h_1}S_1 & & 0 & & & & 0 & & & & 0 \\
\vdots & 0 & 0 & \ddots & 0 & & & & 0 & & & & 0 \\
* & 0 & 0 & & -\frac{1}{h_N}R_N & & & & 0 & & & & 0 \\
* & 0 & 0 & & 0 & & & & 0 & & & & 0 \\
* & 0 & 0 & & 0 & & & & 0 & & & & 0 \\
\vdots & & & & & & & \ddots & 0 & & & & 0 \\
* & 0 & 0 & & & & & & -\frac{1}{h_1}S_N & 0 & 0 & 0 & 0 \\
0 & 0 & 0 & & & & & & 0 & 0 & -T_1 & 0 & 0 \\
0 & 0 & 0 & & & & & & 0 & 0 & 0 & \ddots & 0 \\
0 & 0 & 0 & & 0 & 0 & 0 & 0 & 0 & 0 & 0 & 0 & -T_N
\end{bmatrix}
\begin{bmatrix}
x(t) \\
\int_{t-h_1}^{t} x(\theta)d\theta \\
\int_{t-h_1}^{t} x(\theta - h_1)d\theta \\
\vdots \\
\int_{t-h_1}^{t} x(\theta - h_N)d\theta \\
\int_{t-h_N}^{t} x(\theta)d\theta \\
\int_{t-h_N}^{t} x(\theta - h_1)d\theta \\
\vdots \\
\int_{t-h_N}^{t} x(\theta - h_N)d\theta \\
x(t - h_1) \\
\vdots \\
x(t - h_N)
\end{bmatrix}
$$

$$
-[(k_0-\alpha_0)\|x(t)\|\|s(t)\| + (k_1-\alpha_1)\|x(t-h_1)\|\|s(t)\| + \ldots + (k_N-\alpha_N)\|x(t-h_N)\|\|s(t)\|] - (\delta - f_0)\|s(t)\| \quad (42)
$$

Since (35)-(38) hold, then (42) reduces to:

$$
\dot{V} \le z^T(t)Hz(t) < 0 \tag{43}
$$

Therefore, we can conclude that the perturbed time-delay system (16), (4) is robustly globally asymptotically delay-dependent stable. Theorem 3 is proved.

Special case: Single state-delayed systems: For single state-delayed systems that are frequently encountered in control applications and testing examples equation of motion and control algorithm can be easily found from (1), (4), (16) letting N=1. Therefore, the modified LMI delay-dependent stability conditions for which are significantly reduced and can be summarized in the following Corollary.

Corollary 1: Suppose that Assumption 1 holds. Then the transformed single-delayed sliding system (16) with matched parameter perturbations and external disturbances driven by combined controller (4) for which N=1 and restricted by sliding surface s(t)=0 is robustly globally asymptotically delay-dependent stable with respect to the state variables, if the following LMI conditions and parameter requirements are satisfied:

$$
H = \begin{bmatrix}
\begin{matrix}(\tilde{A}_0+\bar{A}_1)^T P + P(\tilde{A}_0+\bar{A}_1) \\ +h_1(S_1+R_1)+T_1\end{matrix} & -P\bar{A}_1\bar{A}_0 & -P\bar{A}_1^2 & 0 \\
-(P\bar{A}_1\bar{A}_0)^T & -\frac{1}{h_1}R_1 & 0 & 0 \\
-(P\bar{A}_1^2)^T & 0 & -\frac{1}{h_1}S_1 & 0 \\
0 & 0 & 0 & -T_1
\end{bmatrix} < 0 \tag{44}
$$

$$CB = B^T PB > 0 \tag{45}$$

$$k_0 = \alpha_0 ; k_1 = \alpha_1 ; \tag{46}$$

$$\delta \geq f_o \tag{47}$$

Proof: The corollary follows from the proof of the Theorem 3 letting N=1.

5.2 Existence conditions

The final step of the control design is the derivation of the sliding mode existence conditions or the reaching conditions for the perturbed time-delay system states to the sliding manifold in finite time. These results are summarized in the following theorem.

Theorem 4: Suppose that Assumption 1 holds. Then the perturbed multivariable time-delay system (1) states with matched parameter uncertainties and external disturbances driven by controller (4) converge to the siding surface s(t)=0 in finite time, if the following conditions are satisfied:

$$k_0 = \alpha_0 + g ; k_1 = \alpha_1 ;..., k_N = \alpha_N ; \tag{48}$$

$$\delta \geq f_o \tag{49}$$

Proof: Let us choose a modified Lyapunov function candidate as:

$$V = \frac{1}{2} s^T(t)(CB)^{-1} s(t) \tag{50}$$

The time-derivative of (50) along the state trajectories of time-delay system (1), (4) can be calculated as follows:

$$\begin{aligned}
\dot{V} &= s^T(t)(CB)^{-1}\dot{s}(t) = s^T(t)(CB)^{-1}C\dot{x}(t) = s^T(t)(CB)^{-1}C\big[A_0x(t) + A_1x(t-h_1) \\
&\quad + ... + A_Nx(t-h_N) + \Delta A_0x(t) + \Delta A_1x(t-h_1) + ... + \Delta A_Nx(t-h_N) + Bu(t) + Df(t)\big] \\
&= s^T(t)(CB)^{-1}\big[CA_0x(t) + CA_1x(t-h_1) + ... + CA_Nx(t-h_N) + CBE_0x(t) + CBE_1x(t-h_1) \\
&\quad + ... + CBE_Nx(t-h_N) - CB\Big((CB)^{-1}\big[CA_0x(t) + CA_1x(t-h) + ... + CA_Nx(t-h_N)\big] \\
&\quad - \big[k_0\|x(t)\| + k_1\|x(t-h_1)\| + ... + k_N\|x(t-h_N)\|\big]\frac{s(t)}{\|s(t)\|} - Gx(t) - \delta\frac{s(t)}{\|s(t)\|}\Big) + CBEf(t)\big] \qquad (51) \\
&= s^T(t)[E_0x(t) + E_1x(t-h_1) + ... + E_Nx(t-h_N) \\
&\quad - \big[k_0\|x(t)\| + k_1\|x(t-h_1)\| + ... + k_N\|x(t-h_N)\|\big]\frac{s(t)}{\|s(t)\|} - Gx(t) - \delta\frac{s(t)}{\|s(t)\|} + Ef(t)] \\
&\leq -[(k_0 - \alpha_0 - g)\|x(t)\|\|s(t)\| + (k_1 - \alpha_1)\|x(t-h_1)\|\|s(t)\| \\
&\quad + ... + (k_N - \alpha_N)\|x(t-h_N)\|\|s(t)\|] - (\delta - f_0)\|s(t)\|
\end{aligned}$$

Since (48), (49) hold, then (51) reduces to:

$$\dot{V} = s^T(t)(CB)^{-1}\dot{s}(t) \leq -(\delta - f_0)\|s(t)\| \leq -\eta\|s(t)\| \tag{52}$$

where

$$\eta = \delta - f_0 \geq 0 \tag{53}$$

Hence we can evaluate that

$$\dot{V}(t) \leq -\eta \sqrt{\frac{2}{\lambda_{\min}(CB)^{-1}}} \sqrt{V(t)} \tag{54}$$

The last inequality (54) is known to prove the finite-time convergence of system (1),(4) towards the sliding surface s(t)=0 (Utkin, 1977), (Perruquetti & Barbot, 2002). Therefore, Theorem 4 is proved.

5.3. Numerical examples

In order to demonstrate the usefulness of the proposed control design techniques let us consider the following examples.

Example 4: Consider a time-delay system (1),(4) with parameters taken from (Li & De Carlo, 2003):

$$A_0 = \begin{bmatrix} 2 & 0 & 1 \\ 1.75 & 0.25 & 0.8 \\ -1 & 0 & 1 \end{bmatrix}; A_1 = \begin{bmatrix} -1 & 0 & 0 \\ -0.1 & 0.25 & 0.2 \\ -0.2 & 4 & 5 \end{bmatrix}; B = \begin{bmatrix} 0 \\ 0 \\ 1 \end{bmatrix};$$

$$\Delta A_0 = 0.2\sin(t)A_0 \,, \Delta A_1 = 0.2\cos(t)A_1 \,, f = 0.3\sin(t)$$

The LMI delay-dependent stability and sliding mode existence conditions are computed by MATLAB programming (see Appendix 4) where LMI Control Toolbox is used. The computational results are following:

$$\text{Geq} = \begin{bmatrix} 1.2573 & 2.5652 & 1.0000 \end{bmatrix}$$

$$\text{A0hat} = \begin{bmatrix} 2.0000 & 0 & 1.0000 \\ 1.7500 & 0.2500 & 0.8000 \\ -7.0038 & -0.6413 & -3.3095 \end{bmatrix}; \text{A1hat} = \begin{bmatrix} -1.0000 & 0 & 0 \\ -0.1000 & 0.2500 & 0.2000 \\ 1.5139 & -0.6413 & -0.5130 \end{bmatrix}$$

$$\text{eigA0hat} = \begin{bmatrix} -0.5298 + 0.5383i \\ -0.5298 - 0.5383i \\ 0.0000 \end{bmatrix}; \text{eigA1hat} = \begin{bmatrix} -0.2630 & -0.0000 & -1.0000 \end{bmatrix}$$

$$\text{G} = \begin{bmatrix} 3.3240 & 10.7583 & 3.2405 \end{bmatrix}; \quad \text{Geq} = \begin{bmatrix} 1.2573 & 2.5652 & 1.0000 \end{bmatrix}$$

$$\text{A0til} = \begin{bmatrix} 2.0000 & 0 & 1.0000 \\ 1.7500 & 0.2500 & 0.8000 \\ -10.3278 & -11.3996 & -6.5500 \end{bmatrix}; \text{eigA0til} = \begin{bmatrix} -2.7000 \\ -0.8000 + 0.5000i \\ -0.8000 - 0.5000i \end{bmatrix}$$

$$\text{P} = 1.0\text{e+008} * \begin{bmatrix} 1.1943 & -1.1651 & 0.1562 \\ -1.1651 & 4.1745 & 0.3597 \\ 0.1562 & 0.3597 & 0.1248 \end{bmatrix}; \text{R1} = 1.0\text{e+008} * \begin{bmatrix} 1.9320 & 0.2397 & 0.8740 \\ 0.2397 & 1.0386 & 0.2831 \\ 0.8740 & 0.2831 & 0.4341 \end{bmatrix}$$

$$S1 = 1.0e+008 * \begin{bmatrix} 0.8783 & 0.1869 & 0.2951 \\ 0.1869 & 1.0708 & 0.2699 \\ 0.2951 & 0.2699 & 0.1587 \end{bmatrix};\ T1 = 1.0e+007 * \begin{bmatrix} 2.3624 & -0.7303 & 0.7264 \\ -0.7303 & 7.5758 & 1.1589 \\ 0.7264 & 1.1589 & 0.4838 \end{bmatrix}$$

lhs = 1.0e+008 *

$$\begin{bmatrix} -1.1632 & 0.4424 & -0.1828 & 0.1743 & -0.1030 & 0.1181 & -0.4064 & -0.1030 & -0.0824 & 0 & 0 & 0 \\ 0.4424 & -1.6209 & -0.1855 & 0.5480 & 0.2138 & 0.2098 & 0.3889 & 0.2138 & 0.1711 & 0 & 0 & 0 \\ -0.1828 & -0.1855 & -0.0903 & 0.0445 & 0.0026 & 0.0215 & -0.0142 & 0.0026 & 0.0021 & 0 & 0 & 0 \\ 0.1743 & 0.5480 & 0.0445 & -1.9320 & -0.2397 & -0.8740 & 0 & 0 & 0 & 0 & 0 & 0 \\ -0.1030 & 0.2138 & 0.0026 & -0.2397 & -1.0386 & -0.2831 & 0 & 0 & 0 & 0 & 0 & 0 \\ 0.1181 & 0.2098 & 0.0215 & -0.8740 & -0.2831 & -0.4341 & 0 & 0 & 0 & 0 & 0 & 0 \\ -0.4064 & 0.3889 & -0.0142 & 0 & 0 & 0 & -0.8783 & -0.1869 & -0.2951 & 0 & 0 & 0 \\ -0.1030 & 0.2138 & 0.0026 & 0 & 0 & 0 & -0.1869 & -1.0708 & -0.2699 & 0 & 0 & 0 \\ -0.0824 & 0.1711 & 0.0021 & 0 & 0 & 0 & -0.2951 & -0.2699 & -0.1587 & 0 & 0 & 0 \\ 0 & 0 & 0 & 0 & 0 & 0 & 0 & 0 & 0 & -0.2362 & 0.0730 & -0.0726 \\ 0 & 0 & 0 & 0 & 0 & 0 & 0 & 0 & 0 & 0.0730 & -0.7576 & -0.1159 \\ 0 & 0 & 0 & 0 & 0 & 0 & 0 & 0 & 0 & -0.0726 & -0.1159 & -0.0484 \end{bmatrix}$$

$$maxh1 = 1;\quad eigsLHS = 1.0e+008 * \begin{bmatrix} -2.8124 \\ -2.0728 \\ -1.0975 \\ -0.9561 \\ -0.8271 \\ -0.7829 \\ -0.5962 \\ -0.2593 \\ -0.0216 \\ -0.0034 \\ -0.0000 \\ -0.0000 \end{bmatrix};\quad NormP = 4.5946e+008$$

$$G = \begin{bmatrix} 3.3240 & 10.7583 & 3.2405 \end{bmatrix};\ NormG = 11.7171$$

$$invBtPB = 8.0109e-008;\ BtP = 1.0e+007 * \begin{bmatrix} 1.5622 & 3.5970 & 1.2483 \end{bmatrix}$$

$$eigP = 1.0e+008 * \begin{bmatrix} 0.0162 \\ 0.8828 \\ 4.5946 \end{bmatrix};\ eigR1 = 1.0e+008 * \begin{bmatrix} 0.0070 \\ 0.9811 \\ 2.4167 \end{bmatrix}$$

$$eigS1 = 1.0e+008 * \begin{bmatrix} 0.0159 \\ 0.7770 \\ 1.3149 \end{bmatrix};\ eigT1 = 1.0e+007 * \begin{bmatrix} 0.0000 \\ 2.5930 \\ 7.8290 \end{bmatrix}$$

$$\text{k0}= 11.9171;\ \text{k1}=0.2;\ \delta \geq 0.3;\ \text{H}<0$$

Considered time-delay system is delay-dependently robustly asymptotically stable for all constant delays $h \leq 1$.

Example 5: Now, let us consider a networked control time-delay system (1), (4) with parameters taken from (Cao et al., 2007):

$$A_0 = \begin{bmatrix} -4 & 0 \\ -1 & -3 \end{bmatrix}, A_1 = \begin{bmatrix} -1.5 & 0 \\ -1 & -0.5 \end{bmatrix}, B = \begin{bmatrix} 2 \\ 2 \end{bmatrix};$$
$$\Delta A_0 = 0.5\sin(t)A_0\ , \Delta A_1 = 0.5\cos(t)A_1\ , f\ = 0.3\sin(t)$$

The LMI delay-dependent stability and sliding mode existence conditions are computed by MATLAB programming (see Appendix 5) where LMI Control Toolbox is used. The computational results are following:

$$\text{maxh1} = 2.0000;\ \text{Geq} = [\ 0.4762 \quad 0.0238]$$

$$\text{A0hat} = \begin{bmatrix} \text{-0.1429} & \text{0.1429} \\ \text{2.8571} & \text{-2.8571} \end{bmatrix};\ \text{A1hat} = \begin{bmatrix} \text{-0.0238} & \text{0.0238} \\ \text{0.4762} & \text{-0.4762} \end{bmatrix}$$

$$\text{eigA0hat} = \begin{bmatrix} \text{-0.0000} \\ \text{-3.0000} \end{bmatrix};\ \text{eigA1hat} = \begin{bmatrix} \text{-0.0000} \\ \text{-0.5000} \end{bmatrix}$$

$$\text{A0til} = \begin{bmatrix} \text{-4.1429} & \text{-0.0571} \\ \text{-1.1429} & \text{-3.0571} \end{bmatrix};\ \text{eigA0til} = \begin{bmatrix} \text{-4.2000} \\ \text{-3.0000} \end{bmatrix}$$

$$\text{P} = \text{1.0e+004} * \begin{bmatrix} \text{5.7534} & \text{-0.1805} \\ \text{-0.1805} & \text{0.4592} \end{bmatrix};\ \text{R1} = \text{1.0e+004} * \begin{bmatrix} \text{8.4457} & \text{-0.2800} \\ \text{-0.2800} & \text{0.6883} \end{bmatrix}$$

$$\text{S1} = \text{1.0e+004} * \begin{bmatrix} \text{7.7987} & \text{0.2729} \\ \text{0.2729} & \text{0.1307} \end{bmatrix};\ \text{T1} = \text{1.0e+004} * \begin{bmatrix} \text{6.7803} & \text{0.3390} \\ \text{0.3390} & \text{0.0170} \end{bmatrix}$$

lhs = 1.0e+004 *

$$\begin{bmatrix} -8.4351 & 1.2170 & -0.6689 & 0.6689 & -0.1115 & 0.1115 & 0 & 0 \\ 1.2170 & -1.5779 & 0.6689 & -0.6689 & 0.1115 & -0.1115 & 0 & 0 \\ -0.6689 & 0.6689 & -4.2228 & 0.1400 & 0 & 0 & 0 & 0 \\ 0.6689 & -0.6689 & 0.1400 & -0.3442 & 0 & 0 & 0 & 0 \\ -0.1115 & 0.1115 & 0 & 0 & -3.8994 & -0.1364 & 0 & 0 \\ 0.1115 & -0.1115 & 0 & 0 & -0.1364 & -0.0653 & 0 & 0 \\ 0 & 0 & 0 & 0 & 0 & 0 & -6.7803 & -0.3390 \\ 0 & 0 & 0 & 0 & 0 & 0 & -0.3390 & -0.0170 \end{bmatrix}$$

$$\text{eigsLHS} = 1.0\text{e}{+}004 * \begin{bmatrix} -8.8561 \\ -6.7973 \\ -4.1971 \\ -3.9040 \\ -1.4904 \\ -0.0971 \\ -0.0000 \\ -0.0000 \end{bmatrix}; \quad \text{NormP} = 5.7595\text{e}{+}004; \quad G = \begin{bmatrix} 2.0000 & 0.1000 \end{bmatrix}$$

$$\text{NormG} = 2.0025; \ \text{invBtPB} = 4.2724\text{e-}006; \ \text{BtP} = 1.0\text{e}{+}005 * \begin{bmatrix} 1.1146 & 0.0557 \end{bmatrix}$$

$$\text{eigsP} = 1.0\text{e}{+}004 * \begin{bmatrix} 0.4530 \\ 5.7595 \end{bmatrix}; \ \text{eigsR1} = 1.0\text{e}{+}004 * \begin{bmatrix} 0.6782 \\ 8.4558 \end{bmatrix}$$

$$\text{eigsS1} = 1.0\text{e}{+}004 * \begin{bmatrix} 0.1210 \\ 7.8084 \end{bmatrix}; \ \text{eigsT1} = 1.0\text{e}{+}004 * \begin{bmatrix} 0.0000 \\ 6.7973 \end{bmatrix}$$

$$\text{k0} = 2.5025; \ \text{k1} = 0.5; \ \delta \geq 0.3; \ H < 0$$

The networked control time-delay system is robustly asymptotically delay-dependent stable for all constant time-delays $h \leq 2.0000$.
Thus, we have designed all the parameters of the combined sliding mode controller.
Numerical examples show the usefulness of the proposed design approach.

6. Conclusion

The problem of the sliding mode control design for matched uncertain multi-input systems with several fixed state delays by using of LMI approach has been considered. A new combined sliding mode controller has been proposed and designed for the stabilization of uncertain time-delay systems with matched parameter perturbations and external disturbances. Delay-independent and delay-dependent global stability and sliding mode existence conditions have been derived by using Lyapunov-Krasovskii functional method and formulated in terms of linear matrix inequality techniques. The allowable upper bounds on the time-delay are determined from the LMI stability conditions. These bounds are independent in different from existing ones of the parameter uncertainties and external disturbances.
Five numerical examples and simulation results with aircraft control application have illustrated the usefulness of the proposed design approach.
The obtained results of this work are presented in (Jafarov, 2008), (Jafarov, 2009).

7. Appendices

A1

```
clear;
clc;
```

```
A0=[-4  0; -1  -3];
A1=[-1.5 0; -1  -0.5];
B =[ 2;  2];
setlmis([])
P =lmivar(1,[2 1]);
R1=lmivar(1,[2 1]);
Geq=inv(B'*P*B)*B'*P
A0hat=A0-B*G*A0
A1hat=A1-B*G*A1
G= place(A0hat,B,[-4.5 -3])
A0til=A0hat-B*G1
eigA0til=eig(A0til)
eigA0hat=eig(A0hat)
eigA1hat=eig(A1hat)
ii = 1;
lmiterm([-1 1 1 P],ii,ii)
lmiterm([-2 1 1 R1],ii,ii)
lmiterm([4 1 1 P],1,A0til','s')
lmiterm([4 1 1 R1],ii,ii)
lmiterm([4 2 2 R1],-ii,ii)
lmiterm([4 1 2 P],1,A1hat)
LMISYS=getlmis;
[copt,xopt]=feasp(LMISYS);
P=dec2mat(LMISYS,xopt,P);
R1=dec2mat(LMISYS,xopt,R1);
evlmi=evallmi(LMISYS,xopt);
[lhs,rhs]=showlmi(evlmi,4);
lhs
P
eigP=eig(P)
R1
eigR1=eig(R1)
eigsLHS=eig(lhs)
BTP=B'*P
BTPB=B'*P*B
invBTPB=inv(B'*P*B)
% recalculate
Geq=inv(B'*P*B)*B'*P
A0hat=A0-B*G*A0
A1hat=A1-B*G*A1
G= place(A0hat,B,[-4.5 -3])
A0til=A0hat-B*G1
eigA0til=eig(A0til)
eigA0hat=eig(A0hat)
eigA1hat=eig(A1hat)
```

```
ii = 1;
setlmis([])
P =lmivar(1,[2 1]);
R1=lmivar(1,[2 1]);
R2=lmivar(1,[2 1]);
lmiterm([-1 1 1 P],ii,ii)
lmiterm([-2 1 1 R1],ii,ii)
lmiterm([4 1 1 P],1,A0til','s')
lmiterm([4 1 1 R1],ii,ii)
lmiterm([4 2 2 R1],-ii,ii)
lmiterm([4 1 2 P],1,A1hat)
LMISYS=getlmis;
[copt,xopt]=feasp(LMISYS);
P=dec2mat(LMISYS,xopt,P);
R1=dec2mat(LMISYS,xopt,R1);
evlmi=evallmi(LMISYS,xopt);
[lhs,rhs]=showlmi(evlmi,4);
lhs
P
eigP=eig(P)
R1
eigR1=eig(R1)
eigsLHS=eig(lhs)
BTP=B'*P
BTPB=B'*P*B
invBTPB=inv(B'*P*B)
normG1 = norm(G1)
```

A2

```
clear;
clc;
A0=[-1 0.7; 0.3 1];
A1=[-0.1 0.1; 0 0.2];
A2=[0.2 0; 0 0.1];
B=[1; 1]
setlmis([])
P =lmivar(1,[2 1]);
R1=lmivar(1,[2 1]);
R2=lmivar(1,[2 1]);
Geq=inv(B'*P*B)*B'*P
A0hat=A0-B*G*A0
A1hat=A1-B*G*A1
A2hat=A2-B*G*A2
G= place(A0hat,B,[-4.2-.6i -4.2+.6i])
A0til=A0hat-B*G1
```

```
eigA0til=eig(A0til)
eigA0hat=eig(A0hat)
eigA1hat=eig(A1hat)
eigA2hat=eig(A2hat)
ii = 1;
lmiterm([-1 1 1 P],ii,ii)
lmiterm([-2 1 1 R1],ii,ii)
lmiterm([-3 1 1 R2],ii,ii)
lmiterm([4 1 1 P],1,A0til','s')
lmiterm([4 1 1 R1],ii,ii)
lmiterm([4 1 1 R2],ii,ii)
lmiterm([4 2 2 R1],-ii,ii)
lmiterm([4 1 2 P],1,A1hat)
lmiterm([4 1 3 P],1,A2hat)
lmiterm([4 3 3 R2],-ii,ii)
LMISYS=getlmis;
[copt,xopt]=feasp(LMISYS);
P=dec2mat(LMISYS,xopt,P);
R1=dec2mat(LMISYS,xopt,R1);
R2=dec2mat(LMISYS,xopt,R2);
evlmi=evallmi(LMISYS,xopt);
[lhs,rhs]=showlmi(evlmi,4);
lhs
eigsLHS=eig(lhs)
P
eigP=eig(P)
R1
R2
eigR1=eig(R1)
eigR2=eig(R2)
BTP=B'*P
BTPB=B'*P*B
invBTPB=inv(B'*P*B)
% recalculate
Geq=inv(B'*P*B)*B'*P
A0hat=A0-B*G*A0
A1hat=A1-B*G*A1
A2hat=A2-B*G*A2
G= place(A0hat,B,[-4.2-.6i -4.2+.6i])
A0til=A0hat-B*G1
eigA0til=eig(A0til)
eigA0hat=eig(A0hat)
eigA1hat=eig(A1hat)
eigA2hat=eig(A2hat)
ii = 1;
```

```
setlmis([])
P =lmivar(1,[2 1]);
R1=lmivar(1,[2 1]);
R2=lmivar(1,[2 1]);
lmiterm([-1 1 1 P],ii,ii)
lmiterm([-2 1 1 R1],ii,ii)
lmiterm([-3 1 1 R2],ii,ii)
lmiterm([4 1 1 P],1,A0til','s')
lmiterm([4 1 1 R1],ii,ii)
lmiterm([4 1 1 R2],ii,ii)
lmiterm([4 2 2 R1],-ii,ii)
lmiterm([4 1 2 P],1,A1hat)
lmiterm([4 1 3 P],1,A2hat)
lmiterm([4 3 3 R2],-ii,ii)
LMISYS=getlmis;
[copt,xopt]=feasp(LMISYS);
P=dec2mat(LMISYS,xopt,P);
R1=dec2mat(LMISYS,xopt,R1);
R2=dec2mat(LMISYS,xopt,R2);
evlmi=evallmi(LMISYS,xopt);
[lhs,rhs]=showlmi(evlmi,4);
lhs
eigsLHS=eig(lhs)
P
eigP=eig(P)
R1
R2
eigR1=eig(R1)
eigR2=eig(R2)
BTP=B'*P
BTPB=B'*P*B
invBTPB=inv(B'*P*B)
normG1 = norm(G1)
```

A3

```
clear;
clc;
A0=[-0.228  2.148  -0.021 0;  -1    -0.0869  0    0.039;  0.335 -4.424  -1.184 0;    0    0     1    0];
A1=[ 0     0     -0.002 0;   0     0      0   0.004;    0.034 -0.442  0    0;   0     0     0    0];
B =[-1.169  0.065;    0.0223 0;    0.0547 2.120;    0     0];
setlmis([])
P =lmivar(1,[4 1]);
R1=lmivar(1,[4 1]);
G=inv(B'*P*B)*B'*P
A0hat=A0-B*G*A0
```

```
A1hat=A1-B*G*A1
G1= place(A0hat,B,[-.5+.082i -.5-.082i -.2 -.3])
A0til=A0hat-B*G1
eigA0til=eig(A0til)
eigA0hat=eig(A0hat)
eigA1hat=eig(A1hat)
%break
ii = 1;
lmiterm([-1 1 1 P],ii,ii)
lmiterm([-2 1 1 R1],ii,ii)
lmiterm([4 1 1 P],1,A0til','s')
lmiterm([4 1 1 R1],ii,ii)
lmiterm([4 2 2 R1],-ii,ii)
lmiterm([4 1 2 P],1,A1hat)
LMISYS=getlmis;
[copt,xopt]=feasp(LMISYS);
P=dec2mat(LMISYS,xopt,P);
R1=dec2mat(LMISYS,xopt,R1);
evlmi=evallmi(LMISYS,xopt);
[lhs,rhs]=showlmi(evlmi,4);
lhs
P
eigP=eig(P)
R1
eigR1=eig(R1)
eigsLHS=eig(lhs)
BTP=B'*P
BTPB=B'*P*B
invBTPB=inv(B'*P*B)
gnorm=norm(G)
```

A4

```
clear;
clc;
A0=[2 0 1; 1.75 0.25 0.8; -1 0 1]
A1=[-1 0 0; -0.1 0.25 0.2; -0.2 4 5]
B =[0;0;1]
%break
h1=1.0;
setlmis([]);
P=lmivar(1,[3 1]);
Geq=inv(B'*P*B)*B'*P
A0hat=A0-B*Geq*A0
A1hat=A1-B*Geq*A1
eigA0hat=eig(A0hat)
eigA1hat=eig(A1hat)
```

```
DesPol = [-2.7 -.8+.5i -.8-.5i];
G= place(A0hat,B,DesPol)
A0til=A0hat-B*G
eigA0til=eig(A0til)
R1=lmivar(1,[3 1]);
S1=lmivar(1,[3 1]);
T1=lmivar(1,[3 1]);
lmiterm([-1 1 1 P],1,1);
lmiterm([-1 2 2 R1],1,1);
lmiterm([-2 1 1 S1],1,1);
lmiterm([-3 1 1 T1],1,1);
lmiterm([4 1 1 P],(A0til+A1hat)',1,'s');
lmiterm([4 1 1 S1],h1,1);
lmiterm([4 1 1 R1],h1,1);
lmiterm([4 1 1 T1],1,1);
lmiterm([4 1 2 P],-1,A1hat*A0hat);
lmiterm([4 1 3 P],-1,A1hat*A1hat);
lmiterm([4 2 2 R1],-1/h1,1);
lmiterm([4 3 3 S1],-1/h1,1);
lmiterm([4 4 4 T1],-1,1);
LMISYS=getlmis;
[copt,xopt]=feasp(LMISYS);
P=dec2mat(LMISYS,xopt,P);
R1=dec2mat(LMISYS,xopt,R1);
S1=dec2mat(LMISYS,xopt,S1);
T1=dec2mat(LMISYS,xopt,T1);
evlmi=evallmi(LMISYS,xopt);
[lhs,rhs]=showlmi(evlmi,4);
lhs,h1,P,R1,S1,T1
eigsLHS=eig(lhs)
% repeat
clc;
Geq=inv(B'*P*B)*B'*P
A0hat=A0-B*Geq*A0
A1hat=A1-B*Geq*A1
eigA0hat=eig(A0hat)
eigA1hat=eig(A1hat)
G= place(A0hat,B,DesPol)
A0til=A0hat-B*G
eigA0til=eig(A0til)
setlmis([]);
P=lmivar(1,[3 1]);
R1=lmivar(1,[3 1]);
S1=lmivar(1,[3 1]);
T1=lmivar(1,[3 1]);
```

```
lmiterm([-1 1 1 P],1,1);
lmiterm([-1 2 2 R1],1,1);
lmiterm([-2 1 1 S1],1,1);
lmiterm([-3 1 1 T1],1,1);
lmiterm([4 1 1 P],(A0til+A1hat)',1,'s');
lmiterm([4 1 1 S1],h1,1);
lmiterm([4 1 1 R1],h1,1);
lmiterm([4 1 1 T1],1,1);
lmiterm([4 1 2 P],-1,A1hat*A0hat);
lmiterm([4 1 3 P],-1,A1hat*A1hat);
lmiterm([4 2 2 R1],-1/h1,1);
lmiterm([4 3 3 S1],-1/h1,1);
lmiterm([4 4 4 T1],-1,1);
LMISYS=getlmis;
[copt,xopt]=feasp(LMISYS);
P=dec2mat(LMISYS,xopt,P);
R1=dec2mat(LMISYS,xopt,R1);
S1=dec2mat(LMISYS,xopt,S1);
T1=dec2mat(LMISYS,xopt,T1);
evlmi=evallmi(LMISYS,xopt);
[lhs,rhs]=showlmi(evlmi,4);
lhs,h1,P,R1,S1,T1
eigLHS=eig(lhs)
NormP=norm(P)
G
NormG = norm(G)
invBtPB=inv(B'*P*B)
BtP=B'*P
eigP=eig(P)
eigR1=eig(R1)
eigS1=eig(S1)
eigT1=eig(T1)
```

A5

```
clear; clc;
A0=[-4  0; -1  -3];
A1=[-1.5 0; -1  -0.5];
B =[ 2;  2];
h1=2.0000;
setlmis([]);
P=lmivar(1,[2 1]);
Geq=inv(B'*P*B)*B'*P
A0hat=A0-B*Geq*A0
A1hat=A1-B*Geq*A1
eigA0hat=eig(A0hat)
```

```
eigA1hat=eig(A1hat)
% DesPol = [-.8+.5i -.8-.5i]; G= place(A0hat,B,DesPol);
avec = [2 0.1];
G = avec;
A0til=A0hat-B*G1
eigA0til=eig(A0til)
R1=lmivar(1,[2 1]);
S1=lmivar(1,[2 1]);
T1=lmivar(1,[2 1]);
lmiterm([-1 1 1 P],1,1);
lmiterm([-1 2 2 R1],1,1);
lmiterm([-2 1 1 S1],1,1);
lmiterm([-3 1 1 T1],1,1);
lmiterm([4 1 1 P],(A0til+A1hat)',1,'s');
lmiterm([4 1 1 S1],h1,1);
lmiterm([4 1 1 R1],h1,1);
lmiterm([4 1 1 T1],1,1);
lmiterm([4 1 2 P],-1,A1hat*A0hat);
lmiterm([4 1 3 P],-1,A1hat*A1hat);
lmiterm([4 2 2 R1],-1/h1,1);
lmiterm([4 3 3 S1],-1/h1,1);
lmiterm([4 4 4 T1],-1,1);
LMISYS=getlmis;
[copt,xopt]=feasp(LMISYS);
P=dec2mat(LMISYS,xopt,P);
R1=dec2mat(LMISYS,xopt,R1);
S1=dec2mat(LMISYS,xopt,S1);
T1=dec2mat(LMISYS,xopt,T1);
evlmi=evallmi(LMISYS,xopt);
[lhs,rhs]=showlmi(evlmi,4);
lhs,h1,P,R1,S1,T1
eigsLHS=eig(lhs)
% repeat
Geq=inv(B'*P*B)*B'*P
A0hat=A0-B*Geq*A0
A1hat=A1-B*Geq*A1
eigA0hat=eig(A0hat)
eigA1hat=eig(A1hat)
G = avec;
A0til=A0hat-B*G
eigA0til=eig(A0til)
setlmis([]);
P=lmivar(1,[2 1]);
R1=lmivar(1,[2 1]);
S1=lmivar(1,[2 1]);
```

```
T1=lmivar(1,[2 1]);
lmiterm([-1 1 1 P],1,1);
lmiterm([-1 2 2 R1],1,1);
lmiterm([-2 1 1 S1],1,1);
lmiterm([-3 1 1 T1],1,1);
lmiterm([4 1 1 P],(A0til+A1hat)',1,'s');
lmiterm([4 1 1 S1],h1,1);
lmiterm([4 1 1 R1],h1,1);
lmiterm([4 1 1 T1],1,1);
lmiterm([4 1 2 P],-1,A1hat*A0hat);
lmiterm([4 1 3 P],-1,A1hat*A1hat);
lmiterm([4 2 2 R1],-1/h1,1);
lmiterm([4 3 3 S1],-1/h1,1);
lmiterm([4 4 4 T1],-1,1);
LMISYS=getlmis;
[copt,xopt]=feasp(LMISYS);
P=dec2mat(LMISYS,xopt,P);
R1=dec2mat(LMISYS,xopt,R1);
S1=dec2mat(LMISYS,xopt,S1);
T1=dec2mat(LMISYS,xopt,T1);
evlmi=evallmi(LMISYS,xopt);
[lhs,rhs]=showlmi(evlmi,4);
lhs,h1,P,R1,S1,T1
eigsLHS=eig(lhs)
NormP=norm(P)
G
NormG = norm(G)
invBtPB=inv(B'*P*B)
BtP=B'*P
eigsP=eig(P)
eigsR1=eig(R1)
eigsS1=eig(S1)
eigsT1=eig(T1)
```

8. References

Utkin, V. I. (1977), Variable structure system with sliding modes, *IEEE Transactions on Automatic Control*, Vol. 22, pp. 212-222.

Sabanovic, A.; Fridman, L. & Spurgeon, S. (Editors) (2004). *Variable Structure Systems: from Principles to Implementation*, The Institution of Electrical Engineering, London.

Perruquetti, W. & Barbot, J. P. (2002). *Sliding Mode Control in Engineering*, Marcel Dekker, New York.

Richard J. P. (2003). Time-delay systems: an overview of some recent advances and open problems, *Automatica*, Vol. 39, pp. 1667-1694.

Young, K. K. O.; Utkin, V. I. & Özgüner, Ü. (1999). A control engineer's guide to sliding mode control, *Transactions on Control Systems Technology*, Vol. 7, No. 3, pp. 328-342.

Spurgeon, S. K. (1991). Choice of discontinuous control component for robust sliding mode performance, *International Journal of Control*, Vol. 53, No. 1, pp. 163-179.

Choi, H. H. (2002). Variable structure output feedback control design for a class of uncertain dynamic systems, *Automatica*, Vol. 38, pp. 335-341.

Jafarov, E. M. (2009). *Variable Structure Control and Time-Delay Systems*, Prof. Nikos Mastorakis (Ed.), 330 pages, A Series of Reference Books and Textbooks, WSEAS Press, ISBN: 978-960-474-050-5.

Shyu, K. K. & Yan, J. J. (1993). Robust stability of uncertain time-delay systems and it's stabilization by variable structure control, *International Journal of Control*, Vol. 57, pp. 237-246.

Koshkouei, A. J. & Zinober, A. S. I. (1996). Sliding mode time-delay systems, *Proceedings of the IEEE International Workshop on Variable Structure Control*, pp. 97-101, Tokyo, Japan.

Luo, N.; De La Sen N. L. M. & Rodellar, J. (1997). Robust stabilization of a class of uncertain time-delay systems in sliding mode, *International Journal of Robust and Nonlinear Control*, Vol. 7, pp. 59-74.

Li, X. & De Carlo, R. A. (2003). Robust sliding mode control of uncertain time-delay systems, *International Journal of Control*, Vol. 76, No. 1, pp. 1296-1305.

Gouisbaut, F.; Dambrine, M. & Richard, J. P. (2002). Robust control of delay systems: a sliding mode control design via LMI, *Systems and Control Letters*, Vol. 46, pp. 219-230.

Fridman, E.; Gouisbaut, F.; Dambrine, M. & Richard, J. P. (2003). Sliding mode control of systems with time-varying delays via descriptor approach, *International Journal of Systems Science*, Vol. 34, No. 8-9, pp. 553-559.

Cao, J.; Zhong, S. & Hu, Y. (2007). Novel delay-dependent stability conditions for a class of MIMO networked control systems with nonlinear perturbation, *Applied Mathematics and Computation*, doi: 10.1016/j, pp. 1-13.

Jafarov, E. M. (2005). Robust sliding mode controllers design techniques for stabilization of multivariable time-delay systems with parameter perturbations and external disturbances, *International Journal of Systems Science*, Vol. 36, No. 7, pp. 433-444.

Hung, J. Y.; Gao, & Hung, W. J. C. (1993). Variable structure control: a survey, *IEEE Transactions on Industrial Electronics*, Vol. 40, No. 1, pp. 2 – 22.

Xu, J.-X.; Hashimoto, H.; Slotine, J.-J. E.; Arai, Y. & Harashima, F. (1989). Implementation of VSS control to robotic manipulators-smoothing modification, *IEEE Transactions on Industrial Electronics*, Vol. 36, No. 3, pp. 321-329.

Tan, S.-C.; Lai, Y. M.; Tse, C. K.; Martinez-Salamero, L. & Wu, C.-K. (2007). A fast-response sliding-mode controller for boost-type converters with a wide range of operating conditions, *IEEE Transactions on Industrial Electronics*, Vol. 54, No. 6, pp. 3276-3286.

Li, H.; Chen, B.; Zhou, Q. & Su, Y. (2010). New results on delay-dependent robust stability of uncertain time delay systems, *International Journal of Systems Science*, Vol. 41, No. 6, pp. 627-634.

Schmidt, L. V. (1998). *Introduction to Aircraft Flight Dynamics*, AIAA Education Series, Reston, VA.

Jafarov, E. M. (2008). Robust delay-dependent stabilization of uncertain time-delay systems by variable structure control, *Proceedings of the International IEEE Workshop on Variable Structure Systems VSS'08*, pp. 250-255, June 2008, Antalya, Turkey.

Jafarov, E. M. (2009). Robust sliding mode control of multivariable time-delay systems, *Proceedings of the 11th WSEAS International Conference on Automatic Control, Modelling and Simulation*, pp. 430-437, May-June 2009, Istanbul, Turkey.

9

A Robust Reinforcement Learning System Using Concept of Sliding Mode Control for Unknown Nonlinear Dynamical System

Masanao Obayashi, Norihiro Nakahara, Katsumi Yamada, Takashi Kuremoto, Kunikazu Kobayashi and Liangbing Feng
Yamaguchi University
Japan

1. Introduction

In this chapter, a novel control method using a reinforcement learning (RL) (Sutton and Barto (1998)) with concept of sliding mode control (SMC) (Slotine and Li (1991)) for unknown dynamical system is considered.

In designing the control system for unknown dynamical system, there are three approaches. The first one is the conventional model-based controller design, such as optimal control and robust control, each of which is mathematically elegant, however both controller design procedures present a major disadvantage posed by the requirement of the knowledge of the system dynamics to identify and model it. In such cases, it is usually difficult to model the unknown system, especially, the nonlinear dynamical complex system, to make matters worse, almost all real systems are such cases.

The second one is the way to use only the soft-computing, such as neural networks, fuzzy systems, evolutionary systems with learning and so on. However, in these cases it is well known that modeling and identification procedures for the dynamics of the given uncertain nonlinear system and controller design procedures often become time consuming iterative approaches during parameter identification and model validation at each step of the iteration, and in addition, the control system designed through such troubles does not guarantee the stability of the system.

The last one is the way to use the method combining the above the soft-computing method with the model-based control theory, such as optimal control, sliding mode control (SMC), H_∞ control and so on. The control systems designed through such above control theories have some advantages, that is, the good nature which its adopted theory has originally, robustness, less required iterative learning number which is useful for fragile system controller design not allowed a lot of iterative procedure. This chapter concerns with the last one, that is, RL system, a kind of soft-computing method, supported with robust control theory, especially SMC for uncertain nonlinear systems.

RL has been extensively developed in the computational intelligence and machine learning societies, generally to find optimal control policies for Markovian systems with discrete state and action space. RL-based solutions to the continuous-time optimal control problem have been given in Doya (Doya (2000). The main advantage of using RL for solving optimal

control problems comes from the fact that a number of RL algorithms, e.g. Q-learning (Watkins et al. (1992)) and actor-critic learning (Wang et al. (2002)) and Obayashi et al. (2008)), do not require knowledge or identification/learning of the system dynamics. On the other hand, remarkable characteristics of SMC method are simplicity of its design method, good robustness and stability for deviation of control conditions.
Recently, a few researches as to robust reinforcement learning have been found, e.g., Morimoto et al. (2005) and Wang et al. (2002) which are designed to be robust for external disturbances by introducing the idea of H_∞ control theory (Zhau et al. (1996)), and our previous work (Obayashi et al. (2009)) is for deviations of the system parameters by introducing the idea of sliding mode control commonly used in model-based control. However, applying reinforcement learning to a real system has a serious problem, that is, many trials are required for learning to design the control system.
Firstly we introduce an actor-critic method, a kind of RL, to unite with SMC. Through the computer simulation for an inverted pendulum control without use of the inverted pendulum dynamics, it is clarified the combined method mentioned above enables to learn in less trial of learning than the only actor-critic method and has good robustness (Obayashi et al. (2009a)).
In applying the controller design, another problem exists, that is, incomplete observation problem of the state of the system. To solve this problem, some methods have been suggested, that is, the way to use observer theory (Luenberger (1984)), state variable filter theory (Hang (1976), Obayashi et al. 2009b) and both of the theories (Kung and Chen (2005)).
Secondly we introduce a robust reinforcement learning system using the concept of SMC, which uses neural network-type structure in an actor/critic configuration, refer to Fig. 1, to the case of the system state partly available by considering the variable state filter (Hang (1976)).

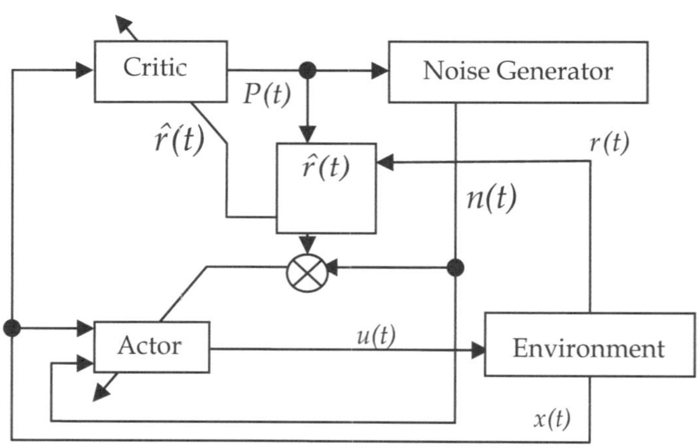

Fig. 1. The construction of the actor-critic system. (symbols in this figure are reffered to section 2)

The rest of this chapter is organized as follows. In Section 2, the conventional actor-critic reinforcement learing system is described. In Section 3, the controlled system, variable filter and sliding mode control are shortly explained. The proposed actor-critic reinforcement learning system with state variable filter using sliding mode control is described in Section 4. Comparison between the proposed system and the conventional system through simulation experiments is executed in Section 5. Finally, the conclusion is given in Section 6.

2. Actor-critic reinforcement learning system

Reinforcement learning (RL, Sutton and Barto (1998)), as experienced learning through trial and error, which is a learning algorithm based on calculation of reward and penalty given through mutual action between the agent and environment, and which is commonly executed in living things. The actor-critic method is one of representative reinforcement learning methods. We adopted it because of its flexibility to deal with both continuous and discrete state-action space environment. The structure of the actor-critic reinforcement learning system is shown in Fig. 1. The actor plays a role of a controller and the critic plays role of an evaluator in control field. Noise plays a part of roles to search the optimal action.

2.1 Structure and learning of critic

2.1.1 Structure of critic

The function of the critic is calculation of $P(t)$: the prediction value of sum of the discounted rewards *r(t)* that will be gotten over the future. Of course, if the value of $P(t)$ becomes bigger, the performance of the system becomes better. These are shortly explained as follows.

The sum of the discounted rewards that will be gotten over the future is defined as $V(t)$.

$$V(t) \equiv \sum_{l=0}^{\infty} \gamma^n \cdot r(t+l), \tag{1}$$

where γ ($0 \le \gamma < 1$) is a constant parameter called discount rate.

Equation (1) is rewritten as

$$V(t) = r(t) + \gamma V(t+1). \tag{2}$$

Here the prediction value of $V(t)$ is defined as $P(t)$. The prediction error $\hat{r}(t)$ is expressed as follows,

$$\hat{r}(t) = \hat{r}_t = r(t) + \gamma P(t+1) - P(t). \tag{3}$$

The parameters of the critic are adjusted to reduce this prediction error $\hat{r}(t)$. In our case the prediction value $P(t)$ is calculated as an output of a radial basis function neural network (RBFN) such as,

$$P(t) = \sum_{j=1}^{J} \omega_j^c y_j^c(t), \tag{4}$$

$$y_j^c(t) = \exp\left[-\sum_{i=1}^{n} (x_i(t) - c_{ij}^c)^2 / (\sigma_{ij}^c)^2 \right]. \tag{5}$$

Here, $y_j^c(t)$: jth node's output of the middle layer of the critic at time t, ω_j^c : the weight of jth output of the middle layer of the critic, x_i : i th state of the environment at time t, c_{ij}^c and σ_{ij}^c : center and dispersion in the i th input of j th basis function, respectively, J : the number of nodes in the middle layer of the critic, n : number of the states of the system (see Fig. 2).

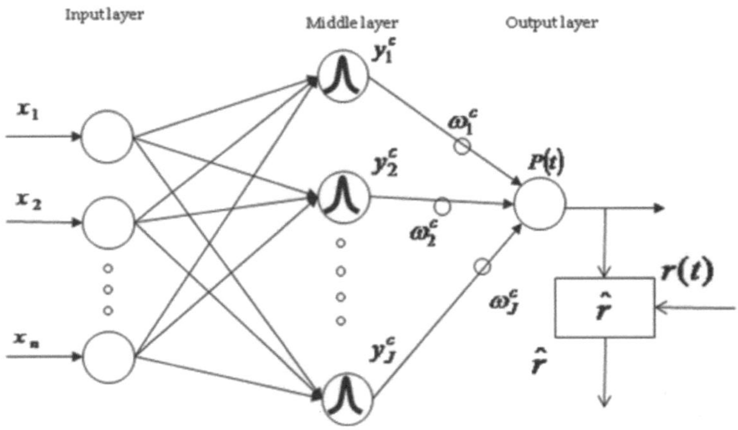

Fig. 2. Structure of the critic.

2.1.2 Learning of parameters of critic

Learning of parameters of the critic is done by back propagation method which makes prediction error $\hat{r}(t)$ go to zero. Updating rule of parameters are as follows,

$$\Delta\omega_i^c = -\eta_c \cdot \frac{\partial \hat{r}_t^2}{\partial \omega_i^c}, \quad (i = 1, \cdots, J). \tag{6}$$

Here η_c is a small positive value of learning coefficient.

2.2 Structure and learning of actor

2.2.1 Structure of actor

Figure 3 shows the structure of the actor. The actor plays the role of controller and outputs the control signal, action $a(t)$, to the environment. The actor basically also consists of radial basis function network. The jth basis function of the middle layer node of the actor is as follows,

$$y_j^a(t) = \exp\left[-\sum_{i=1}^{n}(x_i(t) - c_{ij}^a)^2 / (\sigma_{ij}^a)^2\right], \tag{7}$$

$$u'(t) = \sum_{j=1}^{J} \omega_j^a \cdot y_j^a(t), \tag{8}$$

$$u_1(t) = u_{\max} \cdot \frac{1 + \exp(-u'(t))}{1 - \exp(-u'(t))}, \tag{9}$$

$$u(t) = u_1(t) + n(t) . \tag{10}$$

Here y_j^a : jth node's output of the middle layer of the actor, c_{ij}^a and σ_{ij}^a : center and dispersion in ith input of jth node basis function of the actor, respectively, ω_j^a : connection weight from jth node of the middle layer to the output, $u(t)$: control input, $n(t)$: additive noise.

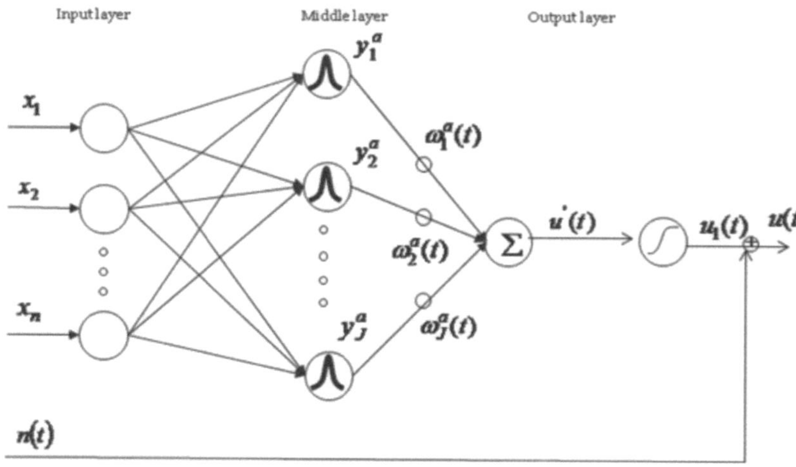

Fig. 3. Structure of the actor.

2.2.2 Noise generator

Noise generator let the output of the actor have the diversity by making use of the noise. It comes to realize the learning of the trial and error according to the results of performance of the system by executing the decided action. Generation of the noise $n(t)$ is as follows,

$$n(t) = n_t = noise_t \cdot \min\left(1, \exp(-P(t)\right), \tag{11}$$

where $noise_t$ is uniformly random number of $[-1, 1]$, min $(\cdot)$: minimum of $\cdot$. As the $P(t)$ will be bigger (this means that the action goes close to the optimal action), the noise will be smaller. This leads to the stable learning of the actor.

2.2.3 Learning of parameters of actor

Parameters of the actor, $\omega_j^a \ (j = 1, \cdots, J)$, are adjusted by using the results of executing the output of the actor, i.e. the prediction error $\hat{r}_t$ and noise.

$$\Delta\omega_j^a = \eta_a \cdot n_t \cdot \hat{r}_t \cdot \frac{\partial u_1(t)}{\partial \omega_j^a}. \tag{12}$$

$\eta_a (> 0)$ is the learning coefficient. Equation (12) means that $(-n_t \cdot \hat{r}_t)$ is considered as an error, ω_j^a is adjusted as opposite to sign of $(-n_t \cdot \hat{r}_t)$. In other words, as a result of executing $u(t)$, e.g. if the sign of the additive noise is positive and the sign of the prediction error is positive, it means that positive additive noise is sucess, so the value of ω_j^a should be increased (see Eqs. (8)-(10)), and vice versa.

3. Controlled system, variable filter and sliding mode control

3.1 Controlled system

This paper deals with next nth order nonlinear differential equation.

$$x^{(n)} = f(\mathbf{x}) + b(\mathbf{x})u, \tag{13}$$

$$y = x\,, \tag{14}$$

where $\mathbf{x} = [x, \dot{x}, \cdots, x^{(n-1)}]^T$ is state vector of the system. In this paper, it is assumed that a part of states, $y(=x)$, is observable, u is control input, $f(\mathbf{x}), b(\mathbf{x})$ are unknown continuous functions.

Object of the control system: To decide control input u which leads the states of the system to their targets **x**. We define the error vector e as follows,

$$\begin{aligned} \mathbf{e} &= [e, \dot{e}, \cdots, e^{(n-1)}]^T, \\ &= [x - x_d, \dot{x} - \dot{x}_d, \cdots, x^{(n-1)} - x_d^{(n-1)}]^T. \end{aligned} \tag{15}$$

The estimate vector of e, $\hat{e}$, is available through the state variable filter (see Fig. 4).

3.2 State variable filter

Usually it is that not all the state of the system are available for measurement in the real system. In this work we only get the state x, that is, e, so we estimate the values of error vector **e**, i.e. $\hat{e}$, through the state variable filter, Eq. (16) (Hang (1976) (see Fig. 4).

$$\hat{e}_i = \frac{\omega_n \cdot p^i}{p^n + \omega_{n-1} p^{n-1} + \cdots + \omega_0} e\,, \quad (i = 0, \cdots, n-1) \tag{16}$$

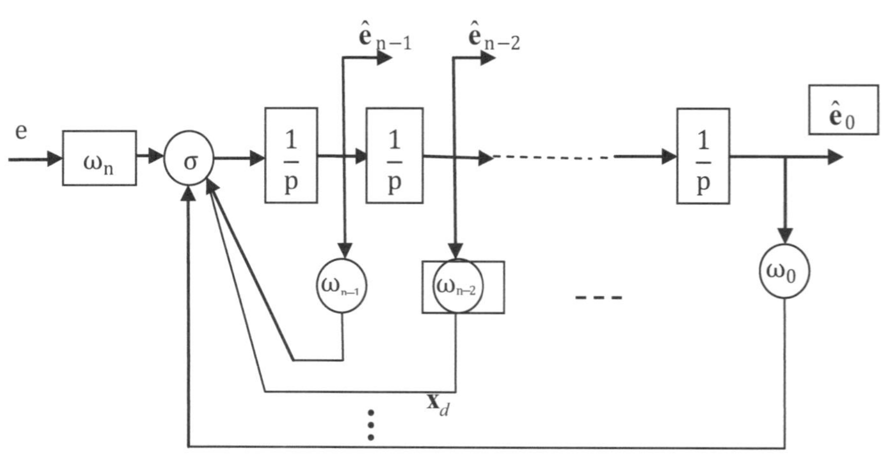

Fig. 4. Internal structure of the state variable filter.

3.3 Sliding mode control

Sliding mode control is described as follows. First it restricts states of the system to a sliding surface set up in the state space. Then it generates a sliding mode s (see in Eq. (18)) on the sliding surface, and then stabilizes the state of the system to a specified point in the state space. The feature of sliding mode control is good robustness.

Sliding time-varying surface H and sliding scalar variable s are defined as follows,

$$H : \{\, \mathbf{e} \mid s(\mathbf{e}) = 0 \}\,, \tag{17}$$

$$s(\mathbf{e}) = \boldsymbol{\alpha}^T \mathbf{e}\,, \tag{18}$$

where $\alpha_{n-1} = 1$ $\boldsymbol{\alpha} = [\alpha_0, \alpha_1, \cdots, \alpha_{n-1}]^T$, and $\alpha_{n-1}p^{n-1} + \alpha_{n-2}p^{n-2} + \cdots + \alpha_0$ is strictly stable in Hurwitz, p is Laplace transformation variable.

4. Actor-critic reinforcement learning system using sliding mode control with state variable filter

In this section, reinforcement learning system using sliding mode control with the state variable filter is explained. Target of this method is enhancing robustness which can not be obtained by conventional reinforcement. The method is almost same as the conventional actor-critic system except using the sliding variable s as the input to it inspite of the system states. In this section, we mainly explain the definition of the reward and the noise generation method.

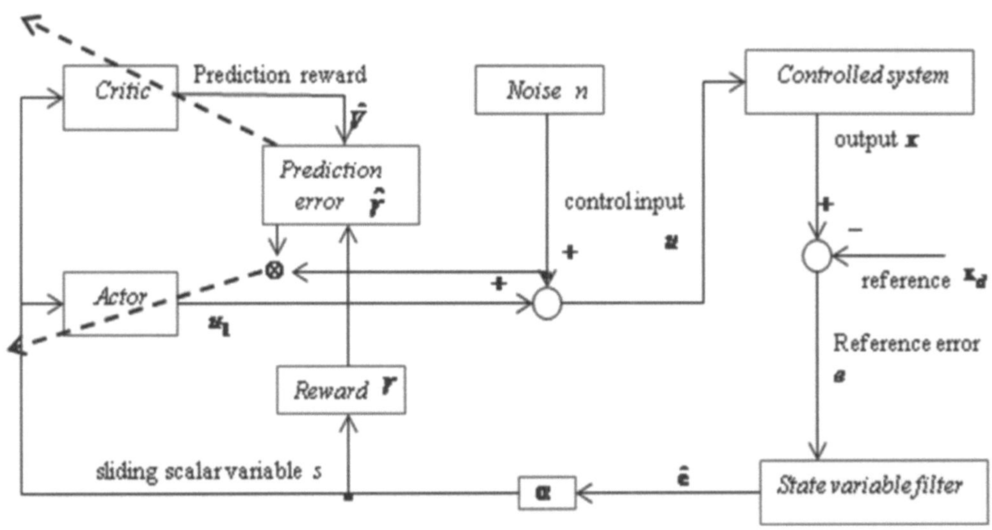

Fig. 5. Proposed reinforcement learning control system using sliding mode control with state variable filter.

4.1 Reward

We define the reward $r(t)$ to realize the sliding mode control as follows,

$$r(t) = \exp\{-s(t)^2\}\,, \tag{19}$$

here, from Eq. (18) if the actor-critic system learns so that the sliding variable s becomes smaller, i.e., error vector $\mathbf{e}$ would be close to zero, the reward $r(t)$ would be bigger.

4.2 Noise

Noise $n(t)$ is used to maintain diversity of search of the optimal input and to find the optimal input. The absolute value of sliding variable s is bigger, $n(t)$ is bigger, and that of s is smaller, it is smaller.

$$n(t) = z \cdot \overline{n} \cdot \exp\left(-\beta \cdot \frac{1}{s^2}\right), \tag{20}$$

where, z is uniform random number of range [-1, 1]. $\overline{n}$ is upper limit of the perturbation signal for searching the optimal input u. β is predefined positive constant for adjusting.

5. Computer simulation

5.1 Controlled object

To verify effectiveness of the proposed method, we carried out the control simulation using an inverted pendulum with dynamics described by Eq. (21) (see Fig. 6).

$$mg\ddot{\theta} = mgl\sin\theta - \mu_v\dot{\theta} + T_q \,. \tag{21}$$

Parameters in Eq. (21) are described in Table 1.

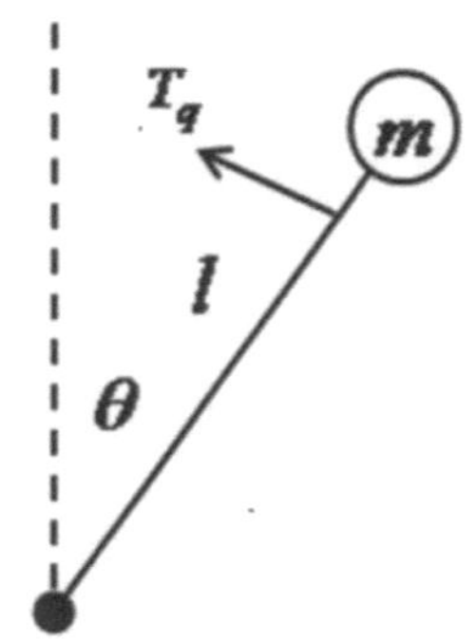

Fig. 6. An inverted pendulum used in the computer simulation.

θ	joint angle	-
m	mass	1.0 [kg]
l	length of the pendulum	1.0 [m]
g	gravity	9.8 [m/sec^2]
μ_V	coefficient of friction	0.02
T_q	input torque	-
$\mathbf{X} = [\theta, \dot{\theta}]$	observation vector	-

Table 1. Parameters of the system used in the computer simulation.

5.2 Simulation procedure

Simulation algorithm is as follows,

Step 1. Initial control input T_{q0} is given to the system through Eq. (21).

Step 2. Observe the state of the system. If the end condition is satisfied, then one trial ends, otherwise, go to Step 3.

Step 3. Calculate the error vector $\mathbf{e}$, Eq. (15). If only $y(=x)$, i.e., e is available, calculate $\hat{\mathbf{e}}$, the estimate value of through the state variable filters, Eq. (16).

Step 4. Calculate the sliding variable s, Eq. (18).
Step 5. Calculate the reward r by Eq. (19).
Step 6. Calculate the prediction reward $P(t)$ and the control input $u(t)$, i.e., torque T_q by Eqs. (4) and (10), respectively.
Step 7. Renew the parameters ω_i^c, ω_j^a of the actor and the critic by Eqs. (6) and (12).
Step 8. Set T_q in Eq. (21) of the system. Go to Step 2.

5.3 Simulation conditions

One trial means that control starts at $(\theta_0, \dot{\theta}_0) = (\pi/18[rad], 0\,[rad/\sec])$ and continues the system control for 20[sec], and sampling time is 0.02[sec]. The trial ends if $|\theta| \geq \pi/4$ or controlling time is over 20[sec]. We set upper limit for output u_1 of the actor. Trial success means that θ is in range $[-\pi/360, \pi/360]$ for last 10[sec]. The number of nodes of the hidden layer of the critic and the actor are set to 15 by trial and error (see Figs. (2)–(3)). The parameters used in this simulation are shown in Table 2.

α_0 : sliding variable parameter in Eq. (18)	5.0
η_c : learning coefficient of the actor in Eqs. (6)-(A6)	0.1
η_a : learning coefficient of the critic in Eqs. (12)-A(7)	0.1
$U_{\max}$: Maximun value of the Torque in Eqs. (9)-(A3)	20
γ : forgetting rate in Eq. (3)	0.9

Table 2. Parameters used in the simulation for the proposed system.

5.4 Simulation results

Using subsection 5.2, simulation procedure, subsection 5.3, simulation conditions, and the proposed method mentioned before, the control simulation of the inverted pendulum Eq. (21) are carried out.

5.4.1 Results of the proposed method

a. The case of complete observation

The results of the proposed method in the case of complete observation, that is, $\theta, \dot{\theta}$ are available, are shown in Fig. 7.

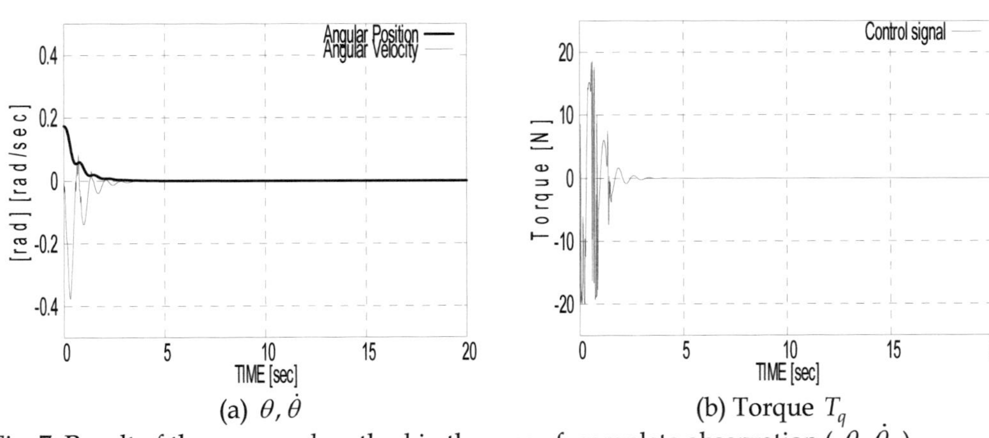

(a) $\theta, \dot{\theta}$ (b) Torque T_q

Fig. 7. Result of the proposed method in the case of complete observation ($\theta, \dot{\theta}$).

b. The case of incomplete observation using the state variable filters .

In the case that only θ is available, we have to estimate $\dot{\theta}$ as $\hat{\dot{\theta}}$. Here, we realize it by use of the state variable filter (see Eqs. (22)-(23), Fig. 8). By trial and error, the parameters, $\omega_0, \omega_1, \omega_2$, of it are set to $\omega_0 = 100$, $\omega_1 = 10$, $\omega_2 = 50$. The results of the proposed method with state variable filter in the case of incomplete observation are shown in Fig. 9.

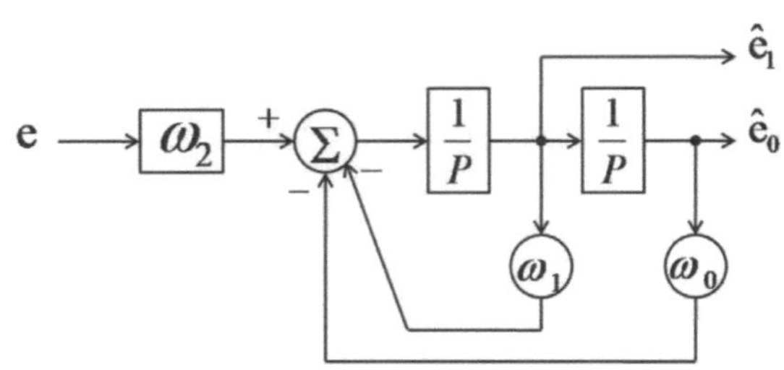

Fig. 8. State variable filter in the case of incomplete observation (θ).

$$\hat{e}_0 = \frac{\omega_2}{p^2 + \omega_1 p + \omega_0} e \tag{22}$$

$$\hat{e}_1 = \frac{\omega_2 p}{p^2 + \omega_1 p + \omega_0} e \tag{23}$$

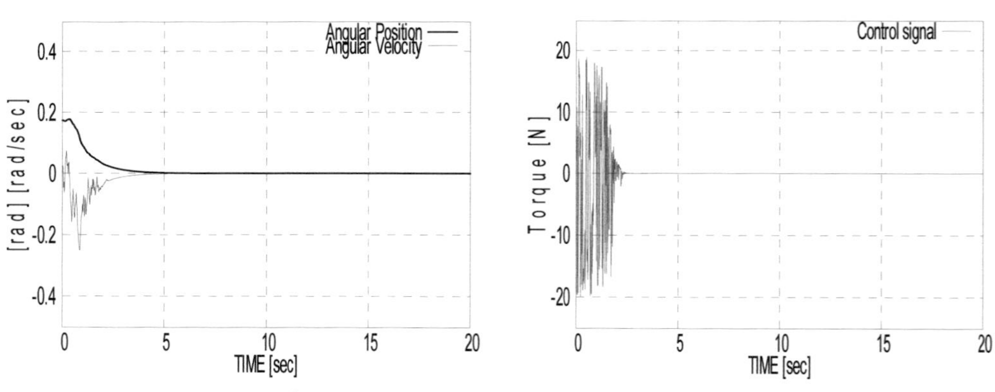

(a) $\theta, \dot{\theta}$ (b) Torque T_q

Fig. 9. Results of the proposed method with the state variable filter in the case of incomplete observation (only θ is available).

c. The case of incomplete observation using the difference method

Instead of the state variable filter in 5.4.1 *B*, to estimate the velocity angle, we adopt the commonly used difference method, like that,

$$\hat{\dot{\theta}}_t = \theta_t - \theta_{t-1} . \tag{24}$$

We construct the sliding variable s in Eq. (18) by using $\theta, \hat{\dot{\theta}}$. The results of the simulation of the proposed method are shown in Fig. 10.

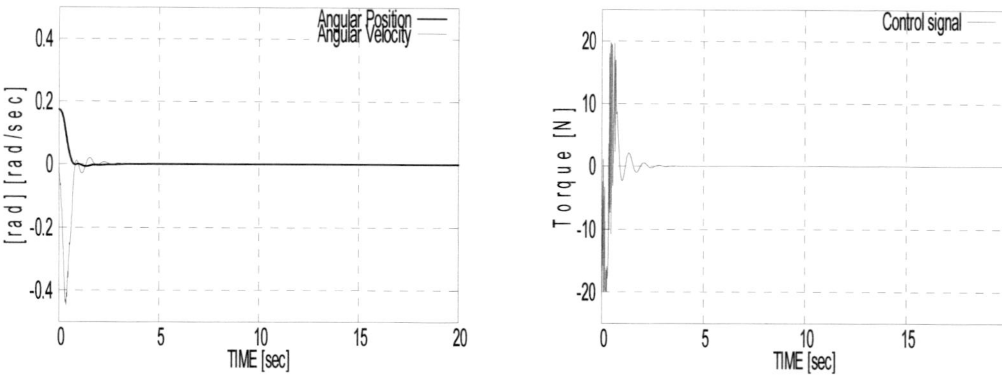

(a) $\theta, \dot{\theta}$ (b) Torque T_q

Fig. 10. Result of the proposed method using the difference method in the case of incomplete observation (only θ is available).

5.4.2 Results of the conventional method.

d. Sliding mode control method

The control input is given as follows,

$$u(t) = \begin{cases} U_{\max}, & if\ \theta \cdot \sigma > 0 \\ -U_{\max}, & if\ \theta \cdot \sigma \le 0 \end{cases}$$
$$\sigma = c\theta + \dot{\theta} \tag{25}$$
$$U_{\max =} = 20.0\,[\mathrm{N}]$$

Result of the control is shown in Fig. 11. In this case, angular, velocity angular, and Torque are all oscillatory because of the bang-bang control.

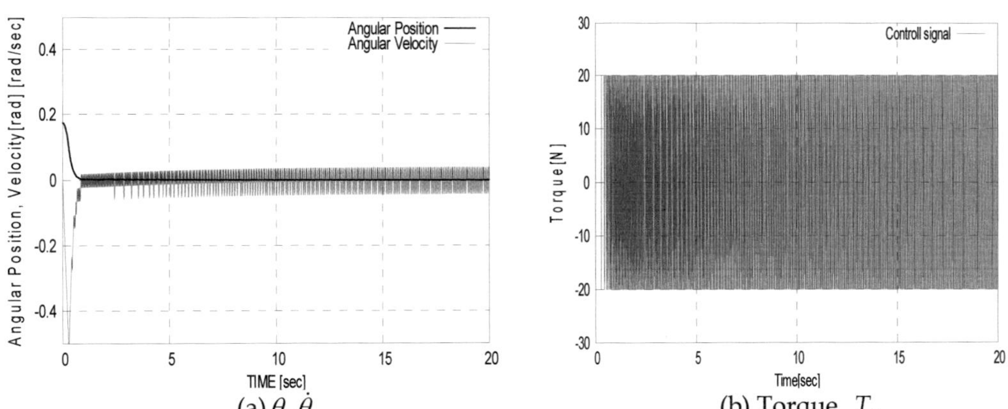

(a) $\theta, \dot{\theta}$ (b) Torque T_q

Fig. 11. Result of the conventional (SMC) method in the case of complete observation ($\theta, \dot{\theta}$).

e. Conventional actor-critic method

The structure of the actor of the conventional actor-critic control method is shown in Fig. 12. The detail of the conventional actor-critic method is explained in Appendix. Results of the simulation are shown in Fig. 13.

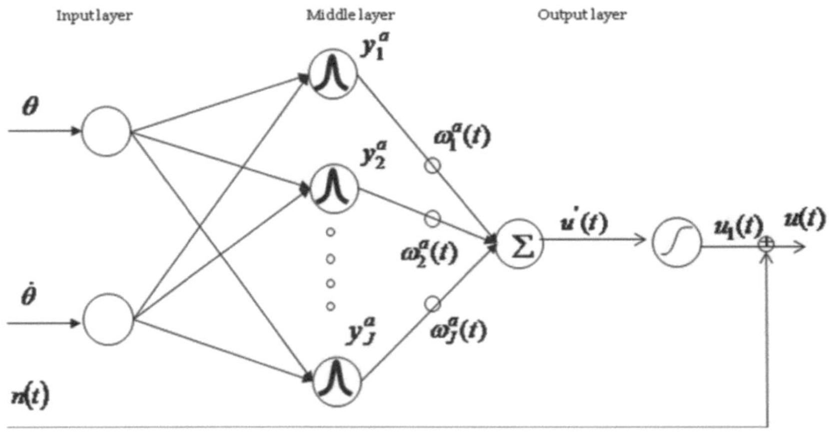

Fig. 12. Structure of the actor of the conventional actor-critic control method.

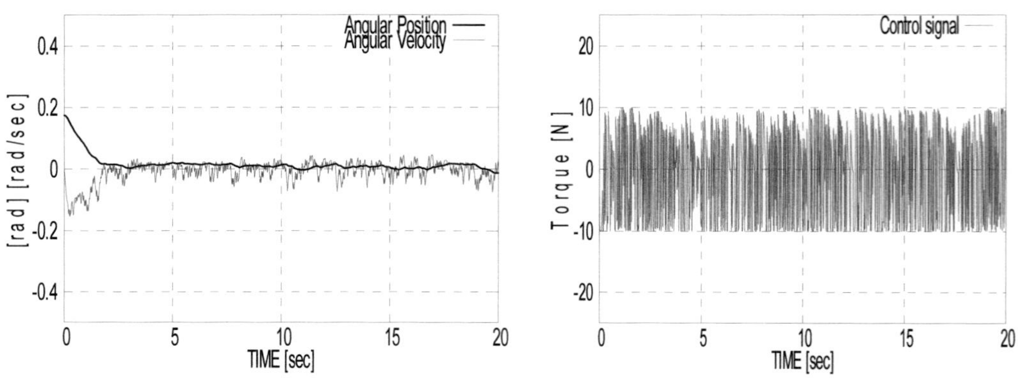

(a) $\theta, \dot{\theta}$ (b) Torque T_q

Fig. 13. Result of the conventional (actor-critic) method in the case of complete observation (θ,$\dot{\theta}$).

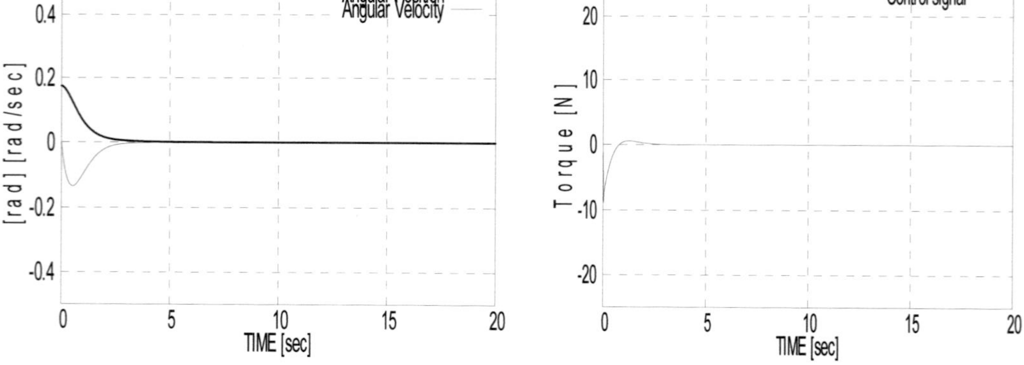

(a) $\theta, \dot{\theta}$ (b) Torque T_q

Fig. 14. Result of the conventional PID control method in the case of complete observation ($\theta, \dot{\theta}$).

f. Conventional PID control method

The control signal $u(t)$ in the PID control is

$$u(t) = -K_p e(t) - K_I \int_0^t e(t) \cdot dt - K_d \cdot \dot{e}(t) , \tag{26}$$

here, $K_p = 45, K_I = 1, K_d \cdot = 10$. Fig. 14 shows the results of the PID control.

5.4.3 Discussion

Table 3 shows the control performance, i.e. average error of $\theta, \dot{\theta}$, through the controlling time when final learning for all the methods the simulations have been done. Comparing the proposed method with the conventional actor-critic method, the proposed method is better than the conventional one. This means that the performance of the conventional actor-critic method hass been improved by making use of the concept of sliding mode control.

Kinds of Average error	Proposed method			Conventional method		
	Actor-Critic + SMC			SMC	PID	Actor-Critic
	Complete observation	Incomplete Observation (θ : available)		Complete observation		
		S.v.f.	Difference			
$\int \theta dt / t$	0.3002	0.6021	0.1893	0.2074	0.4350	0.8474
$\int \dot{\theta} dt / t$	0.4774	0.4734	0.4835	1.4768	0.4350	1.2396

Table 3. Control performance when final learning (S.v.f. : state variable filter, Difference: Difference method).

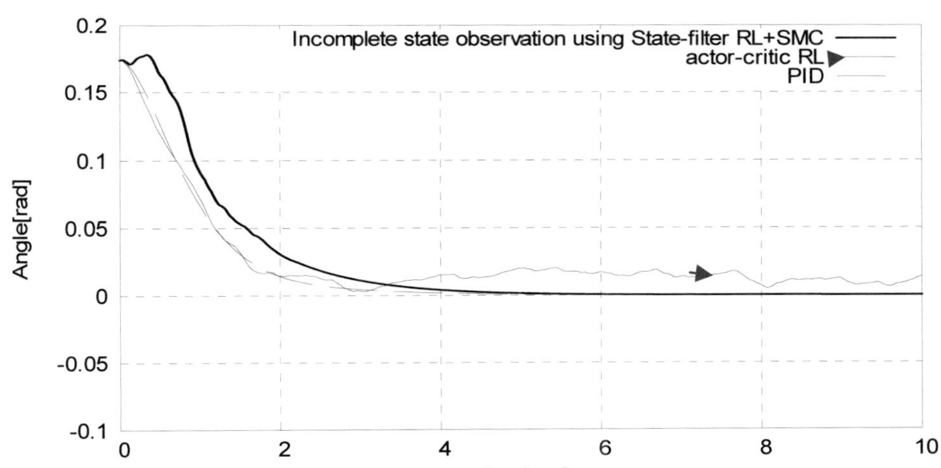

Fig. 15. Comparison of the porposed method with incomplete observation, the conventional actor-critic method and PID method for the angle, θ .

Figure 15 shows the comparison of the porposed method with incomplete observation, the conventional actor-critic method and PID method for the angle, θ. In this figure, the proposed method and PID method converge to zero smoothly, however the conventional actor-critic method does not converge. The comparison of the proposed method with PID control, the latter method converges quickly. These results are corresponding to Fig.16, i.e. the torque of the PID method converges first, the next one is the proposed method, and the conventional one does not converge.

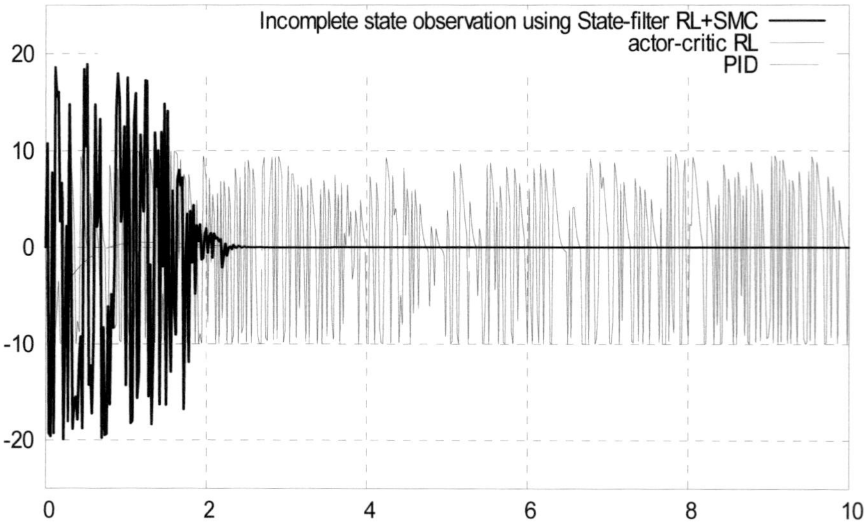

Fig. 16. Comparison of the porposed method with incomplete observation, the conventional actor-critic method and PID method for the Torque, T_q .

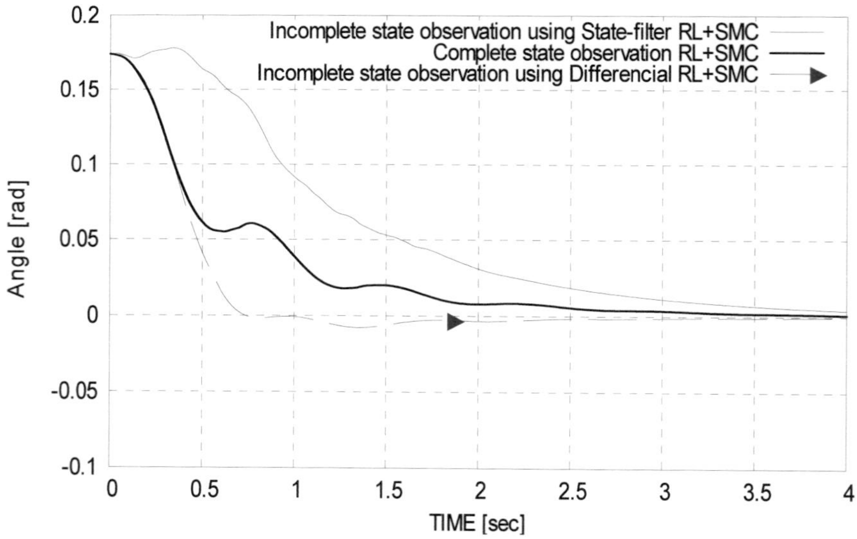

Fig. 17. The comparison of the porposed method among the case of the complete observation, the case with the state variable filter, and with the difference method for the angle, θ .

Fig. 17 shows the comparison of the porposed method among the case of the complete observation, the case with the state variable filter, and with the difference method for the angle, θ. Among them, the incomplete state observation with the difference method is best of three, especially, better than the complete observation. This reason can be explained by Fig. 18. That is, the value of s of the case of the difference method is bigger than that of the observation of the velocity angle, this causes that the input gain becomes bigger and the convergence speed has been accelerated.

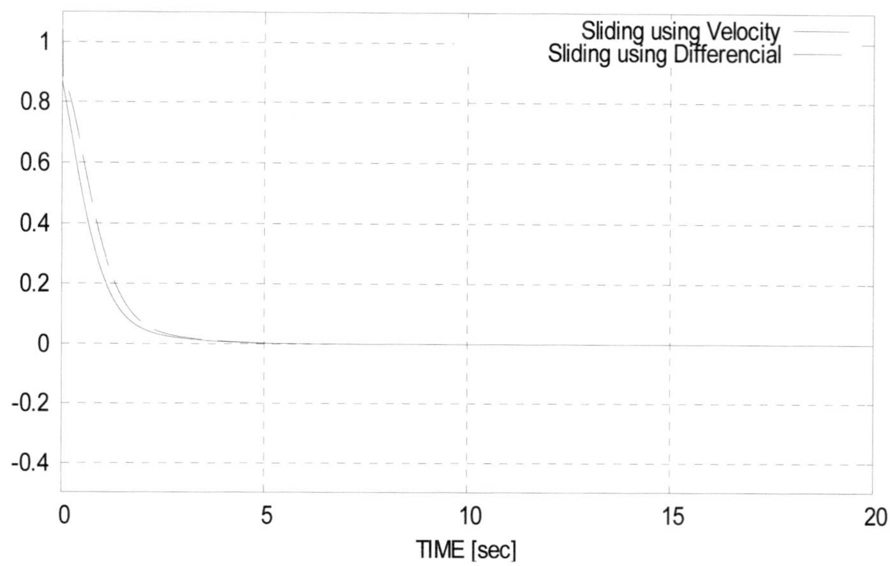

Fig. 18. The values of the sliding variable s for using the velocity and the difference between the angle and 1 sampling past angle.

5.4.4 Verification of the robust performance of each method

At first, as above mentioned, each controller was designed at $m = 1.0\,[\text{kg}]$ in Eq. (21). Next we examined the range of m in which the inverted pendulum control is success. Success is defined as the case that if $|\theta| \leq \pi / 45$ through the last 1[sec]. Results of the robust performance for change of m are shown in Table 4. As to upper/lower limit of m for success, the proposed method is better than the conventional actor-critic method not only for gradually changing m smaller from 1.0 to 0.001, but also for changing m bigger from 1.0 to 2.377. However, the best one is the conventional SMC method, next one is the PID control method.

6. Conclusion

A robust reinforcement learning method using the concept of the sliding mode control was mainly explained. Through the inverted pendulum control simulation, it was verified that the robust reinforcement learning method using the concept of the sliding mode control has good performance and robustness comparing with the conventional actor-critic method, because of the making use of the ability of the SMC method.

The way to improve the control performance and to clarify the stability of the proposed method theoretically has been remained.

	Proposed method		Conventional method		
	Actor-Critic + SMC		SMC	PID	Actor-Critic
	Complete observation	Incomplete observ. + s.v.f.*	Complete observation	Complete observation	Complete observation
m-max [kg]	2.081	2.377	11.788	4.806	1.668
m-min [kg]	0.001	0.001	0.002	0.003	0.021

*(s.v.f.: state variable filter)

Table 4. Robust control performance for change of m in Eq. (21).

7. Acknowledgement

This work has been supported by Japan JSPS-KAKENHI (No.20500207 and No.20500277).

8. Appendix

The structure of the critic of the conventional actor-critic control method is shown in Fig. 2. The number of nodes of the hidden layer of it is 15 as same as that of the proposed method. The prediction reward, $P(t)$, is as follow,

$$P(t) = \sum_{i=1}^{J} \omega_i^c \cdot \exp\left\{ \frac{-(\theta - c_{\theta i}^c)^2}{(\sigma_{\theta i}^c)^2} + \frac{-(\theta - c_{\dot{\theta} i}^c)^2}{(\sigma_{\dot{\theta} i}^c)^2} \right\}, \tag{A1}$$

The structure of actor is also similar with critic shown in Fig. 11. The output of the actor, $u'(t)$, and the control input, $u(t)$, are as follows, respectively,

$$u'(t) = \sum_{i=1}^{J} \omega_i^a \cdot \exp\left\{ \frac{-(\theta - c_{\theta i}^a)^2}{(\sigma_{\theta i}^a)^2} + \frac{-(\theta - c_{\dot{\theta} i}^a)^2}{(\sigma_{\dot{\theta} i}^a)^2} \right\}, \tag{A2}$$

$$u_1(t) = u_{\max} \cdot \frac{1 + \exp(-u'(t))}{1 - \exp(-u'(t))}, \tag{A3}$$

$$u(t) = u_1(t) + n(t). \tag{A4}$$

The center, $c_{\theta i}^c, c_{\dot{\theta} i}^c, c_{\theta i}^a, c_{\dot{\theta} i}^a$ of the critic and actor of the RBF network are set to equivalent distance in the range of $-3 < c < 3$. The variance, $\sigma_{\theta i}^c, \sigma_{\dot{\theta} i}^c, \sigma_{\theta i}^a, \sigma_{\dot{\theta} i}^a$ of the critic and actor of the RBF networks are set to be at equivalent distance in the range of $[0 < \sigma < 1]$. The values mentioned above, particularly, near the original are set to close. The reward $r(t)$ is set as Eq. (A5) in order it to maximize at $(\theta, \dot{\theta}) = (0,0)$,

$$r(t) = \exp\left(-\frac{(\theta_t)^2}{\sigma_1} - \frac{(\dot{\theta}_t)^2}{\sigma_2} \right). \quad \text{(A5)}$$

The learning of parameters of critic and actor are carried out through the back-propagation algorithm as Eqs. (A6)-(A7). $(\eta_c, \eta_a > 0)$

$$\Delta\omega_i^c = -\eta_c \cdot \frac{\partial \hat{r}_t^2}{\partial \omega_i^c}, \quad (i = 1, \cdots, J), \quad \text{(A6)}$$

$$\Delta\omega_j^a = \eta_a \cdot n_t \cdot \hat{r}_t \cdot \frac{\partial u_1(t)}{\partial \omega_j^a}, \quad (j = 1, \cdots, J). \quad \text{(A7)}$$

9. References

K. Doya. (2000). "Reinforcement learning in continuous time and space", *Neural Computation*, 12(1), pp.219-245

C.C. Hang. (1976)."On state variable filters for adaptive system design", *IEEEE Trans. Automatic Control*, Vo.21, No.6,874-876

C.C. Kung & T.H. Chen. (2005)."Observer-based indirect adaptive fuzzy sliding mode control with state variable filters for unknown nonlinear dynamical systems", *Fuzzy Sets and Systems*, Vol.155, pp.292-308

D.G. Luenberger. (1984)."Linear and Nonlinear Programming", *Addison-Wesley Publishing Company*, MA

J. Morimoto & K. Doya. (2005) "Robust Reinforcement Learning", *Neural Computation* 17,335-359

M. Obayashi & T. Kuremoto & K. Kobayashi. (2008)."A Self-Organized Fuzzy-Neuro Reinforcement Learning System for Continuous State Space for Autonomous Robots", *Proc. of International Conference on Computational Intelligence for Modeling, Control and Automation (CIMCA 2008)*, 552–559

M. Obayashi & N. Nakahara & T. Kuremoto & K. Kobayashi. (2009a). "A Robust Reinforcement Learning Using Concept of Slide Mode Control", The Journal of the Artificial Life and Robotics, Vol. 13, No. 2, pp.526-530

M. Obayashi & K. Yamada, T. & Kuremoto & K. Kobayashi. (2009b). "A Robust Reinforcement Learning Using Sliding Mode Control with State Variable Filters", *Proceedings of International Automatic Control Conference (CACS 2009),CDROM*

J.J.E. Slotine & W. Li. (1991). "Applied Nonlinear Control ", *Prentice-Hall*, Englewood Cliffs, NJ

R.S. Sutton & A.G. Barto. (1998). " Reinforcement Learning: An Introduction", *The MIT Press.*

W.Y. Wang & M.L. Chan & C.C. James & T.T. Lee. (2002). " H_∞ Tracking-Based Sliding Mode Control for Uncertain Nonlinear Systems via an Adaptive Fuzzy-Neural Approach", *IEEE Trans. on Systems, Man, and Cybernetics,* Vol.32, No.4, August, pp.483-492

X. S. Wang & Y. H. Cheng & J. Q. Yi. (2007). "A fuzzy Actor-Critic reinforcement learning network", *Information Sciences,* 177, pp.3764-3781

C. Watkins & P. Dayan. (1992)."Q-learning," *Machine learning,* Vol.8, pp.279-292

K. Zhau & J.C.Doyle & K.Glover. (1996). "Robust optimal control", Englewood Cliffs NJ, Prentice Hall

Part 4

Selected Trends in Robust Control Theory

10

Robust Controller Design: New Approaches in the Time and the Frequency Domains

Vojtech Veselý, Danica Rosinová and Alena Kozáková
Slovak University of Technology
Slovak Republic

1. Introduction

Robust stability and robust control belong to fundamental problems in control theory and practice; various approaches have been proposed to cope with uncertainties that always appear in real plants as a result of identification / modelling errors, e.g. due to linearization and approximation, etc. A control system is robust if it is insensitive to differences between the actual plant and its model used to design the controller. To deal with an uncertain plant a suitable uncertainty model is to be selected and instead of a single model, behaviour of a whole class of models is to be considered. Robust control theory provides analysis and design approaches based upon an incomplete description of the controlled process applicable in the areas of non-linear and time-varying processes, including multi input – multi output (MIMO) dynamic systems.

MIMO systems usually arise as interconnection of a finite number of subsystems, and in general, multivariable centralized controllers are used to control them. However, practical reasons often make restrictions on controller structure necessary or reasonable. In an extreme case, the controller is split into several local feedbacks and becomes a decentralized controller. Compared to centralized full-controller systems such a control structure brings about certain performance deterioration; however, this drawback is weighted against important benefits, e.g. hardware, operation and design simplicity, and reliability improvement. Robust approach is one of useful ways to address the decentralized control problem (Boyd et al., 1994; Henrion et al., 2002; de Oliveira et al., 1999; Gyurkovics & Takacs, 2000; Ming Ge et al., 2002; Skogestad & Postlethwaite, 2005; Kozáková and Veselý, 2008; Kozáková et al., 2009a).

In this chapter two robust controller design approaches are presented: in the time domain the approach based on Linear (Bilinear) matrix inequality (LMI, BMI), and in the frequency domain the recently developed Equivalent Subsystem Method (ESM) (Kozáková et al., 2009b). As proportional-integral-derivative (PID) controllers are the most widely used in industrial control systems, this chapter focuses on the time- and frequency domain PID controller design techniques resulting from both approaches.

The development of Linear Matrix Inequality (LMI) computational techniques has provided an efficient tool to solve a large set of convex problems in polynomial time (e.g. Boyd et al., 1994). Significant effort has been therefore made to formulate crucial control problems in

algebraic way (e.g. Skelton et al., 1998), so that the numerical LMI solution can be employed. This approach is advantageously used in solving control problems for linear systems with convex (affine or polytopic) uncertainty domain. However, many important problems in linear control design, such as decentralized control, simultaneous static output feedback (SOF) or more generally - structured linear control problems have been proven as NP hard (Blondel & Tsitsiklis, 1997). Though there exist solvers for bilinear matrix inequalities (BMI), suitable to solve e.g. SOF, they are numerically demanding and restricted to problems of small dimensions. Intensive research has been devoted to overcome nonconvexity and transform the nonconvex or NP-hard problem into convex optimisation problem in LMI framework. Various techniques have been developed using inner or outer convex approximations of the respective nonconvex domains. The common tool in both inner and outer approximation is the use of linearization or convexification. In (Han & Skelton, 2003; de Oliveira et al., 1999), the general convexifying algorithm for the nonconvex function together with potential convexifying functions for both continuous and discrete-time case have been proposed. Linearization approach for continuous and discrete-time system design was independently used in (Rosinová & Veselý, 2003; Veselý, 2003).
When designing a (PID) controller, the derivative part of the controller causes difficulties when uncertainties are considered. In multivariable PID control schemes using LMI developed recently (Zheng et al., 2002), the incorporation of the derivative part requires inversion of the respective matrix, which does not allow including uncertainties. Another way to cope with the derivative part is to assume the special case when output and its derivative are state variables, robust PID controller for first and second order SISO systems are proposed for this case in (Ming Ge et al., 2002).
In Section 2, the state space approach to the design of (decentralized or multi-loop) PID robust controllers is proposed for linear uncertain system with guaranteed cost using a new quadratic cost function. The major contribution is in considering the derivative part in robust control framework. The resulting matrix inequality can be solved either using BMI solver, or using linearization approach and following LMI solution.
The frequency domain design techniques have probably been the most popular among the practitioners due to their insightfulness and link to the classical control theory. In combination with the robust approach they provide a powerful engineering tool for control system analysis and synthesis. An important field of their implementation is control of MIMO systems, in particular the decentralized control (DC) due to simplicity of hardware and information processing algorithms. The DC design proceeds in two main steps: 1) selection of a suitable control configuration (pairing inputs with outputs); 2) design of local controllers for individual subsystems. There are two main approaches applicable in Step 2: sequential (dependent) design, and independent design. When using *sequential design* local controllers are designed sequentially as a series controller, hence information about "lower level" controllers is directly used as more loops are closed. Main drawbacks are lack of failure tolerance when lower level controllers fail, strong dependence of performance on the loop closing order, and a trial-and-error design process.
According to the *independent design*, local controllers are designed to provide stability of each individual loop without considering interactions with other subsystems. The effect of interactions is assessed and transformed into bounds for individual designs to guarantee stability and a desired performance of the full system. Main advantages are direct design of local controllers with no need for trial and error; the limitation consists in that information

about controllers in other loops is not exploited, therefore obtained stability and performance conditions are only sufficient and thus potentially conservative.
Section 3 presents a frequency domain robust decentralized controller design technique applicable for uncertain systems described by a set of transfer function matrices. The core of the technique is the Equivalent Subsystems Method - a Nyquist-based DC design method guaranteeing performance of the full system (Kozáková et al., 2009a; 2009b). To guarantee specified performance (including stability), the effect of interactions is assessed using a selected characteristic locus of the matrix of interactions further used to reshape frequency responses of decoupled subsystems thus generating so-called equivalent subsystems. Local controllers of equivalent subsystems independently tuned to guarantee specified performance measure value in each of them constitute the decentralized (diagonal) controller; when applied to real subsystems, the resulting controller guarantees the same performance measure value for the full system. To guarantee robust stability over the specified operating range of the plant, the *M-Δ* stability conditions are used (Skogestad & Postlethwaite, 2005; Kozáková et al., 2009a, 2009b). Two versions of the robust DC design methodology have been developed: a the two-stage version (Kozáková & Veselý, 2009; Kozáková et al. 2009a), where robust stability is achieved by additional redesign of the DC parameters; in the direct version, robust stability conditions are integrated in the design of local controllers for equivalent subsystems. Unlike standard robust approaches, the proposed technique allows considering full nominal model thus reducing conservatism of robust stability conditions. Further conservatism relaxing is achieved if the additive affine type uncertainty description and the related $M_{af} - Q$ stability conditions are used (Kozáková & Veselý, 2007; 2008).
In the sequel, $X > 0$ denotes positive definite matrix; * in matrices denotes the respective transposed term to make the matrix symmetric; I denotes identity matrix and 0 denotes zero matrix of the respective dimensions.

2. Robust PID controller design in the time domain

In this section the PID control problem formulation via LMI is presented that is appropriate for polytopic uncertain systems. Robust PID control scheme is then proposed for structured control gain matrix, thus enabling decentralized PID control design.

2.1 Problem formulation and preliminaries

Consider the class of linear affine uncertain time-invariant systems described as:

$$\begin{aligned} \delta x(t) &= (A + \delta A)x(t) + (B + \delta B)u(t) \\ y(t) &= Cx(t) \end{aligned} \tag{1}$$

where

$$\begin{aligned} \delta x(t) &= \dot{x}(t) && \text{for continuous-time system} \\ \delta x(t) &= x(t+1) && \text{for discrete-time system} \end{aligned}$$

$x(t) \in R^n, u(t) \in R^m, y(t) \in R^l$ are state, control and output vectors respectively; A, B, C are known constant matrices of the respective dimensions corresponding to the nominal system, $\delta A, \delta B$ are matrices of uncertainties of the respective dimensions. The affine uncertainties are assumed

$$\delta A(t)=\sum_{j=1}^{p}\gamma_j\tilde{A}_j,\ \delta B(t)=\sum_{j=1}^{p}\gamma_j\tilde{B}_j \tag{2}$$

where $\underline{\gamma}_j \le \gamma_j \le \overline{\gamma}_j$ are unknown uncertainty parameters; $\tilde{A}_j,\tilde{B}_j, j=1,2,...,p$ are constant matrices of uncertainties of the respective dimensions and structure. The uncertainty domain for a system described in (1), (2) can be equivalently described by a polytopic model given by its vertices

$$\{(A_1,B_1,C),(A_2,B_2,C),...,(A_N,B_N,C)\},\ N=2^p \tag{3}$$

The (decentralized) feedback control law is considered in the form

$$u(t)=FCx(t) \tag{4}$$

where F is an output feedback gain matrix. The uncertain closed-loop polytopic system is then

$$\delta x(t)=A_C(\alpha)x(t) \tag{5}$$

where

$$\begin{aligned}&A_C(\alpha)\in\left\{\sum_{i=1}^{N}\alpha_i A_{Ci},\quad \sum_{i=1}^{N}\alpha_i=1,\quad \alpha_i\ge 0\right\},\\ &A_{Ci}=A_i+B_iFC.\end{aligned} \tag{6}$$

To assess the performance, a quadratic cost function known from LQ theory is frequently used. In practice, the response rate or overshoot are often limited, therefore we include the additional derivative term for state variable into the cost function to damp the oscillations and limit the response rate.

$$J_c=\int_0^{\infty}[x(t)^TQx(t)+u(t)^TRu(t)+\delta x(t)^TS\delta x(t)]dt \text{ for a continuous-time and} \tag{7}$$

$$J_d=\sum_{k=0}^{\infty}[x(t)^TQx(t)+u(t)^TRu(t)+\delta x(t)^TS\delta x(t)] \text{ for a discrete-time system} \tag{8}$$

where $Q,S\in R^{n\times n}, R\in R^{m\times m}$ are symmetric positive definite matrices. The concept of *guaranteed cost control* is used in a standard way: let there exist a feedback gain matrix F_0 and a constant J_0 such that

$$J\le J_0 \tag{9}$$

holds for the closed loop system (5), (6). Then the respective control (4) is called the guaranteed cost control and the value of J_0 is the guaranteed cost.
The main aim of Section 2 of this chapter is to solve the next problem.
Problem 2.1
Find a (decentralized) robust PID control design algorithm that stabilizes the uncertain system (1) with guaranteed cost with respect to the cost function (7) or (8).

We start with basic notions concerning Lyapunov stability and convexifying functions. In the following we use D-stability concept (Henrion et al., 2002) to receive the respective stability conditions in more general form.

Definition 2.1 (D-stability)

Consider the D-domain in the complex plain defined as

$$D = \{s \text{ is complex number} : \begin{bmatrix} 1 \\ s \end{bmatrix}^* \begin{bmatrix} r_{11} & r_{12} \\ r_{12}^* & r_{22} \end{bmatrix} \begin{bmatrix} 1 \\ s \end{bmatrix} < 0\} .$$

The considered linear system (1) is D-stable if and only if all its poles lie in the D-domain. (For simplicity, we use in Def. 2.1 scalar values of parameters r_{ij}, in general the stability domain can be defined using matrix values of parameters r_{ij} with the respective dimensions.) The standard choice of r_{ij} is $r_{11} = 0$, $r_{12} = 1$, $r_{22} = 0$ for a continuous-time system; $r_{11} = -1$, $r_{12} = 0$, $r_{22} = 1$ for a discrete-time system, corresponding to open left half plane and unit circle respectively.

The *quadratic D-stability* of uncertain system is equivalent to the existence of one Lyapunov function for the whole uncertainty set.

Definition 2.2 (Quadratic D-stability)

The uncertain system (5) is quadratically D-stable if and only if there exists a symmetric positive definite matrix P such that

$$r_{12}PA_C(\alpha) + r_{12}^* A_C^T(\alpha)P + r_{11}P + r_{22}A_C^T(\alpha)PA_C(\alpha) < 0 \tag{10}$$

To obtain less conservative results than using quadratic stability, a robust stability notion is considered based on the parameter dependent Lyapunov function (PDLF) defined as

$$P(\alpha) = \sum_{i=1}^{N} \alpha_i P_i \text{ where } P_i = P_i^T > 0 \tag{11}$$

Definition 2.3 (deOliveira et al., 1999)

System (5) is *robustly D-stable* in the convex uncertainty domain (6) with parameter-dependent Lyapunov function (11) if and only if there exists a matrix $P(\alpha) = P(\alpha)^T > 0$ such that

$$r_{12}P(\alpha)A_C(\alpha) + r_{12}^* A_C^T(\alpha)P(\alpha) + r_{11}P(\alpha) + r_{22}A_C^T(\alpha)P(\alpha)A_C(\alpha) < 0 \tag{12}$$

for all α such that $A_C(\alpha)$ is given by (6).

Now recall the sufficient robust *D*-stability condition proposed in (Peaucelle et al., 2000), proven as not too conservative (Grman et al., 2005).

Lemma 2.1

If there exist matrices $E \in R^{nxn}, G \in R^{nxn}$ and N symmetric positive definite matrices $P_i \in R^{nxn}$ such that for all i = 1,…, N:

$$\begin{bmatrix} r_{11}P_i + A_{Ci}^T E^T + EA_{Ci} & r_{12}P_i - E + A_{Ci}^T G \\ r_{12}^* P_i - E^T + G^T A_{Ci} & r_{22}P_i - (G + G^T) \end{bmatrix} < 0 \tag{13}$$

then uncertain system (5) is robustly D-stable.

Note that matrices E and G are not restricted to any special form; they were included to relax the conservatism of the sufficient condition. To transform nonconvex problem of structured control (e.g. output feedback, or decentralized control) into convex form, the convexifying (linearizing) function can be used (Han&Skelton, 2003; deOliveira et al., 2000; Rosinová&Veselý, 2003; Veselý, 2003). The respective potential convexifying function for X^{-1} and XWX has been proposed in the linearizing form:

- the linearization of $X^{-1} \in R^{nxn}$ about the value $X_k > 0$ is

$$\Phi(X^{-1}, X_k) = X_k^{-1} - X_k^{-1}(X - X_k)X_k^{-1} \tag{14}$$

- the linearization of $XWX \in R^{nxn}$ about X_k is

$$\Psi(XWX, X_k) = -X_k W X_k + XWX_k + X_k WX \tag{15}$$

Both functions defined in (14) and (15) meet one of the basic requirements on convexifying function: to be equal to the original nonconvex term if and only if $X_k = X$. However, the question how to choose the appropriate nice convexifying function remains still open.

2.2 Robust optimal controller design

In this section the new design algorithm for optimal control with guaranteed cost is developed using parameter dependent Lyapunov function and convexifying approach employing iterative procedure. The proposed control design approach is based on sufficient stability condition from Lemma 2.1. The next theorem provides the new form of robust stability condition for linear uncertain system with guaranteed cost.

Theorem 2.1

Consider uncertain linear system (1), (2) with static output feedback (4) and cost function (7) or (8). The following statements are equivalent:

i. Closed loop uncertain system (5) is robustly D-stable with PDLF (11) and guaranteed cost with respect to cost function (7) or (8): $J \le J_0 = x^T(0)P(\alpha)x(0)$.

ii. There exist matrices $P(\alpha) > 0$ defined by (11) such that

$$\begin{aligned} & r_{12}P(\alpha)A_C(\alpha) + r_{12}^{*}A_C^T(\alpha)P(\alpha) + r_{11}P(\alpha) + r_{22}A_C^T(\alpha)P(\alpha)A_C(\alpha) + \\ & +Q + C^T F^T RFC + A_C^T(\alpha)SA_C(\alpha) < 0 \end{aligned} \tag{16}$$

iii. There exist matrices $P(\alpha) > 0$ defined by (11) and matrices H, G and F of the respective dimensions such that

$$\begin{bmatrix} r_{11}P(\alpha) + A_{Ci}^T(\alpha)H^T + HA_{Ci}(\alpha) + Q + C^T F^T RFC & * \\ r_{12}^{*}P(\alpha) - H^T + G^T A_{Ci}(\alpha) & r_{22}P(\alpha) - (G + G^T) + S \end{bmatrix} < 0 \tag{17}$$

$A_{Ci} = (A_i + B_i FC)$ denotes the i-th closed loop system vertex. Matrix F is the guaranteed cost control gain for the uncertain system (5), (6).

Proof. For brevity the detail steps of the proof are omitted where standard tools are applied. (i) ⇔ (ii): the proof is analogous to that in (Rosinová, Veselý, Kučera, 2003). The (ii) ⇒(i) is shown by taking $V(t) = x(t)P(\alpha)x(t)$ as a candidate Lyapunov function for (5) and writing $\delta V(t)$, where

$$\delta V(t) = \dot{V}(t) \quad \text{for continuous-time system}$$
$$\delta V(t) = V(t+1) - V(t) \quad \text{for discrete-time system}$$

$$\delta V(t) = r_{12}^{*}\delta x(t)^T P(\alpha)x(t) + r_{12}x(t)^T P(\alpha)\delta x(t) + r_{11}x(t)^T P(\alpha)x(t) + r_{22}\delta x(t)^T P(\alpha)\delta x(t) \quad (18)$$

Substituting for δx from (5) to (18) and comparing with (16) provides D-stability of the considered system when the latter inequality holds. The guaranteed cost can be proved by summing or integrating both sides of the following inequality for t from 0 to ∞:

$$\delta V(t) < -x(t)^T [Q + C^T F^T RFC + A_C^T(\alpha) S A_C(\alpha)] x(t)$$

The (i) $\Rightarrow$(ii) can be proved by contradiction.
(ii) $\Leftrightarrow$ (iii): The proof follows the same steps to the proof of Lemma 2.1: (iii) $\Rightarrow$(ii) is proved in standard way multiplying both sides of (17) by the full rank matrix (equivalent transformation):

$$\begin{bmatrix} I & A_C^T(\alpha) \end{bmatrix} \{l.h.s.(17)\} \begin{bmatrix} I \\ A_C(\alpha) \end{bmatrix} < 0 \cdot$$

(ii) $\Rightarrow$(iii) follows from applying a Schur complement to (16) rewritten as

$$r_{12}P(\alpha)A_C(\alpha) + r_{12}^{*}A_C^T(\alpha)P(\alpha) + Q + C^T F^T RFC + r_{11}P(\alpha) + A_C^T(\alpha)[r_{22}P(\alpha) + S]A_C(\alpha) < 0$$

Therefore $\begin{bmatrix} X_{11} & X_{12} \\ X_{12}^T & X_{22} \end{bmatrix} < 0$ where

$$X_{11} = r_{11}P(\alpha) + r_{12}P(\alpha)A_C(\alpha) + r_{12}^{*}A_C^T(\alpha)P(\alpha) + Q + C^T F^T RFC$$
$$X_{12} = A_C^T(\alpha)[r_{22}P(\alpha) + S]$$
$$X_{22} = -[r_{22}P(\alpha) + S]$$

which for $H = r_{12}P(\alpha)$, $G = [r_{22}P(\alpha) + S]$ gives (17).
The proposed guaranteed cost control design is based on the robust stability condition (17). Since the matrix inequality (17) is not LMI when both $P(\alpha)$ and F are to be found, we use the inner approximation for the continuous time system applying linearization formula (15) together with using the respective quadratic forms to obtain LMI formulation, which is then solved by iterative procedure.

2.3 PID robust controller design for continuous-time systems

Control algorithm for PID is considered as

$$u(t) = K_P y(t) + K_I \int_0^t y(t)dt + F_d C_d \dot{x}(t) \quad (19)$$

The proportional and integral term can be included into the state vector in the common way defining the auxiliary state $z = \int_0^t y(t)$, i.e. $\dot{z}(t) = y(t) = Cx(t)$. Then the closed-loop system for PI part of the controller is

$$\dot{x}_n = \begin{bmatrix} \dot{x} \\ \dot{z} \end{bmatrix} = \begin{bmatrix} A+\delta A & 0 \\ C & 0 \end{bmatrix} \begin{bmatrix} x \\ z \end{bmatrix} + \begin{bmatrix} B+\delta B \\ 0 \end{bmatrix} u(t) \text{ and } u(t) = FCx(t) + F_d C_d \dot{x}(t) \tag{20}$$

where $FCx(t)$ and $F_d C_d \dot{x}(t)$ correspond respectively to the PI and D term of PID controller. The resulting closed loop system with PID controller (19) is then

$$\dot{x}_n(t) = A_C(\alpha)x_n(t) + B(\alpha)\begin{bmatrix} F_d C_d & 0 \end{bmatrix}\dot{x}_n(t) \tag{21}$$

where the PI controller term is included in $A_C(\alpha)$. (For brevity we omit the argument *t*.) To simplify the denotation, in the following we consider PD controller (which is equivalent to the assumption, that the I term of PID controller has been already included into the system dynamics in the above outlined way) and the closed loop is described by

$$\dot{x}(t) = A_C(\alpha)x(t) + B(\alpha)F_d C_d \dot{x}(t) \tag{22}$$

Let us consider the following performance index

$$J_s = \int_0^{\infty} \begin{bmatrix} x & \dot{x} \end{bmatrix}^T \begin{bmatrix} Q + C^T F^T RFC & 0 \\ 0 & S \end{bmatrix} \begin{bmatrix} x \\ \dot{x} \end{bmatrix} dt \tag{23}$$

which formally corresponds to (7). Then for Lyapunov function (11) we have the necessary and sufficient condition for robust stability with guaranteed cost in the form (16), which for continuous time system can be rewritten as:

$$\begin{bmatrix} x & \dot{x} \end{bmatrix}^T \begin{bmatrix} Q + C^T F^T RFC & P(\alpha) \\ P(\alpha) & S \end{bmatrix} \begin{bmatrix} x \\ \dot{x} \end{bmatrix} < 0 \tag{24}$$

The main result on robust PID control stabilization is summarized in the next theorem.

Theorem 2.2

Consider a continuous uncertain linear system (1), (2) with PID controller (19) and cost function (23). The following statements are equivalent:

i Closed loop system (21) is robustly D-stable with PDLF (11) and guaranteed cost with respect to cost function (23): $J \le J_0 = x^T(0)P(\alpha)x(0)$.

ii There exist matrices $P(\alpha) > 0$ defined by (11), and *H*, *G*, *F* and F_d of the respective dimensions such that

$$\begin{bmatrix} A_{Ci}^T H^T + HA_{Ci} + Q + C^T F^T RFC & * \\ P_i - M_{di}^T H + G^T A_{Ci} & -M_{di}^T G - G^T M_{di} + S \end{bmatrix} < 0 \tag{25}$$

$A_{Ci} = (A_i + B_i FC)$ denotes the i-th closed loop system vertex, M_{di} includes the derivative part of the PID controller: $M_{di} = I - B_i F_d C_d$.

Proof. Owing to (22) for any matrices *H* and *G*:

$$\begin{aligned} &\left(-x^T H - \dot{x}^T G^T\right)(\dot{x} - A_C(\alpha)x - B(\alpha)F_d C_d \dot{x}) + \\ &+(\dot{x} - A_C(\alpha)x - B(\alpha)F_d C_d \dot{x})^T \left(H^T x - G\dot{x}\right) = 0 \end{aligned} \tag{26}$$

Summing up the l.h.s of (26) and (24) and taking into consideration linearity w.r.t. α we get condition (25).
Theorem 2.2 provides the robust stability condition for the linear uncertain system with PID controller. Notice that the derivative term does not appear in the matrix inversion and allows including the uncertainty in control matrix B into the stability condition.
Considering PID control design, there are unknown matrices H, G, F and F_d to be solved from (25). (Recall that $A_{Ci} = (A_i + B_i FC)$, $M_{di} = I - B_i F_d C_d$.) Then, inequality (25) is bilinear with respect to unknown matrices and can be solved either by BMI solver, or by linearization approach using (15) to cope with the respective unknown matrices products. For the latter case the PID iterative control design algorithm based on LMI (4x4 matrix) has been proposed. The resulting closed loop system with PD controller is

$$\dot{x}(t) = \left(I - B_i F_d C_d\right)^{-1}\left(A_i + B_i FC\right)x(t), \quad i=1,\dots,N \tag{27}$$

The extension of the proposed algorithm to decentralized control design is straightforward since the respective F and F_d matrices are assumed as being of the prescribed structure, therefore it is enough to prescribe the decentralized structure for both matrices.

2.4 PID robust controller design for discrete-time systems

Control algorithm for discrete-time PID (often denoted as PSD controller) is considered as

$$u(k) = k_P e(k) + k_I \sum_{i=0}^{k} e(k) + k_D[e(k) - e(k-1)] \tag{28}$$

control error $e(k) = w - y(k)$; discrete time being denoted for clarity as k instead of t. PSD description in state space:

$$\begin{aligned} z(k+1) &= \begin{bmatrix} 0 & 1 \\ 0 & 1 \end{bmatrix} z(k) + \begin{bmatrix} 0 \\ 1 \end{bmatrix} e(k) = A_R z(k) + B_R e(k) \\ u(k) &= \begin{bmatrix} k_D & k_I - k_D \end{bmatrix} z(k) + (k_P + k_I + k_D) e(k) \end{aligned} \tag{29}$$

Combining (1) for $t \approx k$ and (29) the augmented closed loop system is received as

$$\begin{bmatrix} x(k+1) \\ z(k+1) \end{bmatrix} = \begin{bmatrix} A + \delta A & 0 \\ -B_R C & A_R \end{bmatrix} \begin{bmatrix} x(k) \\ z(k) \end{bmatrix} + \begin{bmatrix} B + \delta B \\ 0 \end{bmatrix} \begin{bmatrix} -K_2 & K_1 \end{bmatrix} \begin{bmatrix} C & 0 \\ 0 & I \end{bmatrix} \begin{bmatrix} x(k) \\ z(k) \end{bmatrix} \tag{30}$$

where $K_2 = (k_P + k_I + k_D)$, $K_1 = \begin{bmatrix} k_D & k_I - k_D \end{bmatrix}$.
Note that there is a significant difference between PID (19) and PSD (28) control design problem: for continuous time PID structure results in closed loop system that is not strictly proper which complicates the controller design, while for discrete time PSD structure, the control design is formulated as static output feedback (SOF) problem therefore the respective techniques to SOF design can be applied.
In this section an algorithm for PSD controller design is proposed. Theorem 2.1 provides the robust stability condition for the linear time varying uncertain system, where a constrained control structure can be assumed: considering $A_{Ci} = (A_i + B_i FC)$ we have SOF problem formulation which is also the case of discrete time PSD control structure for

$F=[-(k_P+k_I+k_D)\quad k_D\quad k_I-k_D]$ (see (30)); (taking block diagonal structure of feedback matrix gain F provides decentralized controller). Inequality (17) is LMI for stability analysis for unknown H, G and P_i, however considering control design, having one more unknown matrix F in $A_{Ci}=(A_i+B_iFC)$, the inequality (17) is no more LMI. Then, to cope with the respective unknown matrix products the inner approximation approach can be used, when the resulting LMI is sufficient for the original one to hold.

The next robust output feedback design method is based on (17) using additional constraint on output feedback matrix and the state feedback control design approach proposed respectively in (Crusius and Trofino, 1999; deOliveira et al., 1999). For stabilizing PSD control design (without considering cost function) we have the following algorithm (taking H=0, Q=0, R=0, S=0).

PSD controller design algorithm

Solve the following LMI for unknown matrices F, M, G and P_i of appropriate dimensions, the P_i being symmetric, positive definite, M, G being any matrices with corresponding dimensions:

$$\begin{bmatrix} -P_i & A_iG+B_iKC \\ G^TA_i^T+C^TK^TB_i^T & -G-G^T+P_i+S \end{bmatrix}<0 \tag{31}$$

$$\begin{aligned} &P_i>0, \quad i=1,\dots,N \\ &MC=CG \end{aligned} \tag{32}$$

Compute the corresponding output feedback gain matrix

$$F=KM^{-1} \tag{33}$$

where $F=[-(k_{Pi}+k_{Ii}+k_{Di})\quad k_{Di}\quad k_{Ii}-k_{Di}]$

The algorithm above is quite simple and often provides reasonable results.

2.5 Examples

In this subsection the major contribution of the proposed approach: design of robust controller with derivative feedback is illustrated on the examples. The results obtained using the proposed new iterative algorithm based on (25) to design the PD controller are provided and discussed. The impact of matrix S choice is studied as well. We consider affine models of uncertain system (1), (2) with symmetric uncertainty domain:

$$\underline{\varepsilon}_j=-q,\ \overline{\varepsilon}_j=q$$

Example 2.1

Consider the uncertain system (1), (2) where

$$A=\begin{bmatrix} -4.365 & -0.6723 & -0.3363 \\ 7.0880 & -6.5570 & -4.6010 \\ -2.4100 & 7.5840 & -14.3100 \end{bmatrix} \quad B=\begin{bmatrix} 2.3740 & 0.7485 \\ 1.3660 & 3.4440 \\ 0.9461 & -9.6190 \end{bmatrix} \quad C=C_d=\begin{bmatrix} 0 & 1 & 0 \\ 0 & 0 & 1 \end{bmatrix}$$

uncertainty parameter q=1; uncertainty matrices

$$\bar{A}_1 = \begin{bmatrix} -0.5608 & 0.8553 & 0.5892 \\ 0.6698 & -1.3750 & -0.9909 \\ 3.1917 & 1.7971 & -2.5887 \end{bmatrix} \quad \bar{B}_1 = \begin{bmatrix} 2.3740 & 0.7485 \\ 1.3660 & 3.4440 \\ 0.9461 & -9.6190 \end{bmatrix}$$

$$\bar{A}_2 = \begin{bmatrix} 0.6698 & -1.3750 & -0.9909 \\ -2.8963 & -1.5292 & 10.5160 \\ -3.5777 & 2.8389 & 1.9087 \end{bmatrix} \quad \bar{B}_2 = \begin{bmatrix} 0.1562 & 0.1306 \\ -0.4958 & 4.0379 \\ -0.0306 & 0.8947 \end{bmatrix}$$

The uncertain system can be described by 4 vertices; corresponding maximal eigenvalues in the vertices of open loop system are respectively: -4.0896 ± 2.1956i; -3.9243; 1.5014; -4.9595. Notice, that the open loop uncertain system is unstable (positive eigenvalue in the third vertex). The stabilizing optimal PD controller has been designed by solving matrix inequality (25). Optimality is considered in the sense of guaranteed cost w.r.t. cost function (23) with matrices $R = I_{2\times 2}, Q = 0.001 * I_{3\times 3}$. The results summarized in Tab.2.1 indicate the differences between results obtained for different choice of cost matrix S respective to a derivative of x.

S	Controller matrices F (proportional part) F$_d$ (derivative part)	Max eigenvalues in vertices
1e-6 *I	$F = \begin{bmatrix} -1.0567 & -0.5643 \\ -2.1825 & -1.4969 \end{bmatrix}$	-4.8644 **-2.4074**
	$F_d = \begin{bmatrix} -0.3126 & -0.2243 \\ -0.0967 & 0.0330 \end{bmatrix}$	-3.8368 ± 1.1165 i -4.7436
0.1 *I	$F = \begin{bmatrix} -1.0724 & -0.5818 \\ -2.1941 & -1.4642 \end{bmatrix}$	-4.9546 **-2.2211**
	$F_d = \begin{bmatrix} -0.3227 & -0.2186 \\ -0.0969 & 0.0340 \end{bmatrix}$	-3.7823 ± 1.4723 i -4.7751

Table 2.1 PD controllers from Example 2.1.

Example 2.2
Consider the uncertain system (1), (2) where

$$A = \begin{bmatrix} -2.9800 & 0.9300 & 0 & -0.0340 \\ -0.9900 & -0.2100 & 0.0350 & -0.0011 \\ 0 & 0 & 0 & 1 \\ 0.3900 & -5.5550 & 0 & -1.8900 \end{bmatrix} \quad B = \begin{bmatrix} -0.0320 \\ 0 \\ 0 \\ -1.6000 \end{bmatrix} \quad C = \begin{bmatrix} 0 & 0 & 1 & 0 \\ 0 & 0 & 0 & 1 \end{bmatrix}$$

$$A = \begin{bmatrix} 0 & 1.5 & 0 & 0 \\ 0 & 0 & 0 & 0 \\ 0 & 0 & 0 & 0 \\ 0 & 0 & 0 & 0 \end{bmatrix} \quad B = \begin{bmatrix} 0 \\ 0 \\ 0 \\ 0 \end{bmatrix}.$$

The results are summarized in Tab.2.2 for $R = 1, Q = 0.0005 * I_{4\times4}$ for various values of cost function matrix S. As indicated in Tab.2.2, increasing values of S slow down the response as assumed (max. eigenvalue of closed loop system is shifted to zero).

S	q_{max}	Max. eigenvalue of closed loop system
1e-8 *I	1.1	-0.1890
0.1 *I	1.1	-0.1101
0.2 *I	1.1	-0.0863
0.29 *I	1.02	-0.0590

Table 2.2 Comparison of closed loop eigenvalues (Example 2.2) for various S.

3. Robust PID controller design in the frequency domain

In this section an original frequency domain robust control design methodology is presented applicable for uncertain systems described by a set of transfer function matrices. A two-stage as well as a direct design procedures were developed, both being based on the Equivalent Subsystems Method - a Nyquist-based decentralized controller design method for stability and guaranteed performance (Kozáková et al., 2009a;2009b), and stability conditions for the *M-Δ* structure (Skogestad & Postlethwaite, 2005; Kozáková et al., 2009a, 2009b). Using the additive affine type uncertainty and related M_{af}-Q structure stability conditions, it is possible to relax conservatism of the *M-Δ* stability conditions (Kozáková & Veselý, 2007).

3.1 Preliminaries and problem formulation

Consider a MIMO system described by a transfer function matrix $G(s) \in R^{m\times m}$, and a controller $R(s) \in R^{m\times m}$ in the standard feedback configuration (Fig. 1); *w, u, y, e, d* are respectively vectors of reference, control, output, control error and disturbance of compatible dimensions. Necessary and sufficient conditions for internal stability of the closed-loop in Fig. 1 are given by the Generalized Nyquist Stability Theorem applied to the closed-loop characteristic polynomial

$$\det F(s) = \det[I + Q(s)] \tag{34}$$

where $Q(s) = G(s)R(s) \in R^{m\times m}$ is the open-loop transfer function matrix.

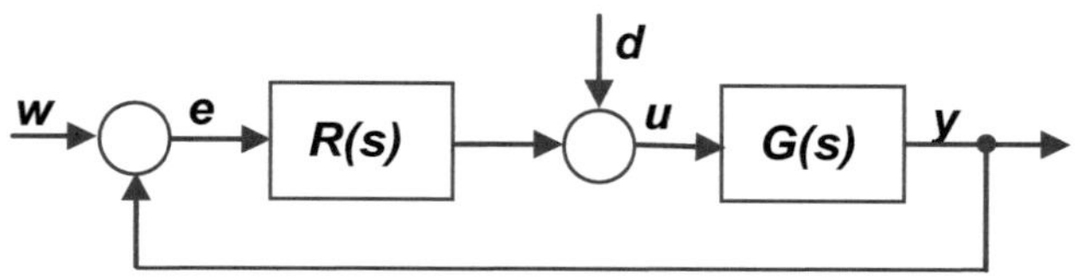

Fig. 1. Standard feedback configuration

The following standard notation is used: D - the standard *Nyquist D-contour* in the complex plane; *Nyquist plot* of $g(s)$ - the image of the Nyquist contour under $g(s)$; $N[k, g(s)]$ - the number of anticlockwise encirclements of the point *(k, j0)* by the Nyquist plot of $g(s)$. *Characteristic functions* of $Q(s)$ are the set of m algebraic functions $q_i(s), i = 1,...,m$ given as

$$\det[q_i(s)I_m - Q(s)] = 0 \quad i = 1,...,m \tag{35}$$

Characteristic loci (CL) are the set of loci in the complex plane traced out by the characteristic functions of $Q(s)$, $\forall s \in D$. The closed-loop characteristic polynomial (34) expressed in terms of characteristic functions of $Q(s)$ reads as follows

$$\det F(s) = \det[I + Q(s)] = \prod_{i=1}^{m}[1 + q_i(s)] \tag{36}$$

Theorem 3.1 (Generalized Nyquist Stability Theorem)
The closed-loop system in Fig. 1 is stable if and only if

$$\begin{aligned} &\text{a.} \quad \det F(s) \neq 0 \quad \forall s \in D \\ &\text{b.} \quad N[0, \det F(s)] = \sum_{i=1}^{m} N\{0, [1 + q_i(s)]\} = n_q \end{aligned} \tag{37}$$

where $F(s) = (I + Q(s))$ and n_q is the number of unstable poles of $Q(s)$.
Let the uncertain plant be given as a set Π of N transfer function matrices

$$\Pi = \{G^k(s)\},\ k = 1,2,...,\ N \ \text{ where } G^k(s) = \left\{G_{ij}^k(s)\right\}_{m\times m} \tag{38}$$

The simplest uncertainty model is the unstructured uncertainty, i.e. a full complex perturbation matrix with the same dimensions as the plant. The set of unstructured perturbations D_U is defined as follows

$$D_U := \{E(j\omega) : \sigma_{\max}[E(j\omega)] \leq \ell(\omega),\ \ \ell(\omega) = \max_k\ \sigma_{\max}[E(j\omega)]\} \tag{39}$$

where $\ell(\omega)$ is a scalar weight function on the norm-bounded perturbation $\Delta(s) \in R^{m\times m}$, $\sigma_{\max}[\Delta(j\omega)] \leq 1$ over given frequency range, $\sigma_{\max}(\cdot)$ is the maximum singular value of (.), i.e.

$$E(j\omega) = \ell(\omega)\Delta(j\omega) \tag{40}$$

For unstructured uncertainty, the set Π can be generated by either additive (E_a), multiplicative input (E_i) or output (E_o) uncertainties, or their inverse counterparts (E_{ia}, E_{ii}, E_{io}), the latter used for uncertainty associated with plant poles located in the closed right half-plane (Skogestad & Postlethwaite, 2005).
Denote $G(s)$ any member of a set of possible plants Π_k, $k = a,i,o,ia,ii,io$; $G_0(s)$ the nominal model used to design the controller, and $\ell_k(\omega)$ the scalar weight on a normalized perturbation. Individual uncertainty forms generate the following related sets Π_k:
Additive uncertainty:

$$\begin{aligned} &\Pi_a := \{G(s) : G(s) = G_0(s) + E_a(s),\ E_a(j\omega) \leq \ell_a(\omega)\Delta(j\omega)\} \\ &\ell_a(\omega) = \max_k\ \sigma_{\max}[G^k(j\omega) - G_0(j\omega)],\ k = 1,2,\ldots,N \end{aligned} \tag{41}$$

Multiplicative input uncertainty:

$$\begin{aligned}&\Pi_i := \{G(s): G(s) = G_0(s)[I + E_i(s)],\ E_i(j\omega) \le \ell_i(j\omega)\Delta(j\omega)\}\\&\ell_i(\omega) = \max_k \sigma_{\max}\{G_0^{-1}(j\omega)[G^k(j\omega) - G_0(j\omega)]\},\ k = 1,2,\ldots,N\end{aligned} \tag{42}$$

Multiplicative output uncertainty:

$$\begin{aligned}&\Pi_o := \{G(s): G(s) = [I + E_o(s)]G_0(s),\ E_0(j\omega) \le \ell_o(j\omega)\Delta(j\omega)\}\\&\ell_o(\omega) = \max_k \sigma_{\max}\{[G^k(j\omega) - G_0(j\omega)]G_0^{-1}(j\omega)\},\ k = 1,2,\ldots,N\end{aligned} \tag{43}$$

Inverse additive uncertainty

$$\begin{aligned}&\Pi_{ia} := \{G(s): G(s) = G_0(s)[I - E_{ia}(s)G_0(j\omega)]^{-1},\ E_{ia}(j\omega) \le \ell_{ia}(\omega)\Delta(j\omega)\}\\&\ell_{ia}(\omega) = \max_k \sigma_{\max}\{[G_0(j\omega)]^{-1} - [G^k(j\omega)]^{-1}\},\ k = 1,2,\ldots,N\end{aligned} \tag{44}$$

Inverse multiplicative input uncertainty

$$\begin{aligned}&\Pi_{ii} := \{G(s): G(s) = G_0(s)[I - E_{ii}(s)]^{-1},\ E_{ii}(j\omega) \le \ell_{ii}(\omega)\Delta(j\omega)\}\\&\ell_{ii}(\omega) = \max_k \sigma_{\max}\{I - [G^k(j\omega)]^{-1}[G_0(j\omega)]\},\ k = 1,2,\ldots,N\end{aligned} \tag{45}$$

Inverse multiplicative output uncertainty:

$$\begin{aligned}&\Pi_{io} := \{G(s): G(s) = [I - E_{io}(s)]^{-1}G_0(s),\ E_{io}(j\omega) \le \ell_{io}(\omega)\Delta(j\omega)\}\\&\ell_{io}(\omega) = \max_k \sigma_{\max}\{I - [G_0(j\omega)][G^k(j\omega)]^{-1}\},\ k = 1,2,\ldots,N\end{aligned} \tag{46}$$

Standard feedback configuration with uncertain plant modelled using any above unstructured uncertainty form can be recast into the $M - \Delta$ structure (for additive perturbation Fig. 2) where *M(s)* is the nominal model and $\Delta(s) \in R^{m \times m}$ is the norm-bounded complex perturbation.
If the nominal closed-loop system is stable then *M(s)* is stable and $\Delta(s)$ is a perturbation which can destabilize the system. The following theorem establishes conditions on *M(s)* so that it cannot be destabilized by $\Delta(s)$ (Skogestad & Postlethwaite, 2005).

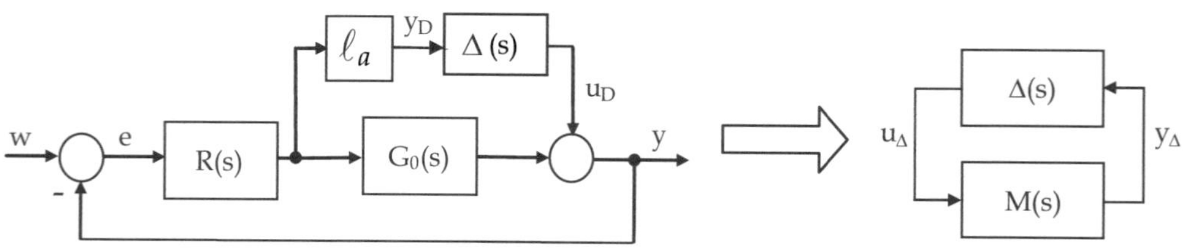

Fig. 2. Standard feedback configuration with unstructured additive uncertainty (left) recast into the $M - \Delta$ structure (right)

Theorem 3.2 (Robust stability for unstructured perturbations)
Assume that the nominal system *M(s)* is stable (nominal stability) and the perturbation $\Delta(s)$ is stable. Then the $M - \Delta$ system in Fig. 2 is stable for all perturbations $\Delta(s): \sigma_{\max}(\Delta) \le 1$ if and only if

$$\sigma_{\max}[M(j\omega)] < 1\,,\ \forall\omega \tag{47}$$

For individual uncertainty forms $M(s) = \ell_k M_k(s)$, $k = a,i,o,ia,ii,io$; the corresponding matrices $M_k(s)$ are given below (disregarding the negative signs which do not affect resulting robustness condition); commonly, the nominal model $G_0(s)$ is obtained as a model of mean parameter values.

$$M(s) = \ell_a(s)R(s)[I + G_0(s)R(s)]^{-1} = \ell_a(s)M_a(s) \text{ additive uncertainty} \tag{48}$$

$$M(s) = \ell_i(s)R(s)[I + G_0(s)R(s)]^{-1}G_0(s) = \ell_i(s)M_i(s) \text{ multiplicative input uncertainty} \tag{49}$$

$$M(s) = \ell_o(s)G_0(s)R(s)[I + G_0(s)R(s)]^{-1} = \ell_o(s)M_o(s) \text{ multiplicative output uncertainty} \tag{50}$$

$$M(s) = \ell_{ia}(s)[I + G_0(s)R(s)]^{-1}G_0(s) = \ell_{ia}(s)M_{ia}(s) \text{ inverse additive uncertainty} \tag{51}$$

$$M(s) = \ell_{ii}(s)[I + R(s)G_0(s)]^{-1} = \ell_{ii}(s)M_{ii}(s) \text{ inverse multiplicative input uncertainty} \tag{52}$$

$$M(s) = \ell_{io}(s)[I + G_0(s)R(s)]^{-1} = \ell_{io}(s)M_{io}(s) \text{ inverse multiplicative output uncertainty} \tag{53}$$

Conservatism of the robust stability conditions can be reduced by structuring the unstructured additive perturbation by introducing the *additive affine-type uncertainty* $E_{af}(s)$ that brings about new way of nominal system computation and robust stability conditions modifiable for the decentralized controller design as (Kozáková & Veselý, 2007; 2008).

$$E_{af}(s) = \sum_{i=1}^{p} G_i(s)q_i \tag{54}$$

where $G_i(s) \in R^{m\times m}$, i=0,1, ..., p are stable matrices, p is the number of uncertainties defining 2^p polytope vertices that correspond to individual perturbed models; q_i are polytope parameters. The set Π_{af} generated by the additive affine-type uncertainty (E_{af}) is

$$\Pi_{af} := \{G(s) : G(s) = G_0(s) + E_{af}\,,\ E_{af} = \sum_{i=1}^{p} G_i(s)q_i\,,\ q_i \in < q_{i\min}, q_{i\max} >,\ q_{i\min} + q_{i\max} = 0\} \tag{55}$$

where $G_0(s)$ is the „afinne“ nominal model. Put into vector-matrix form, individual perturbed plants (elements of the set Π_{af}) can be expressed as follows

$$G(s) = G_0(s) + [I_{q1} \ldots I_{qp}]\begin{bmatrix} G_1(s) \\ \vdots \\ G_p(s) \end{bmatrix} = G_0(s) + QG_u(s) \tag{56}$$

where $Q = [I_{q_1} \ldots I_{q_p}]^T \in R^{m\times(m\times p)}$, $I_{q_i} = q_i I_{m\times m}$, $G_u(s) = [G_1 \ldots G_p]^T \in R^{(m\times p)\times m}$.

Standard feedback configuration with uncertain plant modelled using the additive affine type uncertainty is shown in Fig. 3 (on the left); by analogy with previous cases, it can be recast into the $M_{af} - Q$ structure in Fig. 3 (on the right) where

$$M_{af} = G_u R(I + G_0 R)^{-1} = G_u (I + RG_0)^{-1} R \tag{57}$$

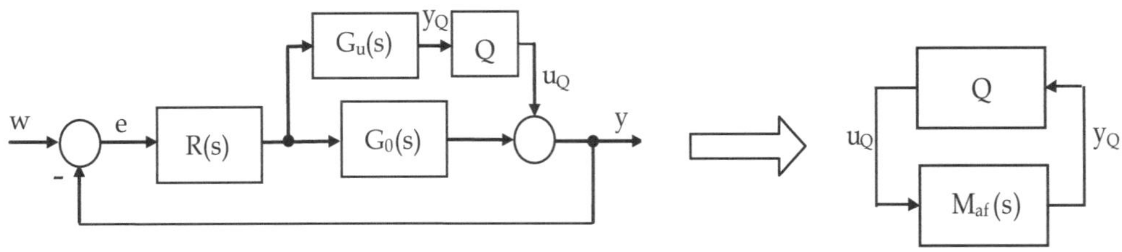

Fig. 3. Standard feedback configuration with unstructured affine-type additive uncertainty (left), recast into the M_{af} -Q structure (right)

Similarly as for the *M-Δ* system, stability condition of the $M_{af} - Q$ system is obtained as

$$\sigma_{\max}(M_{af}Q) < 1 \tag{58}$$

Using singular value properties, the small gain theorem, and the assumptions that $q_0 = |q_{i\min}| = |q_{i\max}|$ and the nominal model $M_{af}(s)$ is stable, (58) can further be modified to yield the robust stability condition

$$\sigma_{\max}(M_{af}) q_0 \sqrt{p} < 1 \tag{59}$$

The main aim of Section 3 of this chapter is to solve the next problem.

Problem 3.1

Consider an uncertain system with m subsystems given as a set of N transfer function matrices obtained in N working points of plant operation, described by a nominal model $G_0(s)$ and any of the unstructured perturbations (41) - (46) or (55).

Let the nominal model $G_0(s)$ can be split into the diagonal part representing mathematical models of decoupled subsystems, and the off-diagonal part representing interactions between subsystems

$$G_0(s) = G_d(s) + G_m(s) \tag{60}$$

where

$$G_d(s) = diag\{G_i(s)\}_{m \times m}, \ \det G_d(s) \neq 0 \quad \forall s \ \ G_m(s) = G_0(s) - G_d(s) \tag{61}$$

A decentralized controller

$$R(s) = diag\{R_i(s)\}_{m \times m}, \det R(s) \neq 0 \quad \forall s \in D \tag{62}$$

is to be designed with $R_i(s)$ being transfer function of the i-th local controller. The designed controller has to guarantee stability over the whole operating range of the plant specified by either (41) - (46) or (55) (robust stability) and a specified performance of the nominal model (nominal performance). To solve the above problem, a frequency domain robust decentralized controller design technique has been developed (Kozáková & Veselý, 2009; Kozáková et. al., 2009b); the core of it is the Equivalent Subsystems Method (ESM).

3.2 Decentralized controller design for performance: equivalent subsystems method

The Equivalent Subsystems Method (ESM) an original Nyquist-based DC design method for stability and guaranteed performance of the full system. According to it, local controller designs are performed independently for so-called equivalent subsystems that are actually Nyquist plots of decoupled subsystems shaped by a selected characteristic locus of the interactions matrix. Local controllers of equivalent subsystems independently tuned for stability and specified feasible performance constitute the decentralized controller guaranteeing specified performance of the full system. Unlike standard robust approaches, the proposed technique considers full mean parameter value nominal model, thus reducing conservatism of resulting robust stability conditions. In the context of robust decentralized controller design, the Equivalent Subsystems Method (Kozáková et. al., 2009b) is applied to design a decentralized controller for the nominal model $G_0(s)$ as depicted in Fig. 4.

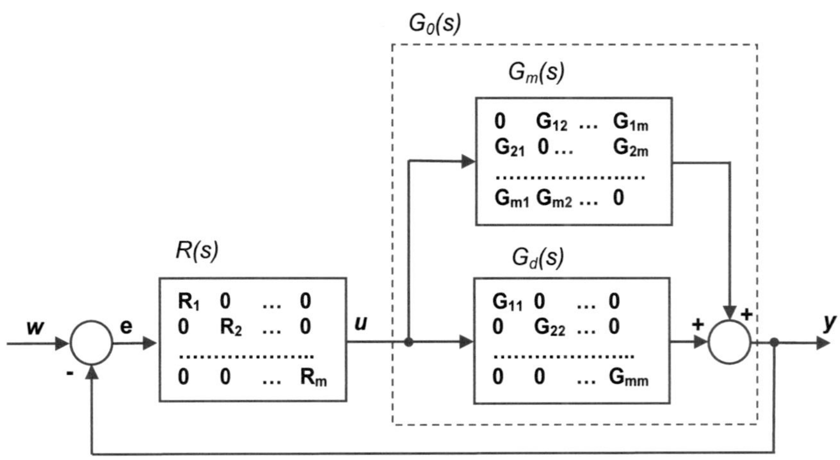

Fig. 4. Standard feedback loop under decentralized controller

The key idea behind the method is factorisation of the closed-loop characteristic polynomial *detF(s)* in terms of the split nominal system (60) under the decentralized controller (62) (existence of $R^{-1}(s)$ is implied by the assumption (62) that $\det R(s) \neq 0$)

$$\det F(s) = \det\{I + [G_d(s) + G_m(s)]R(s)\} = \det[R^{-1}(s) + G_d(s) + G_m(s)]\det R(s) \tag{63}$$

Denote

$$F_1(s) = R^{-1}(s) + G_d(s) + G_m(s) = P(s) + G_m(s) \tag{64}$$

where

$$P(s) = R^{-1}(s) + G_d(s) \tag{65}$$

is a diagonal matrix $P(s) = diag\{p_i(s)\}_{m\times m}$. Considering (63) and (64), the stability condition (37b) in *Theorem 3.1* modifies as follows

$$N\{0, \det[P(s) + G_m(s)]\} + N[0, \det R(s)] = n_q \tag{66}$$

and a simple manipulation of (65) yields

$$I + R(s)[G_d(s) - P(s)] = I + R(s)G^{eq}(s) = 0 \tag{67}$$

where

$$G^{eq}(s) = diag\{G_i^{eq}(s)\}_{m\times m} = diag\{G_i(s) - p_i(s)\}_{m\times m} \quad i = 1,\dots,m \tag{68}$$

is a diagonal matrix of equivalent subsystems $G_i^{eq}(s)$; on subsystems level, (67) yields m equivalent characteristic polynomials

$$CLCP_i^{eq}(s) = 1 + R_i(s)G_i^{eq}(s) \qquad i = 1,2,\dots,m \tag{69}$$

Hence, by specifying *P(s)* it is possible to affect performance of individual subsystems (including stability) through $R^{-1}(s)$. In the context of the independent design philosophy, design parameters $p_i(s), i = 1,2,\dots,m$ represent constraints for individual designs. General stability conditions for this case are given in *Corollary 3.1.*

Corollary 3.1 (Kozáková & Veselý, 2009)

The closed-loop in Fig. 4 comprising the system (60) and the decentralized controller (62) is stable if and only if

1. there exists a diagonal matrix $P(s) = diag\{p_i(s)\}_{i=1,\dots,m}$ such that all equivalent subsystems (68) can be stabilized by their related local controllers $R_i(s)$, i.e. all equivalent characteristic polynomials $CLCP_i^{eq}(s) = 1 + R_i(s)G_i^{eq}(s)$, $i = 1,2,\dots,m$ have roots with $\text{Re}\{s\} < 0$;
2. the following two conditions are met $\forall s \in D$:

$$\begin{aligned} &\text{a.} \quad \det[P(s) + G_m(s)] \neq 0 \\ &\text{b.} \quad N[0, \det F(s)] = n_q \end{aligned} \tag{70}$$

where $\det F(s) = \det(I + G(s)R(s))$ and n_q is the number of open loop poles with $\text{Re}\{s\} > 0$. In general, $p_i(s)$ are to be transfer functions, fulfilling conditions of Corollary 3.1, and the stability condition resulting form the small gain theory; according to it if both $P^{-1}(s)$ and $G_m(s)$ are stable, the necessary and sufficient closed-loop stability condition is

$$\left\| P(s)^{-1} G_m(s) \right\| < 1 \qquad \text{or} \qquad \sigma_{\min}[P(s)] > \sigma_{\max}[G_m(s)] \tag{71}$$

To provide closed-loop stability of the full system under a decentralized controller, $p_i(s), i = 1,2,\dots,m$ are to be chosen so as to appropriately cope with the interactions $G_m(s)$.
A special choice of *P(s)* is addressed in (Kozáková et al.2009a;b): if considering characteristic functions $g_i(s)$ of $G_m(s)$ defined according to (35) for $i = 1,\dots,m$, and choosing P(s) to be diagonal with identical entries equal to any selected characteristic function $g_k(s)$ *of* $[-G_m(s)]$, where $k \in \{1,\dots,m\}$ is fixed, i.e.

$$P(s) = -g_k(s)I, \quad k \in \{1,\dots,m\} \text{ is fixed} \tag{72}$$

then substituting (72) in (70a) and violating the well-posedness condition yields

$$\det[P(s) + G_m(s)] = \prod_{i=1}^{m} [-g_k(s) + g_i(s)] = 0 \qquad \forall s \in D \tag{73}$$

In such a case the full closed-loop system is at the limit of instability with equivalent subsystems generated by the selected $g_k(s)$ according to

$$G_{ik}^{eq}(s) = G_i(s) + g_k(s) \quad i = 1,2,\ldots, m\,, \; \forall s \in D \tag{74}$$

Similarly, if choosing $P(s-\alpha) = -g_k(s-\alpha)I$, $0 \le \alpha \le \alpha_m$ where α_m denotes the maximum feasible degree of stability for the given plant under the decentralized controller $R(s)$, then

$$\det F_1(s-\alpha) = \prod_{i=1}^{m} [-g_k(s-\alpha) + g_i(s-\alpha)] = 0 \quad \forall s \in D \tag{75}$$

Hence, the closed-loop system is stable and has just poles with $\mathrm{Re}\{s\} \le -\alpha$, i.e. its degree of stability is α. Pertinent equivalent subsystems are generated according to

$$G_{ik}^{eq}(s-\alpha) = G_i(s-\alpha) + g_k(s-\alpha) \quad i = 1,2,\ldots, m \tag{76}$$

To guarantee stability, the following additional condition has to be satisfied simultaneously

$$\det F_{1k} = \prod_{i=1}^{m} [-g_k(s-\alpha) + g_i(s)] = \prod_{i=1}^{m} r_{ik}(s) \ne 0 \quad \forall s \in D \tag{77}$$

Simply put, by suitably choosing α: $0 \le \alpha \le \alpha_m$ to generate $P(s-\alpha)$ it is possible to guarantee performance under the decentralized controller in terms of the degree of stability α. Lemma 3.1 provides necessary and sufficient stability conditions for the closed-loop in Fig. 4 and conditions for guaranteed performance in terms of the degree of stability.
Definition 3.1 (Proper characteristic locus)
The characteristic locus $g_k(s-\alpha)$ of $G_m(s-\alpha)$, where fixed $k \in \{1,\ldots,m\}$ and $\alpha > 0$, is called *proper characteristic locus* if it satisfies conditions (73), (75) and (77). The set of all proper characteristic loci of a plant is denoted P_S.
Lemma 3.1
The closed-loop in Fig. 4 comprising the system (60) and the decentralized controller (62) is stable if and only if the following conditions are satisfied $\forall s \in D$, $\alpha \ge 0$ and fixed $k \in \{1,\ldots,m\}$:

1. $g_k(s-\alpha) \in P_S$
2. all equivalent characteristic polynomials (69) have roots with $\mathrm{Re}\, s \le -\alpha$;
3. $N[0, \det F(s-\alpha)] = n_{q\alpha}$

where $F(s-\alpha) = I + G(s-\alpha)R(s-\alpha)$; $n_{q\alpha}$ is the number of open loop poles with $\mathrm{Re}\{s\} > -\alpha$.
Lemma 3.1 shows that local controllers independently tuned for stability and a specified (feasible) degree of stability of equivalent subsystems constitute the decentralized controller guaranteeing the same degree of stability for the full system. The design technique resulting from Corollary 3.1 enables to design local controllers of equivalent subsystems using any SISO frequency-domain design method, e.g. the Neymark D-partition method (Kozáková et al. 2009b), standard Bode diagram design etc. If considering other performance measures in the ESM, the design proceeds according to Corollary 3.1 with $P(s)$ and $G_{ik}^{eq}(s) = G_i(s) + g_k(s)$, $i = 1,2,\ldots, m$ generated according to (72) and (74), respectively.

According to the latest results, guaranteed performance in terms of maximum overshoot is achieved by applying Bode diagram design for specified phase margin in equivalent subsystems. This approach is addressed in the next subsection.

3.3 Robust decentralized controller design

The presented frequency domain robust decentralized controller design technique is applicable for uncertain systems described as a set of transfer function matrices. The basic steps are:

1. *Modelling the uncertain system*

This step includes choice of the nominal model and modelling uncertainty using any unstructured uncertainty (41)-(46) or (55). The nominal model can be calculated either as the mean value parameter model (Skogestad & Postlethwaite, 2005), or the "affine" model, obtained within the procedure for calculating the affine-type additive uncertainty (Kozáková & Veselý, 2007; 2008). Unlike the standard robust approach to decentralized control design which considers diagonal model as the nominal one (interactions are included in the uncertainty), the ESM method applied in the design for nominal performance allows to consider the *full* nominal model.

2. *Guaranteeing nominal stability and performance*

The ESM method is used to design a decentralized controller (62) guaranteeing stability and specified performance of the nominal model (nominal stability, nominal performance).

3. *Guaranteeing robust stability*

In addition to nominal performance, the decentralized controller has to guarantee closed-loop stability over the whole operating range of the plant specified by the chosen uncertainty description (robust stability). Robust stability is examined by means of the *M-Δ* stability condition (47) or the M_{af}-Q stability condition (59) in case of the affine type additive uncertainty (55).

Corollary 3.2 (Robust stability conditions under DC)

The closed-loop in Fig. 3 comprising the uncertain system given as a set of transfer function matrices and described by any type of unstructured uncertainty (41) - (46) or (55) with nominal model fulfilling (60), and the decentralized controller (62) is stable over the pertinent uncertainty region if any of the following conditions hold

1. for any (41)-(46), conditions of *Corollary 3.1* and (47) are simultaneously satisfied where $M(s) = \ell_k M_k(s),\ k = a,i,o,ia,ii,io$ and M_k given by (48)-(53) respectively.
2. for (55), conditions of *Corollary 3.1* and (59) are simultaneously satisfied.

Based on Corollary 3.2, two approaches to the robust decentralized control design have been developed: the two-stage and the direct approaches.

1. *The two stage robust decentralized controller design approach based on the M-Δ structure stability conditions* (Kozáková & Veselý, 2008;, Kozáková & Veselý, 2009; Kozáková et al. 2009a).

In the first stage, the decentralized controller for the nominal system is designed using ESM, afterwards, fulfilment of the *M-Δ* or M_{af}-Q stability conditions (47) or (59), respectively is examined; if satisfied, the design procedure stops, otherwise the second stage follows: either controller parameters are additionally modified to satisfy robust stability conditions in the tightest possible way (Kozáková et al. 2009a), or the redesign is carried out with modified performance requirements (Kozáková & Veselý, 2009).

2. Direct decentralized controller design for robust stability and nominal performance

By direct integration of the robust stability condition (47) or (59) in the ESM, local controllers of equivalent subsystems are designed with regard to robust stability. Performance specification for the full system in terms of the maximum peak of the complementary sensitivity M_T corresponding to maximum overshoot in individual equivalent subsystems is translated into lower bounds for their phase margins according to (78) (Skogestad & Postlethwaite, 2005)

$$PM \geq 2\arcsin\left(\frac{1}{2M_T}\right) \geq \frac{1}{M_T}[rad] \tag{78}$$

where PM is the phase margin, M_T is the maximum peak of the complementary sensitivity

$$T(s) = G(s)R(s)[I + G(s)R(s)]^{-1} \tag{79}$$

As for MIMO systems

$$M_T = \sigma_{\max}(T) \tag{80}$$

the upper bound for M_T can be obtained using the singular value properties in manipulations of the M-Δ condition (47) considering (48)-(53), or the $M_{af}-Q$ condition (58) considering (57) and (59). The following upper bounds $\sigma_{\max}[T_0(j\omega)]$ for the nominal complementary sensitivity $T_0(s) = G_0(s)R(s)[I + G_0(s)R(s)]^{-1}$ have been derived:

$$\sigma_{\max}[T_0(j\omega)] < \frac{\sigma_{\min}[G_0(j\omega)]}{|\ell_a(\omega)|} = L_A(\omega) \qquad \forall\omega \text{ additive uncertainty} \tag{81}$$

$$\sigma_{\max}[T_0(j\omega)] < \frac{1}{|\ell_k(\omega)|} = L_K(\omega),\ k = i,o, \qquad \forall\omega \text{ multiplicative input/output uncertainty} \tag{82}$$

$$\sigma_{\max}[T_0(j\omega)] < \frac{1}{q_0\sqrt{p}}\frac{\sigma_{\min}[G_0(j\omega)]}{\sigma_{\max}[G_u(j\omega)]} = L_{AF}(\omega) \qquad \forall\omega \text{ additive affine-type uncertainty} \tag{83}$$

Using (80) and (78) the upper bounds for the complementary sensitivity of the nominal system (81)-(83) can be directly implemented in the ESM due to the fact that performance achieved in equivalent subsystems is simultaneously guaranteed for the full system. The main benefit of this approach is the possibility to specify maximum overshoot in the full system guaranteeing robust stability in terms of $\sigma_{\max}(T_0)$, translate it into minimum phase margin of equivalent subsystems and design local controllers independently for individual single input - single output equivalent subsystems.
The design procedure is illustrated in the next subsection.

3.4 Example

Consider a laboratory plant consisting of two interconnected DC motors, where each armature voltage (U_1, U_2) affects rotor speeds of both motors (ω_1, ω_2). The plant was identified in three operating points, and is given as a set $\Pi = \{G_1(s), G_2(s), G_3(s)\}$ where

$$G_1(s) = \begin{bmatrix} \dfrac{-0.402s + 2.690}{s^2 + 2.870s + 1.840} & \dfrac{0.006s - 1.680}{s^2 + 11.570s + 3.780} \\ \dfrac{0.003s - 0.720}{s^2 + 9.850s + 1.764} & \dfrac{-0.170s + 1.630}{s^2 + 1.545s + 0.985} \end{bmatrix}$$

$$G_2(s) = \begin{bmatrix} \dfrac{-0.342s + 2.290}{s^2 + 2.070s + 1.840} & \dfrac{0.005s - 1.510}{s^2 + 10.570s + 3.780} \\ \dfrac{0.003s - 0.580}{s^2 + 8.850s + 1.764} & \dfrac{-0.160s + 1.530}{s^2 + 1.045s + 0.985} \end{bmatrix}$$

$$G_3(s) = \begin{bmatrix} \dfrac{-0.423s + 2.830}{s^2 + 4.870s + 1.840} & \dfrac{0.006s - 1.930}{s^2 + 13.570s + 3.780} \\ \dfrac{0.004s - 0.790}{s^2 + 10.850s + 1.764} & \dfrac{-0.200s + 1.950}{s^2 + 1.945s + 0.985} \end{bmatrix}$$

In calculating the affine nominal model $G_0(s)$, all possible allocations of $G_1(s)$, $G_2(s)$, $G_3(s)$ into the $2^2 = 4$ polytope vertices were examined (24 combinations) yielding 24 affine nominal model candidates and related transfer functions matrices $G_4(s)$ needed to complete the description of the uncertainty region. The selected affine nominal model $G_0(s)$ is the one guaranteeing the smallest additive uncertainty calculated according to (41):

$$G_0(s) = \begin{bmatrix} \dfrac{-0.413\, s + 2.759}{s^2 + 3.870s + 1.840} & \dfrac{-0.006s - 1.807}{s^2 + 12.570s + 3.780} \\ \dfrac{0.004s - 0.757}{s^2 + 10.350s + 1.764} & \dfrac{-0.187s + 1.791}{s^2 + 1.745s + 0.985} \end{bmatrix}$$

The upper bound $L_{AF}(\omega)$ for $T_0(s)$ calculated according to (82) is plotted in Fig. 5. Its worst (minimum value) $M_T = \min_{\omega} L_{AF}(\omega) = 1.556$ corresponds to $PM \geq 37.48^{\circ}$ according to (78).

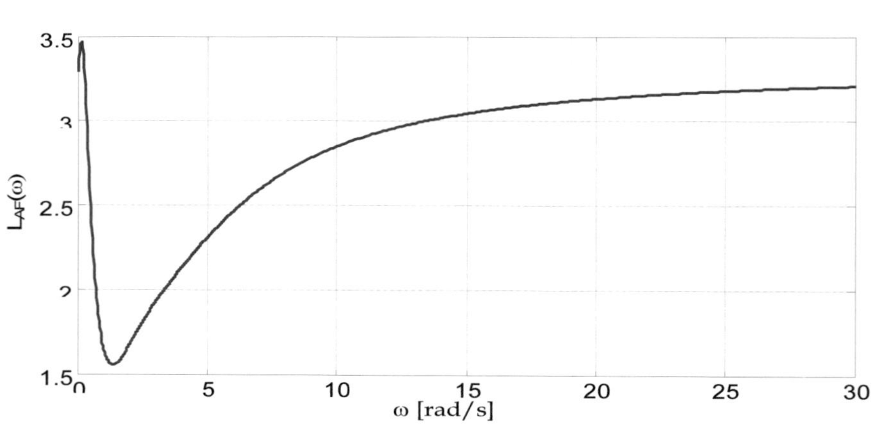

Fig. 5. Plot of $L_{AF}(\omega)$ calculated according to (82)

The Bode diagram design of local controllers for guaranteed PM was carried out for equivalent subsystems generated according to (74) using characteristic locus $g_1(s)$ of the matrix of interactions $G_m(s)$, i.e. $G_{i1}^{eq}(s) = G_i(s) + g_2(s)$ $i = 1,2$. Bode diagrams of equivalent subsystems $G_{11}^{eq}(s), G_{21}^{eq}(s)$ are in Fig. 6. Applying the PI controller design from Bode diagram for required phase margin $PM = 39^{\circ}$ has yielded the following local controllers

$$R_1(s) = \frac{3.367\ s + 1.27}{s} \qquad R_2(s) = \frac{1.803s + 0.491}{s}$$

Bode diagrams of compensated equivalent subsystems in Fig. 8 prove the achieved phase margin. Robust stability was verified using the original M_{af}-Q condition (59) with p=2 and q_0=1; as depicted in Fig. 8, the closed loop under the designed controller is robustly stable.

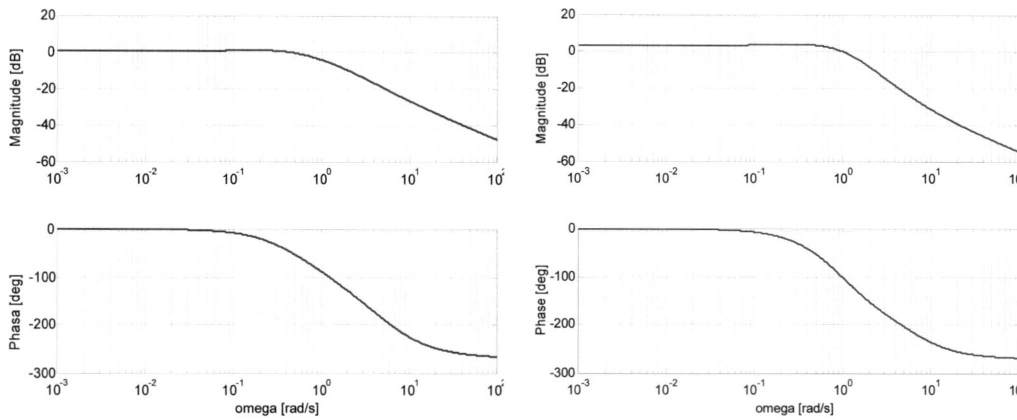

Fig. 6. Bode diagrams of equivalent subsystems $G_{11}^{eq}(s)$ (left), $G_{21}^{eq}(s)$ (right)

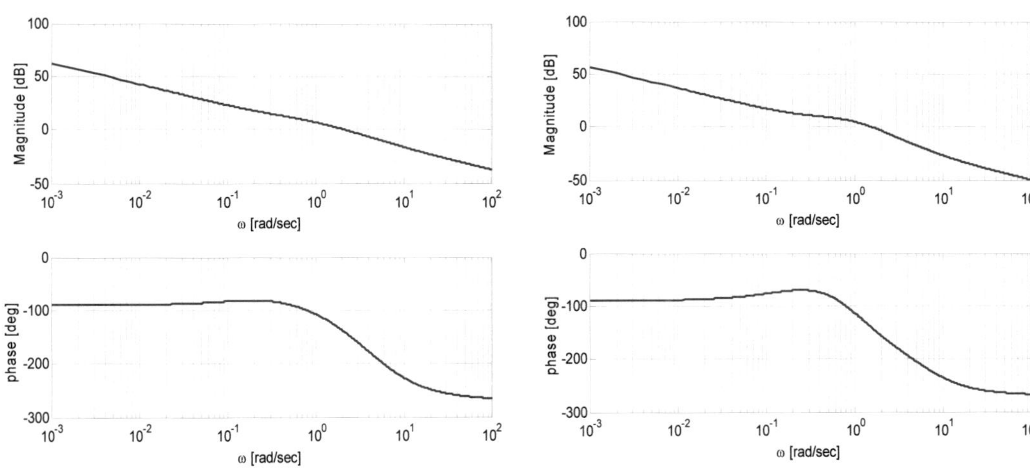

Fig. 7. Bode diagrams of equivalent subsystems $G_{11}^{eq}(s)$ (left), $G_{21}^{eq}(s)$ (right) under designed local controllers $R_1(s)$, $R_2(s)$, respectively.

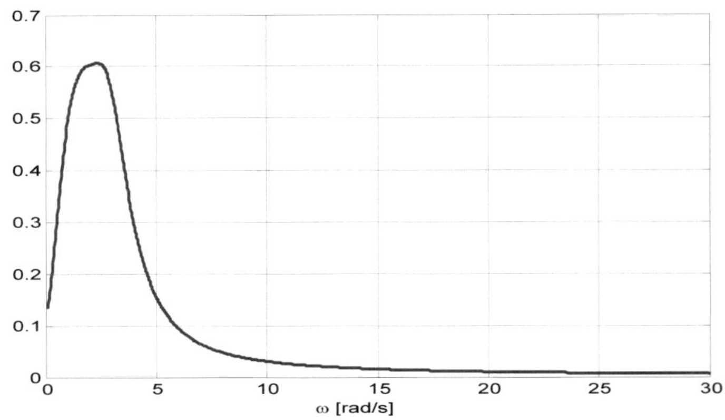

Fig. 8. Verification of robust stability using condition (59) in the form $\sigma_{\max}(M_{af}) < \frac{1}{\sqrt{2}}$

4. Conclusion

The chapter reviews recent results on robust controller design for linear uncertain systems applicable also for decentralized control design.
In the first part of the chapter the new robust PID controller design method based on LMI` is proposed for uncertain linear system. The important feature of this PID design approach is that the derivative term appears in such form that enables to consider the model uncertainties. The guaranteed cost control is proposed with a new quadratic cost function including the derivative term for state vector as a tool to influence the overshoot and response rate.
In the second part of the chapter a novel frequency-domain approach to the decentralized controller design for guaranteed performance is proposed. Its principle consists in including plant interactions in individual subsystems through their characteristic functions, thus yielding a diagonal system of equivalent subsystems. Local controllers of equivalent subsystems independently tuned for specified performance constitute the decentralized controller guaranteeing the same performance for the full system. The proposed approach allows direct integration of robust stability condition in the design of local controllers of equivalent subsystems.
Theoretical results are supported with results obtained by solving some examples.

5. Acknowledgment

This research work has been supported by the Scientific Grant Agency of the Ministry of Education of the Slovak Republic, Grant No. 1/0544/09.

6. References

Blondel, V. & Tsitsiklis, J.N. (1997). NP-hardness of some linear control design problems. *SIAM J. Control Optim.*, Vol. 35, 2118-2127.

Boyd, S.; El Ghaoui, L.; Feron, E. & Balakrishnan, V. (1994). *Linear matrix inequalities in system and control theory*, SIAM Studies in Applied Mathematics, Philadelphia.

Crusius, C.A.R. & Trofino, A. (1999). LMI Conditions for Output Feedback Control Problems. *IEEE Trans. Aut. Control*, Vol. 44, 1053-1057.

de Oliveira, M.C.; Bernussou, J. & Geromel, J.C. (1999). A new discrete-time robust stability condition. *Systems and Control Letters*, Vol. 37, 261-265.

de Oliveira, M.C.; Camino, J.F. & Skelton, R.E. (2000). A convexifying algorithm for the design of structured linear controllers, *Proc. 39nd IEEE CDC*, pp. 2781-2786, Sydney, Australia, 2000.

Ming Ge; Min-Sen Chiu & Qing-Guo Wang (2002). Robust PID controller design via LMI approach. *Journal of Process Control*, Vol.12, 3-13.

Grman, Ľ. ; Rosinová, D. ; Kozáková, A. & Veselý, V. (2005). Robust stability conditions for polytopic systems. *International Journal of Systems Science*, Vol. 36, No. 15, 961-973, ISSN 1464-5319 (electronic) 0020-7721 (paper)

Gyurkovics, E. & Takacs, T. (2000). Stabilisation of discrete-time interconnected systems under control constraints. *IEE Proceedings - Control Theory and Applications*, Vol. 147, No. 2, 137-144

Han, J. & Skelton, R.E. (2003). An LMI optimization approach for structured linear controllers, *Proc. 42nd IEEE CDC*, 5143-5148, Hawaii, USA, 2003

Henrion, D.; Arzelier, D. & Peaucelle, D. (2002). Positive polynomial matrices and improved LMI robustness conditions. 15th *IFAC World Congress*, CD-ROM, Barcelona, Spain, 2002

Kozáková, A. & Veselý, V. (2007). Robust decentralized controller design for systems with additive affine-type uncertainty. *Int. J. of Innovative Computing, Information and Control (IJICIC)*, Vol. 3, No. 5 (2007), 1109-1120, ISSN 1349-4198.

Kozáková, A. & Veselý, V. (2008). Robust MIMO PID controller design using additive affine-type uncertainty. *Journal of Electrical Engineering*, Vol. 59, No.5 (2008), 241-247, ISSN 1335 - 3632

Kozáková, A., Veselý, V. (2009). Design of robust decentralized controllers using the M-Δ structure robust stability conditions. *Int. Journal of Systems Science,* Vol. 40, No.5 (2009), 497-505, ISSN 1464-5319 (electronic) 0020-7721 (paper).

Kozáková, A.; Veselý, V. & Osuský, J. (2009a). A new Nyquist-based technique for tuning robust decentralized controllers, *Kybernetika*, Vol. 45, No.1 (2009), 63-83, ISSN 0023-5954.

Kozáková, A.; Veselý,V. Osuský, J.(2009b). Decentralized Controllers Design for Performance: Equivalent Subsystems Method, *Proceedings of the European Control Conference, ECC'09*, 2295-2300, ISBN 978-963-311-369-1, Budapest, Hungary August 2009, EUCA Budapest.

Peaucelle, D.; Arzelier, D.; Bachelier, O. & Bernussou, J. (2000). A new robust *D*-stability condition for real convex polytopic uncertainty. *Systems and Control Letters*, Vol. 40, 21-30

Rosinová, D.; Veselý, V. & Kučera, V. (2003). A necessary and sufficient condition for static output feedback stabilizability of linear discrete-time systems. *Kybernetika*, Vol. 39, 447-459

Rosinová, D. & Veselý, V. (2003). Robust output feedback design of discrete-time systems – linear matrix inequality methods. *Proceedings* 2th *IFAC Conf. CSD'03* (CD-ROM), Bratislava, Slovakia, 2003

Skelton, R.E.; Iwasaki, T. & Grigoriadis, K. (1998). *A Unified Algebraic Approach to Linear Control Design*, Taylor and Francis, Ltd, London, UK

Skogestad, S. & Postlethwaite, I. (2005). *Multivariable fedback control: analysis and design,* John Wiley & Sons Ltd., ISBN -13978-0-470-01167-6 (H/B), The Atrium, Southern Gate. Chichester, West Sussex, UK

Veselý, V. (2003). Robust output feedback synthesis: LMI Approach, *Proceedings* 2th *IFAC Conference CSD'03* (CD-ROM), Bratislava, Slovakia, 2003

Zheng Feng; Qing-Guo Wang & Tong Heng Lee (2002). On the design of multivariable PID controllers via LMI approach. *Automatica*, Vol. 38, 517-526

11

Robust Stabilization and Discretized PID Control

Yoshifumi Okuyama
Tottori University, Emeritus
Japan

1. Introduction

At present, almost all feedback control systems are realized using discretized (discrete-time and discrete-value, i.e., digital) signals. However, the analysis and design of discretized/quantized control systems has not been entirely elucidated. The first attempt to elucidate the problem was described in a paper by Kalman (1) in 1956. Since then, many researchers have studied this problem, particularly the aspect of understanding and mitigating the quantization effects in quantized feedback control, e.g.,(2–4). However, few results have been obtained for the stability analysis of the nonlinear discrete-time feedback system.

This article describes the robust stability analysis of discrete-time and discrete-value control systems and presents a method for designing (stabilizing) PID control for nonlinear discretized systems. The PID control scheme has been widely used in practice and theory thus far irrespective of whether it is continuous or discrete in time (5; 6) since it is a basic feedback control technique.

In the previous study (7–9), a robust stability condition for nonlinear discretized control systems that accompany discretizing units (quantizers) at equal spaces was examined in a frequency domain. It was assumed that the discretization is executed at the input and output sides of a nonlinear continuous elemet (sensor/actuator) and that the sampling period is chosen such that the size is suitable for discretization in the space. This paper presents a designing problem for discretized control systems on a grid pattern in the time and controller variables space. In this study, the concept of modified Nyquist and Nichols diagrams for nonlinear control systems given in (10; 11) is applied to the designing procedure in the frequency domain.

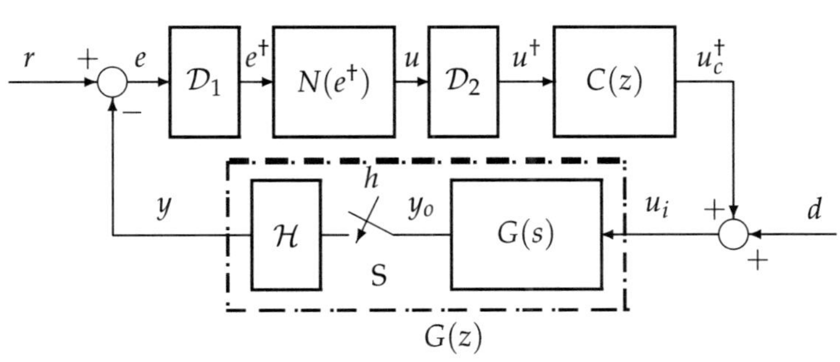

Fig. 1. Nonlinear discretized PID control system.

2. Discretized control system

The discretized control system in question is represented by a sampled-data (discrete-time) feedback system as shown in Fig. 1. In the figure, $G(z)$ is the z-transform of continuous plant $G(s)$ together with the zero-order hold, $C(z)$ is the z-transform of the digital PID controller, and $\mathcal{D}_1$ and $\mathcal{D}_2$ are the discretizing units at the input and output sides of the nonlinear element, respectively.

The relationship between e and $u^{\dagger} = N_d(e)$ is a stepwise nonlinear characteristic on an integer-grid pattern. Figure 2 (a) shows an example of discretized sigmoid-type nonlinear characteristic. For C-language expression, the input/output characteristic can be written as

$$\begin{aligned} e^{\dagger} &= \gamma * (\text{double})(\text{int})(e/\gamma) \\ u &= 0.4 * e^{\dagger} + 3.0 * \text{atan}(0.6 * e^{\dagger}) \\ u^{\dagger} &= \gamma * (\text{double})(\text{int})(u/\gamma), \end{aligned} \tag{1}$$

where (int) and (double) denote the conversion into an integral number (a round-down discretization) and the reconversion into a double-precision real number, respectively. Note that even if the continuous characteristic is linear, the input/output characterisitc becomes nonlinear on a grid pattern as shown in Fig. 2 (b), where the linear continuous characteristic is chosen as $u = 0.85 * e^{\dagger}$.

In this study, a round-down discretization, which is usually executed on a computer, is applied. Therefore, the relationship between $e^{\dagger}$ and $u^{\dagger}$ is indicated by small circles on the stepwise nonlinear characteristic. Here, each signal $e^{\dagger}, u^{\dagger}, \cdots$ can be assigned to an integer number as follows:

$$\begin{aligned} e^{\dagger} &\in \{\cdots, -3\gamma, -2\gamma, -\gamma, 0, \gamma, 2\gamma, 3\gamma, \cdots\}, \\ u^{\dagger} &\in \{\cdots, -3\gamma, -2\gamma, -\gamma, 0, \gamma, 2\gamma, 3\gamma, \cdots\}, \end{aligned}$$

where γ is the resolution of each variable. Without loss of generality, hereafter, it is assumed that $\gamma = 1.0$. That is, the variables $e^{\dagger}, u^{\dagger}, \cdots$ are defined by integers as follows:

$$e^{\dagger}, u^{\dagger} \in Z, \quad Z = \{\cdots -3, -2, -1, 0, 1, 2, 3, \cdots\}.$$

On the other hand, the time variable t is given as $t \in \{0, h, 2h, 3h, \cdots\}$ for the sampling period h. When assuming $h = 1.0$, the following expression can be defined:

$$t \in Z_{+}, \quad Z_{+} = \{0, 1, 2, 3, \cdots\}.$$

Therefore, each signal $e^{\dagger}(t), u^{\dagger}(t), \cdots$ traces on a grid pattern that is composed of integers in the time and controller variables space.

The discretized nonlinear characteristic

$$u^{\dagger} = N_d(e^{\dagger}) = K\, e^{\dagger} + g(e^{\dagger}), \quad 0 < K < \infty, \tag{2}$$

as shown in Fig. 2(a) is partitioned into the following two sections:

$$|g(e^{\dagger})| \leq \bar{g} < \infty, \tag{3}$$

for $|e^{\dagger}| < \varepsilon$, and

$$|g(e^{\dagger})| \leq \beta\, |e^{\dagger}|, \quad 0 \leq \beta \leq K, \tag{4}$$

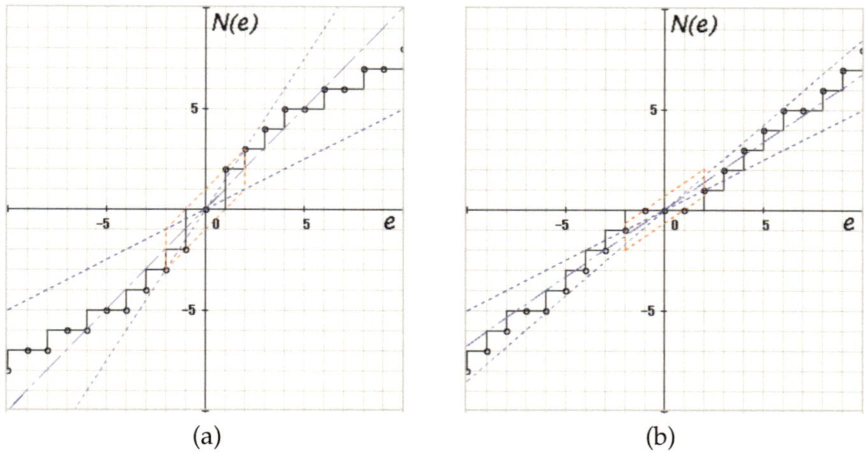

Fig. 2. Discretized nonlinear characteristics on a grid pattern.

for $|e^{\dagger}| \geq \varepsilon$. (In Fig. 2 (a) and (b), the threshold is chosen as $\varepsilon = 2.0$.)
Equation (3) represents a bounded nonlinear characteristic that exists in a finite region. On the other hand, equation (4) represents a sectorial nonlinearity for which the equivalent linear gain exists in a limited range. It can also be expressed as follows:

$$0 \leq g(e^{\dagger})e^{\dagger} \leq \beta e^{\dagger 2} \leq K e^{\dagger 2}. \tag{5}$$

When considering the robust stability in a global sense, it is sufficient to consider the nonlinear term (4) for $|e^{\dagger}| \geq \varepsilon$ because the nonlinear term (3) can be treated as a disturbance signal. (In the stability problem, a fluctuation or an offset of error is assumed to be allowable in $|e^{\dagger}| < \varepsilon$.)

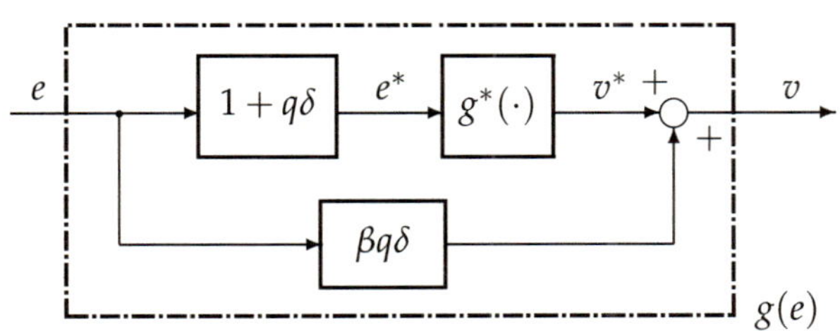

Fig. 3. Nonlinear subsystem $g(e)$.

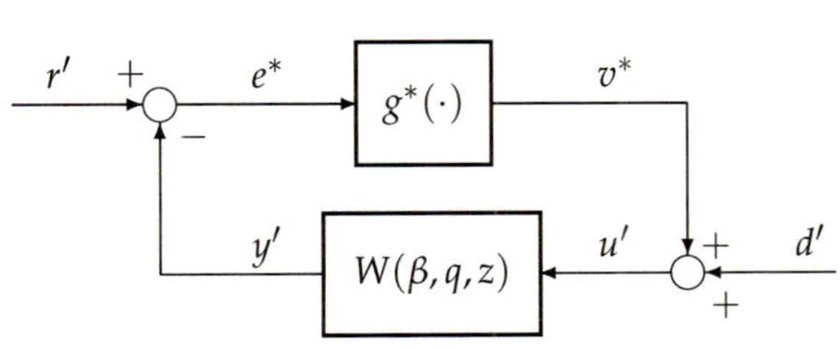

Fig. 4. Equivalent feedback system.

3. Equivalent discrete-time system

In this study, the following new sequences $e_m^{*\dagger}(k)$ and $v_m^{*\dagger}(k)$ are defined based on the above consideration:

$$e_m^{*\dagger}(k) = e_m^{\dagger}(k) + q \cdot \frac{\Delta e^{\dagger}(k)}{h}, \tag{6}$$

$$v_m^{*\dagger}(k) = v_m^{\dagger}(k) - \beta q \cdot \frac{\Delta e^{\dagger}(k)}{h}, \tag{7}$$

where q is a non-negative number, $e_m^{\dagger}(k)$ and $v_m^{\dagger}(k)$ are neutral points of sequences $e^{\dagger}(k)$ and $v^{\dagger}(k)$,

$$e_m^{\dagger}(k) = \frac{e^{\dagger}(k) + e^{\dagger}(k-1)}{2}, \tag{8}$$

$$v_m^{\dagger}(k) = \frac{v^{\dagger}(k) + v^{\dagger}(k-1)}{2}, \tag{9}$$

and $\Delta e^{\dagger}(k)$ is the backward difference of sequence $e^{\dagger}(k)$, that is,

$$\Delta e^{\dagger}(k) = e^{\dagger}(k) - e^{\dagger}(k-1). \tag{10}$$

The relationship between equations (6) and (7) with respect to the continuous values is shown by the block diagram in Fig. 3. In this figure, δ is defined as

$$\delta(z) := \frac{2}{h} \cdot \frac{1 - z^{-1}}{1 + z^{-1}}. \tag{11}$$

Thus, the loop transfer function from v^* to e^* can be given by $W(\beta, q, z)$, as shown in Fig. 4, where

$$W(\beta, q, z) = \frac{(1 + q\delta(z))G(z)C(z)}{1 + (K + \beta q\delta(z))G(z)C(z)}, \tag{12}$$

and r', d' are transformed exogenous inputs. Here, the variables such as v^*, u' and y' written in Fig. 4 indicate the z-transformed ones.
In this study, the following assumption is provided on the basis of the relatively fast sampling and the slow response of the controlled system.

[Assumption] The absolute value of the backward difference of sequence $e(k)$ does not exceed γ, i.e.,

$$|\Delta e(k)| = |e(k) - e(k-1)| \leq \gamma. \tag{13}$$

If condition (13) is satisfied, $\Delta e^{\dagger}(k)$ is exactly $\pm\gamma$ or 0 because of the discretization. That is, the absolute value of the backward difference can be given as

$$|\Delta e^{\dagger}(k)| = |e^{\dagger}(k) - e^{\dagger}(k-1)| = \gamma \text{ or } 0. \quad \square$$

The assumption stated above will be satisfied by the following examples. The phase trace of backward difference $\Delta e^{\dagger}$ is shown in the figures.

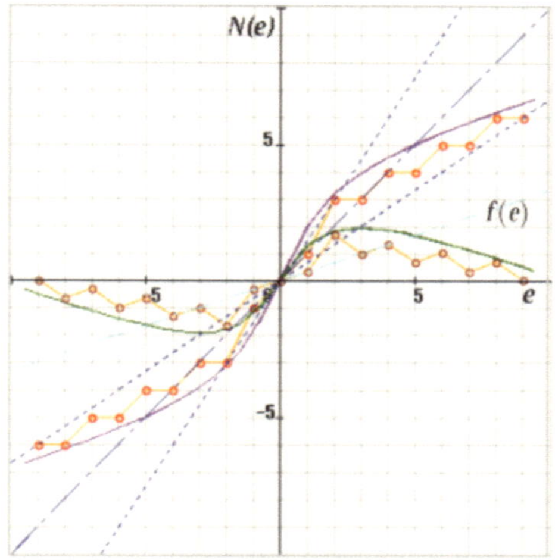

Fig. 5. Nonlinear characteristics and discretized outputs.

4. Norm inequalities

In this section, some lemmas with respect to an ℓ_2 norm of the sequences are presented. Here, we define a new nonlinear function

$$f(e) := g(e) + \beta\, e. \tag{14}$$

When considering the discretized output of the nonlinear characteristic, $v^\dagger = g(e^\dagger)$, the following expression can be given:

$$f(e^\dagger(k)) = v^\dagger(k) + \beta e^\dagger(k). \tag{15}$$

From inequality (4), it can be seen that the function (15) belongs to the first and third quadrants. Figure 5 shows an example of the continuous nonlinear characteristics $u = N(e)$ and $f(e)$, the discretized outputs $u^\dagger = N_d(e^\dagger)$ and $f(e^\dagger)$, and the sector (4) to be considered. When considering the equivalent linear characteristic, the following inequality can be defined:

$$0 \leq \psi(k) := \frac{f(e^\dagger(k))}{e^\dagger(k)} \leq 2\beta. \tag{16}$$

When this type of nonlinearity $\psi(k)$ is used, inequality (4) can be expressed as

$$v^\dagger(k) = g(e^\dagger(k)) = (\psi(k) - \beta)e^\dagger(k). \tag{17}$$

For the neutral points of $e^\dagger(k)$ and $v^\dagger(k)$, the following expression is given from (15):

$$\frac{1}{2}(f(e^\dagger(k)) + f(e^\dagger(k-1))) = v_m^\dagger(k) + \beta e_m^\dagger(k). \tag{18}$$

Moreover, equation (17) is rewritten as $v_m^\dagger(k) = (\psi(k) - \beta)e_m^\dagger(k)$. Since $|e_m^\dagger(k)| \leq |e_m(k)|$, the following inequality is satisfied when a round-down discretization is executed:

$$|v_m^\dagger(k)| \leq \beta |e_m^\dagger(k)| \leq \beta |e_m(k)|. \tag{19}$$

Based on the above premise, the following norm conditions are examined

[Lemma-1] The following inequality holds for a positive integer p:

$$\|v_m^{\dagger}(k)\|_{2,p} \le \beta \|e_m^{\dagger}(k)\|_{2,p} \le \beta \|e_m(k)\|_{2,p}. \tag{20}$$

Here, $\|\cdot\|_{2,p}$ denotes the Euclidean norm, which can be defined by

$$\|x(k)\|_{2,p} := \left(\sum_{k=1}^{p} x^2(k) \right)^{1/2}.$$

(Proof) The proof is clear from inequality (19). □

[Lemma-2] If the following inequality is satisfied with respect to the inner product of the neutral points of (15) and the backward difference:

$$\langle\, v_m^{\dagger}(k) + \beta e_m^{\dagger}(k), \Delta e^{\dagger}(k) \,\rangle_p \ge 0, \tag{21}$$

the following inequality can be obtained:

$$\|v_m^{*\dagger}(k)\|_{2,p} \le \beta \|e_m^{*\dagger}(k)\|_{2,p} \tag{22}$$

for any $q \ge 0$. Here, $\langle \cdot, \cdot \rangle_p$ denotes the inner product, which is defined as

$$\langle\, x_1(k), x_2(k) \,\rangle_p = \sum_{k=1}^{p} x_1(k) x_2(k).$$

(Proof) The following equation is obtained from (6) and (7):

$$\beta^2 \|e_m^{*\dagger}(k)\|_{2,p}^2 - \|v_m^{*\dagger}(k)\|_{2,p}^2 = \beta^2 \|e_m^{\dagger}(k)\|_{2,p}^2 - \|v_m^{\dagger}(k)\|_{2,p}^2 + \frac{2\beta q}{h} \cdot \langle v_m^{\dagger}(k) + \beta e_m^{\dagger}(k), \Delta e^{\dagger}(k) \rangle_p. \tag{23}$$

Thus, (22) is satisfied by using the left inequality of (20). Moreover, as for the input of $g^*(\cdot)$, the following inequality can be obtained from (23) and the right inequality (20):

$$\|v_m^{*\dagger}(k)\|_{2,p} \le \beta \|e_m^{*}(k)\|_{2,p}. \tag{24}$$

□

The left side of inequality (21) can be expressed as a sum of trapezoidal areas.

[Lemma-3] For any step p, the following equation is satisfied:

$$\sigma(p) := \langle\, v_m^{\dagger}(k) + \beta e_m^{\dagger}(k), \Delta e^{\dagger}(k) \,\rangle_p = \frac{1}{2} \sum_{k=1}^{p} (f(e^{\dagger}(k)) + f(e^{\dagger}(k-1))) \Delta e^{\dagger}(k). \tag{25}$$

(Proof) The proof is clear from (18). □

In order to understand easily, an example of the sequences of continuous/discretized signals and the sum of trapezoidal areas is depicted in Fig. 6. The curve e and the sequence of circles $e^{\dagger}$ show the input of the nonlinear element and its discretized signal. The curve u and the sequence of circles $u^{\dagger}$ show the corresponding output of the nonlinear characteristic and its discretized signal, respectively. As is shown in the figure, the sequences of circles $e^{\dagger}$ and $u^{\dagger}$ trace on a grid pattern that is composed of integers. The sequence of circles $v^{\dagger}$ shows

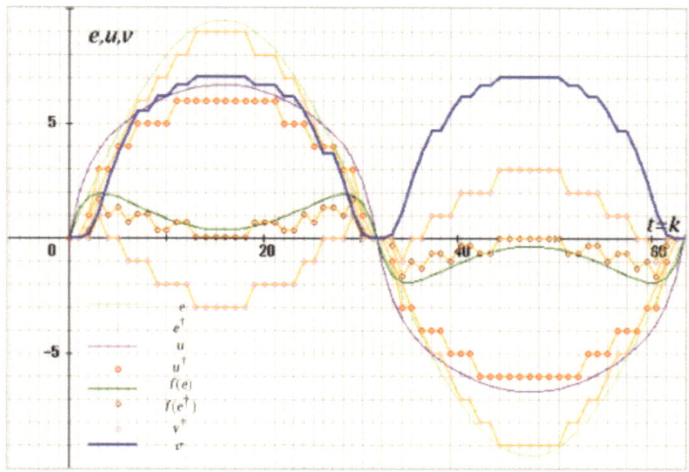

Fig. 6. Discretized input/output signals of a nonlinear element.

the discretized output of the nonlinear characteristic $g(\cdot)$. The curve of shifted nonlinear characteristic $f(e)$ and the sequence of circles $f(e^{\dagger})$ are also shown in the figure.
In general, the sum of trapezoidal areas holds the following property.

[Lemma-4] If inequality (13) is satisfied with respect to the discretization of the control system, the sum of trapezoidal areas becomes non-negative for any p, that is,

$$\sigma(p) \geq 0. \tag{26}$$

(Proof) Since $f(e^{\dagger}(k))$ belongs to the first and third quadrants, the area of each trapezoid

$$\tau(k) := \frac{1}{2}(f(e^{\dagger}(k)) + f(e^{\dagger}(k-1)))\Delta e^{\dagger}(k) \tag{27}$$

On the other hand, the trapezoidal area $\tau(k)$ is non-positive when $e(k)$ decreases (increases) in the first (third) quadrant. Strictly speaking, when ($e(k) \geq 0$ and $\Delta e(k) \geq 0$) or ($e(k) \leq 0$ and $\Delta e(k) \leq 0$), $\tau(k)$ is non-negative for any k. On the other hand, when ($e(k) \geq 0$ and $\Delta e(k) \leq 0$) or ($e(k) \leq 0$ and $\Delta e(k) \geq 0$), $\tau(k)$ is non-positive for any k. Here, $\Delta e(k) \geq 0$ corresponds to $\Delta e^{\dagger}(k) = \gamma$ or 0 (and $\Delta e(k) \leq 0$ corresponds to $\Delta e^{\dagger}(k) = -\gamma$ or 0) for the discretized signal, when inequality (13) is satisfied.
The sum of trapezoidal area is given from (25) as:

$$\sigma(p) = \sum_{k=1}^{p} \tau(k). \tag{28}$$

Therefore, the following result is derived based on the above. The sum of trapezoidal areas becomes non-negative, $\sigma(p) \geq 0$, regardless of whether $e(k)$ (and $e^{\dagger}$(k)) increases or decreases. Since the discretized output traces the same points on the stepwise nonlinear characteristic, the sum of trapezoidal areas is canceled when $e(k)$ (and $e^{\dagger}(k)$ decreases (increases) from a certain point $(e^{\dagger}(k), f(e^{\dagger}(k)))$ in the first (third) quadrant. (Here, without loss of generality, the response of discretized point $(e^{\dagger}(k), f(e^{\dagger}(k)))$ is assumed to commence at the origin.) Thus, the proof is concluded. □

5. Robust stability in a global sense

By applying a small gain theorem to the loop transfer characteristic (12), the following robust stability condition of the discretized nonlinear control system can be derived

[Theorem] If there exists a $q \geq 0$ in which the sector parameter β with respect to nonlinear term $g(\cdot)$ satisfies the following inequality, the discrete-time control system with sector nonlinearity (4) is robust stable in an ℓ_2 sense:

$$\beta < \beta_0 = K \cdot \eta(q_0, \omega_0) = \max_q \min_\omega K \cdot \eta(q, \omega), \tag{29}$$

when the linearized system with nominal gain K is stable.
The η-function is written as follows:

$$\eta(q, \omega) := \frac{-q\Omega \sin\theta + \sqrt{q^2\Omega^2 \sin^2\theta + \rho^2 + 2\rho\cos\theta + 1}}{\rho}, \qquad \forall \omega \in [0, \omega_c], \tag{30}$$

where $\Omega(\omega)$ is the distorted frequency of angular frequency ω and is given by

$$\delta(e^{j\omega h}) = j\Omega(\omega) = j\frac{2}{h}\tan\left(\frac{\omega h}{2}\right), \quad j = \sqrt{-1} \tag{31}$$

and ω_c is a cut-off frequency. In addition, $\rho(\omega)$ and $\theta(\omega)$ are the absolute value and the phase angle of $KG(e^{j\omega h})C(e^{j\omega h})$, respectively.
(Proof) Based on the loop characteristic in Fig. 4, the following inequality can be given with respect to $z = e^{j\omega h}$:

$$\|e_m^*(z)\|_{2,p} \leq c_1\|r_m'(z)\|_{2,p} + c_2\|d_m'(z)\|_{2,p} + \sup_{z=1}|W(\beta, q, z)| \cdot \|w_m^{*\dagger}(z)\|_{2,p}. \tag{32}$$

Here, $r_m'(z)$ and $d_m'(z)$ denote the z-transformation for the neutral points of sequences $r'(k)$ and $d'(k)$, respectively. Moreover, c_1 and c_2 are positive constants.
By applying inequality (24), the following expression is obtained:

$$\left(1 - \beta \cdot \sup_{z=1}|W(\beta, q, z)|\right)\|e_m^*(z)\|_{2,p} \leq c_1\|r_m'(z)\|_{2,p} + c_2\|d_m'(z)\|_{2,p}. \tag{33}$$

Therefore, if the following inequality (i.e., the small gain theorem with respect to ℓ_2 gains) is valid,

$$|W(\beta, q, e^{j\omega h})| = \left|\frac{(1 + jq\Omega(\omega))P(e^{j\omega h})C(e^{j\omega h})}{1 + (K + j\beta q\Omega(\omega))P(e^{j\omega h})C(e^{j\omega h})}\right| = \left|\frac{(1 + jq\Omega(\omega))\rho(\omega)e^{j\theta(\omega)}}{K + (K + j\beta q\Omega(\omega))\rho(\omega)e^{j\theta(\omega)}}\right| < \frac{1}{\beta}. \tag{34}$$

the sequences $e_m^*(k)$, $e_m(k)$, $e(k)$ and $y(k)$ in the feedback system are restricted in finite values when exogenous inputs $r(k)$, $d(k)$ are finite and $p \to \infty$. (The definition of ℓ_2 stable for discrete-time systems was given in (10; 11).)
From the square of both sides of inequality (34),

$$\beta^2\rho^2(1 + q^2\Omega^2) < (K + K\rho\cos\theta - \beta\rho q\Omega\sin\theta)^2 + (K\rho\sin\theta + \beta\rho q\Omega\cos\theta)^2.$$

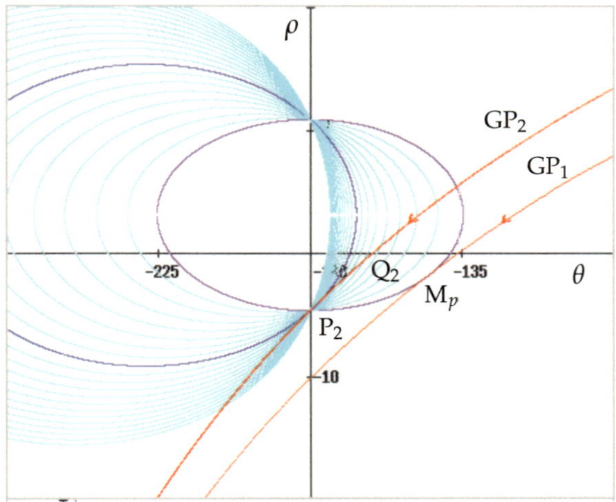

Fig. 7. An example of modified Nichols diagram ($M = 1.4$, $c_q = 0.0, 0.2, \cdots, 4.0$).

Thus, the following quadratic inequality can be obtained:

$$\beta^2\rho^2 < -2\beta K\rho q\Omega \sin\theta + K^2(1 + \rho\cos\theta)^2 + K^2\rho^2\sin^2\theta. \tag{35}$$

Consequently, as a solution of inequality (35),

$$\beta < \frac{-Kq\Omega\sin\theta + K\sqrt{q^2\Omega^2\sin^2\theta + \rho^2 + 2\rho\cos\theta + 1}}{\rho} = K\eta(q,\omega). \tag{36}$$

□

6. Modified Nichols diagram

In the previous papers,the inverse function was used instead of the η-function, i.e.,

$$\xi(q,\omega) = \frac{1}{\eta(q,\omega)}.$$

Using the notation, inequality (29) can be rewritten as follows:

$$M_0 = \xi(q_0,\omega_0) = \min_q \max_\omega \xi(q,\omega) < \frac{K}{\beta}. \tag{37}$$

When $q = 0$, the ξ-function can be expressed as:

$$\xi(0,\omega) = \frac{\rho}{\sqrt{\rho^2 + 2\rho\cos\theta + 1}} = |T(e^{j\omega h})|, \tag{38}$$

where $T(z)$ is the complementary sensitivity function for the discrete-time system. It is evident that the following curve on the gain-phase plane,

$$\xi(0,\omega) = M, \quad (M : \text{const.}) \tag{39}$$

corresponds to the contour of the constant M in the Nichols diagram. In this study, since an arbitrary non-negative number q is considered, the ξ-function that corresponds to (38) and (39) is given as follows:

$$\frac{\rho}{-q\Omega\sin\theta+\sqrt{q^2\Omega^2\sin^2\theta+\rho^2+2\rho\cos\theta+1}}=M. \tag{40}$$

From this expression, the following quadratic equation can be obtained:

$$(M^2-1)\rho^2+2\rho M(M\cos\theta-q\Omega\sin\theta)+M^2=0. \tag{41}$$

The solution of this equation is expressed as follows:

$$\rho=-\frac{M}{M^2-1}(M\cos\theta-q\Omega\sin\theta)\pm\frac{M}{M^2-1}\sqrt{(M\cos\theta-q\Omega\sin\theta)^2-(M^2-1)}. \tag{42}$$

The modified contour in the gain-phase plane (θ, ρ) is drawn based on the equation of (42). Although the distorted frequency Ω is a function of ω, the term $q\Omega = c_q \geq 0$ is assumed to be a constant parameter. This assumption for M contours was also discussed in (11). Figure 7 shows an example of the modified Nichols diagram for $c_q \geq 0$ and $M = 1.4$. Here, GP_1 is a gain-phase curve that touches an M contour at the peak value ($M_p = \xi(0,\omega_p) = 1.4$). On the other hand, GP_2 is a gain-phase curve that crosses the $\theta = -180°$ line and all the M contours at the gain crossover point P_2. That is, the gain margin g_M becomes equal to $-20\log_{10} M/(M+1) = 4.68$[dB]. The latter case corresponds to the discrete-time system in which Aizerman's conjecture is valid (14; 15). At the continuous saddle point P_2, the following equation is satisfied:

$$\left(\frac{\partial\xi(q,\omega)}{\partial q}\right)_{q=q_0,\omega=\omega_0}=0. \tag{43}$$

Evidently, the phase margin p_M is obtained from the phase crossover point Q_2.

7. Controller design

The PID controller applied in this study is given by the following algorithm:

$$u_c(k)=K_pu^\dagger(k)+C_i\sum_{j=0}^{k}u^\dagger(j)+C_d\Delta u^\dagger(k), \tag{44}$$

where $\Delta u^\dagger(k) = u^\dagger(k) - u^\dagger(k-1)$ is a backward difference in integer numbers, and each coefficient is defined as

$$K_p,\ C_i,\ C_d \in Z_+, \quad Z_+=\{0,1,2,3\cdots\}.$$

Here, K_p, C_i, and C_d correspond to K_p, K_ph/T_I, and K_pT_D/h in the following (discrete-time z-transform expression) PID algorithm:

$$C(z)=K_p\left(1+\frac{h}{T_I(1-z^{-1})}+\frac{T_D}{h}(1-z^{-1})\right). \tag{45}$$

We use algorithm (44) without division because the variables $u^\dagger$, u_c, and coefficients K_p, C_i, C_d are integers.

Using the z-transform expression, equation (44) is written as:

$$\begin{aligned} u_c(z) &= C(z)u(z) \\ &= \left(K_p + C_i(1 + z^{-1} + z^{-2} + \cdots) + C_d(1 - z^{-1})\right) u(z). \end{aligned}$$

In the closed form, controller $C(z)$ can be given as

$$C(z) = K_p + C_i \cdot \frac{1}{1 - z^{-1}} + C_d(1 - z^{-1}) \tag{46}$$

for discrete-time systems. When comparing equations (45) and (46), C_i and C_d become equal to $K_p h/T_I$ and $K_p T_D/h$, respectively.
The design method adopted in this paper is based on the classical parameter specifications in the modified Nichols diagram. This method can be conveniently designed, and it is significant in a physical sense (i.e., mechanical vibration and resonance).
Furthermore, in this article, PID-D^2 is considered. The algorithm is written as

$$u_c(k) = K_p u^\dagger(k) + C_i \sum_{j=0}^{k} u^\dagger(j) + C_{d1}\Delta u^\dagger(k) + C_{d2}\Delta^2 u^\dagger(k), \tag{47}$$

where

$$\Delta^2 u^\dagger(k) = \Delta u^\dagger(k) - \Delta u^\dagger(k-1) = u^\dagger(k) - 2u^\dagger(k-1) + u^\dagger(k-2).$$

Thus, the controller $C(z)$ can be given as

$$C(z) = K_p + C_i \cdot \frac{1}{1 - z^{-1}} + C_{d1}(1 - z^{-1}) + C_{d2}(1 - 2z^{-1} + z^{-2}) \tag{48}$$

for discrete-time systems.

8. Numerical examples

[Example-1] Consider the following third order controlled system:

$$G(s) = \frac{K_1}{(s + 0.04)(s + 0.2)(s + 0.4)}, \tag{49}$$

where $K_1 = 0.0002 = 2.0 \times 10^{-4}$.

	K_p	C_i	C_d	β_0	g_M[dB]	p_M[deg]	M_p
(i)	100	0	0	$\overline{\beta}$	7.72	34.9	1.82
(ii)	100	3	0	0.98	5.92	23.8	2.61
(iii)	100	3	120	$\overline{\beta}$	11.1	35.4	1.69
(iv)	50	0	0	$\overline{\beta}$	10.8	48.6	1.29
(v)	50	2	0	1.00	7.92	30.6	1.99
(vi)	50	2	60	$\overline{\beta}$	13.3	40.5	1.45

Table 1. PID parameters for Example-1 (g_M: gain margins, p_M: phase margins, M_p: peak values, β_0: allowable sectors).

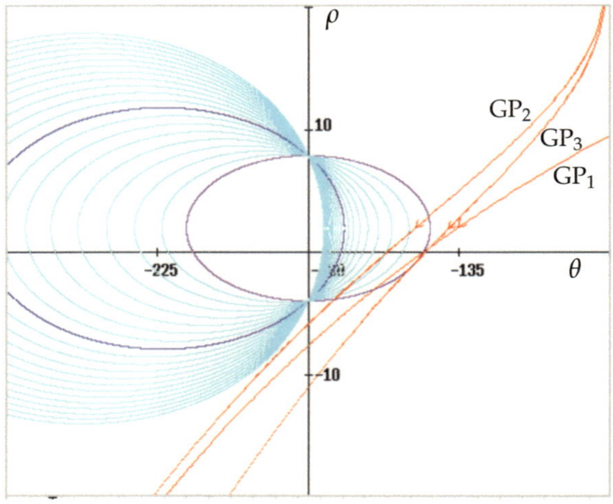

Fig. 8. Modified contours and gain-phase curves for Example-1 ($M = 1.69$, $c_q = 0.0, 0.2, \cdots, 4.0$).

The discretized nonlinear characteristic (discretized sigmoid, i.e. arc tangent (12)) is as shown in Fig. **??** (a). In this article, the resolution value and the sampling period are assumed to be $\gamma = 1.0$ and $h = 1.0$ as described in section 2.
When choosing the nominal gain $K = 1.0$ and the threshold $\varepsilon = 2.0$, the sectorial area of the discretized nonlinear characteristic for $\varepsilon \leq |e|$ can be determined as [0.5, 1.5] drawn by dotted lines in the figure. Figure 8 shows gain-phase curves of $KG(e^{j\omega h})C(e^{j\omega h})$ on the modified Nichols diagram. Here, GP_1, GP_2, and GP_3 are cases (i), (ii), and (iii), respectively. The PID parameters are specified as shown in Table 1. The gain margins g_M, the phase margin p_M and the peak value M_p can be obtained from the gain crossover points P, the phase crossover points Q, and the points of contact with regard to the M contours, respectively.
The max-min value β_0 is calculated from (29) (e.g., (ii)) as follows:

$$\beta_0 = \max_{q} \min_{\omega} K \cdot \eta(q, \omega) = K \cdot \eta(q_0, \omega_0) = 0.98.$$

Therefore, the allowable sector for nonlinear characteristic $g(\cdot)$ is given as [0.0, 1.98]. The stability of discretized control system (ii) (and also systems (i),(iii)) will be guaranteed. In this example, the continuous saddle point (43) appears (i.e., Aizerman's conjecture is satisfied). Thus, the allowable interval of equivalent linear gain K_ℓ can be given as $0 < K_\ell < 1.98$. In the case of (i) and (iii), β_0 becomes not less than K. However, from the definition of (4), $\bar{\beta}$ in the tables should be considered $\beta_0 = \bar{\beta} = 1.0$. Figure 9 shows step responses for the three cases. In this figure, the time-scale line is drawn in $10h$ increments because of avoiding indistinctness. Sequences of the input $u^\dagger(k)$ and the output $u_c^\dagger$ of PID controller are also shown in the figure. Here, $u_c^\dagger(k)$ is drawn to the scale of $1/100$. Figure 10 shows phase traces (i.e., sequences of $(e(k), \Delta e(k))$ and $(e^\dagger((k), \Delta e^\dagger(k)))$. As is obvious from Fig. 10, assumtion (13) is satisfied. The step response (i) remains a sutained oscillation and an off-set. However, as for (ii) and (iii) the responses are improved by using the PID, especially integral (I: a summation in this paper) algorithm.
The discretized linear characteristic as shown in Fig. **??** (b) is also considered here. In the figure, the sectorial area of the discretized characteristic for $\varepsilon \leq |e|$ can be determined as [0.5, 0.85] drawn by dotted lines, and the nominal gain is given as $K = 0.675$. When

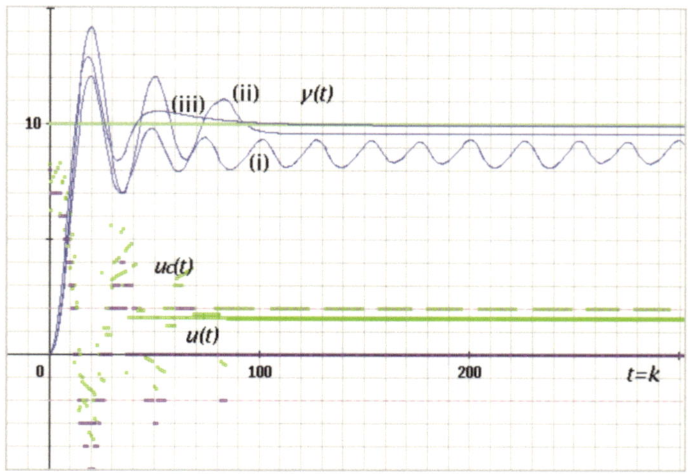

Fig. 9. Step responses for Example-1.

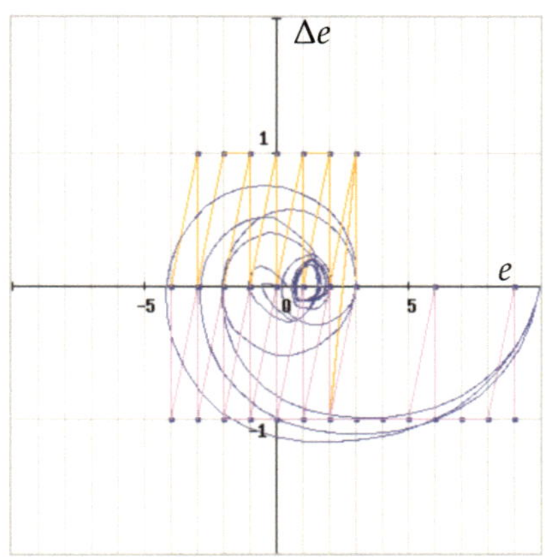

Fig. 10. Phase traces for Example-1.

normalizing the nominal gain for $K = 1.0$ (i.e., choosing the gain constant $K_2 = K_1/0.675$), the sectorial area is determined as [0.74, 1.26]. In this case, an example of step responces is depicted in Fig. 11. The PID parameters used here are also shown in Table 1.

[Example-2] Consider the following fourth order controlled system:

$$G(s) = \frac{K_1}{(s+0.04)(s+0.2)(s+0.4)(s+1.0)}, \tag{50}$$

where $K_1 = 0.0002 = 2.0 \times 10^{-4}$. The same nonlinear characteristic and the nominal gain are chosen as shown in Example-1.

Figure 12 shows gain-phase curves of $KG(e^{j\omega h})C(e^{j\omega h})$ on the modified Nichols diagram. Here, GP_1, GP_2, GP_3 and GP_4 are cases (i), (ii), (iii) and (iv) in Table 2, respectively. In this example, PID-D^2 control scheme is also used. The PID-D^2 parameters are specified as shown

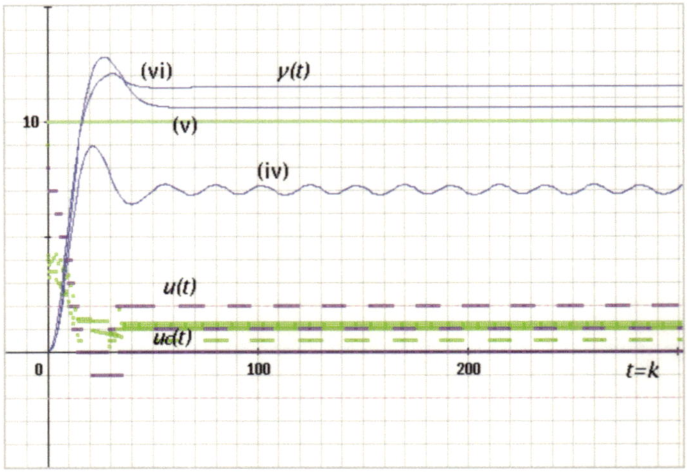

Fig. 11. Step responses for Example-1 (Discretized linear case).

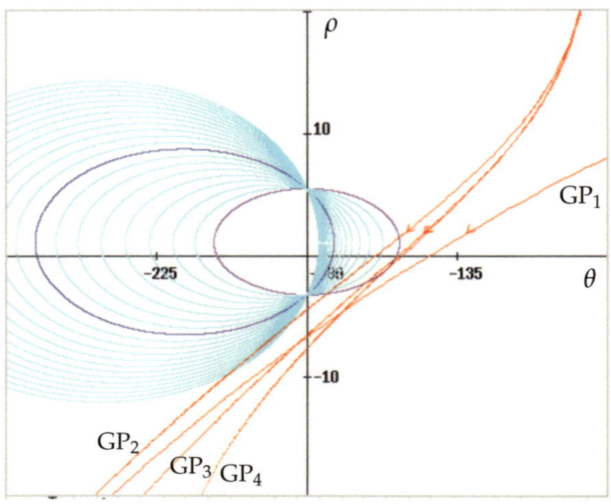

Fig. 12. Modified contours and gain-phase curves for Example-2 ($M = 2.14$, $c_q = 0.0, 0.2, \cdots, 4.0$).

in the table. The max-min value β_0 is calculated from (29) (e.g., (iv)) as follows:

$$\beta_0 = \max_{q} \min_{\omega} K \cdot \eta(q, \omega) = K \cdot \eta(q_0, \omega_0) = 0.69.$$

Therefore, the allowable sector for nonlinear characteristic $g(\cdot)$ is given as [0.0, 1.69]. The stability of discretized control system (ii) (and also systems (i),(iii),(iv)) will be guaranteed. In this example, the continuous saddle point (43) appears (i.e., Aizerman's conjecture is satisfied). Thus, the allowable interval of equivalent gain K_ℓ can be given as $0 < K_\ell < 1.69$.

As is shown in Fig. 13, the step response (i) remains a sutained oscillation and an off-set. However, as for (ii), (iii) and (iv) the responses are improved by using PI, PID and PID-D^2 algorithm (D^2: a second difference).

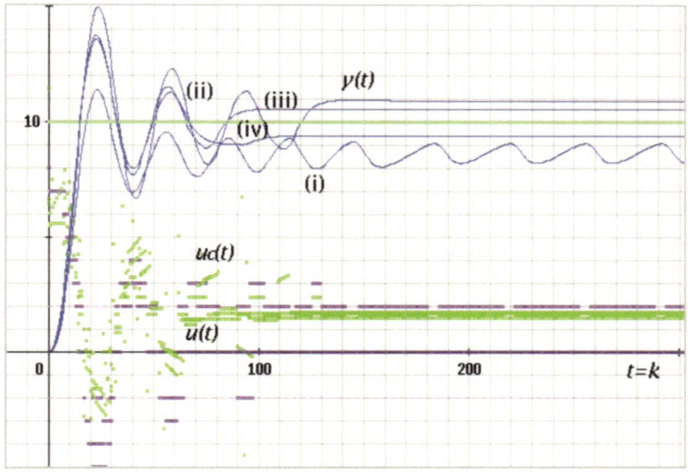

Fig. 13. Step responses for Example-2.

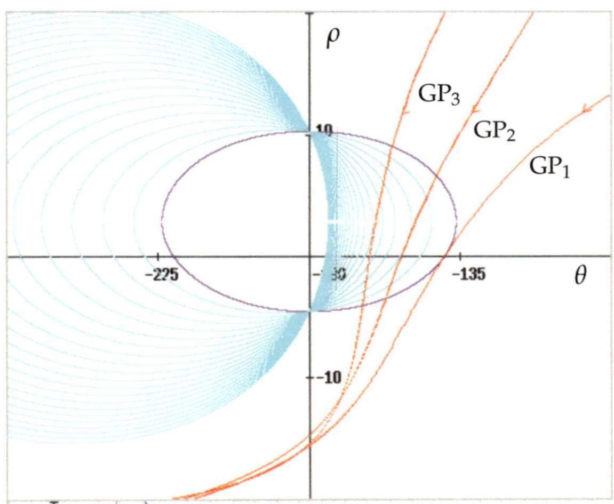

Fig. 14. Modified contours and gain-phase curves for Example-3 ($M = 1.44$, $c_q = 0.0, \cdots, 4.0$).

[Example-3] Consider the following nonminimum phase controlled system:

$$G(s) = \frac{K_2(s+0.2)(-s+0.4)}{(s+0.02)(s+0.04)(s+1.0)}, \tag{51}$$

	K_p	C_i	C_{d1}	C_{d2}	β_0	g_M[dB]	p_M[deg]	M_p
(i)	80	0	0	0	$\bar{\beta}$	6.8	37.2	1.79
(ii)	80	3	0	0	0.69	4.69	20.9	3.10
(iii)	80	3	60	0	1.00	6.63	27.4	2.26
(iv)	80	3	60	120	$\bar{\beta}$	7.76	28.8	2.14

Table 2. PID-D^2 parameters for Example-2.

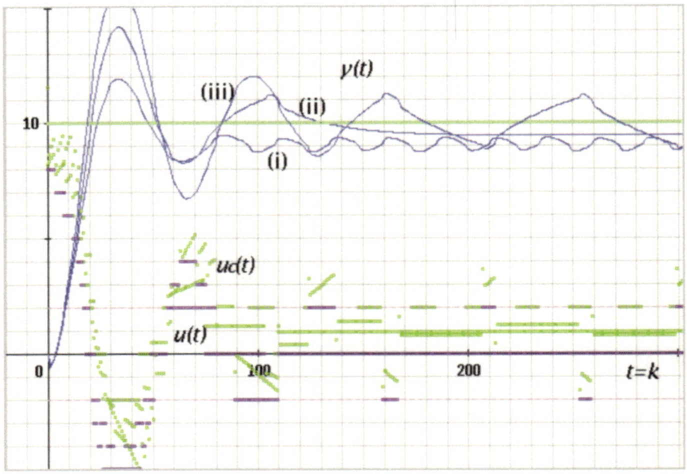

Fig. 15. Step responses for Example-3.

where $K_3 = 0.001 = 1.0 \times 10^{-3}$. Also, in this example, the same nonlinear characteristic and the nominal gain are chosen as shown in Example-1. The modified Nichols diagram with gain-phase curves of $KG(e^{j\omega h})C(e^{j\omega h})$ is as shown in Fig. 14. Here, GP_1, GP_2 and GP_3 are cases (i), (ii), and (iii), and the PID parameters are specified as shown in Table 3. Figure 15 shows time responses for the three cases.

For example, in the case of (iii), although the allowable sector of equivalent linear gain is $0 < K_\ell < 5.9$, the allowable sector for nonlinear characteristic becomes [0.0, 1.44] as shown in Table 3. Since the sectorial area of the discretized nonlinear characteristic is [0.5, 1.5], the stability of the nonlinear control system cannot be guaranteed. The response for (iii) actually fluctuates as shown in Figs. 15 and 16. This is a counter example for Aizerman's conjecture.

9. Conclusion

In this article, we have described robust stabilization and discretized PID control for continuous plants on a grid pattern with respect to controller variables and time elapsed. A robust stability condition for nonlinear discretized feedback systems was presented along with a method for designing PID control. The design procedure employs the modified Nichols diagram and its parameter specifications. The stability margins of the control system are specified directly in the diagram. Further, the numerical examples showed that the time responses can be stabilized for the required performance. The concept described in this article will be applicable to digital and discrete-event control system in general.

	K_p	C_i	C_d	β_0	g_M[dB]	p_M[deg]	M_p
(i)	100	0	0	0.92	15.5	40.6	1.44
(ii)	100	2	0	0.71	14.7	27.7	2.09
(iii)	100	4	40	0.44	15.3	18.1	3.18

Table 3. PID parameters for Example-3.

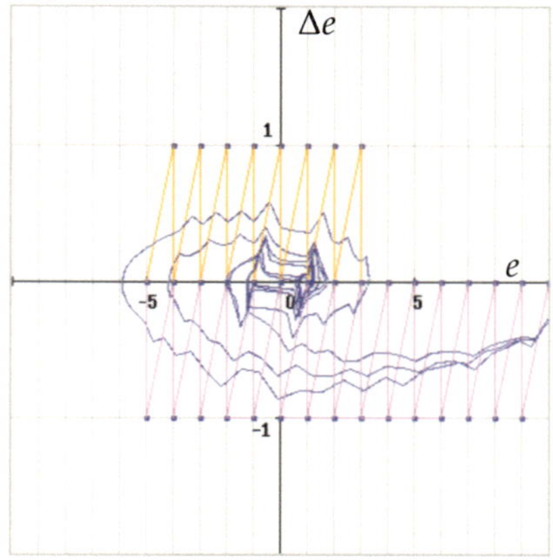

Fig. 16. Phase traces for Example-3.

10. References

[1] R. E. Kalman, "Nonlinear Aspects of Sampled-Data Control Systems", *Proc. of the Symposium on Nonlinear Circuit Analysis*, vol. VI, pp.273-313, 1956.

[2] R. E. Curry, *Estimation and Control with Quantized Measurements*, Cambridge, MIT Press, 1970.

[3] D. F. Delchamps, "Stabilizing a Linear System with Quantized State Feedback", *IEEE Trans. on Automatic Control*, vol. 35, pp. 916-924, 1990.

[4] M. Fu, "Robust Stabilization of Linear Uncertain Systems via Quantized Feedback", *Proc. of IEEE Int. Conf. on Decision and Control*, TuA06-5, 2003.

[5] A. Datta, M.T. Ho and S.P. Bhattacharyya, *Structure and Synthesis of PID Controllers*, Springer-Verlag, 2000.

[6] F. Takemori and Y. Okuyama, "Discrete-Time Model Reference Feedback and PID Control for Interval Plants" *Digital Control 2000:Past, Present and Future of PID Control*, Pergamon Press, pp. 260-265, 2000.

[7] Y. Okuyama, "Robust Stability Analysis for Discretized Nonlinear Control Systems in a Global Sense", *Proc. of the 2006 American Control Conference*, Minneapolis, USA, pp. 2321-2326, 2006.

[8] Y. Okuyama, "Robust Stabilization and PID Control for Nonlinear Discretized Systems on a Grid Pattern", *Proc. of the 2008 American Control Conference*, Seattle, USA, pp. 4746-4751, 2008.

[9] Y. Okuyama, "Discretized PID Control and Robust Stabilization for Continuous Plants", *Proc. of the 17th IFAC World Congress*, Seoul, Korea, pp. 1492-1498, 2008.

[10] Y. Okuyama *et al.*, "Robust Stability Evaluation for Sampled-Data Control Systems with a Sector Nonlinearity in a Gain-Phase Plane" *Int. J. of Robust and Nonlinear Control*, Vol. 9, No. 1, pp. 15-32, 1999.

[11] Y. Okuyama *et al.*, "Robust Stability Analysis for Non-Linear Sampled-Data Control Systems in a Frequency Domain", *European Journal of Control*, Vol. 8, No. 2, pp. 99-108, 2002.

[12] Y. Okuyama *et al.*, "Amplitude Dependent Analysis and Stabilization for Nonlinear Sampled-Data Control Systems", *Proc. of the 15th IFAC World Congress*, T-Tu-M08, 2002.

[13] Y. Okuyama, "Robust Stabilization and for Discretized PID Control Systems with Transmission Delay", *Proc. of IEEE Int. Conf. on Decision and Control*, Shanghai, P. R. China, pp. 5120-5126, 2009.
[14] L. T. Grujic, "On Absolute Stability and the Aizerman Conjecture", *Automatica*, pp. 335-349. 1981.
[15] Y. Okuyama *et al.*, "Robust Stability Analysis for Nonlinear Sampled-Data Control Systems and the Aizerman Conjecture", *Proc. of IEEE Int. Conf. on Decision and Control*, Tampa, USA, pp. 849-852, 1998.

12

Simple Robust Normalized PI Control for Controlled Objects with One-order Modelling Error

Makoto Katoh
Osaka Institute of Technology
Japan

1. Introduction

In this section, the small gain theorem is introduced as a background theory of this chapter. Then, a large mission on safety and a small mission on analytic solutions are introduced after indicating the some problems in discussing robust PI control systems. Moreover, the way how it came to be possible to obtain the analytic solution of PI control adjustment for the concrete robust control problems with uncertain modeling error which is impossible using the space theory for MIMO systems, is shown for a SISO system. The worst lines of closed loop gain margin were shown in a parameter plane. Finally, risk, merit and demerit of the robust control is discussed and the countermeasure for safeness of that is introduced. And some theme, eg., in the lag time system, the MIMO system and a class of non-linear system for expansion of the approach of this chapter is introduced.

- Many researchers have studied on many kinds of robust system recently. The basic robust stability concept is based on the small gain theorem (Zbou K. with Doyle F. C. and Glover K., 1996). The theorem insists that a closed loop system is internal (robust) stable sufficiently and necessary if the H_∞ norm of the nominal closed loop transfer function is smaller than the inverse of H_∞ norm of the any uncertainty of feedback elements. (Fig. 1) Moreover, the expansion of the theorem claims that a closed loop system is stable sufficiently if the product of H_∞ norms of open loop transfer functions is smaller than 1 when the forward and the feedback transfer functions are both stable.

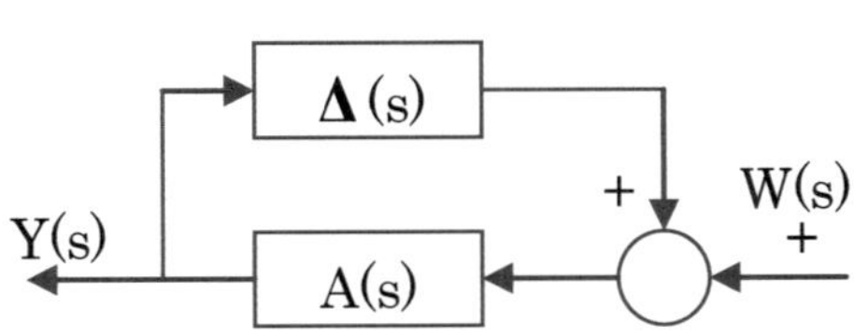

when $\|\Delta\|_\infty \le \gamma$ if $\|A\|_\infty < \dfrac{1}{\gamma}$ then internal stable

Fig. 1. Feed back system configuration with unknown feedback element

- In MIMO state space models (A,B,C,D), a necessary and sufficient condition using LMI (Linear Matrix Inequality) for the above bounded norm of controlled objects is known as the following Bounded Real Lemma (Zhou K. And Khargonekar P.P., 1988) using the Riccati unequality and Shure complement.

$$\exists P = P^T > 0 \quad such \quad that \quad \begin{bmatrix} PA + A^T P & PB & C^T \\ B^T P & -\frac{1}{\gamma} I_m & D^T \\ C & D & -\frac{1}{\gamma} I_p \end{bmatrix} < 0 \Leftrightarrow \|G(s)\|_\infty < \frac{1}{\gamma} \qquad (0)$$

A gain margin between the critical closed loop gain of a dependent type IP controller by the *Furwits* criteria and the analytical closed loop gain solution when closed loop Hardy space norm became 1, and the parametric stability margin (Bhattacharyya S. P., Chapellat H., and Keel L. H., 1994; Katoh 2010) on uncertain time constant and damping coefficient were selected in this chapter for its easiness and robustness although it was expected also using this lemma that internal stable concrete conditions for controlled objects and forward controllers may obtain.

- One of H_∞ control problems is described to obtain a robust controller K(s) when Hardy space norm of closed loop transfer function matrix is bounded like Fig.2 assuming various (additive, multiplicative, left co-prime factor etc.) uncertainty of controlled objects P(s) (Zbou K. with Doyle F. C. and Glover K., 1996).

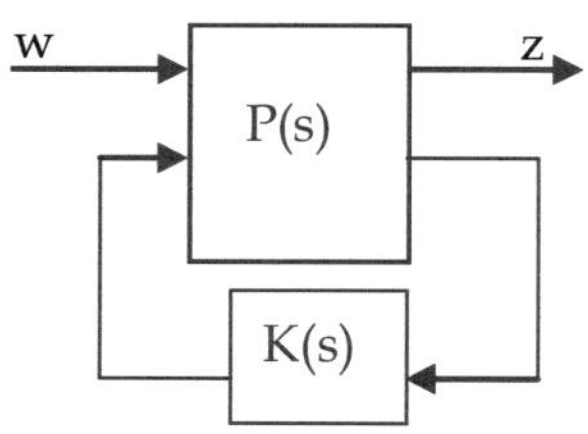

$$\mathbf{z} = \mathbf{\Phi w}$$

$$\mathbf{\Phi} = \mathbf{P}_{11} + \mathbf{P}_{12}\mathbf{K}(\mathbf{I} - \mathbf{P}_{22}\mathbf{K})^{-1}\mathbf{P}_{21}, \quad \|\mathbf{\Phi}\|_\infty < \gamma$$

Fig. 2. Feed back system configuration for obtained robust control K(s) when Hardy space norm of closed loop transfer function matrix is bounded

- The purpose of this chapter for the robust control problem is to obtain analytical solution of closed loop gain of a dependent type IP controller and analyze robustness by closed loop gain margin for 2nd order controlled objects with one-order feedback like (left co-prime factor) uncertainty as Fig.1 in some tuning regions of IP controller when Hardy space norm of closed loop transfer function matrix is bounded less than 1.
- Though another basic robust problem is a cooperation design in frequency region between competitive sensitivity and co-sensitivity function, it was omitted in this chapter because a tuning region of IP control was superior for unknown input disturbance other tuning region was superior for unknown reference disturbance.
- However, there is some one not simple for using higher order controllers with many stable zeros and using the norm with window (Kohonen T., 1995, 1997) for I in Hardy space for evaluating the uncertainty of models. Then, a number of robust PI or PID controller and compensator design methods have recently been proposed. But, they are not considered on the modelling error or parameter uncertainty.
- Our given large mission is to construct safe robust systems using simple controllers and simple evaluating method of the uncertainty of models. Then, we have proposed robust PI controllers for controlled objects without stable zeros (Katoh M., 2008, 2009). Our small mission in this chapter is to obtain analytical solution of controller gain with flat

gain curve in a band width as Butter-worse filter for the 3rd order closed systems with one-order modelling errors and to show the robust property by loop gain margin for damping coefficients of nominal controlled objects and time constants of missing objects (sensor and signal conditioner) using Table Computation Tool (Excel: Microsoft Co. LTD). It is interesting and important historically that infinity time constant is contained in the investing set though it isn't existing actually. Moreover, we confirm the robustness for a parameter change by raising and lowering of step response using CAD Tool (Simulink: Mathworks Co. LTD).

- Risk of Integral term of PI controller when disconnecting the feedback line can be rescued by M/A station used in many industrial applications or by shutdown of the plant in our standing point. Then, we show a simple soft M/A station for simulation with PI controllers in appendix.
- This method is not actually because it becomes complicated to computation for higher order objects contained plants with lag time as pointed out in appendix but useful.

2. System description

In this section, a description of the higher order generalized system for later 2nd order examples with one-order modeling error is presented although they may not computed concretely.

2.1 Normalized transfer function

In this section, how to normalize and why to normalize transfer functions are explained.

The following transfer functions of controlled objects Eq. (1) with multiplicative *one-order modeling error* Eq. (2) are normalized using a general natural angular frequency ω_n^* and gain $K^* = K_o K_s$ as Eq. (3) although the three positions distributed for normalization are different.

$$G(s) = K_o \prod_{i=1}^{r} \frac{\omega_{ni}^{\ 2}}{s^2 + 2\varsigma_i \omega_{ni} s + \omega_{ni}^{\ 2}} \prod_{j=1}^{q} \frac{\alpha_j}{s + \alpha_j} \tag{1}$$

$$H(s) = \frac{K_s}{\varepsilon s + 1} \tag{2}$$

$$\begin{aligned}\bar{G}_l(s) &= \frac{1}{K^*} G(s) H(s) \\ &= \frac{\omega_n^{*2r+q+1}}{\varepsilon s^{2r+q+1} + \beta_{2r+q} \omega_n^* s^{2r+q} + \cdots + \beta_1 \omega_n^{*2r+q} s + \omega_n^{*2r+q+1}}\end{aligned} \tag{3}$$

Moreover, converting the differential operator s to $\bar{s}$ as,

$$\bar{s} \triangleq \frac{s}{\omega_n^*}, \bar{\varepsilon} = \omega_n^* \varepsilon \tag{4}$$

the following normalized open loop transfer function is obtained:

$$\bar{G}_l(\bar{s}) = \frac{1}{\bar{\varepsilon} \bar{s}^{n+1} + \bar{\beta}_{2r+q} \bar{s}^n + \cdots + \bar{\beta}_1 \bar{s} + 1} \quad \text{where } n = 2r + q \tag{5}$$

Neglecting *one-order modeling error*, the following normalized open loop transfer function is obtained:

$$\bar{G}(\bar{s}) = \frac{1}{\bar{s}^n + \gamma_{n-1}\bar{s}^{n-1} \cdots + \gamma_1 \bar{s} + 1} \quad \text{where n = 2r + q} \tag{6}$$

2.2 State space models

In this section, 3 kinds of description on normalized state space models are shown although they may not computed concretely. First shows a continuous realization form of the higher order transfer functions to a SISO system. Second shows a normalized sampled system form for the first continuous realization on sampling points. Third shows a normalized continuously approximated form using logarithm conversion for the second sampled system.

Minimum realization of normalized transfer function: The normalized transfer function, shown in Eq. (6), is converted to the following *SISO* controllable minimum realization:

$$\begin{aligned} \dot{\mathbf{x}}(t) &= \tilde{\mathbf{A}}\mathbf{x}(t) + \tilde{\mathbf{b}}u(t) \\ y(t) &= \tilde{\mathbf{c}}\mathbf{x}(t) + du(t) \end{aligned} \tag{7}$$

Normalized sampled system on sampling points: Integrating the response between two sampling points to the next sampling point, the following precise sampled system is obtained:

$$\begin{aligned} \mathbf{x}((k+1)h) &= e^{\tilde{\mathbf{A}}h}\mathbf{x}(kh) - \tilde{\mathbf{A}}^{-1}[\mathbf{I} - e^{\tilde{\mathbf{A}}h}]\tilde{\mathbf{b}}u(kh) \\ y(kh) &= \tilde{\mathbf{c}}\mathbf{x}(kh) + \tilde{d}u(kh) \end{aligned} \tag{8}$$

Normalized sampled system approximated:

Approximating Eq. (3) by the advanced difference method, the following sampled system is obtained:

$$\begin{aligned} \mathbf{x}(k+1) &= (\mathbf{I} + \tilde{\mathbf{A}}h)\mathbf{x}(k) + \tilde{\mathbf{b}}hu(k) \\ y(k) &= \tilde{\mathbf{c}}\mathbf{x}(k) + du(k) \end{aligned} \tag{9}$$

Normalized System in *continuous region*:

Returning to the continuous region after conversion using the matrix logarithm function, the following system is obtained in *continuous region*:

$$\begin{aligned} \dot{\mathbf{x}}(t) &= \mathbf{A}^*\mathbf{x}(t) + \mathbf{b}^*u(t) \\ y(t) &= \tilde{\mathbf{c}}\mathbf{x}(t) + \tilde{d}u(t) \end{aligned} \tag{10}$$

$$\begin{aligned} \textit{where} \quad \mathbf{A}^* &= \frac{1}{h}\ln(\mathbf{I} + \tilde{\mathbf{A}}h) \\ &= \tilde{\mathbf{A}} - \tilde{\mathbf{A}}^2 h + \frac{1}{2}\tilde{\mathbf{A}}^3 h^2 \cdots \\ \mathbf{b}^* &= (\mathbf{I} - \tilde{\mathbf{A}}h + \frac{1}{2}\tilde{\mathbf{A}}^2 h^2 \cdots)\tilde{\mathbf{b}} \end{aligned} \tag{11}$$

The condition of convergence for logarithm conversion Eq. (11) of controllable accompany description Eq. (7) is not described because it is assumed that the sampled time h is sufficiently small. The approximated order is then selected as the 9th order. Thus, $\tilde{d} = 0$ is assumed for the simplification.

3. Controller and parameter tuning

In this section, an IP controller and a number of parameter tuning methods are presented in order to increase the robustness of the control system.

3.1 Normalized IP controller

In this section, 3 kinds of description on normalized integral lead dependent type IP controller which is not conventional proportional lead dependent type PI controller are shown. First is showing inherent frequency for normalization as magnitudes of integral and proportional in continuous systems. Second is showing that in digital systems. Third is showing again that of digital systems in returning approximated continuous systems.

$$C(\bar{s}) = \bar{K}_i(\frac{1}{\bar{s}} + \bar{p}) = \bar{K}_i\omega_n(\frac{1}{s} + \frac{\bar{p}}{\omega_n}) \tag{12}$$

$$C^*(\bar{z}) = \frac{\bar{K}_i\bar{p}(\bar{z} - 1 + \bar{h}/\bar{p})}{\bar{z} - 1} = \frac{\bar{K}_i\bar{p}(z - 1 + h\omega_n^*/\bar{p})}{z - 1} \tag{13}$$

$$C^*(\bar{s}) = \bar{K}_i(\frac{1}{\bar{s}} + \bar{p}) = \bar{K}_i\omega_n^*(\frac{1}{s} + \frac{\bar{p}}{\omega_n^*}) \tag{14}$$

Note that the digital *IP* controller of Eq. (13) is asymptotic to the proportional control as h approaches zero or $\bar{p}$ becomes larger. This controller is called *IPL tuning*. Then, the stable zero = - 1/ $\bar{p}$ must be placed not in the neighborhood of the system poles for safety.

3.2 Stability of closed loop transfer function

In this section, more higher order systems are processed for consideration generally on three tuning region classified by the amplitude of P control parameter using Hurwits approach in example of a second-order system with one-order modelling error. It is guessed that there may be four elementary tuning regions and six combinatorial tuning regions generally in the aspect of Hurwits stability.

The following normalized loop transfer function is obtained from the normalized controlled object Eq. (5) and the normalized controller Eq. (12):

$$W(\bar{s}) = \frac{\bar{K}_i(1 + \overline{ps})(\overline{\varepsilon s} + 1)}{\overline{\varepsilon s}^{n+2} + \bar{\beta}_{2r+q}\bar{s}^{n+1} + \cdots + \bar{\beta}_1\bar{s}^2 + (\bar{K}_i\bar{p} + 1)\bar{s} + \bar{K}_i} \tag{15}$$

If the original parameters $\forall i, j, \varsigma_i > 0, \alpha_j > 0$ are positive, then $\forall k, \beta_k > 0$.
Assuming $\bar{p} > 0$ and $\bar{K}_i > 0$, and that

$$\varphi(\bar{s}) \triangleq \varepsilon\bar{s}^{n+2} + \beta_{2r+q}\bar{s}^{n+1} + \cdots + \beta_1\bar{s}^2 + (\bar{K}_i\bar{p} + 1)\bar{s} + \bar{K}_i \tag{16}$$

is a *Hurwits* polynomial, the stability limits of $\overline{K}_i$ can be obtained as a region of $\overline{p}$. Then, this region is called a *IPL* region when $\overline{p}$ has a maximum lower bound and an *IP0* region when $\overline{p}$ =0. The region between zero and the minimum upper bound is called the *IPS*. The region between the minimum upper bound and the maximum lower bound is called the *IPM* region. Generally, there are four elementary regions and six combinatorial regions.

3.3 Stationary points investing approach on fraction equation

In this section, Stationary Points Investing approach on Fraction Equation for searching local maximum with equality restriction is shown using Lagrange's undecided multiplier approach. Then, multiple same solutions of the independent variable are solved at the stationary points. *They can be used to check for mistakes in calculation as self-diagnostics approach.*

Here, the common normalized control parameters $\overline{K}_i$ and $\overline{p}$ will be obtained in *continuous region,* which has reduction models reduced from *original region.*

Stationary Points Investing for Fraction Equation approach for searching local maximum with equality restriction:

$$|W(j\overline{\omega})| = [\frac{u(\overline{\omega})}{v(\overline{\omega})}]^{0.5} \tag{17}$$

$$\rightarrow solve\_local\_muximum/minimum \quad for \quad \overline{\omega}=\overline{\omega}_s$$

$$such \quad that \quad |W(j\overline{\omega}_s)|=1$$

This is the design policy of servo control for wide band width. In particular, $|W(o)|=1$ means that steady state error is 0.

Next, *Lagrange*'s undecided multiplier approach is applied to obtain stationary points $\overline{\omega}_s$ with equality restriction using the above *u,v* notations.

Then, the original problem can be converted to the following problem:

$$J^{+}(\overline{\omega},\lambda) = |W(j\overline{\omega})|^{2+} = \frac{u(\overline{\omega})}{v(\overline{\omega})} + \lambda\{u(\overline{\omega})-v(\overline{\omega})\} \tag{18}$$

$$\rightarrow solve\_local\_maximum/minimum$$

where λ is a *Lagrange* multiplier.

The necessary conditions for obtaining the local minimum/maximum of a new function become as followings.

$$\left.\frac{\partial J^{+}(\overline{\omega},\lambda)}{\partial\overline{\omega}}\right|_{\overline{\omega}=\overline{\omega}_s} = \frac{u'(\overline{\omega}_s)v(\overline{\omega}_s)-u(\overline{\omega}_s)v'(\overline{\omega}_s)}{v(\overline{\omega}_s)^2} + \lambda\{u'(\overline{\omega}_s)-v'(\overline{\omega}_s)\}=0 \tag{19}$$

$$\left.\frac{\partial J^{+}(\overline{\omega},\lambda)}{\partial\lambda}\right|_{\overline{\omega}=\overline{\omega}_s} = u(\overline{\omega}_s)-v(\overline{\omega}_s)=0 \tag{20}$$

The following relations are obtained from eqs. (19) and (20):

$$u'(\bar{\omega}_s) = v'(\bar{\omega}_s)\, or \quad \lambda = -\frac{1}{v(\bar{\omega}_s)} \tag{21}$$
$$u(\bar{\omega}_s) = v(\bar{\omega}_s)$$

Solutions of control parameters:
Solving these simultaneous equations, the following functions can be obtained:

$$\begin{aligned} \bar{\boldsymbol{\omega}}_{\mathbf{s}} &= \mathbf{f}(\mathbf{a}, \bar{p}) \\ \bar{K}_i^{\,j} &= g(\mathbf{a}, \bar{p}, \bar{\omega}_{sj}) \qquad (j = 1, 2, \cdots, \alpha) \end{aligned} \tag{22}$$

where $\bar{\omega}_s$ is the stationary points vector.
Multiple solutions of $\bar{K}_i$ can be used to check for mistakes in calculation.

3.4 Example of a second-order system with one-order modelling error

In this section, an IP control system in continuous design for a second-order original controlled object without one-order sensor and signal conditioner dynamics is assumed for simplicity. The closed loop system with uncertain one-order modeling error is normalized and obtained the stable region of the integral gain in the three tuning region classified by the amplitude of P control parameter using Hurwits approach. Then, the safeness of the only I tuning region and the risk of the large P tuning region are discussed. Moreover, the analytic solutions of stationary points and double same integral gains are obtained using the Stationary Points Investing on Fraction Equation approach for the gain curve of a closed loop system.

Here, an *IP* control system for a second-order controlled object without sensor dynamics is assumed.

Closed-loop transfer function:

$$G(s) = \frac{K_o \omega_n^2}{s^2 + 2\varsigma\omega_n s + \omega_n^2} \tag{23}$$

$$H(s) = \frac{K_s}{\varepsilon s + 1} \tag{24}$$

$$\bar{G}_o(s) = \frac{1}{K_s K_o} G(s)H(s) = \frac{\omega_n^2}{(\varepsilon s + 1)(s^2 + 2\varsigma\omega_n s + \omega_n^2)} \tag{25}$$

$$\bar{s} \triangleq \frac{s}{\omega_n}, \bar{\varepsilon} = \omega_n \varepsilon \tag{26}$$

$$W(\bar{s}) = \frac{\bar{K}_i(1 + \overline{ps})(\overline{\varepsilon s} + 1)}{\overline{\varepsilon s}^4 + (2\varsigma\bar{\varepsilon} + 1)\bar{s}^3 + (\bar{\varepsilon} + 2\varsigma)\bar{s}^2 + (\bar{K}_i\bar{p} + 1)\bar{s} + \bar{K}_i} \tag{27}$$

Stable conditions by *Hurwits* approach with four parameters*:*

a. In the case of a certain time constant

***IPL&IPS* Common Region:**

$$0 < \bar{K}_i < \max[0, \min[\bar{k}_2, \bar{k}_3, \infty]] \tag{28}$$

$$\bar{k}_2 \triangleq \frac{2\varsigma(\varepsilon^2 + 2\varsigma\varepsilon + 1)}{\varepsilon \bar{p}} \tag{29}$$

$$[\bar{p}\{4\varsigma^2\varepsilon + 2\varsigma\varepsilon^2 + 2\varsigma - \varepsilon\} - (2\varsigma\varepsilon + 1)^2] \tag{30}$$

$$\bar{k}_3 \triangleq \frac{+\sqrt{\begin{array}{l}[\bar{p}\{4\varsigma^2\varepsilon + 2\varsigma\varepsilon^2 + 2\varsigma - \varepsilon\} - (2\varsigma\varepsilon + 1)^2]^2 \\ +8\varepsilon\varsigma \bar{p}^2(\varepsilon^2 + 2\varsigma\varepsilon + 1)\end{array}}}{2\varepsilon \bar{p}^2} \tag{31}$$

$where \quad \bar{p} > 0 \quad for \quad 4\varsigma^2\varepsilon + 2\varsigma\varepsilon^2 + 2\varsigma \le \varepsilon$

IPL, IPS Separate Region:

The integral gain stability region is given by Eqs. (28)-(30).

$$\begin{array}{ll} 0 < \dfrac{(2\varsigma\bar{\varepsilon} + 1)^2}{4\varsigma^2\bar{\varepsilon} + 2\varsigma\bar{\varepsilon}^2 + 2\varsigma - \bar{\varepsilon}} \le \bar{p} & (PL) \\ 0 < \bar{p} < \dfrac{(2\varsigma\bar{\varepsilon} + 1)^2}{4\varsigma^2\bar{\varepsilon} + 2\varsigma\bar{\varepsilon}^2 + 2\varsigma - \bar{\varepsilon}} & (PS) \\ for \quad 4\varsigma^2\bar{\varepsilon} + 2\varsigma\bar{\varepsilon}^2 + 2\varsigma - \bar{\varepsilon} > 0 & \end{array} \tag{32}$$

It can be proven that $\bar{k}_3$ >0 in the *IPS* region, and

$$\bar{k}_2 \to \infty, \bar{k}_3 \to \infty \quad when \quad \bar{p} \to 0 \tag{33}$$

***IP0* Region:**

$$0 < \bar{K}_i < \frac{2\varsigma(\varepsilon^2 + 2\varsigma\varepsilon + 1)}{(2\varsigma\varepsilon + 1)^2} \quad where \quad \bar{p} = 0 \tag{34}$$

The IP0 region is most safe because it has not zeros.

b. In the case of an uncertain positive time constant

***IPL&IPS* Common Region:**

$$0 < \bar{K}_i < \max[0, \min[\bar{k}_2\big|_{\bar{\varepsilon}=1}, \bar{k}_3\big|_{\bar{\varepsilon}=\bar{\varepsilon}_p}]] \quad when\, \bar{p} > 0 \tag{35}$$

$where \quad \bar{k}_3^{'}(\bar{p}, \varsigma, \bar{\varepsilon}_p) = 0$

$$for \quad 4\varsigma^2\varepsilon + 2\varsigma\varepsilon^2 + 2\varsigma < \varepsilon \tag{36}$$

$$\min_{\varepsilon} \bar{k}_2 = \frac{4\varsigma(\varsigma + 1)}{\bar{p}} \quad when \quad \bar{\varepsilon} = 1 \tag{37}$$

***IPL, IPS* Separate Region:**

This region is given by Eq. (32).

***IP0* Region:**

$$0 < \bar{K}_i < 2\varsigma(1-\varsigma^2)\Big|_{\varepsilon=\bar{\varepsilon}_p, 0<\varsigma<0.707}, \frac{1}{2\varsigma}\Big|_{\varepsilon=+\infty, \varsigma>0.707} \tag{38}$$

when $(\varepsilon_p = \dfrac{\varsigma}{(1-2\varsigma^2)} > 0, 0 < \varsigma < 0.707), p = 0$

c. Robust loop gain margin

The following loop gain margin is obtained from eqs. (28) through (38) in the cases of certain and uncertain parameters:

$$gm \triangleq \frac{\bar{K}_{iUL}}{\bar{K}_i} \tag{39}$$

where $\bar{K}_{iUL}$ is the upper limit of the stable loop gain $\bar{K}_i$.

Stable conditions by Hurwits approach with three parameters:

The stability conditions will be shown in order to determine the risk of *one order modelling error,*

$$0 < \bar{K}_i \quad \textit{where} \quad \bar{p} \geq \frac{1}{2\varsigma} \quad (PL) \tag{40}$$

$$0 < \bar{K}_i < \frac{2\varsigma}{1-2\varsigma\bar{p}} \quad \textit{where} \quad 0 \leq \bar{p} < \frac{1}{2\varsigma} \quad (P0) \tag{41}$$

Hurwits Stability is omitted because $\bar{h}$ is sufficiently small, although it can be checked using the bilinear transform.

Robust loop gain margin:

$$gm = \infty \quad (PL\_region) \tag{42}$$

It is risky to increase the loop gain in the *IPL* region too much, even if the system does not become unstable because a model order error may cause instability in the *IPL* region. In the *IPL* region, the sensitivity of the disturbance from the output rises and the flat property of the gain curve is sacrificed, even if the disturbance from the input can be isolated to the output upon increasing the control gain.

Frequency transfer function:

$$\begin{aligned} &|W(j\bar{\omega})| \triangleq \left[\frac{\bar{K}_i^2\{1+\bar{\omega}^2\bar{p}^2\}}{(\bar{K}_i - 2\varsigma\bar{\omega}^2)^2 + \bar{\omega}^2\{1-\bar{\omega}^2+\bar{K}_i\bar{p}\}^2}\right]^{0.5} \\ &\rightarrow solve\_local\_maximum/minmum \quad for \quad \bar{\omega} = \bar{\omega}_s \\ &such \quad that \quad |W(j\bar{\omega}_s)| = 1 \end{aligned} \tag{43}$$

When the evaluation function is considered to be two variable functions ($\bar{\omega}$ and $\bar{K}_i$) and the stationary point is obtained, the system with the parameters does not satisfy the above stability conditions.

Therefore, only the stationary points in the direction of $\bar{\omega}$ will be obtained without considering the evaluation function on $\bar{K}_i$ alone.

Stationary points and the integral gain:

Using the *Stationary Points Investing for Fraction Equation* approach based on *Lagrange*'s undecided multiplier approach with equality restriction, the following two loop gain equations on *x* are obtained. Both identities can be used to check for miscalculation.

$$\bar{K}_{i1} = 0.5\{x^2 + 2(2\varsigma^2 - 1)x + 1\} / \{2\varsigma + (x-1)\bar{p}\} \tag{44}$$

$$\bar{K}_{i2} = 0.5\{3x^2 + 4(2\varsigma^2 - 1)x + 1\} / \{2\varsigma + (2x-1)\bar{p}\} \tag{45}$$

$$where \quad x = \bar{\omega}_s^2 \geq 0$$

Equating the right-hand sides of these equations, the third-order algebraic equation and the solutions for semi-positive stationary points are obtained as follows:

$$x = 0, x = \sqrt{\frac{2(2\varsigma^2 - 1)(2\varsigma - \bar{p})}{\bar{p}} - 1} \tag{46}$$

These points, which are called the first and second stationary points, call the first and second tuning methods, respectively, which specify the points for gain 1.

4. Numerical results

In this section, the solutions of double same integral gain for a tuning region at the stationary point of the gain curve of the closed system are shown and checked in some parameter tables on normalized proportional gains and normalized damping coefficients. Moreover, loop gain margins are shown in some parameter tables on uncertain time constants of one-order modeling error and damping coefficients of original controlled objects for some tuning regions contained with safest only I region.

	0.9	0.95	1.00	1.05	1.10
0.8	0.8612	1.0496	1.1892	1.3068	1.4116
0.9	0.6999	0.9424	1.0963	1.2197	1.3271
1.0	-99	0.8186	1.0000	1.1335	1.2457
1.1	-99	0.6430	0.8932	1.0446	1.1647
1.2	-99	-99	0.7598	0.3480	1.0812

Table 1. ω_p values for ς and $\bar{p}$ in *IPL* tuning by the first tuning method

	0.9	0.95	1.00	1.05	1.10
0.8	0.7750	1.0063	1.2500	1.5063	1.7750
0.9	0.6889	0.8944	1.1111	1.3389	1.5778
1.0	1.2272	0.8050	1.0000	1.2050	1.4200
1.1	1.1077	0.7318	0.9091	1.0955	1.2909
1.2	1.0149	1.0791	0.8333	1.0042	1.1833

Table 2. $\bar{K}_{i1} = \bar{K}_{i2}$ values for ς and p in *IPL* tuning by the first tuning method

Table 1 lists the stationary points for the first tuning method. Table 2 lists the integration gains ($\overline{K}_{i1} = \overline{K}_{i2}$) obtained by substituting Eq. (46) into Eqs. (44) and (45) for various damping coefficients.
Table 3 lists the integration gains ($\overline{K}_{i1} = \overline{K}_{i2}$) for the second tuning method.

	0.9	0.95	1.00	1.05	1.10
1.3	1.0	0.8333	0.7143	0.6250	0.5556
1.4	1.250	1.0	0.8333	0.7143	0.6250
1.5	1.667	1.250	1.0	0.8333	0.7143
1.6	2.50	1.667	1.250	1.0	0.8333
1.7	5.00	2.50	1.667	1.250	1.0

Table 3. $\overline{K}_{i1} = \overline{K}_{i2}$ values for ς and $\overline{p}$ in *IPL* tuning by the second tuning method

Then, a table of loop gain margins ($gm > 1$) generated by Eq. (39) using the stability limit and the loop gain by the second tuning method on uncertain $\overline{\varepsilon}$ in a given region of $\overline{\varepsilon}$ for each controlled ς by *IPL* ($\overline{p}$ =1.5) control is very useful for analysis of robustness. Then, the unstable region, the unstable region, which does not become unstable even if the loop gain becomes larger, and robust stable region in which uncertainty of the time constant, are permitted in the region of $\overline{\varepsilon}$.
Figure 3 shows a reference step up-down response with unknown input disturbance in the *continuous region*. The gain for the disturbance step of the *IPL tuning* is controlled to be approximately 0.38 and the settling time is approximately 6 sec.
The robustness on indicial response for the damping coefficient change of ±0.1 is an advantageous property. Considering *Zero Order Hold.* with an imperfect dead-time compensator using 1st-order *Pade* approximation, the overshoot in the reference step response is larger than that in the *original region* or that in the *continuous region.*

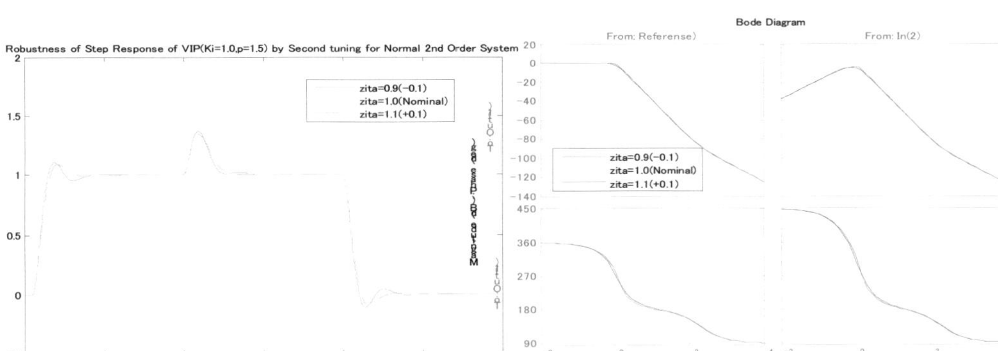

$(\varsigma_0 = 1 \pm 0.1, K_i = 1.0, p = 1.5, \omega_n = 1.005, \varsigma = 1, s_0 = 199.3, k = -0.0050674)$

Fig. 3. Robustness of *IPL* tuning for damping coefficient change.

Then, Table 4 lists robust loop gain margins ($gm > 1$) using the stability limit by Eq.(37) and the loop gain by the second tuning method on uncertain $\overline{\varepsilon}$ in the region of $(0.1 \le \overline{\varepsilon} \le 10)$ for each controlled ς (>0.7) by *IPL*($\overline{p}$ =1.5) control. The first gray row shows the area that is also unstable.

Table 5 does the same for each controlled ς (>0.4) by *IPS*($\bar{p}$ =0.01). Table 6 does the same for each controlled ς (>0.4) by *IP0*($\bar{p}$ =0.0).

eps/zita	0.3	0.7	0.8	0.9	1	1.1	1.2
0.1	-2.042	-1.115	1.404	5.124	10.13	16.49	24.28
0.2	-1.412	-0.631	0.788	2.875	5.7	9.33	13.83
1.5	-0.845	-0.28	0.32	1.08	2	3.08	4.32
2.4	-1.019	-0.3	0.326	1.048	1.846	2.702	3.6
3.2	-1.488	-0.325	0.342	1.06	1.8	2.539	3.26
5	-2.128	-0.386	0.383	1.115	1.778	2.357	2.853
10	-4.596	-0.542	0.483	1.26	1.81	2.187	2.448

Table 4. Robust loop gain margins on uncertain $\bar{\varepsilon}$ in each region for each controlled ς at IPL ($\bar{p}$=1.5)

eps/zita	0.4	0.5	0.6	0.7	0.8
0.1	1.189	1.832	2.599	3.484	4.483
0.6	1.066	1.524	2.021	2.548	3.098
1	1.097	1.492	1.899	2.312	2.729
2.1	1.254	1.556	1.839	2.106	2.362
10	1.717	1.832	1.924	2.003	2.073

Table 5. Robust loop gain margins on uncertain $\bar{\varepsilon}$ in each region for each controlled ς at IPS ($\bar{p}$=0.01)

eps/zita	0.3	0.4	0.5	0.6	0.7	0.8	0.9	1
0.1	0.6857	1.196	1.835	2.594	3.469	4.452	5.538	6.722
0.4	0.6556	1.087	1.592	2.156	2.771	3.427	4.118	4.84
0.5	0.6604	1.078	1.556	2.081	2.645	3.24	3.859	4.5
0.6	0.6696	1.075	1.531	2.025	2.547	3.092	3.655	4.231
1	0.7313	1.106	1.5	1.904	2.314	2.727	3.141	3.556
2.1	0.9402	1.264	1.563	1.843	2.109	2.362	2.606	2.843
10	1.5722	1.722	1.835	1.926	2.004	2.073	2.136	2.195
9999	1.9995	2	2	2	2	2	2	2

Table 6. Robust loop gain margins on uncertain $\bar{\varepsilon}$ in each region for each controlled ς at *IP0* ($\bar{p}$=0.0)

These table data with additional points were converted to the 3D mesh plot as following Fig. 4. As IP0 and IPS with very small $\bar{p}$ are almost equivalent though the equations differ quiet, the number of figures are reduced. It implies validity of both equations.

According to the line of worst loop gain margin as the parameter of attenuation in the controlled objects which are described by gray label, this parametric stability margin (PSM) (Bhattacharyya S. P., Chapellat H., and Keel L. H., 1994) is classified to 3 regions in IPS and IP0 tuning regions and to 4 regions in IPL tuning regions as shown in Fig.5. We may call the larger attenuation region with more than 2 loop gain margin to the strong robust segment region in which region uncertainty time constant of one-order modeling error is allowed in the any region and some change of attenuation is also allowed.

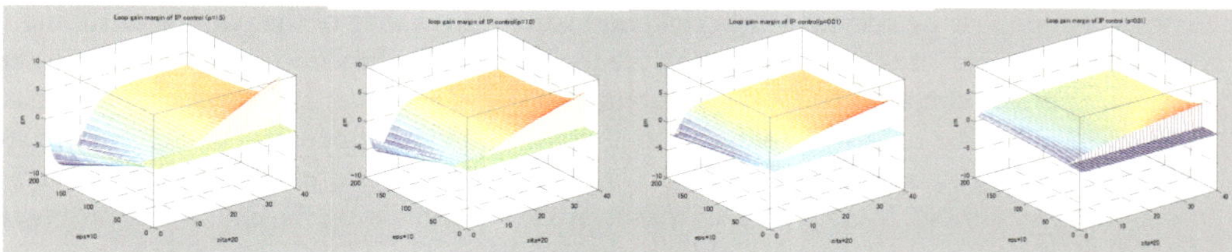

(a) $\overline{p}$ =1.5 (b) $\overline{p}$ =1.0 (c) $\overline{p}$ =0.5 (d) $\overline{p}$ =0.01or 0

Fig. 4. Mesh plot of closed loop gain margin

Next, we call the larger attenuation region with more than $\gamma > 1$ and less than 2 loop gain margin to the weak robust segment region in which region uncertainty time constant of one-order modeling error is only allowed in some region over some larger loop gain margin and some larger change of attenuation is not allowed. The third and the forth segment is almost unstable. Especially, notice that the joint of each segment is large bending so that the sensitivity of uncertainty for loop gain margin is larger more than the imagination.

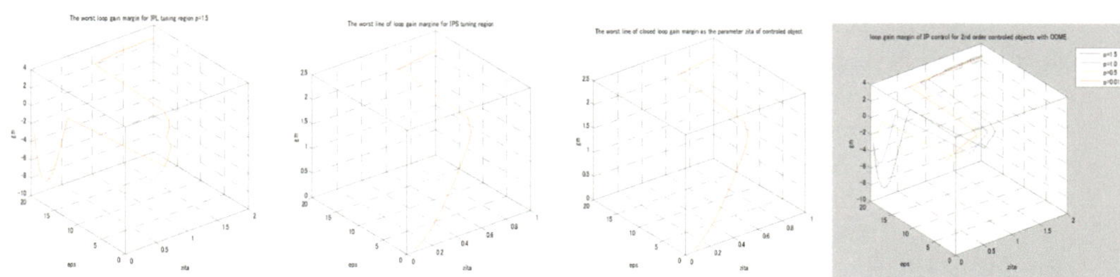

(a) $\overline{p}$ =1.5 (b) $\overline{p}$ =0.01 (c) $\overline{p}$ =0 (d) $\overline{p}$ =1.5, 1.0, 0.5, 0.01

Fig. 5. The various worst lines of loop gain margin in a parameter plane (certain&uncertain)

Moreover, the readers had to notice that the strong robust region and weak robust region of IPL is shift to larger damping coefficient region than ones of IPS and IP0. Then, this is also one of risk on IPL tuning region and change of tuning region from IP0 or IPS to IPL region.

5. Conclusion

In this section, the way to convert this IP control tuning parameters to independent type PI control is presented. Then, parameter tuning policy and the reason adopted the policy on the controller are presented. The good and no good results, limitations and meanings in this chapter are summarized. The closed loop gain curve obtained from the second order example with one-order feedback modeling error implies the butter-worth filter model matching method in higher order systems may be useful. The Hardy space norm with bounded window was defined for I, and robust stability was discussed for MIMO system by an expanssion of small gain theorem under a bounded condition of closed loop systems.

- We have obtained first an integral gain leading type of normalized IP controller to facilitate the adjustment results of tuning parameters explaining in the later. The controller is similar that conventional analog controllers are proportional gain type of PI controller. It can be converted easily to independent type of PI controller as used in recent computer controls by adding some converted gains. The policy of the parameter tuning is to make the norm of the closed loop of frequency transfer function contained one-order modeling error with uncertain time constant to become less than 1. The reason of selected the policy is to be able to be similar to the conventional expansion of the small gain theorem and to be possible in PI control. Then, the controller and uncertainty of the model becomes very simple. Moreover, a simple approach for obtaining the solution is proposed by optimization method with equality restriction using Lagrange's undecided multiplier approach for the closed loop frequency transfer function.
- The stability of the closed loop transfer function was investigated using Hurwits Criteria as the structure of coefficients were known though they contained uncertain time constant.
- The loop gain margin which was defined as the ratio of the upper stable limit of integral gain and the nominal integral gain, was investigated in the parameter plane of damping coefficient and uncertain time constant. Then, the robust controller is safe in a sense if the robust stable region using the loop gain margin is the single connection and changes continuously in the parameter plane even if the uncertain time constant changes larger in a wide region of damping coefficient and even if the uncertain any adjustment is done. Then, IP0 tuning region is most safe and IPL region is most risky.
- Moreover, it is historically and newly good results that the worst loop gain margin as each damping coefficient approaches to 2 in a larger region of damping coefficients.
- The worst loop gain margin line in the uncertainty time constant and controlled objects parameters plane had 3 or 4 segments and they were classified strong robust segment region for more than 2 closed loop gain margin and weak robust segment region for more than $\gamma > 1$ and less than 2 loop gain margin. Moreover, the author was presented also risk of IPL tuning region and the change of tuning region.
- It was not good results that the analytical solution and the stable region were complicated to obtain for higher order systems with higher order modeling error though they were easy and primary. Then, it was unpractical.

6. Appendix

A. Example of a second-order system with lag time and one-order modelling error

In this section, for applying the robust PI control concept of this chapter to systems with lag time, the systems with one-order model error are approximated using Pade approximation and only the simple stability region of the integral gain is shown in the special proportional tuning case for simplicity because to obtain the solution of integral gain is difficult.

Here, a digital *IP* control system for a second-order controlled object with lag time L without sensor dynamics is assumed. For simplicity, only special proportional gain case is shown.

Transfer functions:

$$G(s) = \frac{Ke^{-Ls}}{Ts+1} = \frac{K(1-0.5Ls)}{(Ts+1)(0.5Ls+1)} = \frac{K\omega_n^2(1-0.5Ls)}{s^2+2\varsigma\omega_n s+\omega_n^2} \quad \text{(A.1)}$$

$$\omega_n = \frac{1}{\sqrt{0.5TL}}, \varsigma = 0.5\sqrt{0.5TL}\{(T+0.5L)/(0.5TL)\} \tag{A2}$$

$$H(s) = \frac{K_s}{\varepsilon s + 1} \tag{A3}$$

Normalized operation:

The normalize operations as same as above mentioned are done as follows.

$$\overline{s} \triangleq \frac{s}{\omega_n}, \overline{L} \triangleq L\omega_n \tag{A4}$$

$$\overline{G}(\overline{s}) = \frac{(1-0.5\overline{L}\overline{s})}{\overline{s}^2 + 2\varsigma\overline{s} + 1} \tag{A5}$$

$$\overline{\varepsilon} \triangleq \varepsilon\omega_n \tag{A6}$$

$$\overline{H}(s) = \frac{1}{\overline{\varepsilon}\,\overline{s} + 1} \tag{A7}$$

$$\overline{K}_i \triangleq \frac{K_i}{\omega_n} \qquad p = \frac{1}{\omega_n}\overline{p} \tag{A8}$$

$$C(\overline{s}) = \overline{K}_i(\frac{1}{\overline{s}} + \overline{p}) \qquad C(s) = \overline{K}_i\omega_n(\frac{1}{s} + \frac{1}{\omega_n}\overline{p}) \tag{A9}$$

Closed loop transfer function:

The closed loop transfer function is obtained using above normalization as follows;

$$W(\overline{s}) = \frac{\overline{K}_i(\frac{1}{\overline{s}} + \overline{p})\frac{(1-0.5\overline{L}\overline{s})}{(\overline{s}^2 + 2\varsigma\overline{s} + 1)}}{1 + \overline{K}_i(\frac{1}{\overline{s}} + \overline{p})\frac{(1-0.5\overline{L}\overline{s})}{(\overline{\varepsilon}\,\overline{s} + 1)(\overline{s}^2 + 2\varsigma\overline{s} + 1)}}$$

$$= \frac{\overline{K}_i(1+\overline{p}\overline{s})(1-0.5\overline{L}\overline{s})(\overline{\varepsilon}\,\overline{s}+1)}{\overline{\varepsilon}\overline{s}^4 + (2\varsigma\overline{\varepsilon} + 1)\overline{s}^3 + (\overline{\varepsilon} + 2\varsigma - 0.5\overline{p}\overline{L}\overline{K}_i)\overline{s}^2 + (1 + \overline{K}_i\overline{p} - 0.5\overline{L}\overline{K}_i)\overline{s} + \overline{K}_i} \tag{A.10}$$

if $\overline{p} = 0.5\overline{L}$ *then*

$$W(\overline{s}) = \frac{K_i(1-0.5\overline{L}\overline{s})(\overline{\varepsilon}\overline{s}+1)}{\overline{\varepsilon}\overline{s}^4 + (2\varsigma\overline{\varepsilon}+1)\overline{s}^3 + (\overline{\varepsilon} + 2\varsigma - 0.5^2\overline{L}^2\overline{K}_i)\overline{s}^2 + \overline{s} + \overline{K}_i} \tag{A11}$$

if $\overline{p} = 0$ *then*

$$W(\overline{s}) = \frac{\overline{K}_i(1-0.5\overline{L}\overline{s})(\overline{\varepsilon}\overline{s}+1)}{\overline{\varepsilon}\overline{s}^4 + (2\varsigma\overline{\varepsilon}+1)\overline{s}^3 + (\overline{\varepsilon}+2\varsigma)\overline{s}^2 + (1-0.5\overline{L}\overline{K}_i)\overline{s} + \overline{K}_i} \tag{A.12}$$

Stability analysis by Hurwits Approach

1. $\bar{p} < 0.5\bar{L}, 0 < \bar{K}_i < \min\{\frac{\bar{\varepsilon} + 2\varsigma}{0.5\bar{p}\bar{L}}, \frac{1}{(0.5\bar{L} - \bar{p})}\}, \varsigma > 0, \bar{\varepsilon} > 0$

$$\{(2\varsigma\bar{\varepsilon} + 1)(2\varsigma + \bar{\varepsilon} - 0.5^2\bar{L}^2\bar{K}_i) - \bar{\varepsilon}\} > \bar{K}_i(2\varsigma\bar{\varepsilon} + 1)^2 \quad when \quad \bar{p} = 0.5\bar{L} \tag{A13}$$

$$\frac{2\varsigma(\bar{\varepsilon}^2 + 2\varsigma\bar{\varepsilon} + 1)}{(2\varsigma\bar{\varepsilon} + 1)\{(2\varsigma\bar{\varepsilon} + 1) + 0.5^2\bar{L}^2\}} > \bar{K}_i \quad when \quad \bar{p} = 0.5\bar{L} \tag{A14}$$

$k_3 < k_2$ then

$$0 < \bar{K}_i < \min\{\frac{\bar{\varepsilon} + 2\varsigma}{0.5^2\bar{L}^2}, \frac{2\varsigma(\bar{\varepsilon}^2 + 2\varsigma\bar{\varepsilon} + 1)}{(2\varsigma\bar{\varepsilon} + 1)\{(2\varsigma\bar{\varepsilon} + 1) + 0.5^2\bar{L}^2\}}\} \quad when \quad \bar{p} = 0.5\bar{L} \tag{A15}$$

In *continuous region* with *one order modelling error,*

$$0 < \bar{K}_i < \frac{2\varsigma}{(1 + 0.5^2\bar{L}^2)} \quad when \quad \bar{p} = 0.5\bar{L} \tag{A16}$$

Analytical solution of Ki for flat gain curve using Stationary Points Investing for Fraction Equation approach is complicated to obtain, then it is remained for reader's theme.
In the future, another approach will be developed for safe and simple robust control.

B. Simple soft M/A station

In this section, a configuration of simple soft M/A station and the feedback control system with the station is shown for a simple safe interlock avoiding dangerous large overshoot.

B.1 Function and configuration of simple soft M/A station

This appendix describes a simple interlock plan for an simple soft M/A station that has a parameter-identification mode (manual mode) and a control mode (automatic mode).
The simple soft M/A station is switched from automatic operation mode to manual operation mode for safety when it is used to switch the identification mode and the control mode and when the value of *Pv* exceeds the prescribed range. This serves to protect the plant; for example, in the former case, it operates when the integrator of the PID controller varies erratically and the control system malfunctions. In the latter case, it operates when switching from P control with a large steady-state deviation with a high load to PI or PID control, so that the liquid in the tank spillovers. Other dangerous situations are not considered here because they do not fall under general basic control.
There have several attempts to arrange and classify the control logic by using a case base. Therefore, the M/A interlock should be enhanced to improve safety and maintainability; this has not yet been achieved for a simple M/A interlock plan (Fig. A1).
For safety reasons, automatic operation mode must not be used when changing into manual operation mode by changing the one process value, even if the process value recovers to an appropriate level for automatic operation.
Semiautomatic parameter identification and PID control are driven by case-based data for memory of tuners, which have a nest structure for identification.
This case-based data memory method can be used for reusing information, and preserving integrity and maintainability for semiautomatic identification and control. The semiautomatic approach is adopted not only to make operation easier but also to enhance safety relative to the fully automatic approach.

Notation in computer control (Fig. B1, B3)

Pv : Process value
Av: Actual value
Cv : Control value
Mv : Manipulated value
Sp: Set point
A : Auto
M : Manual
T : Test

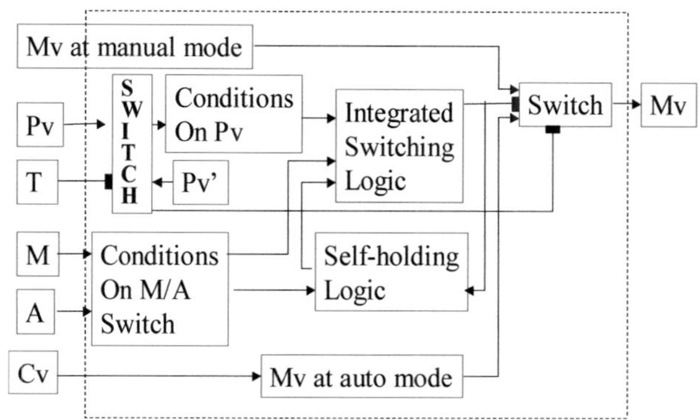

Fig. B1 A Configuration of Simple Soft M/A Station

B.2 Example of a SISO system

Fig. B2 shows the way of using M/A station in a configuration of a SISO control system.

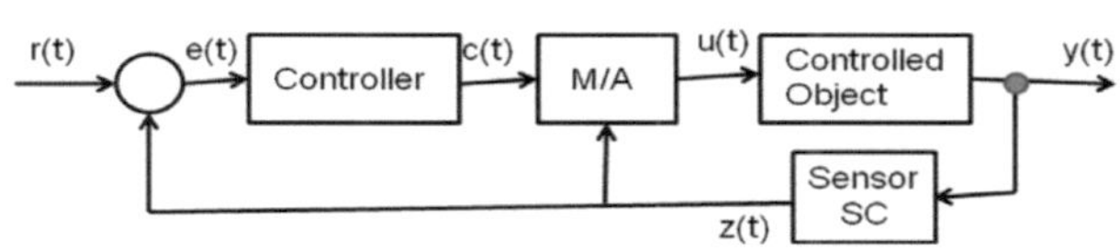

Fig. B2 Configuration of a IP Control System with a M/A Station for a SISO Controlled Object

where the transfer function needed in Fig.B2 is as follows.

1. Controlled Object: $G(s) = \dfrac{K}{Ts+1}e^{-Ls}$
2. Sensor & Signal Conditioner: $G_s(s) = \dfrac{K_s}{T_s s+1}$
3. Controller: $C(s) = 0.5K_{i2}(\dfrac{1}{s} + 0.5L)$
4. Sensor Caribration Gain: $1 / K_s$
5. Normalized Gain before M/A Station: $1 / \sqrt{0.5TL}$
6. Normalized Gain after M/A Station: $1 / K$

Fig. B3 shows examples of simulated results for 2 kinds of switching mode when *Pv* becomes higher than a given threshold. (a) shows one to out of service and (b) does to manual mode.

In former, Mv is down and Cv is almost hold. In latter, Mv is hold and Cv is down.

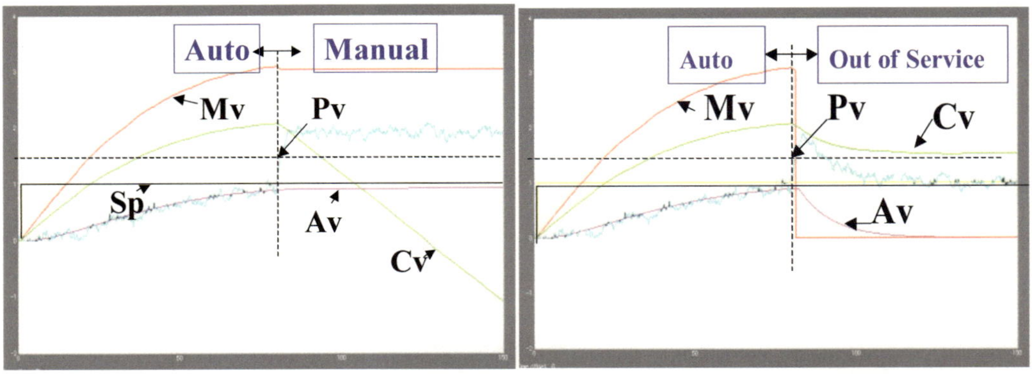

(a) Switching example from auto mode to out of service by Pv High

(b) Switching example from auto mode to manual mode by Pv High

Fig. B3 Simulation results for 2 kinds of switching mode

C. New norm and expansion of small gain theorem

In this section, a new range restricted norm of Hardy space with window(Kohonen T., 1995) H_∞^w is defined for I, of which window is described to notation of norm with superscript w, and a new expansion of small gain theorem based on closed loop system like general H_∞^w control problems and robust sensitivity analysis is shown for applying the robust PI control concept of this chapter to MIMO systems.

The robust control was aims soft servo and requested internal stability for a closed loop control system. Then, it was difficult to apply process control systems or hard servo systems which was needed strong robust stability without deviation from the reference value in the steady state like integral terms.

The method which sets the maximum value of closed loop gain curve to 1 and the results of this numerical experiments indicated the above sections will imply the following new expansion of small gain theorem which indicates the upper limit of Hardy space norm of a forward element using the upper limit of all uncertain feedback elements for robust stability.

For the purpose using unbounded functions in the all real domain on frequency like integral term in the forward element, the domain of Hardy norm of the function concerned on frequency is limited clearly to a section in a positive real one-order space so that the function becomes bounded in the section.

Proposition

Assuming that feedback transfer function H(s) (with uncertainty) are stable and the following inequality is holds,

$$\|H(s)\|_\infty \le \frac{1}{\gamma}, \gamma \ge 1 \tag{C-1}$$

Moreover , if the negative closed loop system as shown in Fig.C-1 is stable and the following inequality holds,

$$\|W(s)\|_\infty = \left\| \frac{G(s)}{1+G(s)H(s)} \right\|_\infty \le 1 \tag{C-2}$$

then the following inequality on the open loop transfer function is hold in a region of frequency.

$$\left\|G(j\omega)H(j\omega)\right\|_{\infty}^{w} \le \frac{1}{\gamma-1}, \gamma \ge 1 \quad for \quad \omega \in [\omega_{\min}, \omega_{\max}] \tag{C-3}$$

In the same feedback system, G(s) holds the following inequality in a region of frequency.

$$\left\|G(j\omega)\right\|_{\infty}^{w} \le \frac{\gamma}{\gamma-1}, \gamma \ge 1 \quad for \quad \omega \in [\omega_{\min}, \omega_{\max}] \tag{C-4}$$

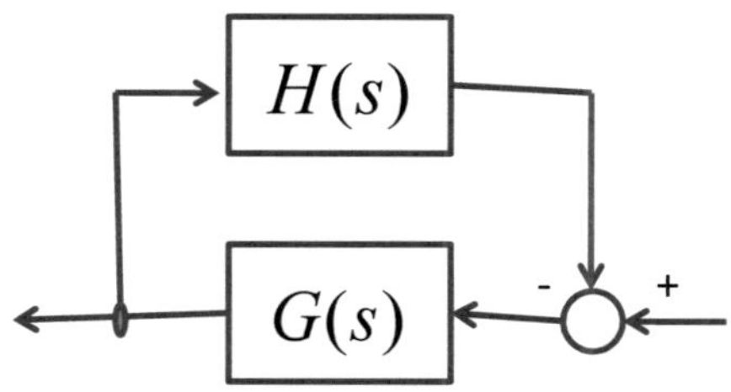

Fig. C-1 Configuration of a negative feed back system

(proof)

Using triangle inequality on separation of norms of summension and inequality on separation of norms of product like Helder's one under a region of frequency $[\omega_{\min}, \omega_{\max}]$, as a domain of the norm of Hardy space with window, the following inequality on the frequency transfer function of $G(j\omega)$ is obtained from the assumption of the proposition.

$$\left\|W(j\omega)\right\|_{\infty} = \left\|\frac{G(j\omega)}{1+G(j\omega)H(j\omega)}\right\|_{\infty} \le 1 \tag{C-5}$$

$$\begin{aligned} &\left\|G(j\omega)\right\|_{\infty}^{w} \le \left\|1+G(j\omega)H(j\omega)\right\|_{\infty}^{w} \le 1+\left\|G(j\omega)\right\|_{\infty}^{w}\left\|H(j\omega)\right\|_{\infty}^{w} \\ &1-\left\|H(j\omega)\right\|_{\infty}^{w} \le \left\|\frac{1}{G(j\omega)}\right\|_{\infty}^{w} \end{aligned} \tag{C-6}$$

$$\begin{aligned} &if \quad \left\|H(j\omega)\right\|_{\infty}^{w} \le \frac{1}{\gamma} \le 1, \gamma \ge 1 \quad then \\ &\frac{\gamma}{\gamma-1} \ge \frac{1}{1-\left\|H(j\omega)\right\|_{\infty}^{w}} \ge \left\|G(j\omega)\right\|_{\infty}^{w} \end{aligned} \tag{C-7}$$

Moreover, the following inequality on open loop frequency transfer function is shown.

$$\frac{1}{\gamma-1} \ge \left\|G(j\omega)\right\|_{\infty}^{w}\left\|H(j\omega)\right\|_{\infty}^{w} \ge \left\|G(j\omega)H(j\omega)\right\|_{\infty}^{w} \tag{C-8}$$

On the inequality of norm, the reverse proposition may be shown though the separation of product of norms in the Hardy space with window are not clear. The sufficient conditions on closed loop stability are not clear. They will remain reader's theme in the future.

D. Parametric robust topics

In this section, the following three topics (Bhattacharyya S. P., Chapellat H., and Keel L. H., 1994.) are introduced at first for parametric robust property in static one, dynamic one and stable one as assumptions after linearizing a class of non-linear system to a quasi linear parametric variable (QLPV) model by Taylar expansion using first order reminder term. (M.Katoh, 2010)

1. Continuity for change of parameter

Boundary Crossing Theorem

1) fixed order polynomials P(λ,s)

2) continuous polynomials with respect to one parameter λ on a fixed interval I=[a,b].

If P(a,s) has all its roots in S, P(b,s) has at least one root in U, then there exists at least one ρ in (a,b] such that:

a) P(ρ,s) has all roots in S U∂S

b) P(ρ,s) has at least one root in ∂S

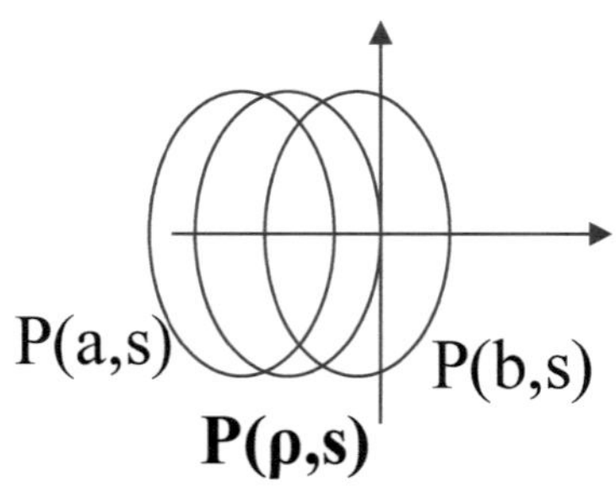

Fig. D-1 Image of boundary crossing theorem

2. Convex for change of parameter

Segment Stable Lemma

Let define a segment using two stable polynomials as follows.

$$\begin{aligned}
&\delta_\lambda(s) \triangleq \lambda\delta_1(s)+(1-\lambda)\delta_2(s)\\
&[\delta_1(s),\delta_2(s)] \triangleq \{\delta_\lambda(s):\lambda\in[0,1]\}\\
&where \quad \delta_1(s),\delta_2(s)_is_polynomials_of_\deg ree_n\\
&\qquad and_stable_with_respect_to_S
\end{aligned} \tag{D-1}$$

Then, the followings are equivalent:

a) The segment $[\delta_1(s),\delta_2(s)]$ is stable with respect to S

$b)\delta_\lambda(s^*) \neq 0, \quad for\_all\_s^* \in \partial S; \quad \lambda \in [0,1]$

3. Worst stability margin for change of parameter

Parametric stability margin (PSM) is defined as the worst case stability margin within the parameter variation. It can be applied to a QLPV system of a class of non-linear system. There are non-linear systems such as becoming worse stability margin than linearized system although there are ones with better stability margin than it. There is a case which is characterized by the one parameter *m* which describes the injection rate of I/O, the interpolation rate of segment or degree of non-linearity.

E. Risk and Merit Analysis

Let show a summary and enhancing of the risk discussed before sections for safety in the following table.

Kinds	Evaluation of influence	Countermeasure
1) Disconnection of feedback line 2) Overshoot over limit value	1) Spill-over threshold 2) Attack to weak material	Auto change to manual mode by M/A station Auto shut down
Change of tuning region from IPS to IPL by making proportional gain to large	Grade down of stability region from strong or weak to weak or un-stability	Use IP0 and not use IPS Not making proportional gain to large in IPS tuning region
Change of damping coefficient or inverse of time constant over weak robust limit	Grade down of stability region from strong or weak to weak or un-stability	Change of tuning region from IPL to IPS or IP0

Table E-1 Risk analysis for safety

It is important to reduce risk as above each one by adequate countermeasures after understanding the property of and the influence for the controlled objects enough.
Next, let show a summary and enhancing of the merit and demerit discussed before sections for robust control in the following table, too.

Kinds	Merit	Demerit
1) Steady state error is vanishing as time by effect of integral	1) It is important property in process control and hard servo area	It is dislike property in soft servo and robot control because of hardness for disturbance
There is a strong robust stability damping region in which the closed loop gain margin for any uncertainty is over 2 and almost not changing.	It is uniform safety for some proportional gain tuning region and changing of damping coefficient. For integral loop gain tuning, it recommends the simple limiting sensitivity approach.	1) Because the region is different by proportional gain, there is a risk of grade down by the gain tuning.
There is a weak robust stability damping region in which the worst closed loop gain margin for any uncertainty is over given constant.	1) It can specify the grade of robust stability for any uncertainty	1) Because the region is different by proportional gain, there is a risk of grade down by the gain tuning. It is different safety for some proportional gain tuning region.

Table E-2 Merit analysis for control

It is important to apply to the best application area which the merit can be made and the demerit can be controlled by the wisdom of everyone.

7. References

Bhattacharyya S. P., Chapellat H., and Keel L. H.(1994). *Robust Control*, The Parametric Approach, Upper Saddle River NJ07458 in USA: Prentice Hall Inc.

Katoh M. and Hasegawa H., (1998). Tuning Methods of 2nd Order Servo by I-PD Control Scheme, *Proceedings of The 41st Joint Automatic Control Conference*, pp. 111-112. (in Japanese)

Katoh M.,(2003). An Integral Design of A Sampled and Continuous Robust Proper Compensator, *Proceedings of CCCT2003*, (pdf000564), Vol. III, pp. 226-229.

Katoh M.,(2008). Simple Robust Normalized IP Control Design for Unknown Input Disturbance, *SICE Annual Conference 2008*, August 20-22, The University Electro-Communication, Japan, pp.2871-2876, No.:PR0001/08/0000-2871

Katoh M., (2009). Loop Gain Margin in Simple Robust Normalized IP Control for Uncertain Parameter of One-Order Model Error, *International Journal of Advanced Computer Engineering*, Vol.2, No.1, January-June, pp.25-31, ISSN:0974-5785, Serials Publications, New Delhi (India)

Katoh M and Imura N., (2009). Double-agent Convoying Scenario Changeable by an Emergent Trigger, Proceedings of the 4th International Conference on Autonomous Robots and Agents, Feb 10-12, Wellington, New Zealand, pp.442-446

Katoh M. and Fujiwara A., (2010). Simple Robust Stability for PID Control System of an Adjusted System with One-Changeable Parameter and Auto Tuning, *International Journal of Advanced Computer Engineering*, Vol.3, No.1, ISSN:0974-5785, Serials Publications, New Delhi (India)

Katoh M.,(2010). Static and Dynamic Robust Parameters and PI Control Tuning of TV-MITE Model for Controlling the Liquid Level in a Single Tank", TC01-2, *SICE Annual Conference 2010*, 18/August TC01-3

Krajewski W., Lepschy A., and Viaro U.,(2004). Designing *PI* Controllers for Robust Stability and Performance, *Institute of Electric and Electronic Engineers Transactions on Control System Technology*, Vol. 12, No. 6, pp. 973- 983.

Kohonen T.,(1995, 1997). *Self-Organizing Maps*, Springer

Kojori H. A., Lavers J. D., and Dewan S. B.,(1993). A Critical Assessment of the Continuous-System Approximate Methods for the Stability Analysis of a Sampled Data System, *Institute of Electric and Electronic Engineers Transactions on Power Electronics*, Vol. 8, No. 1, pp. 76-84.

Miyamoto S.,(1998). Design of *PID* Controllers Based on H_∞-Loop Shaping Method and *LMI* Optimization, *Transactions of the Society of Instrument and Control Engineers*, Vol. 34, No. 7, pp. 653-659. (in Japanese)

Namba R., Yamamoto T., and Kaneda M., (1998). A Design Scheme of Discrete Robust *PID* Control Systems and Its Application, *Transactions on Electrical and Electronic Engineering*, Vol. 118-C, No. 3, pp. 320-325. (in Japanese)

Olbrot A. W. and Nikodem M.,(1994) . Robust Stabilization: Some Extensions of the Gain Margin Maximization Problem, *Institute of Electric and Electronic Engineers Transactions on Automatic Control*, Vol. 39, No. 3, pp. 652- 657.

Zbou K. with Doyle F. C. and Glover K.,(1996). *Robust and Optimal Control*, Prentice Hall Inc.

Zhau K. and Khargonekar P.P., (1988). An Algebraic Riccati Equation Approach to H Optimization, Systems & Control Letters, 11, pp.85-91.

13

Passive Fault Tolerant Control

M. Benosman
Mitsubishi Electric Research Laboratories 201 Broadway street, Cambridge, MA 02139, USA*

1. Introduction

Today, as a result of increasing complexity of industrial automation technologies, fault handling of such automatic systems has become a challenging task. Indeed, although industrial systems are usually designed to perform optimally over time, performance degradation occurs inevitably. These are due, for example, to aging of system components, which have to be monitored to prevent system-wide failures. Fault handling is also necessary to allow redesign of the control in such a way to recover, as much as possible, an optimal performance. To this end, researchers in the systems control community have focused on a specific control design strategy, called Fault tolerant control (FTC). Indeed, FTC is aimed at achieving acceptable performance and stability for the safe, i.e. fault-free system as well as for the faulty system. Many methods have been proposed to deal with this problem. For survey papers on FTC, the reader may refer to (5; 38; 53). While the available schemes can be classified into two types, namely *passive* and *active* FTC (53), the work presented here falls into the first category of passive FTC. Indeed, active FTC is aimed at ensuring the stability and some performance, possibly degraded, for the post-fault model, and this by reconfiguring on-line the controller, by means of a fault detection and diagnosis (FDD) component that detects, isolates and estimates the current fault (53). Contrary to this active approach, the passive solution consists in using a unique robust controller that, will deal with *all the expected faults*. The passive FTC approach has the drawback of being reliable only for the class of faults expected and taken into account in the design. However, it has the advantage of avoiding the time delay required in active FTC for on-line fault diagnosis and control reconfiguration (42; 54), which is very important in practical situations where the time window during which the faulty system stay stabilizable is very short, e.g. the unstable double inverted pendulum example (37). In fact, in practical applications, passive FTC complement active FTC schemes. Indeed, passive FTC schemes are necessary during the fault detection and estimation phase (50), to ensure the stability of the faulty system, before switching to active FTC. Several passive FTC methods have been proposed, mainly based on robust theory, e.g. multi-objective linear optimization and LMIs techniques (25), QFT method (47; 48), H_∞ (36; 37), absolute stability theory (6), nonlinear regulation theory (10; 11), Lyapunov reconstruction (9) and passivity-based FTC (8). As for active FTC, many methods have been proposed for active linear FTC, e.g. (19; 29; 43; 46; 51; 52), as well as for nonlinear FTC, e.g. (4; 7; 13; 14; 20; 21; 28; 32–35; 39; 41; 49).
We consider in this work the problem of fault tolerant control for failures resulting from *loss of*

*E-mail: benosman@merl.com

actuator effectiveness. FTCs dealing with actuator faults are relevant in practical applications and have already been the subject of many publications. For instance, in (43), the case of uncertain linear time-invariant models was studied. The authors treated the problem of actuators stuck at unknown constant values at unknown time instants. The active FTC approach they proposed was based on an output feedback adaptive method. Another active FTC formulation was proposed in (46), where the authors studied the problem of loss of actuator effectiveness in linear discrete-time models. The loss of control effectiveness was estimated via an adaptive Kalman filter. The estimation was complemented by a fault reconfiguration based on the LQG method. In (30), the authors proposed a multiple-controller based FTC for linear uncertain models. They introduced an active FTC scheme that ensured the stability of the system regardless of the decision of FDD.

However, as mentioned earlier and as presented for example in (50), the aforementioned active schemes will incur a delay period during which the associate FDD component will have to converge to a best estimate of the fault. During this time period of FDD response delay, it is essential to control the system with a passive fault tolerant controller which is robust against actuator faults so as to ensure at least the stability of the system, before switching to another controller based on the estimated post-fault model, that ensures optimal post-fault performance. In this context, we propose here passive FTC schemes against actuator loss of effectiveness. The results presented here are based on the work of the author introduced in (6; 8). We first consider linear FTC and present some results on passive FTC for loss of effectiveness faults based on absolute stability theory. Next we present an extension of the linear results to some nonlinear models and use passivity theory to write nonlinear fault tolerant controllers. In this chapter several controllers are proposed for different problem settings: a) Linear time invariant (LTI) certain plants, b) uncertain LTI plants, c) LTI models with input saturations, d) nonlinear plants affine in the control with single input, e) general nonlinear models with constant as well as time-varying faults and with input saturation. We underline here that we focus in this chapter on the theoretical developments of the controllers, readers interested in numerical applications should refer to (6; 8).

2. Preliminaries

Throughout this chapter we will use the L_2 norm denoted $||.||$, i.e. for $x \in \mathbb{R}^n$ we define $||x|| = \sqrt{x^T x}$. The notation $L_f h$ denotes the standard Lie derivative of a scalar function $h(.)$ along a vector function $f(.)$. Let us introduce now some definitions from (40), that will be frequently used in the sequel.

Definition 1 ((40), p.45): The solution $x(t, x_0)$ of the system $\dot{x} = f(x)$, $x \in \mathbb{R}^n$, f locally Lipschitz, is stable conditionally to Z, if $x_0 \in Z$ and for each $\epsilon > 0$ there exists $\delta(\epsilon) > 0$ such that

$$||\tilde{x}_0 - x_0|| < \delta \textit{ and } \tilde{x}_0 \in Z \Rightarrow ||x(t, \tilde{x}_0) - x(t, x_0)|| < \epsilon, \ \forall t \geq 0.$$

If furthermore, there exist $r(x_0) > 0$, s.t. $||x(t, \tilde{x}_0) - x(t, x_0)|| \Rightarrow 0$, $\forall ||\tilde{x}_0 - x_0|| < r(x_0)$ *and* $\tilde{x}_0 \in Z$, the solution is asymptotically stable conditionally to Z. If $r(x_0) \to \infty$, the stability is global.

Definition 2 ((40), p.48): Consider the system $H : \dot{x} = f(x, u)$, $y = h(x, u)$, $x \in \mathbb{R}^n$, $u, y \in \mathbb{R}^m$, with zero inputs, i.e. $\dot{x} = f(x, 0)$, $y = h(x, 0)$ and let $Z \subset \mathbb{R}^n$ be its largest positively invariant set contained in $\{x \in \mathbb{R}^n | y = h(x, 0) = 0\}$. We say that H is globally zero-state detectable (GZSD) if $x = 0$ is globally asymptotically stable conditionally to Z. If $Z = \{0\}$, the system H is zero-state observable (ZSO).

Definition 3 ((40), p.27): We say that H is dissipative in $X \subset \mathbb{R}^n$ containing $x = 0$, if there exists a function $S(x)$, $S(0) = 0$ such that for all $x \in X$

$$S(x) \geq 0 \text{ and } S(x(T)) - S(x(0)) \leq \int_0^T \omega(u(t), y(t))dt,$$

for all $u \in U \subset \mathbb{R}^m$ and all $T > 0$ such that $x(t) \in X$, $\forall\ t \in [0, T]$. Where the function $\omega : \mathbb{R}^m \times \mathbb{R}^m \rightarrow \mathbb{R}$ called the supply rate, is locally integrable for every $u \in U$, i.e. $\int_{t_0}^{t_1} |\omega(u(t), y(t))|dt < \infty$, $\forall\ t_0 \leq t_1$. S is called the storage function. If the storage function is differentiable the previous conditions writes as

$$\dot{S}(x(t)) \leq \omega(u(t), y(t)).$$

The system H is said to be passive if it is dissipative with the supply rate $w(u, y) = u^T y$.
Definition 4 ((40), p.36): We say that H is output feedback passive (OFP(ρ)) if it is dissipative with respect to $\omega(u, y) = u^T y - \rho y^T y$ for some $\rho \in \mathbb{R}$.
We will also need the following definition to study the case of time-varying faults in Section 8.
Definition 5 (24): A function $\overline{x} : [0, \infty) \rightarrow \mathbb{R}^n$ is called a limiting solution of the system $\dot{x} = f(t, x)$, f a smooth vector function, with respect to an unbounded sequence t_n in $[0, \infty)$, if there exist a compact $\kappa \subset \mathbb{R}^n$ and a sequence $\{x_n : [t_n, \infty) \rightarrow \kappa\}$ of solutions of the system such that the associated sequence $\{\hat{x}_n :\rightarrow x_n(t + t_n)\}$ converges uniformly to $\overline{x}$ on every compact subset of $[0, \infty)$.
Also, throughout this paper it is said that a statement $P(t)$ holds almost everywhere (a.e.) if the Lebesgue measure of the set $\{t \in [0, \infty)\ |P(t)\ is\ false\}$ is zero. We denote by df the differential of the function $f : \mathbb{R}^n \rightarrow \mathbb{R}$. We also mean by semiglobal stability of the equilibrium point x^0 for the autonomous system $\dot{x} = f(x)$, $x \in \mathbb{R}^n$ with f a smooth function, that for each compact set $K \subset \mathbb{R}^n$ containing x^0, there exist a locally Lipschitz state feedback, such that x^0 is asymptotically stable, with a basin of attraction containing K ((44), Definition 3, p. 1445).

3. FTC for known LTI plants

First, let us consider linear systems of the form

$$\dot{x} = Ax + B\alpha u, \tag{1}$$

where, $x \in \mathbb{R}^n$, $u \in \mathbb{R}^m$ are the state and input vector, respectively, and $\alpha \in \mathbb{R}^{m \times m}$ is a diagonal time variant fault matrix, with diagonal elements $\alpha_{ii}(t)$, $i = 1, ..., m$ s.t., $0 < \epsilon_1 \leq \alpha_{ii}(t) \leq 1$. The matrices A, B have appropriate dimensions and satisfy the following assumption.
Assumption(1): The pair (A, B) is controllable.

3.1 Problem statement

Find a state feedback controller $u(x)$ such that the closed-loop controlled system (1) admits $x = 0$ as a globally uniformly asymptotically (GUA) stable equilibrium point $\forall \alpha(t)$ (s.t. $0 < \epsilon_1 \leq \alpha_{ii}(t) \leq 1$).

3.2 Problem solution

Hereafter, we will re-write the problem of stabilizing (1), for $\forall \alpha(t)$ s.t., $0 < \epsilon_1 \leq \alpha_{ii}(t) \leq 1$, as an absolute stability problem or Lure's problem (2). Let us first recall the definition of sector nonlinearities.

Definition 6 ((22), p. 232): A static function $\psi : [0,\infty) \times \mathbb{R}^m \to \mathbb{R}^m$, s.t. $[\psi(t,y) - K_1 y]^T [\psi(t,y) - K_2 y] \leq 0, \ \forall (t,y)$, with $K = K_2 - K_1 = K^T > 0$, where $K_1 = diag(k1_1, ..., k1_m)$, $K_2 = diag(k2_1, ..., k2_m)$, is said to belong to the sector $[K_1, \ K_2]$.
We can now recall the definition of absolute stability or Lure's problem.
Definition 7 (Absolute stability or Lure's problem (22), p. 264): We assume a linear system of the form

$$\begin{aligned} \dot{x} &= Ax + Bu \\ y &= Cx + Du \\ u &= -\psi(t,y), \end{aligned} \tag{2}$$

where, $x \in \mathbb{R}^n$, $u \in \mathbb{R}^m$, $y \in \mathbb{R}^m$, (A,B) controllable, (A,C) observable and $\psi : [0,\infty) \times \mathbb{R}^m \to \mathbb{R}^m$ is a static nonlinearity, piecewise continuous in t, locally Lipschitz in y and satisfies a sector condition as defined above. Then, the system (2) is absolutely stable if the origin is GUA stable for any nonlinearity in the given sector. It is absolutely stable within a finite domain if the origin is uniformly asymptotically (UA) stable within a finite domain.
We can now introduce the idea used here, which is as follows:
Let us associate with the faulty system (1) a virtual output vector $y \in \mathbb{R}^m$

$$\begin{aligned} \dot{x} &= Ax + B\alpha u \\ y &= Kx, \end{aligned} \tag{3}$$

and let us write the controller as an output feedback

$$u = -y. \tag{4}$$

From (3) and (4), we can write the closed-loop system as

$$\begin{aligned} \dot{x} &= Ax + Bv \\ y &= Kx \\ v &= -\alpha(t)y. \end{aligned} \tag{5}$$

We have thus transformed the problem of stabilizing (1), for all bounded matrices $\alpha(t)$, to the problem of stabilizing the system (5) for all $\alpha(t)$. It is clear that the problem of GUA stabilizing (5) is a Lure's problem in (2), with the linear time varying stationarity $\psi(t,y) = \alpha(t)y$, and where the 'nonlinearities' admit the sector bounds $K_1 = diag(\epsilon_1, ..., \epsilon_1)$, $K_2 = I_{m \times m}$.
Based on this formulation we can now solve the problem of passive fault tolerant control of (1) by applying the absolute stability theory (26).
We can first write the following result:
Proposition 1: Under Assumption 1, the closed-loop of (1) with the static state feedback

$$u = -Kx, \tag{6}$$

where K is solution of the optimal problem

$$\begin{aligned} & \min_{k_{ij}} \left(\textstyle\sum_{i=1}^{i=m} \sum_{j=1}^{j=n} k_{ij}^2 \right) \\ & \begin{bmatrix} P\hat{A}(K) + \hat{A}^T(K)P & (\hat{C}^T - P\hat{B})W^{-1} \\ ((\hat{C}^T - P\hat{B})W^{-1})^T & -I \end{bmatrix} < 0 \\ & P > 0 \\ & rank \begin{bmatrix} K \\ KA \\ \vdots \\ KA^{n-1} \end{bmatrix} = n, \end{aligned} \tag{7}$$

for $P = P^T > 0$, $W = (\hat{D} + \hat{D}^T)^{0.5}$ and $\{\hat{A}(K),\ \hat{B}(K),\ \hat{C}(K),\ \hat{D}(K)\}$ is a minimal realization of the transfer matrix

$$\hat{G} = [I + K(sI - A)^{-1}B][I + \epsilon_1 \times I_{m\times m}K(sI - A)^{-1}B]^{-1}, \tag{8}$$

admits the origin $x = 0$ as GUA stable equilibrium point.

Proof: We saw that the problem of stabilizing (1) with a static state feedback $u = -Kx$ is equivalent to the stabilization of (5). Studying the stability of (5) is a particular case of Lure's problem defined by (2), with the 'nonlinearity' function $\psi(t,y) = -\alpha(t)y$ associated with the sector bounds $K_1 = \epsilon_1 \times I_{m\times m}$, $K_2 = I_{m\times m}$ (introduced in Definition 1). Then based on Theorem 7.1, in ((22), p. 265), we can write that under Assumption1 and the constraint of observability of the pair (A, K), the origin $x = 0$ is GUA stable equilibrium point for (5), if the matrix transfer function

$$\hat{G} = [I + G(s)][I + \epsilon_1 \times I_{m\times m}G(s)]^{-1},$$

where $G(s) = K(sI - A)^{-1}B$, is strictly positive real (SPR). Now, using the KYP lemma as presented in (Lemma 6.3, (22), p. 240), we can write that a sufficient condition for the GUA stability of $x = 0$ along the solution of (1) with $u = -Kx$ is the existence of $P = P^T > 0$, L and W, s.t.

$$\begin{array}{l} P\hat{A}(K) + \hat{A}^T(K)P = -L^TL - \epsilon P,\ \epsilon > 0 \\ P\hat{B}(K) = \hat{C}^T(K) - L^TW \\ W^TW = \hat{D}(K) + \hat{D}^T(K), \end{array} \tag{9}$$

where, $\{\hat{A},\ \hat{B},\ \hat{C},\ \hat{D}\}$ is a minimal realization of $\hat{G}$. Finally, adding to equation (9), the observability condition of the pair (A, K), we arrive at the condition

$$\begin{array}{l} P\hat{A}(K) + \hat{A}^T(K)P = -L^TL - \epsilon P,\ \epsilon > 0 \\ P\hat{B}(K) = \hat{C}^T(K) - L^TW \\ W^TW = \hat{D}(K) + \hat{D}^T(K) \\ rank \begin{bmatrix} K \\ KA \\ \vdots \\ KA^{n-1} \end{bmatrix} = n. \end{array} \tag{10}$$

Next, if we choose $W = W^T$ we can write $W = (\hat{D} + \hat{D}^T)^{0.5}$. The second equation in (10) leads to $L^T = (\hat{C}^T - P\hat{B})W^{-1}$. Finally, from the first equation in (10), we arrive at the following condition on P

$$P\hat{A}(K) + \hat{A}^T(K)P + (\hat{C}^T - P\hat{B})W^{-1}((\hat{C}^T - P\hat{B})W^{-1})^T < 0,$$

which is in turn equivalent to the LMI

$$\begin{bmatrix} P\hat{A}(K) + \hat{A}^T(K)P & (\hat{C}^T - P\hat{B})W^{-1} \\ ((\hat{C}^T - P\hat{B})W^{-1})^T & -I \end{bmatrix} < 0. \tag{11}$$

Thus, to solve equation (10) we can solve the constrained optimal problem

$$\min_{k_{ij}}(\sum_{i=1}^{i=m}\sum_{j=1}^{j=n}k_{ij}^2)$$
$$\begin{bmatrix} P\hat{A}(K)+\hat{A}^T(K)P & (\hat{C}^T-P\hat{B})W^{-1} \\ ((\hat{C}^T-P\hat{B})W^{-1})^T & -I \end{bmatrix}<0$$
$$P>0$$
$$rank\begin{bmatrix} K \\ KA \\ \vdots \\ KA^{n-1} \end{bmatrix}=n. \qquad \square \tag{12}$$

Note that the inequality constraints in (7) can be easily solved by available LMI algorithms, e.g. *feasp* under Matlab. Furthermore, to solve equation (10), we can propose two other different formulations:

1. Through nonlinear algebraic equations: Choose $W = W^T$ which implies by the third equation in (10) that $W=(\hat{D}(K)+\hat{D}^T(K))^{0.5}$, for any K s.t.

$$\begin{array}{l} P\hat{A}(K)+\hat{A}^T(K)P=-L^TL-\epsilon P,\ \epsilon>0,\ P=P^T>0 \\ P\hat{B}(K)=\hat{C}^T(K)-L^TW \\ rank\begin{bmatrix} K \\ KA \\ \vdots \\ KA^{n-1} \end{bmatrix}=n. \end{array} \tag{13}$$

   To solve (13) we can choose $\epsilon=\tilde{\epsilon}^2$ and $P=\tilde{P}^T\tilde{P}$, which leads to the nonlinear algebraic equation

$$F(k_{ij},\tilde{p}_{ij},l_{ij},\tilde{\epsilon})=0, \tag{14}$$

   where $k_{ij},\ i=1,...,m,\ j=1,...n,\ \tilde{p}_{ij},\ i=1,...,\tilde{n}\ (\hat{A}\in\mathbb{R}^{\tilde{n}\times\tilde{n}}),\ j=1,...\tilde{n}$ and $l_{ij},\ i=1,...,m,\ j=1,...\tilde{n}$ are the elements of K, $\tilde{P}$ and L, respectively. Equation (14) can then be resolved by any nonlinear algebraic equations solver, e.g. *fsolve* under Matlab.

2. Through Algebraic Riccati Equations (ARE): It is well known that the positive real lemma equations, i.e. the first three equations in (10) can be transformed to the following ARE ((3), pp. 270-271):

$$P(\hat{\hat{A}}-\hat{B}R^{-1}\hat{C})+(\hat{\hat{A}}^T-\hat{C}^TR^{-1}\hat{B}^T)P+P\hat{B}R^{-1}\hat{B}^TP+\hat{C}^TR^{-1}\hat{C}=0, \tag{15}$$

   where $\hat{\hat{A}}=\hat{A}+0.5\epsilon.I_{\tilde{n}\times\tilde{n}}$, $R=\hat{D}(K)+\hat{D}^T(K)>0$. Then, if a solution $P=P^T>0$ is found for (15) it is also a solution for the first three equation in (10), together with

$$W=-VR^{1/2},\ L=(P\hat{B}-\hat{C}^T)R^{-1/2}V^T,\ VV^T=I.$$

   To solve equation (10), we can then solve the constrained optimal problem

$$\begin{array}{l} \min_{k_{ij}}(\sum_{i=1}^{i=m}\sum_{j=1}^{j=n}k_{ij}^2) \\ P > 0 \\ rank\begin{bmatrix} K \\ KA \\ \vdots \\ KA^{n-1} \end{bmatrix} = n, \end{array} \tag{16}$$

where P is the symmetric solution of the ARE (15), that can be directly computed by available solvers, e.g. *care* under Matlab.

There are other linear controllers for LPV system, that might solve the problem stated in Section 3.1, e.g. (1). However, the solution proposed here benefits from the simplicity of the formulation based on the absolute stability theory, and allows us to design FTCs for uncertain and saturated LTI plants, as well as nonlinear affine models, as we will see in the sequel. Furthermore, reformulating the FTC problem in the absolute stability theory framework may be applied to solve the FTC problem for several other systems, like infinite dimensional systems, i.e. PDEs models, stochastic systems and systems with delays (see (26) and the references therein). Furthermore, compared to optimal controllers, e.g. LQR, the proposed solution offers greater robustness, since it compensates for the loss of effectiveness over $[\epsilon_1,\ 1]$. Indeed, it is well known that *in the time invariant case,* optimal controllers like LQR compensates for a loss of effectiveness over $[1/2,\ 1]$ ((40), pp. 99-102). A larger loss of effectiveness can be covered but at the expense of higher control amplitude ((40), Proposition 3.32, p.100), which is not desirable in practical situations.
Let us consider now the more practical case of LTI plants with parameter uncertainties.

4. FTC for uncertain LTI plants

We consider here models with structured uncertainties of the form

$$\dot{x} = (A + \Delta A)x + (B + \Delta B)\alpha u, \tag{17}$$

where $\Delta A \in \circ\mathcal{A} = \{\Delta A \in \mathbb{R}^{n\times n} | \Delta A_{min} \leq \Delta A \leq \Delta A_{max},\ \Delta A_{min}, \Delta A_{max} \in \mathbb{R}^{n\times n}\}$,
$\Delta B \in \circ\mathcal{B} = \{\Delta B \in \mathbb{R}^{n\times m} | \Delta B_{min} \leq \Delta B \leq \Delta B_{max},\ \Delta B_{min}, \Delta B_{max} \in \mathbb{R}^{n\times m}\}$,
$\alpha = diag(\alpha_{11}, ..., \alpha_{mm}),\ 0 < \epsilon_1 \leq \alpha_{ii} \leq 1\ \forall i \in \{1, ..., m\}$, and $A,\ B,\ x,\ u$ as defined before.

4.1 Problem statement

Find a state feedback controller $u(x)$ such that the closed-loop controlled system (17) admits $x = 0$ as a globally asymptotically (GA) stable equilibrium point $\forall \alpha (s.t.\ 0 < \epsilon_1 \leq \alpha_{ii} \leq 1),\ \forall \Delta A \in \circ\mathcal{A},\ \Delta B \in \circ\mathcal{B}$.

4.2 Problem solution

We first re-write the model (17) as follows:

$$\begin{array}{l} \dot{x} = (A + \Delta A)x + (B + \Delta B)v \\ y = Kx \\ v = -\alpha y. \end{array} \tag{18}$$

The formulation given by (18), is an uncertain Lure's problem (as defined in (15) for example). We can write the following result:

Proposition 2: Under Assumption 1, the system (17) admits $x = 0$ as GA stable equilibrium point, with the static state feedback $u = -\tilde{K}\tilde{H}^{-1}x$, where $\tilde{K}$, $\tilde{H}$ are solutions of the LMIs

$$\begin{array}{l} \tilde{Q} + \tilde{H}A^T - \tilde{K}^T L^T B^T + A\tilde{H} - BL\tilde{K} \leq 0\ \forall L \in L^v,\ \tilde{Q} = \tilde{Q}^T > 0,\ \tilde{H} > 0 \\ -\tilde{Q} + \tilde{H}\Delta A^T - \tilde{K}^T L^T \Delta B^T + \Delta A\tilde{H} - \Delta BL\tilde{K} < 0,\ \forall(\Delta A, \Delta B, Ł) \in \circ\mathcal{A}^v \times \circ\mathcal{B}^v \times L^v, \end{array} \tag{19}$$

where, L^v is the set containing the vertices of $\{\epsilon_1 I_{m\times m},\ I_{m\times m}\}$, and $\circ\mathcal{A}^v$, $\circ\mathcal{B}^v$ are the set of vertices of $\circ\mathcal{A}$, $\circ\mathcal{B}$ respectively.
Proof: Under Assumption 1, and using Theorem 5 in ((15), p. 330), we can write the stabilizing static state feedback $u = -Kx$, where K is such that, for a given $H > 0$, $Q = Q^T > 0$ we have

$$\begin{cases} Q + (A - BLK)^T H + H(A - BLK) \leq 0\ \forall L \in L^v \\ -Q + ((\Delta A - \Delta BLK)^T H + H(\Delta A - \Delta BLK)) < 0\ \forall(\Delta A, \Delta B, Ł) \in \circ\mathcal{A}^v \times \circ\mathcal{B}^v \times L^v, \end{cases} \tag{20}$$

where, L^v is the set containing the vertices of $\{\epsilon_1 I_{m\times m},\ I_{m\times m}\}$, and $\circ\mathcal{A}^v$, $\circ\mathcal{B}^v$ are the set of vertices of $\circ\mathcal{A}$, $\circ\mathcal{B}$ respectively. Next, inequalities (20) can be transformed to LMIs by defining the new variables $\tilde{K} = KH^{-1}$, $\tilde{H} = H^{-1}$, $\tilde{Q} = H^{-1}QH^{-1}$ and multiplying both sides of the inequalities in (20) by H^{-1}, we can write finally (20) as

$$\begin{array}{l} \tilde{Q} + \tilde{H}A^T - \tilde{K}^T L^T B^T + A\tilde{H} - BL\tilde{K} \leq 0\ \forall L \in L^v,\ \tilde{Q} = \tilde{Q}^T > 0,\ \tilde{H} > 0 \\ -\tilde{Q} + \tilde{H}\Delta A^T - \tilde{K}^T L^T \Delta B^T + \Delta A\tilde{H} - \Delta BL\tilde{K} < 0\ \forall(\Delta A, \Delta B, Ł) \in \circ\mathcal{A}^v \times \circ\mathcal{B}^v \times L^v, \end{array} \tag{21}$$

the controller gain will be given by $K = \tilde{K}\tilde{H}^{-1}$.□
Let us consider now the practical problem of input saturation. Indeed, in practical applications the available actuators have limited maximum amplitudes. For this reason, it is more realistic to consider bounded control amplitudes in the design of the fault tolerant controller.

5. FTC for LTI plants with control saturation

We consider here the system (1) with input constraints $|u_i| \leq u_{max_i}$, $i = 1, ..., m$, and study the following FTC problem.

5.1 Problem statement

Find a bounded feedback controller, i.e. $|u_i| \leq u_{max_i}$, $i = 1, ..., m$, *such that the closed-loop controlled system (1) admits* $x = 0$ *as a uniformly asymptotically (UA) stable equilibrium point* $\forall\alpha(t)$ (s.t. $0 < \epsilon_1 \leq \alpha_{ii}(t) \leq 1$), $i = 1, ..., m$, *within an estimated domain of attraction.*

5.2 Problem solution

Under the actuator constraint $|u_i| \leq u_{max_i}$, $i = 1, ..., m$, the system (1) can be re-written as

$$\begin{array}{l} \dot{x} = Ax + BU_{max}v \\ y = Kx \\ v = -\alpha(t)sat(y), \end{array} \tag{22}$$

where $U_{max} = diag(u_{max_1}, ..., u_{max_m})$, $sat(y) = (sat(y_1), ..., sat(y_m))^T$, $sat(y_i) = sign(y_i)min\{1, |y_i|\}$.
Thus we have rewritten the system (1) as a MIMO Lure's problem with a generalized sector condition, which is a generalization of the SISO case presented in (16).
Next, we define the two functions $\psi_1 : \mathbb{R}^n \to \mathbb{R}^m$, $\psi_1(x) = -\epsilon_1 I_{m\times m} sat(Kx)$ and

$\psi_2 : \mathbb{R}^n \to \mathbb{R}^m,\ \psi_2(x) = -sat(Kx)$.
We can then write that v is spanned by the two functions $\psi_1,\ \psi_2$:

$$v(x,t) \in co\{\psi_1(x), \psi_2(x)\},\ \forall x \in \mathbb{R}^n, t \in \mathbb{R}, \tag{23}$$

where $co\{\psi_1(x), \psi_2(x)\}$ denotes the convex hull of $\psi_1,\ \psi_2$, i.e.

$$co\{\psi_1(x), \psi_2(x)\} := \{\sum_{i=1}^{i=2} \gamma_i(t)\psi_i(x),\ \sum_{i=1}^{i=2} \gamma_i(t) = 1,\ \gamma_i(t) \geq 0\ \forall t\}.$$

Note that in the SISO case, the problem of analyzing the stability of $x = 0$ for the system (22) under the constraint (23) is a Lure's problem with a generalized sector condition as defined in (16).
Let us recall now some material from (16; 17), that we will use to prove Proposition 4.
Definition 8 ((16), p.538): The ellipsoid level set $\varepsilon(P,\rho) := \{x \in \mathbb{R}^n :\ V(x) = x^T P x \leq \rho\},\ \rho > 0,\ P = P^T > 0$ is said to be contractive invariant for (22) if

$$\dot{V} = 2x^T P(Ax - BU_{max}\alpha sat(Kx)) < 0,$$

for all $x \in \varepsilon(P,\rho)\backslash\{0\},\ \forall t \in \mathbb{R}$.
Proposition 3 *((16), P. 539):* An ellipsoid $\varepsilon(P,\rho)$ is contractively invariant for

$$\dot{x} = Ax + Bsat(Fx),\ B \in \mathbb{R}^{n\times 1}$$

if and only if

$$(A + BF)^T P + P(A + BF) < 0,$$

and there exists an $H \in \mathbb{R}^{1\times n}$ such that

$$(A + BH)^T P + P(A + BH) < 0,$$

and $\varepsilon(P,\rho) \subset \{x \in \mathbb{R}^N :\ |Fx| \leq 1\}$.
Fact 1 ((16), p.539): Given a level set $L_V(\rho) = \{x \in \mathbb{R}^n/\ V(x) \leq \rho\}$ and a set of functions $\psi_i(u),\ i \in \{1,...,N\}$. Suppose that for each $i \in \{1,...,N\}$, $L_V(\rho)$ is contractively invariant for $\dot{x} = Ax + B\psi_i(u)$. Let $\psi(u,t) \in co\{\psi_i(u),\ i \in \{1,...,N\}\}$ for all $u, t \in \mathbb{R}$, then $L_V(\rho)$ is contractively invariant for $\dot{x} = Ax + B\psi(u,t)$.
Theorem 1((17), p. 353): Given an ellipsoid level set $\varepsilon(P,\rho)$, if there exists a matrix $H \in \mathbb{R}^{m\times n}$ such that

$$(A + BM(v,K,H))^T P + P(A + BM(v,K,H)) < 0,$$

for all[1] $v \in \mathcal{V} := \{v \in \mathbb{R}^n | v_i = 1\ or\ 0\}$, and $\varepsilon(P,\rho) \subset \mathcal{L}(H) := \{x \in \mathbb{R}^N :\ |h_i x| \leq 1,\ i = 1,...,m\}$, where

$$M(v,K,H) = \begin{bmatrix} v_1 k_1 + (1 - v_1)h_1 \\ \vdots \\ v_m k_m + (1 - v_m)h_m \end{bmatrix}, \tag{24}$$

then $\varepsilon(P,\rho)$ is a contractive domain for $\dot{x} = Ax + Bsat(Kx)$.
We can now write the following result:
Proposition 4: Under Assumption 1, the system (1) admits $x = 0$ as a UA stable equilibrium

[1] Hereafter, $h_i,\ k_i$ denote the ith line of $H,\ K$, respectively.

point, within the estimated domain of attraction $\varepsilon(P,\rho)$, with the static state feedback $u = Kx = YQ^{-1}x$, where Y, Q solve the LMI problem

$$\begin{array}{l} inf_{Q>0,Y,G}J \\ \begin{bmatrix} JR & I \\ I & Q \end{bmatrix} \geq 0,\ J>0 \\ QA^T + AQ + M(v,Y,G)^T(BU_{max}\alpha_\epsilon)^T + (BU_{max}\alpha_\epsilon)M(v,Y,G) < 0,\ \forall v \in \mathcal{V} \\ QA^T + AQ + M(v,Y,G)^T(BU_{max})^T + (BU_{max})M(v,Y,G) < 0,\ \forall v \in \mathcal{V} \\ \begin{bmatrix} 1 & g_i \\ g_i^T & Q \end{bmatrix} \geq 0,\ i=1,...,m \end{array} \tag{25}$$

where $g_i \in \mathbb{R}^{1\times n}$ is the ith line of G, $\alpha_\epsilon = \epsilon_1 \times I_{m\times m}$, M given by (24), $P = \rho Q^{-1}$, and $R > 0$, $\rho > 0$ are chosen.

Proof: Based on Theorem 1 recalled above, the following inequalities

$$(A + BU_{max}\alpha_\epsilon M(v,-K,H))^T P + P(A + BU_{max}\alpha_\epsilon M(v,-K,H)) < 0, \tag{26}$$

together with the condition $\varepsilon(P,\rho) \subset \mathcal{L}(H)$ are sufficient to ensure that $\varepsilon(P,\rho)$ is contractive invariant for (1) with $\alpha = \epsilon_1 I_{m\times m}$, $u = -u_{max}sat(Kx)$.

Again based on Theorem 1, the following inequalities

$$(A + BU_{max}M(v,-K,H))^T P + P(A + BU_{max}M(v,-K,H)) < 0, \tag{27}$$

together with $\varepsilon(P,\rho) \subset \mathcal{L}(H)$ are sufficient to ensure that $\varepsilon(P,\rho)$ is contractive invariant for (1) with $\alpha = I_{m\times m}$, $u = -u_{max}sat(Kx)$. Now based on the direct extension to the MIMO case, of Fact 1 recalled above, we conclude that $\varepsilon(P,\rho)$ is contractive invariant for (1) with $u = -u_{max}sat(Kx)$, $\forall \alpha_{ii}(t)$, $i = 1,...,m$, $s.t.,0 < \epsilon_1 \leq \alpha_{ii}(t) \leq 1$.

Next, the inequalities conditions (26), (27) under the constraint $\varepsilon(P,\rho) \subset \mathcal{L}(H)$ can be transformed to LMI conditions ((17), p. 355) as follows: To find the control gain K such that we have the bigger estimation of the attraction domain, we can solve the LMI problem

$$\begin{array}{l} inf_{Q>0,Y,G}J \\ \begin{bmatrix} JR & I \\ I & Q \end{bmatrix} \geq 0,\ J>0 \\ QA^T + AQ + M(v,Y,G)^T(BU_{max}\alpha_\epsilon)^T + (BU_{max}\alpha_\epsilon)M(v,Y,G) < 0,\ \forall v \in \mathcal{V} \\ QA^T + AQ + M(v,Y,G)^T(BU_{max})^T + (BU_{max})M(v,Y,G) < 0,\ \forall v \in \mathcal{V} \\ \begin{bmatrix} 1 & g_i \\ g_i^T & Q \end{bmatrix} \geq 0,\ i=1,...,m, \end{array} \tag{28}$$

where $Y = -KQ$, $Q = (P/\rho)^{-1}$, $G = H(P/\rho)^{-1}$, $M(v,Y,G) = M(v,-K,H)Q$, $g_i = h_i(P/\rho)^{-1}$, $h_i \in \mathbb{R}^{1\times n}$ is the ith line of H and $R > 0$ is chosen. □

Remark 1: To solve the problem (25) we have to deal with $2^{m+1} + m + 1$ LMIs, to reduce the number of LMIs we can force $Y = G$, which means $K = -H(P/\rho)^{-1}Q^{-1}$. Indeed, in this case the second and third conditions in (25) reduce to the two LMIs

$$\begin{array}{l} QA^T + AQ + G^T(BU_{max}\alpha_\epsilon)^T + (BU_{max}\alpha_\epsilon)G < 0 \\ QA^T + AQ + G^T(BU_{max})^T + (BU_{max})G < 0, \end{array} \tag{29}$$

which reduces the total number of LMIs in (25) to $m+3$. ♦

In the next section, we report some results in the extension of the previous linear controllers to single input nonlinear affine plants.

6. FTC for nonlinear single input affine plants

Let us consider now the nonlinear affine system

$$\dot{x} = f(x) + g(x)\alpha u, \tag{30}$$

where $x \in \mathbb{R}^n, u \in \mathbb{R}$ represent, respectively, the state vector and the scalar input. The vector fields f, columns of g are supposed to satisfy the classical smoothness assumptions, with $f(0) = 0$. The fault coefficient is such that $0 < \epsilon_1 \leq \alpha \leq 1$.

6.1 Problem statement

Find a state feedback controller $u(x)$ such that the closed-loop controlled system (44) admits $x = 0$ as a local (global) asymptotically stable equilibrium point $\forall \alpha$ (s.t. $0 < \epsilon_1 \leq \alpha \leq 1$).

6.2 Problem solution

We follow here the same idea used above for the linear case, and associate with the faulty system (44) a virtual scalar output, the corresponding system writes as

$$\begin{aligned} \dot{x} &= f(x) + g(x)\alpha u \\ y &= k(x), \end{aligned} \tag{31}$$

where $k : \mathbb{R} \to \mathbb{R}$ is a continuous function.
Let us chose now the controller as the simple output feedback

$$u = -k(x). \tag{32}$$

We can then write from (31) and (32) the closed-loop system as

$$\begin{aligned} \dot{x} &= f(x) + g(x)v \\ y &= k(x) \\ v &= -\alpha y. \end{aligned} \tag{33}$$

As before we have cast the problem of stabilizing (44), for all α as an absolute stability problem (33) as defined in ((40), p.55). We can then use the absolute stability theory to solve the problem.
Proposition 5: The closed-loop system (44) with the static state feedback

$$u = -k(x), \tag{34}$$

where k is such that there exist a C^1 function $S : \mathbb{R}^n \to \mathbb{R}$ positive semidefinite, radially unbounded, i.e. $S(x) \to +\infty$, $||x|| \to +\infty$, that satisfies the PDEs

$$\begin{aligned} L_f S(x) &= -0.5 q^T(x) q(x) + \left(\tfrac{\epsilon_1}{1-\epsilon_1}\right) k^2(x) \\ L_g S(x) &= \left(\tfrac{1+\epsilon_1}{1-\epsilon_1}\right) k(x) - q^T w, \end{aligned} \tag{35}$$

where the function $w : \mathbb{R}^n \to \mathbb{R}^l$ is s.t. $w^T w = \frac{2}{1-\epsilon_1}$, and $q : \mathbb{R}^n \to \mathbb{R}^l$, $l \in \mathbb{N}$, under the condition of local (global) detectability of the system

$$\begin{aligned} \dot{x} &= f(x) + g(x)v \\ y &= k(x), \end{aligned} \tag{36}$$

admits the origin $x = 0$ as a local (global) asymptotically stable equilibrium point.
Proof: We saw the equivalence between the problem of stabilizing (44), and the absolute stability problem (33), with the 'nonlinearities' sector bounds ϵ_1 and 1. Based on this, we can use the sufficient condition provided in Proposition 2.38 in ((40), p. 55) to ensure the absolute stability of the origin $x = 0$ of (33), for all $\alpha \in [\epsilon_1, 1]$.
First we have to ensure that the parallel interconnection of the system

$$\begin{aligned} \dot{x} &= f(x) + g(x)v \\ y &= k(x), \end{aligned} \tag{37}$$

with the trivial unitary gain system

$$y = v, \tag{38}$$

is OFP($-\tilde{k}$), with $\tilde{k} = \frac{\epsilon_1}{1-\epsilon_1}$ and with a C^1 radially unbounded storage function S.
Based on Definition 4, this is true if the parallel interconnection of (37) and (38) is dissipative with respect to the supply rate

$$\omega(v, \tilde{y}) = v^T \tilde{y} + \left(\frac{\epsilon_1}{1-\epsilon_1}\right) \tilde{y}^T \tilde{y}, \tag{39}$$

where $\tilde{y} = y + v$. This means, based on Definition 3, that it exists a C^1 function $S : \mathbb{R}^n \to \mathbb{R}$, with $S(0) = 0$ and $S(x) \geq 0, \ \forall x$, s.t.

$$\begin{aligned} \dot{S}(x(t)) &\leq \omega(v, \tilde{y}) \\ &\leq v^T y + ||v||^2 + \left(\tfrac{\epsilon_1}{1-\epsilon_1}\right) ||y + v||^2. \end{aligned} \tag{40}$$

Furthermore, S should be radially unbounded.
From the condition (40) and Theorem 2.39 in ((40), p. 56), we can write the following condition on S, k for the dissipativity of the parallel interconnection of (37) and (38) with respect to the supply rate (39):

$$\begin{aligned} L_f S(x) &= -0.5 q^T(x) q(x) + \left(\tfrac{\epsilon_1}{1-\epsilon_1}\right) k^2(x) \\ L_g S(x) &= k(x) + 2\left(\tfrac{\epsilon_1}{1-\epsilon_1}\right) k(x) - q^T w, \end{aligned} \tag{41}$$

where the function $w : \mathbb{R}^n \to \mathbb{R}^l$ is s.t. $w^T w = \frac{2}{1-\epsilon_1}$, and $q : \mathbb{R}^n \to \mathbb{R}^l$, $l \in \mathbb{N}$. Finally, based on Proposition 2.38 in ((40), p. 55), to ensure the local (global) asymptotic stability of $x = 0$, the system (37) has to be locally (globally) ZSD, which is imposed by the local (global) detectability of (36). □
Solving the condition (41) might be computationally demanding, since it requires to solve a system of PDEs. We can simplify the static state feedback controller, by considering a lower bound of the condition (40). Indeed, condition (40) is true if the inequality

$$\dot{S} \leq v^T y, \tag{42}$$

is satisfied. Thus, it suffices to ensure that the system (37) is passive with the storage function S. Now, based again on the necessary and sufficient condition given in Theorem 2.39 ((40), p.56), the storage function and the feedback gain have to satisfy the condition

$$\begin{aligned} L_f S(x) &\leq 0 \\ L_g S(x) &= k(x). \end{aligned} \tag{43}$$

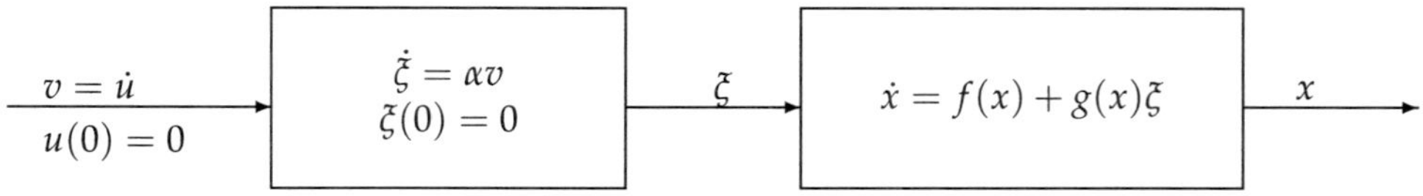

Fig. 1. The model (45) in cascade form

However, by considering a lower bound of (40), we are considering the extreme case where $\epsilon_1 \to 0$, which may result in a conservative feedback gain (refer to (6)). It is worth noting that in that case the controller given by (52), (41), reduces to the classical damping or Jurdjevic-Quinn control $u = -L_g S(x)$, e.g. ((40), p. 111), but based on a semidefinite function S.

7. FTC for nonlinear multi-input affine plants with constant loss of effectiveness actuator faults

We consider here affine nonlinear models of the form

$$\dot{x} = f(x) + g(x)u, \tag{44}$$

where $x \in \mathbb{R}^n, u \in \mathbb{R}^m$ represent respectively the state and the input vectors. The vector fields f, and the columns of g are assumed to be C^1, with $f(0) = 0$.
We study actuator's faults modelled by a multiplicative constant coefficient, i.e. a loss of effectiveness, which implies the following form for the faulty model[2]

$$\dot{x} = f(x) + g(x)\alpha u, \tag{45}$$

where $\alpha \in \mathbb{R}^{m\times m}$ is a diagonal constant matrix, with the diagonal elements $\alpha_{ii},\ i = 1, ..., m$ s.t., $0 < \epsilon_1 \leq \alpha_{ii} \leq 1$. We write then the FTC problem as follows.
Problem statement: *Find a feedback controller such that the closed-loop controlled system (45) admits $x = 0$ as a globally asymptotically stable (GAS) equilibrium point $\forall \alpha$ (s.t. $0 < \epsilon_1 \leq \alpha_{ii} \leq 1$).*

7.1 Problem solution

Let us first rewrite the faulty model (45) in the following cascade form (see figure 1)

$$\begin{aligned} \dot{x} &= f(x) + g(x)h(\xi) \\ \dot{\xi} &= \alpha v,\ \xi(0) = 0 \\ y &= h(\xi) = \xi, \end{aligned} \tag{46}$$

where we define the virtual input $v = \dot{u}$ with $u(0) = 0$. This is indeed, a cascade form where the controlling subsystem, i.e. ξ dynamics, is linear (40). Using this cascade form, it is possible to write a stabilizing controller for the faulty model (45), as follows.

[2] Hereafter, we will denote by x the states of the faulty system (45) to avoid cumbersome notations. However, we remind the reader that the solutions of the healthy system (44) and the faulty system (45) are different.

Theorem 2: Consider the closed-loop system that consists of the faulty system (45) and the dynamic state feedback

$$\begin{aligned} \dot{u} &= -L_g W(x)^T - k\xi,\ u(0) = 0 \\ \dot{\xi} &= \epsilon_1(-(L_g W(x))^T - k\xi),\ \xi(0) = 0, \end{aligned} \tag{47}$$

where W is a C^1 radially unbounded, positive semidefinite function, s.t. $L_f W \leq 0$, and $k > 0$. Consider the fictitious system

$$\begin{aligned} \dot{x} &= f(x) + g(x)\xi \\ \dot{\xi} &= \epsilon_1(-(L_g W)^T + \tilde{v}) \\ y &= h(\xi) = \xi. \end{aligned} \tag{48}$$

If the system (48) is (G)ZSD with the input $\tilde{v}$ and the output y, then the closed-loop system (45) with (47) admits the origin $(x,\xi) = (0,0)$ as (G)AS equilibrium point.
Proof: We first prove that the cascade system (48) is passive from $\tilde{v}$ to $y = \xi$.
To do so, let us first consider the linear part of the cascade system

$$\begin{aligned} \dot{\xi} &= \epsilon_1 v,\ \xi(0) = 0 \\ y &= h(\xi) = \xi. \end{aligned} \tag{49}$$

The system (49) is passive, with the C^1 positive definite, radially unbounded, storage function $U(\xi) = \frac{1}{2}\xi^T\xi$. Indeed, we can easily see that $\forall\ T > 0$

$$\begin{aligned} U(\xi(T)) = \frac{1}{2}\xi^T(T)\xi(T) &\leq \int_0^T v^T y dt \\ &\leq \frac{1}{\epsilon_1}\int_0^T \dot{\xi}^T\xi dt \\ \leq \frac{1}{\epsilon_1}\int_{\xi(0)}^{\xi(T)} \xi^T d\xi &\leq \frac{1}{2}\frac{1}{\epsilon_1}\xi^T(T)\xi(T), \end{aligned}$$

which is true for $0 < \epsilon_1 \leq 1$
Next, we can verify that the nonlinear part of the cascade

$$\begin{aligned} \dot{x} &= f(x) + g(x)\xi \\ y &= L_g W(x), \end{aligned} \tag{50}$$

is passive, with the C^1 radially unbounded, positive semidefinite storage function W. Since, $\dot{W} = L_f W + L_g W\xi \leq L_g W\xi$. Thus we have proved that both the linear and the nonlinear parts of the cascade are passive , we can then conclude that the feedback interconnection (48) of (49) and (50) (see figure 2) is passive from the new input $\tilde{v} = L_g W + v$ to the output ξ, with the storage function $S(x,\xi) = W(x) + U(\xi)$ (see Theorem 2.10 in (40), p. 33).
Finally, the passivity associated with the (G)ZSD implies that the control $\tilde{v} = -k\xi,\ k > 0$ achieves (G)AS (Theorem 2.28 in (40), p. 49).
Up to now, we proved that the negative feedback output feedback $\tilde{v} = -k\xi,\ k > 0$ achieves the desired AS for $\alpha = \epsilon_1 I_{m\times m}$. We have to prove now that the result holds for all α s.t. $0 < \epsilon_1 \leq \alpha_{ii} \leq 1$, even if ξ is fed back from the fault's model (46) with $\alpha = \epsilon_1 I_{m\times m}$, since we do not know the actual value of α. If we multiply the control law (47) by a constant gain matrix $\tilde{k} = diag(\tilde{k}_1, ..., \tilde{k}_m)$, $1 \leq \tilde{k}_i \leq \frac{1}{\epsilon_1}$, we can write the new control as

$$\begin{aligned} \dot{u} &= -\tilde{k}(L_g W(x)^T - k\xi),\ k > 0,\ u(0) = 0 \\ \dot{\xi} &= \epsilon_1\tilde{k}(-(L_g W(x))^T - k\xi),\ \xi(0) = 0. \end{aligned} \tag{51}$$

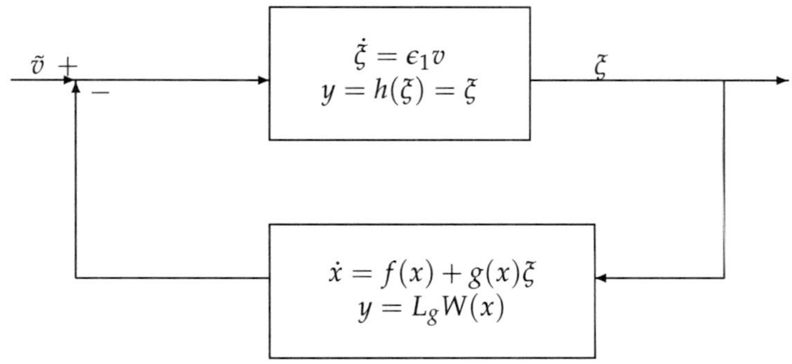

Fig. 2. Feedback interconnection of (49) and (50)

It is easy to see that this gain does not change the stability result, since we can define for the nonlinear cascade part (50) the new storage function $\tilde{W} = \tilde{k}W$ and the passivity is still satisfied from its input ξ to the new output $\tilde{k}L_gW(x)$. Next, since the ZSD property remains unchanged, we can chose the new stabilizing output feedback $\tilde{v} = -\tilde{k}k\xi$, which is still a stabilizing feedback with the new gain $\tilde{k}k > 0$, and thus the stability result holds for all α $s.t.$ $0 < \epsilon_1 \leq \alpha_{ii} \leq 1,\ i = 1, ..., m$. □

The stability result obtained in Theorem 2, depends on the ZSD property. Indeed, if the ZSD is global, the stability obtained is global otherwise only local stability is ensured. Furthermore, we note here that with the dynamic controller (47) we ensure that the initial control is zero, regardless of the initial value of the states. This might be important for practical applications, where an abrupt switch from zero to a non zero initial value of the control is not tolerated by the actuators.

In Theorem 2, one of the necessary conditions is the existence of $W \geq 0$, s.t. the uncontrolled part of (45) satisfies $L_fW \leq 0$. To avoid this condition that may not be satisfied for some practical systems, we propose the following Theorem.

Theorem 3: Consider the closed-loop system that consists of the faulty system (45) and the dynamic state feedback

$$\begin{array}{c}\dot{u} = \frac{1}{\epsilon_1}(-k(\xi - \beta K(x)) - \beta L_g W^T + \beta \frac{\partial K}{\partial x}(f + g\xi)),\ \beta = diag(\beta_{11}, ..., \beta_{mm}),\ 0 < \frac{\tilde{\epsilon}_1}{\epsilon_1} \leq \beta_{ii} \leq 1,\\ \dot{\xi} = -k(\xi - \beta K(x)) - \beta L_g W^T + \beta \frac{\partial K}{\partial x}(f + g\xi),\ \xi(0) = 0,\ u(0) = 0,\end{array} \tag{52}$$

where $k > 0$ and the C^1 function $K(x)$ is s.t. there exists a C^1 radially unbounded, positive semidefinite function W satisfying

$$\frac{\partial W}{\partial x}(f(x) + g(x)\beta K(x)) \leq 0,\ \forall x \in \mathbb{R}^n,\ \forall \beta = diag(\beta_{11}, ..., \beta_{mm}),\ 0 < \tilde{\epsilon}_1 \leq \beta_{ii} \leq 1. \tag{53}$$

Consider the fictitious system

$$\begin{array}{c}\dot{x} = f(x) + g(x)\xi\\ \dot{\xi} = \beta \frac{\partial K}{\partial x}(f + g\xi) - \beta L_g W^T + \tilde{\tilde{v}}\\ \tilde{y} = \xi - \beta K(x).\end{array} \tag{54}$$

If (54) is (G)ZSD with the input $\tilde{\tilde{v}}$ and the output $\tilde{y}$, for for all β s.t. $\beta_{ii},\ i = 1, ..., m,\ 0 < \tilde{\epsilon}_1 \leq \beta_{ii} \leq 1$. Then, the closed-loop system (45) with (52) admits the origin $(x, \xi) = (0, 0)$ as (G)AS

equilibrium point.
Proof: We will first prove that the controller (52) achieves the stability results for a faulty model with $\alpha = \epsilon_1 I_{m\times m}$ and then we will prove that the stability result holds the same for all α s.t. $\alpha_{ii},\ i = 1,...,m, 0 < \epsilon_1 \leq \alpha_{ii} \leq 1$.
First, let us define the virtual output $\tilde{y} = \xi - \beta K(x)$, we can write the model (46) with $\alpha = \epsilon_1 I_{m\times m}$ as

$$\begin{aligned} \dot{x} &= f(x) + g(x)(\tilde{y} + \beta K(x)) \\ \dot{\xi} &= \epsilon_1 I_{m\times m} v \\ \tilde{y} &= \xi - \beta K(x), \end{aligned} \tag{55}$$

we can then write

$$\dot{\tilde{y}} = \epsilon_1 I_{m\times m} v - \beta \frac{\partial K}{\partial x}(f + g(\tilde{y} + \beta K(x))) = \tilde{v}.$$

To study the passivity of (55), we define the positive semidefinite storage function

$$V = \beta W(x) + \frac{1}{2}\tilde{y}^T \tilde{y},$$

and write

$$\dot{V} = \beta L_{f+g\beta K} W + \beta L_g W \tilde{y} + \tilde{y}^T \tilde{v},$$

and using the condition (53), we can write

$$\dot{V} \leq \tilde{y}^T(\beta L_g W^T + \tilde{v}),$$

which establishes the passivity of (55) from the new input $\tilde{\tilde{v}} = \tilde{v} + \beta L_g W^T$ to the output $\tilde{y}$. Finally, using the (G)ZSD condition for $\alpha = \epsilon_1 I_{m\times m}$, we conclude about the (G)AS of (46) for $\alpha = \epsilon_1 I_{m\times m}$, with the controller (52) (Theorem 2.28 in (40), p. 49). Now, remains to prove that the same result holds for all α s.t. $\alpha_{ii},\ i = 1,...,m, 0 < \epsilon_1 \leq \alpha_{ii} \leq 1$, i.e. the controller (52) has the appropriate gain margin. In our particular case, it is straightforward to analyse the gain margin of (52), since if we multiply the controller in (52) by a matrix α, s.t. $\alpha_{ii},\ i = 1,...,m$, $0 < \epsilon_1 \leq \alpha_{ii} \leq 1$, the new control writes as

$$\begin{aligned} \dot{u} &= \tfrac{1}{\epsilon_1} I_{m\times m}(-\alpha k(\xi - \beta K(x)) - \alpha\beta L_g W^T + \alpha\beta \tfrac{\partial K}{\partial x}(f + g\xi)), \\ & k > 0,\ \beta = diag(\beta_1, ..., \beta_m),\ 0 < \beta_i \leq 1, \\ \dot{\xi} &= \alpha k(\xi - \beta K(x)) - \alpha\beta L_g W^T + \alpha\beta \tfrac{\partial K}{\partial x}(f + g\xi),\ \xi(0) = 0,\ u(0) = 0. \end{aligned} \tag{56}$$

We can see that this factor will not change the structure of the initial control (52), since it will be directly absorbed by the gains, i.e. we can write $\tilde{k} = \alpha k$, with all the elements of diagonal matrix $\tilde{k}$ positive, we can also define $\tilde{\beta} = \alpha\beta$ which is still a diagonal matrix with bounded elements in $[\tilde{\epsilon}_1, 1]$, s.t. (53) and (54) are is still satisfied. Thus the stability result remains unchanged. □
The previous theorems may guaranty *global* AS. However, the conditions required may be difficult to satisfy for some systems. We present below a control law ensuring, under less demanding conditions, semiglobal stability instead of global stability.
Theorem 4: Consider the closed-loop system that consists of the faulty system (45) and the dynamic state feedback

$$\begin{aligned} \dot{u} &= -k(\xi - u_{nom}(x)),\ k > 0, \\ \dot{\xi} &= -k\epsilon_1(\xi - u_{nom}(x)),\ \xi(0) = 0,\ u(0) = 0, \end{aligned} \tag{57}$$

where the nominal controller $u_{nom}(x)$ achieves semiglobal asymptotic and local exponential stability of $x = 0$ for the safe system (44). Then, the closed-loop (45) with (57) admits the origin $(x, \xi) = (0, 0)$ as semiglobal AS equilibrium point.
Proof: The prove is a direct application of the Proposition 6.5 in ((40), p. 244), to the system (46), with $\alpha = \epsilon_1 I_{m \times m}$. Any positive gain α, s.t. $1 \leq \alpha_{ii} \leq \frac{1}{\epsilon_1}$, $i = 1, ..., m$, will be absorbed by $k > 0$, keeping the stability results unchanged. Thus the control law (57) stabilize (46) and equivalently (45) for all α s.t. α_{ii}, $i = 1, ..., m$, $0 < \epsilon_1 \leq \alpha_{ii} \leq 1$. □
Let us consider now the practical problem of input saturation. Indeed, in practical systems the actuator powers are limited, and thus the control amplitude bounds should be taken into account in the controller design. To solve this problem, we consider a more general model than the affine model (44). In the following we first study the problem of FTC with input saturation, on the general model

$$\dot{x} = f(x) + g(x, u)u, \tag{58}$$

where, x, u, f are defined as before, g is now function of both the states and the inputs, and is assumed to be C^1 w.t.r. to x, u.
The actuator faut model, writes as

$$\dot{x} = f(x) + g(x, \alpha u)\alpha u, \tag{59}$$

with the loss of effectiveness matrix α defined as before. This problem is treated in the following Theorem, for the scalar case where $\alpha \in [\epsilon_1, \ 1]$, i.e. when the same fault occurs on all the actuators.
Theorem 5: Consider the closed-loop system that consists of the faulty system (59), for $\alpha \in [\epsilon_1, \ 1]$, and the static state feedback

$$\begin{aligned}
u(x) &= -\lambda(x) G(x, 0)^T \\
G(x, 0) &= \frac{\partial W(x)}{\partial x} \epsilon_1 g(x, 0) \\
\lambda(x) &= \frac{2\overline{u}}{(1+\gamma_1(|x|^2+4\overline{u}^2|G(x,0)|^2))(1+|G(x,0)|^2)} > 0 \\
\gamma_1 &= \int_0^{2s} \frac{\overline{\gamma}_1(s)}{1+\overline{\gamma}_1(1)} ds \\
\overline{\gamma}_1(s) &= \frac{1}{s} \int_s^{2s} (\tilde{\gamma}_1(t) - 1) dt + s \\
\tilde{\gamma}_1(s) &= max_{\{(x,u)||x|^2+|u|^2 \leq s\}} \{1 + \int_0^1 \frac{\partial W(x)}{\partial x} \frac{\partial g(x, \tau\epsilon_1 u)}{\partial u} d\tau\},
\end{aligned} \tag{60}$$

where W is a C^2 radially unbounded, positive semidefinite function, s.t. $L_f W \leq 0$. Consider the fictitious system

$$\begin{aligned}
\dot{x} &= f(x) + g(x, \epsilon_1 u)\epsilon_1 u \\
y &= \frac{\partial W(x)}{\partial x} \epsilon_1 g(x, \epsilon_1 u).
\end{aligned} \tag{61}$$

If (61) is (G)ZSD, then the closed-loop system (59) with (60) admits the origin as (G)AS equilibrium point. Furthermore $|u(x)| \leq \overline{u}$, $\forall x$.
Proof: Let us first consider the faulty model (59) with $\alpha = \epsilon_1$. For this model, we can compute the derivative of W as

$$\dot{W}(x) = L_f W + \frac{\partial W(x)}{\partial x} \epsilon_1 g(x, \epsilon_1 u) u$$

$$\dot{W}(x) \leq \frac{\partial W(x)}{\partial x} \epsilon_1 g(x, \epsilon_1 u) u.$$

Now, using Lemma II.4 (p.1562 in (31)), we can directly write the controller (60), s.t.

$$\dot{W} \leq -\frac{1}{2} \lambda(x) |G(x, 0)|^2.$$

Furthermore $|u(x)| \leq \overline{u},\ \forall x$.
We conclude then that the trajectories of the closed-loop equations converge to the invariant set $\{x|\ \lambda(x)|G(x,0)|^2 = 0\}$ which is equivalent to the set $\{x|\ G(x,0) = 0\}$. Based on Theorem 2.21 (p. 43, (40)), and the assumption of (G)ZSD for (61), we conclude about the (G)AS of the origin of (59), (60), with $\alpha = \epsilon_1$. Now multiplying u by any positive coefficient α, s.t. $0 < \epsilon_1 \leq \alpha \leq 1$ does not change the stability result. Furthermore, if $|u(x)| \leq \overline{u},\ \forall x$, then $|\alpha u(x)| \leq \overline{u},\ \forall x$, which completes the proof. □
Remark 2: In Theorem 5, we consider only the case of scalar fault $\alpha \in [\epsilon_1,\ 1]$, i.e. the case of uniform fault, since we need this assumption to be able to apply the result of Lemma II.4 in (31). However, this assumption can be satisfied in practice by a class of actuators, namely pneumatically driven diaphragm-type actuators (23), for which the failure of the pressure supply system might lead to a uniform fault of all the actuators. Furthermore, in Proposition 6 below we treat for the case of systems affine in the control, i.e. $g(x,u) = g(x)$, the general case of any diagonal matrix of loss of effectiveness coefficients.
Proposition 6: Consider the closed-loop system that consists of the faulty system (45), and the static state feedback

$$\begin{aligned} u(x) &= -\lambda(x)G(x)^T \\ G(x) &= \tfrac{\partial W(x)}{\partial x}\epsilon_1 g(x) \\ \lambda(x) &= \tfrac{2\overline{u}}{1+|G(x)|^2}. \end{aligned} \tag{62}$$

where W is a C^2 radially unbounded, positive semidefinite function, s.t. $L_f W \leq 0$. Consider the fictitious system

$$\begin{aligned} \dot{x} &= f(x) + g(x)\epsilon_1 u \\ y &= \tfrac{\partial W(x)}{\partial x}\epsilon_1 g(x). \end{aligned} \tag{63}$$

If (63) is (G)ZSD, then the closed-loop system (45) with (62) admits the origin as (G)AS equilibrium point. Furthermore $|u(x)| \leq \overline{u},\ \forall x$.
Proof: The proof follows the same steps as in the proof of Theorem 5, except that in this case the constraint of considering that the same fault occurs on all the actuators, i.e. for a scalar α, is relaxed. Indeed, in this case we can directly ensure the negativeness of $\dot{W}$, since if u is such that $\dot{W} \leq -\lambda(x)L_g W(x)\epsilon_1 L_g W(x)^T \leq 0$, then in the case of a diagonal fault matrix, the derivative writes as $\dot{W} \leq -\lambda(x)L_g W(x)\epsilon_1\alpha L_g W(x)^T \leq -\lambda(x)\epsilon_1^2 L_g W(x)L_g W(x)^T \leq -\epsilon_1^2\lambda(x)|G(x)|^2$. Thus, the stability result remains unchanged. □
Up to now we have considered the case of abrupt faults, modelled with constant loss of effectiveness matrices. However, in practical applications, the faults are usually time-varying or incipient, modelled with time-varying loss of effectiveness coefficients, e.g. (50). We consider in the following section this case of time-varying loss of effectiveness matrices.

8. FTC for nonlinear multi-input affine plants with time-varying loss of effectiveness actuator faults

We consider here faulty models of the form

$$\dot{x} = f(x) + g(x)\alpha(t)u, \tag{64}$$

where $\alpha(t)$ is a diagonal time-varying matrix, with C^1 diagonal elements $\alpha_{ii}(t),\ i = 1,...,m$ s.t., $0 < \epsilon_1 \leq \alpha_{ii}(t) \leq 1,\ \forall t$. We write then the FTC problem as follows.
Problem statement: *Find a feedback controller such that the closed-loop controlled system (64) admits $x = 0$ as a uniformly asymptotically stable (UAS) equilibrium point $\forall\alpha(t)$ (s.t. $0 < \epsilon_1 \leq \alpha_{ii}(t) \leq 1$).*

8.1 Problem solution

To solve this problem we use some of the tools introduced in (24), where a generalization of Krasovskii-LaSalle theorem, has been proposed for nonlinear time-varying systems.
We can first write the following result.
Theorem 6: Consider the closed-loop system that consists of the faulty system (64) with the dynamic state feedback

$$\begin{aligned} \dot{u} &= -L_g W(x)^T - k\xi,\ k > 0,\ u(0) = 0 \\ \dot{\xi} &= \tilde{\alpha}(t)(-(L_g W(x))^T - k\xi),\ \xi(0) = 0, \end{aligned} \tag{65}$$

where $\tilde{\alpha}(t)$ is a C^1 function, s.t. $0 < \epsilon_1 \leq \tilde{\alpha}(t) \leq 1,\ \forall t$, and W is a C^1, positive semidefinite function, such that:
1- $L_f W \leq 0$,
2- The system $\dot{x} = f(x)$ is AS conditionally to the set $M = \{x \mid W(x) = 0\}$,
3- $\forall(\overline{x}, \overline{\xi})$ limiting solutions for the system

$$\begin{aligned} \dot{x} &= f(x) + g(x)\xi \\ \dot{\xi} &= \alpha(t)(-(L_g W)^T - k\xi) \\ y &= h(x, \xi) = \xi, \end{aligned} \tag{66}$$

w.r.t. unbounded sequence $\{t_n\}$ in $[0, \infty)$, then if $h(\overline{x}, \overline{\xi}) = 0$, a.e., then either $(\overline{x}, \overline{\xi})(t_0) = (0, 0)$ for some $t_0 \geq 0$ or $(0, 0)$ is a ω-limit point of $(\overline{x}, \overline{\xi})$, i.e. $\lim_{t \to \infty}(\overline{x}, \overline{\xi})(t) \to (0, 0)$.
Then the closed-loop system (64) with (65) admits the origin $(x, \xi) = (0, 0)$ as UAS equilibrium point.
Proof: Let us first rewrite the system (64) for $\alpha(t) = \tilde{\alpha}(t)$, in the cascade form

$$\begin{aligned} \dot{x} &= f(x) + g(x)h(\xi) \\ \dot{\xi} &= \tilde{\alpha}(t)v,\ v = \dot{u},\ \xi(0) = 0,\ u(0) = 0 \\ y &= h(\xi) = \xi. \end{aligned} \tag{67}$$

Replacing $v = \dot{u}$ by its value in (65) gives the feedback system

$$\begin{aligned} \dot{x} &= f(x) + g(x)h(\xi) \\ \dot{\xi} &= \tilde{\alpha}(t)(-L_g W(x)^T + \tilde{v}),\ \xi(0) = 0,\ u(0) = 0 \\ y &= h(\xi) = \xi. \end{aligned} \tag{68}$$

We prove that (68) is passive from the input $\tilde{v}$ to the output ξ. We consider first the linear part of (67)

$$\begin{aligned} \dot{\xi} &= \tilde{\alpha}(t)v,\ \xi(0) = 0 \\ y &= h(\xi) = \xi, \end{aligned} \tag{69}$$

which is passive with the storage function $U(\xi) = \frac{1}{2}\xi^T\xi$, i.e. $\dot{U}(t, \xi) = \xi^T\dot{\xi} = \xi^T\tilde{\alpha}(t)v \leq \xi^T v = v^T\xi$.
Next, we consider the nonlinear part

$$\begin{aligned} \dot{x} &= f(x) + g(x)\xi \\ y &= L_g W(x), \end{aligned} \tag{70}$$

which is passive with the storage function $W(x)$, s.t. $\dot{W} = L_f W + L_g W\xi \leq L_g W\xi$.
We conclude that the feedback interconnection (68) of (69) and (70) is passive from $\tilde{v}$ to ξ, with

the storage function $S(x,\xi) = W(x) + U(\xi)$ (see Theorem 2.10, p. 33 in (40)).
This implies that the derivative of S along (68) with $\tilde{v} = -k\xi,\ k > 0$, writes

$$\dot{S}(t,x,\xi) \leq \tilde{v}^T\xi \leq 0.$$

Now we define for (68) with $\tilde{v} = -k\xi,\ k > 0$, the positive invariant set

$$M = \{(x,\xi)|W(x) + U(\xi) = 0\}$$

$$M = \{(x,0)|W(x) = 0\}.$$

We note that the restriction of (68) with $\tilde{v} = -k\xi,\ k > 0$ on M is $\dot{x} = f(x)$, then applying Theorem 5 in (18), we conclude that, under Condition 2 of Theorem 6, the origin $(x,\xi) = (0,0)$ is US for the system (64) for $\alpha = \tilde{\alpha}$ and the dynamic controller (65). Now, multiplying u by any $\alpha(t)$, s.t. $0 < \epsilon_1 \leq \alpha_{ii}(t) \leq 1,\ \forall t$, does not change neither the passivity property, nor the AS condition of $\dot{x} = f(x)$ on M, which implies the US of $(x,\xi) = (0,0)$ for (64), (65) $\forall\alpha(t)$, s.t. $0 < \epsilon_1 \leq \alpha_{ii}(t) \leq 1,\ \forall t$.
Now we first note the following fact: for any $\sigma > 0$ and any $t \geq t_0$ we can write

$$S(t,x(t),\xi(t)) - S(t_0,x(t_0),\xi(t_0)) \leq -\int_{t_0}^{t} \mu(h(\xi(\tau)))d\tau = -\int_{t_0}^{t} k|\xi(\tau)|^2 d\tau,$$

thus we have

$$\int_{t_0}^{t} (\mu(h(\xi(\tau))) - \sigma)d\tau \leq \int_{t_0}^{t} \mu(h(\xi(\tau)))d\tau \leq S(t_0,x(t_0),\xi(t_0)) < \tilde{M};\tilde{M} > 0.$$

Finally, using Theorem 1 in (24), under Condition 3 of Theorem 6, we conclude that $(x,\xi) = (0,0)$ is UAS for (64), (65). □
Remark 3: The function $\tilde{\alpha}$ in (65) has been chosen to be any C^1 time varying function, s.t. $0 < \epsilon_1 \leq \tilde{\alpha}(t) \leq 1,\ \forall t$. The general time-varying nature of the function was necessary in the proof to be able to use the results of Theorem 5 in (18) to prove the US of the faulty system's equilibrium point. However, in practice one can simply chose $\tilde{\alpha}(t) = 1,\ \forall t$ ♦.
Remark 4: Condition 3 in Theorem 6 is general and has been used to properly prove the stability results in the time-varying case. However, in practical application it can be further simplified, using the notion of reduced limiting system. Indeed, using Theorem 3 and Lemma 7 in (24), Condition 3 simplifies to:
$\forall(\overline{x},\overline{\xi})$ solutions for the reduced limiting system

$$\begin{aligned} \dot{\overline{x}} &= f(\overline{x}) + g(\overline{x})\overline{\xi} \\ \dot{\overline{\xi}} &= \alpha_\gamma(t)(-(L_g W(\overline{x}))^T - k\overline{\xi}) \\ y &= h(\overline{x},\overline{\xi}) = \overline{\xi}, \end{aligned} \tag{71}$$

where the limiting function $\alpha_\gamma(t)$ is defined us $\alpha_\gamma(t) \triangleq \lim_{n\to\infty}\alpha(t+t_n)$ w.r.t. unbounded sequence $\{t_n\}$ in $[0,\infty)$. Then, if $h(\overline{x},\overline{\xi}) = 0$, a.e., then either $(\overline{x},\overline{\xi})(t_0) = (0,0)$ for some $t_0 \geq 0$ or $(0,0)$ is a ω-limit point of $(\overline{x},\overline{\xi})$. Now, since in our case the diagonal matrix-valued function α is s.t. $0 < \epsilon_1 \leq \alpha_{ii}(t) \leq 1,\ \forall t$, then it obviously satisfies a permanent excitation (PE) condition of the form

$$\int_{t}^{t+T} \alpha(\tau)\alpha(\tau)^T d\tau \geq rI,\ T > 0,\ r > 0,\ \forall t,$$

which implies, based on Lemma 8 in (24), that to check Condition 3 we only need to check the classical ZSD condition:
$\forall \overline{x}$ solutions for the system

$$\begin{array}{c} \dot{\overline{x}} = f(\overline{x}) \\ L_g W(\overline{x}) = 0, \end{array} \tag{72}$$

either $\overline{x}(t_0) = 0$ for some $t_0 \geq 0$ or 0 is a ω-limit point of $\overline{x}$. ♦
Let us consider again the problem of input saturation. We consider here again the more general model (58), and study the problem of FTC with input saturation for the time-varying faulty model

$$\dot{x} = f(x) + g(x, \alpha(t)u)\alpha(t)u, \tag{73}$$

with the diagonal loss of effectiveness matrix $\alpha(t)$ defined as before. This problem is treated in the following Theorem, for the scalar case where $\alpha(t) \in [\epsilon_1,\ 1],\ \forall t$, i.e. when the same fault occurs on all the actuators.
Theorem 7: Consider the closed-loop system that consists of the faulty system (73) for $\alpha \in [\epsilon_1,\ 1],\ \forall t$, with the static state feedback

$$\begin{array}{c} u(x) = -\lambda(x)G(x,0)^T \\ G(x,0) = \frac{\partial W(x)}{\partial x} g(x,0) \\ \lambda(x) = \frac{2\overline{u}}{(1+\gamma_1(|x|^2+4\overline{u}^2|G(x,0)|^2))(1+|G(x,0)|^2)} > 0 \\ \gamma_1 = \int_0^{2s} \frac{\overline{\gamma}_1(s)}{1+\overline{\gamma}_1(1)} ds \\ \overline{\gamma}_1(s) = \frac{1}{s}\int_s^{2s} (\tilde{\gamma}_1(t) - 1)dt + s \\ \tilde{\gamma}_1(s) = max_{\{(x,u)||x|^2+|u|^2 \leq s\}}\{1 + \int_0^1 \frac{\partial W(x)}{\partial x} \frac{\partial g(x, \tau\epsilon_1 u)}{\partial u} d\tau\}, \end{array} \tag{74}$$

where W is a C^2, positive semidefinite function, such that:
1- $L_f W \leq 0$,
2- The system $\dot{x} = f(x)$ is AS conditionally to the set $M = \{x \mid W(x) = 0\}$,
3- $\forall \overline{x}$ limiting solutions for the system

$$\begin{array}{c} \dot{x} = f(x) + g(x, \epsilon_1 u(x))(-\lambda(x)\alpha(t)\frac{\partial W}{\partial x}(x)g(x,0))^T \\ y = h(x) = \lambda(x)^{0.5}|\frac{\partial W}{\partial x}(x)g(x,0)|, \end{array} \tag{75}$$

w.r.t. unbounded sequence $\{t_n\}$ in $[0, \infty)$, then if $h(\overline{x}) = 0$, a.e., then either $\overline{x}(t_0) = 0$ for some $t_0 \geq 0$ or 0 is a ω-limit point of $\overline{x}$.
Then the closed-loop system (73) with (74) admits the origin $x = 0$ as UAS equilibrium point. Furthermore $|u(x)| \leq \overline{u},\ \forall x$.
Proof: We first can write, based on Condition 1 in Theorem 7

$$\dot{W} \leq \frac{\partial W}{\partial x} g(x, \alpha(t)u)\alpha(t)u,$$

using Lemma II.4 in (31), and considering the controller (74), we have

$$\dot{W} \leq -\frac{\epsilon_1}{2}\lambda(x)|G(x,0)|^2,\ |u(x)| \leq \overline{u}\ \forall x.$$

Next, we define for (73) and the controller (74) the positive invariant set $M = \{x|\ W(x) = 0\}$. Note that we can also write

$$M = \{x|\ \dot{W}(x) = 0\} \Leftrightarrow \{x|\ G(x,0) = 0\} \Leftrightarrow \{x|\ u(x) = 0\}.$$

Thus, the restriction of (73) on M is the system $\dot{x} = f(x)$. Finally, using Theorem 5 in (18), and under Condition 2 in Theorem 7, we conclude that $x = 0$ is US for (73) and the controller (74). Furthermore if $|u(x)| \leq \bar{u}$ then $|\alpha(t)u(x)| \leq \bar{u}\ \forall t, x$.
Now we note that for the virtual output $y = h(x) = \lambda(x)^{0.5}|\frac{\partial W}{\partial x}(x)g(x,0)|$, and $\sigma > 0$ we can write

$$W(t, x(t)) - W(t_0, x(t_0)) \leq -\frac{\epsilon_1}{2}\int_{t_0}^{t} |y(\tau)|^2 d\tau = -\int_{t_0}^{t} \mu(y(\tau))d\tau,$$

thus we have

$$\int_{t_0}^{t} (\mu(y(\tau)) - \sigma)d\tau \leq \int_{t_0}^{t} \mu(y(\tau))d\tau \leq W(t_0, x(t_0)) \leq \tilde{M},\ \tilde{M} > 0.$$

Finally, based on this last inequality and under Condition 3 in Theorem 7, using Theorem 1 in (24), we conclude that $x = 0$ is UAS equilibrium point for (73), (74). □
Remark 5: Here again we can simplify Condition 3 of Theorem 7, as follows. Based on Proposition 3 and Lemma 7 in (24), this condition is equivalent to: $\forall \bar{x}$ solutions for the reduced limiting system

$$\begin{array}{c} \dot{\bar{x}} = f(\bar{x}) + g(\bar{x}, \epsilon_1 u(\bar{x}))(-\lambda(\bar{x})\alpha_\gamma(t)\frac{\partial W}{\partial x}(\bar{x})g(\bar{x},0))^T \\ y = h(\bar{x}) = \lambda(\bar{x})^{0.5}|\frac{\partial W}{\partial x}(\bar{x})g(\bar{x},0)|, \end{array} \quad (76)$$

where the limiting function $\alpha_\gamma(t)$ is defined us $\alpha_\gamma(t) \triangleq \lim_{n\to\infty} \alpha(t + t_n)$ w.r.t. unbounded sequence $\{t_n\}$ in $[0, \infty)$. Then, if $h(\bar{x}) = 0$, a.e., then either $\bar{x}(t_0) = 0$ for some $t_0 \geq 0$ or 0 is a ω-limit point of $\bar{x}$. Which writes directly as the ZSD condition:
$\forall \bar{x}$ solutions for the system

$$\begin{array}{c} \dot{\bar{x}} = f(\bar{x}) \\ \frac{\partial W}{\partial x}(\bar{x})g(\bar{x},0) = 0, \end{array} \quad (77)$$

either $\bar{x}(t_0) = 0$ for some $t_0 \geq 0$ or 0 is a ω-limit point of $\bar{x}$. ♦
Theorem 7 deals with the case of the general nonlinear model (73). For the particular case of affine nonlinear models, i.e. $g(x, u) = g(x)$, we can directly write the following Proposition.
Proposition 7: Consider the closed-loop system that consists of the faulty system (64) with the static state feedback

$$\begin{array}{rl} u(x) = & -\lambda(x)G(x)^T \\ G(x) = & \frac{\partial W(x)}{\partial x}g(x) \\ \lambda(x) = & \frac{2\bar{u}}{1+|G(x)|^2}. \end{array} \quad (78)$$

where W is a C^2, positive semidefinite function, such that:
1- $L_f W \leq 0$,
2- The system $\dot{x} = f(x)$ is AS conditionally to the set $M = \{x \mid W(x) = 0\}$,
3- $\forall \bar{x}$ limiting solutions for the system

$$\begin{array}{c} \dot{x} = f(x) + g(x)(-\lambda(x)\alpha(t)\frac{\partial W}{\partial x}(x)g(x))^T \\ y = h(x) = \lambda(x)^{0.5}|\frac{\partial W}{\partial x}(x)g(x)|, \end{array} \quad (79)$$

w.r.t. unbounded sequence $\{t_n\}$ in $[0, \infty)$, then if $h(\bar{x}) = 0$, a.e., then either $\bar{x}(t_0) = 0$ for some $t_0 \geq 0$ or 0 is a ω-limit point of $\bar{x}$.
Then the closed-loop system (64) with (78) admits the origin $x = 0$ as UAS equilibrium point. Furthermore $|u(x)| \leq \bar{u},\ \forall x$.
Proof: The proof is a direct consequence of Theorem 7. However in this case the constraint of

considering that the same fault occurs on all the actuators, is relaxed. Indeed, in this case we can directly write $\forall \alpha(t) \in \mathbb{R}^{m\times m}$, *s.t.* $0 < \epsilon_1 \leq \alpha_{ii}(t) \leq 1$, $\forall t$:

$$\dot{W} \leq -\lambda(x) L_g W(x) \alpha(t) L_g W(x)^T$$

$$\dot{W} \leq -\lambda(x) \epsilon_1 L_g W(x) L_g W(x)^T \leq -\epsilon_1 |G(x)|^2.$$

The rest of the proof remains unchanged. □

If we compare the dynamic controllers proposed in the Theorems 2, 3, 4, 6 and the static controllers of Theorems 5, 7, we can see that the dynamic controllers ensure that the control at the initialization time is zero, whereas this is not true for the static controllers. In the opposite, the static controllers have the advantage to ensure that the feedback control amplitude stays within the desired bound $\overline{u}$. We can also notice that, except for the controller in Theorem 3, all the remaining controllers proposed here do not involve the vector field f in there computation. This implies that these controllers are robust with respect to any uncertainty Δf as long as the conditions on f, required in the different theorems are still satisfied by the uncertain vector field $f + \Delta f$. Furthermore, the dynamic controller of Theorem 4 inherits the same robustness properties of the nominal controller u_{nom} used to write equation (57) (refer to Proposition 6.5, (40), p. 244).

9. Conclusion and future work

In this chapter we have presented different passive fault tolerant controllers for linear as well as for nonlinear models. Firstly, we have formulated the FTC problem in the context of the absolute stability theory, which has led to direct solutions to the passive FTC problem for LTI systems with uncertainties as well as input saturations. Open problems to which this formulation may be applied include infinite dimension models, stochastic models as well as time-delay models. Secondly, we have proposed several fault tolerant controllers for nonlinear models, by formulating the FTC problem as a cascade passivity-based control. Although, the proposed formulation has led to solutions for a large class of loss of actuator effectiveness faults for nonlinear systems, a more general result treating component faults entering the system through the vector field f plus additive faults on g, as well as the complete loss of some actuators is still missing and should be the subject of future work.

10. References

[1] F. Amato. *Robust control of linear systems subject to uncertain time-varying parameters.* Lecture Notes in Control and Information Sciences. Springer edition, 2006.

[2] M. Aizerman and F. Gantmacher. *Absolute stability of regulator systems.* Holden-Day, INC. 1964.

[3] B. Anderson and S. Vongpanitlerd. *Network analysis and synthesis.* Network series. Prentice-hall edition, 1973.

[4] M. Basin and M. Pinsky. Stability impulse control of faulted nonlinear systems. *IEEE, Transactions on Automatic Control*, 43(11):1604–1608, 1998.

[5] M. Benosman. A survey of some recent results on nonlinear fault tolerant control. *Mathematical Problems in Engineering*, 2010. Article ID 586169, 25 pages, doi:10.1155/2010/586169.

[6] M. Benosman and K.-Y. Lum. Application of absolute stability theory to robust control against loss of actuator effectiveness. *IET Control Theory and Applications*, 3(6):772–788, June 2009.

[7] M. Benosman and K.-Y. Lum. On-line references reshaping and control re-allocation for nonlinear fault tolerant control. *IEEE, Transactions on Control Systems Technology*, 17(2):366–379, March 2009.

[8] M. Benosman and K.-Y. Lum. Application of passivity and cascade structure to robust control against loss of actuator effectiveness. *Int. Journal of Robust and Nonlinear Control*, 20:673–693, 2010.

[9] M. Benosman and K.-Y. Lum. A passive actuators' fault tolerant control for affine nonlinear systems. *IEEE, Transactions on Control Systems Technology*, 18(1):152–163, 2010.

[10] C. Bonivento, L. Gentili, and A. Paoli. Internal model based fault tolerant control of a robot manipulator. In *IEEE, Conference on Decision and Control*, 2004.

[11] C. Bonivento, A. Isidori, L. Marconi, and A. Paoli. Implicit fault-tolerant control: application to induction motors. *Automatica*, 40(3):355–371, 2004.

[12] C. Byrnes, A. Isidori, and J.C.Willems. Passivity, feedback equivalence, and the global stabilization of minimum phase nonlinear systems. *IEEE, Transactions on Automatic Control*, 36(11):1228–1240, November 1991.

[13] M. Demetriou, K. Ito, and R. Smith. Adaptive monitoring and accommodtion of nonlinear actuator faults in positive real infinite dimensional systems. *IEEE, Transactions on Automatic Control*, 52(12):2332–2338, 2007.

[14] A. Fekih. Effective fault tolerant control design for nonlinear systems: application to a class of motor control system. *IET Control Theory and Applications*, 2(9):762–772, 2008.

[15] T. Gruijic and D. Petkovski. On robustness of Lurie systems with multiple non-linearities. *Automatica*, 23(3):327–334, 1987.

[16] T. Hu, B. Huang, and Z. Lin. Absolute stability with a generalized sector condition. *IEEE, Transactions on Automatic Control*, 49(4):535–548, 2004.

[17] T. Hu, Z. Lin, and B. Chen. An analysis and design method for linear systems subject to actuator saturation and disturbance. *Automatica*, 38:351–359, 2002.

[18] A. Iggidr and G. Sallet. On the stability of noautonomous systems. *Automatica*, 39:167–171, 2003.

[19] B. Jiang and F. Chowdhury. Fault estimation and accommodation for linear MIMO discret-time systems. *IEEE, Transactions on Control Systems Technology*, 13(3):493–499, 2005.

[20] B. Jiang and F. Chowdhury. Parameter fault detection and estimation of a class of nonlinear systems using observers. *Journal of Franklin Institute*, 342(7):725–736, 2005.

[21] B. Jiang, M. Staroswiecki, and V. Cocquempot. Fault accomodation for nonlinear dynamics systems. *IEEE, Transactions on Automatic Control*, 51(9):1578–1583, 2006.

[22] H. Khalil. *Nonlinear systems*. Prentice-Hall, third edition, 2002.

[23] M. Karpenkoa, N. Sepehri, and D. Scuseb. Diagnosis of process valve actuator faults using a multilayer neural network. *Control Engineering Practice*, 11(11):1289–1299, November 2003.

[24] T.-C. Lee and Z.-P. Jiang. A generalization of Krasovskii-LaSalle theorem for nonlinear time-varying systems: converse results and applications. *IEEE, Transactions on Automatic Control*, 50(8):1147–1163, August 2005.

[25] F. Liao, J. Wang, and G. Yang. Reliable robust flight tracking control: An LMI approach. *IEEE, Transactions on Control Systems Technology*, 10(1):76–89, 2002.

[26] M. Liberzon. Essays on the absolute stability theory. *Automation and remote control*, 67(10):1610–1644, 2006.

[27] X. Mao. Exponential stability of stochastic delay interval systems with markovian switching. *IEEE, Transactions on Automatic Control*, 47(10):1604–1612, 2002.
[28] H.-J. Ma and G.-H. Yang. FTC synthesis for nonlinear systems: Sum of squares optimization approach. In *IEEE, Conference on Decision and Control*, pages 2645–2650, New Orleans, USA,, 2007.
[29] M. Mahmoud, J. Jiang, and Y. Zhang. Stabilization of active fault tolerant control systems with imperfect fault detection and diagnosis. *Stochastic Analysis and Applications*, 21(3):673–701, 2003.
[30] M. Maki, J. Jiang, and K. Hagino. A stability guaranteed active fault-tolerant control system against actuator failures. *Int. Journal of Robust and Nonlinear Control*, 14:1061–1077, 2004.
[31] F. Mazenc and L. Praly. Adding integrations, saturated controls, and stabilization for feedforward systems. *IEEE, Transactions on Automatic Control*, 41(11):1559–1578, November 1996.
[32] P. Mhaskar. Robust model predictive control design for fault tolrant control of process systems. *Ind. Eng. Chem. Res.*, 45:8565–8574, 2006.
[33] P. Mhaskar, A. Gani, and P. Christofides. Fault-tolrant control of nonlinear processes: Performance-based reconfiguration and robustness. *Int. Journal of Robust and Nonlinear Control*, 16:91–111, 2006.
[34] P. Mhaskar, A. Gani, N. El-Farra, C. McFall, P. Christofides, and J. Davis. Integrated fault detection and fault-tolerant control of nonlinear process systems. *AIChE J.*, 52:2129–2148, 2006.
[35] P. Mhaskar, C. McFall, A. Gani, P. Christofides, and J. Davis. Isolation and handling of actuator faults in nonlinear systems. *Automatica*, 44(1):53–62, January 2008.
[36] H. Niemann and J. Stoustrup. Reliable control using the primary and dual Youla parametrization. In *IEEE, Conference on Decision and Control*, pages 4353–4358, 2002.
[37] H. Niemann and J. Stoustrup. Passive fault tolerant control of a double inverted pendulium-a case study. *Control Engineering Practice*, 13:1047–1059, 2005.
[38] R. Patton. Fault-tolerant control systems: The 1997 situation. In *IFAC Symposium Safe-Process' 97*, pages 1033–1055, U.K., 1997.
[39] P.G. de Lima and G.G. Yen. Accommodation controller malfunctions through fault tolerant control architecture. *IEEE Transactions on Aerospace and Electronic Systems*, 43(2):706–722, 2007.
[40] R. Sepulchre, M. Jankovic, and P. Kokotovic. *Constructive nonlinear control*. Springer edition, 1997.
[41] A. Shumsky. Algebraic approach to the problem of fault accommodation in nonlinear systems. In *IFAC 17th Word Congress*, pages 1884–1889, 2008.
[42] M. Staroswiecki, H. Yang, and B. Jiang. Progressive accommodation of aircraft actuator faults. In *6th IFAC Symposium of Fault Detection Supervision and Safety for Technical Processes*, pages 877–882, 2006.
[43] G. Tao, S. Chen, and S. Joshi. An adaptive actuator failure compensation controller using output feedback. *IEEE, Transactions on Automatic Control*, 47(3):506–511, 2002.
[44] A. Teel and L. Praly. Tools for semiglobal stabilization by partial state and output feedback. 33(5):1443–1488, 1995.
[45] M. Vidyasagar. *Nonlinear systems analysis*. Prentice-Hall, second edition, 1993.

[46] N. Wu, Y. Zhang, and K. Zhou. Detection, estimation, and accomodation of loss of control effectiveness. *Int. Journal of Adaptive Control and Signal Processing*, 14:775–795, 2000.

[47] S. Wu, M. Grimble, and W. Wei. QFT based robust/fault tolerant flight control design for a remote pilotless vehicle. In *IEEE International Conference on Control Applications*, China, August 1999.

[48] S. Wu, M. Grimble, and W. Wei. QFT based robust/fault tolerant flight control design for a remote pilotless vehicle. *IEEE, Transactions on Control Systems Technology*, 8(6):1010–1016, 2000.

[49] H. Yang, V. Cocquempot, and B. Jiang. Robust fault tolerant tracking control with application to hybrid nonlinear systems. *IET Control Theory and Applications*, 3(2):211–224, 2009.

[50] X. Zhang, T. Parisini, and M. Polycarpou. Adaptive fault-tolerant control of nonlinear uncertain systems: An information-based diagnostic approach. *IEEE, Transactions on Automatic Control*, 49(8):1259–1274, 2004.

[51] Y. Zhang and J. Jiang. Design of integrated fault detection, diagnosis and reconfigurable control systems. In *IEEE, Conference on Decision and Control*, pages 3587–3592, 1999.

[52] Y. Zhang and J. Jiang. Integrated design of reconfigurable fault-tolerant control systems. *Journal of Guidance, Control, and Dynamics*, 24(1):133–136, 2000.

[53] Y. Zhang and J. Jiang. Bibliographical review on reconfigurable fault-tolerant control systems. In *Proceeding of the 5th IFAC symposium on fault detection, supervision and safety for technical processes*, pages 265–276, Washington DC, 2003.

[54] Y. Zhang and J. Jiang. Issues on integration of fault diagnosis and reconfigurable control in active fault-tolerant control systems. In *6th IFAC Symposium on fault detection supervision and safety of technical processes*, pages 1513–1524, China, August 2006.

14

Design Principles of Active Robust Fault Tolerant Control Systems

Anna Filasová and Dušan Krokavec
Technical University of Košice
Slovakia

1. Introduction

The complexity of control systems requires the fault tolerance schemes to provide control of the faulty system. The fault tolerant systems are that one of the more fruitful applications with potential significance for those domains in which control must proceed while the controlled system is operative and testing opportunities are limited by given operational considerations. The real problem is usually to fix the system with faults so that it can continue its mission for some time with some limitations of functionality. These large problems are known as the fault detection, identification and reconfiguration (FDIR) systems. The practical benefits of the integrated approach to FDIR seem to be considerable, especially when knowledge of the available fault isolations and the system reconfigurations is used to reduce the cost and to increase the control reliability and utility. Reconfiguration can be viewed as the task to select these elements whose reconfiguration is sufficient to do the acceptable behavior of the system. If an FDIR system is designed properly, it will be able to deal with the specified faults and maintain the system stability and acceptable level of performance in the presence of faults.

The essential aspect for the design of fault-tolerant control requires the conception of diagnosis procedures that can solve the fault detection and isolation problem. The fault detection is understood as a problem of making a binary decision either that something has gone wrong or that everything is in order. The procedure composes residual signal generation (signals that contain information about the failures or defects) followed by their evaluation within decision functions, and it is usually achieved designing a system which, by processing input/output data, is able generating the residual signals, detect the presence of an incipient fault and isolate it.

In principle, in order to achieve fault tolerance, some redundancy is necessary. So far direct redundancy is realized by redundancy in multiple hardware channels, fault-tolerant control involve functional redundancy. Functional (analytical) redundancy is usually achieved by design of such subsystems, which functionality is derived from system model and can be realized using algorithmic (software) redundancy. Thus, analytical redundancy most often means the use of functional relations between system variables and residuals are derived from implicit information in functional or analytical relationships, which exist between measurements taken from the process, and a process model. In this sense a residual is a fault indicator, based on a deviation between measurements and model-equation-based computation and model based diagnosis use models to obtain residual signals that are as a rule zero in the fault free case and non-zero otherwise.

A fault in the fault diagnosis systems can be detected and isolated when has to cause a residual change and subsequent analyze of residuals have to provide information about faulty component localization. From this point of view the fault decision information is capable in a suitable format to specify possible control structure class to facilitate the appropriate adaptation of the control feedback laws. Whereas diagnosis is the problem of identifying elements whose abnormality is sufficient to explain an observed malfunction, reconfiguration can be viewed as a problem of identifying elements whose in a new structure are sufficient to restore acceptable behavior of the system.

1.1 Fault tolerant control

Main task to be tackled in achieving fault-tolerance is design a controller with suitable reconfigurable structure to guarantee stability, satisfactory performance and plant operation economy in nominal operational conditions, but also in some components malfunction. Generally, fault-tolerant control is a strategy for reliable and highly efficient control law design, and includes fault-tolerant system requirements analysis, analytical redundancy design (fault isolation principles) and fault accommodation design (fault control requirements and reconfigurable control strategy). The benefits result from this characterization give a unified framework that should facilitate the development of an integrated theory of FDIR and control (fault-tolerant control systems (FTCS)) to design systems having the ability to accommodate component failures automatically.

FTCS can be classified into two types: passive and active. In passive FTCS, fix controllers are used and designed in such way to be robust against a class of presumed faults. To ensure this a closed-loop system remains insensitive to certain faults using constant controller parameters and without use of on-line fault information. Because a passive FTCS has to maintain the system stability under various component failures, from the performance viewpoint, the designed controller has to be very conservative. From typical relationships between the optimality and the robustness, it is very difficult for a passive FTCS to be optimal from the performance point of view alone.

Active FTCS react to the system component failures actively by reconfiguring control actions so that the stability and acceptable (possibly partially degraded, graceful) performance of the entire system can be maintained. To achieve a successful control system reconfiguration, this approach relies heavily on a real-time fault detection scheme for the most up-to-date information about the status of the system and the operating conditions of its components. To reschedule controller function a fixed structure is modified to account for uncontrollable changes in the system and unanticipated faults. Even though, an active FTCS has the potential to produce less conservative performance.

The critical issue facing any active FTCS is that there is only a limited amount of reaction time available to perform fault detection and control system reconfiguration. Given the fact of limited amount of time and information, it is highly desirable to design a FTCS that possesses the guaranteed stability property as in a passive FTCS, but also with the performance optimization attribute as in an active FTCS.

Selected useful publications, especially interesting books on this topic (Blanke et al.,2003), (Chen and Patton,1999), (Chiang et al.,2001), (Ding,2008), (Ducard,2009), (Simani et al.,2003) are presented in References.

1.2 Motivation

A number of problems that arise in state control can be reduced to a handful of standard convex and quasi-convex problems that involve matrix inequalities. It is known that the optimal solution can be computed by using interior point methods (Nesterov and Nemirovsky,1994) which converge in polynomial time with respect to the problem size and efficient interior point algorithms have recently been developed for and further development of algorithms for these standard problems is an area of active research. For this approach, the stability conditions may be expressed in terms of linear matrix inequalities (LMI), which have a notable practical interest due to the existence of powerful numerical solvers. Some progres review in this field can be found e.g. in (Boyd et al.,1994), (Herrmann et al.,2007), (Skelton et al.,1998), and the references therein.

In contradiction to the standard pole placement methods application in active FTCS design there don't exist so much structures to solve this problem using LMI approach (e.g. see (Chen et al.,1999), (Filasova and Krokavec,2009), (Liao et al.,2002), (Noura et al.,2009)). To generalize properties of non-expansive systems formulated as H_∞ problems in the bounded real lemma (BRL) form, the main motivation of this chapter is to present reformulated design method for virtual sensor control design in FTCS structures, as well as the state estimator based active control structures for single actuator faults in the continuous-time linear MIMO systems. To start work with this formalism structure residual generators are designed at first to demonstrate the application suitability of the unified algebraic approach in these design tasks. LMI based design conditions are outlined generally to posse the sufficient conditions for a solution. The used structure is motivated by the standard ones (Dong et al.,2009), and in this presented form enables to design systems with the reconfigurable controller structures.

2. Problem description

Through this chapter the task is concerned with the computation of reconfigurable feedback $\mathbf{u}(t)$, which control the observable and controllable faulty linear dynamic system given by the set of equations

$$\dot{\mathbf{q}}(t) = \mathbf{A}\mathbf{q}(t) + \mathbf{B}_u\mathbf{u}(t) + \mathbf{B}_f\mathbf{f}(t) \tag{1}$$

$$\mathbf{y}(t) = \mathbf{C}\mathbf{q}(t) + \mathbf{D}_u\mathbf{u}(t) + \mathbf{D}_f\mathbf{f}(t) \tag{2}$$

where $\mathbf{q}(t) \in \mathbf{R}^n$, $\mathbf{u}(t) \in \mathbf{R}^r$, $\mathbf{y}(t) \in \mathbf{R}^m$, and $\mathbf{f}(t) \in \mathbf{R}^l$ are vectors of the state, input, output and fault variables, respectively, matrices $\mathbf{A} \in \mathbf{R}^{n\times n}$, $\mathbf{B}_u \in \mathbf{R}^{n\times r}$, $\mathbf{C} \in \mathbf{R}^{m\times n}$, $\mathbf{D}_u \in \mathbf{R}^{m\times r}$, $\mathbf{B}_f \in \mathbf{R}^{n\times l}$, $\mathbf{D}_f \in \mathbf{R}^{m\times l}$ are real matrices. Problem of the interest is to design the asymptotically stable closed-loop systems with the linear memoryless state feedback controllers of the form

$$\mathbf{u}(t) = -\mathbf{K}_o\mathbf{y}_e(t) \tag{3}$$

$$\mathbf{u}(t) = -\mathbf{K}\mathbf{q}_e(t) - \mathbf{L}\mathbf{f}_e(t) \tag{4}$$

respectively. Here $\mathbf{K}_o \in \mathbf{R}^{r\times m}$ is the output controller gain matrix, $\mathbf{K} \in \mathbf{R}^{r\times n}$ is the nominal state controller gain matrix, $\mathbf{L} \in \mathbf{R}^{r\times l}$ is the compensate controller gain matrix, $\mathbf{y}_e(t)$ is by virtual sensor estimated output of the system, $\mathbf{q}_e(t) \in \mathbf{R}^n$ is the system state estimate vector, and $\mathbf{f}_e(t) \in \mathbf{R}^l$ is the fault estimate vector. Active compensate method can be applied for such systems, where

$$\begin{bmatrix} \mathbf{B}_f \\ \mathbf{D}_f \end{bmatrix} = \begin{bmatrix} \mathbf{B}_u \\ \mathbf{D}_u \end{bmatrix} \mathbf{L} \tag{5}$$

and the additive term $\mathbf{B}_f\mathbf{f}(t)$ is compensated by the term

$$-\mathbf{B}_f\mathbf{f}_e(t) = -\mathbf{B}_u\mathbf{L}\mathbf{f}_e(t) \tag{6}$$

which implies (4). The estimators are then given by the set of the state equations

$$\dot{\mathbf{q}}_e(t) = \mathbf{A}\mathbf{q}_e(t) + \mathbf{B}_u\mathbf{u}(t) + \mathbf{B}_f\mathbf{f}_e(t) + \mathbf{J}(\mathbf{y}(t) - \mathbf{y}_e(t)) \tag{7}$$

$$\dot{\mathbf{f}}_e(t) = \mathbf{M}\mathbf{f}_e(t) + \mathbf{N}(\mathbf{y}(t) - \mathbf{y}_e(t)) \tag{8}$$

$$\mathbf{y}_e(t) = \mathbf{C}\mathbf{q}_e(t) + \mathbf{D}_u\mathbf{u}(t) + \mathbf{D}_f\mathbf{f}_e(t) \tag{9}$$

where $\mathbf{J} \in \boldsymbol{R}^{n\times m}$ is the state estimator gain matrix, and $\mathbf{M} \in \boldsymbol{R}^{l\times l}$, $\mathbf{N} \in \boldsymbol{R}^{l\times m}$ are the system and input matrices of the fault estimator, respectively or by the set of equation

$$\dot{\mathbf{q}}_{fe}(t) = \mathbf{A}\mathbf{q}_{fe}(t) + \mathbf{B}_u\mathbf{u}_f(t) + \mathbf{J}(\mathbf{y}_f(t) - \mathbf{D}_u\mathbf{u}_f(t) - \mathbf{C}_f\mathbf{q}_{fe}(t)) \tag{10}$$

$$\mathbf{y}_e(t) = \mathbf{E}(\mathbf{y}_f(t) + (\mathbf{C} - \mathbf{E}\mathbf{C}_f)\mathbf{q}_{fe}(t) \tag{11}$$

where $\mathbf{E} \in \boldsymbol{R}^{m\times m}$ is a switching matrix, generally used in such a way that $\mathbf{E} = \mathbf{0}$, or $\mathbf{E} = \mathbf{I}_m$.

3. Basic preliminaries

Definition 1 (Null space) *Let* $\mathbf{E}$, $\mathbf{E} \in \boldsymbol{R}^{h\times h}$, *rank*$(\mathbf{E}) = k < h$ *be a rank deficient matrix. Then the null space* $\mathcal{N}_{\mathbf{E}}$ *of* $\mathbf{E}$ *is the orthogonal complement of the row space of* $\mathbf{E}$.

Proposition 1 (Orthogonal complement) *Let* $\mathbf{E}$, $\mathbf{E} \in \boldsymbol{R}^{h\times h}$, *rank*$(\mathbf{E}) = k < h$ *be a rank deficient matrix. Then an orthogonal complement* $\mathbf{E}^{\perp}$ *of* $\mathbf{E}$ *is*

$$\mathbf{E}^{\perp} = \mathbf{E}^{\circ}\mathbf{U}_2^T \tag{12}$$

where $\mathbf{U}_2^T$ *is the null space of* $\mathbf{E}$ *and* $\mathbf{E}^{\circ}$ *is an arbitrary matrix of appropriate dimension.*

Proof. The singular value decomposition (SVD) of $\mathbf{E}$, $\mathbf{E} \in \boldsymbol{R}^{h\times h}$, rank$(\mathbf{E}) = k < h$ gives

$$\mathbf{U}^T\mathbf{E}\mathbf{V} = \begin{bmatrix}\mathbf{U}_1^T \\ \mathbf{U}_2^T\end{bmatrix} \mathbf{E}\begin{bmatrix}\mathbf{V}_1 & \mathbf{V}_2\end{bmatrix} = \begin{bmatrix}\boldsymbol{\Sigma}_1 & \mathbf{0}_{12} \\ \mathbf{0}_{21} & \mathbf{0}_{22}\end{bmatrix} \tag{13}$$

where $\mathbf{U}^T \in \boldsymbol{R}^{h\times h}$ is the orthogonal matrix of the left singular vectors, $\mathbf{V} \in \boldsymbol{R}^{h\times h}$ is the orthogonal matrix of the right singular vectors of $\mathbf{E}$ and $\boldsymbol{\Sigma}_1 \in \boldsymbol{R}^{k\times k}$ is the diagonal positive definite matrix of the form

$$\boldsymbol{\Sigma}_1 = \operatorname{diag}\begin{bmatrix}\sigma_1 & \cdots & \sigma_k\end{bmatrix}, \quad \sigma_1 \geq \cdots \geq \sigma_k > 0 \tag{14}$$

which diagonal elements are the singular values of $\mathbf{E}$. Using orthogonal properties of $\mathbf{U}$ and $\mathbf{V}$, i.e. $\mathbf{U}^T\mathbf{U} = \mathbf{I}_h$, as well as $\mathbf{V}^T\mathbf{V} = \mathbf{I}_h$, and

$$\begin{bmatrix}\mathbf{U}_1^T \\ \mathbf{U}_2^T\end{bmatrix}\begin{bmatrix}\mathbf{U}_1 & \mathbf{U}_2\end{bmatrix} = \begin{bmatrix}\mathbf{I}_1 & \mathbf{0} \\ \mathbf{0} & \mathbf{I}_2\end{bmatrix}, \quad \mathbf{U}_2^T\mathbf{U}_1 = \mathbf{0} \tag{15}$$

respectively, where $\mathbf{I}_h \in \boldsymbol{R}^{h\times h}$ is the identity matrix, then $\mathbf{E}$ can be written as

$$\mathbf{E} = \mathbf{U}\boldsymbol{\Sigma}\mathbf{V}^T = \begin{bmatrix}\mathbf{U}_1 & \mathbf{U}_2\end{bmatrix}\begin{bmatrix}\boldsymbol{\Sigma}_1 & \mathbf{0}_{12} \\ \mathbf{0}_{21} & \mathbf{0}_{22}\end{bmatrix}\begin{bmatrix}\mathbf{V}_1^T \\ \mathbf{V}_2^T\end{bmatrix} = \begin{bmatrix}\mathbf{U}_1 & \mathbf{U}_2\end{bmatrix}\begin{bmatrix}\mathbf{S}_1 \\ \mathbf{0}_2\end{bmatrix} = \mathbf{U}_1\mathbf{S}_1 \tag{16}$$

where $\mathbf{S}_1 = \mathbf{\Sigma}_1 \mathbf{V}_1^T$. Thus, (15) and (16) implies

$$\mathbf{U}_2^T \mathbf{E} = \mathbf{U}_2^T \begin{bmatrix} \mathbf{U}_1 & \mathbf{U}_2 \end{bmatrix} \begin{bmatrix} \mathbf{S}_1 \\ \mathbf{0}_2 \end{bmatrix} = \mathbf{0} \tag{17}$$

It is evident that for an arbitrary matrix $\mathbf{E}^{\circ}$ is

$$\mathbf{E}^{\circ} \mathbf{U}_2^T \mathbf{E} = \mathbf{E}^{\perp} \mathbf{E} = \mathbf{0} \tag{18}$$

$$\mathbf{E}^{\perp} = \mathbf{E}^{\circ} \mathbf{U}_2^T \tag{19}$$

respectively, which implies (12). This concludes the proof. ■

Proposition 2. (Schur Complement) *Let* $\mathbf{Q} > \mathbf{0}$, $\mathbf{R} > 0$, $\mathbf{S}$ *are real matrices of appropriate dimensions, then the next inequalities are equivalent*

$$\begin{bmatrix} \mathbf{Q} & \mathbf{S} \\ \mathbf{S}^T & -\mathbf{R} \end{bmatrix} < 0 \Leftrightarrow \begin{bmatrix} \mathbf{Q} + \mathbf{S}\mathbf{R}^{-1}\mathbf{S}^T & \mathbf{0} \\ \mathbf{0} & -\mathbf{R} \end{bmatrix} < 0 \Leftrightarrow \mathbf{Q} + \mathbf{S}\mathbf{R}^{-1}\mathbf{S}^T < 0,\ \mathbf{R} > 0 \tag{20}$$

Proof. Let the linear matrix inequality takes form

$$\begin{bmatrix} \mathbf{Q} & \mathbf{S} \\ \mathbf{S}^T & -\mathbf{R} \end{bmatrix} < 0 \tag{21}$$

then using Gauss elimination principle it yields

$$\begin{bmatrix} \mathbf{I} & \mathbf{S}\mathbf{R}^{-1} \\ \mathbf{0} & \mathbf{I} \end{bmatrix} \begin{bmatrix} \mathbf{Q} & \mathbf{S} \\ \mathbf{S}^T & -\mathbf{R} \end{bmatrix} \begin{bmatrix} \mathbf{I} & \mathbf{0} \\ \mathbf{R}^{-1}\mathbf{S}^T & \mathbf{I} \end{bmatrix} = \begin{bmatrix} \mathbf{Q} + \mathbf{S}\mathbf{R}^{-1}\mathbf{S}^T & \mathbf{0} \\ \mathbf{0} & -\mathbf{R} \end{bmatrix} \tag{22}$$

Since

$$\det \begin{bmatrix} \mathbf{I} & \mathbf{S}\mathbf{R}^{-1} \\ \mathbf{0} & \mathbf{I} \end{bmatrix} = 1 \tag{23}$$

and it is evident that this transform doesn't change negativity of (21), and so (22) implies (20). This concludes the proof. ■

Note that in the next the matrix notations **E**, **Q**, **R**, **S**, **U**, and **V** be used in another context, too.

Proposition 3 (Bounded real lemma) *For given* $\gamma \in \boldsymbol{R}$ *and the linear system (1), (2) with* $\mathbf{f}(t) = \mathbf{0}$ *if there exists symmetric positive definite matrix* $\mathbf{P} > 0$ *such that*

$$\begin{bmatrix} \mathbf{A}^T\mathbf{P} + \mathbf{P}\mathbf{A} & \mathbf{P}\mathbf{B}_u & \mathbf{C}^T \\ * & -\gamma^2 \mathbf{I}_r & \mathbf{D}_u^T \\ * & * & -\mathbf{I}_m \end{bmatrix} < 0 \tag{24}$$

where $\mathbf{I}_r \in \boldsymbol{R}^{r \times r}$, $\mathbf{I}_m \in \boldsymbol{R}^{m \times m}$ *are the identity matrices, respectively then given system is asymptotically stable.*

Hereafter, $*$ denotes the symmetric item in a symmetric matrix.

Proof. Defining Lyapunov function as follows

$$v(\mathbf{q}(t)) = \mathbf{q}^T(t)\mathbf{P}\mathbf{q}(t) + \int_0^t \left(\mathbf{y}^T(r)\mathbf{y}(r) - \gamma^2 \mathbf{u}^T(r)\mathbf{u}(r)\right) \mathrm{d}r > 0 \tag{25}$$

where $\mathbf{P} = \mathbf{P}^T > 0$, $\mathbf{P} \in \boldsymbol{R}^{n \times n}$, $\gamma \in \boldsymbol{R}$, and evaluating the derivative of $v(\mathbf{q}(t))$ with respect to t then it yields

$$\dot{v}(\mathbf{q}(t)) = \dot{\mathbf{q}}^T(t)\mathbf{P}\mathbf{q}(t) + \mathbf{q}^T(t)\mathbf{P}\dot{\mathbf{q}}(t) + \mathbf{y}^T(t)\mathbf{y}(t) - \gamma^2 \mathbf{u}^T(t)\mathbf{u}(t) < 0 \tag{26}$$

Thus, substituting (1), (2) with $\mathbf{f}(t) = \mathbf{0}$ it can be written

$$\begin{aligned}\dot{v}(\mathbf{q}(t)) = (\mathbf{A}\mathbf{q}(t) + \mathbf{B}_u\mathbf{u}(t))^T\mathbf{P}\mathbf{q}(t) + \mathbf{q}^T(t)\mathbf{P}(\mathbf{A}\mathbf{q}(t) + \mathbf{B}_u\mathbf{u}(t)) + \\ +(\mathbf{C}\mathbf{q}(t) + \mathbf{D}_u\mathbf{u}(t))^T(\mathbf{C}\mathbf{q}(t) + \mathbf{D}_u\mathbf{u}(t)) - \gamma^2\mathbf{u}^T(t)\mathbf{u}(t) < 0\end{aligned} \tag{27}$$

and with notation

$$\mathbf{q}_c^T(t) = \left[\, \mathbf{q}^T(t)\, \mathbf{u}^T(t) \,\right] \tag{28}$$

it is obtained

$$\dot{v}(\mathbf{q}(t)) = \mathbf{q}_c^T(t)\mathbf{P}_c\mathbf{q}_c(t) < 0 \tag{29}$$

where

$$\mathbf{P}_c = \begin{bmatrix} \mathbf{A}^T\mathbf{P} + \mathbf{P}\mathbf{A} & \mathbf{P}\mathbf{B}_u \\ * & -\gamma^2\mathbf{I}_r \end{bmatrix} + \begin{bmatrix} \mathbf{C}^T\mathbf{C} & \mathbf{C}^T\mathbf{D}_u \\ * & \mathbf{D}_u^T\mathbf{D}_u \end{bmatrix} < 0 \tag{30}$$

Since

$$\begin{bmatrix} \mathbf{C}^T\mathbf{C} & \mathbf{C}^T\mathbf{D}_u \\ * & \mathbf{D}_u^T\mathbf{D}_u \end{bmatrix} = \begin{bmatrix} \mathbf{C}^T \\ \mathbf{D}_u^T \end{bmatrix} \left[\, \mathbf{C} \; \mathbf{D}_u \,\right] \geq 0 \tag{31}$$

Schur complement property implies

$$\begin{bmatrix} \mathbf{0} & \mathbf{0} & \mathbf{C}^T \\ * & \mathbf{0} & \mathbf{D}_u^T \\ * & * & -\mathbf{I}_m \end{bmatrix} \geq 0 \tag{32}$$

then using (32) the LMI (30) can now be written compactly as (24). This concludes the proof. ■

Remark 1 (Lyapunov inequality) Considering Lyapunov function of the form

$$v(\mathbf{q}(t)) = \mathbf{q}^T(t)\mathbf{P}\mathbf{q}(t) > 0 \tag{33}$$

where $\mathbf{P} = \mathbf{P}^T > 0$, $\mathbf{P} \in R^{n\times n}$, and the control law

$$\mathbf{u}(t) = -\mathbf{K}_o\big(\mathbf{y}(t) - \mathbf{D}_u\mathbf{u}(t)\big) = -\mathbf{K}_o\mathbf{C}\mathbf{q}(t) \tag{34}$$

where $\mathbf{K}_o \in R^{r\times m}$ is a gain matrix. Because in this case (27) gives

$$\dot{v}(\mathbf{q}(t)) = (\mathbf{A}\mathbf{q}(t) + \mathbf{B}_u\mathbf{u}(t))^T\mathbf{P}\mathbf{q}(t) + \mathbf{q}^T(t)\mathbf{P}(\mathbf{A}\mathbf{q}(t) + \mathbf{B}_u\mathbf{u}(t)) < 0 \tag{35}$$

then inserting (34) into (35) it can be obtained

$$\dot{v}(\mathbf{q}(t)) = \mathbf{q}^T(t)\mathbf{P}_{cb}\mathbf{q}(t) < 0 \tag{36}$$

where

$$\mathbf{P}_{cb} = \mathbf{A}^T\mathbf{P} + \mathbf{P}\mathbf{A} - \mathbf{P}\mathbf{B}_u\mathbf{K}_o\mathbf{C} - (\mathbf{P}\mathbf{B}_u\mathbf{K}_o\mathbf{C})^T < 0 \tag{37}$$

Especially, if all system state variables are measurable the control policy can be defined as follows

$$\mathbf{u}(t) = -\mathbf{K}\mathbf{q}(t) \tag{38}$$

and (37) can be written as

$$\mathbf{A}^T\mathbf{P} + \mathbf{P}\mathbf{A} - \mathbf{P}\mathbf{B}_u\mathbf{K} - (\mathbf{P}\mathbf{B}_u\mathbf{K})^T < 0 \tag{39}$$

Note that in a real physical dynamic plant model usually $\mathbf{D}_u = \mathbf{0}$. ■

***Proposition 4** Let for given real matrices* $\mathbf{F}$, $\mathbf{G}$ *and* $\mathbf{\Theta} = \mathbf{\Theta}^T > 0$ *of appropriate dimension a matrix* $\mathbf{\Lambda}$ *has to satisfy the inequality*

$$\mathbf{F\Lambda G}^T + \mathbf{G\Lambda}^T\mathbf{F}^T - \mathbf{\Theta} < 0 \tag{40}$$

then any solution of $\mathbf{\Lambda}$ *can be generated using a solution of inequality*

$$\begin{bmatrix} -\mathbf{FHF}^T - \mathbf{\Theta} & \mathbf{FH} + \mathbf{G\Lambda}^T \\ * & -\mathbf{H} \end{bmatrix} < 0 \tag{41}$$

where $\mathbf{H} = \mathbf{H}^T > 0$ *is a free design parameter.*

Proof. If (40) yields then there exists a matrix $\mathbf{H}^{-1} = \mathbf{H}^{-T} > 0$ such that

$$\mathbf{F\Lambda G}^T + \mathbf{G\Lambda}^T\mathbf{F}^T - \mathbf{\Theta} + \mathbf{G\Lambda}^T\mathbf{H}^{-1}\mathbf{\Lambda G}^T < 0 \tag{42}$$

Completing the square in (42) it can be obtained

$$(\mathbf{FH} + \mathbf{G\Lambda}^T)\mathbf{H}^{-1}(\mathbf{FH} + \mathbf{G\Lambda}^T)^T - \mathbf{FHF}^T - \mathbf{\Theta} < 0 \tag{43}$$

and using Schur complement (43) implies (41). ■

4. Fault isolation

4.1 Structured residual generators of sensor faults

4.1.1 Set of the state estimators

To design structured residual generators of sensor faults based on the state estimators, all actuators are assumed to be fault-free and each estimator is driven by all system inputs and all but one system outputs. In that sense it is possible according with given nominal fault-free system model (1), (2) to define the set of structured estimators for $k = 1, 2, \ldots, m$ as follows

$$\dot{\mathbf{q}}_{ke}(t) = \mathbf{A}_{ke}\mathbf{q}_{ke}(t) + \mathbf{B}_{uke}\mathbf{u}(t) + \mathbf{J}_{sk}\mathbf{T}_{sk}(\mathbf{y}(t) - \mathbf{D}_u\mathbf{u}(t)) \tag{44}$$

$$\mathbf{y}_{ke}(t) = \mathbf{C}\mathbf{q}_{ke}(t) + \mathbf{D}_u\mathbf{u}(t) \tag{45}$$

where $\mathbf{A}_{ke} \in R^{n\times n}$, $\mathbf{B}_{uke} \in R^{n\times r}$, $\mathbf{J}_{sk} \in R^{n\times(m-1)}$, and $\mathbf{T}_{sk} \in R^{(m-1)\times m}$ takes the next form

$$\mathbf{T}_{sk} = \mathbf{I}_{m\oslash k} = \begin{bmatrix} 1\,0 \cdots 0\,0\,0\,0\,0 \cdots 0\,0 \\ \vdots \qquad\qquad \vdots \\ 0\,0 \cdots 0\,1\,0\,0\,0 \cdots 0\,0 \\ 0\,0 \cdots 0\,0\,0\,1\,0 \cdots 0\,0 \\ \vdots \qquad\qquad \vdots \\ 0\,0 \cdots 0\,0\,0\,0\,0 \cdots 0\,1 \end{bmatrix} \tag{46}$$

Note that $\mathbf{T}_{sk}$ can be obtained by deleting the k-th row in identity matrix $\mathbf{I}_m$.
Since the state estimate error is defined as $\mathbf{e}_k(t) = \mathbf{q}(t) - \mathbf{q}_{ke}(t)$ then

$$\begin{aligned} \dot{\mathbf{e}}_k(t) &= \mathbf{A}\mathbf{q}(t) + \mathbf{B}_u\mathbf{u}(t) - \mathbf{A}_{ke}\mathbf{q}_{ke}(t) - \mathbf{B}_{uke}\mathbf{u}(t) - \mathbf{J}_{sk}\mathbf{T}_{sk}(\mathbf{y}(t) - \mathbf{D}_u\mathbf{u}(t)) = \\ &= (\mathbf{A} - \mathbf{A}_{ke} - \mathbf{J}_{sk}\mathbf{T}_{sk}\mathbf{C})\mathbf{q}(t) + (\mathbf{B}_u - \mathbf{B}_{uke})\mathbf{u}(t) + \mathbf{A}_{ke}\mathbf{e}_k(t) \end{aligned} \tag{47}$$

To obtain the state estimate error autonomous it can be set

$$\mathbf{A}_{ke} = \mathbf{A} - \mathbf{J}_{sk}\mathbf{T}_{sk}\mathbf{C}, \qquad \mathbf{B}_{uke} = \mathbf{B}_u \tag{48}$$

It is obvious that (48) implies

$$\dot{\mathbf{e}}_k(t) = \mathbf{A}_{ke}\mathbf{e}_k(t) = (\mathbf{A} - \mathbf{J}_{sk}\mathbf{T}_{sk}\mathbf{C})\mathbf{e}_k(t) \tag{49}$$

(44) can be rewritten as

$$\begin{aligned}\dot{\mathbf{q}}_{ke}(t) &= (\mathbf{A} - \mathbf{J}_{sk}\mathbf{T}_{sk}\mathbf{C})\mathbf{q}_{ke}(t) + \mathbf{B}_u\mathbf{u}(t) + \mathbf{J}_{sk}\mathbf{T}_{sk}(\mathbf{y}(t) - \mathbf{D}_u\mathbf{u}(t)) = \\ &= \mathbf{A}\mathbf{q}_{ke}(t) + \mathbf{B}_u\mathbf{u}(t) + \mathbf{J}_{sk}\mathbf{T}_{sk}(\mathbf{y}(t) - (\mathbf{C}\mathbf{q}_{ke}(t) + \mathbf{D}_u\mathbf{u}(t)))\end{aligned} \tag{50}$$

and (44), (45) can be rewritten equivalently as

$$\dot{\mathbf{q}}_{ke}(t) = \mathbf{A}\mathbf{q}_{ke}(t) + \mathbf{B}_u\mathbf{u}(t) + \mathbf{J}_{sk}\mathbf{T}_{sk}(\mathbf{y}(t) - \mathbf{y}_{ke}(t)) \tag{51}$$

$$\mathbf{y}_{ke}(t) = \mathbf{C}\mathbf{q}_{ke}(t) + \mathbf{D}_u\mathbf{u}(t) \tag{52}$$

Theorem 1 *The k-th state-space estimator (52), (53) is stable if there exist a positive definite symmetric matrix $\mathbf{P}_{sk} > 0$, $\mathbf{P}_{sk} \in R^{n\times n}$ and a matrix $\mathbf{Z}_{sk} \in R^{n\times(m-1)}$ such that*

$$\mathbf{P}_{sk} = \mathbf{P}_{sk}^T > 0 \tag{53}$$

$$\mathbf{A}^T\mathbf{P}_{sk} + \mathbf{P}_{sk}\mathbf{A} - \mathbf{Z}_{sk}\mathbf{T}_{sk}\mathbf{C} - \mathbf{C}^T\mathbf{T}_{sk}^T\mathbf{Z}_{sk}^T < 0 \tag{54}$$

Then $\mathbf{J}_{sk}$ can be computed as

$$\mathbf{J}_{sk} = \mathbf{P}_{sk}^{-1}\mathbf{Z}_{sk} \tag{55}$$

Proof. Since the estimate error is autonomous Lyapunov function of the form

$$v(\mathbf{e}_k(t)) = \mathbf{e}_k^T(t)\mathbf{P}_{sk}\mathbf{e}_k(t) > 0 \tag{56}$$

where $\mathbf{P}_{sk} = \mathbf{P}_{sk}^T > 0$, $\mathbf{P}_{sk} \in R^{n\times n}$ can be considered. Thus,

$$\dot{v}(\mathbf{e}_k(t)) = \mathbf{e}_k^T(t)(\mathbf{A} - \mathbf{J}_{sk}\mathbf{T}_{sk}\mathbf{C})^T\mathbf{P}_{sk}\mathbf{e}_k(t) + \mathbf{e}_k^T(t)\mathbf{P}_{sk}(\mathbf{A} - \mathbf{J}_{sk}\mathbf{T}_{sk}\mathbf{C})\mathbf{e}_k(t) < 0 \tag{57}$$

$$\dot{v}(\mathbf{e}_k(t)) = \mathbf{e}_k^T(t)\mathbf{P}_{skc}\mathbf{e}_k(t) < 0 \tag{58}$$

respectively, where

$$\mathbf{P}_{skc} = \mathbf{A}^T\mathbf{P}_{sk} + \mathbf{P}_{sk}\mathbf{A} - \mathbf{P}_{sk}\mathbf{J}_{sk}\mathbf{T}_{sk}\mathbf{C} - (\mathbf{P}_{sk}\mathbf{J}_{sk}\mathbf{T}_{sk}\mathbf{C})^T < 0 \tag{59}$$

Using notation $\mathbf{P}_{sk}\mathbf{J}_{sk} = \mathbf{Z}_{sk}$ (59) implies (54). This concludes the proof. ■

4.1.2 Set of the residual generators

Exploiting the model-based properties of state estimators the set of residual generators can be considered as

$$\mathbf{r}_{sk}(t) = \mathbf{X}_{sk}\mathbf{q}_{ke}(t) + \mathbf{Y}_{sk}(\mathbf{y}(t) - \mathbf{D}_u\mathbf{u}(t)), \quad k = 1,2,\ldots,m \tag{60}$$

Subsequently

$$\mathbf{r}_{sk}(t) = \mathbf{X}_{sk}(\mathbf{q}(t) - \mathbf{e}_k(t)) + \mathbf{Y}_{sk}\mathbf{C}\mathbf{q}(t) = (\mathbf{X}_{sk} + \mathbf{Y}_{sk}\mathbf{C})\mathbf{q}(t) - \mathbf{X}_{sk}\mathbf{e}_k(t) \tag{61}$$

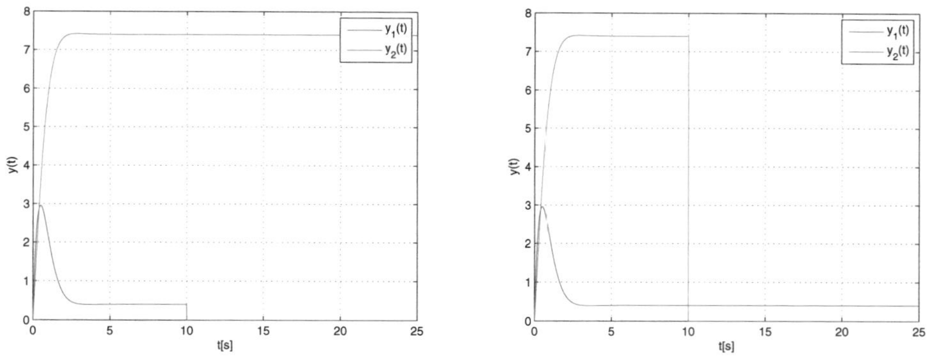

Fig. 1. Measurable outputs for single sensor faults

To eliminate influences of the state variable vector it is necessary in (61) to consider

$$\mathbf{X}_{sk} + \mathbf{Y}_{sk}\mathbf{C} = \mathbf{0} \tag{62}$$

Choosing $\mathbf{X}_{sk} = -\mathbf{T}_{sk}\mathbf{C}$ (62) implies

$$\mathbf{X}_{sk} = -\mathbf{T}_{sk}\mathbf{C}, \qquad \mathbf{Y}_{sk} = \mathbf{T}_{sk} \tag{63}$$

Thus, the set of residuals (60) takes the form

$$\mathbf{r}_{sk}(t) = \mathbf{T}_{sk}\big(\mathbf{y}(t) - \mathbf{D}_u\mathbf{u}(t) - \mathbf{C}\mathbf{q}_{ke}(t)\big), \quad k = 1, 2, \ldots, m \tag{64}$$

When all actuators are fault-free and a fault occurs in the l-th sensor the residuals will satisfy the isolation logic

$$\|\mathbf{r}_{sk}(t)\| \leq h_{sk},\ k = l, \qquad \|\mathbf{r}_{sk}(t)\| > h_{sk},\ k \neq l \tag{65}$$

This residual set can only isolate a single sensor fault at the same time. The principle can be generalized based on a regrouping of faults in such way that each residual will be designed to be sensitive to one group of sensor faults and insensitive to others.

Illustrative example

To demonstrate algorithm properties it was assumed that the system is given by (1), (2) where the nominal system parameters are given as

$$\mathbf{A} = \begin{bmatrix} 0 & 1 & 0 \\ 0 & 0 & 1 \\ -5 & -9 & -5 \end{bmatrix}, \quad \mathbf{B}_u = \begin{bmatrix} 1 & 3 \\ 2 & 1 \\ 1 & 5 \end{bmatrix}, \quad \mathbf{C} = \begin{bmatrix} 1 & 2 & 1 \\ 1 & 1 & 0 \end{bmatrix}, \quad \mathbf{D}_u = \begin{bmatrix} 0 & 0 \\ 0 & 0 \end{bmatrix}$$

and it is obvious that

$$\mathbf{T}_{s1} = \mathbf{I}_{2\oslash 1} = \begin{bmatrix} 0 & 1 \end{bmatrix}, \mathbf{T}_{s2} = \mathbf{I}_{2\oslash 2} = \begin{bmatrix} 1 & 0 \end{bmatrix}, \mathbf{T}_{s1}\mathbf{C} = \begin{bmatrix} 1 & 1 & 0 \end{bmatrix}, \mathbf{T}_{s2}\mathbf{C} = \begin{bmatrix} 1 & 2 & 1 \end{bmatrix}$$

Solving (53), (54) with respect to the LMI matrix variables $\mathbf{P}_{sk}$, and $\mathbf{Z}_{sk}$ using Self-Dual-Minimization (SeDuMi) package for Matlab, the estimator gain matrix design problem was feasible with the results

$$\mathbf{P}_{s1} = \begin{bmatrix} 0.8258 & -0.0656 & 0.0032 \\ -0.0656 & 0.8541 & 0.0563 \\ 0.0032 & 0.0563 & 0.2199 \end{bmatrix}, \quad \mathbf{Z}_{s1} = \begin{bmatrix} 0.6343 \\ 0.2242 \\ -0.8595 \end{bmatrix}, \quad \mathbf{J}_{s1} = \begin{bmatrix} 0.8312 \\ 0.5950 \\ -4.0738 \end{bmatrix}$$

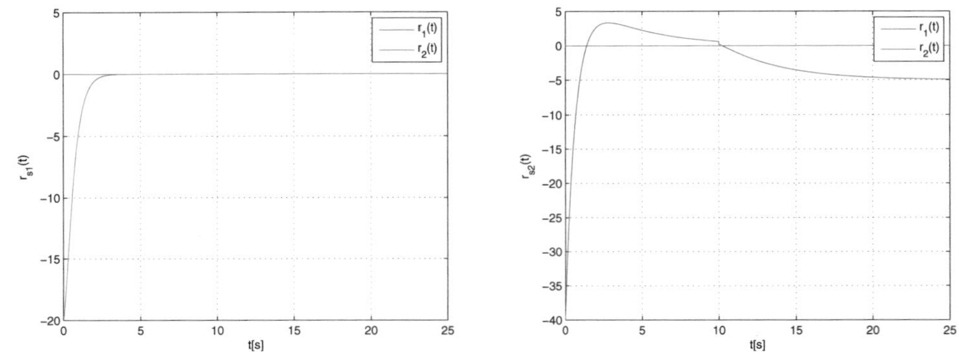

Fig. 2. Residuals for the 1st sensor fault

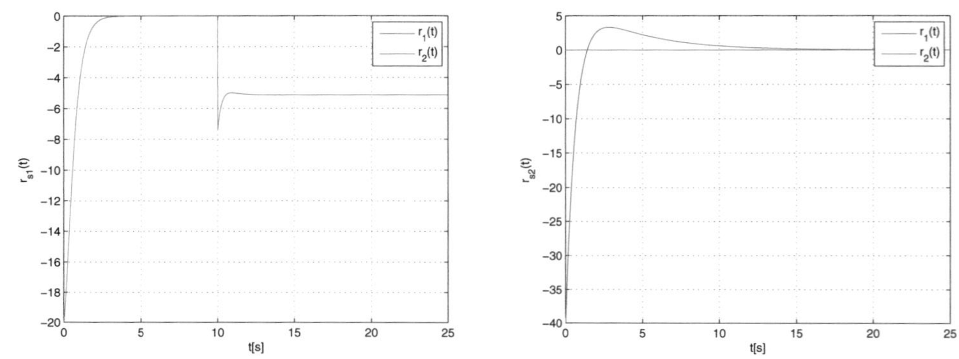

Fig. 3. Residuals for the 2nd sensor fault

$$\mathbf{P}_{s2} = \begin{bmatrix} 0.8258 & -0.0656 & 0.0032 \\ -0.0656 & 0.8541 & 0.0563 \\ 0.0032 & 0.0563 & 0.2199 \end{bmatrix}, \quad \mathbf{Z}_{s2} = \begin{bmatrix} 0.0335 \\ 0.6344 \\ -0.9214 \end{bmatrix}, \quad \mathbf{J}_{s2} = \begin{bmatrix} 0.1412 \\ 1.0479 \\ -4.4614 \end{bmatrix}$$

respectively. It is easily verified that the system matrices of state estimators are stable with the eigenvalue spectra

$$\rho(\mathbf{A} - \mathbf{J}_{s1}\mathbf{T}_{s1}\mathbf{C}) = \{-1.0000 \ \ -2.3459 \ \ -3.0804\}$$

$$\rho(\mathbf{A} - \mathbf{J}_{s2}\mathbf{T}_{s2}\mathbf{C}) = \{-1.5130 \ \ -1.0000 \ \ -0.2626\}$$

respectively, and the set of residuals takes the form

$$\mathbf{r}_{s1}(t) = \begin{bmatrix} 0 & 1 \end{bmatrix} \left(\mathbf{y}(t) - \begin{bmatrix} 1 & 2 & 1 \\ 1 & 1 & 0 \end{bmatrix} \mathbf{q}_{ke}(t) \right)$$

$$\mathbf{r}_{s2}(t) = \begin{bmatrix} 1 & 0 \end{bmatrix} \left(\mathbf{y}(t) - \begin{bmatrix} 1 & 2 & 1 \\ 1 & 1 & 0 \end{bmatrix} \mathbf{q}_{ke}(t) \right)$$

Fig. 1-3 plot the residuals variable trajectories over the duration of the system run. The results show that one residual profile remain about the same through the entire run while the second shows step changes, which can be used in the fault isolation stage.

4.2 Structured residual generators of actuator faults

4.2.1 Set of the state estimators

To design structured residual generators of actuator faults based on the state estimators, all sensors are assumed to be fault-free and each estimator is driven by all system outputs and all but one system inputs. To obtain this a congruence transform matrix $\mathbf{T}_{ak} \in \mathbf{R}^{n\times n}$, $k = 1, 2, \ldots, r$ be introduced, and so it is natural to write

$$\mathbf{T}_{ak}\dot{\mathbf{q}}(t) = \mathbf{T}_{ak}\mathbf{A}\mathbf{q}(t) + \mathbf{T}_{ak}\mathbf{B}_u\mathbf{u}(t) \tag{66}$$

$$\dot{\mathbf{q}}_k(t) = \mathbf{A}_k\mathbf{q}(t) + \mathbf{B}_{uk}\mathbf{u}(t) \tag{67}$$

respectively, where

$$\mathbf{A}_k = \mathbf{T}_{ak}\mathbf{A}, \qquad \mathbf{B}_{uk} = \mathbf{T}_{ak}\mathbf{B}_u \tag{68}$$

as well as

$$\mathbf{y}_k(t) = \mathbf{C}\mathbf{T}_{ak}\mathbf{q}(t) = \mathbf{C}\mathbf{q}_k(t) \tag{69}$$

The set of state estimators associated with (67), (69) for $k = 1, 2, \ldots, r$ can be defined in the next form

$$\dot{\mathbf{q}}_{ke}(t) = \mathbf{A}_k\mathbf{q}_{ke}(t) + \mathbf{B}_{uke}\mathbf{u}(t) + \mathbf{L}_k\mathbf{y}(t) - \mathbf{J}_k\mathbf{y}_{ke}(t) \tag{70}$$

$$\mathbf{y}_{ke}(t) = \mathbf{C}\mathbf{q}_{ke}(t) \tag{71}$$

$\mathbf{A}_{ke} \in \mathbf{R}^{n\times n}$, $\mathbf{B}_{uke} \in \mathbf{R}^{n\times r}$, $\mathbf{J}_k, \mathbf{L}_k \in \mathbf{R}^{n\times m}$. Denoting the estimate error as $\mathbf{e}_k(t) = \mathbf{q}_k(t) - \mathbf{q}_{ke}(t)$ the next differential equations can be written

$$\begin{aligned}
&\dot{\mathbf{e}}_k(t) = \dot{\mathbf{q}}_k(t) - \dot{\mathbf{q}}_{ke}(t) = \\
&= \mathbf{A}_k\mathbf{q}(t) + \mathbf{B}_{uk}\mathbf{u}(t) - \mathbf{A}_k\mathbf{q}_{ke}(t) - \mathbf{B}_{uke}\mathbf{u}(t) - \mathbf{L}_k\mathbf{y}(t) + \mathbf{J}_k\mathbf{y}_{ke}(t) = \\
&= \mathbf{A}_k\mathbf{q}(t) + \mathbf{B}_{uk}\mathbf{u}(t) - \mathbf{A}_k\big(\mathbf{q}_k(t) - \mathbf{e}_k(t)\big) - \mathbf{B}_{uke}\mathbf{u}(t) - \\
&- \mathbf{L}_k\mathbf{C}\mathbf{q}(t) + \mathbf{J}_k\mathbf{C}\big(\mathbf{q}_k(t) - \mathbf{e}_k(t)\big) = \\
&= (\mathbf{A}_k - \mathbf{A}_k\mathbf{T}_{ak} + \mathbf{J}_k\mathbf{C}\mathbf{T}_{ak} - \mathbf{L}_k\mathbf{C})\mathbf{q}(t) + (\mathbf{B}_{uk} - \mathbf{B}_{uke})\mathbf{u}(t) + (\mathbf{A}_k - \mathbf{J}_k\mathbf{C})\mathbf{e}_k(t)
\end{aligned} \tag{72}$$

$$\dot{\mathbf{e}}_k(t) = (\mathbf{T}_{ak}\mathbf{A} - \mathbf{A}_{ke}\mathbf{T}_{ak} - \mathbf{L}_k\mathbf{C})\mathbf{q}(t) + (\mathbf{B}_{uk} - \mathbf{B}_{uke})\mathbf{u}(t) + \mathbf{A}_{ke}\mathbf{e}_k(t) \tag{73}$$

respectively, where

$$\mathbf{A}_{ke} = \mathbf{A}_k - \mathbf{J}_k\mathbf{C} = \mathbf{T}_{ak}\mathbf{A} - \mathbf{J}_k\mathbf{C}, \qquad k = 1, 2, \ldots, r \tag{74}$$

are elements of the set of estimators system matrices. It is evident, to make estimate error autonomous that it have to be satisfied

$$\mathbf{L}_k\mathbf{C} = \mathbf{T}_{ak}\mathbf{A} - \mathbf{A}_{ke}\mathbf{T}_{ak}, \qquad \mathbf{B}_{uke} = \mathbf{B}_{uk} = \mathbf{T}_{ak}\mathbf{B}_u \tag{75}$$

Using (75) the equation (73) can be rewritten as

$$\dot{\mathbf{e}}_k(t) = \mathbf{A}_{ke}\mathbf{e}_k(t) = (\mathbf{A}_k - \mathbf{J}_k\mathbf{C})\mathbf{e}_k(t) = (\mathbf{T}_{ak}\mathbf{A} - \mathbf{J}_k\mathbf{C})\mathbf{e}_k(t) \tag{76}$$

and the state equation of estimators are then

$$\dot{\mathbf{q}}_{ke}(t) = (\mathbf{T}_{ak}\mathbf{A} - \mathbf{J}_k\mathbf{C})\mathbf{q}_{ke}(t) + \mathbf{B}_{uk}\mathbf{u}(t) + \mathbf{L}_k\mathbf{y}(t) - \mathbf{J}_k\mathbf{y}_{ke}(t) \tag{77}$$

$$\mathbf{y}_{ke}(t) = \mathbf{C}\mathbf{q}_{ke}(t) \tag{78}$$

4.2.2 Congruence transform matrices

Generally, the fault-free system equations (1), (2) can be rewritten as

$$\dot{\mathbf{q}}(t) = \mathbf{A}\mathbf{q}(t) + \mathbf{b}_{uk}\mathbf{u}_k(t) + \sum_{h=1,h\neq k}^{r} \mathbf{b}_{uh}\mathbf{u}_h(t) \tag{79}$$

$$\dot{\mathbf{y}}(t) = \mathbf{C}\dot{\mathbf{q}}(t) + \mathbf{D}_u\dot{\mathbf{u}}(t) = \mathbf{C}\mathbf{A}\mathbf{q}(t) + \mathbf{C}\mathbf{b}_{uk}\mathbf{u}_k(t) + \mathbf{D}_u\dot{\mathbf{u}}(t) + \sum_{h=1,h\neq k}^{r} \mathbf{C}\mathbf{b}_{uh}\mathbf{u}_h(t) \tag{80}$$

$$\mathbf{C}\mathbf{b}_{uk}\mathbf{u}_k(t) = \dot{\mathbf{y}}(t) - \mathbf{C}\mathbf{A}\mathbf{q}(t) - \mathbf{D}_u\dot{\mathbf{u}}(t) - \sum_{h=1,h\neq k}^{r} \mathbf{C}\mathbf{b}_{uh}\mathbf{u}_h(t) \tag{81}$$

respectively. Thus, using matrix pseudoinverse it yields

$$\mathbf{u}_k(t) \doteq (\mathbf{C}\mathbf{b}_{uk})^{\ominus 1}\Big(\dot{\mathbf{y}}(t) - \mathbf{C}\mathbf{A}\mathbf{q}(t) - \mathbf{D}_u\dot{\mathbf{u}}(t) - \sum_{h=1,h\neq k}^{r} \mathbf{C}\mathbf{b}_{uh}\mathbf{u}_h(t)\Big) \tag{82}$$

and substituting (81)

$$\mathbf{b}_{uk}\mathbf{u}_k(t) \doteq \mathbf{b}_{uk}(\mathbf{C}\mathbf{b}_{uk})^{\ominus 1}\mathbf{C}\mathbf{b}_{uk}\mathbf{u}_k(t) \tag{83}$$

$$\big(\mathbf{I}_n - \mathbf{b}_{uk}(\mathbf{C}\mathbf{b}_{uk})^{\ominus 1}\mathbf{C}\big)\mathbf{b}_{uk}\mathbf{u}_k(t) \doteq \mathbf{0} \tag{84}$$

respectively. It is evident that if

$$\mathbf{T}_{ak} = \mathbf{I}_n - \mathbf{b}_{uk}(\mathbf{C}\mathbf{b}_{uk})^{\ominus 1}\mathbf{C}\,, \qquad k = 1, 2, \ldots, r \tag{85}$$

influence of $\mathbf{u}_k(t)$ in (77) be suppressed (the k-th column in $\mathbf{B}_{uk} = \mathbf{T}_{ak}\mathbf{B}_u$ is the null column, approximatively).

4.2.3 Estimator stability

***Theorem* 2** *The k-th state-space estimator (77), (78) is stable if there exist a positive definite symmetric matrix* $\mathbf{P}_{ak} > 0$, $\mathbf{P}_{ak} \in R^{n\times n}$ *and a matrix* $\mathbf{Z}_{ak} \in R^{n\times m}$ *such that*

$$\mathbf{P}_{ak} = \mathbf{P}_{ak}^T > 0 \tag{86}$$

$$\mathbf{A}^T\mathbf{T}_{ak}\mathbf{P}_{ak} + \mathbf{P}_{ak}\mathbf{T}_{ak}\mathbf{A} - \mathbf{Z}_{ak}\mathbf{C} - \mathbf{C}^T\mathbf{Z}_{ak}^T < 0 \tag{87}$$

Then $\mathbf{J}_k$ *can be computed as*

$$\mathbf{J}_k = \mathbf{P}_{ak}^{-1}\mathbf{Z}_{ak} \tag{88}$$

Proof. Since the estimate error is autonomous Lyapunov function of the form

$$v(\mathbf{e}_k(t)) = \mathbf{e}_k^T(t)\mathbf{P}_{ak}\mathbf{e}_k(t) > 0 \tag{89}$$

where $\mathbf{P}_{ak} = \mathbf{P}_{ak}^T > 0$, $\mathbf{P}_{ak} \in R^{n\times n}$ can be considered. Thus,

$$\dot{v}(\mathbf{e}_k(t)) = \mathbf{e}_k^T(t)(\mathbf{T}_{ak}\mathbf{A} - \mathbf{J}_k\mathbf{C})^T\mathbf{P}_{ak}\mathbf{e}_k(t) + \mathbf{e}_k^T(t)\mathbf{P}_{ak}(\mathbf{T}_{ak}\mathbf{A} - \mathbf{J}_k\mathbf{C})\mathbf{e}_k(t) < 0 \tag{90}$$

$$\dot{v}(\mathbf{e}_k(t)) = \mathbf{e}_k^T(t)\mathbf{P}_{akc}\mathbf{e}_k(t) < 0 \tag{91}$$

respectively, where

$$\mathbf{P}_{akc} = \mathbf{A}^T\mathbf{T}_{ak}^T\mathbf{P}_{ak} + \mathbf{P}_{ak}\mathbf{T}_{ak}\mathbf{A} - \mathbf{P}_{ak}\mathbf{J}_k\mathbf{C} - (\mathbf{P}_{ak}\mathbf{J}_k\mathbf{C})^T < 0 \tag{92}$$

Using notation $\mathbf{P}_{ak}\mathbf{J}_k = \mathbf{Z}_{ak}$ (92) implies (87). This concludes the proof. ■

4.2.4 Estimator gain matrices

Knowing $\mathbf{J}_k$, $k = 1, 2, \ldots, r$ elements of this set can be inserted into (75). Thus

$$\begin{aligned}\mathbf{L}_k\mathbf{C} &= \mathbf{A}_k - \mathbf{A}_{ke}\mathbf{T}_{ak} = \mathbf{A}_k - \left(\mathbf{A}_k - \mathbf{J}_k\mathbf{C}\right)\left(\mathbf{I} - \mathbf{b}_{uk}(\mathbf{C}\mathbf{b}_{uk})^{\ominus 1}\mathbf{C}\right) = \\ &= \left(\mathbf{J}_k + \left(\mathbf{A}_k - \mathbf{J}_k\mathbf{C}\right)\mathbf{b}_{uk}(\mathbf{C}\mathbf{b}_{uk})^{\ominus 1}\right)\mathbf{C} = \left(\mathbf{J}_k + \mathbf{A}_{ke}\mathbf{b}_{uk}(\mathbf{C}\mathbf{b}_{uk})^{\ominus 1}\right)\mathbf{C}\end{aligned} \tag{93}$$

and

$$\mathbf{L}_k = \mathbf{J}_k + \mathbf{A}_{ke}\mathbf{b}_{uk}(\mathbf{C}\mathbf{b}_{uk})^{\ominus 1}, \quad k = 1, 2, \ldots, r \tag{94}$$

4.2.5 Set of the residual generators

Exploiting the model-based properties of state estimators the set of residual generators can be considered as

$$\mathbf{r}_{ak}(t) = \mathbf{X}_{ak}\mathbf{q}_{ke}(t) + \mathbf{Y}_{ak}(\mathbf{y}(t) - \mathbf{D}_u\mathbf{u}(t)), \quad k = 1, 2, \ldots, m \tag{95}$$

Subsequently

$$\mathbf{r}_{ak}(t) = \mathbf{X}_{ak}\left(\mathbf{T}_{ak}\mathbf{q}(t) - \mathbf{e}_k(t)\right) + \mathbf{Y}_{ak}\mathbf{C}\mathbf{q}(t) = (\mathbf{X}_{ak}\mathbf{T}_{ak} + \mathbf{Y}_{ak}\mathbf{C})\mathbf{q}(t) - \mathbf{X}_{ak}\mathbf{e}_k(t) \tag{96}$$

To eliminate influences of the state variable vector it is necessary to consider

$$\mathbf{X}_{ak}\mathbf{T}_{ak} + \mathbf{Y}_{ak}\mathbf{C} = \mathbf{0} \tag{97}$$

$$\mathbf{X}_{ak}\left(\mathbf{I}_n - \mathbf{b}_{uk}(\mathbf{C}\mathbf{b}_{uk})^{\ominus 1}\mathbf{C}\right) + \mathbf{Y}_{ak}\mathbf{C} = \mathbf{0} \tag{98}$$

respectively. Choosing $\mathbf{X}_{ak} = -\mathbf{C}$ (98) gives

$$-\left(\mathbf{C} - \mathbf{C}\mathbf{b}_{uk}(\mathbf{C}\mathbf{b}_{uk})^{\ominus 1}\mathbf{C}\right) + \mathbf{Y}_{ak}\mathbf{C} = -(\mathbf{I}_m - \mathbf{C}\mathbf{b}_{uk}(\mathbf{C}\mathbf{b}_{uk})^{\ominus 1})\mathbf{C} + \mathbf{Y}_{ak}\mathbf{C} = \mathbf{0} \tag{99}$$

i.e.

$$\mathbf{Y}_{ak} = \mathbf{I}_m - \mathbf{C}\mathbf{b}_{uk}(\mathbf{C}\mathbf{b}_{uk})^{\ominus 1} \tag{100}$$

Thus, the set of residuals (95) takes the form

$$\mathbf{r}_{ak}(t) = (\mathbf{I}_m - \mathbf{C}\mathbf{b}_{uk}(\mathbf{C}\mathbf{b}_{uk})^{\ominus 1})\mathbf{y}(t) - \mathbf{C}\mathbf{q}_{ke}(t) \tag{101}$$

When all sensors are fault-free and a fault occurs in the l-th actuator the residuals will satisfy the isolation logic

$$\|\mathbf{r}_{sk}(t)\| \le h_{sk},\ k = l, \qquad \|\mathbf{r}_{sk}(t)\| > h_{sk},\ k \ne l \tag{102}$$

This residual set can only isolate a single actuator fault at the same time. The principle can be generalized based on a regrouping of faults in such way that each residual will be designed to be sensitive to one group of actuator faults and insensitive to others.

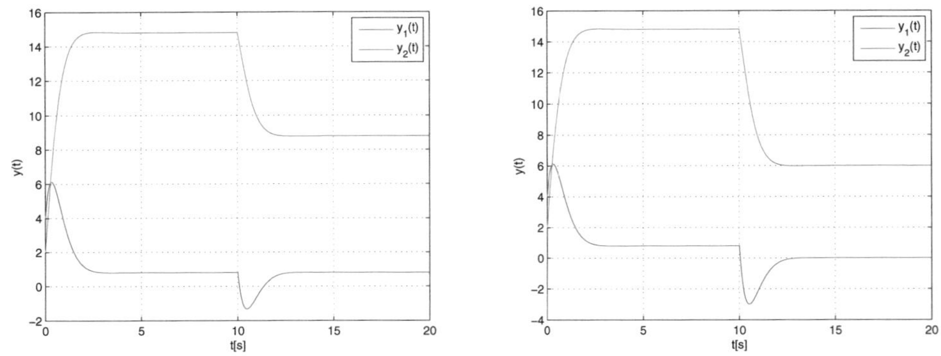

Fig. 4. System outputs for single actuator faults

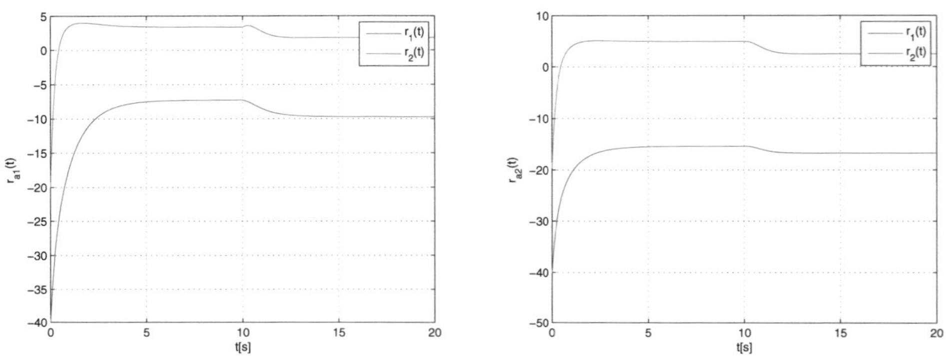

Fig. 5. Residuals for the 1st actuator fault

Illustrative example

Using the same system parameters as that given in the example in Subsection 4.1.2 the next design parameters be computed

$$\mathbf{b}_{u1} = \begin{bmatrix} 1 \\ 2 \\ 1 \end{bmatrix}, \ (\mathbf{C}\mathbf{b}_{u1})^{\ominus 1} = \begin{bmatrix} 0.1333 & 0.0667 \end{bmatrix}, \ \mathbf{T}_{a1} = \begin{bmatrix} 0.8000 & -0.3333 & -0.1333 \\ -0.4000 & 0.3333 & -0.2667 \\ -0.2000 & -0.3333 & 0.8667 \end{bmatrix}$$

$$\mathbf{b}_{u2} = \begin{bmatrix} 3 \\ 1 \\ 5 \end{bmatrix}, \ (\mathbf{C}\mathbf{b}_{u2})^{\ominus 1} = \begin{bmatrix} 0.0862 & 0.0345 \end{bmatrix}, \ \mathbf{T}_{a2} = \begin{bmatrix} 0.6379 & -0.6207 & -0.2586 \\ -0.1207 & 0.7931 & -0.0862 \\ -0.6034 & -1.0345 & 0.5690 \end{bmatrix}$$

$$\mathbf{A}_1 = \begin{bmatrix} 0.6667 & 2.0000 & 0.3333 \\ 1.3333 & 2.0000 & 1.6667 \\ -4.3333 & -8.0000 & -4.6667 \end{bmatrix}, \quad \mathbf{A}_2 = \begin{bmatrix} 1.2931 & 2.9655 & 0.6724 \\ 0.4310 & 0.6552 & 1.2241 \\ -2.8448 & -5.7241 & -3.8793 \end{bmatrix}$$

$$\mathbf{Y}_{a1} = \begin{bmatrix} 0.2 & -0.4 \\ -0.4 & 0.8 \end{bmatrix}, \quad \mathbf{Y}_{a2} = \begin{bmatrix} 0.1379 & -0.3448 \\ -0.3448 & 0.8621 \end{bmatrix}$$

Solving (86), (87) with respect to the LMI matrix variables $\mathbf{P}_{ak}$, and $\mathbf{Z}_{ak}$ using

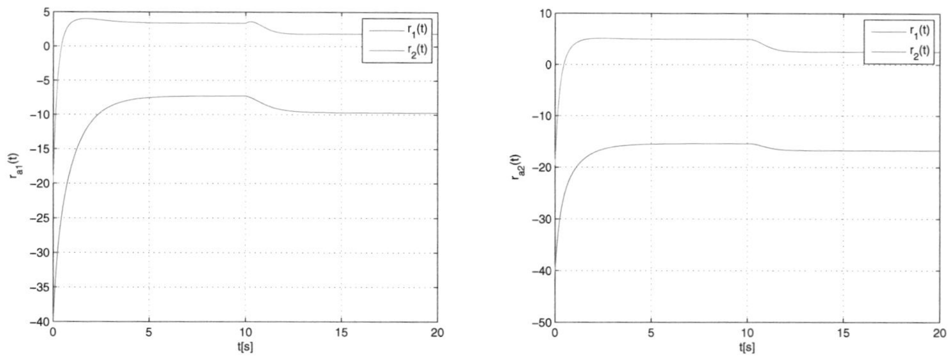

Fig. 6. Residuals for the 2nd actuator fault

Self-Dual-Minimization (SeDuMi) package for Matlab, the estimator gain matrix design problem was feasible with the results

$$\mathbf{P}_{a1} = \begin{bmatrix} 0.7555 & -0.0993 & 0.0619 \\ -0.0993 & 0.7464 & 0.1223 \\ 0.0619 & 0.1223 & 0.3920 \end{bmatrix}$$

$$\mathbf{Z}_{a1} = \begin{bmatrix} 0.0257 & 0.7321 \\ 0.4346 & 0.2392 \\ -0.7413 & -0.7469 \end{bmatrix}, \ \mathbf{J}_1 = \begin{bmatrix} 0.3504 & 1.2802 \\ 0.9987 & 0.8810 \\ -2.2579 & -2.3825 \end{bmatrix}, \ \mathbf{L}_1 = \begin{bmatrix} 0.2247 & 1.2173 \\ 0.7807 & 0.7720 \\ -2.8319 & -2.6695 \end{bmatrix}$$

$$\mathbf{P}_{a2} = \begin{bmatrix} 0.6768 & -0.0702 & 0.0853 \\ -0.0702 & 0.7617 & 0.0685 \\ 0.0853 & 0.0685 & 0.4637 \end{bmatrix}$$

$$\mathbf{Z}_{a2} = \begin{bmatrix} 0.2127 & 0.9808 \\ 0.3382 & 0.0349 \\ -0.6686 & -0.4957 \end{bmatrix}, \ \mathbf{J}_2 = \begin{bmatrix} 0.5888 & 1.6625 \\ 0.6462 & 0.3270 \\ -1.6457 & -1.4233 \end{bmatrix}, \ \mathbf{L}_2 = \begin{bmatrix} 0.3878 & 1.5821 \\ 0.6720 & 0.3373 \\ -2.6375 & -1.8200 \end{bmatrix}$$

respectively. It is easily verified that the system matrices of state estimators are stable with the eigenvalue spectra

$$\rho(\mathbf{T}_{a1}\mathbf{A} - \mathbf{J}_1\mathbf{C}) = \{-1.0000 \ \ -1.6256 \pm 0.3775\,\mathrm{i}\}$$

$$\rho(\mathbf{T}_{a2}\mathbf{A} - \mathbf{J}_2\mathbf{C}) = \{-1.0000 \ \ -1.5780 \pm 0.4521\,\mathrm{i}\}$$

respectively, and the set of residuals takes the form

$$\mathbf{r}_{a1}(t) = \begin{bmatrix} 0.2 & -0.4 \\ -0.4 & 0.8 \end{bmatrix} \mathbf{y}(t) - \begin{bmatrix} 1 & 2 & 1 \\ 1 & 1 & 0 \end{bmatrix} \mathbf{q}_{1e}(t)$$

$$\mathbf{r}_{a2}(t) = \begin{bmatrix} 0.1379 & -0.3448 \\ -0.3448 & 0.8621 \end{bmatrix} \mathbf{y}(t) - \begin{bmatrix} 1 & 2 & 1 \\ 1 & 1 & 0 \end{bmatrix} \mathbf{q}_{2e}(t)$$

Fig. 4-6 plot the residuals variable trajectories over the duration of the system run. The results show that both residual profile show changes through the entire run, therefore a fault isolation has to be more sophisticated.

5. Control with virtual sensors

5.1 Stability of the system

Considering a sensor fault then (1), (2) can be written as

$$\dot{\mathbf{q}}_f(t) = \mathbf{A}\mathbf{q}_f(t) + \mathbf{B}_u\mathbf{u}_f(t) \tag{103}$$

$$\mathbf{y}_f(t) = \mathbf{C}_f\mathbf{q}_f(t) + \mathbf{D}_u\mathbf{u}_f(t) \tag{104}$$

where $\mathbf{q}_f(t) \in \mathbb{R}^n$, $\mathbf{u}_f(t) \in \mathbb{R}^r$ are vectors of the state, and input variables of the faulty system, respectively, $\mathbf{C}_f \in \mathbb{R}^{m\times n}$ is the output matrix of the system with a sensor fault, and $\mathbf{y}_f(t) \in \mathbb{R}^m$ is a faulty measurement vector. This interpretation means that one row of $\mathbf{C}_f$ is null row.

Problem of the interest is to design a stable closed-loop system with the output controller

$$\mathbf{u}_f(t) = -\mathbf{K}_o\mathbf{y}_e(t) \tag{105}$$

where

$$\mathbf{y}_e(t) = \mathbf{E}\mathbf{y}_f(t) + (\mathbf{C} - \mathbf{E}\mathbf{C}_f)\mathbf{q}_{fe}(t) \tag{106}$$

$\mathbf{K}_o \in \mathbb{R}^{r\times m}$ is the controller gain matrix, and $\mathbf{E} \in \mathbb{R}^{m\times m}$ is a switching matrix, generally used in such a way that $\mathbf{E} = \mathbf{0}$, or $\mathbf{E} = \mathbf{I}_m$. If $\mathbf{E} = \mathbf{0}$ full state vector estimation is used for control, if $\mathbf{E} = \mathbf{I}_m$ the outputs of the fault-free sensors are combined with the estimated state variables to substitute a missing output of the faulty sensor.

Generally, the controller input is generated by the virtual sensor realized in the structure

$$\dot{\mathbf{q}}_{fe}(t) = \mathbf{A}\mathbf{q}_{fe}(t) + \mathbf{B}_u\mathbf{u}_f(t) + \mathbf{J}(\mathbf{y}_f(t) - \mathbf{D}_u\mathbf{u}_f(t) - \mathbf{C}_f\mathbf{q}_{fe}(t)) \tag{107}$$

The main idea is, instead of adapting the controller to the faulty system virtually adapt the faulty system to the nominal controller.

***Theorem* 3** *Control of the faulty system with virtual sensor defined by (103) – (107) is stable in the sense of bounded real lemma if there exist positive definite symmetric matrices* $\mathbf{Q}$, $\mathbf{R} \in \mathbb{R}^{n\times n}$, *and matrices* $\mathbf{K}_o \in \mathbb{R}^{r\times m}$, $\mathbf{J} \in \mathbb{R}^{n\times m}$ *such that*

$$\begin{bmatrix} \mathbf{\Phi}_1 & \mathbf{Q}\mathbf{B}_u\mathbf{K}_o(\mathbf{C}-\mathbf{E}\mathbf{C}_f) & -\mathbf{Q}\mathbf{B}_u\mathbf{K}_o\mathbf{E} & \left(\mathbf{C}_f - \mathbf{D}_u\mathbf{K}_o(\mathbf{C}-\mathbf{E}\mathbf{C}_f)\right)^T \\ * & \mathbf{\Phi}_2 & \mathbf{0} & \left(\mathbf{D}_u\mathbf{K}_o(\mathbf{C}-\mathbf{E}\mathbf{C}_f)\right)^T \\ * & * & -\gamma^2\mathbf{I}_r & -\left(\mathbf{D}_u\mathbf{K}_o\mathbf{E}\right)^T \\ * & * & * & -\mathbf{I}_m \end{bmatrix} < 0 \tag{108}$$

where

$$\mathbf{\Phi}_1 = \mathbf{Q}\left(\mathbf{A} - \mathbf{B}_u\mathbf{K}_o(\mathbf{C}-\mathbf{E}\mathbf{C}_f)\right) + \left(\mathbf{A} - \mathbf{B}_u\mathbf{K}_o(\mathbf{C}-\mathbf{E}\mathbf{C}_f)\right)^T\mathbf{Q} \tag{109}$$

$$\mathbf{\Phi}_2 = \mathbf{R}(\mathbf{A} - \mathbf{J}\mathbf{C}_f) + (\mathbf{A} - \mathbf{J}\mathbf{C}_f)^T\mathbf{R} \tag{110}$$

Proof. Assembling (103), (104), and (107) gives

$$\begin{bmatrix} \dot{\mathbf{q}}_f(t) \\ \dot{\mathbf{q}}_{fe}(t) \end{bmatrix} = \begin{bmatrix} \mathbf{A} & \mathbf{0} \\ \mathbf{J}\mathbf{C}_f & \mathbf{A}-\mathbf{J}\mathbf{C}_f \end{bmatrix} \begin{bmatrix} \mathbf{q}_f(t) \\ \mathbf{q}_{fe}(t) \end{bmatrix} + \begin{bmatrix} \mathbf{B}_u \\ \mathbf{B}_u \end{bmatrix} \mathbf{u}_f(t) \tag{111}$$

$$\mathbf{y}_f(t) = \mathbf{C}_f\mathbf{q}_f(t) + \mathbf{D}_u\mathbf{u}_f(t) \tag{112}$$

Thus, defining the estimation error vector

$$\mathbf{e}_{qf}(t) = \mathbf{q}_f(t) - \mathbf{q}_{fe}(t) \tag{113}$$

as well as the congruence transform matrix

$$\mathbf{T} = \mathbf{T}^{-1} = \begin{bmatrix} \mathbf{I} & \mathbf{0} \\ \mathbf{I} & -\mathbf{I} \end{bmatrix} \tag{114}$$

and then multiplying left-hand side of (111) by (114) results in

$$\mathbf{T}\begin{bmatrix} \dot{\mathbf{q}}_f(t) \\ \dot{\mathbf{q}}_{fe}(t) \end{bmatrix} = \mathbf{T}\begin{bmatrix} \mathbf{A} & \mathbf{0} \\ \mathbf{JC}_f & \mathbf{A}-\mathbf{JC}_f \end{bmatrix}\mathbf{T}^{-1}\mathbf{T}\begin{bmatrix} \mathbf{q}_f(t) \\ \mathbf{q}_{fe}(t) \end{bmatrix} + \mathbf{T}\begin{bmatrix} \mathbf{B}_u \\ \mathbf{B}_u \end{bmatrix}\mathbf{u}_f(t) \tag{115}$$

$$\begin{bmatrix} \dot{\mathbf{q}}_f(t) \\ \dot{\mathbf{e}}_{qf}(t) \end{bmatrix} = \begin{bmatrix} \mathbf{A} & \mathbf{0} \\ \mathbf{0} & \mathbf{A}-\mathbf{JC}_f \end{bmatrix}\begin{bmatrix} \mathbf{q}_f(t) \\ \mathbf{e}_{qf}(t) \end{bmatrix} + \begin{bmatrix} \mathbf{B}_u \\ \mathbf{0} \end{bmatrix}\mathbf{u}_f(t) \tag{116}$$

respectively. Subsequently, inserting (105), (106) into (116), (112) gives

$$\begin{bmatrix} \dot{\mathbf{q}}_f(t) \\ \dot{\mathbf{e}}_{qf}(t) \end{bmatrix} = \begin{bmatrix} \mathbf{A}-\mathbf{B}_u\mathbf{K}_o(\mathbf{C}-\mathbf{EC}_f) & \mathbf{B}_u\mathbf{K}_o(\mathbf{C}-\mathbf{EC}_f) \\ \mathbf{0} & \mathbf{A}-\mathbf{JC}_f \end{bmatrix}\begin{bmatrix} \mathbf{q}_f(t) \\ \mathbf{e}_{qf}(t) \end{bmatrix} + \begin{bmatrix} -\mathbf{B}_u\mathbf{K}_o\mathbf{E} \\ \mathbf{0} \end{bmatrix}\mathbf{y}_e(t) \tag{117}$$

together with

$$\mathbf{y}_f(t) = \begin{bmatrix} \mathbf{C}_f - \mathbf{D}_u\mathbf{K}_o(\mathbf{C}-\mathbf{EC}_f) & \mathbf{D}_u\mathbf{K}_o(\mathbf{C}-\mathbf{EC}_f) \end{bmatrix}\begin{bmatrix} \mathbf{q}_f(t) \\ \mathbf{e}_{qf}(t) \end{bmatrix} - \mathbf{D}_u\mathbf{K}_o\mathbf{E}\mathbf{y}_e(t) \tag{118}$$

and it is evident, that the separation principle yields.
Denoting

$$\mathbf{q}_\varepsilon^T(t) = \begin{bmatrix} \mathbf{q}_f^T(t) & \mathbf{e}_{qf}^T(t) \end{bmatrix}, \qquad \mathbf{w}_\varepsilon(t) = \mathbf{y}_e(t) \tag{119}$$

$$\mathbf{A}_\varepsilon = \begin{bmatrix} \mathbf{A}-\mathbf{B}_u\mathbf{K}_o(\mathbf{C}-\mathbf{EC}_f) & \mathbf{B}_u\mathbf{K}_o(\mathbf{C}-\mathbf{EC}_f) \\ \mathbf{0} & \mathbf{A}-\mathbf{JC}_f \end{bmatrix}, \qquad \mathbf{B}_\varepsilon = \begin{bmatrix} -\mathbf{B}_u\mathbf{K}_o\mathbf{E} \\ \mathbf{0} \end{bmatrix} \tag{120}$$

$$\mathbf{C}_\varepsilon = \begin{bmatrix} \mathbf{C}_f - \mathbf{D}_u\mathbf{K}_o(\mathbf{C}-\mathbf{EC}_f) & \mathbf{D}_u\mathbf{K}_o(\mathbf{C}-\mathbf{EC}_f) \end{bmatrix}, \qquad \mathbf{D}_\varepsilon = -\mathbf{D}_u\mathbf{K}_o\mathbf{E} \tag{121}$$

To accept the separation principle a block diagonal symmetric matrix $\mathbf{P}_\varepsilon > 0$ is chosen, i.e.

$$\mathbf{P}_\varepsilon = \mathrm{diag}\begin{bmatrix} \mathbf{Q} & \mathbf{R} \end{bmatrix} \tag{122}$$

where $\mathbf{Q} = \mathbf{Q}^T > 0$, $\mathbf{R} = \mathbf{R}^T > 0$, $\mathbf{Q},\ \mathbf{R} \in R^{n\times n}$ Thus, with (109), (110) it yields

$$\mathbf{P}_\varepsilon\mathbf{A}_\varepsilon + \mathbf{A}_\varepsilon^T\mathbf{P}_\varepsilon = \begin{bmatrix} \mathbf{\Phi}_1 & \mathbf{QB}_u\mathbf{K}_o(\mathbf{C}-\mathbf{EC}_f) \\ * & \mathbf{\Phi}_2 \end{bmatrix}, \qquad \mathbf{P}_\varepsilon\mathbf{B}_\varepsilon = \begin{bmatrix} -\mathbf{QB}_u\mathbf{K}_o\mathbf{E} \\ \mathbf{0} \end{bmatrix} \tag{123}$$

and inserting (121), (123), into (24) gives (108). This concludes the proof. ■

It is evident that there are the cross parameter interactions in the structure of (108). Since the separation principle pre-determines the estimator structure (error vectors are independent on the state as well as on the input variables), the controller, as well as estimator have to be designed independent.

5.2 Output feedback controller design

Theorem 4 (Unified algebraic approach) *A system (103), (104) with control law (105) is stable if there exist positive definite symmetric matrices* $\mathbf{P} > 0$, $\mathbf{\Pi} = \mathbf{P}^{-1} > 0$ *such that*

$$\begin{bmatrix} \mathbf{B}_u^{\perp}(\mathbf{A\Pi} + \mathbf{\Pi A}^T)\mathbf{B}_u^{\perp T} & \mathbf{B}_u^{\perp}\mathbf{\Pi C}_{fi}^T \\ * & -\mathbf{I}_m \end{bmatrix} < 0, \qquad i = 0, 1, 2, \ldots, m \tag{124}$$

$$\begin{bmatrix} \mathbf{C}_{fi}^{\bullet T\perp} \begin{bmatrix} \mathbf{PA} + \mathbf{A}^T\mathbf{P} & \mathbf{0} \\ * & -\gamma^2\mathbf{I}_r \end{bmatrix} \mathbf{C}_{fi}^{\bullet T\perp T} & \mathbf{C}_{fi}^{\bullet T\perp} \begin{bmatrix} \mathbf{C}_{fi}^T \\ \mathbf{0} \end{bmatrix} \\ * & -\mathbf{I}_m \end{bmatrix} < 0 \; i = 1, 2, \ldots, m, \; \mathbf{E} = \mathbf{I}_m \tag{125}$$

$$\begin{bmatrix} \mathbf{C}^{T\perp}(\mathbf{PA} + \mathbf{A}^T\mathbf{P})\mathbf{C}^{T\perp T} & \mathbf{C}^{T\perp}\mathbf{C}_{fi}^T \\ * & -\mathbf{I}_m \end{bmatrix} < 0 \quad i = 0, 1, 2, \ldots m, \quad \mathbf{E} = 0 \tag{126}$$

where

$$\mathbf{C}_{fi}^{\bullet T\perp} = \begin{bmatrix} (\mathbf{C} - \mathbf{EC}_{fi})^T \\ \mathbf{E} \end{bmatrix}^{\perp} \tag{127}$$

and $\mathbf{B}_u^{\perp}$ *is the orthogonal complement to* $\mathbf{B}_u$. *Then the control law gain matrix* $\mathbf{K}_o$ *exists if for obtained* $\mathbf{P}$ *there exist a symmetric matrices* $\mathbf{H} > 0$ *such that*

$$\begin{bmatrix} -\mathbf{FHF}^T - \mathbf{\Theta}_i & \mathbf{FH} + \mathbf{G}_i\mathbf{K}_o^T \\ * & -\mathbf{H} \end{bmatrix} < 0 \tag{128}$$

where $i = 0, 1, 2, \ldots, m$, *and*

$$\mathbf{\Theta}_i = -\begin{bmatrix} \mathbf{PA} + \mathbf{A}^T\mathbf{P} & \mathbf{0} & \mathbf{C}_{fi}^T \\ * & -\gamma^2\mathbf{I}_r & \mathbf{0} \\ * & * & -\mathbf{I}_m \end{bmatrix} < 0, \quad \mathbf{F} = -\begin{bmatrix} \mathbf{PB}_u \\ \mathbf{0} \\ \mathbf{0} \end{bmatrix}, \quad \mathbf{G} = \begin{bmatrix} (\mathbf{C} - \mathbf{EC}_{fi})^T \\ \mathbf{E} \\ \mathbf{0} \end{bmatrix} \tag{129}$$

Proof. Considering $\mathbf{e}_q(t) = \mathbf{0}$ then inserting $\mathbf{Q} = \mathbf{P}$ (108) implies

$$\begin{bmatrix} \mathbf{\Phi}_1 & -\mathbf{PB}_u\mathbf{K}_o\mathbf{E} & (\mathbf{C}_f - \mathbf{D}_u\mathbf{K}_o(\mathbf{C} - \mathbf{EC}_f))^T \\ * & -\gamma^2\mathbf{I}_r & -(\mathbf{D}_u\mathbf{K}_o\mathbf{E})^T \\ * & * & -\mathbf{I}_m \end{bmatrix} < 0 \tag{130}$$

where

$$\mathbf{\Phi}_1 = \mathbf{P}\big(\mathbf{A} - \mathbf{B}_u\mathbf{K}_o(\mathbf{C} - \mathbf{EC}_f)\big) + \big(\mathbf{A} - \mathbf{B}_u\mathbf{K}_o(\mathbf{C} - \mathbf{EC}_f)\big)^T\mathbf{P} \tag{131}$$

For the simplicity it is considered in the next that $\mathbf{D}_u = \mathbf{0}$ (in real physical systems this condition is satisfied) and subsequently (130), (131) can now be rewritten as

$$\begin{bmatrix} \mathbf{PA} + \mathbf{A}^T\mathbf{P} & \mathbf{0} & \mathbf{C}_f^T \\ * & -\gamma^2\mathbf{I}_r & \mathbf{0} \\ * & * & -\mathbf{I}_m \end{bmatrix} - \\ -\begin{bmatrix} \mathbf{PB}_u \\ \mathbf{0} \\ \mathbf{0} \end{bmatrix} \mathbf{K}_o \begin{bmatrix} \mathbf{C} - \mathbf{EC}_f & \mathbf{E} & \mathbf{0} \end{bmatrix} - \begin{bmatrix} (\mathbf{C} - \mathbf{EC}_f)^T \\ \mathbf{E} \\ \mathbf{0} \end{bmatrix} \mathbf{K}_o^T \begin{bmatrix} \mathbf{B}_u^T\mathbf{P} & \mathbf{0} & \mathbf{0} \end{bmatrix} < 0 \tag{132}$$

Defining the congruence transform matrix

$$\mathbf{T}_v = \text{diag}\left[\, \mathbf{P}^{-1}\ \mathbf{I}_r\ \mathbf{I}_m \,\right] \tag{133}$$

then pre-multiplying left-hand side and right-hand side of (132) by (133) gives

$$\begin{gathered}\begin{bmatrix} \mathbf{A}\mathbf{P}^{-1} + \mathbf{P}^{-1}\mathbf{A}^T & \mathbf{0} & \mathbf{P}^{-1}\mathbf{C}_f^T \\ * & -\gamma^2\mathbf{I}_r & \mathbf{0} \\ * & * & -\mathbf{I}_m \end{bmatrix} - \\ -\begin{bmatrix} \mathbf{B}_u \\ \mathbf{0} \\ \mathbf{0} \end{bmatrix} \mathbf{K}_o \left[(\mathbf{C}-\mathbf{E}\mathbf{C}_f)\mathbf{P}^{-1}\ \mathbf{E}\ \mathbf{0} \right] - \begin{bmatrix} \mathbf{P}^{-1}(\mathbf{C}-\mathbf{E}\mathbf{C}_f)^T \\ \mathbf{E} \\ \mathbf{0} \end{bmatrix} \mathbf{K}_o^T \left[\mathbf{B}_u^T\ \mathbf{0}\ \mathbf{0} \right] < 0 \end{gathered} \tag{134}$$

Since it yields

$$\mathbf{B}_u^{\circ\perp} = \begin{bmatrix} \mathbf{B}_u \\ \mathbf{0} \\ \mathbf{0} \end{bmatrix}^{\perp} = \begin{bmatrix} \mathbf{B}_u^{\perp} & \mathbf{0} & \mathbf{0} \\ \mathbf{0} & \mathbf{I}_r & \mathbf{0} \\ \mathbf{0} & \mathbf{0} & \mathbf{I}_m \end{bmatrix} \tag{135}$$

pre-multiplying left hand side of (134) by (135) as well as right-hand side of (134) by transposition of (135) leads to inequalities

$$\begin{bmatrix} \mathbf{B}_u^{\perp}(\mathbf{A}\mathbf{P}^{-1} + \mathbf{P}^{-1}\mathbf{A}^T)\mathbf{B}_u^{\perp T} & \mathbf{0} & \mathbf{B}_u^{\perp}\mathbf{P}^{-1}\mathbf{C}_f^T \\ * & -\gamma^2\mathbf{I}_r & \mathbf{0} \\ * & * & -\mathbf{I}_m \end{bmatrix} < 0 \tag{136}$$

$$\begin{bmatrix} \mathbf{B}_u^{\perp}(\mathbf{A}\mathbf{P}^{-1} + \mathbf{P}^{-1}\mathbf{A}^T)\mathbf{B}_u^{\perp T} & \mathbf{B}_u^{\perp}\mathbf{P}^{-1}\mathbf{C}_f^T \\ * & -\mathbf{I}_m \end{bmatrix} < 0 \tag{137}$$

respectively. Considering all possible structures $\mathbf{C}_{fi}$, $i = 1, 2, \ldots, m$ associated with simple sensor faults, as well as fault-free regime associated with the nominal matrix $\mathbf{C} = \mathbf{C}_{f0}$, then using the substitution $\mathbf{P}^{-1} = \mathbf{\Pi}$ the inequality (136) implies (124).
Analogously, using orthogonal complement

$$\mathbf{C}_f^{\circ T\perp} = \begin{bmatrix} (\mathbf{C}-\mathbf{E}\mathbf{C}_f)^T \\ \mathbf{E} \\ \mathbf{0} \end{bmatrix}^{\perp} = \begin{bmatrix} \begin{bmatrix} (\mathbf{C}-\mathbf{E}\mathbf{C}_f)^T \\ \mathbf{E} \end{bmatrix}^{\perp} & \mathbf{0} \\ \mathbf{0} & \mathbf{I}_m \end{bmatrix} = \begin{bmatrix} \mathbf{C}_f^{\bullet T\perp} & \mathbf{0} \\ * & \mathbf{I}_m \end{bmatrix} \tag{138}$$

and pre-multiplying left-hand side of (132) by (138) and its right-hand side by transposition of (138) results in

$$\begin{bmatrix} \mathbf{C}_f^{\bullet T\perp} \begin{bmatrix} \mathbf{P}\mathbf{A} + \mathbf{A}^T\mathbf{P} & \mathbf{0} \\ * & -\gamma^2\mathbf{I}_r \end{bmatrix} \mathbf{C}_f^{\bullet T\perp T} & \mathbf{C}_f^{\bullet T\perp} \begin{bmatrix} \mathbf{C}_f^T \\ \mathbf{0} \end{bmatrix} \\ * & -\mathbf{I}_m \end{bmatrix} < 0 \tag{139}$$

Considering all possible structures $\mathbf{C}_{fi}$, $i = 1, 2, \ldots, m$ (139) implies (125).
Inequality (125) takes a simpler form if $\mathbf{E} = \mathbf{0}$. Thus, now

$$\mathbf{C}_f^{\circ T\perp} = \begin{bmatrix} \mathbf{C}^T \\ \mathbf{0} \\ \mathbf{0} \end{bmatrix}^{\perp} = \begin{bmatrix} \mathbf{C}^{T\perp} & \mathbf{0} & \mathbf{0} \\ \mathbf{0} & \mathbf{I}_r & \mathbf{0} \\ \mathbf{0} & \mathbf{0} & \mathbf{I}_m \end{bmatrix} \tag{140}$$

and pre-multiplying left-hand sides of (132) by (140) and its right-hand side by transposition of (140) results in

$$\begin{bmatrix} \mathbf{C}^{T\perp}(\mathbf{PA}+\mathbf{A}^T\mathbf{P})\mathbf{C}^{T\perp T} & \mathbf{0} & \mathbf{C}^{T\perp}\mathbf{C}_f^T \\ * & -\gamma^2\mathbf{I}_r & \mathbf{0} \\ * & * & -\mathbf{I}_m \end{bmatrix} < 0 \tag{141}$$

which implies (126). This concludes the proof. ■

Solving LMI problem (124), (125), (126) with respect to LMI variable $\mathbf{P}$, then it is possible to construct (128), and subsequently to solve (127) defining the feedback control gain $\mathbf{K}_o$, and $\mathbf{H}$ as LMI variables.

Note, (124), (125), (126) have to be solved iteratively to obtain any approximation $\mathbf{P}^{-1} = \mathbf{\Pi}$. This implies that these inequalities together define only the sufficient condition of a solution, and so one from $(\mathbf{P}, \mathbf{\Pi}^{-1})$ can be used in design independently while verifying solution using the latter. Since of an approximative solution the matrix Θ defined in (129) need not be negative definite, and so it is necessary to introduce into (128) a negative definite matrix $\mathbf{\Theta}_{fi}^{\circ}$ as follows

$$\mathbf{\Theta}_{fi}^{\circ} = \mathbf{\Theta}_{fi} - \mathbf{\Delta} < 0 \tag{142}$$

where $\mathbf{\Delta} > 0$.

If (124), (125), (126) is infeasible the principle can be modified based on inequalities regrouping e.g. in such way that solving (124), (125), and (124), (126) separatively and obtaining two virtual sensor structures (one for $\mathbf{E} = \mathbf{0}$ and other for $\mathbf{E} = \mathbf{I}_m$). It is evident that virtual sensor switching be more sophisticated in this case.

5.3 Virtual sensor design

Theorem 5 *Virtual sensor (107) associated with the system (103), (104) is stable if there exist symmetric positive definite matrix $\mathbf{R} \in \mathbb{R}^{n\times n}$, and a matrix $\mathbf{Z} \in \mathbb{R}^{n\times m}$, such that*

$$\mathbf{R} = \mathbf{R}^T > 0 \tag{143}$$

$$\mathbf{RA} + \mathbf{A}^T\mathbf{R} - \mathbf{ZC}_{fi} + \mathbf{C}_{fi}^T\mathbf{Z}^T < 0, \qquad i = 0, 1, 2, \ldots, m \tag{144}$$

The virtual sensor matrix parameter is then given as

$$\mathbf{J} = \mathbf{R}^{-1}\mathbf{Z} \tag{145}$$

Proof. Supposing that $\mathbf{q}(t) = \mathbf{0}$ and $\mathbf{D}_u = \mathbf{0}$ then (108), (110) is reduced as follows

$$\begin{bmatrix} \mathbf{\Phi}_2 & \mathbf{0} & \mathbf{0} \\ * & -\gamma^2\mathbf{I}_r & \mathbf{0} \\ * & * & -\mathbf{I}_m \end{bmatrix} < 0 \tag{146}$$

$$\mathbf{R}(\mathbf{A} - \mathbf{JC}_f) + (\mathbf{A} - \mathbf{JC}_f)^T\mathbf{R} < 0 \tag{147}$$

respectively. Thus, with the notation

$$\mathbf{Z} = \mathbf{RJ} \tag{148}$$

(147) implies (144). This concludes the proof. ■

Illustrative example

Using for $\mathbf{E} = \mathbf{0}$ the same system parameters as that given in the example in Subsection 4.1.2 then the next design parameters were computed

$$\mathbf{B}_u^{\perp} = \begin{bmatrix} -0.8581 & 0.1907 & 0.4767 \end{bmatrix}, \; \mathbf{C}^{T\perp} = \begin{bmatrix} 0.5774 & -0.5774 & 0.5774 \end{bmatrix}$$

$$\mathbf{C}_{f0} = \begin{bmatrix} 1 & 2 & 1 \\ 1 & 1 & 0 \end{bmatrix}, \; \mathbf{C}_{f1} = \begin{bmatrix} 0 & 0 & 0 \\ 1 & 1 & 0 \end{bmatrix}, \; \mathbf{C}_{f2} = \begin{bmatrix} 1 & 2 & 1 \\ 0 & 0 & 0 \end{bmatrix}$$

Solving (124) and the set of polytopic inequalities (126) with respect to $\mathbf{P}, \mathbf{\Pi}$ using the SeDuMi package the problem was feasible and the matrices

$$\mathbf{P} = \begin{bmatrix} 0.6836 & 0.0569 & -0.0569 \\ 0.0569 & 0.6836 & 0.0569 \\ -0.0569 & 0.0569 & 0.6836 \end{bmatrix}$$

as well as $\mathbf{H} = 0.1\mathbf{I}_2$ was used to construct the next ones

$$\mathbf{\Theta}_0 = \begin{bmatrix} 0.5688 & 0.9111 & -3.0769 & & & 1 & 1 \\ 0.9111 & -0.9100 & -5.8103 & & & 2 & 1 \\ -3.0769 & -5.8103 & -6.7225 & & & 1 & 0 \\ & & & -0.1 & & & \\ & & & & -0.1 & & \\ 1.0000 & 2.0000 & 1.0000 & & & -1 & \\ 1.0000 & 1.0000 & 0 & & & & -1 \end{bmatrix}, \quad \Theta_1, \quad \Theta_2$$

$$\mathbf{F}^T = -\begin{bmatrix} 0.7405 & 1.4810 & 0.7405 & 0 & 0 & 0 & 0 \\ 1.8234 & 1.1386 & 3.3044 & 0 & 0 & 0 & 0 \end{bmatrix}, \; \mathbf{G}^T = \begin{bmatrix} 1 & 2 & 1 & 0 & 0 & 0 & 0 \\ 1 & 1 & 0 & 0 & 0 & 0 & 0 \end{bmatrix}$$

To obtain negativity of $\mathbf{\Theta}_{fi}^{\circ}$ the matrix $\mathbf{\Delta} = 4.94\mathbf{I}_7$ was introduced. Solving the set of polytopic inequalities (128) with respect to $\mathbf{K}_o$ the problem was also feasible and it gave the result

$$\mathbf{K}_o = \begin{bmatrix} -0.0734 & -0.0008 \\ -0.1292 & 0.1307 \end{bmatrix}$$

which secure robustness of control stability with respect to all structures of output matrices $\mathbf{C}_{fi}$, $i = 0, 1, 2$. In this sense

$$\rho(\mathbf{A} - \mathbf{B}_u\mathbf{K}_o\mathbf{C}) = \begin{bmatrix} -1.0000 & -1.3941 \pm 2.3919\,\mathrm{i} \end{bmatrix}$$

$$\rho(\mathbf{A} - \mathbf{B}_u\mathbf{K}_o\mathbf{C}_{f1}) = \begin{bmatrix} -1.0000 & -2.2603 \pm 1.6601\,\mathrm{i} \end{bmatrix}$$

$$\rho(\mathbf{A} - \mathbf{B}_u\mathbf{K}_o\mathbf{C}_{f2}) = \begin{bmatrix} -1.0000 & -1.1337 \pm 1.8591\,\mathrm{i} \end{bmatrix}$$

Solving the set of polytopic inequalities (144) with respect to $\mathbf{R}, \mathbf{Z}$ the feasible solution was

$$\mathbf{R} = \begin{bmatrix} 0.7188 & 0.0010 & 0.0016 \\ 0.0010 & 0.7212 & 0.0448 \\ 0.0016 & 0.0448 & 0.1299 \end{bmatrix}, \quad \mathbf{Z} = \begin{bmatrix} -0.0006 & 0.4457 \\ 0.0117 & 0.0701 \\ -0.0629 & -0.5894 \end{bmatrix}$$

Thus, the virtual sensor gain matrix $\mathbf{J}$ was computed as

$$\mathbf{J} = \begin{bmatrix} 0.0002 & 0.6296 \\ 0.0473 & 0.3868 \\ -0.5003 & -4.6799 \end{bmatrix}$$

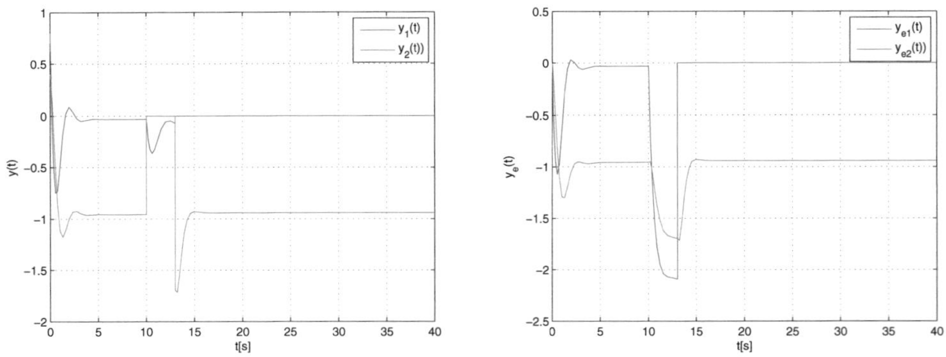

Fig. 7. System output and its estimation

which secure robustness of virtual sensor stability with respect to all structures of output matrices $\mathbf{C}_{fi}$, $i = 0, 1, 2$. In this sense

$$\rho(\mathbf{A} - \mathbf{J}\mathbf{C}) = \begin{bmatrix} -1.0000 \\ -1.1656 \\ -3.4455 \end{bmatrix}, \quad \rho(\mathbf{A} - \mathbf{J}\mathbf{C}_{f1}) = \begin{bmatrix} -1.0000 \\ -1.2760 \\ -3.7405 \end{bmatrix}$$

$$\rho(\mathbf{A} - \mathbf{B}_u\mathbf{K}_o\mathbf{C}_{f2}) = \begin{bmatrix} -1.0000 \\ -1.1337 + 1.8591\,\mathrm{i} \\ -1.1337 - 1.8591\,\mathrm{i} \end{bmatrix}$$

As was mentioned above the simulation results were obtained by solving the semi-definite programming problem under Matlab with SeDuMi package 1.2, where the initial conditions were set to

$$\mathbf{q}(0) = \begin{bmatrix} 0.2\ 0.2\ 0.2 \end{bmatrix}^T, \quad \mathbf{q}_e(0) = \begin{bmatrix} 0\ 0\ 0 \end{bmatrix}^T$$

respectively, and the control law in forced mode was

$$\mathbf{u}_f(t) = -\mathbf{K}_o\mathbf{y}_e(t) + \mathbf{w}(t), \quad \mathbf{w}(t) = \begin{bmatrix} -0.2\ -0.2 \end{bmatrix}^T$$

Fig. 7 shows the trajectory of the system outputs and the trajectory of the estimate system outputs using virtual sensor structure. It can be seen there a reaction time available to perform fault detection and isolation in the trajectory of the estimate system outputs, as well as a reaction time of control system reconfiguration in the system output trajectory. The results confirm that the true signals and their estimation always reside between limits given by static system error of the closed-loop structure.

6. Active control structures with a single actuator fault

6.1 Stability of the system

Theorem 6 *Fault tolerant control system defined by (1) – (9) is stable in the sense of bounded real lemma if there exist positive definite symmetric matrices* $\mathbf{Q}$, $\mathbf{R} \in R^{n \times n}$, $\mathbf{S} \in R^{l \times l}$, *and matrices*

$\mathbf{K} \in \mathbb{R}^{r \times n}, \mathbf{L} \in \mathbb{R}^{r \times l}, \mathbf{J} \in \mathbb{R}^{n \times m}, \mathbf{M} \in \mathbb{R}^{l \times l}, \mathbf{N} \in \mathbb{R}^{l \times m}$ *such that*

$$\begin{bmatrix} \mathbf{\Phi}_{11} & \mathbf{QB}_u\mathbf{K} & \mathbf{QB}_u\mathbf{L} & \mathbf{0} & \mathbf{0} & (\mathbf{C}-\mathbf{D}_u\mathbf{K})^T \\ * & \mathbf{\Phi}_{22} & \mathbf{R}(\mathbf{B}_f-\mathbf{JD}_f)-(\mathbf{SNC})^T & \mathbf{0} & \mathbf{0} & (\mathbf{D}_u\mathbf{K})^T \\ * & * & \mathbf{\Phi}_{33} & -\mathbf{SM} & \mathbf{S} & (\mathbf{D}_u\mathbf{L})^T \\ * & * & * & -\gamma^2\mathbf{I}_l & \mathbf{0} & \mathbf{0} \\ * & * & * & * & -\gamma^2\mathbf{I}_l & \mathbf{0} \\ * & * & * & * & * & -\mathbf{I}_m \end{bmatrix} < 0 \tag{149}$$

where

$$\mathbf{\Phi}_{11} = \mathbf{Q}(\mathbf{A}-\mathbf{B}_u\mathbf{K}) + (\mathbf{A}-\mathbf{B}_u\mathbf{K})^T\mathbf{Q}, \quad \mathbf{\Phi}_{22} = \mathbf{R}(\mathbf{A}-\mathbf{JC}) + (\mathbf{A}-\mathbf{JC})^T\mathbf{R} \tag{150}$$

$$\mathbf{\Phi}_{33} = \mathbf{S}(\mathbf{M}-\mathbf{ND}_f) + (\mathbf{M}-\mathbf{ND}_f)^T\mathbf{S} \tag{151}$$

Proof. Considering equality $\dot{\mathbf{f}}(t) = \dot{\mathbf{f}}(t)$ and assembling this equality with (1) – (4), and with (7) – (9) gives the result

$$\begin{bmatrix} \dot{\mathbf{q}}(t) \\ \dot{\mathbf{q}}_e(t) \\ \dot{\mathbf{f}}(t) \\ \dot{\mathbf{f}}_e(t) \end{bmatrix} = \begin{bmatrix} \mathbf{A} & -\mathbf{B}_u\mathbf{K} & \mathbf{B}_f & -\mathbf{B}_u\mathbf{L} \\ \mathbf{JC} & \mathbf{A}-\mathbf{B}_u\mathbf{K}-\mathbf{JC} & \mathbf{JD}_f & \mathbf{B}_f-\mathbf{JD}_f-\mathbf{B}_u\mathbf{L} \\ \mathbf{0} & \mathbf{0} & \mathbf{0} & \mathbf{0} \\ \mathbf{NC} & -\mathbf{NC} & \mathbf{ND}_f & \mathbf{M}-\mathbf{ND}_f \end{bmatrix} \begin{bmatrix} \mathbf{q}(t) \\ \mathbf{q}_e(t) \\ \mathbf{f}(t) \\ \mathbf{f}_e(t) \end{bmatrix} + \begin{bmatrix} \mathbf{0} \\ \mathbf{0} \\ \dot{\mathbf{f}}(t) \\ \mathbf{0} \end{bmatrix} \tag{152}$$

$$\mathbf{y} = \begin{bmatrix} \mathbf{C} & -\mathbf{D}_u\mathbf{K} & \mathbf{D}_f & -\mathbf{D}_u\mathbf{L} \end{bmatrix} \begin{bmatrix} \mathbf{q}(t) \\ \mathbf{q}_e(t) \\ \mathbf{f}(t) \\ \mathbf{f}_e(t) \end{bmatrix} \tag{153}$$

which can be written in a compact form as

$$\dot{\mathbf{q}}_\alpha(t) = \mathbf{A}_\alpha\mathbf{q}_\alpha(t) + \mathbf{f}_\alpha(t) \tag{154}$$

$$\mathbf{y} = \mathbf{C}_\alpha\mathbf{q}_\alpha(t) \tag{155}$$

where

$$\mathbf{q}_\alpha^T(t) = \begin{bmatrix} \mathbf{q}^T(t) & \mathbf{q}_e^T(t) & \mathbf{f}^T(t) & \mathbf{f}_e^T(t) \end{bmatrix}, \quad \mathbf{f}_\alpha^T(t) = \begin{bmatrix} \mathbf{0}^T & \mathbf{0}^T & \dot{\mathbf{f}}^T(t) & \mathbf{0}^T \end{bmatrix} \tag{156}$$

$$\mathbf{A}_\alpha = \begin{bmatrix} \mathbf{A} & -\mathbf{B}_u\mathbf{K} & \mathbf{B}_f & -\mathbf{B}_u\mathbf{L} \\ \mathbf{JC} & \mathbf{A}-\mathbf{B}_u\mathbf{K}-\mathbf{JC} & \mathbf{JD}_f & \mathbf{B}_f-\mathbf{JD}_f-\mathbf{B}_u\mathbf{L} \\ \mathbf{0} & \mathbf{0} & \mathbf{0} & \mathbf{0} \\ \mathbf{NC} & -\mathbf{NC} & \mathbf{ND}_f & \mathbf{M}-\mathbf{ND}_f \end{bmatrix} \tag{157}$$

$$\mathbf{C}_\alpha = \begin{bmatrix} \mathbf{C} & -\mathbf{D}_u\mathbf{K} & \mathbf{D}_f & -\mathbf{D}_u\mathbf{L} \end{bmatrix} \tag{158}$$

Using notations

$$\mathbf{e}_q(t) = \mathbf{q}(t) - \mathbf{q}_e(t), \qquad \mathbf{e}_f(t) = \mathbf{f}(t) - \mathbf{f}_e(t) \tag{159}$$

where $\mathbf{e}_q(t)$ is the error between the actual state and the estimated state, and $\mathbf{e}_f(t)$ is the error between the actual fault and the estimated fault, respectively then it is possible to define the state transformation

$$\mathbf{q}_\beta(t) = \mathbf{T}\mathbf{q}_\alpha(t) = \begin{bmatrix} \mathbf{q}(t) \\ \mathbf{e}_q(t) \\ \mathbf{f}(t) \\ \mathbf{e}_f(t) \end{bmatrix}, \ \mathbf{f}_\beta(t) = \mathbf{T}\mathbf{f}_\alpha(t) = \begin{bmatrix} \mathbf{0} \\ \mathbf{0} \\ \dot{\mathbf{f}}(t) \\ \dot{\mathbf{f}}(t) \end{bmatrix}, \ \mathbf{T} = \mathbf{T}^{-1} = \begin{bmatrix} \mathbf{I} & \mathbf{0} & \mathbf{0} & \mathbf{0} \\ \mathbf{I} & -\mathbf{I} & \mathbf{0} & \mathbf{0} \\ \mathbf{0} & \mathbf{0} & \mathbf{I} & \mathbf{0} \\ \mathbf{0} & \mathbf{0} & \mathbf{I} & -\mathbf{I} \end{bmatrix} \tag{160}$$

and to rewrite (154), (155) as follows

$$\dot{\mathbf{q}}_\beta(t) = \mathbf{A}_\beta \mathbf{q}_\beta(t) + \mathbf{f}_\beta(t) \tag{161}$$

$$\mathbf{y} = \mathbf{C}_\beta \mathbf{q}_\beta(t) \tag{162}$$

where

$$\mathbf{A}_\beta = \mathbf{T}\mathbf{A}_\alpha\mathbf{T}^{-1} = \begin{bmatrix} \mathbf{A} & \mathbf{B}_u\mathbf{K} & \mathbf{B}_f - \mathbf{B}_u\mathbf{L} & \mathbf{B}_u\mathbf{L} \\ \mathbf{0} & \mathbf{A}-\mathbf{J}\mathbf{C} & \mathbf{0} & \mathbf{B}_f - \mathbf{J}\mathbf{D}_f \\ \mathbf{0} & \mathbf{0} & \mathbf{0} & \mathbf{0} \\ \mathbf{0} & -\mathbf{N}\mathbf{C} & -\mathbf{M} & \mathbf{M}-\mathbf{N}\mathbf{D}_f \end{bmatrix} \tag{163}$$

$$\mathbf{C}_\beta = \mathbf{C}_\alpha\mathbf{T}^{-1} = \begin{bmatrix} \mathbf{C}-\mathbf{D}_u\mathbf{K} & \mathbf{D}_u\mathbf{K} & \mathbf{D}_f - \mathbf{D}_u\mathbf{L} & \mathbf{D}_u\mathbf{L} \end{bmatrix} \tag{164}$$

Since (5) implies

$$\mathbf{B}_f - \mathbf{B}_u\mathbf{L} = \mathbf{0}, \qquad \mathbf{D}_f - \mathbf{D}_u\mathbf{L} = \mathbf{0} \tag{165}$$

it obvious that (163), (165) can be simplified as

$$\mathbf{A}_\beta = \mathbf{T}\mathbf{A}_\alpha\mathbf{T}^{-1} = \begin{bmatrix} \mathbf{A} & \mathbf{B}_u\mathbf{K} & \mathbf{0} & \mathbf{B}_u\mathbf{L} \\ \mathbf{0} & \mathbf{A}-\mathbf{J}\mathbf{C} & \mathbf{0} & \mathbf{B}_f - \mathbf{J}\mathbf{D}_f \\ \mathbf{0} & \mathbf{0} & \mathbf{0} & \mathbf{0} \\ \mathbf{0} & -\mathbf{N}\mathbf{C} & -\mathbf{M} & \mathbf{M}-\mathbf{N}\mathbf{D}_f \end{bmatrix} \tag{166}$$

$$\mathbf{C}_\beta = \mathbf{C}_\alpha\mathbf{T}^{-1} = \begin{bmatrix} \mathbf{C}-\mathbf{D}_u\mathbf{K} & \mathbf{D}_u\mathbf{K} & \mathbf{0} & \mathbf{D}_u\mathbf{L} \end{bmatrix} \tag{167}$$

Eliminating out equality $\dot{\mathbf{f}}(t) = \dot{\mathbf{f}}(t)$ it can be written

$$\dot{\mathbf{q}}_\delta(t) = \mathbf{A}_\delta \mathbf{q}_\delta(t) + \mathbf{B}_\delta \mathbf{w}_\delta(t) \tag{168}$$

$$\mathbf{y} = \mathbf{C}_\delta \mathbf{q}_\delta(t) + \mathbf{D}_\delta \mathbf{w}_\delta(t) \tag{169}$$

where

$$\mathbf{q}_\delta^T(t) = \begin{bmatrix} \mathbf{q}^T(t) & \mathbf{e}_q^T(t) & \mathbf{e}_f^T(t) \end{bmatrix}, \qquad \mathbf{w}_\delta^T(t) = \begin{bmatrix} \mathbf{f}^T(t) & \dot{\mathbf{f}}^T(t) \end{bmatrix} \tag{170}$$

$$\mathbf{A}_\delta = \begin{bmatrix} \mathbf{A} & \mathbf{B}_u\mathbf{K} & \mathbf{B}_u\mathbf{L} \\ \mathbf{0} & \mathbf{A}-\mathbf{J}\mathbf{C} & \mathbf{B}_f - \mathbf{J}\mathbf{D}_f \\ \mathbf{0} & -\mathbf{N}\mathbf{C} & \mathbf{M}-\mathbf{N}\mathbf{D}_f \end{bmatrix}, \qquad \mathbf{B}_\delta = \begin{bmatrix} \mathbf{0} & \mathbf{0} \\ \mathbf{0} & \mathbf{0} \\ -\mathbf{M} & \mathbf{I} \end{bmatrix} \tag{171}$$

$$\mathbf{C}_\delta = \begin{bmatrix} \mathbf{C}-\mathbf{D}_u\mathbf{K} & \mathbf{D}_u\mathbf{K} & \mathbf{D}_u\mathbf{L} \end{bmatrix}, \qquad \mathbf{D}_\delta = \begin{bmatrix} \mathbf{0} & \mathbf{0} \end{bmatrix} \tag{172}$$

To apply the separation principle a block diagonal symmetric matrix $\mathbf{P}_\delta > 0$ has to be chosen, i.e.

$$\mathbf{P}_\delta = \text{diag}\begin{bmatrix} \mathbf{Q} & \mathbf{R} & \mathbf{S} \end{bmatrix} \tag{173}$$

where $\mathbf{Q}$, $\mathbf{R} \in R^{n\times n}$, $\mathbf{S} \in R^{l\times l}$. Thus, with (150), (151) it yields

$$\mathbf{P}_\delta\mathbf{A}_\delta + \mathbf{A}_\delta^T\mathbf{P}_\delta = \begin{bmatrix} \mathbf{\Phi}_{11} & \mathbf{Q}\mathbf{B}_u\mathbf{K} & \mathbf{Q}\mathbf{B}_u\mathbf{L} \\ * & \mathbf{\Phi}_{22} & \mathbf{R}(\mathbf{B}_f - \mathbf{J}\mathbf{D}_f) - (\mathbf{S}\mathbf{N}\mathbf{C})^T \\ * & * & \mathbf{\Phi}_{33} \end{bmatrix}, \quad \mathbf{P}_\delta\mathbf{B}_\delta = \begin{bmatrix} \mathbf{0} & \mathbf{0} \\ \mathbf{0} & \mathbf{0} \\ -\mathbf{S}\mathbf{M} & \mathbf{S} \end{bmatrix} \tag{174}$$

and inserting (171), (172), and (174) into (24) gives (149). This concludes the proof. ∎

6.2 Feedback controller gain matrix design

It is evident that there are the cross parameter interactions in the structure of (149). Since the separation principle pre-determines the estimator structure (error vectors are independent on the state as well as on the input variables), at the first design step can be computed a feedback controller gain matrix $\mathbf{K}$, and at the next step be designed the estimators gain matrices $\mathbf{J} \in R^{n\times m}$, $\mathbf{M} \in R^{l\times l}$, $\mathbf{N} \in R^{l\times m}$, including obtained $\mathbf{K}$.

Theorem 7. *For a fault-free system (1), (2) exists a stable nominal control (4) if there exist a positive definite symmetric matrix* $\mathbf{X} > 0$, $\mathbf{X} \in R^{n\times n}$, *a matrix* $\mathbf{Y} \in R^{r\times n}$, *and a positive scalar* $\gamma > 0$, $\gamma \in R$ *such that*

$$\mathbf{X} = \mathbf{X}^T > 0 \tag{175}$$

$$\begin{bmatrix} \mathbf{AX} + \mathbf{XA}^T - \mathbf{Y}^T\mathbf{B}_u^T - \mathbf{B}_u\mathbf{Y} & \mathbf{B}_u\mathbf{L} & \mathbf{XC}^T - \mathbf{Y}^T\mathbf{D}_u^T \\ * & -\gamma^2\mathbf{I}_l & \mathbf{L}^T\mathbf{D}_u^T \\ * & * & -\mathbf{I}_m \end{bmatrix} < 0 \tag{176}$$

The control law gain matrix is then given as

$$\mathbf{K} = \mathbf{YX}^{-1} \tag{177}$$

Proof. Considering $\mathbf{e}_q(t) = \mathbf{0}$, then separating $\mathbf{q}(t)$ from (168)-(169) gives

$$\dot{\mathbf{q}}(t) = \mathbf{A}^\circ\mathbf{q}(t) + \mathbf{B}^\circ\mathbf{w}^\circ(t) \tag{178}$$

$$\mathbf{y}(t) = \mathbf{C}^\circ\mathbf{q}(t) + \mathbf{D}^\circ\mathbf{w}^\circ(t) \tag{179}$$

where

$$\mathbf{w}^\circ(t) = \mathbf{e}_f(t) \tag{180}$$

$$\mathbf{A}^\circ = \mathbf{A} - \mathbf{B}_u\mathbf{K}, \quad \mathbf{B}^\circ = \mathbf{B}_u\mathbf{L}, \quad \mathbf{C}^\circ = \mathbf{C} - \mathbf{D}_u\mathbf{K}, \quad \mathbf{D}^\circ = \mathbf{D}_u\mathbf{L} \tag{181}$$

and with (181), and $\mathbf{P} = \mathbf{Q}$ inequality (24) can be written as

$$\begin{bmatrix} \mathbf{QA} + \mathbf{A}^T\mathbf{Q} - \mathbf{QB}_u\mathbf{K} - \mathbf{K}^T\mathbf{B}_u^T\mathbf{Q} & \mathbf{QB}_u\mathbf{L} & \mathbf{C}^T - \mathbf{K}^T\mathbf{D}_u^T \\ * & -\gamma^2\mathbf{I}_l & \mathbf{L}^T\mathbf{D}_u^T \\ * & * & -\mathbf{I}_m \end{bmatrix} < 0 \tag{182}$$

Introducing the congruence transform matrix

$$\mathbf{H} = \operatorname{diag}\begin{bmatrix} \mathbf{Q}^{-1} & \mathbf{I}_l & \mathbf{I}_m \end{bmatrix} \tag{183}$$

then multiplying left-hand side, as well right-hand side of (182) by (183) gives

$$\begin{bmatrix} \mathbf{AQ}^{-1} + \mathbf{Q}^{-1}\mathbf{A}^T - \mathbf{B}_u\mathbf{KQ}^{-1} - \mathbf{QK}^T\mathbf{B}_u^T & \mathbf{B}_u\mathbf{L} & \mathbf{Q}^{-1}(\mathbf{C}^T - \mathbf{K}^T\mathbf{D}_u^T) \\ * & -\gamma^2\mathbf{I}_l & \mathbf{L}^T\mathbf{D}_u^T \\ * & * & -\mathbf{I}_m \end{bmatrix} < 0 \tag{184}$$

With notation

$$\mathbf{Q}^{-1} = \mathbf{X} > 0, \qquad \mathbf{KQ}^{-1} = \mathbf{Y} \tag{185}$$

(184) implies (176). This concludes the proof. ■

6.3 Estimator system matrix design

Theorem 8 *For given scalar $\gamma > 0$, $\gamma \in R$, and matrices $\mathbf{Q} = \mathbf{Q}^T > 0$, $\mathbf{Q} \in R^{n\times n}$, $\mathbf{K} \in R^{r\times n}$, $\mathbf{L} \in R^{r\times l}$ estimators (7) – (9) associated with the system (1), (2) are stable if there exist symmetric positive definite matrices $\mathbf{R} \in R^{n\times n}$, $\mathbf{S} \in R^{l\times l}$, and matrices $\mathbf{Z} \in R^{n\times m}$, $\mathbf{V} \in R^{l\times l}$, $\mathbf{W} \in R^{l\times m}$ such that*

$$\mathbf{R} = \mathbf{R}^T > 0 \tag{186}$$

$$\mathbf{S} = \mathbf{S}^T > 0 \tag{187}$$

$$\begin{bmatrix} \mathbf{\Phi}_{22} & \mathbf{RB}_f - \mathbf{ZD}_f - (\mathbf{WC})^T & \mathbf{0} & \mathbf{0} & (\mathbf{D}_u\mathbf{K})^T \\ * & \mathbf{\Phi}_{33} & -\mathbf{V} & \mathbf{S} & (\mathbf{D}_u\mathbf{L})^T \\ * & * & -\gamma^2\mathbf{I}_l & \mathbf{0} & \mathbf{0} \\ * & * & * & -\gamma^2\mathbf{I}_l & \mathbf{0} \\ * & * & * & * & -\mathbf{I}_m \end{bmatrix} \tag{188}$$

where

$$\mathbf{\Phi}_{22} = \mathbf{RA} - \mathbf{ZC} + \mathbf{A}^T\mathbf{R} - \mathbf{C}^T\mathbf{Z}^T, \quad \mathbf{\Phi}_{33} = \mathbf{V} - \mathbf{WD}_f + \mathbf{V}^T - \mathbf{D}_f^T\mathbf{W}^T \tag{189}$$

The estimators matrix parameters are then given as

$$\mathbf{M} = \mathbf{S}^{-1}\mathbf{V}, \quad \mathbf{N} = \mathbf{S}^{-1}\mathbf{W}, \quad \mathbf{J} = \mathbf{R}^{-1}\mathbf{Z} \tag{190}$$

Proof. Supposing that $\mathbf{q}(t) = \mathbf{0}$ then (149) is reduced as follows

$$\begin{bmatrix} \mathbf{\Phi}_{22} & \mathbf{R}(\mathbf{B}_f - \mathbf{JD}_f) - (\mathbf{SNC})^T & \mathbf{0} & \mathbf{0} & (\mathbf{D}_u\mathbf{K})^T \\ * & \mathbf{\Phi}_{33} & -\mathbf{SM} & \mathbf{S} & (\mathbf{D}_u\mathbf{L})^T \\ * & * & -\gamma^2\mathbf{I}_l & \mathbf{0} & \mathbf{0} \\ * & * & * & -\gamma^2\mathbf{I}_l & \mathbf{0} \\ * & * & * & * & -\mathbf{I}_m \end{bmatrix} < 0 \tag{191}$$

where

$$\mathbf{\Phi}_{22} = \mathbf{R}(\mathbf{A} - \mathbf{JC}) + (\mathbf{A} - \mathbf{JC})^T\mathbf{R}, \quad \mathbf{\Phi}_{33} = \mathbf{S}(\mathbf{M} - \mathbf{ND}_f) + (\mathbf{M} - \mathbf{ND}_f)^T\mathbf{S} \tag{192}$$

Thus, with notation

$$\mathbf{SM} = \mathbf{V}, \qquad \mathbf{SN} = \mathbf{W}, \qquad \mathbf{RJ} = \mathbf{Z} \tag{193}$$

(191), (192) implies (188), (189). This concludes the proof. ■

It is obvious that $\mathbf{F}_e = \mathbf{A} - \mathbf{JC}$, as well as $\mathbf{M}$ have to be stable matrices.

6.4 Illustrative example

To demonstrate algorithm properties it was assumed that the system is given by (1), (2) where

$$\mathbf{A} = \begin{bmatrix} 0 & 1 & 0 \\ 0 & 0 & 1 \\ -5 & -9 & -5 \end{bmatrix}, \quad \mathbf{B}_u = \begin{bmatrix} 1 & 3 \\ 2 & 1 \\ 1 & 5 \end{bmatrix}, \quad \mathbf{B}_f = \begin{bmatrix} 1 \\ 2 \\ 1 \end{bmatrix}, \quad \mathbf{L} = \begin{bmatrix} -1 \\ 0 \end{bmatrix}$$

$$\mathbf{C} = \begin{bmatrix} 1 & 2 & 1 \\ 1 & 1 & 0 \end{bmatrix}, \quad \mathbf{D}_u = \begin{bmatrix} 0 & 0 \\ 0 & 0 \end{bmatrix}, \quad \mathbf{D}_f = \begin{bmatrix} 0 \\ 0 \end{bmatrix}$$

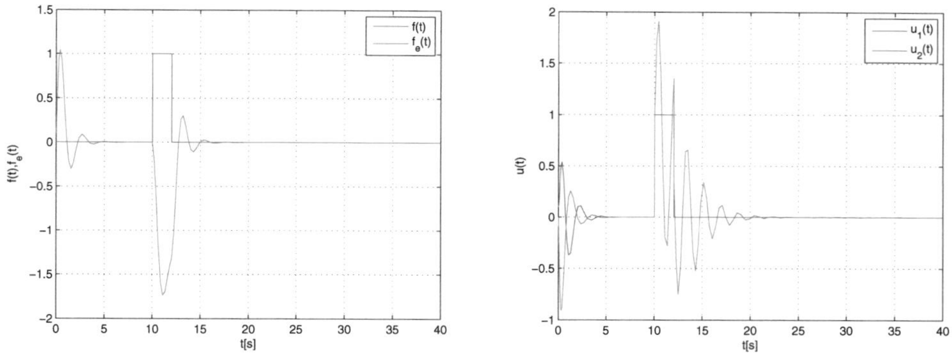

Fig. 8. The first actuator fault as well as its estimation and system input variables

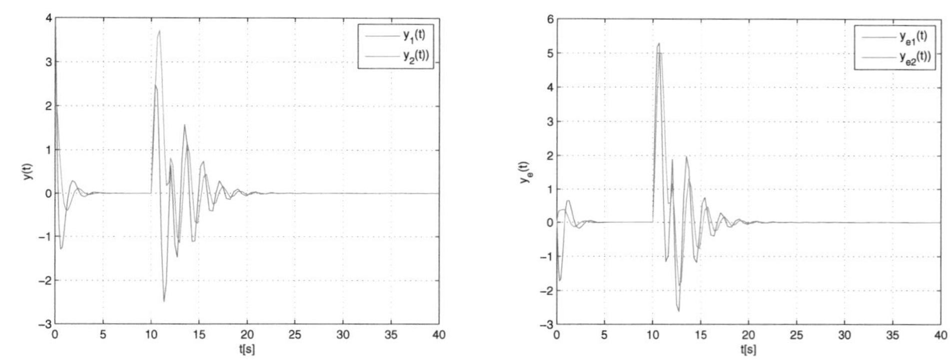

Fig. 9. System output and its estimation

Solving (175), (176) with respect to the LMI matrix variables γ, $\mathbf{X}$, and $\mathbf{Y}$ using Self-Dual-Minimization (SeDuMi) package for Matlab, the feedback gain matrix design problem was feasible with the result

$$\mathbf{X} = \begin{bmatrix} 1.7454 & -0.8739 & 0.0393 \\ -0.8739 & 1.3075 & -0.5109 \\ 0.0393 & -0.5109 & 2.0436 \end{bmatrix}, \quad \mathbf{Y} = \begin{bmatrix} 0.9591 & 1.2907 & -0.1049 \\ -0.1950 & -0.5166 & -0.4480 \end{bmatrix}, \quad \gamma = 1.8509$$

$$\mathbf{K} = \begin{bmatrix} 1.2524 & 1.7652 & 0.0436 \\ -0.0488 & -0.2624 & -0.3428 \end{bmatrix}$$

In the next step the solution to (186) – (188) using design parameters $\gamma = 1.8509$ was also feasible giving the LMI variables

$$\mathbf{V} = -1.3690, \quad \mathbf{S} = 1.1307, \quad \mathbf{W} = \begin{bmatrix} 0.9831 & 0.7989 \end{bmatrix}$$

$$\mathbf{R} = \begin{bmatrix} 1.7475 & 0.0013 & 0.0128 \\ 0.0013 & 1.4330 & 0.0709 \\ 0.0128 & 0.0709 & 0.6918 \end{bmatrix}, \quad \mathbf{Z} = \begin{bmatrix} -0.0320 & 1.0384 \\ 0.1972 & 0.1420 \\ -2.0509 & -1.1577 \end{bmatrix}$$

which gives

$$\mathbf{J} = \begin{bmatrix} 0.0035 & 0.6066 \\ 0.2857 & 0.1828 \\ -2.9938 & -1.7033 \end{bmatrix}, \quad \mathbf{N} = \begin{bmatrix} 0.8694 & 0.7066 \end{bmatrix}, \quad \mathbf{M} = -1.2108$$

Since $\mathbf{M} < 0$ it is evident that the fault estimator is stable and verifying the rest subsystem stability it can see that

$$\mathbf{A}_c = \mathbf{A} - \mathbf{B}_u\mathbf{K} = \begin{bmatrix} -1.1062 & 0.0282 & 0.9847 \\ -2.4561 & -3.2659 & 1.2555 \\ -6.0087 & -9.4430 & -3.3297 \end{bmatrix}, \quad \varrho(\mathbf{A}_c) = \{-0.7110 \quad -3.4954 \pm \mathrm{i}\,4.3387\}$$

$$\mathbf{A}_{qe} = \mathbf{A} - \mathbf{JC} = \begin{bmatrix} -0.6101 & 0.3864 & -0.0035 \\ -0.4684 & -0.7541 & 0.7143 \\ -0.3029 & -1.3092 & -2.0062 \end{bmatrix}, \quad \varrho(\mathbf{A}_{qe}) = \{-1.0000 \quad -1.1852 \pm \mathrm{i}\,0.7328\}$$

where $\varrho(\cdot)$ is eigenvalue spectrum of a real square matrix. It is evident that the designed observer-based control structure results the stable system.
The example is shown of the closed-loop system response in the autonomous mode where Fig. 8 represents the first actuator fault as well as its estimation, and the system input variables, respectively, and Fig. 9 is concerned with the system outputs and its estimation, respectively.

7. Concluding remarks

This chapter provides an introduction to the aspects of reconfigurable control design method with emphasis on the stability conditions and related system properties. Presented viewpoint has been that non-expansive system properties formulated in the H_∞ design conditions underpins the nature of dynamic and feedback properties. Sufficient conditions of asymptotic stability of systems have thus been central to this approach. Obtained closed-loop eigenvalues express the internal dynamics of the system and they are directly related to aspects of system performance as well as affected by the different types of faults. On the other hand, control structures alternation achieved under virtual sensors, or by design or re-design of an actuator fault estimation can be done robust with respect of unaccepted faults. The role and significance of another reconfiguration principles may be found e.g. in the literature (Blanke et al.,2003), (Krokavec and Filasova,2007), (Noura et al.,2009), and references therein.

8. Acknowledgment

The main results presented in this chapter was derived from works supported by VEGA, Grant Agency of Ministry of Education and Academy of Science of Slovak Republic, under Grant No. 1/0256/11. This support is very gratefully acknowledged.

9. References

[Blanke et al.,2003] Blanke, M.; Kinnaert, M.; Lunze, J. & Staroswiecki, M. (2003). *Diagnosis and Fault-tolerant Control.* Springer, ISBN 3-540-01056-4, Berlin.

[Boyd et al.,1994] Boyd, D.; El Ghaoui, L.; Peron, E. & and Balakrishnan, V. (1994). *Linear Matrix Inequalities in System and Control Theory.* SIAM Society for Industrial and Applied Mathematics, ISBN, 0-89871-334-X, Philadelphia

[Chen and Patton,1999] Chen, J. & Patton, R.J. (1999). *Robust Model-Based Fault Diagnosis for Dynamic Systems.* Kluwer Academic Publishers, ISBN 0-7923-8411-3, Norwell.

[Chen et al.,1999] Chen, J.; Patton, R.J. & Chen, Z. (1999). Active fault-tolerant flight control systems design using the linear matrix inequality method. *Transactions of the Institute of Measurement and Control*, ol. 21, No. 2, (1999), pp. 77-84, ISSN 0142-3312.

[Chiang et al.,2001] Chiang, L.H.; Russell, E.L. & Braatz, R.D. (2001). *Fault Detection and Diagnosis in Industrial Systems.* Springer, ISBN 1-85233-327-8, London.

[Ding,2008] Ding, S.X. (2008). *Model-based Fault Diagnosis Techniques: Design Schemes, Alghorithms, and Tools.* Springer, ISBN 978-3-540-76304-8, Berlin.

[Dong et al.,2009] Dong, Q.; Zhong, M. & Ding, S.X. (2009). On active fault tolerant control for a class of time-delay systems. *Preprints of 7th IFAC Symposium on Fault Detection, Supervision and Safety of Technical Processes, SAFEPROCESS 2009*, pp. 882-886, Barcelona, Spain, June 30, - July 3, 2009.

[Ducard,2009] Ducard, G.J.J. (2009). *Fault-tolerant Flight Control and Guidance Systems. Practical Methods for Small Unmanned Aerial Vehicles.* Springer, ISBN 978-1-84882-560-4, London.

[Filasova and Krokavec,2009] Filasová, A. & Krokavec, D. (2009). LMI-supported design of residual generators based on unknown-input estimator scheme. *Preprints ot the 6th IFAC Symposium on Robust Control Design ROCOND '09*, pp. 313-319, Haifa, Israel, June 16-18, 2009,

[Gahinet et al.,1995] P. Gahinet, P.; Nemirovski, A.; Laub, A.J. & Chilali, M. (1995). *LMI Control Toolbox User's Guide,* The MathWorks, Natick.

[Herrmann et al.,2007] Herrmann, G.; Turner, M.C. & Postlethwaite, I. (2007). Linear matrix inequalities in control. *Mathematical Methods for Robust and Nonlinear Control*, Turner, M.C. and Bates, D.G. (Eds.), pp. 123-142, Springer, ISBN 978-1-84800-024-7, Berlin.

[Jiang,2005] Jiang, J. (2005). Fault-tolerant Control Systems. An Introductory Overview. *Acta Automatica Sinica*, Vol. 31, No. 1, (2005), pp. 161-174, ISSN 0254-4156.

[Krokavec and Filasova,2007] Krokavec, D. & Filasová, A. (2007). *Dynamic Systems Diagnosis.* Elfa, ISBN 978-80-8086-060-8, Košice (in Slovak).

[Krokavec and Filasova,2008] Krokavec, D. & Filasová, A. (2008) Diagnostics and reconfiguration of control systems. *Advances in Electrical and Electronic Engineering*, Vol. 7, No. 1-2, (2008), pp. 15-20, ISSN 1336-1376.

[Krokavec and Filasova,2008] Krokavec D. & A. Filasová, A. (2008). Performance of reconfiguration structures based on the constrained control. *Proceedigs of the 17th IFAC World Congress 2008*, pp. 1243-1248, Seoul, Korea, July 06-11, 2008.

[Krokavec and Filasova,2009] Krokavec, D. & Filasová, A. (2009). Control reconfiguration based on the constrained LQ control algorithms. *Preprints of 7th IFAC Symposium on Fault Detection, Supervision and Safety of Technical Processes SAFEPROCESS 2009*, pp. 686-691, Barcelona, Spain, June 30, - July 3, 2009.

[Liao et al.,2002] Liao, F.; Wang, J.L. & Yang, G.H. (2002). Reliable robust flight tracking control. An LMI approach. *IEEE Transactions on Control Systems Technology*, Vol. 10, No. 1, (2002), pp. 76-89, ISSN 1063-6536.

[Nesterov and Nemirovsky,1994] Nesterov, Y.; & Nemirovsky, A. (1994). *Interior Point Polynomial Methods in Convex Programming. Theory and Applications,* SIAM, ISBN 0-89871-319-6, Philadelphia.

[Nobrega et al.,2000] Nobrega, E.G.; Abdalla, M.O. & Grigoriadis, K.M. (2000). LMI-based filter design for fault detection and isolation. *Proceedings of the 39th IEEE Conference Decision and Control 2000*, Vol. 5, pp. 4329-4334, Sydney, Australia, December 12-15, 2000.

[Noura et al.,2009] Noura, H.; Theilliol, D.; Ponsart, J.C. & Chamseddine, A. (2009). *Fault-tolerant Control Systems. Design and Practical Applications.* Springer, ISBN 978-1-84882-652-6, Berlin.

[Patton,1997] Patton. R.J. (1997). Fault-tolerant control. The 1997 situation. *Proceedings of the 3rd IFAC Symposium on Fault Detection, Supervision and Safety for Technical Processes SAFEPROCESSŠ97*, Vol. 2, pp. 1033–1054, Hull, England, August 26-28, 1997.

[Peaucelle et al.,2002] Peaucelle, D.; Henrion, D.; Labit, Y. & Taitz, K. (2002). *User's Guide for Sedumi Interface 1.04.* LAAS-CNRS, Toulouse.

[Simani et al.,2003] Simani, S.; Fantuzzi, C. & Patton, R.J. (2003). *Model-based Fault Diagnosis in Dynamic Systems Using Identification Techniques.* Springer, ISBN 1-85233-685-4, London.

[Skelton et al.,1998] Skelton, R.E., Iwasaki, T. & Grigoriadis, K. (1998) A Unified Algebraic Approach to Linear Control Design. Taylor & Francis, ISBN 0-7484-0592-5, London.

[Staroswiecki,2005] Staroswiecki, M. (2005). Fault tolerant control. The pseudo-inverse method revisited. *Proceedings of the 16th IFAC World Congress 2005*, Prag, Czech Repulic, July 4-8, 2005.

[Theilliol et al.2008] Theilliol, D.; Join, C. & Zhang, Y.M. (2008). Actuator fault tolerant control design based on a reconfigurable reference input. *International Journal of Applied Mathematics and Computer Science*, Vol.18, No.4, (2008), pp. 553-560, ISSN 1641-876X.

[Zhang and Jiang,2008] Zhang, Y.M. & Jiang, J. (2008). Bibliographical review on reconfigurable fault-tolerant control systems. *Annual Reviews in Control*, Vol. 32, (2008), pp. 229Ű-252, ISSN 1367-5788.

[Zhou and Ren,2001] Zhou, K.M. & Ren, Z. (2001). A new controller architecture for high performance, robust and fault tolerant control. *IEEE Transactions on Automatic Control*, Vol. 40, No. 10, (2001), pp. 1613-1618, ISSN 0018-9286.

15

Robust Model Predictive Control for Time Delayed Systems with Optimizing Targets and Zone Control

Alejandro H. González[1] and Darci Odloak[2]
[1]*Institute of Technological Development for the Chemical Industry (INTEC), CONICET - Universidad Nacional del Litoral (U.N.L.). Güemes 3450, (3000) Santa Fe,*
[2]*Department of Chemical Engineering, University of São Paulo, Av. Prof. Luciano Gualberto, trv 3 380, 61548 São Paulo,*
[1]*Argentina*
[2]*Brazil*

1. Introduction

Model Predictive Control (MPC) is frequently implemented as one of the layers of a control structure where a Real Time Optimization (RTO) algorithm - laying in an upper layer of this structure - defines optimal targets for some of the inputs and/or outputs (Kassmann et al., 2000). The main scope is to reach the most profitable operation of the process system while preserving safety and product specification constraints. The model predictive controller is expected to drive the plant to the optimal operating point, while minimizing the dynamic error along the input and output paths. Since in the control structure considered here the model predictive controller is designed to track the optimal targets, it is expected that for nonlinear process systems, the linear model included in the controller will become uncertain as we move from the design condition to the optimal condition. The robust MPC presented in this chapter explicitly accounts for model uncertainty of open loop stable systems, where a different model corresponds to each operating point of the process system. In this way, even in the presence of model uncertainty, the controller is capable of maintaining all outputs within feasible zones, while reaching the desired optimal targets. In several other process systems, the aim of the MPC layer is not to guide all the controlled variables to optimal targets, but only to maintain them inside appropriate ranges or zones. This strategy is designated as zone control (Maciejowski, 2002). The zone control may be adopted in some systems, where there are highly correlated outputs to be controlled, and there are not enough inputs to control all the outputs. Another class of zone control problems relates to using the surge capacity of tanks to smooth out the operation of a process unit. In this case, it is desired to let the level of the tank to float between limits, as necessary, to buffer disturbances between sections of a plant. The paper by Qin and Badgwell (2003), which surveys the existing industrial MPC technology, describes a variety of industrial controllers and mention that they always provide a zone control option. Other example of zone control can be found in Zanin et al, (2002), where the authors exemplify the application of this strategy in the real time optimization of a FCC system. Although this strategy shows to have an acceptable performance, stability is not usually proved, even when an infinite horizon is used, since the control system keeps switching from one controller to another throughout the continuous operation of the process.

There are several research works that treat the problem of how to obtain a stable MPC with fixed output set points. Although stability of the closed loop is commonly achieved by means of an infinite prediction horizon, the problem of how to eliminate output steady state offset when a supervisory layer produces optimal economic set points, and how to explicitly incorporate the model uncertainty into the control problem formulation for this case, remain an open issue. For the nominal model case, Rawlings (2000), Pannochia and Rawlings (2003), Muske and Badgwell (2002), show how to include disturbance models in order to assure that the inputs and states are led to the desired values without offset. Muske and Badgwell (2002) and Pannochia and Rawlings (2003) develop rank conditions to assure the detectability of the augmented model.

For the uncertain system, Odloak (2004) develops a robust MPC for the multi-plant uncertainty (that is, for a finite set of possible models) that uses a non-increasing cost constraint (Badgwell, 1997). In this strategy, the MPC cost function to be minimized is computed using a nominal model, but the non-increasing cost constraint is settled for each of the models belonging to the set. The stability is then achieved by means of the recursive feasibility of the optimization problem, instead of the optimality. On the other hand, there exist some recent MPC formulations that are based on the existence of a control Lyapunov function (CLF), which is independent of the control cost function. Although the construction of the CFL may not be a trivial task, these formulations also allow the explicit characterization of the stability region subject to constraints and they do not need an infinite output horizon. Mashkar et al. (2006) explore this approach for the control of nominal nonlinear systems, and Mashkar (2006) extends the approach for the case of model uncertainty and control actuator fault. More recently, González et al. (2009) extended the infinite horizon approach to stabilize the closed loop with the MPC controller for the case of multi-model uncertainty and optimizing targets. They developed a robust MPC by adapting the non-increasing cost constraint strategy to the case of zone control of the outputs and it is desirable to guide some of the manipulated inputs to the targets given by a supervisory stationary optimization stage, while maintaining the controlled output in their corresponding zones, taking into account a finite set of possible models. This problem, that seems to interchange an output tracking by an input-tracking formulation, is not trivial, since once the output lies outside the corresponding zone (because of a disturbance, or a change in the output zones), the priority of the controller is again to control the outputs, even if this implies that the input must be settled apart from its targets.

Since in many process systems, mainly from the chemical and petrochemical industries, the process model shows significant time delays, the main contribution of this chapter is to extend the approach of González et al. (2009) to the case of input delayed multi-model systems by introducing minor modifications in the state space model, in such a way that the structure of the control algorithm is preserved. Simulation of a process system of the oil refining industry illustrates the performance of the proposed strategy.

2. System representation

Consider a system with nu inputs and ny outputs, and assume for simplicity that the poles relating any input u_j to any output y_i are non-repeated. To account for the implementation of an intuitive MPC formulation, an output prediction oriented model (OPOM) originally presented in Odloak (2004) is adapted here to the case of time delayed systems. Let us designate $\theta_{i,j}$ the time delay between input u_j and output y_i, and define $p > \max_{i,j} \theta_{i,j}$. Then, the state space model considered here is defined as follows:

$$\begin{aligned} x(k+1) &= Ax(k) + B\Delta u(k) \\ y(k) &= Cx(k) \end{aligned} \tag{1}$$

where $x(k) = \begin{bmatrix} y(k)^T & y(k+1)^T & \cdots & y(k+p)^T & x^s(k)^T & x^d(k)^T \end{bmatrix}^T$

$$A = \begin{bmatrix} 0 & I_{ny} & 0 & \cdots & 0 & 0 & 0 \\ 0 & 0 & I_{ny} & \cdots & 0 & 0 & 0 \\ \vdots & \vdots & & \ddots & \vdots & \vdots & \vdots \\ 0 & 0 & 0 & \cdots & I_{ny} & 0 & 0 \\ 0 & 0 & 0 & \cdots & 0 & I_{ny} & \Psi((p+1)T) \\ 0 & 0 & 0 & \cdots & 0 & I_{ny} & 0 \\ 0 & 0 & 0 & \cdots & 0 & 0 & F \end{bmatrix}, \quad B = \begin{bmatrix} S_1 \\ S_2 \\ \vdots \\ S_p \\ S_{p+1} \\ B^s \\ B^d \end{bmatrix} \tag{2}$$

$$C = \begin{bmatrix} I_{ny} & 0 & \cdots & 0 \end{bmatrix}$$

$x^s \in \Re^{ny}$, $x^d \in C^{nd}$, $F \in C^{nd \times nd}$, $\mathbf{\Psi} \in \Re^{ny \times nd}$, $I_{ny} = diag(\begin{bmatrix} 1 & \cdots & 1 \end{bmatrix}) \in \Re^{ny \times ny}$.

The advantage of using the structure of the transition matrix A is that the state vector is divided into components that are associated to the system modes. In the state equation (1), the state components x^s correspond to the (predicted) output steady state, which are in addition the integrating modes of the system (the integrating modes are induced by the incremental form of the inputs), and the components x^d correspond to the stable modes of the system. Naturally, when the system approaches steady state these last components tend to zero. For the case of non-repeated pole, F is a diagonal matrix with components of the form $e^{r_i T}$ where r_i is a pole of the system and T is the sampling period. It is assumed that the system has nd stable poles and B^s is the gain matrix of the system. The upper left block of matrix A is included to account for the time delay of the system. $S_1, \ldots, S_{p+1}$ are the step response coefficients of the system. Matrix Ψ, which appears in the extended state matrix, is defined as follows

$$\Psi(t) = \begin{bmatrix} \phi_1(t) & 0 & \cdots & 0 \\ 0 & \phi_2(t) & \cdots & 0 \\ \vdots & \vdots & \ddots & \vdots \\ 0 & 0 & \cdots & \phi_{ny}(t) \end{bmatrix}$$

where

$$\phi_i(t) = \begin{bmatrix} e^{r_{i,1,1}(t-\theta_{i,1})} & \cdots & e^{r_{i,1,na}(t-\theta_{i,1})} & \cdots & e^{r_{i,nu,1}(t-\theta_{i,nu})} & \cdots & e^{r_{i,nu,na}(t-\theta_{i,nu})} \end{bmatrix},$$

$r_{i,j,k}$, with k=1,…,na, are the poles of the transfer function that relates input u_j and output y_i and na is the order of this transfer function. It is assumed that na is the same for any pair (u_j, y_i). The time delay affects the dimension of the state matrix A through parameter p and the components of matrix Ψ. Input matrix B is also affected by the value of the time delay as the step response coefficients S_n will be equal to zero for any n smaller than the time delay.

2.1 Model uncertainty

With the model structure presented in (1), model uncertainty is related to uncertainty in matrices F, B^s, B^d and the matrix of time delays θ. The uncertainty in these parameters also reflects in the uncertainty of the step response coefficients, which appear in (2). There are several practical ways to represent model uncertainty in model predictive control. One simple way to represent model uncertainty is to consider the multi-plant system (Badgwell, 1997), where we have a discrete set Ω of plants, and the real plant is unknown, but it is assumed to be one of the components of this set. With this representation of model uncertainty, we can define the set of possible plants as $\Omega = \{\Theta_1, ..., \Theta_L\}$ where each Θ_n corresponds to a particular plant: $\Theta_n = \left(F, B^s, B^d, \theta\right)_n$, $n = 1, ..., L$.
Also, let us assume that the true plant, which lies within the set Ω is designated as θ_T and there is a most likely plant that also lies in Ω and is designated as Θ_N. In addition, it is assumed that the current estimated state corresponds to the true plant.
Badgwell (1997) developed a robust linear quadratic regulator for stable systems with the multi-plant uncertainty. Later, Odloak (2004) extended the method of Badgwell to the output tracking of stable systems considering the same kind of model uncertainty. These strategies include a new constraint corresponding to each of the models lying in Ω, that prevents an increase in the true plant cost function at successive time steps. More recently, González and Odloak (2009) presented an extension of the method by combining the approach presented in Odloak (2004) with the idea of including the output set point as a new restricted optimization variable to develop a robust MPC for systems where the control objective is to maintain the outputs into their corresponding feasible zone, while reaching the desired optimal input target given by the supervisory stationary optimization. In this work the controller proposed by González et al. (2009) is extended to the case of uncertain systems with time delays.

2.2. System steady state

As was already said, one of the advantages of the model defined in (1) and (2) is that the state component $x^s(k)$ represents the predicted output at steady state, and furthermore this component concentrates the integrating modes of the system. Observe that for the model defined in (1) and (2), if $\Delta u(k+j) = 0$ for $j \geq 0$, then the future states can be computed as follows

$$x(k+j) = A^j x(k)$$

Assuming that F has all the eigenvalues inside the unit circle (i.e. the system is open loop stable), it is easy to show that

$$\lim_{j\to\infty} A^j = \begin{bmatrix} 0 & 0 & \cdots & I_{ny} & 0 \\ 0 & 0 & \cdots & I_{ny} & 0 \\ \vdots & \vdots & \cdots & \vdots & \vdots \\ 0 & 0 & \cdots & I_{ny} & 0 \\ 0 & 0 & \cdots & 0 & 0 \end{bmatrix}$$

Then, it becomes clear that $\lim_{j\to\infty} x(k+j) = \begin{bmatrix} x^s(k)^T & \cdots & x^s(k)^T & 0 \end{bmatrix}^T$ and consequently $\lim_{j\to\infty} y(k+j) = C \lim_{j\to\infty} x(k+j) = x^s(k)$. Therefore, $x^s(k)$ can be interpreted as the prediction of the output at steady state. The state component $x^s(k)$ is assumed to be known or estimated through a stable state observer. A stable observer for the model defined in (1) and (2) is given by

$$\hat{x}(k+1|k+1) = \hat{x}(k+1|k) + K\left(y(k+1) - C\hat{x}(k+1|k)\right)$$

where $K = \begin{bmatrix} I_{ny} & \cdots & I_{ny} & 0 \end{bmatrix}^T$ is the observer gain, and

$$\hat{x}(k+1|k+1) = \hat{x}(k+1|k) + K\left(y(k+1) - C\hat{x}(k+1|k)\right)$$

$$\hat{x}(k+1|k+1) = \left(I - KC\right)A\hat{x}(k|k) + \left(I - KC\right)B\Delta u(k)$$

For open loop stable systems this is a stable observer as matrix $\left(I - KC\right)A$ has the eigenvalues of F and the remaining eigenvalues are equal to zero.

3. Control structure

In this work, we consider the control structure shown in Figure 1. In this structure, the economic optimization stage is dedicated to the calculation of the (stationary) desired target, $u_{des,k}$, for the input manipulated variables. This stage may be based on a rigorous stationary model and takes into account the process measurements and some economic parameters. In addition, this stage works with a smaller frequency than the low-level control stage, which allows a separation between the two stages. In the zone control framework the low-level control stage, given by the MPC controller, is devoted to guide the manipulated input from the current stationary value u_{ss} to the desired value given by the supervisory economic stage, $u_{des,k}$, while keeping the outputs within specified zones. In general, the target $u_{des,k}$ will vary whenever the plant operation or the economic parameters change. If it is assumed that the system is currently at a stationary value given by (u_{ss}, y_{ss}), the desired target $u_{des,k}$ should satisfy not only the input constraints

$$u_{\min} \le u_{des,k} \le u_{\max}$$

but also the output zone condition

$$y_{\min} \le B^s\left(\Theta_n\right)\left(u_{des,k} - u_{ss}\right) + \hat{x}_n^s\left(k\right) \le y_{\max}\,,\ n = 1,\ldots,L \tag{3}$$

where $u_{\min}$ and $u_{\max}$ represent the lower and upper bounds of the input, $y_{\min}$ and $y_{\max}$ represent the lower and upper limits of the output, $B^s(\Theta_n)$ is the gain corresponding to a given model Θ_n, and $\hat{x}_n^s(k)$ is the estimated steady-state values of the output corresponding to model Θ_n. Note that in the control structure depicted in Figure 1, as the model structure adopted here has integral action, the estimation of component $x_n^s(k)$ tends to the measured output at steady state for all the models lying in Ω, which means that $\hat{x}_n^s(k) = y_{ss}$ if the system is at steady state (See González and Odloak 2009 for details). Taking into account this fact, equation (3) can be rewritten as

$$y_{\min} \le B^s(\Theta_n) u_{des,k} + d_{n,ss} \le y_{\max} \qquad n = 1, \cdots, L, \tag{4}$$

where $d_{n,ss} = \hat{x}_n^s(k) - B^s(\Theta_n) u_{ss} = y_{ss} - B^s(\Theta_n) u_{ss}$ is the output bias based on the comparison between the current actual output at steady state and the current predicted output at steady state for each model. In other words, $B^s(\Theta_n) u_{des,k} + d_{n,ss}$ can be interpreted as the corrected output steady state. Note that, since $u_{ss} = \sum_{j=0}^{\bar{k}} \Delta u(j)$, for a large $\bar{k}$, the term $B^s(\Theta_n) u_{ss}$ represents the output prediction based only on the past inputs.

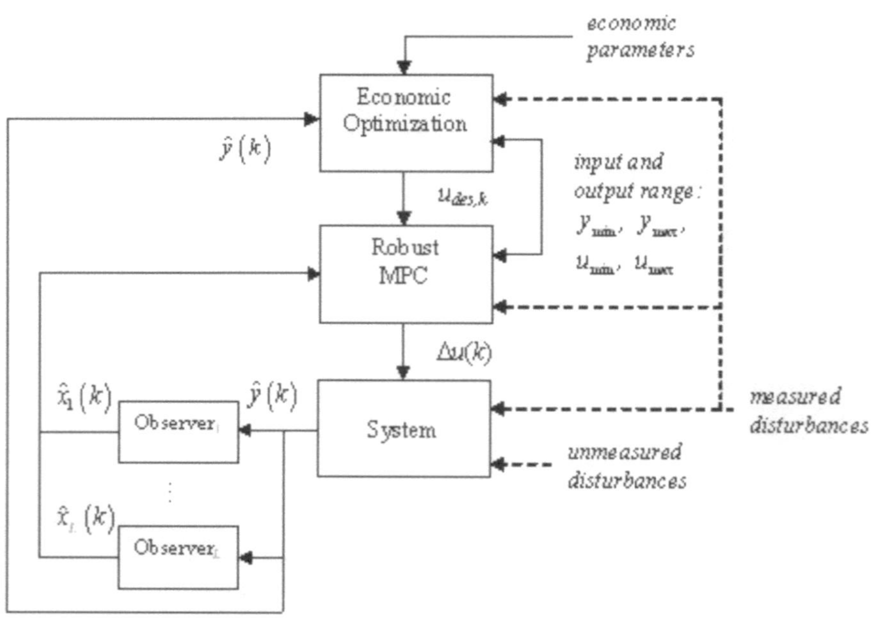

Fig. 1. Control structure.

Based on the later concepts, it is possible to define two input feasible sets for the stationary desired target $u_{des,k}$. The first one is the global input feasible set $\vartheta_o = \{u : u_{\min} \le u \le u_{\max}\}$, which represents a box-type set. In addition, it is possible to define the more restricted input feasible set ϑ_u, which is computed taking into account both, the input constraints and the output limits:

$$\vartheta_u = \left\{ u : u_{\min} \le u \le u_{\max} \text{ and } y_{\min} \le B^s(\Theta_n)u - \overbrace{B^s(\Theta_n)u_{ss} + y_{ss}}^{d_{n,ss}} \le y_{\max},\ n = 1,\cdots,L \right\}. \quad (5)$$

This set, which depends on the current stationary point given by (u_{ss}, y_{ss}), is the intersection of several sets, each one corresponding to a model lying in set Ω. When the output zones are narrow, the restricted input feasible set is smaller than the global feasible set, defined solely by the input constraints. An intuitive diagram of the input feasible set is shown in Figure 4, where three models are used to represent the uncertainty set. In the following sections it will be shown that the proposed controller remains stable and feasible even when the desired input target $u_{des,k}$ is outside the set ϑ_u, or the set ϑ_u itself is null.

4. Nominal MPC with zone control and input target

One way to handle the zone control strategy, that is, to maintain the controlled output inside its corresponding range, is by means of an appropriate choice of the output error penalization in the conventional MPC cost function. In this case the output weight is made equal to zero when the system output is inside the range, and the output weight is different from zero if the output prediction is violating any of the constraints, so that the output variable is strictly controlled only if it is outside the feasible range. In this way, the closed loop is guided to a feasible steady state. In Zanin et al. (2002), an algorithm assigns three possible values to the output set points used in the MPC controller: the upper bound of the output feasible range if the predicted output is larger than the upper bound; the lower bound of the output feasible range if the predicted output is smaller than this lower bound; and the predicted output itself, if the predicted output is inside the feasible range. However, a rigorous analysis of the stability of this strategy is not possible even when using an infinite output horizon. González et al. (2006) describe a stable MPC based on the incremental model defined in (1) and (2), that takes into account a stationary optimization of the plant operation. The controller was designed specifically for a heat exchanger network with a number of degrees of freedom larger than zero. In that work, the mismatch between the stationary and the dynamic model was treated by means of an appropriate choice of the weighting matrices in the control cost. However, stability and offset elimination was assured only when the model was perfect.
Based on the work of González et al (2006), we consider the following nominal cost function:

$$\begin{aligned} V_k = \sum_{j=0}^{\infty} \Big\{ & \left(y(k+j\,|\,k) - y_{sp,k}\right)^T Q_y \left(y(k+j\,|\,k) - y_{sp,k}\right) + \left(u(k+j\,|\,k) - u_{des,k}\right)^T Q_u \\ & \left(u(k+j\,|\,k) - u_{des,k}\right) \Big\} + \sum_{j=0}^{m-1} \Delta u(k+j\,|\,k)^T R\, \Delta u(k+j\,/\,k) \end{aligned} \quad (6)$$

where $\Delta u(k+j\,|\,k)$ is the control move computed at time k to be applied at time $k+j$, m is the control or input horizon, Q_y, Q_u, R are positive weighting matrices of appropriate dimension, $y_{sp,k}$ and $u_{des,k}$ are the output and input targets, respectively. The output target $y_{sp,k}$ becomes a computed set point when the output has no optimizing target and consequently the output is controlled by zone. This cost explicitly incorporates an input deviation penalty that tries to accommodate the system at an optimal economic stationary point.

In the case of systems without time delay the term corresponding to the infinite output error in the cost V_k is divided in two parts: the first goes from the current time k to the end of the control horizon, $k+m$-1; while the second one goes from time $k+m$ to infinity. This is so because beyond the control horizon no control actions are implemented and so, considering only the state at time $k+m$, the infinite series can be reduced to a single terminal cost. In the case of time delayed systems, however, the horizon beyond which the entire output evolution can be predicted by a terminal cost is given by $k+p$. As a result, the cost defined in (6) can be developed as follows

$$\begin{aligned} V_k =& \sum_{j=0}^{p}\left\{\left(y(k+j\,|\,k)-y_{sp,k}\right)^T Q_y\left(y(k+j\,|\,k)-y_{sp,k}\right)\right. \\ &+\sum_{j=0}^{\infty}\left\{\left(y(k+p+j\,|\,k)-y_{sp,k}\right)^T Q_y\left(y(k+p+j\,|\,k)-y_{sp,k}\right)\right. \\ &\qquad \left.\left(u(k+j/k)-u_{des,k}\right)^T Q_u\left(u(k+j/k)-u_{des,k}\right)\right\} \\ &+\sum_{j=0}^{m-1}\Delta u(k+j/k)^T R\,\Delta u(k+j/k) \end{aligned} \tag{7}$$

The first term on the right hand side of (7) can be developed as follows

$$V_{k,1} = \sum_{j=0}^{p}\left\{\left(y(k+j\,|\,k)-y_{sp,k}\right)^T Q_y\left(y(k+j\,|\,k)-y_{sp,k}\right)\right.$$

$$V_{k,1} = \left(\tilde{y}_k - \tilde{I}_y y_{sp,k}\right)^T \tilde{Q}_y\left(\tilde{y}_k - \tilde{I}_y y_{sp,k}\right)$$

where

$$\tilde{y}_k = N_x x(k) + \tilde{S}\Delta u_k \tag{8}$$

$$\tilde{y}_k = \begin{bmatrix} y(k\,|\,k) \\ y(k+1\,|\,k) \\ \vdots \\ y(k+p\,|\,k) \end{bmatrix}, \quad N_x = \begin{bmatrix} I_{(p+1)ny} & 0 \end{bmatrix} \in \Re^{(p+1)ny\times nx}; \; \tilde{S} = \begin{bmatrix} 0 & 0 & \cdots & 0 \\ S_1 & 0 & \cdots & 0 \\ S_2 & S_1 & \cdots & 0 \\ \vdots & \vdots & & \vdots \\ S_p & S_{p-1} & \cdots & S_{p-m+1} \end{bmatrix},$$

$$\tilde{I}_y = \begin{bmatrix} I_{ny} & \cdots & I_{ny} \end{bmatrix}^T, \; \tilde{I}_y \in \Re^{(p+1)ny\times ny}$$

$$\tilde{Q}_y = diag\left(\underbrace{Q_y \quad \cdots \quad Q_y}_{p+1}\right), \quad nx = (p+1)ny + ny + nd$$

Consequently, considering (8), the term $V_{k,1}$ can be written as follows

$$V_{k,1}=\left[N_x x(k)+\tilde{S}\Delta u_k-\tilde{I}_y y_{sp,k}\right]^T\tilde{Q}_y\left[N_x x(k)+\tilde{S}\Delta u_k-\tilde{I}_y y_{sp,k}\right] \quad (9)$$

The term corresponding to the infinite horizon error on the system output in (7) can be written as follows

$$V_{k,2}=\sum_{j=1}^{\infty}\left(y(k+p+j\,|\,k)-y_{sp,k}\right)^T Q_y\left(y(k+p+j\,|\,k)-y_{sp,k}\right)$$

$$V_{k,2}=\sum_{j=1}^{\infty}\left(x^s(k+m\,|\,k)+\Psi(p+j-m)x^d(k+m\,|\,k)-y_{sp,k}\right)^T Q_y \left(x^s(k+m\,|\,k)+\Psi(p+j-m)x^d(k+m\,|\,k)-y_{sp,k}\right) \quad (10)$$

where, $x^s(k+m\,|\,k)=x^s(k)+\tilde{B}^s\Delta u_k$, $\tilde{B}^s=\left[\underbrace{B^s \ \cdots \ B^s}_{m}\right]$,

$$\Delta u_k=\left[\Delta u(k/k)^T \ \cdots \ \Delta u(k+m-1/k)^T\right]^T\in\Re^{m.nu}$$

Also,

$$x^d(k+m\,|\,k)=F^m x^d(k)+\tilde{B}^d\Delta u_k,\ \tilde{B}^d=\left[F^{m-1}B^d \quad F^{m-2}B^d \quad \cdots \quad B^d\right]$$

$$\Psi(p+j-m)=\Psi(p-m)F^j \quad (11)$$

In order to force $V_{k,2}$ to be bounded, we include the following constraint in the control problem

$$x^s(k+m\,|\,k)-y_{sp,k}=0 \ \text{ or } \ x^s(k)+\tilde{B}^s\Delta u_k-y_{sp,k}=0$$

With the above equation and (11), Eq. (10) becomes

$$V_{k,2}=\sum_{j=1}^{\infty}\left(\Psi(p-m)F^j x^d(k+m\,|\,k)\right)^T Q_y\left(\Psi(p-m)F^j x^d(k+m\,|\,k)\right)$$

$$V_{k,2}=\left(F^m x^d(k)+\tilde{B}^d\Delta u_k\right)^T Q_d\left(F^m x^d(k)+\tilde{B}^d\Delta u_k\right)$$

where

$$Q_d=\sum_{j=1}^{\infty}\left(\Psi(p-m)F^j\right)^T Q_y\left(\Psi(p-m)F^j\right)$$

Finally, the infinite term corresponding to the error on the input along the infinite horizon in (7) can be written as follows

$$V_{k,3}=\sum_{j=1}^{\infty}\left(u(k+j\,|\,k)-u_{des,k}\right)^T Q_u\left(u(k+j\,|\,k)-u_{des,k}\right) \quad (12)$$

Then, it is clear that in order to force (12) to be bounded one needs the inclusion of the following constraint

$$u(k+m\,|\,k)-u_{des,k}=0$$

or

$$u(k-1)+\tilde{I}_u^T \Delta u_k - u_{des,k}=0 \qquad (13)$$

where $\tilde{I}_u^T = \left[\underbrace{I_{nu} \quad \cdots \quad I_{nu}}_{m}\right]$

Then, assuming that (13) is satisfied, (12) can be written as follows

$$V_{k,3}=\left(\tilde{I}_u u(k-1)+M\Delta u_k-\tilde{I}_u u_{des,k}\right)^T \tilde{Q}_u \left(\tilde{I}_u u(k-1)+M\Delta u_k-\tilde{I}_u u_{des,k}\right)$$

where $M=\begin{bmatrix} I_{nu} & 0 & \cdots & 0 \\ I_{nu} & I_{nu} & \cdots & 0 \\ \vdots & \vdots & \ddots & \vdots \\ I_{nu} & I_{nu} & \cdots & I_{nu} \end{bmatrix}$; $\tilde{Q}_u = diag\left(\underbrace{Q_u \quad \cdots \quad Q_u}_{m}\right)$

Now, taking into account the proposed terminal constraints, the control cost defined in (7) can be written as follows

$$\begin{aligned} V_k &= \left[N_x x(k)+\tilde{S}\Delta u_k-\tilde{I}_y y_{sp,k}\right]^T \tilde{Q}_y \left[N_x x(k)+\tilde{S}\Delta u_k-\tilde{I}_y y_{sp,k}\right] \\ &+\left(F^m x^d(k)+\tilde{B}^d \Delta u_k\right)^T Q_d \left(F^m x^d(k)+\tilde{B}^d \Delta u_k\right) \\ &+\left(\tilde{I}_u u(k-1)+M\Delta u_k-\tilde{I}_u u_{des,k}\right)^T \tilde{Q}_u \left(\tilde{I}_u u(k-1)+M\Delta u_k-\tilde{I}_u u_{des,k}\right)+\Delta u_k^T \tilde{R}\Delta u_k . \end{aligned}$$

To formulate the IHMPC with zone control and input target for the time delayed nominal system, it is convenient to consider the output set point as an additional decision variable of the control problem and the controller results from the solution to the following optimization problem:

$$\min_{\Delta u_k, y_{sp,k}} V_k = \Delta u_k^T H \Delta u_k + 2c_f^T \Delta u_k$$

subject to

$$u(k-1)+\tilde{I}_u^T \Delta u_k - u_{des,k}=0 \qquad (14)$$

$$x^s(k)+\tilde{B}^s \Delta u_k - y_{sp,k}=0 \qquad (15)$$

$$y_{\min} \le y_{sp,k} \le y_{\max} \qquad (16)$$

$$-\Delta u_{\max} \le \Delta u(k+j\,|\,k) \le \Delta u_{\max} \qquad j=0,1,\cdots,m-1$$

$$u_{\min} \le u(k-1)+\sum_{i=0}^{j} \Delta u(k+i\,|\,k) \le u_{\max}; \qquad j=0,1,\cdots,m-1$$

where

$$H = \tilde{S}^T \tilde{Q}_y \tilde{S} + \tilde{B}^{dT} Q_d \tilde{B}^d + M^T \tilde{Q}_u M + \tilde{R}$$

$$c_f^T = x(k)^T N_x^{\ T} \tilde{Q}_y \tilde{S} + x^d(k)^T (F^m)^T Q_d \tilde{B}^d + \left(u(k-1) - u_{des}\right)^T \tilde{I}_u^T \tilde{Q}_u M$$

Constraints (14) and (15) are terminal constraints, and they mean that both, the input and the integrating component of the output errors will be null at the end of the control horizon m. Constraint (16), on the other hand, forces the new decision variable $y_{sp,k}$ to be inside the zone given by $y_{\min}$ and $y_{\max}$. So, as $y_{sp,k}$ is a set point variable, constraint (16) means that the effective output set point of the proposed controller is now the complete feasible zone. Notice that if the output bounds are settled so that the upper bound equals the lower bound, then the problem becomes the traditional set point tracking problem.

4.1 Enlarging the feasible region

The set of constraints added to the optimization problem in the last section may produce a severe reduction in the feasible region of the resulting controller. Specifically, since the input increments are usually bounded, the terminal constraints frequently result in infeasible problems, which means that it is not possible for the controller to achieve the constraints in m time steps, given that m is frequently small to reduce the computational cost. A possible solution to this problem is to incorporate slack variables in the terminal constraints. So, assuming that the slack variables are unconstrained, it is possible to guarantee that the control problem will be feasible. Besides, these slack variables must be penalized in the cost function with large weights to assure the constraint violation will be minimized by the control actions. Thus, the cost function can be written as follows

$$\begin{aligned} V_k = & \sum_{j=0}^{p} \left(y(k+j\,|\,k) - y_{sp,k} - \delta_{y,k}\right)^T Q_y \left(y(k+j\,|\,k) - y_{sp,k} - \delta_{y,k}\right) \\ & + \sum_{j=1}^{\infty} \left(y(k+p+j\,|\,k) - y_{sp,k} - \delta_{y,k}\right)^T Q_y \left(y(k+p+j\,|\,k) - y_{sp,k} - \delta_{y,k}\right) + \\ & + \sum_{j=0}^{m-1} \left(u(k+j\,|\,k) - u_{des,k} - \delta_{u,k}\right)^T Q_u \left(u(k+j\,|\,k) - u_{des,k} - \delta_{u,k}\right) \\ & + \sum_{j=0}^{\infty} \left(u(k+m+j\,|\,k) - u_{des,k} - \delta_{u,k}\right)^T Q_u \left(u(k+m+j\,|\,k) - u_{des,k} - \delta_{u,k}\right) + \\ & + \sum_{j=0}^{m-1} \Delta u(k+j\,|\,k)^T R \,\Delta u(k+j\,|\,k) + \delta_{y,k}^T S_y \delta_{y,k} + \delta_{u,k}^{\ T} S_u \delta_{u,k} \end{aligned} \tag{17}$$

where S_y, S_u are positive definite matrices of appropriate dimension and $\delta_{y,k} \in \Re^{ny}$, $\delta_{u,k} \in \Re^{nu}$ are the slack variables (new decision variables) that eliminate any infeasibility of the control problem. Following the same steps as in the controller where slacks are not considered, it can be shown that the cost defined in (17) will be bounded if the following constraints are included in the control problem:

$$x^s(k) + \tilde{B}^s \Delta u_k - y_{sp,k} - \delta_{y,k} = 0$$

$$u(k-1)+\tilde{I}_u^T \Delta u_k - u_{des,k} - \delta_{u,k} = 0 \tag{18}$$

In this case, the cost defined in (17) can be reduced to the following quadratic function

$$V_k = \begin{bmatrix} \Delta u_k^T & y_{sp,k}^T & \delta_{y,k}^T & \delta_{u,k}^T \end{bmatrix} \begin{bmatrix} H_{11} & H_{12} & H_{13} & H_{14} \\ H_{21} & H_{22} & H_{23} & 0 \\ H_{31} & H_{32} & H_{33} & 0 \\ H_{41} & 0 & 0 & H_{44} \end{bmatrix} \begin{bmatrix} \Delta u_k \\ y_{sp,k} \\ \delta_{y,k} \\ \delta_{u,k} \end{bmatrix} +$$

$$+2\begin{bmatrix} c_{f,1} & c_{f,2} & c_{f,3} & c_{f,4} \end{bmatrix} \begin{bmatrix} \Delta u_k \\ y_{sp,k} \\ \delta_{y,k} \\ \delta_{u,k} \end{bmatrix} + c$$

where

$$H_{11} = \tilde{S}^T \tilde{Q}_y \tilde{S} + (\tilde{B}^d)^T Q_d \tilde{B}^d + M^T \tilde{Q}_u M + \tilde{R}$$

$$H_{12} = H_{21}^T = -\tilde{S}^T \tilde{Q}_y \tilde{I}_y, \quad H_{13} = H_{31}^T = -\tilde{S}^T \tilde{Q}_y \tilde{I}_y, \quad H_{14} = H_{41}^T = -M^T \tilde{Q}_u \tilde{I}_u$$

$$H_{22} = \tilde{I}_u^T \tilde{Q}_u \tilde{I}_u, \quad H_{23} = H_{32}^T = \tilde{I}_y^T \tilde{Q}_y \tilde{I}_y, \quad H_{33} = \tilde{I}_y^T \tilde{Q}_y \tilde{I}_y + S_y, \quad H_{44} = \tilde{I}_u^T \tilde{Q}_u \tilde{I}_u + S_u$$

$$H_{24} = H_{42}^T = H_{34} = H_{43}^T = 0$$

$$c_{f,1} = x(k)^T N_x^T \tilde{Q}_y \tilde{S} + x^d(k)^T (F^m)^T Q_d B_m^d + \left(u(k-1) - u_{des}\right)^T \tilde{I}_u^T \tilde{Q}_u M$$

$$c_{f,2} = -x(k)^T N_x^T \tilde{Q}_y \tilde{I}_y, \quad c_{f,3} = -x(k)^T N_x^T \tilde{Q}_y \tilde{I}_y$$

$$c_{f,4} = -\left(u(k-1) - u_{des,k}\right)^T \tilde{I}_u^T \tilde{Q}_u \tilde{I}_u$$

$$c = x(k)^T N_x^T \tilde{Q}_y N_x x(k) + x^d(k)^T (F^m)^T Q_d F^m x^d(k) + \left(u(k-1) - u_{des,k}\right)^T \tilde{I}_u^T \tilde{Q}_u \tilde{I}_u \left(u(k-1) - u_{des,k}\right)$$

Then, the nominally stable MPC controller with guaranteed feasibility for the case of output zone control of time delayed systems with input targets results from the solution to the following optimization problem:
Problem P1

$$\min_{\substack{\Delta u_k, y_{sp,k}, \\ \delta_{y,k}, \delta_{u,k}}} V_k$$

subject to:

$$-\Delta u_{\max} \le \Delta u(k+j\,|\,k) \le \Delta u_{\max} \qquad j = 0,1,\cdots,m-1$$

$$u_{\min} \le u(k-1) + \sum_{i=0}^{j} \Delta u(k+i \mid k) \le u_{\max}; \qquad j = 0,1,\cdots,m-1$$

$$y_{\min} \le y_{sp,k} \le y_{\max} \tag{19}$$

$$x^s(k) + \tilde{B}^s \, \Delta u_k - y_{sp,k} - \delta_{y,k} = 0 \qquad \left(x^s(k+m \mid k) - y_{sp,k} - \delta_{y,k} = 0 \right)$$

$$u(k-1) + \tilde{I}_u^T \, \Delta u_k - u_{des,k} - \delta_{u,k} = 0 \quad \left(u(k+m-1 \mid k) - u_{des,k} - \delta_{u,k} = 0 \right)$$

It must be noted that the use of slack variables is not only convenient to avoid dynamic feasibility problems, but also to prevent stationary feasibility problems. Stationary feasibility problems are usually produced by the supervisory optimization level shown in the control structure defined in Figure 1. In such a case, for instance, the slack variable $\delta_{y,k}$ allows the predicted output to be different from the set point variable $y_{sp,k}$ at steady state (notice that only $y_{sp,k}$ is constrained to be inside the desired zone). So, the slacked problem formulation allows the system output to remain outside the desired zone, if no stationary feasible solution can be found.

It can be shown that the controller produced through the solution of problem P1 results in a stable closed loop system for the nominal system. However, the aim here is to extend this formulation to the case of multi model uncertainty.

5. Robust MPC with zone control and input target

In the model formulation presented in (1) and (2) for the time delayed system, uncertainty concentrates not only on matrices F, B^s and B^d as in the system without time delay, but also on matrix $\theta \in \Re^{ny \times nu}$ that contains all the time delays between the system inputs and outputs. Observe that the step response coefficients $S_1,\ldots,S_{p+1}$, which appears in the input matrix and $\Psi(p+1)$, which appears in the state matrix of the model defined in (1) and (2) are also uncertain, but can be computed from F, B^s, B^d and θ. Now, considering the multi-model uncertainty, assume that each model is designated by a set of parameters defined as $\Theta_n = \left\{ B_n^s, B_n^d, F_n, \theta_n \right\}$, $n = 1,\ldots,L$. Also, assume that in this case $p > \max_{i,j,n} \theta_n(i,j) + m$ (this condition guarantees that the state vector of all models have the same dimension). Then, for each model Θ_n, we can define a cost function as follows

$$\begin{aligned}
V_k(\Theta_n) = & \sum_{j=0}^{p} \left(y_n(k+j \mid k) - y_{sp,k}(\Theta_n) - \delta_{y,k}(\Theta_n) \right)^T Q_y \left(y_n(k+j \mid k) - y_{sp,k}(\Theta_n) - \delta_{y,k}(\Theta_n) \right) \\
& + \sum_{j=1}^{\infty} \left(y_n(k+p+j \mid k) - y_{sp,k}(\Theta_n) - \delta_{y,k}(\Theta_n) \right)^T Q_y \left(y_n(k+p+j \mid k) - y_{sp,k}(\Theta_n) - \delta_{y,k}(\Theta_n) \right) \\
& + \sum_{j=0}^{m-1} \left(u(k+j \mid k) - u_{des,k} - \delta_{u,k} \right)^T Q_u \left(u(k+j \mid k) - u_{des,k} - \delta_{u,k} \right) \\
& + \sum_{j=0}^{\infty} \left(u(k+m+j \mid k) - u_{des,k} - \delta_{u,k} \right)^T Q_u \left(u(k+m+j \mid k) - u_{des,k} - \delta_{u,k} \right) \\
& + \sum_{j=0}^{m-1} \Delta u(k+j \mid k)^T R \, \Delta u(k+j \mid k) + \delta_{y,k}^T(\Theta_n) S_y \delta_{y,k}(\Theta_n) + \delta_{u,k}{}^T S_u \delta_{u,k}
\end{aligned} \tag{20}$$

Following the same steps as in case of the nominal system, we can conclude that the cost defined in (20) will be bounded if the control actions, set points and slack variables are such that (18) is satisfied and

$$x^s(k)+\tilde{B}^s(\Theta_n)\Delta u_k - y_{sp,k}(\Theta_n)-\delta_{y,k}(\Theta_n)=0$$

Then, if these conditions are satisfied, (20) can be written as follows

$$\begin{aligned} V_k(\Theta_n) &= \left(N_x x(k)+\tilde{S}(\Theta_n)\Delta u_k - \tilde{I}_y y_{sp,k}(\Theta_n)-\tilde{I}_y\delta_{y,k}(\Theta_n)\right)^T \tilde{Q}_y \\ &\qquad \left(N_x x(k)+\tilde{S}(\Theta_n)\Delta u_k - \tilde{I}_y y_{sp,k}(\Theta_n)-\tilde{I}_y\delta_{y,k}(\Theta_n)\right) \\ &+\left(F(\Theta_n)^m x^d(k)+B^d_m(\Theta_n)\Delta u_k\right)^T Q_d(\Theta_n)\left(F(\Theta_n)^m x^d(k)+B^d_m(\Theta_n)\Delta u_k\right) \\ &+\left(\tilde{I}_u u(k-1)+M\Delta u_k-\tilde{I}_u u_{des,k}-\tilde{I}_u\delta_{u,k}\right)^T Q_u\left(\tilde{I}_u u(k-1)+M\Delta u_k-\tilde{I}_u u_{des,k}-\tilde{I}_u\delta_{u,k}\right) \\ &+\Delta u_k^T\tilde{R}\Delta u_k+\delta_{y,k}(\Theta_n)^T S_y\delta_{y,k}(\Theta_n)+\delta_{u,k}{}^T S_u\delta_{u,k} \end{aligned} \tag{21}$$

or

$$V_k(\Theta_n)=\begin{bmatrix}\Delta u_k^T & y_{sp,k}^T(\Theta_n) & \delta_{y,k}^T(\Theta_n) & \delta_{u,k}^T\end{bmatrix}\begin{bmatrix}H_{11}(\Theta_n) & H_{12}(\Theta_n) & H_{13}(\Theta_n) & H_{14}\\ H_{21}(\Theta_n) & H_{22} & H_{23} & 0\\ H_{31}(\Theta_n) & H_{32} & H_{33} & 0\\ H_{41} & 0 & 0 & H_{44}\end{bmatrix}\begin{bmatrix}\Delta u_k\\ y_{sp,k}(\Theta_n)\\ \delta_{y,k}(\Theta_n)\\ \delta_{u,k}\end{bmatrix}$$

$$+2\begin{bmatrix}c_{f,1}(\Theta_n) & c_{f,2} & c_{f,3} & c_{f,4}\end{bmatrix}\begin{bmatrix}\Delta u_k\\ y_{sp,k}(\Theta_n)\\ \delta_{y,k}(\Theta_n)\\ \delta_{u,k}\end{bmatrix}+c(\Theta_n)$$

$$H_{11}(\Theta_n)=\tilde{S}(\Theta_n)^T\tilde{Q}_y\tilde{S}(\Theta_n)+(\tilde{B}^d(\Theta_n))^T Q_d(\Theta_n)\tilde{B}^d(\Theta_n)+M^T\tilde{Q}_u M+\tilde{R}$$

$$H_{12}=H_{21}^T=-\tilde{S}(\Theta_n)^T\tilde{Q}_y\tilde{I}_y,\quad H_{13}=H_{31}^T=-\tilde{S}(\Theta_n)^T\tilde{Q}_y\tilde{I}_y,\quad H_{14}=H_{41}^T=-M^T\tilde{Q}_u\tilde{I}_u$$

$$H_{22}=\tilde{I}_y^T\tilde{Q}_y\tilde{I}_y,\quad H_{23}=H_{32}^T=\tilde{I}_y^T\tilde{Q}_y\tilde{I}_y,\quad H_{33}=\tilde{I}_y^T\tilde{Q}_y\tilde{I}_y$$

$$H_{24}=H_{42}^T=H_{34}=H_{43}^T=0$$

$$c_{f,1}=x(k)^T N_x^T\tilde{Q}_y\tilde{S}(\Theta_n)+x^d(k)^T(F(\Theta_n)^m)^T Q_{xd}\tilde{B}^d(\Theta_n)+\left(u(k-1)-u_{des}\right)^T\tilde{I}_u^T\tilde{Q}_u M$$

$$c_{f,2}=-x(k)^T N_x^T\tilde{Q}_y\tilde{I}_y,\quad c_{f,3}=-x(k)^T N_x^T\tilde{Q}_y\tilde{I}_y$$

$$c_{f,4}=-\left(u(k-1)-u_{des}\right)^T\tilde{I}_u^T\tilde{Q}_u\tilde{I}_u$$

$$c = x(k)^T N_x^T \tilde{Q}_y N_x x(k) + x^d(k)^T (F(\Theta_n)^m)^T Q_{xd} F(\Theta_n)^m x^d(k) +$$
$$+\left(u(k-1) - u_{des,k}\right)^T \tilde{I}_u^T \tilde{Q}_u \tilde{I}_u \left(u(k-1) - u_{des,k}\right)$$

Then, the robust MPC for the system with time delay and multi-model uncertainty is obtained from the solution to the following problem:
Problem P2

$$\min_{\substack{\Delta u_k, y_{sp,k}(\Theta_n), \delta_{y,k}(\Theta_n), \delta_{u,k} \\ n=1,\ldots,L}} V_k(\Theta_N) \tag{22}$$

subject to

$$-\Delta u_{\max} \le \Delta u(k+j \mid k) \le \Delta u_{\max} \qquad j = 0,1,\cdots,m-1$$

$$u_{\min} \le u(k-1) + \sum_{i=0}^{j} \Delta u(k+i \mid k) \le u_{\max}; \qquad j = 0,1,\cdots,m-1$$

$$y_{\min} \le y_{sp,k}(\Theta_n) \le y_{\max}; \qquad n = 1,\cdots,L$$

$$x^s(k) + \tilde{B}^s(\Theta_n)\Delta u_k - y_{sp,k}(\Theta_n) - \delta_{y,k}(\Theta_n) = 0; \quad n = 1,\cdots,L \tag{23}$$

$$u(k-1) + \tilde{I}_u^T \Delta u_k - u_{des,k} - \delta_{u,k} = 0 \tag{24}$$

$$V_k\left(\Delta u_k, \delta_{y,k}(\Theta_n), \delta_{u,k}, y_{sp,k}(\Theta_n), \Theta_n\right) \le \tilde{V}_k\left(\Delta \tilde{u}_k, \tilde{\delta}_{y,k}(\Theta_n), \tilde{\delta}_{u,k}, y_{sp,k}(\Theta_n), \Theta_n\right), \quad n = 1,\cdots,L \tag{25}$$

where, assuming that $\left(\Delta u_{k-1}^*, y_{sp,k-1}(\Theta_n), \delta_{u,k-1}^*, \delta_{y,k-1}^*(\Theta_n)\right)$ is the optimal solution to Problem P2 at time step k-1, we define
$\Delta \tilde{u}_k = \left[\Delta u^*(k \mid k-1)^T \quad \cdots \quad \Delta u^*(k+m-2 \mid k-1)^T \quad 0\right]^T$; $\tilde{y}_{sp,k}(\Theta_n) = y_{sp,k-1}^*(\Theta_n)$ and $\tilde{\delta}_{u,k}$ such that

$$u(k-1) + \tilde{I}_u^T \Delta \tilde{u}_k - u_{des,k} - \tilde{\delta}_{u,k} = 0 \tag{26}$$

and define $\tilde{\delta}_{y,k}(\Theta_n)$ such that

$$x^s(k) + \tilde{B}^s(\Theta_n)\Delta \tilde{u}_k - \tilde{y}_{sp,k}(\Theta_n) - \tilde{\delta}_{y,k}(\Theta_n) = 0 \tag{27}$$

In (20), Θ_N corresponds to the nominal or most probable model of the system.

Remark 1: The cost to be minimized in problem P2 corresponds to the nominal model. However, constraints (23) and (24) are imposed considering the estimated state of each model $\Theta_n \in \Omega$. Constraint (25) is a non-increasing cost constraint that assures the convergence of the true state cost to zero.

Remark 2: The introduction of L set-point variables allows the simultaneous zeroing of all the output slack variables. In that case, whenever possible, the set-point variable $y_{sp,k}(\Theta_n)$

will be equal to the output prediction at steady state (represented by $x_n^s(k+m)$), and so the corresponding output penalization will be removed from the cost. As a result, the controller gains some flexibility that allows achieving the other control objectives.

Remark 3: Note that by hypothesis, one of the observers is based on the actual plant model, and if the initial and the final steady states are known, then the estimated state $\hat{x}_T(k)$ will be equal to the actual plant state at each time *k*.

Remark 4: Conditions (26) and (27) are used to update the pseudo variables of constraint (25), by taking into account the current state estimation $\hat{x}_n^s(k)$ for each of the models lying in Ω, and the last value of the input target.

One important feature that should have a constrained controller is the recursive feasibility (i.e. if the optimization problem is feasible at a given time step, it should remain feasible at any subsequent time step). The following lemma shows how the proposed controller achieves this property.

Lemma. If problem P2 is feasible at time step *k*, it will remain feasible at any subsequent time step *k+j, j*=1,2,...

Proof:

Assume that the output zones remain fixed, and also assume that

$$\Delta u_k^* = \left[\Delta u^*(k|k)^T \quad \cdots \quad \Delta u^*(k+m-1|k)^T \right]^T \in \Re^{m.nu}, \tag{28}$$

$$y_{sp,k}^*(\Theta_1),\cdots,y_{sp,k}^*(\Theta_L),\ \delta_{y,k}^*(\Theta_1),\cdots,\delta_{y,k}^*(\Theta_L) \text{ and } \delta_{u,k}^* \tag{29}$$

correspond to the optimal solution to problem P2 at time *k*.

Consider now the pseudo variables $\left(\Delta\tilde{u}_{k+1},\tilde{y}_{sp,k+1}(\Theta_1),\cdots,\tilde{y}_{sp,k+1}(\Theta_L),\ \tilde{\delta}_{y,k+1}(\Theta_1),\ldots,\tilde{\delta}_{y,k+1}(\Theta_L),\tilde{\delta}_{u,k+1}\right)$ where

$$\Delta\tilde{u}_{k+1} = \left[\Delta u^*(k+1|k)^T \quad \cdots \quad \Delta u^*(k+m-1|k)^T \quad 0 \right]^T \tag{30}$$

$$\tilde{y}_{sp,k+1}(\Theta_n) = y_{sp,k}^*(\Theta_n),\ n=1,\cdots,L, \tag{31}$$

Also, the slacks $\tilde{\delta}_{u,k+1}$ and $\tilde{\delta}_{y,k+1}(\Theta_n)$ are such that

$$u(k) + \tilde{I}_u^T\,\Delta\tilde{u}_{k+1} - u_{des,k} - \tilde{\delta}_{u,k+1} = 0 \tag{32}$$

and

$$\hat{x}_n^s(k+1) + \tilde{B}^s(\Theta_n)\Delta\tilde{u}_{k+1} - \tilde{y}_{sp,k+1}(\Theta_n) - \tilde{\delta}_{y,k+1}(\Theta_n) = 0, \qquad n=1,\ldots,L \tag{33}$$

We can show that the solution defined through (30) to (33) represent a feasible solution to problem P2 at time *k*+1, which proves the recursive feasibility. This means that if problem P2 is feasible at time step *k*, then, it will remain feasible at all the successive time steps *k*+1, *k*+2, ... □

Now, the convergence of the closed loop system with the robust controller resulting from the later optimization problem can be stated as follows:

Theorem. Suppose that the undisturbed system starts at a known steady state and one of the state observers is based on the actual model of the plant. Consider also that the input target is moved to a new value, or the boundaries of the output zones are modified. Then, if condition (3) is satisfied for each model $\Theta_n \in \Omega$, the cost function of the undisturbed true system in closed loop with the controller defined through the solution to problem P2 will converge to zero.

Proof:

Suppose that, at time k the uncertain system starts from a steady state corresponding to output $y(k)=y_{ss}$ and input $u(k-1)=u_{ss}$. We have already shown that, with the model structure considered in (1) and (2), the model states corresponding to this initial steady state can be represented as follows:

$$\hat{x}_n(k)=\begin{bmatrix} \underbrace{y_{ss} \quad \cdots \quad y_{ss}}_{p} & y_{ss} & 0 \end{bmatrix}, \quad n=1,\cdots,L$$

and consequently, $\hat{x}_n^s(k)=y_{ss},\ \ \hat{x}_n^d(k)=0,\ \ n=1,\cdots,L$.

At time k, the cost corresponding to the solution defined in (28) and (29) for the true model is given by

$$\begin{aligned} V_k^*(\Theta_T) = \sum_{j=0}^{\infty} \Big\{ & \left(y_T^*(k+j\,|\,k)-y_{sp,k}^*(\Theta_T)-\delta_{y,k}^*(\Theta_T)\right)^T Q_y \left(y_T^*(k+j\,|\,k)-y_{sp,k}^*(\theta_T)-\delta_{y,k}^*(\theta_T)\right) \\ & +\left(u^*(k+j\,|\,k)-u_{des,k}-\delta_{u,k}^*\right)^T Q_u \left(u^*(k+j\,|\,k)-u_{des,k}-\delta_{u,k}^*\right)\Big\} \\ & +\sum_{j=0}^{m-1} \Delta u^*(k+j\,|\,k)^T R\, \Delta u^*(k+j\,|\,k) + \delta_{y,k}^{*\ T}(\Theta_T) S_y \delta_{y,k}^*(\Theta_T) + \delta_{u,k}^{*\ T} S_u \delta_{u,k}^* \end{aligned} \tag{34}$$

At time step k+1, the cost corresponding to the pseudo variables defined in (30) to (33) for the true model is given by

$$\begin{aligned} & \tilde{V}_{k+1}(\Theta_T) = \\ & \sum_{j=0}^{\infty} \Big\{ \left(y_T^*(k+j+1\,|\,k)-y_{sp,k}^*(\Theta_T)-\delta_{y,k}^*(\Theta_T)\right)^T Q_y \left(y_T^*(k+j+1\,|\,k)-y_{sp,k}^*(\Theta_T)-\delta_{y,k}^*(\Theta_T)\right) \\ & \qquad +\left(u^*(k+j+1/k)-u_{des,k}-\delta_{u,k}^*\right)^T Q_u \left(u^*(k+j+1/k)-u_{des,k}-\delta_{u,k}^*\right)\Big\} \\ & +\sum_{j=0}^{m-1} \Delta u^*(k+j+1\,|\,k)^T R\, \Delta u^*(k+j+1\,|\,k) + \delta_{y,k}^{*\ T}(\Theta_T) S_y \delta_{y,k}^*(\Theta_T) + \delta_{u,k}^{*\ T} S_u \delta_{u,k}^* \end{aligned} \tag{35}$$

Observe that, since the same input sequence is used and the current estimated state corresponding to the actual model of the plant is equal to the actual state, then the predicted state and output trajectory will be the same as the optimal predicted trajectories at time step k. That is, for any $j \geq 1$, we have

$$x_T(k+j\,|\,k+1)=x_T(k+j\,|\,k)$$

and

$$y_T(k+j\,|\,k+1) = y_T(k+j\,|\,k)$$

In addition, for the true model we have $\tilde{\delta}_{y,k+1}(\Theta_T) = \delta^*_{y,k}(\Theta_T)$ and $\tilde{\delta}_{u,k+1} = \delta^*_{u,k}$. However, the first of these equalities is not true for the other models, as for these models we have $\hat{x}_n(k+1\,|\,k+1) \neq x_n(k+1\,|\,k)$, for $\Theta_n \neq \Theta_T$.
Now, subtracting (35) from (34) we have

$$\begin{aligned} V^*_k(\Theta_T) - \tilde{V}_{k+1}(\Theta_T) &= \left(y^*_T(k\,|\,k) - y^*_{sp,k}(\Theta_T) - \delta^*_{y,k}(\Theta_T)\right)^T Q_y \left(y^*_T(k\,|\,k) - y^*_{sp,k}(\Theta_T) - \delta^*_{y,k}(\Theta_T)\right) \\ &+ \left(u^*(k\,|\,k) - u_{des,k} - \delta^*_{u,k}\right)^T Q_u \left(u^*(k\,|\,k) - u_{des,k} - \delta^*_{u,k}\right) + \Delta u^*(k)^T R \Delta u^*(k) \end{aligned}$$

and, from constraint (25), the following relation is obtained

$$V^*_{k+1}(\Theta_T) \leq \tilde{V}_{k+1}(\Theta_T),$$

which finally implies

$$\begin{aligned} V^*_k(\Theta_T) - V^*_{k+1}(\Theta_T) &\geq \left(y^*_T(k\,|\,k) - y^*_{sp,k}(\Theta_T) - \delta^*_{y,k}(\Theta_T)\right)^T Q_y \left(y^*_T(k\,|\,k) - y^*_{sp,k}(\Theta_T) - \delta^*_{y,k}(\Theta_T)\right) \\ &+ \left(u^*(k\,|\,k) - u_{des,k} - \delta^*_{u,k}\right)^T Q_u \left(u^*(k\,|\,k) - u_{des,k} - \delta^*_{u,k}\right) + \Delta u^*(k)^T R \Delta u^*(k) \end{aligned} \quad (36)$$

Since the right hand side of (36) is positive definite, the successive values of the cost will be strictly decreasing and for a large enough time $\bar{k}$, we will have $\left(V^*_{\bar{k}}(\Theta_T) - V^*_{\bar{k}+1}(\Theta_T)\right) = 0$, which proves the convergence of the cost.
The convergence of $V^*_k(\Theta_T)$ means that, at steady state, the following relations should hold

$$y^*_T(\bar{k}\,|\,\bar{k}) - y^*_{sp,\bar{k}}(\Theta_T) = \delta^*_{y,\bar{k}}(\Theta_T)$$

$$u^*(\bar{k}\,|\,\bar{k}) - u_{des,\bar{k}} = \delta^*_{u,\bar{k}}$$

$$\Delta u^*(\bar{k}) = 0$$

At steady state, the state is such that

$$\hat{x}_n(\bar{k}) = \begin{bmatrix} y(\bar{k}) \\ \vdots \\ y(\bar{k}+p) \\ \hat{x}^s_n(\bar{k}) \\ \hat{x}^d_n(\bar{k}) \end{bmatrix} = \begin{bmatrix} y(\bar{k}) \\ \vdots \\ y(\bar{k}) \\ y(\bar{k}) \\ 0 \end{bmatrix}$$

where $y(\bar{k})$ is the actual plant output. Note that the state component $\hat{x}^d_n(\bar{k})$ is null as it corresponds to the stable modes of the system and the input increment is null at steady state. Then, constraint (23) can be written as follows:

$$\delta^{*}_{y,\bar{k}}(\Theta_n)=y^{*}_n(\bar{k}\,|\,\bar{k})-y^{*}_{sp,\bar{k}}(\Theta_n)=y(\bar{k})-y^{*}_{sp,\bar{k}}(\Theta_n), \qquad n=1,\ldots,L. \tag{37}$$

This means that, if the output of the true system is stabilized inside the output zone, then the set point corresponding to each particular model will be placed by the optimizer exactly at the output predicted values. As a result, all the output slacks will be null. On the other hand, if the output of the true system is stabilized at a value outside the output zone, then the set-point variable corresponding to any particular model will be placed by the optimizer at the boundary of the zone. In this case, the output slack variables will be different from zero, but they will all have the same numerical value as can be seen from (37).

Now, to strictly prove the convergence of the input and output to their corresponding targets, we must show that slacks $\delta_{u,\bar{k}}$ and $\delta_{y,\bar{k}}(\Theta_T)$ will converge to zero. It is necessary at this point to notice that in the case of zone control the degrees of freedom of the system are no longer the same as in the fixed set-point problem. So, the desired input values may be exactly achieved by the true system, even in the presence of some bounded disturbances. Let us now assume that the system is stabilized at a point where, $\delta^{*}_{y,\bar{k}}(\Theta_1)=\cdots=\delta^{*}_{y,\bar{k}}(\Theta_L)\neq 0$, and $\delta_{u,\bar{k}}\neq 0$. In addition, assume that the desired input value is constant at $u_{des,k}$. Then, at time $\bar{k}$ large enough, the cost corresponding to model Θ_n will be reduced to

$$V^{*}_{\bar{k}}(\Theta_n)=\delta^{T}_{y,\bar{k}}(\Theta_n)S_y\delta_{y,\bar{k}}(\Theta_{n1})+\delta^{T}_{u,\bar{k}}S_u\delta_{u,\bar{k}}, \quad n=1,\ldots,L, \tag{38}$$

and constraints (21) and (22) become,

$$\hat{x}^{s}_n(\bar{k})-y_{sp,\bar{k}}(\Theta_n)=\delta_{y,\bar{k}}(\Theta_n), \qquad n=1,\cdots,L \tag{39}$$

and $u(\bar{k}-1)-u_{des,\bar{k}}=\delta_{u,\bar{k}}$.

Since $\hat{x}^{s}_n(\bar{k})=y(\bar{k}), \quad n=1,\cdots,L$, Eq. (39) can be written as

$$y(\bar{k})-y_{sp,\bar{k}}(\Theta_n)=\delta_{y,\bar{k}}(\Theta_n), \qquad n=1,\cdots,L.$$

Now, we want to show that if $u(\bar{k}-1)$ and $u_{des,\bar{k}}$ are not on the boundary of the input operating range, then it is possible to guide the system toward a point in which the slack variables $\delta_{y,k}(\theta_n)$ and $\delta_{u,k}$ are null, and this point have a smaller cost than the steady state defined above. Assume also for simplicity that m=1. Let us consider a candidate solution to problem P2 defined by:

$$\Delta\bar{u}(\bar{k}/\bar{k})=u_{des,\bar{k}}-u(\bar{k}-1)=-\delta_{u,\bar{k}} \tag{40}$$

and

$$\bar{y}_{sp,\bar{k}}(\theta_n)=y(\bar{k})-B^{s}(\theta_n)\delta_{u,\bar{k}} \quad n=1,\ldots,L \tag{41}$$

Now, consider the cost function defined in (21), written for time step $\bar{k}$ and the control move defined in (40) and the output set point defined in (41):

$$\bar{V}_k(\Theta_n) = \left(\tilde{I}_y y(\bar{k}) - \tilde{S}_1(\Theta_n)\delta_{u,\bar{k}} - \tilde{I}_y y_{sp,\bar{k}}(\Theta_n) - \tilde{I}_y \delta_{y,\bar{k}}(\Theta_n)\right)^T \tilde{Q}_y$$
$$\left(\tilde{I}_y y(\bar{k}) - \tilde{S}_1(\Theta_n)\delta_{u,\bar{k}} - \tilde{I}_y y_{sp,\bar{k}}(\Theta_n) - \tilde{I}_y \delta_{y,\bar{k}}(\Theta_n)\right)$$
$$+\left(F(\Theta_n)^m x^d(\bar{k}) - \tilde{B}^d(\Theta_n)\delta_{u,\bar{k}}\right)^T Q_d(\Theta_n)\left(F(\Theta_n)^m x^d(\bar{k}) - \tilde{B}^d(\Theta_n)\delta_{u,\bar{k}}\right)$$
$$+\left(\tilde{I}_u u(k-1) - M\delta_{u,\bar{k}} - \tilde{I}_u u_{des,k} - \tilde{I}_u \bar{\delta}_{u,k}\right)^T Q_u\left(\tilde{I}_u u(k-1) - M\delta_{u,\bar{k}} - \tilde{I}_u u_{des,k} - \tilde{I}_u \bar{\delta}_{u,k}\right)$$
$$+(-\delta_{u,\bar{k}})^T \tilde{R}(-\delta_{u,\bar{k}}) + \bar{\delta}_{y,\bar{k}}(\Theta_n)^T S_y \bar{\delta}_{y,\bar{k}}(\Theta_n) + \bar{\delta}_{u,\bar{k}}{}^T S_u \bar{\delta}_{u,\bar{k}}$$

Now, since the solution defined by $\left(\Delta\bar{u}(\bar{k}/\bar{k}), \bar{\delta}_{y,\bar{k}}(\Theta_n), \bar{\delta}_{u,\bar{k}}\right)$ satisfies constraint (23) and (24), the above cost can be reduced to

$$\bar{V}_{\bar{k}}(\Theta_n) = \delta_{u,\bar{k}}^T S_{\min}^u(\Theta_n)\delta_{u,\bar{k}}$$

where

$$S_{\min}^u(\Theta_n) = \left[\tilde{I}_y B^s(\Theta_n) - \tilde{S}_1(\Theta_n)\right]^T \tilde{Q}_y\left[\tilde{I}_y B^s(\Theta_n) - \tilde{S}_1(\Theta_n)\right] + \tilde{B}^d(\Theta_n)^T Q_d \tilde{B}^d(\Theta_n) + \tilde{R}$$

Then, if

$$S_u > S_u^{\min}(\Theta_n), \quad n = 1, \ldots, L, \tag{42}$$

the cost corresponding to the decision variables defined in (40) and (41) will be smaller than the cost obtained in (38). This means that it is not possible for the system to remain at a point in which the slack variables $\delta_{y,k}(\Theta_n)$, $n = 1, \cdots, L$ and $\delta_{u,k}$ are different from zero.
Thus, as long as the system remains controllable, condition (42) is sufficient to guarantee the convergence of the system inputs to their target while the system output will remain within the output zones.□
Observe that only matrix S_u is involved in condition (42) because condition (3) assures that the corrected output prediction, i.e. the one corresponding to the desired input values, lies in the feasible zone. In this case, for all positive matrices S_y, the total cost can be reduced by making the set point variable equal to the steady-state output prediction, which is a feasible solution and produces no additional cost. However, matrix S_y is suggested to be large enough to avoid any numerical problem in the optimization solution.
***Remark** 5*: We can prove the stability of the proposed zone controller under the same assumptions considered in the proof of the convergence. Output tracking stability means that for every $\gamma > 0$, there exists a $\rho(\gamma)$ such that if $\|\bar{x}_T(0)\| < \rho$, then $\|\bar{x}_T(k)\| < \gamma$ for all $k \geq 0$; where the extended state of the true system $\bar{x}_T(k)$ may be defined as follows

$$\bar{x}_T(k) = \begin{bmatrix} y_T(k|k) - x_T^s(k) \\ \vdots \\ y_T(k+p|k) - x_T^s(k) \\ x_T^s(k) - y_{sp,k-1}^*(\Theta_T) \\ x_T^d(k) \\ u(k|k) - u_{des,k} \end{bmatrix}$$

To simplify the proof, we still assume that m=1, and suppose that the optimal solution obtained at step k-1 is given by $\Delta u_{k-1}^{*}=\Delta u^{*}(k-1/k-1)$, $y_{sp,k-1}^{*}(\Theta_1),\cdots,y_{sp,k-1}^{*}(\Theta_L)$, $\delta_{y,k-1}^{*}(\Theta_1),\cdots,\delta_{y,k-1}^{*}(\Theta_L)$ and $\delta_{u,k-1}^{*}$.
A feasible solution to problem P2 at time k is given by:
$\Delta\tilde{u}_k=0$, $\tilde{y}_{sp,k}(\Theta_n)=y_{sp,k-1}^{*}(\Theta_n)$, and $\tilde{\delta}_{u,k}$ and $\tilde{\delta}_{y,k}(\Theta_n)$ are such that

$$u(k-1)+\tilde{I}_u^T\overbrace{\Delta\tilde{u}_k}^{=0}-u_{des,k}-\tilde{\delta}_{u,k}=0 \tag{43}$$

$$\hat{x}_n^s(k)+\tilde{B}^s(\Theta_n)\overbrace{\Delta\tilde{u}_k}^{=0}-\tilde{y}_{sp,k}(\Theta_n)-\tilde{\delta}_{y,k}(\Theta_n)=0, \qquad n=1,\ldots,L\,. \tag{44}$$

Since $\Delta\tilde{u}(k\,|\,k)=0$, we have $u(k\,|\,k)=u(k-1)$ and from (43) we can write

$$\tilde{\delta}_{u,k}=u(k\,|\,k)-u_{des,k} \tag{45}$$

For the true system, (44) can be written as follows

$$x_T^s(k)-y_{sp,k-1}^{*}(\Theta_T)-\tilde{\delta}_{y,k}(\Theta_T)=0$$

and consequently, we have the following relations

$$\tilde{\delta}_{y,k}(\Theta_T)=x_T^s(k)-y_{sp,k-1}^{*}(\Theta_T) \tag{46}$$

and

$$x_T^s(k)=y_{sp,k-1}^{*}(\Theta_T)+\tilde{\delta}_{y,k}(\Theta_T) \tag{47}$$

For the feasible solution defined above, the cost defined in (21) can be written for the actual model Θ_T as follows

$$\begin{aligned}\tilde{V}_k(\Theta_T)=&\left(N_x x_T(k)-\tilde{I}_y y_{sp,k-1}^{*}(\Theta_T)-\tilde{I}_y\tilde{\delta}_{y,k}(\Theta_T)\right)^T\tilde{Q}_y\left(N_x x_T(k)-\tilde{I}_y y_{sp,k-1}^{*}(\Theta_T)-\tilde{I}_y\tilde{\delta}_{y,k}(\Theta_T)\right)\\&+\left(F(\Theta_T)^m x_T^d(k)\right)^T Q_{xd}(\Theta_T)\left(F(\Theta_T)^m x_T^d(k)\right)\\&+\left(\tilde{I}_u u(k-1)-\tilde{I}_u u_{des,k}-\tilde{I}_u\tilde{\delta}_{u,k}\right)^T Q_u\left(\tilde{I}_u u(k-1)-\tilde{I}_u u_{des,k}-\tilde{I}_u\tilde{\delta}_{u,k}\right)\\&+\left(x_T^s(k)-y_{sp,k-1}^{*}(\Theta_T)\right)^T S_y\left(x_T^s(k)-y_{sp,k-1}^{*}(\Theta_T)\right)+\left(u(k\,|\,k)-u_{des,k}\right)^T S_u\left(u(k\,|\,k)-u_{des,k}\right)\end{aligned} \tag{48}$$

Now, using (45), (46) and (47) the cost defined in (48) can be reduced to the following expression

$$\tilde{V}_k(\Theta_T)=\overline{x}_T(k)^T\left\{C_1^T\tilde{Q}_y C_1+C_2^T\left(F(\Theta_T)^m\right)^T Q_d(\Theta_T)\left(F(\Theta_T)^m\right)C_2+C_3^T S_y C_3+C_4^T S_u C_4\right\}$$

where

$$C_1=\begin{bmatrix}I_{(p+1)ny} & 0_{(p+1)ny\times ny} & 0_{(p+1)ny\times nd} & 0_{(p+1)ny\times nu}\end{bmatrix}$$

$$C_2 = \left[0_{nd\times(p+1)ny} \quad 0_{nd\times ny} \quad I_{nd} \quad 0_{nd\times nu} \right]$$

$$C_3 = \left[0_{ny\times(p+1)ny} \quad I_{ny} \quad 0_{ny\times nd} \quad 0_{ny\times nu} \right]$$

$$C_4 = \left[0_{nu\times(p+1)ny} \quad 0_{nu\times ny} \quad 0_{nu\times nd} \quad I_{nu} \right]$$

Thus, the cost defined in (48) can be written as follows:

$$\tilde{V}_{2,k}(\Theta_T) = \overline{x}_T(k)^T H_1(\Theta_T)\overline{x}_T(k), \tag{49}$$

where $H_1 = C_1^T \tilde{Q}_y C_1 + C_2^T \left(F(\Theta_T)^m\right)^T Q_{xd}(\Theta_T)\left(F(\Theta_T)^m\right)C_2 + C_3^T S_y C_3 + C_4^T S_u C_4$.

Because of constraint (25), the optimal true cost (that is, the cost based on the true model, considering the optimal solution that minimizes the nominal cost at time k) will satisfy

$$V_k^*(\Theta_T) \le \tilde{V}_k(\Theta_T). \tag{50}$$

and

$$V_{k+n}^*(\Theta_T) \le V_k^*(\Theta_T) \text{ for any } n > 1. \tag{51}$$

By a similar procedure as above and based on the optimal solution at time $k+n$, we can find a feasible solution to Problem P2 at time $k + n + 1$, for any $n>1$, such that

$$\tilde{V}_{k+n+1}(\Theta_T) \le V_{k+n}^*(\Theta_T) \tag{52}$$

and from the definition of $\tilde{V}_{k+n+1}$ we have

$$\tilde{V}_{2,k+n+1}(\Theta_T) = \overline{x}_T(k+n+1)^T H_1(\Theta_T)\overline{x}_T(k+n+1)$$

Therefore, combining inequalities (49) to (52) results

$$\overline{x}_T(k+n+1)^T H_1(\Theta_T)\overline{x}_T(k+n+1) \le \overline{x}_T(k)^T H_1(\Theta_T)\overline{x}_T(k), \qquad \forall n > 1.$$

As $H_1(\Theta_T)$ is positive definite, it follows that

$$\left\|\overline{x}_T(k+n+1)\right\| \le \alpha(\Theta_T)\left\|\overline{x}_T(k)\right\|, \qquad \forall n > 1$$

where

$$\alpha(\Theta_T) = \left[\frac{\lambda_{\max}(H_1(\Theta_T))}{\lambda_{\min}(H_1(\Theta_T))}\right]^{1/2} \le \max_j \left[\frac{\lambda_{\max}(H_1(\Theta_j))}{\lambda_{\min}(H_1(\Theta_j))}\right]^{1/2}$$

If we restrict the state at time k to the set defined by $\left\|\overline{x}_T(k)\right\| < \rho$, then, the state at tine $k+n+1$ will be inside the set defined by

$$\left\|\overline{x}_T(k+n+1)\right\| < \alpha(\Theta_T)\rho, \qquad \forall n > 1 .$$

Which proves stability of the closed loop system, as $\overline{x}_T$ will remain inside the ball $\left\|\overline{x}_T\right\| < \alpha(\Theta_T)\rho$, where $\alpha(\Theta_T)$ is limited, as long as the closed loop starts from a state inside the ball $\left\|\overline{x}_T\right\| < \rho$. Therefore, as we have already proved the convergence of the closed loop, we can now assure that under the assumption of state controllability at the final equilibrium point, the proposed MPC is asymptotically stable.□

Remark 6: It is important to observe that even if condition (3) cannot be satisfied by the input target, or the input target is such that one or more outputs need to be kept outside their zones, the proposed controller will still be stable. This is a consequence of the decreasing property of the cost function (inequality (36)) and the inclusion of appropriate slack variables into the optimization problem. When no feasible solution exists, the system will evolve to an operating point in which the slack variables, which at steady state are the same for all the models, are as small as possible, but different from zero. This is an important aspect of the controller, as in practical applications a disturbance may move the system to a point from which it is not possible to reach a steady state that satisfies (3). When this happens, the controller will do the best to compensate the disturbance, while maintaining the system under control.

Remark 7: We may consider the case when the desired input target $u_{des,k}$ is outside the feasible set ϑ_u and the case where the set ϑ_u itself is null. If ϑ_u is not null, the input target $u_{des,k}$ could be located within the global input feasible set ϑ_o, but outside the restricted input feasible set ϑ_u. In this case, the slack variables at steady state, $\delta_{u,ss}$ and $\delta_{y,ss}(\Theta_n)$, cannot be simultaneously zeroed, and the relative magnitude of matrices S_y and S_u will define the equilibrium point. If the priority is to maintain the output inside the corresponding range, the choice must be $S_y >> S_u$, while preserving $S_u > S_u^{\min}$. Then, the controller will guide the system to a point in which $\delta_{y,ss}(\Theta_n) \approx 0,\ n = 1,\cdots,L$ and $\delta_{u,ss} \neq 0$. On the other hand, if ϑ_u is null, that is, there is no input belonging to the global input feasible set ϑ_o that simultaneously satisfies all the zones for the models lying in Ω, then, the slack variables $\delta_{y,ss}(\Theta_n), n = 1,\cdots,L$, cannot be zeroed, no matter the value of $\delta_{u,ss}$. In this case (assuming that $S_y >> S_u$), the slack variables $\delta_{y,ss}(\Theta_n), n = 1,\cdots,L$, will be made as small as possible, independently of the value of $\delta_{u,ss}$. Then, once the output slack is established, the input slack will be accommodated to satisfy these values of the outputs.

6. Simulation results for the system with time delay

The system adopted to test the performance of the robust controller presented here is based on the FCC system presented in Sotomayor and Odloak (2005) and González et al. (2009). It is a typical example of the chemical process industry, and instead of output set points, this system has output zones. The objective of the controller is then to guide the manipulated inputs to the corresponding targets and to maintain the outputs (that are more numerous than the inputs) within the corresponding feasible zones. The system considered here has 2 inputs and 3 outputs. Three models constitute the multi-model set Ω on which the robust controller is based. In two of these models, time delays were included to represent a possible degradation of the process conditions along an operation campaign. The third model corresponds to the process at the design conditions. The parameters corresponding to each of these models can be seen in the following transfer functions:

$$G(\Theta_1)=\begin{bmatrix} \dfrac{0.4515e^{-2s}}{2.9846s+1} & \dfrac{0.2033e^{-4s}}{1.7187s+1} \\ \dfrac{1.5e^{-6s}}{20s+1} & \dfrac{(0.1886s-3.8087)e^{-3s}}{17.7347s^2+10.8348s+1} \\ \dfrac{1.7455e^{-6s}}{9.1085s+1} & \dfrac{-6.1355e^{-5s}}{10.9088s+1} \end{bmatrix},$$

$$G(\Theta_2)=\begin{bmatrix} \dfrac{0.25e^{-2s}}{3.5s+1} & \dfrac{0.135e^{-5s}}{2.77s+1} \\ \dfrac{0.9e^{-3s}}{25s+1} & \dfrac{(0.1886s-2.8)e^{-4s}}{19.7347s^2+10.8348s+1} \\ \dfrac{1.25e^{-5s}}{11.1085s+1} & \dfrac{-5e^{-6s}}{12.9088s+1} \end{bmatrix},$$

$$G(\Theta_3)=\begin{bmatrix} \dfrac{0.7}{1.98s+1} & \dfrac{0.5}{2.7s+1} \\ \dfrac{2.3}{25s+1} & \dfrac{0.1886s-4.8087}{15.7347s^2+10.8348s+1} \\ \dfrac{3}{7s+1} & \dfrac{-8.1355}{7.9088s+1} \end{bmatrix}.$$

In this reduced system, the manipulated input variables correspond to: u_1 air flow rate to the catalyst regenerator, u_2 opening of the regenerated catalyst valve, and the controlled outputs are the following: y_1 riser temperature, y_2 regenerator dense phase temperature, y_3: regenerator dilute phase temperature.

In the simulations considered here, model Θ_1 is assumed to be the true model, while model Θ_3 represents the nominal model that is used into the MPC cost. In the discussion that follows, unless explicitly mentioned, the adopted tuning parameters of the controller are $m=3$, $T=1$, $Q_y=0.5*diag(1\ \ 1\ \ 1)$, $Q_u=diag(1\ \ 1)$, $R=diag(1\ \ 1)$, $S_y=10^3*diag(1\ \ 1\ \ 1)$ and $S_u=10^5*diag(1\ \ 1)$. The input and output constraints, as well as the maximum input increments, are shown in Tables 1 and 2.

Output	y_{min}	y_{max}
y_1 (ºC)	510	530
y_2 (ºC)	600	610
y_3 (ºC)	530	590

Table 1. Output zones of the FCC system

Input	Δu_{max}	u_{min}	u_{max}
u_1 (ton/h)	25	75	250
u_2 (%)	25	25	101

Table 2. Input constraints of the FCC system

Before starting the detailed analysis of the properties of the proposed robust controller, we find it useful to justify the need for a robust controller for this specific system. We compare, the performance of the proposed robust controller defined through Problem P2, with the performance of the nominal MPC defined through Problem P1. We consider the same scenario described above except for the input targets that are not fully included in the control problem (we consider a target only to input u_1 by simply making $Q_u = diag(1 \quad 0)$ and $S_u = 10^5 * diag(1 \quad 0)$. This is a possible situation that may happen in practice when the process optimizer is sending a target to one of the outputs. Figures 2 and 3 show the output and input responses respectively for the two controllers when the system starts from a steady state where the outputs are outside their zones. It is clear that the conventional MPC cannot stabilize the plant corresponding to model Θ_1 when the controller uses model Θ_3 to calculate the output predictions. However, the proposed robust controller performs quite well and is able to bring the three outputs to their zones

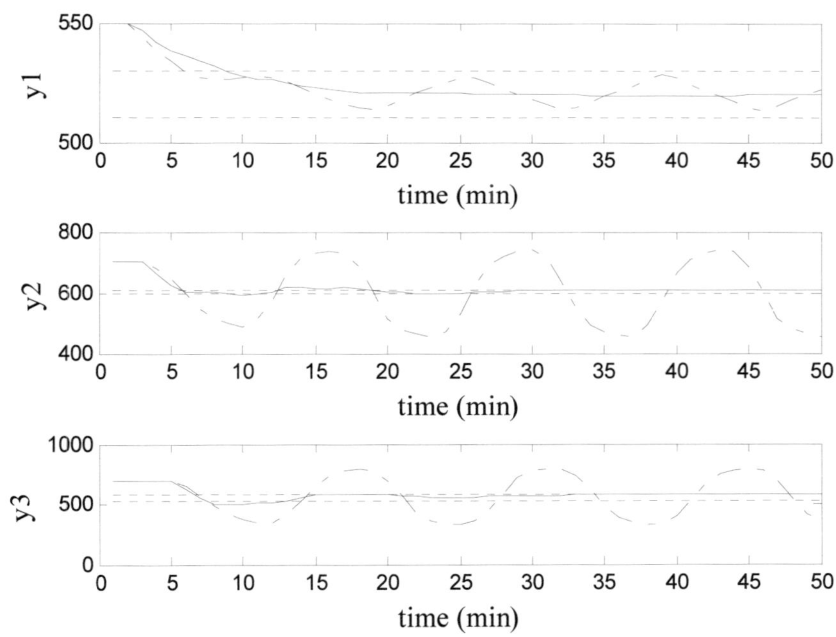

Fig. 2. Controlled outputs for the nominal (- - -) and robust (——) MPC.

We now concentrate our analysis on the application of the proposed controller to the FCC system. As was defined in Eq. (5), each of the three models produces an input feasible set, whose intersection constitutes the restricted input feasible set of the controller. These sets have different shapes and sizes for different stationary operating points (since the disturbance $d_n(k)$ is included into Eq. (5), except for the true model case, where the input feasible set remains unmodified as the estimated states exactly match the true states. The closed loop simulation begins at u_{ss}=[230.5977 60.2359] and y_{ss}=[549.5011 704.2756 690.6233], which are values taken from the real FCC system. For such an operating point, the input feasible set corresponding to models 1, 2 and 3 are depicted in Figure 4. These sets are quite distinct from each other, which results in an empty restricted feasible input set for the controller ($\vartheta_u = \vartheta_u(\Theta_1) \cap \vartheta_u(\Theta_2) \cap \vartheta_u(\Theta_3)$). This means that, we cannot find an input that, taking into account the gains of all the models and all the estimated states, satisfies the output constraints.

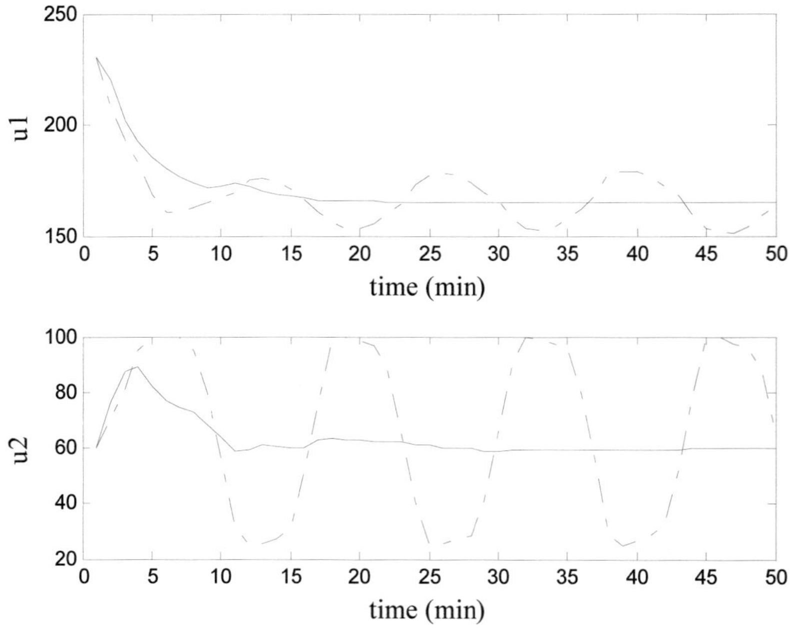

Fig. 3. Manipulated inputs for the nominal (- - -) and robust (——) MPC.

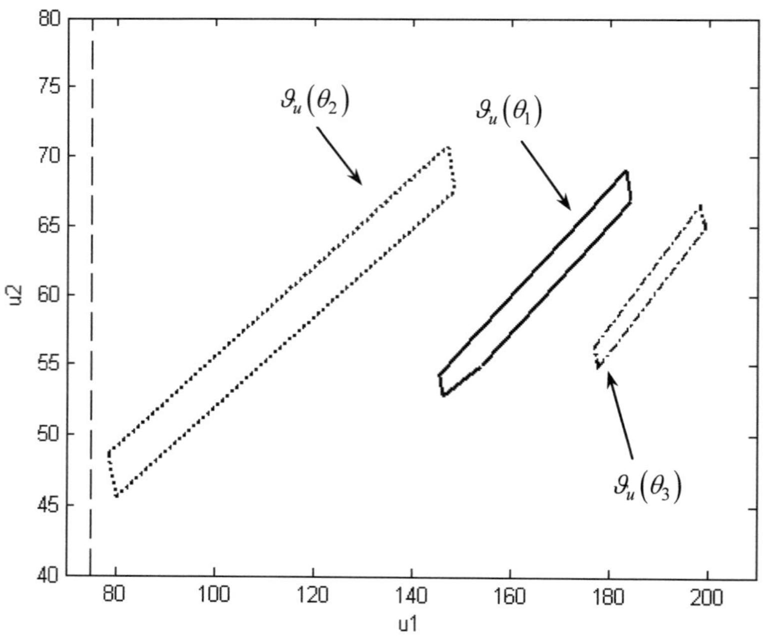

Fig. 4. Input feasible sets of the FCC system

The first objective of the control simulation is to stabilize the system input at $u_{des}^{a} = [165 \;\; 60]$. This input corresponds to the output $y = [520 \; 606.8 \; 577.6]$ for the true system (Θ_1), which results in the input feasible sets shown in Figure 5a. In this figure, it can be seen that the input feasible set corresponding to model 1 is the same as in Fig. 4, while the sets corresponding to the other models adapt their shape and size to the new steady state. Once the system is stabilized at this new steady state, we simulate a step change in the target of the input (at time step *k*=50 min). The new target is given by $u_{des}^{b} = [175 \;\; 64]$, and the corresponding input feasible sets are shown in Figure 5b. In this case, it can be seen that the new target remains inside the new input feasible set ϑ_u^b, which means that the cost can be guided to zero for the true model. Finally, at time step *k*=100 min, when the system reaches the steady state, a different input target is introduced ($u_{des}^{c} = [175 \;\; 58]$). Differently from the previous targets, this new target is outside the input feasible set ϑ_u^c, as can be seen in Figure 5c. Since in this case, the cost cannot be guided to zero and the output requirements are more important than the input ones, the inputs are stabilized in a feasible point as close as possible to the desired target. This is an interesting property of the controller as such a change in the target is likely to occur in the real plant operation.

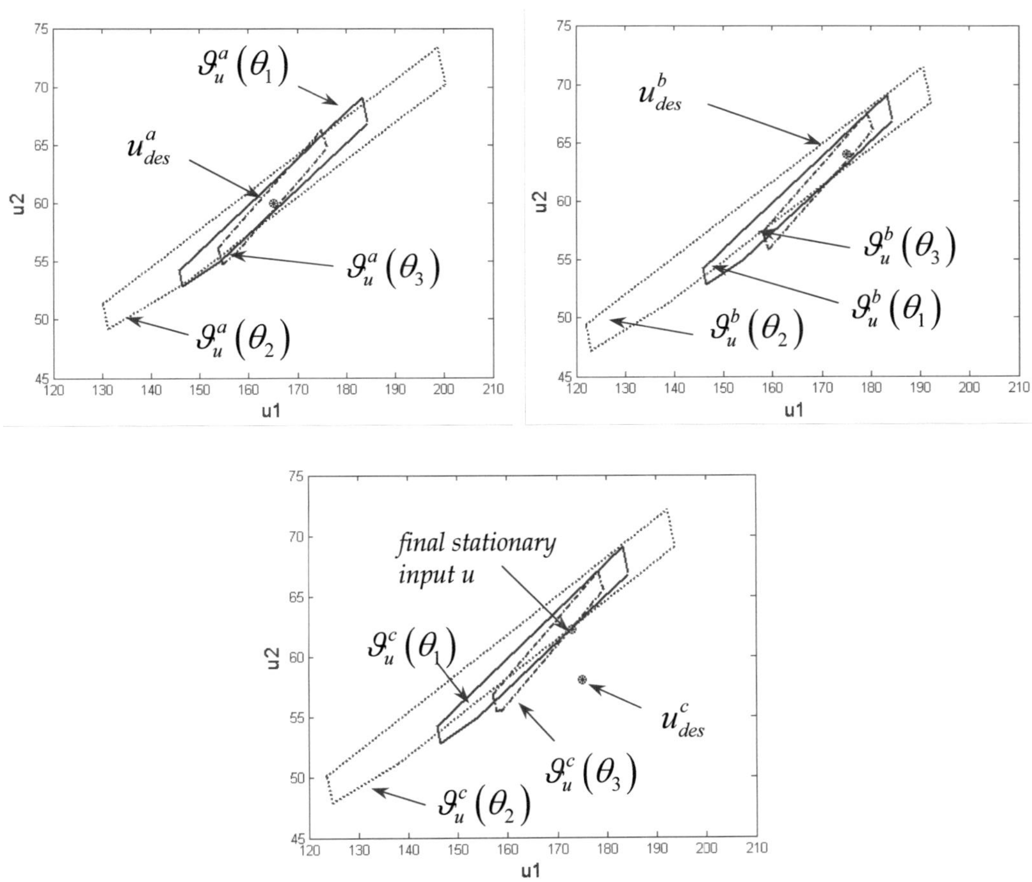

Fig. 5. (a): Initial input feasible sets; (b): Input feasible sets when the first input target is changed; and (c): Input feasible sets when the second input target is changed.

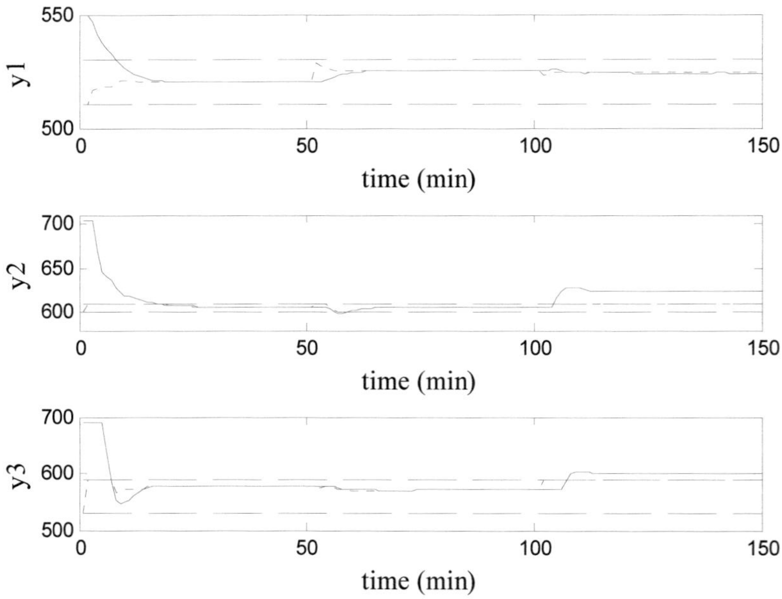

Fig. 6. Controlled outputs and set points for the FCC subsystem with modified input target.

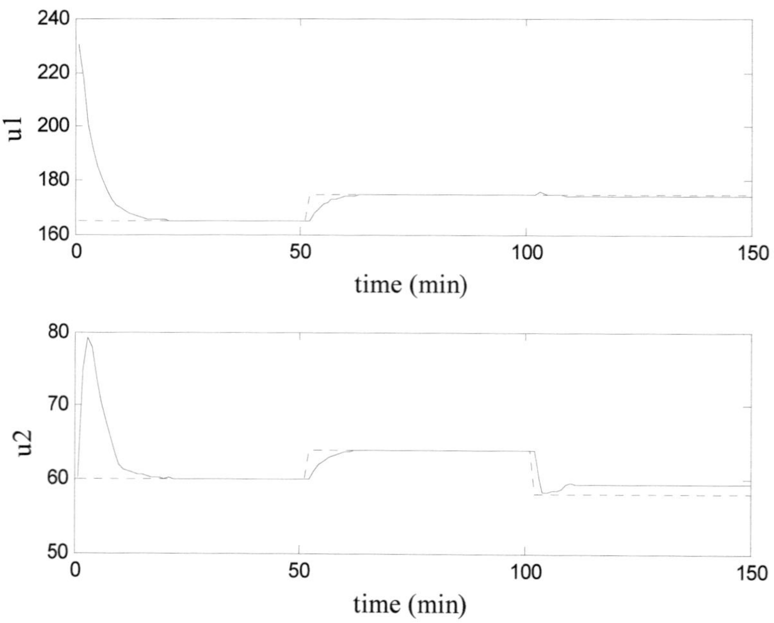

Fig. 7. Manipulated inputs for the FCC subsystem with different input target.

Figure 6 shows the true system outputs (solid line), the set point variables (dotted line) and the output zones (dashed line) for the complete sequence of changes. Figure 7, on the other hand, shows the inputs (solid line), and the input targets (dotted line) for the same sequence. As was established in Theorem 1, the cost function corresponding to the true system is strictly decreasing, and this can be seen in Figure 8. In this figure, the solid line represents the true cost function, while the dotted line represents the cost corresponding to model 3. It is interesting to observe that this last cost function is not decreasing, since the estimated state does not match exactly the true state. Note also that in the last period of time, the cost does not reach zero, as the new target is not inside the input feasible set.

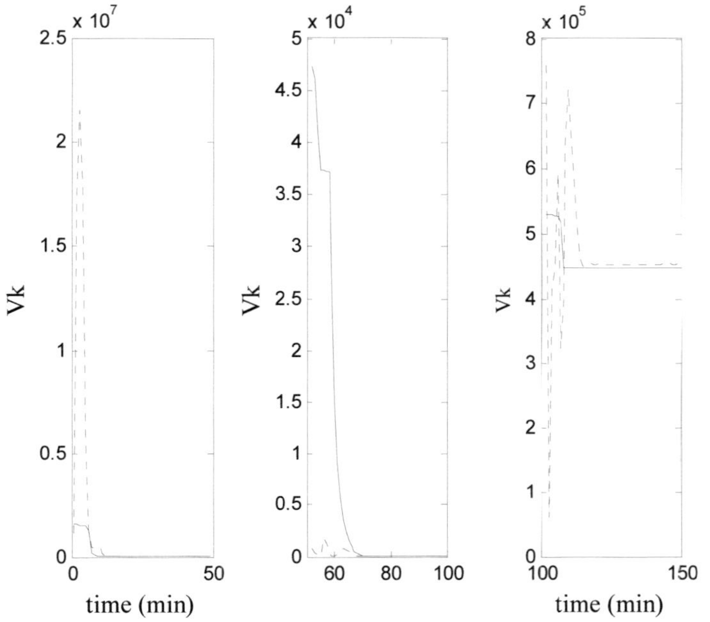

Fig. 8. Cost function corresponding to the true system (solid line) and cost corresponding to model 3 (dotted line).

Output	y_{min}	y_{max}
y_1 (°C)	510	550
y_2 (°C)	400	500
y_3 (°C)	350	500

Table 3. New output zones for the FCC subsystem

Next, we simulate a change in the output zones. The new bounds are given in Table 3. Corresponding to the new control zones, the input feasible set changes its dimension and shape significantly. In Figure 9, $\vartheta_u^a(\Theta_1)$ corresponds to the initial feasible set for the true model, and $\vartheta_u^d(\Theta_1)$, $\vartheta_u^d(\Theta_2)$ and $\vartheta_u^d(\Theta_3)$ represent the new input feasible sets for the three models considered in the robust controller. Since the input target is outside the input feasible set $\vartheta_u^d = \vartheta_u^d(\Theta_1) \cap \vartheta_u^d(\Theta_2) \cap \vartheta_u^d(\Theta_3)$, it is not possible to guide the system to a point in which the control cost is null at the end of the simulation time. When the output weight S_y is as large as the input weight S_u, all the outputs are guided to their corresponding zones, while the inputs show a steady state offset with respect to the target u_{des}^a. The complete behavior of the outputs and inputs of the FCC subsystem, as well as the output set-points, can be seen in Figures 10 and 11, respectively when $S_y = 10^3 * diag(1 \;\; 1 \;\; 1)$ and $S_u = 10^3 * diag(1 \;\; 1)$. The final stationary value of the input is u= [155 84], which represents the closest feasible input value to the target u_{des}^a. Finally, Figure 12 shows the control cost of the two simulated time periods. Observe that in the last period of time (from 51min to 100 min) the true cost function does not reach zero since the change in the operating point prevents the input and output constraints to be satisfied simultaneously.

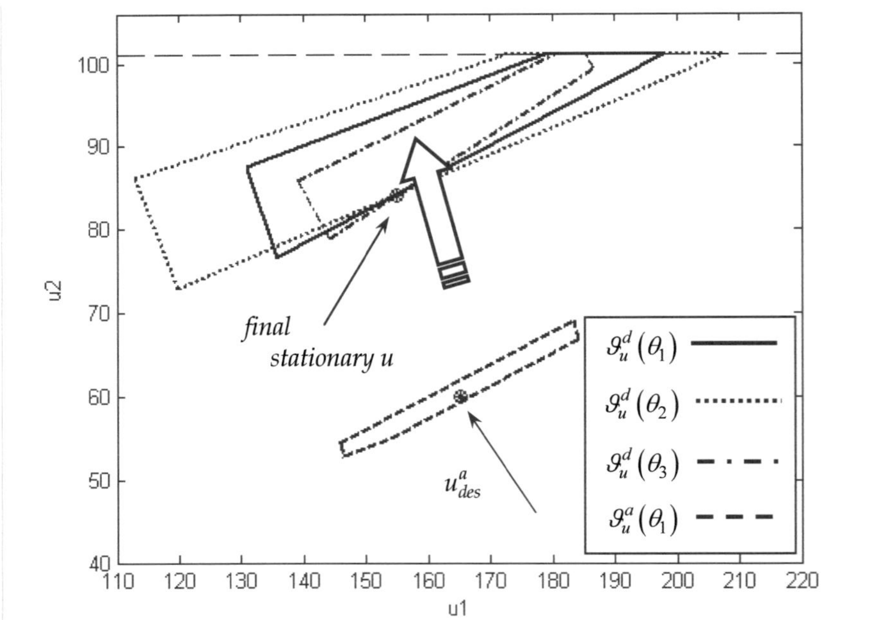

Fig. 9. Input feasible sets for the FCC subsystem when a change in the output zones is introduced.

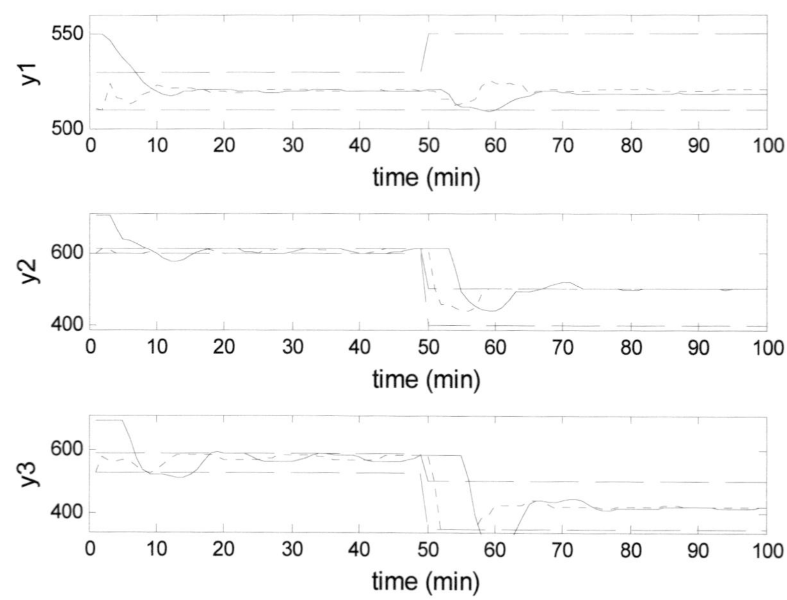

Fig. 10. Controlled outputs and set points for the FCC subsystem with modified zones.

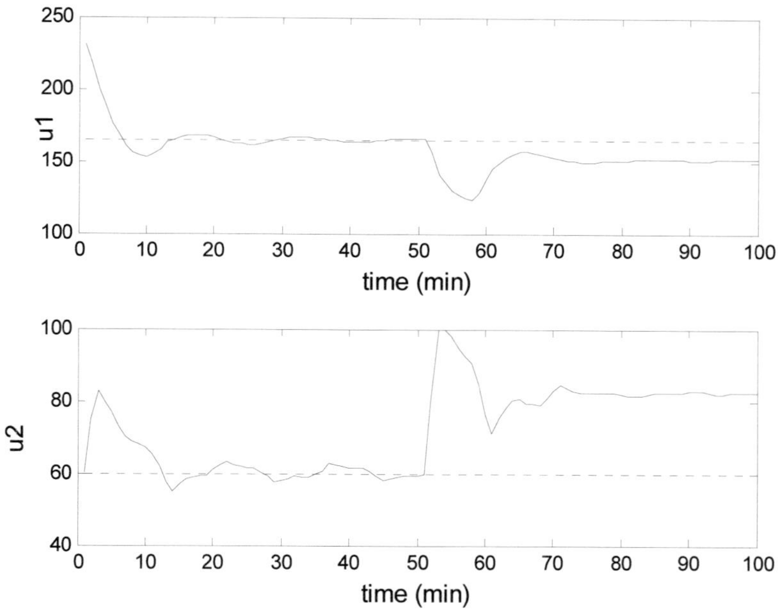

Fig. 11. Manipulated inputs for the FCC subsystem with modified output zones.

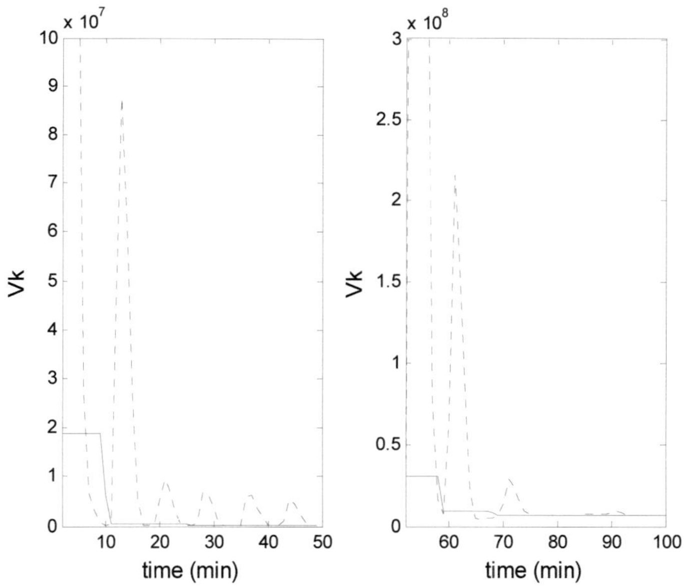

Fig. 12. Cost function for the FCC subsystem with modified zones. True cost function (solid line); Cost function of Model 3 (dotted line).

7. Conclusion

In this chapter, a robust MPC previously presented in the literature was extended to the output zone control of time delayed system with input targets. To this end an extended model that incorporates additional states to account for the time delay is presented. The control structure assumes that model uncertainty can be represented as a discrete set of models (multi-model uncertainty). The proposed approach assures both, recursive feasibility and stability of the closed loop system. The main idea consists in using an extended set of variables in the control optimization problem, which includes the set point to each predicted output. This approach introduces additional degrees of freedom in the zone control problem. Stability is achieved by imposing non-increasing cost constraints that prevent the cost corresponding to the true plant to increase. The strategy was shown, by simulation, to have an adequate performance for a 2*x*3 subsystem of a typical industrial system.

8. References

Badgwell T. A. (1997). Robust model predictive control of stable linear systems. *International Journal of Control*, 68, 797-818.

González A. H.; Odloak D.; Marchetti J. L. & Sotomayor O. (2006). IHMPC of a Heat-Exchanger Network. *Chemical Engineering Research and Design*, 84 (A11), 1041-1050.

González A. H. & Odloak D. (2009). Stable MPC with zone control. Journal of Process Control, 19, 110-122.

González A. H.; Odloak D. & Marchetti J. L. (2009) Robust Model Predictive Control with zone control. IET Control Theory Appl., 3, (1), 121–135.

González A. H.; Odloak D. & Marchetti J. L. (2007). Extended robust predictive control of integrating systems. *AIChE Journal,* 53 1758-1769.

Kassmann D. E.; Badgwell T. & Hawkings R. B. (2000). Robust target calculation for model predictive control. *AIChE Journal*, 45 (5), 1007-1023.

Muske K.R. & Badgwell T. A. (2002). Disturbance modeling for offset free linear model predictive control. *Journal of Process Control,* 12, 617-632.

Odloak D. (2004). Extended robust model predictive control. *AIChE Journal*, 50 (8) 1824-1836.

Pannochia G. & Rawlings J. B. (2003). Disturbance models for offset-free model-predictive control. *AIChE Journal*, 49, 426-437.

Qin S.J. & Badgwell T. A. (2003). A Survey of Industrial Model Predictive Control Technology, *Control Engineering Practice*, 11 (7), 733-764.

Rawlings J. B. (2000). Tutorial overview of model predictive control. IEEE Control Systems Magazine, 38-52.

Sotomayor O. A. Z. & Odloak D. (2005). Observer-based fault diagnosis in chemical plants. *Chemical Engineering Journal*, 112, 93-108.

Zanin A. C.; Gouvêa M. T. & Odloak D. (2002). Integrating real time optimization into the model predictive controller of the FCC system. *Contr. Eng. Pract.*, 10, 819-831.

16

Robust Fuzzy Control of Parametric Uncertain Nonlinear Systems Using Robust Reliability Method

Shuxiang Guo
Faculty of Mechanics, College of Science, Air Force Engineering University Xi'an 710051, P R China

1. Introduction

Stability is of primary importance for any control systems. Stability of both linear and nonlinear uncertain systems has received a considerable attention in the past decades (see for example, Tanaka & Sugeno, 1992; Tanaka, Ikeda, & Wang, 1996; Feng, Cao, Kees, et al. 1997; Teixeira & Zak, 1999; Lee, Park, & Chen, 2001; Park, Kim, & Park, 2001; Chen, Liu, & Tong, 2006; Lam & Leung, 2007, and references therein). Fuzzy logical control (FLC) has proved to be a successful control approach for a great many complex nonlinear systems. Especially, the well-known Takagi-Sugeno (T-S) fuzzy model has become a convenient tool for dealing with complex nonlinear systems. T-S fuzzy model provides an effective representation of nonlinear systems with the aid of fuzzy sets, fuzzy rules and a set of local linear models. Once the fuzzy model is obtained, control design can be carried out via the so called parallel distributed compensation (PDC) approach, which employs multiple linear controllers corresponding to the locally linear plant models (Hong & Langari, 2000). It has been shown that the problems of controller synthesis of nonlinear systems described by the T-S fuzzy model can be reduced to convex problems involving linear matrix inequalities (LMIs) (Park, Kim, & Park, 2001). Many significant results on the stability and robust control of uncertain nonlinear systems using T-S fuzzy model have been reported (see for example, Hong, & Langari, 2000; Park, Kim, & Park, 2001; Xiu & Ren, 2005; Wu & Cai, 2006; Yoneyama, 2006; 2007), and considerable advances have been made. However, as stated in Guo (2010), many approaches for stability and robust control of uncertain systems are often characterized by conservatism when dealing with uncertainties. In practice, uncertainty exists in almost all engineering systems and is frequently a source of instability and deterioration of performance. So, uncertainty is one of the most important factors that have to be taken into account rationally in system analysis and synthesis. Moreover, it has been shown (Guo, 2010) that the increasing in conservatism in dealing with uncertainties by some traditional methods does not mean the increasing in reliability. So, it is significant to deal with uncertainties by means of reliability approach and to achieve a balance between reliability and performance/control-cost in design of uncertain systems.

In fact, traditional probabilistic reliability methods have ever been utilized as measures of stability, robustness, and active control effectiveness of uncertain structural systems by Spencer et al. (1992,1994); Breitung et al. (1998) and Venini & Mariani (1999) to develop

robust control strategies which maximize the overall reliability of controlled structures. Robust control design of systems with parametric uncertainties have also been studied by Mengali and Pieracci (2000); Crespo and Kenny (2005). These works are meaningful in improving the reliability of uncertain controlled systems, and it has been shown that the use of reliability analysis may be rather helpful in evaluating the inherent uncertainties in system design. However, these works are within the probabilistic framework.
In Guo (2007,2010), a non-probabilistic robust reliability method for dealing with bounded parametric uncertainties of linear controlled systems has been presented. The non-probabilistic procedure can be implemented more conveniently than probabilistic one whether in dealing with the uncertainty data or in controller design of uncertain systems, since complex computations are often associated with in controller design of uncertain systems. In this chapter, following the basic idea developed in Guo (2007, 2010), we focus on developing a robust reliability method for robust fuzzy controller design of uncertain nonlinear systems.

2. Problem statements and preliminary knowledge

Consider a nonlinear uncertain system represented by the following T-S fuzzy model with parametric uncertainties:

Plant Rule *i*:

$$\begin{aligned}&\text{IF } x_1(t) \text{ is } F_{i1} \text{ and } \dots \text{ and } x_n(t) \text{ is } F_{in},\\ &\text{THEN } \dot{\mathbf{x}}(t) = \mathbf{A}_i(\boldsymbol{\rho})\mathbf{x}(t) + \mathbf{B}_i(\boldsymbol{\rho})\mathbf{u}(t),\ (i = 1,\dots,r)\end{aligned} \tag{1}$$

Where F_{ij} is a fuzzy set, $\boldsymbol{x}(t) \in R^n$ is the state vector, $\boldsymbol{u}(t) \in R^m$ is the control input vector. r is the number of rules of the T-S fuzzy model. The system matrices $\boldsymbol{A}(\rho)$ and $\boldsymbol{B}(\rho)$ depend on the uncertain parameters $\boldsymbol{\rho} = \{\rho_1, \rho_2, \cdots, \rho_p\}$.
The defuzzified output of the fuzzy system can be represented by

$$\dot{\boldsymbol{x}}(t) = \sum_{i=1}^{r} \mu_i(\boldsymbol{x}(t))[\boldsymbol{A}_i(\rho)\boldsymbol{x}(t) + \boldsymbol{B}_i(\rho)\boldsymbol{u}(t)] \tag{2}$$

In which

$$\mu_i(\boldsymbol{x}(t)) = \omega_i(\boldsymbol{x}(t)) \Big/ \sum_{i=1}^{r} \omega_i(\boldsymbol{x}(t)),\quad \omega_i(\boldsymbol{x}(t)) = \prod_{j=1}^{n} F_{ij}(x_j(t)) \tag{3}$$

Where $F_{ij}(x_j(t))$ is the grade of membership of $x_j(t)$ in the fuzzy set F_{ij}, $\omega_i(\boldsymbol{x}(t))$ satisfies $\omega_i(\boldsymbol{x}(t)) \geq 0$ for all i ($i = 1,\dots,r$). Therefore, there exist the following relations

$$\mu_i(\boldsymbol{x}(t)) \geq 0\ (i = 1,\dots,r),\ \sum_{i=1}^{r} \mu_i(\boldsymbol{x}(t)) = 1 \tag{4}$$

If the system (1) is local controllable, a fuzzy model of state feedback controller can be stated as follows:

Control Rule *i*:

$$\text{IF } x_1(t) \text{ is } F_{i1} \text{ and } \dots \text{ and } x_n(t) \text{ is } F_{in},\ \text{THEN } \boldsymbol{u}(t) = \boldsymbol{K}_i\boldsymbol{x}(t),\ (i = 1,\dots,r) \tag{5}$$

Where $K_i \in R^{m\times n}$ $(i=1,\dots,r)$ are gain matrices to be determined. The final output of the fuzzy controller can be obtained by

$$u(t)=\sum_{i=1}^{r}\mu_i(x(t))K_i x(t) \tag{6}$$

By substituting the control law (6) into (2), we obtain the closed-loop system as follows

$$\dot{x}(t)=\sum_{i=1}^{r}\sum_{j=1}^{r}\mu_i(x(t))\mu_j(x(t))[A_i(\rho)+B_i(\rho)K_j]x(t) \tag{7}$$

When the parameterized notation (Tuan, Apkarian, and Narikiyo 2001) is used, equations (6) and (7) can be rewritten respectively as

$$u(t)=K(\mu)x(t) \tag{8}$$

$$\dot{x}(t)=\big(A(\rho,\mu)+B(\rho,\mu)K(\mu)\big)x(t) \tag{9}$$

Where

$$\mu=\big(\mu_1(x(t)),\dots,\mu_r(x(t))\big)\in\Omega:=\left\{\mu\in R^r:\sum_{i=1}^{r}\mu_i(x(t))=1,\mu_i(x(t))\ge 0\right\} \tag{10}$$

$$K(\mu)=\sum_{i=1}^{r}\mu_i(x(t))K_i\ ,\ A(\rho,\mu)=\sum_{i=1}^{r}\mu_i(x(t))A_i(\rho)\ ,\ B(\rho,\mu)=\sum_{i=1}^{r}\mu_i(x(t))B_i(\rho) \tag{11}$$

Note that the uncertain parameters $\rho=\{\rho_1,\rho_2,\cdots,\rho_p\}$ are involved in the expressions of (9) and (11). Following the basic idea developed by Guo (2007,2010), the uncertain-but-bounded parameters $\rho=\{\rho_1,\rho_2,\cdots,\rho_p\}$ involved in the problem can be considered as interval variables and expressed in the following normalized form

$$\rho_i=\rho_{i0}+\rho_{id}\delta_i\ \ (i=1,\dots,p) \tag{12}$$

where ρ_{i0} and ρ_{id} are respectively the nominal and deviational values of the uncertain parameter ρ_i, $\delta_i\in[-1,1]$ is a standard interval variable. Furthermore, the system matrices are expressed in a corresponding form of that depend on the standard interval variables $\delta=[\delta_1,\delta_2,\cdots,\delta_p]$. Suppose that the stability of the control system can be reduced to solving a matrix inequality as follows

$$M(\delta,P_1,P_2,\dots,P_l)<0 \tag{13}$$

where, $P_1,P_2,\dots,P_l$ are feasible matrices to be determined. The sign " <0 " indicates that the matrix is negative-definite.

If the performance function (it may also be referred to as limit-state function) used for reliability analysis is defined in terms of the criterion (13) and represented by $M=M(\delta,P_1,P_2,\dots,P_l)$, and the reliable domain in the space built by the standard variables

$\boldsymbol{\delta}=[\delta_1,\delta_2,\cdots,\delta_p]$ is indicated by $\Omega_r=\{\boldsymbol{\delta}:\boldsymbol{M}(\boldsymbol{\delta},\boldsymbol{P}_1,\boldsymbol{P}_2,\ldots,\boldsymbol{P}_l)<0\}$, then the robust reliability can be given as follows

$$\eta_r = \sup_{\boldsymbol{\delta}\in \boldsymbol{R}^p} \left\{\|\boldsymbol{\delta}\|_\infty : \boldsymbol{M}(\boldsymbol{\delta},\boldsymbol{P}_1,\boldsymbol{P}_2,\ldots,\boldsymbol{P}_l)<0\right\}-1 \tag{14}$$

Where, $\|\boldsymbol{\delta}\|_\infty$ denote the infinity norm of the vector $\boldsymbol{\delta}=[\delta_1,\delta_2,\cdots,\delta_p]$. Essentially, the robust reliability η_r defined by (14) represents the admissible maximum degree of expansion of the uncertain parameters in the infinity topology space built by the standard interval variables under the condition of that (13) is satisfied. If $\eta_r>0$ holds, the system is reliable for all admissible uncertainties. The larger the value of η_r, the more robust the system with respect to uncertainties and the system is more reliable for this reason. So it is referred to as robust reliability in the paper as that in Ben-Haim (1996) and Guo (2007,2010).

The main objective of this chapter is to develop a method based on the robust reliability idea to deal with bounded parametric uncertainties of the system (1) and to obtain reliability-based robust fuzzy controller (6) for stabilizing the nonlinear system.

Before deriving the main results, the following lemma is given to provide a basis.

Lemma 1 (Guo, 2010). Given real matrices $\boldsymbol{Y}$, $\boldsymbol{E}_1,\boldsymbol{E}_2,\cdots,\boldsymbol{E}_n$, $\boldsymbol{F}_1,\boldsymbol{F}_2,\ldots$, and $\boldsymbol{F}_n$ with appropriate dimensions and $\boldsymbol{Y}=\boldsymbol{Y}^T$, then for any uncertain matrices $\boldsymbol{\Delta}_1=diag\{\delta_{11},\cdots,\delta_{1m_1}\}$, $\boldsymbol{\Delta}_2=diag\{\delta_{21},\cdots,\delta_{2m_2}\}$, $\cdots$, and $\boldsymbol{\Delta}_n=diag\{\delta_{n1},\cdots,\delta_{nm_n}\}$ satisfying $|\delta_{ij}|\le\alpha$ ($i=1,\ldots,n$, $j=1,\ldots,m_n$), the following inequality holds for all admissible uncertainties

$$\boldsymbol{Y}+\sum_{i=1}^{n}\left(\boldsymbol{E}_i\boldsymbol{\Delta}_i\boldsymbol{F}_i+\boldsymbol{F}_i^T\boldsymbol{\Delta}_i^T\boldsymbol{E}_i^T\right)<0 \tag{15}$$

if and only if there exist n constant positive-definite diagonal matrices $\boldsymbol{H}_1$, $\boldsymbol{H}_2$, …, and $\boldsymbol{H}_n$ with appropriate dimensions such that

$$\boldsymbol{Y}+\sum_{i=1}^{n}\left(\boldsymbol{E}_i\boldsymbol{H}_i\boldsymbol{E}_i^T+\alpha^2\boldsymbol{F}_i^T\boldsymbol{H}_i^{-1}\boldsymbol{F}_i\right)<0 \tag{16}$$

3. Methodology and main results

3.1 Basic theory

The following commonly used Lyapunov function is considered

$$V(\boldsymbol{x}(t))=\boldsymbol{x}(t)^T\boldsymbol{P}\boldsymbol{x}(t) \tag{17}$$

where $\boldsymbol{P}$ is a symmetric positive definite matrix. The time derivative of $\boldsymbol{V}(\boldsymbol{x}(t))$ is

$$\dot{V}(\boldsymbol{x}(t))=\dot{\boldsymbol{x}}(t)^T\boldsymbol{P}\boldsymbol{x}(t)+\boldsymbol{x}(t)^T\boldsymbol{P}\dot{\boldsymbol{x}}(t) \tag{18}$$

Substituting (9) into (18), we can obtain

$$\dot{V}(\boldsymbol{x}(t))=\boldsymbol{x}^T(t)\left\{(\boldsymbol{A}(\rho,\mu)+\boldsymbol{B}(\rho,\mu)\boldsymbol{K}(\mu))^T\boldsymbol{P}+\boldsymbol{P}(\boldsymbol{A}(\rho,\mu)+\boldsymbol{B}(\rho,\mu)\boldsymbol{K}(\mu))\right\}\boldsymbol{x}(t) \tag{19}$$

So, $\dot{V}(\boldsymbol{x}(t))<0$ is equivalent to (20) and further equivalent to (21) that are represented as follows

$$\left(A(\rho,\mu)+B(\rho,\mu)K(\mu)\right)^T P+P\left(A(\rho,\mu)+B(\rho,\mu)K(\mu)\right)<0\,,\ \mu\in\Omega \tag{20}$$

$$\left(A(\rho,\mu)X+B(\rho,\mu)Y(\mu)\right)^T+\left(A(\rho,\mu)X+B(\rho,\mu)Y(\mu)\right)<0\,,\ \mu\in\Omega \tag{21}$$

In which, $X=P^{-1}$, $Y(\mu)=K(\mu)X$ possess the following form

$$Y(\mu)=\sum_{i=1}^{r}\mu_i(x(t))Y_i=\sum_{i=1}^{r}\mu_i(x(t))K_iX\ \ (i=1,\ldots,r) \tag{22}$$

Let

$$Q_{ij}(\rho,X,Y_j)=(A_i(\rho)X+B_i(\rho)Y_j)+(A_i(\rho)X+B_i(\rho)Y_j)^T\ \ (i,j=1,\ldots,r) \tag{23}$$

then (21) can be written as

$$\sum_{i=1}^{r}\sum_{j=1}^{r}\mu_i\mu_jQ_{ij}(\rho,X,Y_j)<0 \tag{24}$$

Some convex relaxations for (24) have been developed to make it tractable. Two type of relaxation are adopted here to illustrate the presented method.

3.1.1 A simple relaxation of (24) represented as follows is often used by authors (Lee, Park, & Chen 2001)

$$Q_{ii}(\rho,X,Y_i)<0\,,\ Q_{ij}(\rho,X,Y_j)+Q_{ji}(\rho,X,Y_i)<0\ \ (1\le i<j\le r) \tag{25}$$

These expressions can be rewritten respectively as

$$(A_i(\rho)X+B_i(\rho)Y_i)+(A_i(\rho)X+B_i(\rho)Y_i)^T<0\ \ (i=1,\ldots,r) \tag{26}$$

$$\begin{aligned}(\mathbf{A}_i(\boldsymbol{\rho})\mathbf{X}+\mathbf{B}_i(\boldsymbol{\rho})\mathbf{Y}_j)+(\mathbf{A}_i(\boldsymbol{\rho})\mathbf{X}+\mathbf{B}_i(\boldsymbol{\rho})\mathbf{Y}_j)^T+(\mathbf{A}_j(\boldsymbol{\rho})\mathbf{X}+\mathbf{B}_j(\boldsymbol{\rho})\mathbf{Y}_i)+\\+(\mathbf{A}_j(\boldsymbol{\rho})\mathbf{X}+\mathbf{B}_j(\boldsymbol{\rho})\mathbf{Y}_i)^T<0\quad(1\le i<j\le r)\end{aligned} \tag{27}$$

Expressing all the uncertain parameters $\rho=\{\rho_1,\rho_2,\cdots,\rho_p\}$ as the standard form of (12), furthermore, the system matrices are expressed as a corresponding form of that depend on the standard interval variables $\delta=[\delta_1,\delta_2,\cdots,\delta_p]$. Without loss of generality, suppose that all the uncertain matrices $A_i(\rho)$ and $B_i(\rho)$ can be expressed as

$$A_i(\rho)=A_{i0}+\sum_{j=1}^{p}A_{ij}\delta_{ij}\,,\ B_i(\rho)=B_{i0}+\sum_{k=1}^{q}B_{ik}\delta_{ik}\ \ (i=1,\ldots,r) \tag{28}$$

In which, A_{i0}, B_{i0}, A_{ij}, and B_{ik} are known real constant matrices determined by the nominal and deviational values of the basic variables. To reduce the conservatism caused by dealing with uncertainties as far as possible, representing all the matrices A_{ij} and B_{ik} as the form of the vector products as follows

$$A_{ij} = V_{ij1}V_{ij2}^T,\ B_{ik} = U_{ij1}U_{ik2}^T\ \ (i=1,\dots,r,\ j=1,\dots,p,\ k=1,\dots,q) \tag{29}$$

In which, V_{ij1}, V_{ij2}, U_{ik1}, and U_{ik2} are all column vectors. Denoting

$$\begin{aligned}
&\mathbf{V}_{i1} = \begin{bmatrix}\mathbf{V}_{i11} & \mathbf{V}_{i21} & \cdots & \mathbf{V}_{ip1}\end{bmatrix},\ \ \mathbf{V}_{i2} = \begin{bmatrix}\mathbf{V}_{i12} & \mathbf{V}_{i22} & \cdots & \mathbf{V}_{ip2}\end{bmatrix}^T;\\
&\mathbf{U}_{i1} = \begin{bmatrix}\mathbf{U}_{i11} & \mathbf{U}_{i21} & \cdots & \mathbf{U}_{iq1}\end{bmatrix};\ \ \mathbf{U}_{i2} = \begin{bmatrix}\mathbf{U}_{i12} & \mathbf{U}_{i22} & \cdots & \mathbf{U}_{iq2}\end{bmatrix}^T,\\
&\mathbf{\Delta}_{i1} = diag\{\delta_{i1},\cdots,\delta_{ip}\};\ \ \mathbf{\Delta}_{i2} = diag\{\delta_{i1},\cdots,\delta_{iq}\};\ \ (i=1,\dots,r)
\end{aligned} \tag{30}$$

where, the first four matrices are constructed by the column vectors involved in (29). Then, the expressions in (28) can be further written as

$$A_i(\rho) = A_{i0} + V_{i1}\Delta_{i1}V_{i2},\ B_i(\rho) = B_{i0} + U_{i1}\Delta_{i2}U_{i2}\ \ (i=1,\dots,r) \tag{31}$$

Substituting (31) into equations (26) and (27), we can obtain

$$\begin{aligned}
&(\mathbf{A}_{i0}\mathbf{X} + \mathbf{B}_{i0}\mathbf{Y}_i) + (\mathbf{A}_{i0}\mathbf{X} + \mathbf{B}_{i0}\mathbf{Y}_i)^T + (\mathbf{V}_{i1}\mathbf{\Delta}_{i1}\mathbf{V}_{i2}\mathbf{X}) + (\mathbf{V}_{i1}\mathbf{\Delta}_{i1}\mathbf{V}_{i2}\mathbf{X})^T\\
&+(\mathbf{U}_{i1}\mathbf{\Delta}_{i2}\mathbf{U}_{i2}\mathbf{Y}_i) + (\mathbf{U}_{i1}\mathbf{\Delta}_{i2}\mathbf{U}_{i2}\mathbf{Y}_i)^T < 0\ \ (i=1,\dots,r)
\end{aligned} \tag{32}$$

$$\begin{aligned}
&(\mathbf{A}_{i0}\mathbf{X} + \mathbf{B}_{i0}\mathbf{Y}_j) + (\mathbf{A}_{i0}\mathbf{X} + \mathbf{B}_{i0}\mathbf{Y}_j)^T + (\mathbf{V}_{i1}\mathbf{\Delta}_{i1}\mathbf{V}_{i2}\mathbf{X}) + (\mathbf{V}_{i1}\mathbf{\Delta}_{i1}\mathbf{V}_{i2}\mathbf{X})^T\\
&\quad + (\mathbf{U}_{i1}\mathbf{\Delta}_{i2}\mathbf{U}_{i2}\mathbf{Y}_j) + (\mathbf{U}_{i1}\mathbf{\Delta}_{i2}\mathbf{U}_{i2}\mathbf{Y}_j)^T +\\
&+(\mathbf{A}_{j0}\mathbf{X} + \mathbf{B}_{j0}\mathbf{Y}_i) + (\mathbf{A}_{j0}\mathbf{X} + \mathbf{B}_{j0}\mathbf{Y}_i)^T + (\mathbf{V}_{j1}\mathbf{\Delta}_{j1}\mathbf{V}_{j2}\mathbf{X}) + (\mathbf{V}_{j1}\mathbf{\Delta}_{j1}\mathbf{V}_{j2}\mathbf{X})^T\\
&\quad + (\mathbf{U}_{j1}\mathbf{\Delta}_{j2}\mathbf{U}_{j2}\mathbf{Y}_i) + (\mathbf{U}_{j1}\mathbf{\Delta}_{j2}\mathbf{U}_{j2}\mathbf{Y}_i)^T < 0\ \ \ (1 \le i < j \le r)
\end{aligned} \tag{33}$$

In terms of Lemma 1, the matrix inequality (32) holds for all admissible uncertainties if and only if there exist diagonal positive-definite matrices E_{i1} and E_{i2} with appropriate dimensions such that

$$\begin{aligned}
&(\mathbf{A}_{i0}\mathbf{X} + \mathbf{B}_{i0}\mathbf{Y}_i) + (\mathbf{A}_{i0}\mathbf{X} + \mathbf{B}_{i0}\mathbf{Y}_i)^T + \mathbf{V}_{i1}\mathbf{E}_{i1}\mathbf{V}_{i1}^T + \alpha^2(\mathbf{V}_{i2}\mathbf{X})^T\mathbf{E}_{i1}^{-1}(\mathbf{V}_{i2}\mathbf{X}) +\\
&+\mathbf{U}_{i1}\mathbf{E}_{i2}\mathbf{U}_{i1}^T + \alpha^2(\mathbf{U}_{i2}\mathbf{Y}_i)^T\mathbf{E}_{i2}^{-1}(\mathbf{U}_{i2}\mathbf{Y}_i) < 0\ \ \ \ (i=1,\dots,r)
\end{aligned} \tag{34}$$

Similarly, (33) holds for all admissible uncertainties if and only if there exist constant diagonal positive-definite matrices H_{ij1}, H_{ij2}, H_{ij3}, and H_{ij4} such that

$$\begin{aligned}
&(\mathbf{A}_{i0}\mathbf{X} + \mathbf{B}_{i0}\mathbf{Y}_j) + (\mathbf{A}_{i0}\mathbf{X} + \mathbf{B}_{i0}\mathbf{Y}_j)^T + \mathbf{V}_{i1}\mathbf{H}_{ij1}\mathbf{V}_{i1}^T + \mathbf{V}_{j1}\mathbf{H}_{ij2}\mathbf{V}_{j1}^T +\\
&+\mathbf{U}_{i1}\mathbf{H}_{ij3}\mathbf{U}_{i1}^T + \mathbf{U}_{j1}\mathbf{H}_{ij4}\mathbf{U}_{j1}^T + (\mathbf{A}_{j0}\mathbf{X} + \mathbf{B}_{j0}\mathbf{Y}_i) +\\
&(\mathbf{A}_{j0}\mathbf{X} + \mathbf{B}_{j0}\mathbf{Y}_i)^T + \alpha^2(\mathbf{V}_{i2}\mathbf{X})^T\mathbf{H}_{ij1}^{-1}(\mathbf{V}_{i2}\mathbf{X}) + \alpha^2(\mathbf{V}_{j2}\mathbf{X})^T\mathbf{H}_{ij2}^{-1}(\mathbf{V}_{j2}\mathbf{X}) +\\
&+\alpha^2(\mathbf{U}_{i2}\mathbf{Y}_j)^T\mathbf{H}_{ij3}^{-1}(\mathbf{U}_{i2}\mathbf{Y}_j) + \alpha^2(\mathbf{U}_{j2}\mathbf{Y}_i)^T\mathbf{H}_{ij4}^{-1}(\mathbf{U}_{j2}\mathbf{Y}_i) < 0\ \ (1 \le i < j \le r)
\end{aligned} \tag{35}$$

Applying the well-known Schur complement, (34) and (35) can be written respectively as

$$\begin{bmatrix}
\Xi_i & \alpha X V_{i2}^T & (\alpha U_{i2}Y_i)^T\\
\alpha V_{i2}X & -E_{i1} & 0\\
\alpha U_{i2}Y & 0 & -E_{i2}
\end{bmatrix} < 0\ \ (i=1,\dots,r) \tag{36}$$

$$\begin{bmatrix} \Gamma_{ij} & * & * & * & * \\ \alpha V_{i2}X & -H_{ij1} & * & * & * \\ \alpha V_{j2}X & 0 & -H_{ij2} & * & * \\ \alpha U_{i2}Y_j & 0 & 0 & -H_{ij3} & * \\ \alpha U_{j2}Y_i & 0 & 0 & 0 & -H_{ij4} \end{bmatrix} < 0 \ (1 \le i < j \le r) \tag{37}$$

In which,

$\Xi_i = (A_{i0}X + B_{i0}Y_i) + (A_{i0}X + B_{i0}Y_i)^T + V_{i1}E_{i1}V_{i1}^T + U_{i1}E_{i2}U_{i1}^T$,

$\Gamma_{ij} = (\mathbf{A}_{i0}\mathbf{X} + \mathbf{B}_{i0}\mathbf{Y}_j) + (\mathbf{A}_{i0}\mathbf{X} + \mathbf{B}_{i0}\mathbf{Y}_j)^T + (A_{j0}X + B_{j0}Y_i) + (A_{j0}X + B_{j0}Y_i)^T + V_{i1}H_{ij1}V_{i1}^T + V_{j1}H_{ij2}V_{j1}^T + U_{i1}H_{ij3}U_{i1}^T + U_{j1}H_{ij4}U_{j1}^T$. "*" denotes the transposed matrices in the symmetric positions.

Consequently, the following theorem can be obtained.

Theorem 3.1. For the dynamic system (2) with the uncertain matrices represented by (31) and $|\delta_m| \le \alpha$ ($m = 1,\ldots,p$), it is asymptotically stabilizable with the state feedback controller (6) if there exist a symmetric positive-definite matrix X, matrices Y_i, and constant diagonal positive-definite matrices E_{i1}, E_{i2}, H_{ij1}, H_{ij2}, H_{ij3}, and H_{ij4} ($1 \le i < j \le r$) such that the LMIs represented by (36) and (37) hold for all admissible uncertainties. If the feasible matrices X and Y_i are found out, then the feedback gain matrices deriving the fuzzy controller (6) can be obtained by

$$K_i = Y_i X^{-1} \quad (i = 1,\ldots,r) \tag{38}$$

It should be stated that the condition of (25) is restrictive in practice. It is adopted yet here is merely to show the proposed reliability method and for comparison.

3.1.2 Some improved relaxation for (24) have also been proposed in literatures. A relaxation provided by Tuan, Apkarian, and Narikiyo (2001) is as follows

$$Q_{ii}(\rho, X, Y_i) < 0 \quad (i = 1,\ldots,r) \tag{39}$$

$$Q_{ii}(\rho, X, Y_i) + \frac{r-1}{2}\big(Q_{ij}(\rho, X, Y_j) + Q_{ji}(\rho, X, Y_i)\big) < 0 \quad (1 \le i \ne j \le r) \tag{40}$$

The expression (39) is the same completely with the first expression of (25). So, only (40) is investigated further. It can be rewritten as

$$\begin{aligned} &(\mathbf{A}_i(\boldsymbol{\rho})\mathbf{X} + \mathbf{B}_i(\boldsymbol{\rho})\mathbf{Y}_i) + (\mathbf{A}_i(\boldsymbol{\rho})\mathbf{X} + \mathbf{B}_i(\boldsymbol{\rho})\mathbf{Y}_i)^T + \\ &\frac{r-1}{2}\{(\mathbf{A}_i(\boldsymbol{\rho})\mathbf{X} + \mathbf{B}_i(\boldsymbol{\rho})\mathbf{Y}_j) + (\mathbf{A}_i(\boldsymbol{\rho})\mathbf{X} + \mathbf{B}_i(\boldsymbol{\rho})\mathbf{Y}_j)^T + \\ &\quad + (\mathbf{A}_j(\boldsymbol{\rho})\mathbf{X} + \mathbf{B}_j(\boldsymbol{\rho})\mathbf{Y}_i) + (\mathbf{A}_j(\boldsymbol{\rho})\mathbf{X} + \mathbf{B}_j(\boldsymbol{\rho})\mathbf{Y}_i)^T\} < 0 \quad (1 \le i \ne j \le r) \end{aligned} \tag{41}$$

On substituting the expression (31) into (41), we obtain

$$\begin{aligned}&\{(\mathbf{A}_{i0}\mathbf{X}+\mathbf{B}_{i0}\mathbf{Y}_i)+(\mathbf{A}_{i0}\mathbf{X}+\mathbf{B}_{i0}\mathbf{Y}_i)^T\}+\frac{r-1}{2}\{(\mathbf{A}_{i0}\mathbf{X}+\mathbf{B}_{i0}\mathbf{Y}_j)+(\mathbf{A}_{i0}\mathbf{X}+\mathbf{B}_{i0}\mathbf{Y}_j)^T\}+\\&+\frac{r-1}{2}\{(\mathbf{A}_{j0}\mathbf{X}+\mathbf{B}_{j0}\mathbf{Y}_i)+(\mathbf{A}_{j0}\mathbf{X}+\mathbf{B}_{j0}\mathbf{Y}_i)^T\}+\frac{r+1}{2}\{(\mathbf{V}_{i1}\mathbf{\Delta}_{i1}\mathbf{V}_{i2}\mathbf{X})+(\mathbf{V}_{i1}\mathbf{\Delta}_{i1}\mathbf{V}_{i2}\mathbf{X})^T\}+\\&+(\mathbf{U}_{i1}\mathbf{\Delta}_{i2}\mathbf{U}_{i2}\mathbf{Y}_i)+(\mathbf{U}_{i1}\mathbf{\Delta}_{i2}\mathbf{U}_{i2}\mathbf{Y}_i)^T+(\frac{r-1}{2}\mathbf{U}_{i1}\mathbf{\Delta}_{i2}\mathbf{U}_{i2}\mathbf{Y}_j)+(\frac{r-1}{2}\mathbf{U}_{i1}\mathbf{\Delta}_{i2}\mathbf{U}_{i2}\mathbf{Y}_j)^T\\&+\frac{r-1}{2}\{(\mathbf{V}_{j1}\mathbf{\Delta}_{j1}\mathbf{V}_{j2}\mathbf{X})+(\mathbf{V}_{j1}\mathbf{\Delta}_{j1}\mathbf{V}_{j2}\mathbf{X})^T+(\mathbf{U}_{j1}\mathbf{\Delta}_{j2}\mathbf{U}_{j2}\mathbf{Y}_i)+(\mathbf{U}_{j1}\mathbf{\Delta}_{j2}\mathbf{U}_{j2}\mathbf{Y}_i)^T\}<0\\&\qquad (1\le i\ne j\le r)\end{aligned}\tag{42}$$

In terms of Lemma 1, the matrix inequality (42) hold for all admissible uncertainties if and only if there exist constant diagonal positive-definite matrices F_{i1}, F_{i2}, F_{i3}, H_{ij1}, and H_{ij2} such that

$$\begin{aligned}&\{(\mathbf{A}_{i0}\mathbf{X}+\mathbf{B}_{i0}\mathbf{Y}_i)+(\mathbf{A}_{i0}\mathbf{X}+\mathbf{B}_{i0}\mathbf{Y}_i)^T\}+\frac{r-1}{2}\{(\mathbf{A}_{i0}\mathbf{X}+\mathbf{B}_{i0}\mathbf{Y}_j)+(\mathbf{A}_{i0}\mathbf{X}+\mathbf{B}_{i0}\mathbf{Y}_j)^T\}\\&+\frac{r-1}{2}\{(\mathbf{A}_{j0}\mathbf{X}+\mathbf{B}_{j0}\mathbf{Y}_i)+(\mathbf{A}_{j0}\mathbf{X}+\mathbf{B}_{j0}\mathbf{Y}_i)^T\}+\left(\frac{r+1}{2}\mathbf{V}_{i1}\right)\mathbf{F}_{i1}\left(\frac{r+1}{2}\mathbf{V}_{i1}\right)^T+\alpha^2(\mathbf{V}_{i2}\mathbf{X})^T\mathbf{F}_{i1}^{-1}(\mathbf{V}_{i2}\mathbf{X})\\&+\mathbf{U}_{i1}\mathbf{F}_{i2}\mathbf{U}_{i1}^T+\alpha^2(\mathbf{U}_{i2}\mathbf{Y}_i)^T\mathbf{F}_{i2}^{-1}(\mathbf{U}_{i2}\mathbf{Y}_i)\}+\left(\frac{r-1}{2}\mathbf{U}_{i1}\right)\mathbf{H}_{ij1}\left(\frac{r-1}{2}\mathbf{U}_{i1}\right)^T+\alpha^2(\mathbf{U}_{i2}\mathbf{Y}_j)^T\mathbf{H}_{ij1}^{-1}(\mathbf{U}_{i2}\mathbf{Y}_j)\\&+\left(\frac{r-1}{2}\mathbf{V}_{j1}\right)\mathbf{F}_{j3}\left(\frac{r-1}{2}\mathbf{V}_{j1}\right)^T+\left(\frac{r-1}{2}\mathbf{U}_{j1}\right)\mathbf{H}_{ij2}\left(\frac{r-1}{2}\mathbf{U}_{j1}\right)^T+\alpha^2(\mathbf{V}_{j2}\mathbf{X})^T\mathbf{F}_{j3}^{-1}(\mathbf{V}_{j2}\mathbf{X})\\&+\alpha^2(\mathbf{U}_{j2}\mathbf{Y}_i)^T\mathbf{H}_{ij2}^{-1}(\mathbf{U}_{j2}\mathbf{Y}_i)\}<0\quad(1\le i\ne j\le r)\end{aligned}\tag{43}$$

Applying the Schur complement, (43) is equivalent to

$$\begin{bmatrix}\boldsymbol{\Psi}_{ij} & * & * & * & * & *\\ \alpha V_{i2}X & -F_{i1} & * & * & * & *\\ \alpha U_{i2}Y_i & 0 & -F_{i2} & * & * & *\\ \alpha V_{j2}X & 0 & 0 & -F_{j3} & * & *\\ \alpha U_{i2}Y_j & 0 & 0 & 0 & -H_{ij1} & *\\ \alpha U_{j2}Y_i & 0 & 0 & 0 & 0 & -H_{ij2}\end{bmatrix}<0\ \ (1\le i\ne j\le r)\tag{44}$$

In which,
$$\begin{aligned}\boldsymbol{\Psi}_{ij}=&\{(\mathbf{A}_{i0}\mathbf{X}+\mathbf{B}_{i0}\mathbf{Y}_i)+(\mathbf{A}_{i0}\mathbf{X}+\mathbf{B}_{i0}\mathbf{Y}_i)^T\}+\frac{r-1}{2}\{(\mathbf{A}_{i0}\mathbf{X}+\mathbf{B}_{i0}\mathbf{Y}_j)+(\mathbf{A}_{i0}\mathbf{X}+\mathbf{B}_{i0}\mathbf{Y}_j)^T\}\\&+\frac{r-1}{2}\{(\mathbf{A}_{j0}\mathbf{X}+\mathbf{B}_{j0}\mathbf{Y}_i)+(\mathbf{A}_{j0}\mathbf{X}+\mathbf{B}_{j0}\mathbf{Y}_i)^T\}+\left(\frac{r+1}{2}\mathbf{V}_{i1}\right)\mathbf{F}_{i1}\left(\frac{r+1}{2}\mathbf{V}_{i1}\right)^T+\mathbf{U}_{i1}\mathbf{F}_{i2}\mathbf{U}_{i1}^T+\\&\left(\frac{r-1}{2}\mathbf{V}_{j1}\right)\mathbf{F}_{j3}\left(\frac{r-1}{2}\mathbf{V}_{j1}\right)^T+\left(\frac{r-1}{2}\mathbf{U}_{i1}\right)\mathbf{H}_{ij1}\left(\frac{r-1}{2}\mathbf{U}_{i1}\right)^T+\left(\frac{r-1}{2}\mathbf{U}_{j1}\right)\mathbf{H}_{ij2}\left(\frac{r-1}{2}\mathbf{U}_{j1}\right)^T.\end{aligned}$$

This can be summarized as follows.

Theorem 3.2. For the dynamic system (2) with the uncertain matrices represented by (31) and $|\delta_m|\le\alpha$ ($m=1,\dots,p$), it is asymptotically stabilizable with the state feedback controller (6) if there exist a symmetric positive-definite matrix X, matrices Y_i, and constant diagonal positive-definite matrices E_{i1}, E_{i2}, F_{i1}, F_{i2}, F_{i3}, H_{ij1}, and H_{ij2} ($1 \le i \ne j \le r$) such that the LMIs (36) and (44) hold for all admissible uncertainties. If the feasible matrices X and Y_i are found out, the feedback gain matrices deriving the fuzzy controller (6) can then be given by (38).

3.2 Robust reliability based stabilization

In terms of Theorem 3.1, the closed-loop fuzzy system (7) is stable if all the matrix inequalities (36) and (37) hold for all admissible uncertainties. So, the performance functions used for calculation of reliability of that the uncertain system to be stable can be taken as

$$M_i(\alpha, X, Y_i, E_{i1}, E_{i2}) = \begin{bmatrix} \Xi_i & \alpha X V_{i2}^T & (\alpha U_{i2} Y_i)^T \\ \alpha V_{i2} X & -E_{i1} & 0 \\ \alpha U_{i2} Y_i & 0 & -E_{i2} \end{bmatrix} \quad (i = 1, \ldots, r) \tag{45}$$

$$M_{ij}(\alpha, X, Y_i, Y_j, H_{ij1}, H_{ij2}, H_{ij3}, H_{ij4}) = \begin{bmatrix} \Gamma_{ij} & * & * & * & * \\ \alpha V_{i2} X & -H_{ij1} & * & * & * \\ \alpha V_{j2} X & 0 & -H_{ij2} & * & * \\ \alpha U_{i2} Y_j & 0 & 0 & -H_{ij3} & * \\ \alpha U_{j2} Y_i & 0 & 0 & 0 & -H_{ij4} \end{bmatrix} \quad (1 \le i < j \le r) \tag{46}$$

in which, the expressions of Ξ_i and Γ_{ij} are in the same form respectively as that in (36) and (37).

Therefore, the robust reliability of the uncertain nonlinear system in the sense of stability can be expressed as

$$\eta_r = \sup_{\alpha \in \mathbf{R}^+} \left\{ \alpha : \mathbf{M}_i(\alpha, \mathbf{X}, \mathbf{Y}_i, \mathbf{E}_{i1}, \mathbf{E}_{i2}) < 0, \mathbf{M}_{ij}(\alpha, \mathbf{X}, \mathbf{Y}_i, \mathbf{Y}_j, \mathbf{H}_{ij1}, \mathbf{H}_{ij2}, \mathbf{H}_{ij3}, \mathbf{H}_{ij4}) < 0,\ 1 \le i < j \le r \right\} - 1 \tag{47}$$

where, R^+ denotes the set of all positive real numbers. The robust reliability of that the uncertain closed-loop system (7) is stable may be obtained by solving the following optimization problem

$$\begin{aligned} &\text{Maximize } \alpha \\ &\text{Subject to } \mathbf{M}_i(\alpha, \mathbf{X}, \mathbf{Y}_i, \mathbf{E}_{i1}, \mathbf{E}_{i2}) < 0 \ \ \mathbf{M}_{ij}(\alpha, \mathbf{X}, \mathbf{Y}_i, \mathbf{Y}_j, \mathbf{H}_{ij1}, \mathbf{H}_{ij2}, \mathbf{H}_{ij3}, \mathbf{H}_{ij4}) < 0 \\ &\mathbf{E}_{i1} > 0, \mathbf{E}_{i2} > 0 \ \ \mathbf{X} > 0, \mathbf{H}_{ij1} > 0, \mathbf{H}_{ij2} > 0, \mathbf{H}_{ij3} > 0, \mathbf{H}_{ij4} > 0 \quad (1 \le i < j \le r) \end{aligned} \tag{48}$$

From the viewpoint of robust stabilizing controller design, if inequalities (36) and (37) hold for all admissible uncertainties, then there exists a fuzzy controller (6) such that the closed-loop system (7) to be asymptotically stable. Therefore, the performance functions used for reliability-based design of control to stabilize the uncertain system (2) can also be taken as that of (45) and (46). So, a possible stabilizing controller satisfying the robust reliability requirement can be given by a feasible solution of the following matrix inequalities

$$\begin{aligned} &\mathbf{M}_i(\alpha^*, \mathbf{X}, \mathbf{Y}_i, \mathbf{E}_{i1}, \mathbf{E}_{i2}) < 0, \ \mathbf{M}_{ij}(\alpha^*, \mathbf{X}, \mathbf{Y}_i, \mathbf{Y}_j, \mathbf{H}_{ij1}, \mathbf{H}_{ij2}, \mathbf{H}_{ij3}, \mathbf{H}_{ij4}) < 0; \ \mathbf{E}_{i1} > 0 \mathbf{E}_{i2} > 0, \\ &\mathbf{X} > 0, \mathbf{H}_{ij1} > 0, \mathbf{H}_{ij2} > 0, \mathbf{H}_{ij3} > 0, \mathbf{H}_{ij4} > 0 \quad (1 \le i < j \le r) \ \ \alpha^* = \eta_{cr} + 1 \end{aligned} \tag{49}$$

In which, $M_i(\cdot)$ and $M_{ij}(\cdot)$ are functions of some matrices and represented by (45) and (46) respectively. η_{cr} is the allowable robust reliability.
If the control cost is taken into account, the robust reliability based design optimization of stabilization controller can be carried out by solving the following optimization problem

$$\begin{aligned}&\text{Minimize } Trace(\mathbf{N});\\&\text{Subject to } \mathbf{M}_i(\alpha^*,\mathbf{X},\mathbf{Y}_i,\mathbf{E}_{i1},\mathbf{E}_{i2})<0\ \mathbf{M}_{ij}(\alpha^*,\mathbf{X},\mathbf{Y}_i,\mathbf{Y}_j,\mathbf{H}_{ij1},\mathbf{H}_{ij2},\mathbf{H}_{ij3},\mathbf{H}_{ij4})<0\\&\mathbf{E}_{i1}>0,\ \mathbf{E}_{i2}>0,\ \mathbf{H}_{ij1}>0,\ \mathbf{H}_{ij2}>0,\ \mathbf{H}_{ij3}>0,\ \mathbf{H}_{ij4}>0,\quad (1\le i<j\le r)\\&\begin{bmatrix}\mathbf{N} & \mathbf{I}\\ \mathbf{I} & \mathbf{X}\end{bmatrix}>0,\ \mathbf{X}>0,\mathbf{N}>0,\ \alpha^*=\eta_{cr}+1\end{aligned}\tag{50}$$

In which, the introduced additional matrix N is symmetric positive-definite and with the same dimension as X. When the feasible matrices X and Y_i are found out, the optimal fuzzy controller could be obtained by using (6) together with (38).
If Theorem 3.2 is used, the expression of $M_{ij}(\cdot)$ corresponding to (46) becomes

$$M_{ij}(\alpha,X,Y_i,Y_j,F_{i1},F_{i2},F_{i3},H_{ij1},H_{ij2})=\begin{bmatrix}\Psi_{ij} & * & * & * & * & *\\ \alpha V_{i2}X & -F_{i1} & * & * & * & *\\ \alpha U_{i2}Y_i & 0 & -F_{i2} & * & * & *\\ \alpha V_{j2}X & 0 & 0 & -F_{j3} & * & *\\ \alpha U_{i2}Y_j & 0 & 0 & 0 & -H_{ij1} & *\\ \alpha U_{j2}Y_i & 0 & 0 & 0 & 0 & -H_{ij2}\end{bmatrix}\ (1\le i\ne j\le r)\tag{51}$$

where, Ψ_{ij} is the same with that in (44). Correspondingly, (47) and (48) become respectively as follows.

$$\begin{aligned}\eta_r=\sup_{\alpha\in\mathbf{R}^+}\{&\alpha:\mathbf{M}_i(\alpha,\mathbf{X},\mathbf{Y}_i,\mathbf{E}_{i1},\mathbf{E}_{i2})<0,\mathbf{M}_{ij}(\alpha,\mathbf{X},\mathbf{Y}_i,\mathbf{Y}_j,\mathbf{F}_{i1},\mathbf{F}_{i2},\mathbf{F}_{i3},\mathbf{H}_{ij1},\mathbf{H}_{ij2})<0,\\&\mathbf{E}_{i1}>0,\ \mathbf{E}_{i2}>0,\ \mathbf{X}>0,\mathbf{F}_{i1}>0,\mathbf{F}_{i2}>0,\mathbf{F}_{i3}>0,\mathbf{H}_{ij1}>0,\mathbf{H}_{ij2}>0,\ 1\le i\ne j\le r\}-1\end{aligned}\tag{52}$$

$$\begin{aligned}&\text{Maximize } \alpha\\&\text{Subject to } \mathbf{M}_i(\alpha,\mathbf{X},\mathbf{Y}_i,\mathbf{E}_{i1},\mathbf{E}_{i2})<0\ \ \mathbf{M}_{ij}(\alpha,\mathbf{X},\mathbf{Y}_i,\mathbf{Y}_j,\mathbf{F}_{i1},\mathbf{F}_{i2},\mathbf{F}_{i3},\mathbf{H}_{ij1},\mathbf{H}_{ij2})<0\\&\mathbf{E}_{i1}>0,\mathbf{E}_{i2}>0,\mathbf{X}>0,\mathbf{F}_{i1}>0,\mathbf{F}_{i2}>0,\mathbf{F}_{i3}>0,\mathbf{H}_{ij1}>0,\mathbf{H}_{ij2}>0,\ \ (1\le i\ne j\le r)\end{aligned}\tag{53}$$

Similarly, (50) becomes

$$\begin{aligned}&\text{Minimize } Trace(\mathbf{N})\\&\text{Subject to } \mathbf{M}_i(\alpha^*,\mathbf{X},\mathbf{Y}_i,\mathbf{E}_{i1},\mathbf{E}_{i2})<0,\ \mathbf{M}_{ij}(\alpha^*,\mathbf{X},\mathbf{Y}_i,\mathbf{Y}_j,\mathbf{F}_{i1},\mathbf{F}_{i2},\mathbf{F}_{i3},\mathbf{H}_{ij1},\mathbf{H}_{ij2})<0\\&\mathbf{E}_{i1}>0,\mathbf{E}_{i2}>0,\mathbf{F}_{i1}>0,\mathbf{F}_{i2}>0,\mathbf{F}_{i3}>0,\mathbf{H}_{ij1}>0,\mathbf{H}_{ij2}>0,\quad (1\le i\ne j\le r)\\&\begin{bmatrix}\mathbf{N} & \mathbf{I}\\ \mathbf{I} & \mathbf{X}\end{bmatrix}>0,\ \mathbf{X}>0,\mathbf{N}>0,\ \alpha^*=\eta_{cr}+1\end{aligned}\tag{54}$$

3.3 A special case

Now, we consider a special case in which the matrices of (30) is expressed as

$$V_{i1} = V_1,\ V_{i2} = V_2,\ \Delta_{i1} = \Delta,\ U_{i1} = U_{i2} = 0\ \ (i = 1,\ldots,r) \tag{55}$$

This means that the matrices $A_i(\rho)$ in all the rules have the same uncertainty structure and the matrices $B_i(\rho)$ become certain. In this case, (31) can be written as

$$A_i(\rho) = A_{i0} + V_1\Delta\ V_2,\ B_i(\rho) = B_{i0}\ \ (i = 1,\ldots,r) \tag{56}$$

and the expressions involved in Theorem 3.1 can be simplified further. This is summarized in the following.

Theorem 3.3. For the dynamic system (2) with the matrices represented by (56) and $|\delta_m| \le \alpha$ ($m = 1,\ldots,p$), it is asymptotically stabilizable with the state feedback controller (6) if there exist a symmetric positive-definite matrix X, matrices Y_i, and constant diagonal positive-definite matrices E_i and H_{ij} ($1 \le i < j \le r$) with appropriate dimensions such that the following LMIs hold for all admissible uncertainties

$$\begin{bmatrix} \Xi_i & \alpha X V_2^T \\ \alpha V_2 X & -E_i \end{bmatrix} < 0,\ \begin{bmatrix} \Gamma_{ij} & \alpha X V_2^T \\ \alpha V_2 X & -H_{ij} \end{bmatrix} < 0\ \ (1 \le i < j \le r) \tag{57}$$

In which, $\Xi_i = (A_{i0}X + B_{i0}Y_i) + (A_{i0}X + B_{i0}Y_i)^T + V_1 E_{i1} V_1^T$,

$\Gamma_{ij} = (A_{i0}X + B_{i0}Y_j) + (A_{i0}X + B_{i0}Y_j)^T + (A_{j0}X + B_{j0}Y_i) + (A_{j0}X + B_{j0}Y_i)^T +$ $(2V_1)H_{ij}(2V_1)^T$. If the feasible matrices X and Y_i are found out, the feedback gain matrices deriving the fuzzy controller (6) can then be given by (38).

Proof. In the case, inequalities (32) and (33) become, respectively,

$$(A_{i0}X + B_{i0}Y_i) + (A_{i0}X + B_{i0}Y_i)^T + (V_1\Delta\ V_2X) + (V_1\Delta\ V_2X)^T < 0\ \ (i = 1,\ldots,r) \tag{58}$$

$$\begin{aligned}&(\mathbf{A}_{i0}\mathbf{X} + \mathbf{B}_{i0}\mathbf{Y}_j) + (\mathbf{A}_{i0}\mathbf{X} + \mathbf{B}_{i0}\mathbf{Y}_j)^T + 2(\mathbf{V}_1\Delta\ \mathbf{V}_2\mathbf{X}) + 2(\mathbf{V}_1\Delta\ \mathbf{V}_2\mathbf{X})^T \\ &+(\mathbf{A}_{j0}\mathbf{X} + \mathbf{B}_{j0}\mathbf{Y}_i) + (\mathbf{A}_{j0}\mathbf{X} + \mathbf{B}_{j0}\mathbf{Y}_i)^T < 0\ \ (1 \le i < j \le r)\end{aligned} \tag{59}$$

In terms of Lemma 1, (58) holds for all admissible uncertainties if and only if there exist diagonal positive-definite matrices E_i ($i = 1,\ldots,r$) with appropriate dimensions such that

$$(A_{i0}X + B_{i0}Y_i) + (A_{i0}X + B_{i0}Y_i)^T + V_1 E_i V_1^T + \alpha^2 (V_2X)^T E_i^{-1}(V_2X) < 0\ \ (i = 1,\ldots,r) \tag{60}$$

Similarly, (59) holds for all admissible uncertainties if and only if there exist constant diagonal positive-definite matrices H_{ij} such that

$$\begin{aligned}&(\mathbf{A}_{i0}\mathbf{X} + \mathbf{B}_{i0}\mathbf{Y}_j) + (\mathbf{A}_{i0}\mathbf{X} + \mathbf{B}_{i0}\mathbf{Y}_j)^T + (\mathbf{A}_{j0}\mathbf{X} + \mathbf{B}_{j0}\mathbf{Y}_i) + \\ &+(\mathbf{A}_{j0}\mathbf{X} + \mathbf{B}_{j0}\mathbf{Y}_i)^T + (2\mathbf{V}_1)\mathbf{H}_{ij}(2\mathbf{V}_1)^T + \alpha^2(\mathbf{V}_2\mathbf{X})^T\mathbf{H}_{ij}^{-1}(\mathbf{V}_2\mathbf{X}) < 0\ \ (1 \le i < j \le r)\end{aligned} \tag{61}$$

Applying Schur complement, (57) can be obtained. So, the theorem holds.

By Theorem 3.3, the performance functions used for reliability calculation can be taken as

$$
\begin{aligned}
&\mathbf{M}_i(\alpha,\mathbf{X},\mathbf{Y}_i,\mathbf{E}_i)=\begin{bmatrix}\Xi_i & \alpha\mathbf{X}\mathbf{V}_2^T\\ \alpha\mathbf{V}_2\mathbf{X} & -\mathbf{E}_i\end{bmatrix},\ (i=1,\ldots,r)\\
&\mathbf{M}_{ij}(\alpha,\mathbf{X},\mathbf{Y}_i,\mathbf{Y}_j,\mathbf{H}_{ij})=\begin{bmatrix}\mathbf{\Gamma}_{ij} & \alpha\mathbf{X}\mathbf{V}_2^T\\ \alpha\mathbf{V}_2\mathbf{X} & -\mathbf{H}_{ij}\end{bmatrix},\ (1\le i<j\le r)
\end{aligned} \tag{62}
$$

Accordingly, a possible stabilizing controller satisfying robust reliability requirement can be obtained by a feasible solution of the following matrix inequalities

$$
\begin{aligned}
&\mathbf{M}_i(\alpha^*,\mathbf{X},\mathbf{Y}_i,\mathbf{E}_i)<0,\ \mathbf{M}_{ij}(\alpha^*,\mathbf{X},\mathbf{Y}_i,\mathbf{Y}_j,\mathbf{H}_{ij})<0,\ \ \mathbf{X}>0,\ \ \mathbf{E}_i>0;\ \ \mathbf{H}_{ij}>0\quad(1\le i<j\le r)\\
&\alpha^*=\eta_{cr}+1
\end{aligned} \tag{63}
$$

The optimum stabilizing controller based on the robust reliability and control cost can be obtained by solving the following optimization problem

Minimize $Trace(\mathbf{N})$

Subject to $\mathbf{M}_i(\alpha^*,\mathbf{X},\mathbf{Y}_i,\mathbf{E}_i)<0,\ \mathbf{M}_{ij}(\alpha^*,\mathbf{X},\mathbf{Y}_i,\mathbf{Y}_j,\mathbf{H}_{ij})<0;\ \mathbf{E}_i>0,\mathbf{H}_{ij}>0,\ (1\le i<j\le r)$ (64)

$\begin{bmatrix}\mathbf{N} & \mathbf{I}\\ \mathbf{I} & \mathbf{X}\end{bmatrix}>0,\ \mathbf{X}>0,\mathbf{N}>0,\ \alpha^*=\eta_{cr}+1$

Similarly, the expressions involved in Theorem 3.2 can also be simplified and the corresponding result is summarized in the following.

Theorem 3.4. For the dynamic system (2) with the matrices represented by (56) and $|\delta_m|\le\alpha$ ($m=1,\ldots,p$), it is asymptotically stabilizable with the state feedback controller (6) if there exist a symmetric positive-definite matrix X, matrices Y_i, and constant diagonal positive-definite matrices E_i and H_{ij} ($1\le i\ne j\le r$) with appropriate dimensions such that the following LMIs hold for all admissible uncertainties

$$
\begin{bmatrix}\Xi_i & \alpha XV_2^T\\ \alpha V_2X & -E_i\end{bmatrix}<0,\ \begin{bmatrix}\Psi_{ij} & (\alpha V_2X)^T\\ (\alpha V_2X) & -H_{ij}\end{bmatrix}<0\ \ (1\le i\ne j\le r) \tag{65}
$$

In which, $\Xi_i=(A_{i0}X+B_{i0}Y_i)+(A_{i0}X+B_{i0}Y_i)^T+V_{i1}E_{i1}V_{i1}^T$,

$\Psi_{ij}=(rV_1)H_{ij}(rV_1)^T+\{(A_{i0}X+B_{i0}Y_i)+(A_{i0}X+B_{i0}Y_i)^T\}$

$+\frac{r-1}{2}\{(A_{i0}X+B_{i0}Y_j)+(A_{i0}X+B_{i0}Y_j)^T+(A_{j0}X+B_{j0}Y_i)+(A_{j0}X+B_{j0}Y_i)^T\}$. If the feasible matrices X and Y_i are found out, the feedback gain matrices deriving the fuzzy controller (6) can then be obtained by (38).

Proof. (42) can be rewritten as

$$
\begin{aligned}
&\{(\mathbf{A}_{i0}\mathbf{X}+\mathbf{B}_{i0}\mathbf{Y}_i)+(\mathbf{A}_{i0}\mathbf{X}+\mathbf{B}_{i0}\mathbf{Y}_i)^T\}+\frac{r-1}{2}\{(\mathbf{A}_{i0}\mathbf{X}+\mathbf{B}_{i0}\mathbf{Y}_j)+(\mathbf{A}_{i0}\mathbf{X}+\mathbf{B}_{i0}\mathbf{Y}_j)^T\\
&+(\mathbf{A}_{j0}\mathbf{X}+\mathbf{B}_{j0}\mathbf{Y}_i)+(\mathbf{A}_{j0}\mathbf{X}+\mathbf{B}_{j0}\mathbf{Y}_i)^T\}+r\{(\mathbf{V}_1\Delta\ \mathbf{V}_2\mathbf{X})+(\mathbf{V}_1\Delta\ \mathbf{V}_2\mathbf{X})^T\}<0\\
&\qquad(1\le i\ne j\le r)
\end{aligned} \tag{66}
$$

In terms of Lemma 1, (66) holds for all admissible uncertainties if and only if there exist diagonal positive-definite matrices $\boldsymbol{H}_{ij}$ such that

$$\begin{aligned}&\left\{(\mathbf{A}_{i0}\mathbf{X}+\mathbf{B}_{i0}\mathbf{Y}_i)+(\mathbf{A}_{i0}\mathbf{X}+\mathbf{B}_{i0}\mathbf{Y}_i)^T\right\}+\frac{r-1}{2}\left\{(\mathbf{A}_{i0}\mathbf{X}+\mathbf{B}_{i0}\mathbf{Y}_j)+(\mathbf{A}_{i0}\mathbf{X}+\mathbf{B}_{i0}\mathbf{Y}_j)^T+(\mathbf{A}_{j0}\mathbf{X}+\mathbf{B}_{j0}\mathbf{Y}_i)\right.\\&\left.+(\mathbf{A}_{j0}\mathbf{X}+\mathbf{B}_{j0}\mathbf{Y}_i)^T\right\}+(r\mathbf{V}_1)\mathbf{H}_{ij}(r\mathbf{V}_1)^T+\alpha^2(\mathbf{V}_2\mathbf{X})^T\mathbf{H}_{ij}^{-1}(\mathbf{V}_2\mathbf{X})<0\quad 1\le i\ne j\le r\end{aligned}\tag{67}$$

Applying Schur complement, (67) is equivalent to the second expression of (65). So, the theorem holds.

By Theorem 3.4, the performance functions used for reliability calculation can be taken as

$$\begin{aligned}&\mathbf{M}_i(\alpha,\mathbf{X},\mathbf{Y}_i,\mathbf{E}_i)=\begin{bmatrix}\Xi_i & (\alpha\mathbf{V}_2\mathbf{X})^T\\ (\alpha\mathbf{V}_2\mathbf{X}) & -\mathbf{E}_i\end{bmatrix},\quad (i=1,\ldots,r)\\&\mathbf{M}_{ij}(\alpha,\mathbf{X},\mathbf{Y}_i,\mathbf{Y}_j,\mathbf{H}_{ij})=\begin{bmatrix}\mathbf{\Psi}_{ij} & (\alpha\mathbf{V}_2\mathbf{X})^T\\ (\alpha\mathbf{V}_2\mathbf{X}) & -\mathbf{H}_{ij}\end{bmatrix},\quad (1\le i\ne j\le r)\end{aligned}\tag{68}$$

So, design of the optimal controller based on the robust reliability and control cost could be carried out by solving the following optimization problem

$$\begin{aligned}&\text{Minimize } Trace(\mathbf{N})\\&\text{Subject to } \mathbf{M}_i(\alpha^*,\mathbf{X},\mathbf{Y}_i,\mathbf{E}_i)<0,\ \mathbf{M}_{ij}(\alpha^*,\mathbf{X},\mathbf{Y}_i,\mathbf{Y}_j,\mathbf{H}_{ij})<0,\ \mathbf{E}_i>0,\mathbf{H}_{ij}>0\ (1\le i\ne j\le r)\\&\begin{bmatrix}\mathbf{N} & \mathbf{I}\\ \mathbf{I} & \mathbf{X}\end{bmatrix}>0,\ \mathbf{X}>0,\mathbf{N}>0\ (\alpha^*=\eta_{cr}+1)\end{aligned}\tag{69}$$

4. Numerical examples

Example 1. Consider a simple uncertain nonlinear mass-spring-damper mechanical system with the following dynamic equation (Tanaka, Ikeda, & Wang 1996)

$$\ddot{x}(t)+\dot{x}(t)+c(t)x(t)=(1+0.13\dot{x}^3(t))u(t)$$

Where $c(t)$ is the uncertain term satisfying $c(t)\in[0.5,1.81]$.

Assume that $x(t)\in[-1.5,1.5]$, $\dot{x}(t)\in[-1.5,1.5]$. Using the following fuzzy sets

$$F_1(\dot{x}(t))=0.5+\frac{\dot{x}^3(t)}{6.75},\quad F_2(\dot{x}(t))=0.5-\frac{\dot{x}^3(t)}{6.75}$$

The uncertain nonlinear system can be represented by the following fuzzy model

Plant Rule 1: IF $\dot{x}(t)$ is about F_1, THEN $\dot{\boldsymbol{x}}(t)=\boldsymbol{A}_1\boldsymbol{x}(t)+\boldsymbol{B}_1\boldsymbol{u}(t)$

Plant Rule 2: IF $\dot{x}(t)$ is about F_2, THEN $\dot{\boldsymbol{x}}(t)=\boldsymbol{A}_2\boldsymbol{x}(t)+\boldsymbol{B}_2\boldsymbol{u}(t)$

Where

$$\mathbf{x}(t)=\begin{bmatrix}\dot{x}(t)\\ x(t)\end{bmatrix},\ \mathbf{A}_1=\mathbf{A}_2=\begin{bmatrix}-1 & -c\\ 1 & 0\end{bmatrix},\ \mathbf{B}_1=\begin{bmatrix}1.43875\\ 0\end{bmatrix},\ \mathbf{B}_2=\begin{bmatrix}0.56125\\ 0\end{bmatrix}$$

Expressing the uncertain parameter c as the normalized form, $c = 1.155 + 0.655\delta$, furthermore, the system matrices are expressed as

$$A_1 = A_{10} + V_1 \Delta V_2, \ A_2 = A_{20} + V_1 \Delta V_2, \ B_1 = B_{10}, \ B_2 = B_{20}.$$

In which

$$A_{10} = A_{20} = \begin{bmatrix} -1 & -1.155 \\ 1 & 0 \end{bmatrix}, V_1 = \begin{bmatrix} 1 \\ 0 \end{bmatrix}, \ V_2 = \begin{bmatrix} 0 & -0.655 \end{bmatrix}, \ \Delta = \delta.$$

By solving the optimization problem of (69) with $\alpha^* = 1$ and $\alpha^* = 3$ respectively, the gain matrices are obtained as follows

$$K_1 = \begin{bmatrix} -0.0567 & -0.1446 \end{bmatrix}, \ K_2 = \begin{bmatrix} -0.0730 & -0.1768 \end{bmatrix} \ (\alpha^* = 1);$$

$$K_1 = \begin{bmatrix} -0.3645 & -1.0570 \end{bmatrix}, \ K_2 = \begin{bmatrix} -0.9191 & -2.4978 \end{bmatrix} \ (\alpha^* = 3).$$

When the initial value of the state is taken as $x(0) = \begin{bmatrix} -1 & -1.3 \end{bmatrix}^T$, the simulation results of the controlled system with the uncertain parameter generated randomly within the allowable range $c(t) \in [0.5, 1.81]$ are shown in Fig. 1.

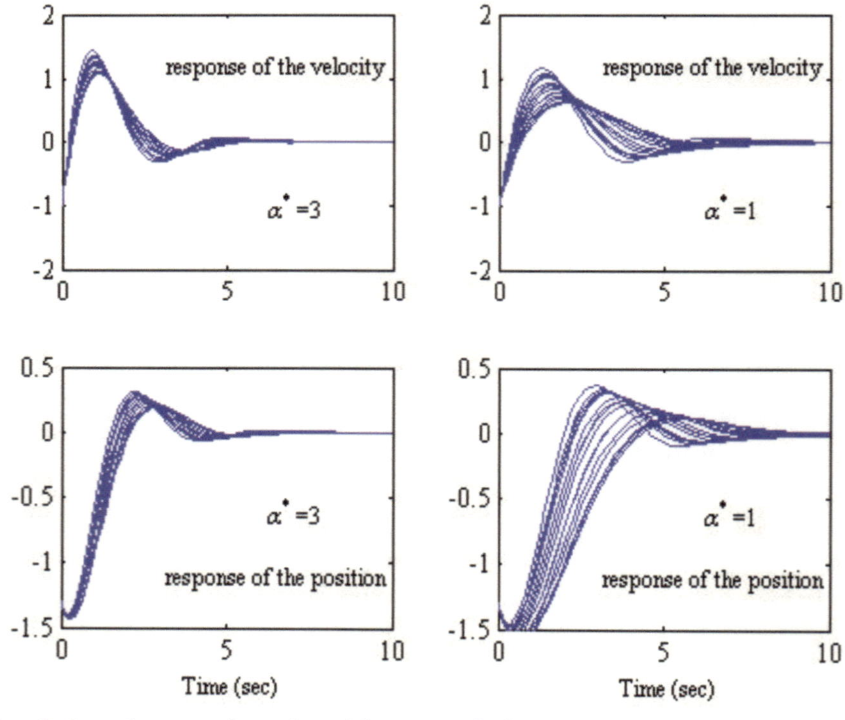

Fig. 1. Simulation of state trajectories of the controlled system (The uncertain parameter c is generated randomly within [0.5, 1.81])

Example 2. Consider the problem of stabilizing the chaotic Lorenz system with parametric uncertainties as follows (Lee, Park, & Chen, 2001)

$$\begin{bmatrix} \dot{x}_1(t) \\ \dot{x}_2(t) \\ \dot{x}_3(t) \end{bmatrix} = \begin{bmatrix} -\sigma x_1(t) + \sigma x_2(t) \\ r x_1(t) - x_2(t) - x_1(t) x_3(t) \\ x_1(t) x_2(t) - b x_3(t) \end{bmatrix}$$

For the purpose of comparison, the T-S fuzzy model of the chaotic Lorenz system is constructed as

Plant Rule 1: IF $x_1(t)$ is about M_1 THEN $\dot{\mathbf{x}}(t) = \mathbf{A}_1\mathbf{x}(t) + \mathbf{B}_1\mathbf{u}(t)$

Plant Rule 2: IF $x_1(t)$ is about M_2 THEN $\dot{\mathbf{x}}(t) = \mathbf{A}_2\mathbf{x}(t) + \mathbf{B}_2\mathbf{u}(t)$

Where

$$\mathbf{A}_1 = \begin{bmatrix} -\sigma & \sigma & 0 \\ r & -1 & -M_1 \\ 0 & M_1 & -b \end{bmatrix}, \ \mathbf{A}_2 = \begin{bmatrix} -\sigma & \sigma & 0 \\ r & -1 & -M_2 \\ 0 & M_2 & -b \end{bmatrix}$$

The input matrices $\boldsymbol{B}_1$ and $\boldsymbol{B}_2$, and the membership functions are respectively

$$\mathbf{B}_1 = \mathbf{B}_2 = \begin{bmatrix} 1 & 0 & 0 \end{bmatrix}^T, \ \mu_1(\mathbf{x}(t)) = \frac{-x_1(t) + M_2}{M_2 - M_1}, \ \mu_2(\mathbf{x}(t)) = \frac{x_1(t) - M_1}{M_2 - M_1}.$$

The nominal values of (σ, r, b) are (10, 28, 8/3), and choosing $[M_1, M_2] = [-20, 30]$. All system parameters are uncertain-but-bounded within 30% of their nominal values.
The gain matrices for deriving the stabilizing controller (6) given by Lee, Park, and Chen (2001) are

$$\boldsymbol{K}_{1L} = \begin{bmatrix} -295.7653 & -137.2603 & -8.0866 \end{bmatrix}, \ \boldsymbol{K}_{2L} = \begin{bmatrix} -443.0647 & -204.8089 & 12.6930 \end{bmatrix}.$$

(1) Reliability-based feasible solutions

In order to apply the presented method, all the uncertain parameters (σ, r, b) are expressed as the following normalized form

$$\sigma = 10 + 3\delta_1, \ r = 28 + 8.4\delta_2, \ b = 8/3 + 0.8\delta_3.$$

Furthermore, the system matrices can be expressed as

$$\boldsymbol{A}_1 = \boldsymbol{A}_{10} + \boldsymbol{V}_1\boldsymbol{\Delta}\boldsymbol{V}_2, \ \boldsymbol{A}_2 = \boldsymbol{A}_{20} + \boldsymbol{V}_1\boldsymbol{\Delta}\boldsymbol{V}_2, \ \boldsymbol{B}_1 = \boldsymbol{B}_{10}, \ \boldsymbol{B}_2 = \boldsymbol{B}_{20}.$$

In which

$$\mathbf{A}_{10} = \begin{bmatrix} -10 & 10 & 0 \\ 28 & -1 & 20 \\ 0 & -20 & -8/3 \end{bmatrix}, \ \mathbf{A}_{20} = \begin{bmatrix} -10 & 10 & 0 \\ 28 & -1 & -30 \\ 0 & 30 & -8/3 \end{bmatrix}, \mathbf{V}_1 = \begin{bmatrix} 1 & 0 & 0 \\ 0 & 1 & 0 \\ 0 & 0 & 1 \end{bmatrix}, \mathbf{V}_2 = \begin{bmatrix} -3 & 3 & 0 \\ 8.4 & 0 & 0 \\ 0 & 0 & -0.8 \end{bmatrix},$$

$$\boldsymbol{\Delta} = diag\{\delta_1, \delta_2, \delta_3\}, \ \mathbf{B}_{10} = \mathbf{B}_{20} = \begin{bmatrix} 1 & 0 & 0 \end{bmatrix}^T$$

By solving the matrix inequalities corresponding to (63) with $\alpha^* = 1$, the gain matrices are found to be

$$\mathbf{K}_1 = \begin{bmatrix} -84.2940 & -23.7152 & -2.4514 \end{bmatrix}, \ \mathbf{K}_2 = \begin{bmatrix} -84.4669 & -23.6170 & 3.8484 \end{bmatrix}$$

The common positive definite matrix $\boldsymbol{X}$ and other feasible matrices obtained are as follows

$$\boldsymbol{X}=\begin{bmatrix} 0.0461 & -0.0845 & 0.0009 \\ -0.0845 & 0.6627 & -0.0027 \\ 0.0009 & -0.0027 & 0.6685 \end{bmatrix},\ \boldsymbol{E}_1 = diag\{3.4260,2.3716,1.8813\},$$
$$\boldsymbol{E}_2 = diag\{3.4483,2.2766,1.9972\},\ \boldsymbol{H} = diag\{2.6535,1.9734,1.3318\}.$$

Again, by solving the matrix inequalities corresponding to (63) with $\alpha^* = 2$, which means that the allowable variation of all the uncertain parameters are within 60% of their nominal values, we obtain

$$\boldsymbol{K}_1 = [-123.6352 \quad -42.4674 \quad -4.4747],\ \boldsymbol{K}_2 = [-125.9081 \quad -42.8129 \quad 6.8254],$$
$$\boldsymbol{X}=\begin{bmatrix} 1.0410 & -1.7704 & 0.0115 \\ -1.7704 & 7.7159 & -0.0360 \\ 0.0115 & -0.0360 & 7.7771 \end{bmatrix},\ \boldsymbol{E}_1 = diag\{98.7271,44.0157,22.7070\},$$
$$\boldsymbol{E}_2 = diag\{101.9235,42.7451,24.7517\},\ \boldsymbol{H} = diag\{68.8833,31.0026,13.8173\}.$$

Clearly, the control inputs of the controllers obtained in the paper in the two cases are all lower than that of Lee, Park, and Chen (2001).

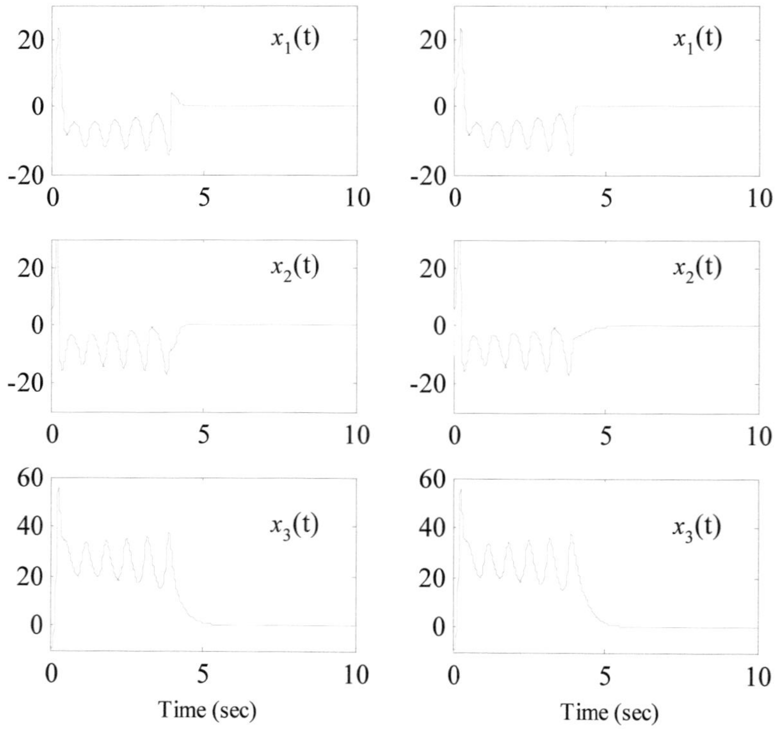

Fig. 2. State trajectories of the controlled nominal chaotic Lorenz system (On the left- and right-hand sides are results respectively of the controller of Lee, Park, and Chen (2001) and of the controller obtained in this paper)

(2) Robust reliability based design of optimal controller

Firstly, if Theorem 3.3 is used, by solving a optimization problem corresponding to (64) with $\alpha^* = 1$, the gain matrices as follows for deriving the controller are obtained

$$\boldsymbol{K}_{1G} = [-20.8512 \quad -13.5211 \quad -3.2536],\ \boldsymbol{K}_{2G} = [-21.2143 \quad -13.1299 \quad 4.3799].$$

The norm of the gain matrices are respectively $\|\boldsymbol{K}_{1G}\| = 25.0635$ and $\|\boldsymbol{K}_{2G}\| = 25.3303$. So, there exist relations

$$\|\boldsymbol{K}_{1L}\| = 326.1639 = 13.0135\|\boldsymbol{K}_{1G}\|,\ \|\boldsymbol{K}_{2L}\| = 488.2767 = 19.2764\|\boldsymbol{K}_{2G}\|.$$

To examine the effect of the controllers, the initial values of the states of the Lorenz system are taken as $\boldsymbol{x}(0) = [10 \quad -10 \quad -10]^T$, the control input is activated at t=3.89s, all as that of Lee, Park, and Chen (2001), the simulated state trajectories of the controlled Lorenz system without uncertainty are shown in Fig. 2. In which, on the left- and right-hand sides are results of the controller of Lee, Park, and Chen (2001) and of the controller obtained in this paper respectively. Simulations of the corresponding control inputs are shown in Fig. 3, in which, the dash-dot line and the solid line represent respectively the input of the controller of Lee, Park, and Chen (2001) and of the controller in the paper.

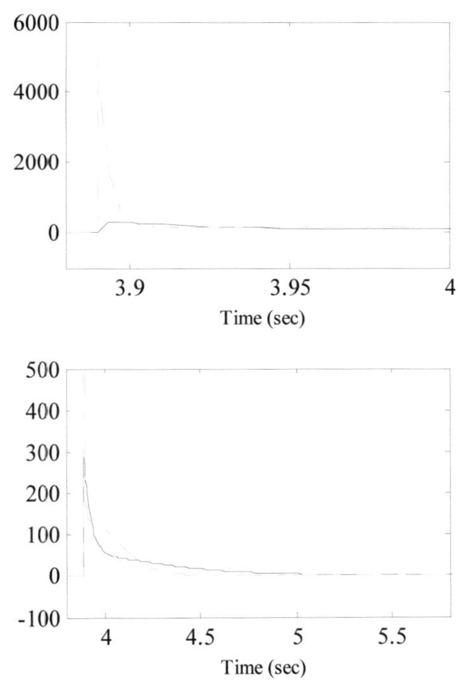

Fig. 3. Control input of the two controllers (dash-dot line and solid line represent respectively the result of Lee, Park, and Chen (2001) and the result of the paper)

The simulated state trajectories and phase trajectory of the controlled Lorenz system are shown respectively in Figs. 4 and 5, in which, all the uncertain parameters are generated randomly within the allowable ranges.

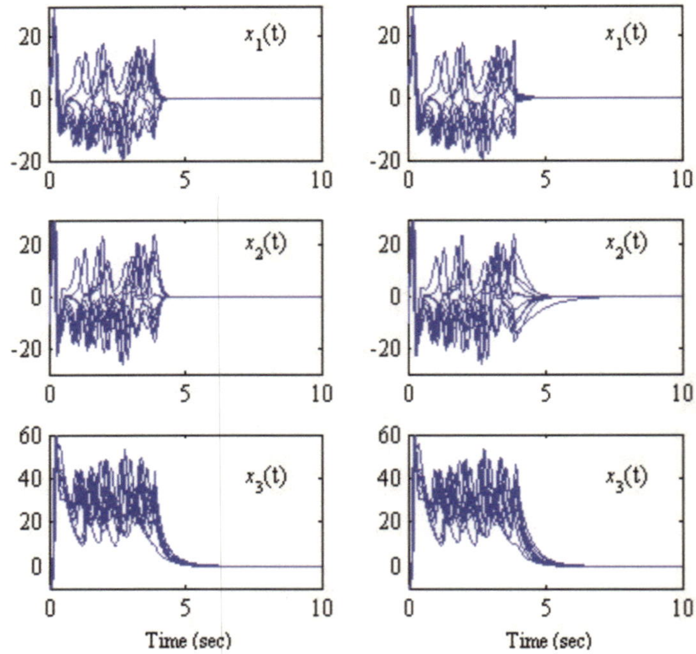

Fig. 4. Ten-times simulated state trajectories of the controlled chaotic Lorenz system with parametric uncertainties (all uncertain parameters are generated randomly within the allowable ranges, and on the left- and right-hand sides are respectively the results of controllers in Lee, Park, and Chen (2001) and in the paper)

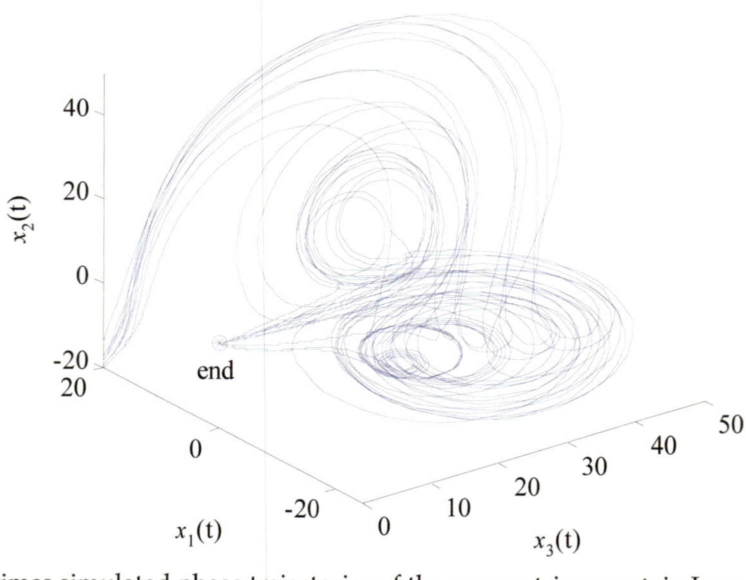

Fig. 5. Ten-times simulated phase trajectories of the parametric uncertain Lorenz system controlled by the presented method (all parameters are generated randomly within their allowable ranges)

It can be seen that the controller obtained by the presented method is effective, and the control effect has no evident difference with that of the controller in Lee, Park, and Chen (2001), but the control input of it is much lower. This shows that the presented method is much less conservative.

Taking $\alpha = 3$, which means that the allowable variation of all the uncertain parameters are within 90% of their nominal values, by applying Theorem 3.3 and solving a corresponding optimization problem of (64) with $\alpha^* = 3$, the gain matrices for deriving the fuzzy controller obtained by the presented method become

$$\mathbf{K}_{1G} = [-54.0211 \quad -32.5959 \quad -6.5886],\ \mathbf{K}_{2G} = [-50.0340 \quad -30.6071 \quad 10.4215].$$

Obviously, the input of the controller in this case is also much lower than that of the controller obtained by Lee, Park, and Chen (2001).

Secondly, when Theorem 3.4 is used, by solving two optimization problems corresponding to (69) with $\alpha^* = 1$ and $\alpha^* = 3$ respectively, the gain matrices for deriving the controller are found to be

$$\mathbf{K}_{1G} = [-20.8198 \quad -13.5543 \quad -3.2560],\ \mathbf{K}_{2G} = [-21.1621 \quad -13.1451 \quad 4.3928]\ (\alpha^* = 1).$$

$$\mathbf{K}_{1G} = [-54.0517 \quad -32.6216 \quad -6.6078],\ \mathbf{K}_{2G} = [-50.0276 \quad -30.6484 \quad 10.4362]\ (\alpha^* = 3)$$

Note that the results based on Theorem 3.4 are in agreement, approximately, with those based on Theorem 3.3.

5. Conclusion

In this chapter, stability of parametric uncertain nonlinear systems was studied from a new point of view. A robust reliability procedure was presented to deal with bounded parametric uncertainties involved in fuzzy control of nonlinear systems. In the method, the T-S fuzzy model was adopted for fuzzy modeling of nonlinear systems, and the parallel-distributed compensation (PDC) approach was used to control design. The stabilizing controller design of uncertain nonlinear systems were carried out by solving a set of linear matrix inequalities (LMIs) subjected to robust reliability for feasible solutions, or by solving a robust reliability based optimization problem to obtain optimal controller. In the optimal controller design, both the robustness with respect to uncertainties and control cost can be taken into account simultaneously. Formulations used for analysis and synthesis are within the framework of LMIs and thus can be carried out conveniently. It is demonstrated, via numerical simulations of control of a simple mechanical system and of the chaotic Lorenz system, that the presented method is much less conservative and is effective and feasible. Moreover, the bounds of uncertain parameters are not required strictly in the presented method. So, it is applicable for both the cases that the bounds of uncertain parameters are known and unknown.

6. References

Ben-Haim, Y. (1996). *Robust Reliability in the Mechanical Sciences*, Berlin: Spring-Verlag

Breitung, K.; Casciati, F. & Faravelli, L. (1998). Reliability based stability analysis for actively controlled structures. *Engineering Structures*, Vol. 20, No. 3, 211–215

Chen, B.; Liu, X. & Tong, S. (2006). Delay-dependent stability analysis and control synthesis of fuzzy dynamic systems with time delay. *Fuzzy Sets and Systems*, Vol. 157, 2224–2240

Crespo, L. G. & Kenny, S. P. (2005). Reliability-based control design for uncertain systems. *Journal of Guidance, Control, and Dynamics*, Vol. 28, No. 4, 649-658

Feng, G.; Cao, S. G.; Kees, N. W. & Chak, C. K. (1997). Design of fuzzy control systems with guaranteed stability. *Fuzzy Sets and Systems*, Vol. 85, 1–10

Guo, S. X. (2010). Robust reliability as a measure of stability of controlled dynamic systems with bounded uncertain parameters. *Journal of Vibration and Control*, Vol. 16, No. 9, 1351-1368

Guo, S. X. (2007). Robust reliability method for optimal guaranteed cost control of parametric uncertain systems. *Proceedings of IEEE International Conference on Control and Automation*, 2925-2928, Guangzhou, China

Hong, S. K. & Langari, R. (2000). An LMI-based H_∞ fuzzy control system design with TS framework. *Information Sciences*, Vol. 123, 163-179

Lam, H. K. & Leung, F. H. F. (2007). Fuzzy controller with stability and performance rules for nonlinear systems. *Fuzzy Sets and Systems*,Vol. 158, 147–163

Lee, H. J.; Park, J. B. & Chen, G. (2001). Robust fuzzy control of nonlinear systems with parametric uncertainties. *IEEE Transactions on Fuzzy Systems*, Vol. 9, 369–379

Park, J.; Kim, J. & Park, D. (2001). LMI-based design of stabilizing fuzzy controllers for nonlinear systems described by Takagi-Sugeno fuzzy model. *Fuzzy Sets and Systems*, Vol. 122, 73–82

Spencer, B. F.; Sain, M. K.; Kantor, J. C. & Montemagno, C. (1992). Probabilistic stability measures for controlled structures subject to real parameter uncertainties. *Smart Materials and Structures*, Vol. 1, 294–305

Spencer, B. F.; Sain, M. K.; Won C. H.; et al. (1994). Reliability-based measures of structural control robustness. *Structural Safety*, Vol. 15, No. 2, 111–129

Tanaka, K.; Ikeda, T. & Wang, H. O. (1996). Robust stabilization of a class of uncertain nonlinear systems via fuzzy control: quadratic stabilizability, H_∞ control theory, and linear matrix inequalities. *IEEE Transactions on Fuzzy Systems*, Vol. 4, No. 1, 1–13

Tanaka, K. & Sugeno, M. (1992). Stability analysis and design of fuzzy control systems. *Fuzzy Sets and Systems*, Vol. 45, 135–156

Teixeira, M. C. M. & Zak, S. H. (1999). Stabilizing controller design for uncertain nonlinear systems using fuzzy models. *IEEE Transactions on Fuzzy Systems*, Vol. 7, 133–142

Tuan, H. D. & Apkarian, P. (1999). Relaxation of parameterized LMIs with control applications. *International Journal of Nonlinear Robust Control*, Vol. 9, 59-84

Tuan, H. D.; Apkarian, P. & Narikiyo, T. (2001). Parameterized linear matrix inequality techniques in fuzzy control system design. *IEEE Transactions on Fuzzy Systems*, Vol. 9, 324–333

Venini, P. & Mariani, C. (1999). Reliability as a measure of active control effectiveness. *Computers and Structures*, Vol. 73, 465-473

Wu, H. N. & Cai, K. Y. (2006). H_2 guaranteed cost fuzzy control design for discrete-time nonlinear systems with parameter uncertainty. *Automatica*, Vol. 42, 1183–1188

Xiu, Z. H. & Ren, G. (2005). Stability analysis and systematic design of Takagi-Sugeno fuzzy control systems. *Fuzzy Sets and Systems*, Vol. 151, 119–138

Yoneyama, J. (2006). Robust H_∞ control analysis and synthesis for Takagi-Sugeno general uncertain fuzzy systems. *Fuzzy Sets and Systems*, Vol. 157, 2205–2223

Yoneyama, J. (2007). Robust stability and stabilization for uncertain Takagi-Sugeno fuzzy time-delay systems. *Fuzzy Sets and Systems*, Vol. 158, 115–134

17

A Frequency Domain Quantitative Technique for Robust Control System Design

José Luis Guzmán[1], José Carlos Moreno[2], Manuel Berenguel[3], Francisco Rodríguez[4], Julián Sánchez-Hermosilla[5]
[1,2,3,4]*Departamento de Lenguajes y Computación;*
[5]*Departamento de Ingeniería Rural, University of Almería*
Spain

1. Introduction

Most control techniques require the use of a plant model during the design phase in order to tune the controller parameters. The mathematical models are an approximation of real systems and contain imperfections by several reasons: use of low-order descriptions, unmodelled dynamics, obtaining linear models for a specific operating point (working with poor performance outside of this working point), etc. Therefore, control techniques that work without taking into account these modelling errors, use a fixed-structure model and known parameters (nominal model) supposing that the model exactly represents the real process, and the imperfections will be removed by means of feedback. However, there exist other control methods called robust control techniques which use these imperfections implicity during the design phase. In the robust control field such imperfections are called uncertainties, and instead of working only with one model (nominal model), a family of models is used forming the *nominal model + uncertainties*. The uncertainties can be classified in parametric or structured and non-parametric or non-structured. The first ones allow representing the uncertainties into the model coefficients (e.g. the value of a pole placed between maximum and minimum limits). The second ones represent uncertainties as unmodelled dynamics (e.g. differences in the orders of the model and the real system) (Morari and Zafiriou, 1989).
The robust control technique which considers more exactly the uncertainties is the Quantitative Feedback Theory (QFT). It is a methodology to design robust controllers based on frequency domain, and was developed by Prof. Isaac Horowitz (Horowitz, 1982; Horowitz and Sidi, 1972; Horowitz, 1993). This technique allows designing robust controllers which fulfil some minimum quantitative specifications considering the presence of uncertainty in the plant model and the existence of perturbations. With this theory, Horowitz showed that the final aim of any control design must be to obtain an open-loop transfer function with the suitable bandwidth (cost of feedback) in order to sensitize the plant and reduce the perturbations. The Nichols plane is used to achieve a desired robust design over the specified region of plant uncertainty where the aim is to design a compensator $C(s)$ and a prefilter $F(s)$ (if it is necessary) (see Figure 1), so that performance and stability specifications are achieved for the family of plants.

This chapter presents for SISO (Single Input Single Output) LTI (Linear Time Invariant) systems, a detailed description of this robust control technique and two real experiences where QFT has successfully applied at the University of Almería (Spain). It starts with a QFT description from a theoretical point of view, afterwards section 3. 1 is devoted to present two well-known software tools for QFT design, and after that two real applications in agricultural spraying tasks and solar energy are presented. Finally, the chapter ends with some conclusions.

2. Synthesis of SISO LTI uncertain feedback control systems using QFT

QFT is a methodology to design robust controllers based on frequency domain (Horowitz, 1993; Yaniv, 1999). This technique allows designing robust controllers which fulfil some quantitative specifications. The Nichols plane is the key tool for this technique and is used to achieve a robust design over the specified region of plant uncertainty. The aim is to design a compensator $C(s)$ and a prefilter $F(s)$ (if it is necessary), as shown in Figure 1, so that performance and stability specifications are achieved for the family of plants $\wp(s)$ describing a plant $P(s)$. Here, the notation $\hat{a}$ is used to represent the Laplace transform for a time domain signal $a(t)$.

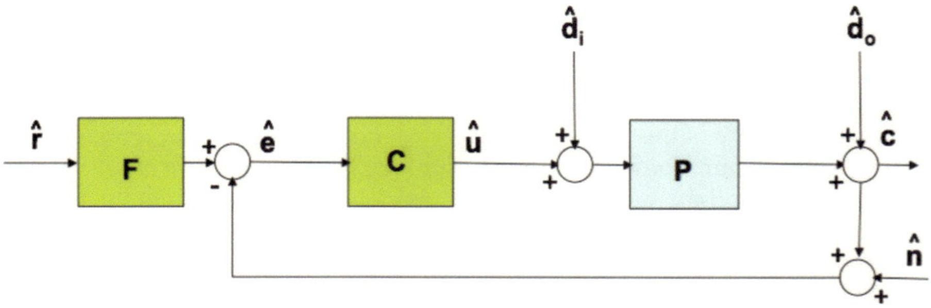

Fig. 1. Two degrees of freedom feedback system.

The QFT technique uses the information of the plant uncertainty in a quantitative way, imposing robust tracking, robust stability, and robust attenuation specifications (among others). The 2DoF compensator $\{F, C\}$, from now onwards the s argument will be omitted when necessary for clarity, must be designed in such a way that the plant behaviour variations due to the uncertainties are inside of some specific tolerance margins in closed-loop. Here, the family $\wp(s)$ is represented by the following equation

$$\begin{aligned}\wp(s) = \Big\{P(s) = k\frac{\prod_{i=1}^{n}(s+z_i)\prod_{z=1}^{m}(s^2+2\xi_z\omega_{0z}+\omega_{0z}^2)}{s^N\prod_{r=1}^{a}(s+p_r)\prod_{t=1}^{b}(s^2+2\xi_t\omega_{0t}+\omega_{0t}^2)} : \\ k \in [k_{min}, k_{max}],\ z_i \in [z_{i,min}, z_{i,max}],\ p_r \in [p_{r,min}, p_{r,max}], \\ \xi_z \in [\xi_{z,min}, \xi_{z,max}],\ \omega_{0z} \in [\omega_{0z,min}, \omega_{0z,max}], \\ \xi_t \in [\xi_{t,min}, \xi_{t,max}],\ \omega_{0t} \in [\omega_{0t,min}, \omega_{0t,max}], \\ n+m < a+b+N\Big\}\end{aligned} \tag{1}$$

A typical QFT design involves the following steps:

1. *Problem specification*. The plant model with uncertainty is identified, and a set of working frequencies is selected based on the system bandwidth, $\Omega = \{\omega_1, \omega_2, ..., \omega_k\}$. The specifications (stability, tracking, input disturbances, output disturbances, noise, and control effort) for each frequency are defined, and the nominal plant P_0 is selected.
2. *Templates*. The quantitative information of the uncertainties is represented by a set of points on the Nichols plane. This set of points is called *template* and it defines a graphical representation of the uncertainty at each design frequency ω. An example is shown in Figure 2, where templates of a second-order system given by $P(s) = k/s(s+a)$, with $k \in [1, 10]$ and $a \in [1, 10]$ are displayed for the following set of frequencies $\Omega = \{0.5, 1, 2, 4, 8, 15, 30, 60, 90, 120, 180\}$ rad/s.
3. *Bounds*. The specifications settled at the first step are translated, for each frequency ω in Ω set, into prohibited zones on the Nichols plane for the loop transfer function $L_0(j\omega) = C(j\omega)P_0(j\omega)$. These zones are defined by limits that are known as *bounds*. There exist so many *bounds* for each frequency as specifications are considered. So, all these *bounds* for each frequency are grouped showing an unique prohibited *boundary*. Figure 3 shows an example for stability and tracking specifications.

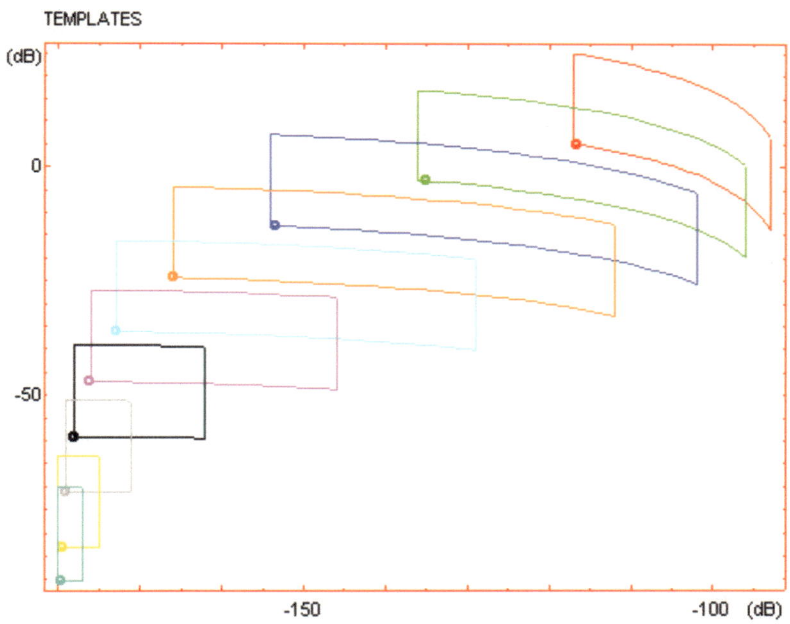

Fig. 2. QFT Template example.

4. *Loop shaping*. This phase consists in designing the C controller in such a way that the nominal loop transfer function $L_0(j\omega) = C(j\omega)P_0(j\omega)$ fulfils the *bounds* calculated in the previous phase. Figure 3 shows the design of L_0 where the *bounds* are fulfilled at each design frequency.
5. *Prefilter*. The prefilter F is designed so that the closed-loop transfer function from reference to output follows the robust tracking specifications, that is, the closed-loop system variations must be inside of a desired tolerance range, as Figure 4 shows.

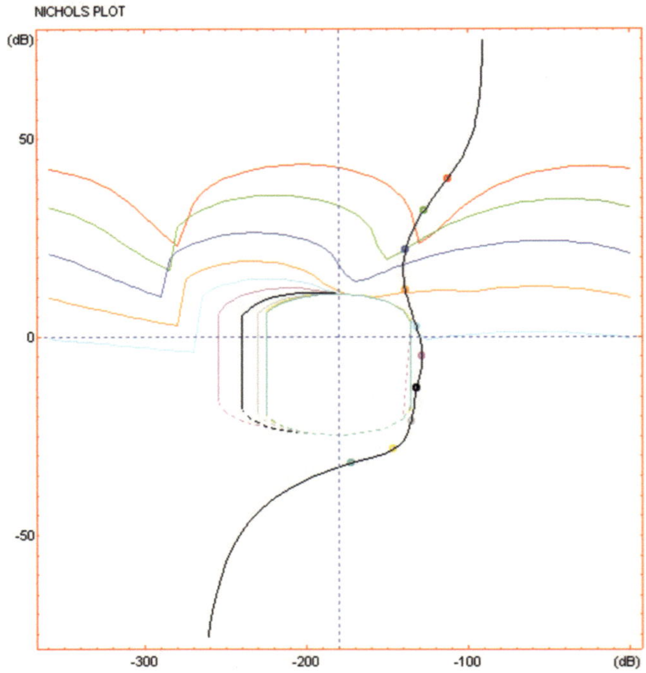

Fig. 3. QFT Bound and Loop Shaping example.

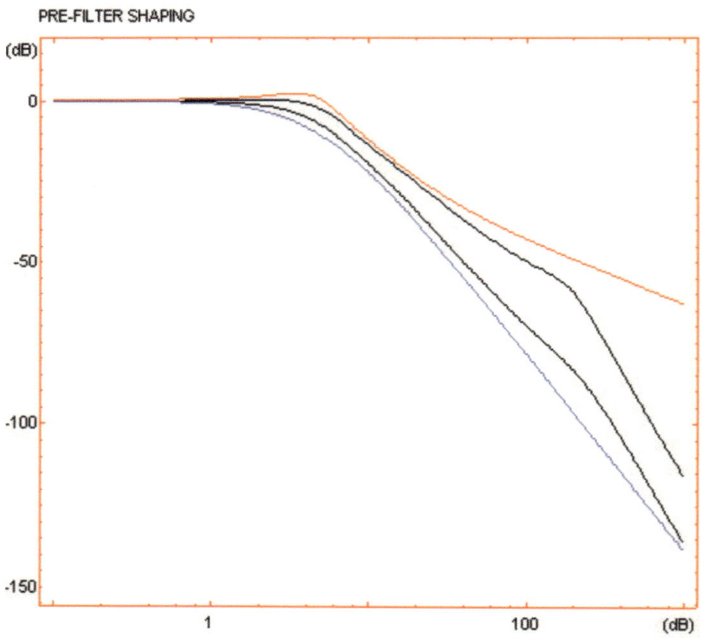

Fig. 4. QFT Prefilter example.

6. *Validation*. This step is devoted to verify that the closed-loop control system fulfils, for the whole family of plants, and for all frequencies in the bandwith of the system, all the specifications given in the first step. Otherwise, new frequencies are added to the set Ω, so that the design is repeated until such specifications are reached.

The closed-loop specifications for system in Figure 1 are typically defined in time domain and/or in the frequency domain. The time domain specifications define the desired outputs for determined inputs, and the frequency domain specifications define in terms of frequency the desired characteristics for the system output for those inputs.
In the following, these types of specifications are described and the specifications translation problem from time domain to frequency domain is considered.

2.1 Time domain specifications

Typically, the closed-loop specifications for system in Figure 1 are defined in terms of the system inputs and outputs. Both of them must be delimited, so that the system operates in a predetermined region. For example:

1. In a regulation problem, the aim is to achieve a plant output close to zero (or nearby a determined operation point). For this case, the time domain specifications could define allowed operation regions as shown in Figures 5*a* and 5*b*, supposing that the aim is to achieve a plant output close to zero.
2. In a reference tracking problem, the plant output must follow the reference input with determined time domain characteristics. In Figure 5*c* a typical specified region is shown, in which the system output must stay. The unit step response is a very common characterization, due to it combines a fast signal (an infinite change in velocity at $t = 0^+$) with a slow signal (it remains in a constant value after transitory).

The classical specifications such as rise time, settling time and maximum overshoot, are special cases of examples in Figure 5. All these cases can be also defined in frequency domain.

2.2 Frequency domain specifications

The closed-loop specifications for system in Figure 1 are typically defined in terms of inequalities on the closed-loop transfer functions for the system, as shown in Equations (2)-(7).

1. Disturbance rejection at the plant output:

$$\left|\frac{\hat{c}}{\hat{d}_o}\right| = \left|\frac{1}{1+P(j\omega)C(j\omega)}\right| \leq \delta_{po}(\omega)\ \forall\omega > 0,\ \forall P \in \wp \tag{2}$$

2. Disturbance rejection at the plant input:

$$\left|\frac{\hat{c}}{\hat{d}_i}\right| = \left|\frac{P(j\omega)}{1+P(j\omega)C(j\omega)}\right| \leq \delta_{pi}(\omega)\ \forall\omega > 0,\ \forall P \in \wp \tag{3}$$

3. Stability:

$$\left|\frac{\hat{c}}{\hat{r}F}\right| = \left|\frac{P(j\omega)C(j\omega)}{1+P(j\omega)C(j\omega)}\right| \leq \lambda\ \forall\omega > 0,\ \forall P \in \wp \tag{4}$$

4. References Tracking:

$$B_l(\omega) \leq \left|\frac{\hat{c}}{\hat{r}}\right| = \left|\frac{F(j\omega)P(j\omega)C(j\omega)}{1+P(j\omega)C(j\omega)}\right| \leq B_u(\omega)\ \forall\omega > 0,\ \forall P \in \wp \tag{5}$$

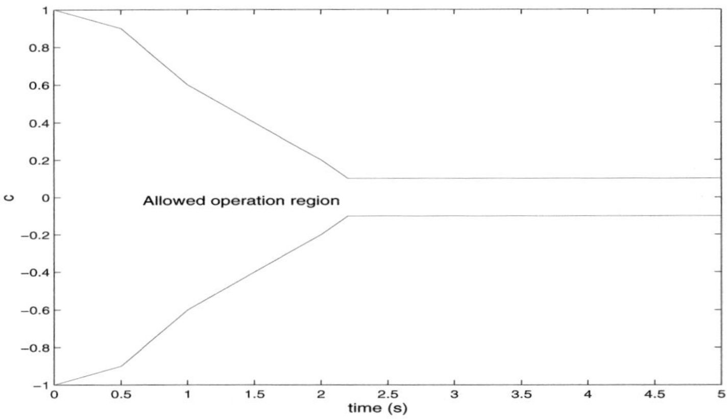

(a) Regulation problem

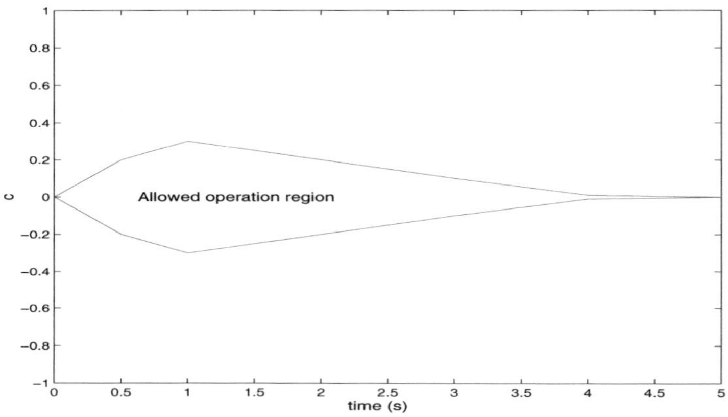

(b) Regulation problem for other initial conditions

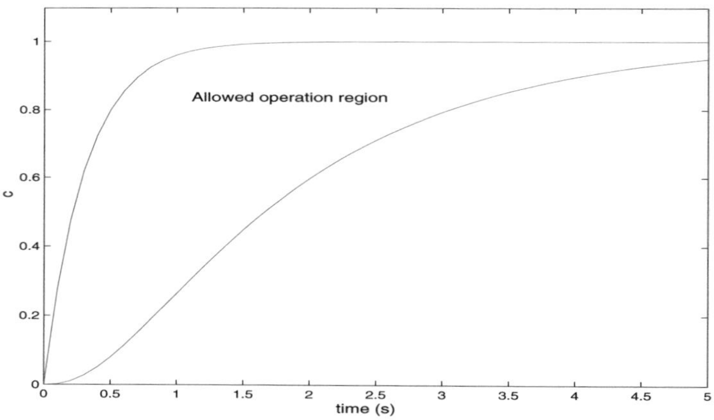

(c) Tracking problem

Fig. 5. Specifications examples in time domain.

5. Noise rejection:

$$\left|\frac{\hat{c}}{\hat{n}}\right| = \left|\frac{P(j\omega)C(j\omega)}{1+P(j\omega)C(j\omega)}\right| \leq \delta_n(\omega)\ \forall\omega > 0,\ \forall P \in \wp \tag{6}$$

6. Control effort:

$$\left|\frac{\hat{u}}{\hat{n}}\right| = \left|\frac{C(j\omega)}{1+P(j\omega)C(j\omega)}\right| \leq \delta_{ce}(\omega)\ \forall\omega > 0,\ \forall P \in \wp \tag{7}$$

For specifications in Eq. (2), (3) and (5), arbitrarily small specifications can be achieved designing C so that $|C(j\omega)| \to \infty$ (due to the appearance of the M-circle in the Nichols plot). So, with an arbitrarily small deviation from the steady state, due to the disturbance, and with a sensibility close to zero, the control system is more independent of the plant uncertainty. Obviously, in order to achieve an increase in $|C(j\omega)|$ is necessary to increase the crossover frequency[1] for the system. So, to achieve arbitrarily small specifications implies to increase the bandwidth[2] of the system. Note that the control effort specification is defined, in this context, from the sensor noise n to the control signal u. In order to define this specification from the reference, only the closed-loop transfer function from the n signal to u signal must be multiplied by F precompensator. However, in QFT, it is not defined in this form because of F must be used with other purposes.

On the other hand, to increase the value of $|C(j\omega)|$ implies a problem in the case of the control effort specification and in the case of the sensor noise rejection, since, as was previously indicated, the bandwidth of the system is increased (so the sensor noise will affect the system performance a lot). A compromise must be achieved among the different specifications.

The stability specification is related to the relative stability margins: phase and gain margins. Hence, supposing that λ is the stability specification in Eq. (4), the phase margin is equal to $2 \cdot \arcsin(0.5\lambda)$ degrees, and the gain margin is equal to $20log_{10}(1+1/\lambda)$ dB.

The output disturbance rejection specification limits the distance from the open-loop transfer function $L(j\omega)$ to the point $(-1,0)$ in Nyquist plane, and it sets an upper limit on the amplification of the disturbances at the plan output. So, this type of specification is also adequated for relative stability.

2.3 Translation of quantitative specifications from time to frequency domain

As was previously indicated, QFT is a frequency domain design technique, so, when the specifications are given in the time domain (typically in terms of the unit step response), it is necessary to translate them to frequency domain. One way to do it is to assume a model for the transfer function T_{cr}, closed-loop transfer function from reference r to the output c, and to find values for its parameters so that the defined time domain limits over the system output are satisfied.

2.3.1 A first-order model

Lets consider the simplest case, a first-order model given by $T_{cr}(s) = K/(s+a)$, so that when $r(t)$ is an unit step the system output is given by $c(t) = (K/a)(1-e^{-at})$. Then, in order to reach $c(t) = r(t)$ for a time t large enough, K should be $K = a$.

[1] The crossover frequency for a system is defined as the frequency in rad/s such that the magnitude of the open-loop transfer function $L(j\omega) = P(j\omega)C(j\omega)$ is equal to zero decibels (dB).

[2] The bandwith of a system is defined as the value of the frequency ω_b in rad/s such that $|T_{cr}(j\omega_b)/T_{cr}(0)|_{dB}$= -3 dB, where T_{cr} is the closed-loop transfer function from the reference r to the output c.

For a first-order model $\tau_c = 1/a = 1/\omega_b$ is the time constant (represents the time it takes the system step response to reach 63.2% of its final value). In general, the greater the bandwith is, the faster the system output will be.
One important difficulty for a first-order model considered is that the first derivative for the output (in time infinitesimaly after zero, $t = 0^+$) is $c = K$, when it would be desirable to be 0. So, problems appear at the neighborhood of time $t = 0$. In Figure 6 typical specified time limits (from Eq. (5) B_l and B_u are the magnitudes of the frequency response for these time domain limits) and the system output are shown when a first-order model is used. As observed, problems appear at the neighborhood of time $t = 0$. On the other hand the first-order model does not allow any overshoot, so from the specified time limits the first order model would be very conservative. Hence, a more complex model must be used for the closed-loop transfer function T_{cr}.

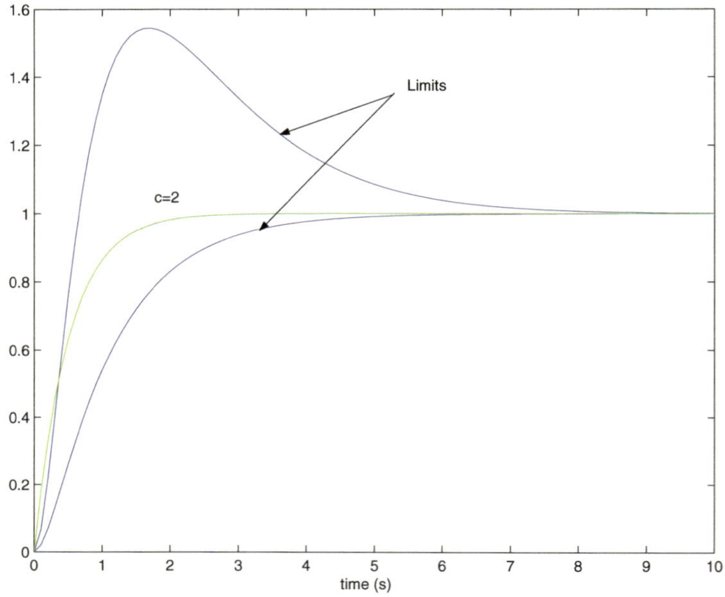

Fig. 6. Inadequate first-order model.

2.3.2 A second-order model

In this case, two free parameters are available (assuming unit static gain): the damping factor ξ and the natural frequency ω_n (rad/s). The model is given by

$$T(s) = \frac{\omega_n^2}{s^2 + 2\xi\omega_n s + \omega_n^2} \tag{8}$$

The unit step response, depending on the value of ξ, is given by

$$c(t) = \begin{cases} 1 - e^{-\xi\omega_n t}(\cos(\omega_n\sqrt{1-\xi^2}t) + \frac{\xi\omega_n}{\omega_n\sqrt{1-\xi^2}}\sin(\omega_n\sqrt{1-\xi^2}t)) & \text{if } \xi < 1 \\ 1 - e^{-\xi\omega_n t}(\cosh(\omega_n\sqrt{\xi^2-1}t) + \frac{\xi\omega_n}{\omega_n\sqrt{\xi^2-1}}\sinh(\omega_n\sqrt{1-\xi^2}t)) & \text{if } \xi > 1 \\ 1 - e^{-\xi\omega_n t}(1+\omega_n t) & \text{if } \xi = 1 \end{cases}$$

In practice, the step response for a system usually has more terms, but normally it contains a dominant second-order component with $\xi < 1$. The second-order model is very popular in control system design in spite of its simplicity, because of it is applicable to a large number of systems. The most important time domain indexes for a second-order model are: overshoot, settling time, rise time, damping factor and natural frequency. In frequency domain, its most important indexes are: resonance peak (related with the damping factor and the overshoot), resonance frequency (related with the natural frequency), and the bandwidth (related with the rise time). The resonance peak is defined as $\max_{\omega} |T_{cr}(j\omega)| \triangleq M_p$. The resonance frequency ω_p is defined as the frequency at which $|T_{cr}(j\omega_p)| = M_p$. One way to control the overshoot is setting an upper limit over M_p. For example, if this limit is fixed on 3 dB, and the practical $T_{cr}(j\omega)$ for ω in the frequency range of interest is ruled by a pair of complex conjugated poles, then this constrain assures an overshoot lower than 27%.
In (Horowitz, 1993) tables with these relations are proposed, where, based on the experience of Professor Horowitz, makes to set a second-order model to be located inside the allowed zone defined by the possible specifications. As Horowitz suggested in his book, if the magnitude of the closed-loop transfer function T_{cr} is located between frequency domain limits $B_u(\omega)$ and $B_l(\omega)$ in Eq. (5), then the time domain response is located between the corresponding time domain specifications, or at most it would be satisfied them in a very approximated way.

2.3.3 A third-order model with a zero

A third-order model with a unit static gain is given by

$$T(s) = \frac{\mu\omega_n^3}{(s^2 + 2\xi\omega_n s + \omega_n^2)(s + \mu\omega_n)} \tag{9}$$

For values of μ less than 5, a similar behaviour as if the pole is not added to the second-order model is obtained . So, the model in Eq. (8) would must be used.
If a zero is added to Eq. (9), it results

$$T(s) = \frac{(1 + s/\lambda\xi\omega_n)\mu\omega_n^3}{(s^2 + 2\xi\omega_n s + \omega_n^2)(s + \mu\omega_n)} \tag{10}$$

The unit responses obtained in this case are shown in Figure 7 for different values of λ.
As shown in Figure 7, this model implies an improvement with respect to that in Eq. (8), because of it is possible to reduce the rise time without increasing the overshoot. Obviously, if $\omega_n > 1$, then the response is ω_n times faster than the case with $\omega_n = 1$ (slower for $\omega_n < 1$). In (Horowitz, 1993), several tables are proposed relating parameters in Eq. (10) with time domain parameters as overshoot, rise time and settling time.

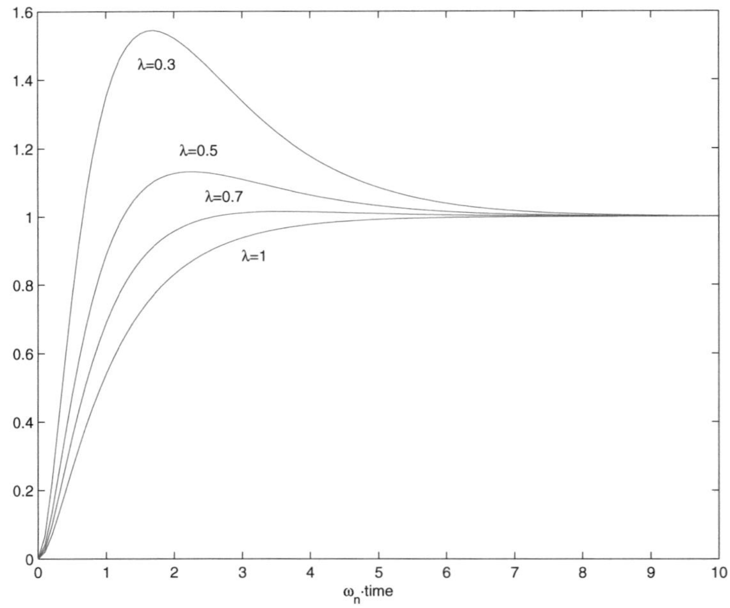

Fig. 7. Third-order model with a zero for $\mu = 5$ and $\xi = 1$.

There exist other techniques to translate specifications from time domain to frequency domain, such as model-based techniques, where based on the structures of the plant and the controller, a set of allowed responses is defined. Another technique is that presented in (Krishnan and Cruickshanks, 1977), where the time domain specifications are formulated as $\int_0^t |c(\tau) - m(\tau)|^2 d\tau \leq \int_0^t v^2(\tau) d\tau$, with $m(t)$ and $v(t)$ specified time domain functions, and where it is established that the energy of the signal, difference between the system output and the specification $m(t)$, must be enclosed by the energy of the signal $v(t)$, for each instant t, and with a translation to the frequency domain given by the inequality $|\hat{c}(j\omega) - \hat{m}(j\omega)| \leq |\hat{v}(j\omega)|$. In (Pritchard and Wigdorowitz, 1996) and (Pritchard and Wigdorowitz, 1997), the relation time-frequency is studied when uncertainty is included in the system, so that it is possible to know the time domain limits for the system response from frequency response of a set of closed-loop transfer functions from reference to the output. This technique may be used to solve the time-frequency translation problem. However, the results obtained in translation from frequency to time and from time to frequency are very conservative.

2.4 Controller design

Now, the procedure previously introduced is explained more in detail. The aim is to design the 2DoF controller $\{F, G\}$ in Figure 1, so that a subset of specifications introduced in section 2.2is satisfied, and the stability of the closed-loop system for all plant P in $\wp$ is assured.
The specifications in section 2. 2are translated in circles on Nyquist plane defining allowed zones for the function $L(j\omega) = P(j\omega)C(j\omega)$. The allowed zone is the outside of the circle for specifications in Eq. (2)-(6), and the inside one for the specification in Eq. (7). Combining the allowed zones for each function L corresponding to each plant P in $\wp$, a set of restrictions for controller C for each frequency ω is obtained. The limits of these zones represented in Nichols

plane are called *bounds* or *boundaries*. These constrains in frequency domain can be formulated over controller C or over function $L_0 = P_0 C$, for any plant P_0 in $\wp$ (so-called nominal plant). In order to explain the detailed design process, the following example, from (Horowitz, 1993), is used. Lets suppose the plant in Figure 1 given by

$$\wp = \left\{ P(s) = \frac{k}{s(s+a)} \text{ with } k \in [1,20] \text{ and } a \in [1,5] \right\} \tag{11}$$

corresponding to a range of motors and loads, where the equation modeling the motor dynamic is $J\ddot{c} + B\dot{c} = Ku$, with $k = K/J$ and $a = B/J$ in Eq. (11). Lets suppose the tracking specifications given by

$$B_l(\omega) \le |T_{cr}(j\omega)|_{dB} = \left| \frac{F(j\omega)P(j\omega)C(j\omega)}{1 + P(j\omega)C(j\omega)} \right|_{dB} \le B_u(\omega) \; \forall P \in \wp \; \forall \omega > 0 \tag{12}$$

shown in Figure 8. In Figure 9, the difference $\delta(\omega) = B_u(\omega) - B_l(\omega)$ is shown for each frequency ω. It is easy to see that in order to satisfy the specifications in Eq. (12), the following inequality must be satisfied

$$\begin{aligned} \Delta|T_{cr}(j\omega)|_{dB} = &\max_{P \in \wp} \left| \frac{P(j\omega)C(j\omega)}{1 + P(j\omega)C(j\omega)} \right|_{dB} - \min_{P \in \wp} \left| \frac{P(j\omega)C(j\omega)}{1 + P(j\omega)C(j\omega)} \right|_{dB} \le \\ &\le \delta(\omega) = B_u(\omega) - B_l(\omega) \; \forall P \in \wp \; \forall \omega > 0 \end{aligned} \tag{13}$$

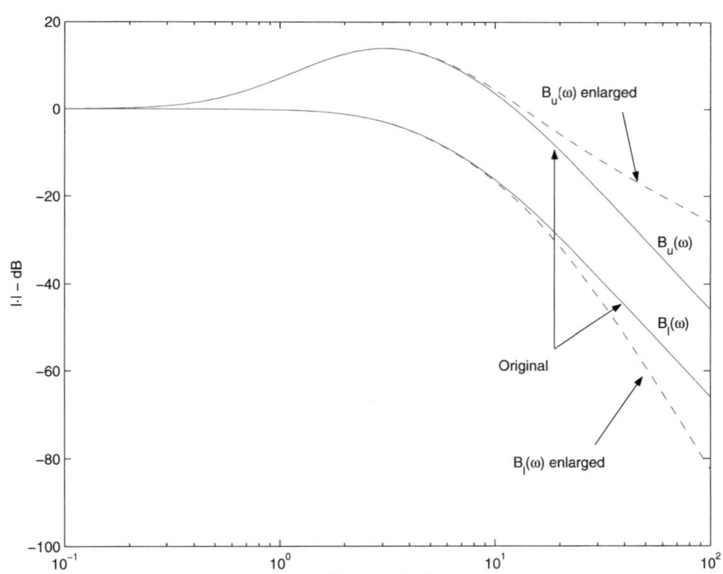

Fig. 8. Tracking specifications (variations over a nominal).

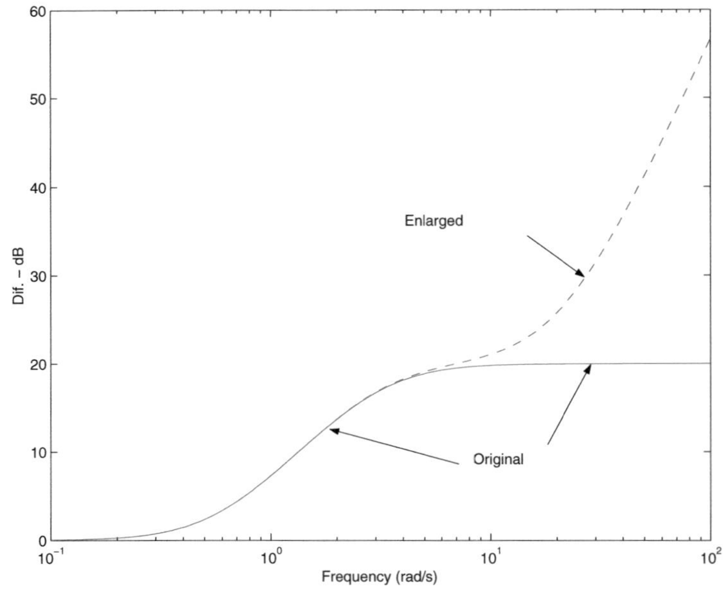

Fig. 9. Specifications on the magnitude variations for the tracking problem.

Making $L = PC$ large enough, for each plant P in $\wp$, and for a frequency ω, it is possible to achieve an arbitrarily small specification $\delta(\omega)$. However, this is not possible in practice, since the system bandwidth must be limited in order to minimize the influence of the sensor noise at the plant input. When C has been designed to satisfy the specifications in Eq. (13), the second degree of freedom, F, is used to locate those variations inside magnitude limits $B_l(\omega)$ and $B_u(\omega)$.

In order to design the first degree of freedom, C, it is necessary to define a set of constrains on C or on L_0 in the frequency domain, what guarantee that if C (respectively L_0) satisfies those restrictions then the specifications are satisfied too. As commented above, these constrains are called *bounds* or *boundaries* in QFT, and in order to compute them it is necessary to take into account:

(i) A set of specifications in frequency domain, that in the case of tracking problem, are given by Eq. (13), and that in other cases (disturbance rejection, control effort, sensor noise,...) are similar as shown in section 2.2

(ii) An object (representation) modeling the plant uncertainty in frequency domain, so-called *template*.

The following sections explain more in detail the meaning of the *templates* and the *bounds*.

Computation of basic graphical elements to deal with uncertainties: *templates*

If there is no uncertainty in plant, the set $\wp$ would contain only one transfer function, P, and for a frequency, ω, $P(j\omega)$ would be a point in the Nichols plane. Due to the uncertainty, a set of points, for each frequency, appears in the Nichols plane. One point for each plant P in $\wp$. These sets are called *templates*. For example, Figure 10 shows the *template* for $\omega = 2$ rad/s, corresponding to the set:

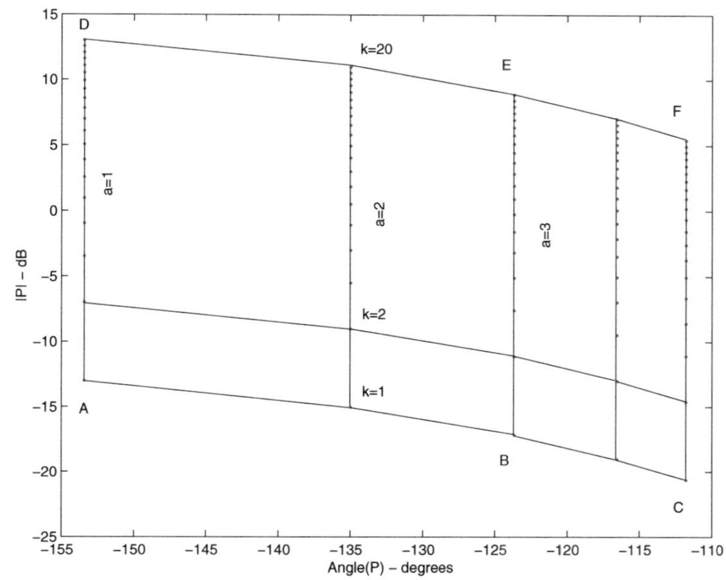

Fig. 10. *Template* for frequency $\omega = 2$ rad/s and the plant given by Eq. (11).

$$\Im(\omega = 2) = \left\{ \frac{k}{2j(2j + a)} : k \in [1, 20] \text{ and } a \in [1, 5] \right\}.$$

For $k = 1$ and driving a from 1 to 5, the segment ABC is obtained in Figure 10. For $a = 3$ and driving k from 1 to 20, the segment BE is calculated. For $k = 20$ and driving a from 1 to 5, the segment DEF is obtained.

Choosing a plant P_0 belonging to the set $\wp$, the nominal open-loop transfer function is defined as $L_0 = P_0 C$. In order to shift a *template* in the Nichols plane, a quantity must be added in phase (degrees) and other quantity in magnitude (decibels) to all points. Using the nominal point $P_0(j\omega)$ as representative of the full *template* at frequency ω and shaping the value of the nominal $L_0(j\omega) = P_0(j\omega)C(j\omega)$ using $C(j\omega)$, it is equivalent to add $|C(j\omega)|_{dB}$ in magnitude and $Angle(C(j\omega))$ degrees in phase to each point $P(j\omega)$ (with magnitude in decibels and phase in degrees) inside the *template* at frequency ω. So, the shaping of the nominal open-loop transfer function at frequency ω (using the degree of freedom C), is equivalent to shift the *template* at that frequency ω to a specific location in the Nichols plane.

The choice of the nominal plant for a *template* is totally free. The design method is valid independently of this choice. However, there exist rules for the more adequate choice in specific situations (Horowitz, 1993).

As was previously indicated, there exists a *template* for each frequency, so that after the definition of the specifications for the control problem, the following step is to define a set of design frequencies Ω. Then, the *templates* would be computed for each frequency ω in Ω.

Once the specifications have been defined and the templates have been computed, the third step is the computation of *boundaries* using these graphical objects and the specifications.

Derivation of boundaries from templates and specifications

Now, zones on Nichols plane are defined for each frequency ω in Ω, so that if the nominal of the *template* shifted by $C(j\omega)$ is located inside that zone, then the specifications are satisfied. For each specification in section 2.2and for each frequency ω in Ω, using the *template* and the corresponding specification, the *boundary* must be computed. Details about the different types of *bounds* and the most important algorithms to compute them can be found in (Moreno et al., 2006). In general, a *boundary* at frequency ω defines a limit of a zone on Nichols plane so that if the nominal $L_0(j\omega)$ of the shifted *template* is located inside that zone, then some specifications are satisfied. So, the most single appearance of a *boundary* defines a threshold value in magnitude for each phase ϕ in the Nichols plane, so that if $Angle(L_0(j\omega)) = \phi$, then $|L_0(j\omega)|_{dB}$ must be located above (or below depending on the type of specification used to compute the *boundary*) that threshold value.

It is important to note that sometimes redefinition of the specifications is necessary. For example, for system in Eq. (11), for $\omega \geq 10$ rad/s the *templates* have similar dimensions, and the specifications from Eq. (13) in Figure 9 are identical. Then, the *boundaries* for $\omega \geq 10$ rad/s will be almost identical. The function $L_0(j\omega)$ must be above the *boundaries* for all frequencies, including $\omega \geq 10$ rad/s, but this is unviable due to it must be satisfied that $L_0(j\omega) \to 0$ when $\omega \to \infty$. Therefore, it is necessary to open the tracking specifications for high frequency (where furthermore the uncertainty is greater), such as it is shown in Figure 8. On the other hand, it must be also taken into account that for a large enough frequency ω, the specification $\delta(\omega)$ in Eq. (13) must be greater or equal than $\max_{P\in\wp} |P(j\omega)|_{dB} - \min_{P\in\wp} |P(j\omega)|_{dB}$ such that, for a small value of $L_0(j\omega)$ for these frequencies, the specifications are also satisfied. The effect of this enlargement for the specifcations is negligible when the modifications are introduced at a frequency large enough. These effects are notable in the response at the neighborhood of $t = 0$.

Considering the tracking bounds as negligible from a specific frequency (in the sense that the specification is large enough), it implies that the stability boundaries are the dominant ones at these frequencies. As was mentioned above, since the templates are almost identical at high frequencies and the stability specification λ is independent of the frequency, the stability bounds are also identical and only one of them can be used as representative of the rest. In QFT, this boundary is usually called high frequency bound, and it is denoted by B_h.

Notice that the use of a discrete set of design frequencies Ω does not imply any problem. The variation of the specifications and the variation of the appearance of the *templates* from a frequency ω^- to a frequency ω^+, with $\omega^- < \omega < \omega^+$, is smooth. Anyway, the methodology let us discern the specific cases in which the number of elements of Ω is insufficient, and let us iterate in the design process to incorporate the *boundaries* for those new frequencies, then reshaping again the compensator $\{F, C\}$.

Design of the nominal open-loop transfer function fulfilling the *boundaries*

In this stage, the function $L_0(j\omega)$ must be shaped fulfilling all the *boundaries* for each frequency. Furthermore, It must assure that the transfer function $1 + L(s)$ has no zeros in the right half plane for any plant P in $\wp$. So, initially $L_0 = P_0$ $(C = 1)$ and poles and zeros are added to this function (poles and zeros of the controller C) in order to satisfy all of these restrictions on the Nichols plane. In this stage, only using the function L_0, it is possible to assure the fulfillment of the specifications for all of the elements in the set $\wp$ when $L_0(j\omega)$ is located inside the allowed zones defined by the *boundary* at frequency ω (computed from the corresponding *template* at that frequency, and from the specifications).

Obviously, there exists an infinite number of acceptable functions L_0 satisfying the *boundaries* and the stability condition. In order to choose among all of these functions, an important factor to be considered is the sensor noise effect at the plant input. The closed-loop transfer function from noise n to the plant input u is given by

$$T_{un}(s) = \frac{-C(s)}{1+P(s)C(s)} = \frac{-L(s)/P(s)}{1+L(s)}.$$

In the range of frequencies in which $|L(j\omega)|$ is large (generally low frequency), $|T_{un}(j\omega)| \rightarrow |1/P(j\omega)|$, so that the value of $|T_{un}(j\omega)|$ at low frequency is independent on the design chosen for L. In the range of frequencies where $|L(j\omega)|$ is small (generally high frequency), $|T_{un}(j\omega)| \rightarrow |G(j\omega)|$. These two asymptotes cross between themselves at the crossover frequency.
In order to reduce the influence of the sensor noise at the plant input, $|C(j\omega)| \rightarrow 0$ when $\omega \rightarrow \infty$ must be guaranteed. It is equivalent to say that $|L_0(j\omega)|$ must be reduced as fast as possible at high frequency. A conditionally stable[3] design for L_0 is especially adequate to achieve this objective. However, as it is shown in (Moreno et al., 2010) this type of designs supposes a problem when there exists a saturation non-linearity type in the system.

Design of the prefilter
At this point, only the second degree of freedom, F, must be shaped. The controller C, designed in the previous step, only guarantees that the specifications in Eq. (13) are satisfied, but not the specifications in Eq. (12). Using F, it is possible to guarantee that the specifications in Eq. (12) are satisfied when with C the specifications in Eq. (13) are assured.
In order to design F, the most common method consists of computing for each frequency ω the following limits

$$F_u(\omega) = \left| \max_{P \in \wp} \left| \frac{P(j\omega)C(j\omega)}{1+P(j\omega)C(j\omega)} \right|_{dB} - B_u(\omega) \right|$$

and

$$F_l(\omega) = \left| \min_{P \in \wp} \left| \frac{P(j\omega)C(j\omega)}{1+P(j\omega)C(j\omega)} \right|_{dB} - B_l(\omega) \right|$$

and shaping F adding poles and zeros until $F_l(\omega) \leq |F(j\omega)| \leq F_u(\omega)$ for all frequency ω in Ω.

Validation of the design
This is the last step in the design process and consists in studying the magnitude of the different closed-loop transfer functions, checking if the specifications for frequencies outside of the set Ω are satisfied. If any specification is not satisfied for a specific frequency, ω_p, then this frequency is added to the set Ω, and the corresponding *template* and *boundary* are

[3] A system is conditionally stable if a gain reduction of the open-loop transfer function L drives the closed-loop poles to the right half plane.

computed for that frequency ω_p. Then, the function L_0 is reshaped, so that the new restriction is satisfied. Afterwards, the precompensator F is reshaped, and finally the new design is validated. So, an iterative procedure is followed until the validation result is satisfactory.

3. Computer-based tools for QFT

As it has been described in the previous section, the QFT framework evolves several stages, where a continuous re-design process must be followed. Furthermore, there are some steps requiring the use of algorithms to calculate the corresponding parameters. Therefore, computer-based tools as support for the QFT methodology are highly valuable to help in the design procedure. This section briefly describes the most well-known tools available in the literature, The Matlab QFT Toolbox (Borghesani et al., 2003) and SISO-QFTIT (Díaz et al., 2005a),(Díaz et al., 2005b).

3.1 Matlab QFT toolbox

The QFT Frequency Domain Control Design Toolbox is a commercial collection of Matlab functions for designing robust feedback systems using QFT, supported by the company Terasoft, Inc (Borghesani et al., 2003). The QFT Toolbox includes a convenient GUI that facilitates classical loop shaping of controllers to meet design requirements in the face of plant uncertainty and disturbances. The interactive GUI for shaping controllers provides a point-click interface for loop shaping using classical frequency domain concepts. The toolbox also includes powerful *bound* computation routines which help in the conversion of closed-loop specifications into *boundaries* on the open-loop transfer function (Borghesani et al., 2003).

The toolbox is used as a combination of Matlab functions and graphical interfaces to perform a complete QFT design. The best way to do that is to create a Matlab script including all the required calls to the corresponding functions. The following lines briefly describe the main steps and functions to use, where an example presented in (Borghesani et al., 2003) is followed for a better understanding (a more detailed description can be found in (Borghesani et al., 2003)).

The example to follow is described by:

$$\wp = \left\{ P(s) = \frac{k}{(s+a)(s+b)} : k = [1,2,5,8,10], a = [1,3,5], b = [20,25,30] \right\}. \tag{14}$$

Once the process and the associated uncertainties are defined, the different steps, explained in section 2, to design the robust control scheme using the QFT toolbox are described in the following:

- *Template computation*. First, the transfer function models representing the process uncertainty must be written. The following code calculates a matrix of 40 plant elements which is stored in the variable P and represents the system defined by Eq. (14).

```
» c = 1; k = 10; b = 20;
» for a = linspace(1,5,10),
»         P(1,1,c) = tf(k,[1,a+b,a*b]); c = c + 1;
» end
» k = 1; b = 30;
» for a = linspace(1,5,10),
```

```
»          P(1,1,c) = tf(k,[1,a+b,a*b]); c = c + 1;
» end
» b = 30; a = 5;
» for k = linspace(1,10,10),
»          P(1,1,c) = tf(k, [1,a+b,a*b]); c = c + 1;
» end
» b = 20; a = 1;
» for k = linspace(1,10,10),
»          P(1,1,c) = tf(k, [1,a+b,a*b]); c = c + 1;
» end
```

Then, the nominal element is selected:

```
» nompt=21;
```

and the frequency array is set:

```
» w = [0.1, 5, 10, 100];
```

Finally, the templates are calculated and visualized using the *plottmpl* function (see (Borghesani et al., 2003) for a detailed explanation):

```
» plottmpl(w,P,nompt);
```

obtaining the *templates* shown in Figure 11.

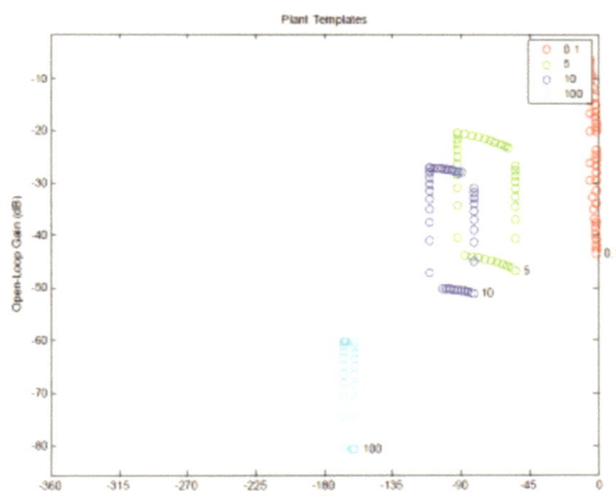

Fig. 11. Matlab QFT Toolbox. Templates for example in Eq. (14)

- *Specifications*. In this step, the system specifications must be defined according to Eq. (2)-(7). Once the specifications are determined, the corresponding bounds on the Nichols plane are computed. The following source code shows the use of specifications in Eq. (2)-(4) for this example.

  A stability specification of $\lambda = 1.2$ in Eq. (4) corresponding to a gain margin (GM) ≥ 5.3 dB and a phase margin (PM) $= 49.25$ degrees is given:

  ```
  » Ws1 = 1.2;
  ```

  Then, the stability *bounds* are computed using the function *sisobnds* (see (Borghesani et al., 2003) for a detailed explanation) and its value is stored in the variable *bdb*1:

» bdb1 = sisobnds(1,w,Ws1,P,0,nompt);

Lets now consider the specifications for output and input disturbance rejection cases, from Eq. (2)-(3). For the case of the output disturbance specification, the performance weight for the bandwidth [0,10] is defined as

» Ws2 = tf(0.02*[1,64,748,2400],[1,14.4,169]);

and the *bounds* are computed in the following way

» bdb2 = sisobnds(2,w(1:3),Ws2,P,0,nompt);

For the input disturbance case, the specification is defined as constant for

» Ws3 = 0.01;

calculating the *bounds* as

» bdb3 = sisobnds(3,w(1:3),Ws3,P,0,nompt);

also for the bandwidth [0,10].

Once the specifications are defined and the corresponding *bounds* are calculated. For each frequency they can be combined using the following functions:

» bdb = grpbnds(bdb1,bdb2,bdb3); // Making a global structure

» ubdb = sectbnds(bdb); // Combining bounds

The resulting *bounds* which will be used for the loop-shaping stage are shown in Figure 12. This figure is obtained using the *plotbnds* function:

» plotbnds(ubdb);

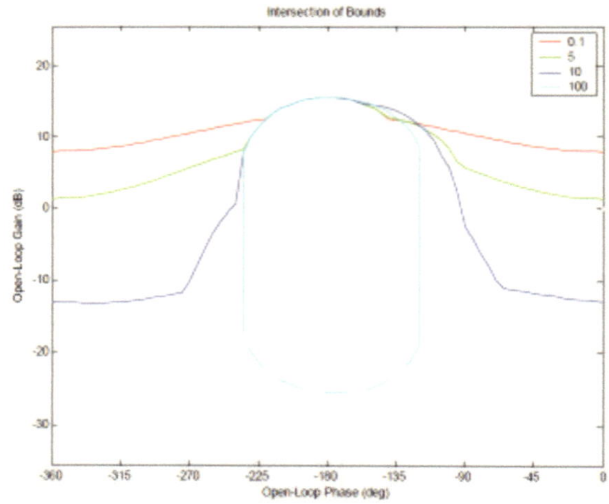

Fig. 12. Matlab QFT Toolbox. Boundaries for example (14)

- *Loop-shaping*. After obtaining the stability and performance *bounds*, the next step consists in designing (loop shaping) the controller. The QFT toolbox includes a graphical interactive GUI, *lpshape*, which helps to perform this task in an straightforward way. Before using this function, it is necessary to define the frequency array for loop shaping, the nominal plant, and the initial controller transfer function. Therefore, these variables must be set previously, where for this example are given by:

  » wl = logspace(-2,3,100); // frequency array for loop shaping

  » C0 = tf(1,1); // Initial Controller

» L0=P(1,1,nompt)*C0; // Nominal open-loop transfer function

Having defined these variables, the graphical interface is opened using the following line:

» lpshape(wl,ubdb,L0,C0);

obtaining the window shown in Figure 13. As shown from this figure, the GUI allows to modify the control transfer functions adding, modifying, and removing poles and zeros. This task can be done from the options available at the right area of the windows or dragging interactively on the loop $L_0(s) = P_0(s)C(s)$ represented by the black line on the Nichols plane.

For this example, the final controller is given by (Borghesani et al., 2003)

$$C(s) = \frac{379(\frac{s}{42}+1)}{\frac{s^2}{247^2}+\frac{s}{247}+1} \quad (15)$$

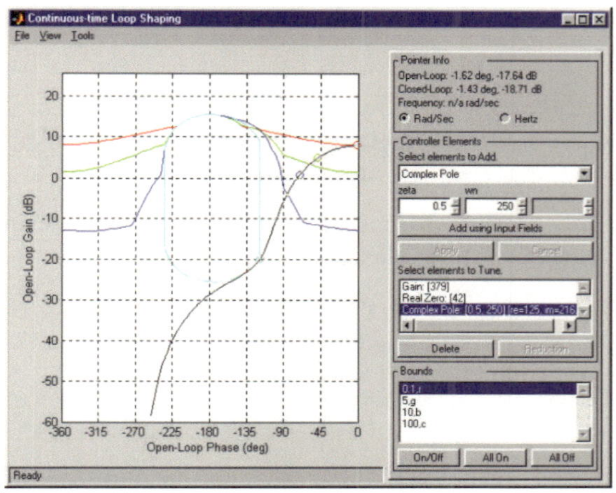

Fig. 13. Matlab QFT Toolbox. Loop shaping for example in Eq. (14)

- *Pre-filter design*. When the control design requires tracking of reference signals, although this is not the case for this example, a pre-filter $F(s)$ must be used in addition to the controller $C(s)$ such as discussed in section 2.. The prefilter can be also designed interactively using a graphical interface similar to that described for the loop shaping stage. To run this option, the *pfshape* function must be used (see (Borghesani et al., 2003) for more details).
- *Validation*. The control system validation can be done testing the resulting robust controller for all uncertain plants defined by Eq. (14) and checking that the different specifications are fulfilled for all of them. This task can be performed directly programming in Matlab or using the *chksiso* function from the QFT toolbox.

3.2 An interactive tool based in Sysquake: SISO-QFTIT

SISO-QFTIT is a free software interactive tool for robust control design using the QFT methodology (Díaz et al., 2005a;b). The main advantages of SISO-QFTIT compared to other existing tools are its easiness of use and its interactive nature. In the tool described in the previous section, a combination between code and graphical interfaces must be used, where

some interactive features are also provided for the loop shaping and filter design stages. However, with SISO-QFTIT all the stages are available from an interactive point of view.
As commented above, the tool has been implemented in Sysquake, a Matlab-like language with fast execution and excellent facilities for interactive graphics (Piguet, 2004). Windows, Mac, and Linux operating systems are supported. Since this tool is completely interactive, one consideration that must be kept in mind is that the tool's main feature -interactivity- cannot be easily illustrated in a written text. Thus, the reader is cordially invited to experience the interactive features of the tool.
The users mainly should operate with only mouse operations on different elements in the window of the application or text insertion in dialog boxes. The actions that they carry out are reflected instantly in all the graphics in the screen. In this way the users take aware visually of the effects that produce their actions on the design that they are carrying out. This tool is specially conceived as much as for beginner users that want to learn the QFT methodology, as for expert users (Díaz et al., 2005b).
The user can work with SISO-QFTIT in two different but not excluding ways (Díaz et al., 2005b):

- Interactive mode. In this work form, the user selects an element in the window and drags it to take it to a certain value, their actions on this element are reflected simultaneously on all the present figures in the window of the tool.
- Dialogue mode. In this work form, the user should simply go selecting entrances of the *Settings* menu and correctly fill the blanks of dialog boxes.

Such as commented in the manual of this interactive software tool, its main interactive advantages and options are the following (Díaz et al., 2005b):

- Variations that take place in the *templates* when modifying the uncertainty of the different elements of the plant or in the value of the *template* calculation frequency.
- Individual or combined variation on the bounds as a result of the configuration of specifications, i.e., by adding zeros and poles to the different specifications.
- The movement of the controller zeros and poles over the complex plane and the modification of its symbolic transfer function when the open loop transfer function is modified in the Nichols plane.
- The change of shape of the open loop transfer function in the Nichols plane and the variation of the expression of the controller transfer function when any movement, addition or suppression of its zeros or poles in the complex plane.
- The changes that take place in the time domain representation of the manipulated and controlled variables due to the modification of the nominal values of the different elements of the plant.
- The changes that take place in the time domain representation of the manipulated and controlled variables due to the introduction of a step perturbation at the input of the plant. The magnitude and the occurrence instant of the perturbation is configured by the user by means of the mouse.

Such as pointed out above, the interactive capabilities of the tool cannot be shown in a written text. However, some screenshots for the example used with the Matlab QFT toolbox are provided. Figure 14*a* shows the resulting templates for the process defined by Eq. (14).

Notice that with this tool, the frequencies, the process uncertainties and the nominal plant can be interactively modified. The stability bounds are shown in Figure 14*b*. The radiobuttons available at the top-right side of the tool allow to choose the desired specification. Once the specification is selected, the rest of the screen is changed to include the specification values in an interactive way. Figure 15*a* displays the loop shaping stage with the combination of the different bounds (same result than in Figure 13). The figure also shows the resulting loop shaping for controller (15). Then, the validation screen is shown in Figure 15*b*, where it is possible to check interactively if the robust control design satisfies the specifications for all uncertain cases. Although for this example it is not necessary to design the pre-filter for the tracking specifications, this tool also provides a screen where it is possible to perform this task (see an example in Figure 16).

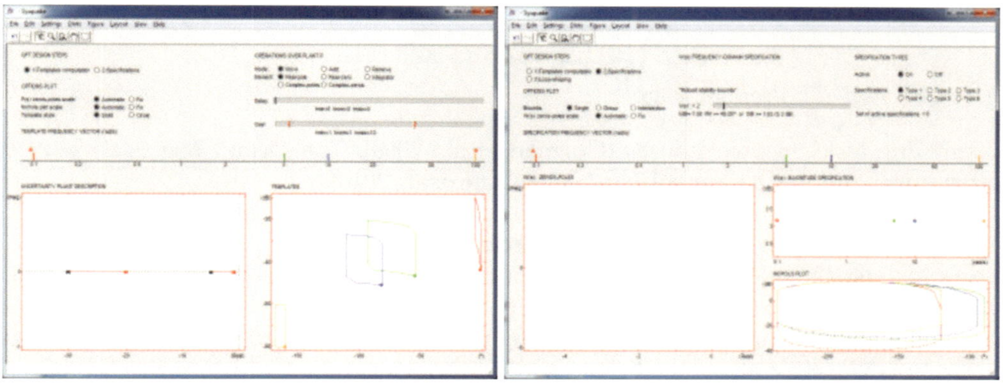

(a) QFT Templates (b) Stability bounds

Fig. 14. SISO-QFTIT. Templates and bounds for the example described in Eq. (14)

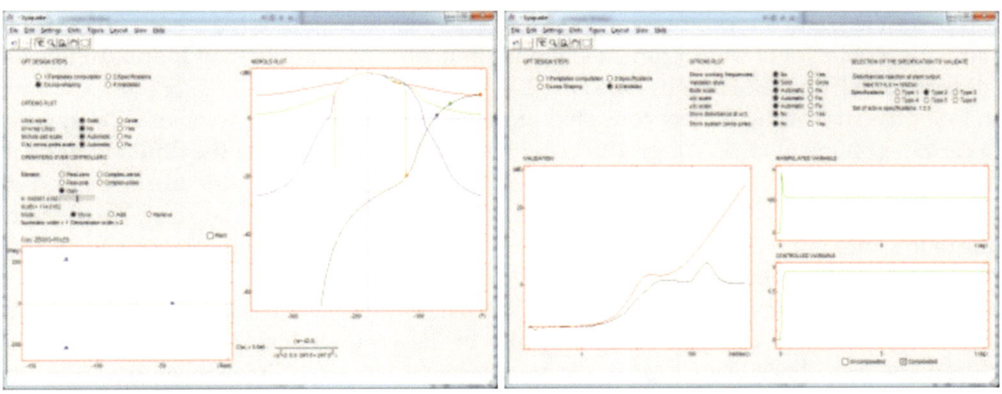

(a) Loop shaping (b) Validation

Fig. 15. SISO-QFTIT. Loop shaping and validation for the example described in Eq. (14)

4. Practical applications

This section presents two industrial projects where the QFT technique has been successfully used. The first one is focused on the pressure control of a mobile robot which was design

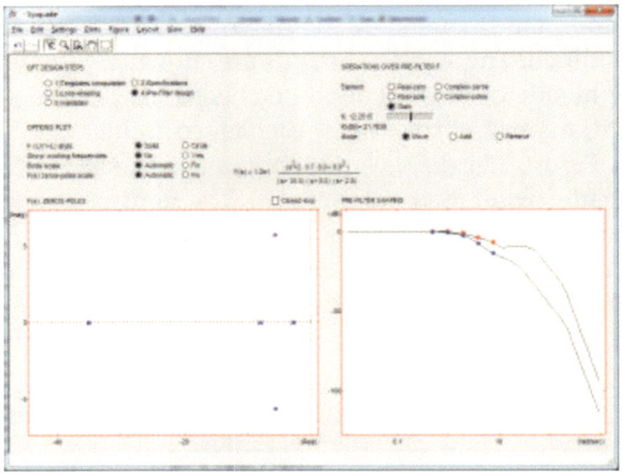

Fig. 16. SISO-QFTIT. Prefilter stage

for spraying tasks in greenhouses (Guzmán et al., 2008). The second one deals with the temperature control of a solar collector field (Cirre et al., 2010).

4.1 In agricultural and robotics context: Fitorobot

During the last six years, the Automatic Control, Robotics and Electronics research group and the Agricultural Engineering Department, both from the University of Almería (Spain), have been working in a project aimed at designing, implementation, and testing a multi-use autonomous vehicle with safe, efficient, and economic operation which moves through the crop lines of a greenhouse and which performs tasks that are tedious and/or hazardous for people. This robot has been called *Fitorobot*. The first version of this vehicle has been equipped for spraying activities, but other configurations have also been designed, such as: a lifting platform to reach high zones to perform tasks (staking, cleaning leaves, harvesting, manual pollination, etc.), and a forklift to transport and raise heavy materials (Sánchez-Gimeno et al., 2006). This mobile robot was designed and built following the paradigm of Mechatronics such as described in (Sánchez-Hermosilla. et al., 2010).

The first objective of the project consisted of developing a prototype to enable the spraying of a certain volume of chemical products per hectare while controlling the different variables that affect the spraying system (pressure, flow, and travel speed). The pressure is selected and the control signal keeps the spraying conditions constant (mainly droplet size). The reference value of the pressure is calculated based on the mobile robot speed and the volume of pesticide to apply, where the pressure working range is between 5 and 15 bar.

There are some circumstances where it is impossible to maintain a constant velocity due to the irregularities of the soil, different slopes of the ground, and the turning movements between the crop lines. Thus, for work at a variable velocity (Guzmán et al., 2008), it is necessary to spray using a variable-pressure system based on the vehicle velocity, which is the proposal adopted and implemented in this work. This system presents some advantages, such as the higher quality of the process, because the product sprayed over each plant is optimal. Furthermore, this system saves chemical products because an optimal quantity is sprayed, reducing the environmental impact and pollution as the volume sprayed to the air is minimized.

The robot prototype (Figure 17) consists of an autonomous mobile platform with a rubber tracked system and differential guidance mechanism (to achieve a more homogeneous distribution of soil-compaction pressure, thus disturbing less the sandy soil typical of Mediterranean greenhouses (Sánchez-Gimeno et al., 2006)). The robot is driven by hydraulic motors fed by two variable displacement pumps powered by a 20-HP gasoline motor, allowing a maximum velocity of 2.9 m/s. Due to the restrictions imposed by the narrow greenhouse lanes, the vehicle dimensions are 70 cm width, 170 cm length, and 180 cm height at the top of the nozzles.

Fig. 17. Mobile robot for agricultural tasks

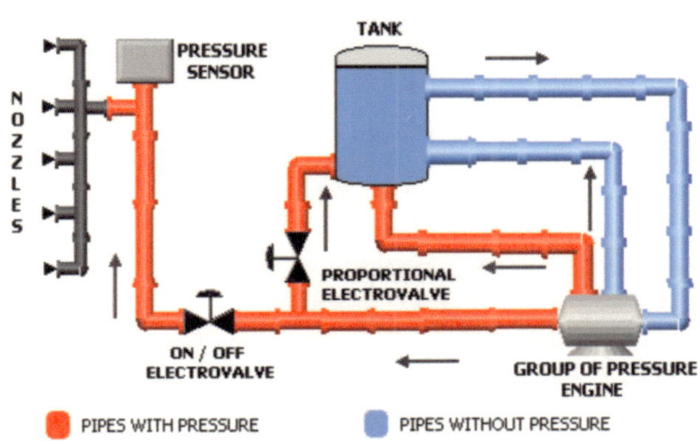

Fig. 18. Scheme of the spraying system

The spraying system carried out by the mobile robot is composed with a 300 l tank used to store the chemical products, a vertical boom sprayer with 10 nozzles, an on/off electrovalve to activate the spraying, a proportional electrovalve to regulate the output pressure, a double-membrane pump with pressure accumulator providing a maximum flow of 30 l/min and a maximum pressure of 30 bar, and a pressure sensor to close the control loop as shown in Figure 18.
In this case, the control problem was focused on regulating the output pressure of the spraying system mounted on the mobile robot despite changes in the vehicle velocity and the nonlinearities of the process.
For an adequate control system design, it was necessary to model the plant by obtaining its associated parameters. Several open-loop step-based tests were performed varying the valve aperture around a particular operating point. The results showed that the system dynamics can be approximated by a first-order system with delay. Thus, it can be modelled using the following transfer function

$$P(s) = \frac{k}{\tau s + 1} e^{-t_r s} \tag{16}$$

where k is the static gain, t_r is the delay time, and τ is the time constant.
Then, several experiments in open loop were performed to design the dynamic model of the spraying system using different amplitude opening steps (5% and 10%) over the same operating points (see Figure 19*a*). The analysis of the results showed that the output-pressure behavior changes when different valve-amplitude steps are produced around the same working point, and also when the same valve opening steps are produced at several operating points, confirming the uncertainty and nonlinear characteristics of the system.

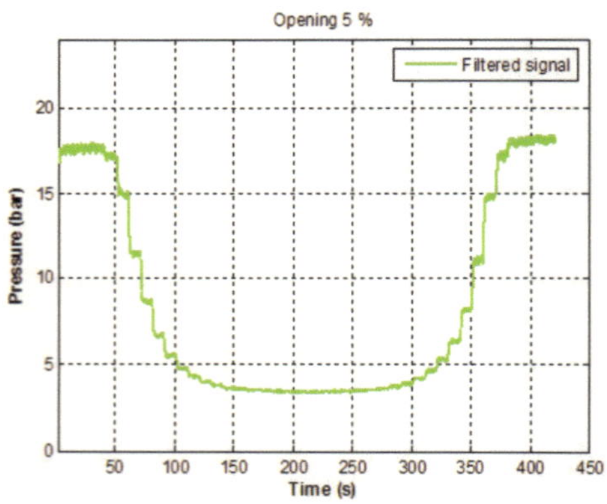

(a) Time domain

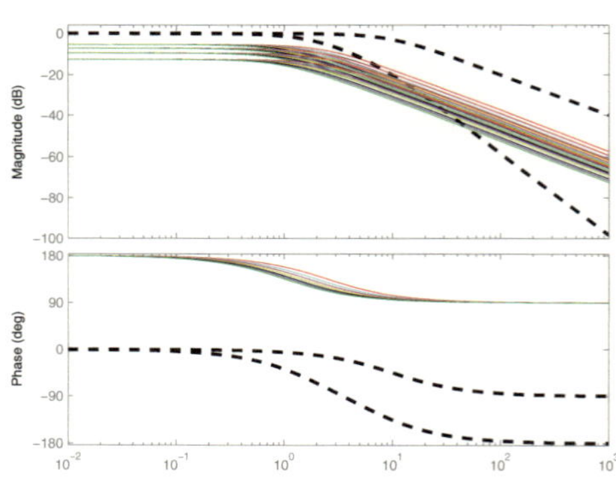

(b) Frequency domain

Fig. 19. System uncertainties from the time and frequency domains

After analyzing the results (see Figure 19*a*), the system was modelled as a first-order dynamical system with uncertain parameters, where the reaction curve method has been used at the different operating points. Therefore, the resulting uncertain model is given by the following transfer function (see Figure 19*b*):

$$\wp = \left\{ P(s) = \frac{k}{\tau s + 1} : k \in [-0.572, -0.150], \tau \in [0.4, 1] \right\} \tag{17}$$

where the gain, k, is given in bar/% aperture and the constant time, τ , in seconds.
Once the system was characterized, the robust control design using QFT was performed considering specifications on stability and tracking.
First, the specifications for each frequency were defined, and the nominal plant P_0 was selected. The set of frequencies and the nominal plant were set to $\Omega = \{0.1, 1, 2, 10\}$ rad/s and $P_0 = \frac{-0.3}{0.7s+1}$, respectively. The stability specification was set to $\lambda = 1.2$ corresponding to a $GM \geq 5.3$ dB and a $PM = 49.25$, and for the tracking specifications the maximum and minimum values for the magnitude have been described by the following transfer functions (frequency response for tracking specifications are shown in Figure 19*b* in dashed lines)

$$B_l(s) = \frac{10}{s+10}, \quad B_u(s) = \frac{12.25}{s^2 + 8.75s + 12.25} \tag{18}$$

Figure 20*a* shows the different *templates* of the plant for the set of frequencies determined above.

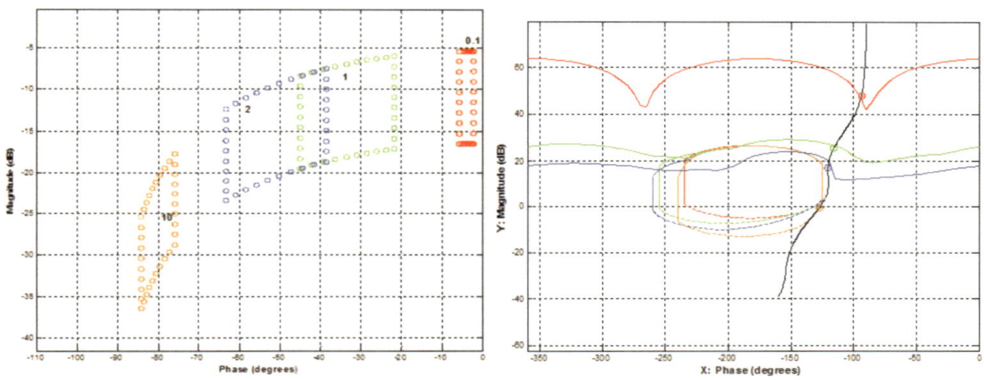

(a) QFT Templates (b) Loopshaping stage

Fig. 20. Templates and feedback controller design by QFT

The specifications are translated to the *boundaries* on the Nichols plane for the loop-transfer function $L(j\omega) = C(j\omega)P(j\omega)$. Figure 20*b* shows the different *bounds* for stability and tracking specifications set previously.
Then, the loop shaping stage was performed in such a way that the nominal loop-transfer function $L_0(j\omega) = C(j\omega)P_0(j\omega)$ was adjusted to make the templates fulfil the *bounds* calculated in the previous phase. Figure 20*b* shows the design of L_0 where the *bounds* are fulfilled at each design frequency. This figure shows the optimal controller using QFT to lie on the *boundaries* at each frequency design. However, a simpler controller fulfilling the specifications was preferred for practical reasons. The resulting controller was the following:

$$C(s) = \frac{27.25(s+1)}{s} \tag{19}$$

To conclude the design process, the prefilter F is determined so that the closed-loop transfer function matches the robust tracking specifications, that is, the closed-loop system variations must be inside of a desired tolerance range:

$$F(s) = \frac{1}{0.1786s + 1} \tag{20}$$

Once the robust design was performed, the system was validated by simulation. Figure 21 shows the validation results where the specifications are clearly satisfied for the whole family of plants described by Eq. (17) for the time domain and frequency domain, respectively.

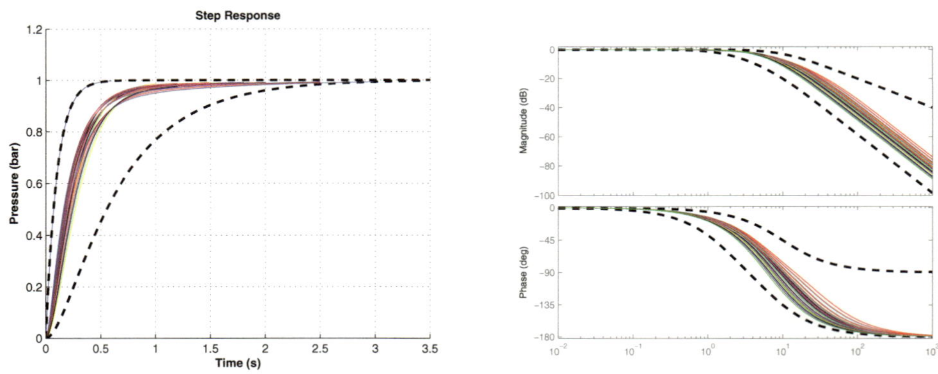

(a) Time domain (b) Frequency domain

Fig. 21. Validation for the QFT design of the pressure system

If the results shown in Figure 19*b* are compared with those shown in Figure 21*b*, a considerable uncertainty reduction can be appreciated, especially in the gain system. Notice that Figure 19 shows the responses of the open-loop system against step inputs for the time and frequency domains, respectively. From these figures, the system uncertainties can be observed by deviations in the static gain and in the time constant of the system, such as described in equation (17).

Finally, the proposed control scheme was tested on the spraying system. The robust control system is characterized by the ability of the closed-loop system to reach desired specifications satisfactorily despite of large variations in the (open-loop) plant dynamics. As commented above, in the pressure system presented in this work such variations appear along the different operating points of the process. Therefore, the system was initially tested through a group of different steps in order to verify that the control system fulfills the robust specifications. Figure 22 shows the results for a sequence of typical steps. It can be observed that the system faithfully follows the proposed reference, reaching the same performance for the different operating points.

4.2 In solar energy field: ACUREX

This section presents a robust control scheme for a distributed solar collector (DSC) field. As DSC are systems subjected to strong disturbances (mainly in solar radiation and inlet oil temperature), a series feedforward was used as a part of the plant, so that the system to be controlled has one input (fluid flow) and one output (outlet temperature) as the disturbances are partially compensated by the series feedforward term, so that the nonlinear plant is transformed into an uncertain linear system. The QFT technique (QFT) was used to design a control structure that guarantee desired control specifications, as settling time and maximum overshoot, under different operating conditions despite system uncertainties and disturbances (Cirre et al., 2010).

The main difference between a conventional power plant and a solar plant is that the primary energy source, while being variable, cannot be manipulated. The objective of the

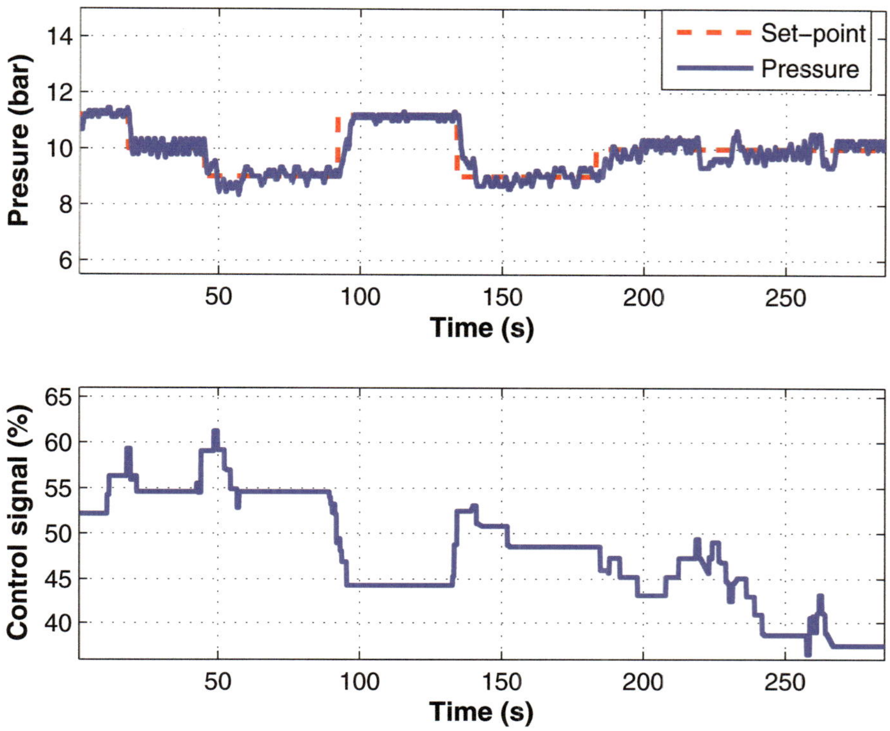

Fig. 22. Experimental tests for the spraying system

control system in a distributed solar collector field (DCS) is to maintain the outlet oil temperature of the loop at a desired level in spite of disturbances such as changes in the solar irradiance level (caused by clouds), mirror reflectivity, or inlet oil temperature. The means available for achieving this is via the adjustment of the fluid flow and the daily solar power cycle characteristics are such that the oil flow has to change substantially during operation. This leads to significant variations in the dynamic characteristics of the field, which cause difficulties in obtaining adequate performance over the operating range with a fixed parameter controller (Camacho et al., 1997; 2007a;b). For that reason, this section summarizes a work developed by the authors where a robust PID controller is designed to control the outlet oil temperature of a DSC loop using the QFT technique.

In this work, the ACUREX thermosolar plant was used, which is located at the Plataforma Solar de Almería (PSA), a research centre of the Spanish Centro de Investigaciones Energéticas Medioambientales y Tecnológicas (CIEMAT), in Almería, Spain. The plant is schematically composed of a distributed collector field, a recirculation pump, a storage tank and a three-way valve, as shown in Figures 23 and 24. The distributed collector field consists of 480 east-west-aligned single-axis-tracking parabolic trough collectors, with a total mirror aperture area of 2672 m2, arranged in 20 rows forming 10 parallel loops (see Figure 23). The parabolic mirrors in each collector concentrate the solar irradiation on an absorber tube through which Santotherm 55 heat transfer oil is flowing. For the collector to concentrate sunlight on its focus, the direct solar radiation must be perpendicular to the mirror plane. Therefore, a sun-tracking algorithm causes the mirrors to revolve around an axis parallel to the tube. Oil is recirculated through the field by a pump that under nominal conditions supplies the field at a flow rate of between 2 l/s (in some applications 3 l/s) and 12 l/s. As it passes through the field, the oil is heated and then the hot oil enters a thermocline storage tank, as shown in Figure 24. A complete detailed description of the ACUREX plant can be found in (Camacho et al., 1997).

Fig. 23. ACUREX solar plant

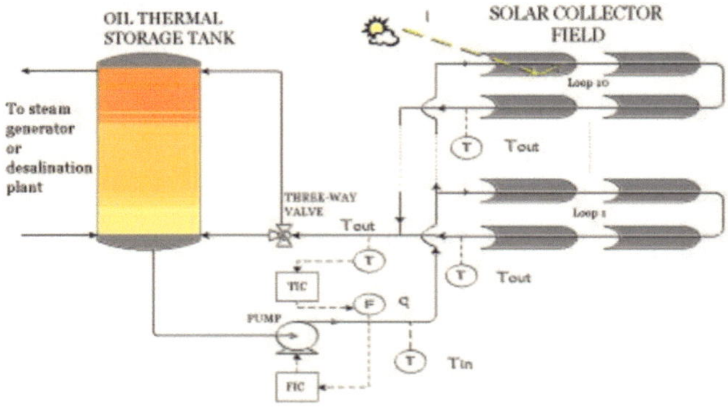

Fig. 24. Simplified layout of the ACUREX plant

As described in (Camacho et al., 1997), DSC dynamics can be approximated by low-order linear descriptions of the plant (as is usually done in the process industry) to model the system around different operating conditions and to design diverse control strategies without accounting for system resonances (Álvarez et al., 2007; Camacho et al., 1997). Thus, different low-order models are found for different operating points mainly due to fluid velocity and system disturbances. Using the series feedforward controller (presented in (Camacho et al., 1997) and improved in (Roca et al., 2008)), a nonlinear plant subjected to disturbances is treated as an uncertain linear plant with only one input (the reference temperature to the feedforward controller, T_{rff}).
After performing an analysis of the frequency response (Berenguel et al., 1994), it was observed that the characteristics of the system (time constants, gains, resonance modes, ...) depend on the fluid flow rate as expected (Álvarez et al., 2007; Camacho et al., 1997). Therefore, in order to control the system with a fixed-parameter controller, the following model has been used

$$\wp = \left\{ P(s) = \frac{k\omega_n^2}{s^2 + 2\xi\omega_n s + \omega_n^2} e^{-\tau_d s} : \xi = 0.8, \right. \tag{21}$$

$$\left. \tau_d = 39s,\ \omega_n \in [0.0038, 0.014]\text{rad/s},\ k \in [0.7, 1.05] \right\},$$

where the chosen nominal plant is $P_0(s)$ with $\omega_n = 0.014$ rad/s and $k = 0.7$.

Thus, once the uncertain model has been obtained, the specifications were determined on time domain and translated into the frequency domain for the QFT design. In this case, the tracking and stability specifications were established (Horowitz, 1993). For tracking specifications only is necessary to impose the minimum and maximum values for the magnitude of the closed-loop transfer function from the reference input to the output in all frequencies. With respect to the stability specification, the desired gain (*GM*) and phase (*PM*) margins are set.
The tracking specifications were required to fulfill a settling time between 5 and 35 minutes and an overshoot less than 30% after 10-20°C setpoint changes for all operating conditions (realistic specifications, see (Camacho et al., 2007a;b)).
For stability specification, $\lambda = 3.77$ in Eq. (4) is selected in order to guarantee at least a phase margin of 35 degrees for all operating conditions.
To design the compensator $C(s)$, the tracking specifications in Eq. (13), shown in Table 1 for each frequency in the set of design frequencies Ω, are used

Table 1. Tracking specifications for the C compensator design

ω (rad/s)	0.0006	0.001	0.003	0.01
$\delta(\omega)$	0.55	1.50	9.01	19.25

The resulting compensator $C(s)$, synthesized in order to achieve the stability specifications and the tracking specifications previously indicated, is the following PID-type controller

$$C(s) = 0.75\left(1 + \frac{1}{180s} + 40s\right) \tag{22}$$

which represents the resulting loop shaping in Figure 25.
Then, in order to satisfy the tracking specifications, the prefilter $F(s)$ must be designed, where the synthesized prefilter is given by

$$F(s) = \frac{0.1}{s + 0.1} \tag{23}$$

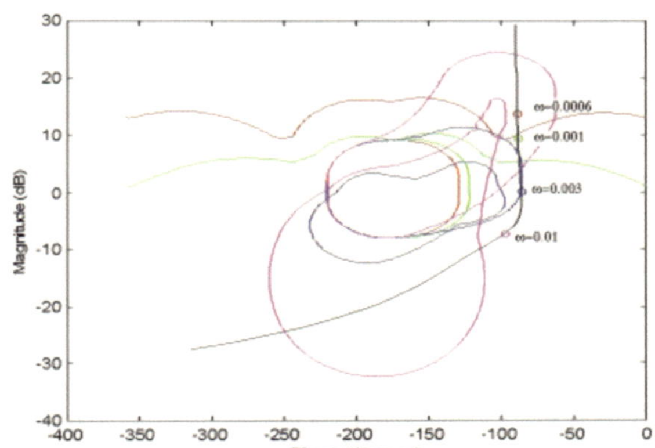

Fig. 25. Tracking and stability boundaries with the designed $L_0(j\omega)$

Figure 26 shows that the tracking specifications are fulfilled for all uncertain cases. Note that the different appearance of Bode diagrams in closed loop for five operating conditions is due to the changing root locus of $L(s)$ when the PID is introduced.

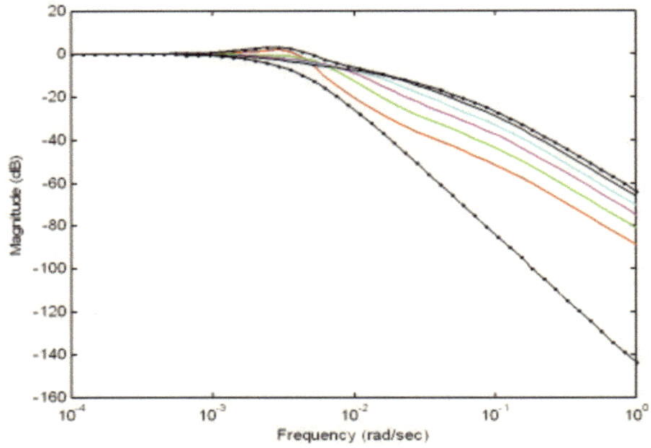

Fig. 26. Tracking specifications (dashed-dotted) and magnitude Bode diagram of some closed loop transfer functions

In order to prove the fulfillment of the tracking and stability specifications of the control structure, experiments were performed under several operating points and under different conditions of disturbances (Cirre et al., 2010), although only representative results are shown in this work.

Figure 27 shows an experiment with the robust controller. At the beginning of the experiment, the flow is saturated until the outlet temperature is higher than the inlet one (the normal

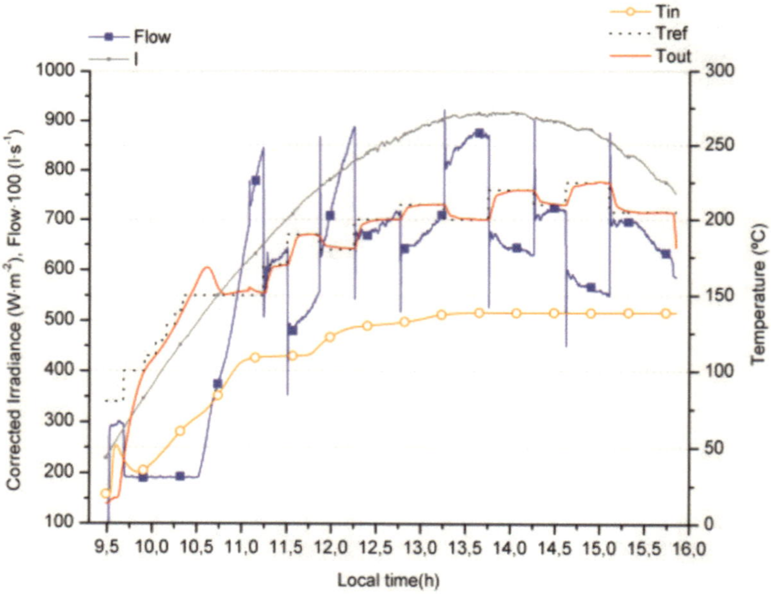

Fig. 27. QTF control results for the ACUREX plant (24/03/2009) (Cirre et al., 2010)

situation during the operation). This situation always appears due to the oil resident inside the pipes is cooler than the oil from the tank. Once the oil is mixed in the pipes, the outlet temperature reaches a higher temperature than the inlet one. During the start up, steps in the reference temperature are made until reaching the nominal operating point. The overshoot at the end of this phase is 18 ^{0}C approximately, and thus the specifications are fulfilled. Analyzing the time responses, a settling time between 11 and 15 minutes is observed at the different operating points. Therefore, both time specifications, overshoot and settling time are properly fulfilled. Disturbances in the inlet temperature (from the beginning until $t = 12.0$ h), due to the temperature variation of the stratified oil inside the tank, are observed during this experiment and correctly rejected by the feedforward action (Cirre et al., 2010).

5. Conclusions

This chapter has introduced the Quantitative Feedback Theory as a robust control technique based on the frequency domain. QFT is a powerful tool which allows to design robust controllers considering the plant uncertainty, disturbances, noise and the desired specifications. It is very versatile tool and has been used in multiple control problems including linear (Horowitz, 1963), non-linear (Moreno et al., 2010), (Moreno et al., 2003), (Moreno, 2003), MIMO (Horowitz, 1979) and non-minimum phase (Horowitz and Sidi, 1978). After describing the theoretical aspects, the most well-known software tools to work with QFT have been described using simple examples. Then, results from two experimental applications were presented, where QFT were successfully used to compensate for the uncertainties in the processes.

6. References

J.D. Álvarez, L. Yebra, and M. Berenguel. Repetitive control of tubular heat exchangers. *Journal of Process Control*, 17:689–701, 2007.

M. Berenguel, E.F. Camacho, and F.R. Rubio. Simulation software package for the acurex field. Technical report, Dep. Ingeniería de Sistemas y Automática, University of Seville (Spain), 1994. www.esi2.us.es/ rubio/libro2.html.

C. Borghesani, Y. Chait, and O. Yaniv. *The QFT Frequency Domain Control Design Toolbox*. Terasoft, Inc., http://www.terasoft.com/qft/QFTManual.pdf, 2003.

E.F. Camacho, M. Berenguel, and F.R. Rubio. *Advanced Control of Solar Plants (1st edn)*. Springer, London, 1997.

E.F. Camacho, F.R. Rubio, M. Berenguel, and L. Valenzuela. A survey on control schemes for distributed solar collector fields. part i: modeling and basic control approaches. *Solar Energy*, 81:1240–1251, 2007a.

E.F. Camacho, F.R. Rubio, M. Berenguel, and L. Valenzuela. A survey on control schemes for distributed solar collector fields. part ii: advances control approaches. *Solar Energy*, 81:1252–1272, 2007b.

M.C. Cirre, J.C. Moreno, M. Berenguel, and J.L. Guzmán. Robust control of solar plants with distributed collectors. In *IFAC International Symposium on Dynamics and Control of Process Systems, DYCOPS*, Leuven, Belgium, 2010.

J. M. Díaz, S. Dormido, and J. Aranda. Interactive computer-aided control design using quantitative feedback theory: The problem of vertical movement stabilization on a high-speed ferry. *International Journal of Control*, 78:813–825, 2005a.

J. M. Díaz, S. Dormido, and J. Aranda. *SISO-QFTIT. An interactive software tool for the design of robust controllers using the QFT methodology*. UNED, http://ctb.dia.uned.es/asig/qftit/, 2005b.

J.L. Guzmán, Rodríguez F., Sánchez-Hermosilla J., and M. Berenguel. Robust pressure control in a mobile robot for spraying tasks. *Transactions of the ASABE*, 51(2):715–727, 2008.

I. Horowitz. *Synthesis of Feedback Systems*. Academic Press, New York, 1963.

I. Horowitz. Quantitative feedback theory. *IEEE Proc.*, 129 (D-6):215–226, 1982.

I. Horowitz and M. Sidi. Synthesis of feedback systems with large plant ignorance for prescribed time-domain tolerances. *International Journal of Control*, 16 (2):287–309, 1972.

I. M. Horowitz. *Quantitative Feedback Design Theory (QFT)*. QFT Publications, Boulder, Colorado, 1993.

I.M. Horowitz. Quantitative synthesis of uncertain multiple input-output feedback systems. *International Journal of Control*, 30:81–106, 1979.

I.M. Horowitz and M. Sidi. Optimum synthesis of non-minimum phase systems with plant uncertainty. *International Journal of Control*, 27(3):361–386, 1978.

K. R. Krishnan and A. Cruickshanks. Frequency domain design of feedback systems for specified insensitivity of time-domain response to parameter variations. *International Journal of Control*, 25 (4):609–620, 1977.

M. Morari and E. Zafiriou. *Robust Process Control*. Prentice Hall, 1989.

J. C. Moreno. *Robust control techniques for systems with input constrains, (in Spanish, Control Robusto de Sistemas con Restricciones a la Entrada)*. PhD thesis, University of Murcia, Spain (Universidad de Murcia, España), 2003.

J. C. Moreno, A. Baños, and M. Berenguel. A synthesis theory for uncertain linear systems with saturation. In *Proceedings of the 4th IFAC Symposium on Robust Control Design*, Milan, Italy, 2003.

J. C. Moreno, A. Baños, and M. Berenguel. Improvements on the computation of boundaries in qft. *International Journal of Robust and Nonlinear Control*, 16(12):575–597, May 2006.

J. C. Moreno, A. Baños, and M. Berenguel. A qft framework for anti-windup control systems design. *Journal of Dynamic Systems, Measurement and Control*, 132(021012):15 pages, 2010.

Y. Piguet. *Sysquake 3 User Manual*. Calerga Sàrl, Lausanne, Switzerland, 2004.

C. J. Pritchard and B. Wigdorowitz. Mapping frequency response bounds to the time domain. *International Journal of Control*, 64 (2):335–343, 1996.

C. J. Pritchard and B. Wigdorowitz. Improved method of determining time-domain transient performance bounds from frequency response uncertainty regions. *International Journal of Control*, 66 (2):311–327, 1997.

L. Roca, M. Berenguel, L.J. Yebra, and D. Alarcón. Solar field control for desalination plants. *Solar Energy*, 82:772–786, 2008.

A. Sánchez-Gimeno, Sánchez-Hermosilla J., Rodríguez F., M. Berenguel, and J.L. Guzmán. Self-propelled vehicle for agricultural tasks in greenhouses. In *World Congress - Agricultural Engineering for a better world*, Bonn, Germany, 2006.

J. Sánchez-Hermosilla, Rodríguez F., González R., J.L. Guzmán, and M. Berenguel. A mechatronic description of an autonomous mobile robot for agricultural tasks in greenhouses. In Alejandra Barrera, editor, *Mobile Robots Navigation*, pages 583–608. In-Tech, 2010. ISBN 978-953-307-076-6.

O. Yaniv, Quantitative Feedback Design of Linear and Nonlinear Control Systems. Kluwer Academic Publishers, 1999.

18

Consensuability Conditions of Multi Agent Systems with Varying Interconnection Topology and Different Kinds of Node Dynamics

Dr. Sabato Manfredi
Faculty of Engineering, University of Naples Federico II, Via Claudio 21, Napoli 80120. Italy

1. Introduction

Many systems in nature and of practical interest can be modeled as large collections of interacting subsystems. Such systems are referred as "Multi Agent Systems" (briefly MASs) and some examples include electrical power distribution networks (P. Kundur, 1994), communication (F. Paganini, 2001), and collections of vehicles traveling in formation (J.K. Hedrick et al., 1990). Several practical issues concern the design of decentralized controllers and the stability analysis of MASs in the presence of uncertainties in the subsystem interconnection topology (i.e. due in practical applications to failures of transmission lines). The analysis and control of collections of interconnected systems have been widely studied in the literature. Early work on stability analysis and decentralized control of large-scale interconnected systems is found in (D. Limebeer & Y.S. Hung, 1983; A. Michel & R. Miller, 1977; P.J. Moylan & D.J. Hill, 1978; Siljak, 1978; J.C. Willems, 1976). Some of the more widely notable stability criteria are based on the passivity conditions (M. Vidyasagar, 1977) and on the well-known notion of connective stability introduced in (Siljak, 1978).
More recently, MASs have appeared broadly in several applications including formation flight, sensor networks, swarms, collective behavior of flocks (Savkin, 2004; C.C. Cheaha et al., 2009; W. Ren, 2009) motivating the recent significative attention of the scientific community to distributed control and consensus problems (i.e. (R.O. Saber & R. Murray, 2004; Z. Lin et al., 2004; V. Blondel et al., 2005; J. N. Tsitsiklis et al., 1986)). One common feature of the consensus algorithm is to allow every agent automatically converge to a common *consensus state* using only local information received from its neighboring agents. "Consensusability" of MASs is a fundamental problem concerning with the existence conditions of the consensus state and it is of great importance in both theoretical and practical features of cooperative protocol (i.e. flocking, rendezvous problem, robot coordination). Results about consensuability of MASs are related to first and second order systems and are based on the assumption of jointly-connected interaction graphs (i.e. in (R.O. Saber & R. Murray, 2004; J. N. Tsitsiklis et al., 1986)). Extension to more general linear MASs whose agents are described by LTI (Linear Time Invariant) systems can be found in (Tuna, 2008) where the closed-loop MASs were shown to be asymptotic consensus stable if the topology had a spanning tree. In (L. Scardovi & R. Sepulchre, 2009) it is investigated the synchronization of a

network of identical linear state-space models under a possibly time-varying and directed interconnection structure. Many investigations are carried out when the dynamic structure is fixed and the communication topology is time varying (i.e. in (R.O. Saber & R. Murray, 2004; W. Ren & R. W. Beard, 2005; Ya Zhanga & Yu-Ping Tian, 2009)). One of main appealing field of research is the investigation of the MASs consensusability under both the dynamic agent structure and communication topology variations. In particular, it is worth analyzing the joint impact of the agent dynamic and the communication topology on the MASs consensusability. The aim of the chapter is to give consensusability conditions of LTI MASs as function of the agent dynamic structure, communication topology and coupling strength parameters. The theoretical results are derived by transferring the consensusability problem into the robust stability analysis of LTI-MASs. Differently from the existing works, here the consensuability conditions are given in terms of the adjacency matrix rather than Laplacian matrix. Moreover, it is shown that the interplay among consensusability, node dynamic and topology must be taken into account for MASs stabilization: specifically, consensuability of MASs is assessed for all topologies, dynamic and coupling strength satisfying a pre-specified bound. From the practical point of view the consensuability conditions can be used for both the analysis and planning of MASs protocols to guarantee robust stability for a wide range of possible interconnection topologies, coupling strength and node dynamics. Also, the number of subsystems affecting the overall system stability is taken into account as it is analyzed the robustness of multi agent systems if the number of subsystems changes. Finally, simulation examples are given to illustrate the theoretical analysis.

2. Problem statement

We consider a network composed of linear systems interconnected by a specific topological structure. The dynamical system at each node is of m-*th* order and described by the matrices (A, B, C). Let $G(V, E, U)$ be a directed weighted graph (digraph) with the set of nodes $V = 1..n$, set of edges $E \subseteq n \times n$, and the associated weighted adjacency matrix $U = \{u_{ij}\}$ with $u_{ij} > 0$ if there is a directed edge of weight u_{ij} from vertex j (node parent) into vertex i (node child). The linear systems are interconnected by a directed weighted graph $G(V, E, U)$. Each node dynamical is described by:

$$\begin{aligned} \dot{x}_i(t) &= Ax_i(t) + Bv_i(t) \\ y_i(t) &= Cx_i(t) \end{aligned} \tag{1}$$

with $v_i(t)$ is the input to the i-*th* node of the form

$$v_i(t) = \sum_{j=1}^{n} u_{ij} y_j(t). \tag{2}$$

In this way, each node dynamic is influenced by the sum of its neighbors' outputs. This yields to the MAS network equation:

$$\dot{x}_i(t) = Ax_i(t) + \sum_{j=1}^{n} u_{ij} BCx_j(t) \tag{3}$$

with $1 \leq i \leq n$, and its compact form:

$$\dot{x}(t) = A_g x(t) \tag{4}$$

with $A_g = (I_n \otimes A) + (U \otimes BC)$, with $\otimes$ denotes the matrix Kronecker product. Notice that the above equation can be associated to the main model used in the literature for describing the synchronization phenomena, energy distribution, tanks network (e.g. in (R. Cogill & S. Lall, 2004)). Moreover the system at each node can be MIMO or SISO type, and the matrix product BC takes into account the coupling strength and the coupling interaction among the state system variables. Observing the MAS model (3) we point out as the overall network dynamic is affected by the node system dynamic matrix A, the coupling matrix BC, and by the adjacency matrix U of the topological structure.

Consider a network with n agents whose topology information exchange is described by a graph $G(V, E, U)$ and let x_i the state of agent-node i-*th*, *consensus* corresponds to the network condition such that the state of the agents as a whole asymptotically converges to an equilibrium state with identical elements (i.e. $x_i = x_j$ for all $(i, j) \in n \times n$). The common value $\bar{x}$ is named *consensus value*. Consensusability of MASs is a fundamental problem concerning with the conditions for assessing network consensus equilibrium. Under the assumption of the existence of a network equilibrium, then consensuability deals with the research of analytical conditions such that the network equilibrium corresponds to a consensus state. In this way, without loss of generality, the consensuability problem can be reduced to the problem of assessing stabilization conditions of the MAS network (3) with respect to the 0 equilibrium point (i.e. $x_i = x_j = 0$ for all $(i, j) \in n \times n$).

Hence, we are interested in solving the following problem:

Problem Given a multi agent network described by (3), to determinate the MAS consensuability conditions as function of node dynamic, topology and coupling strength.

Specifically, consensuability of MASs is assessed for all topologies, dynamic and coupling strength satisfying a pre-specified bound.

In the follows we will present analytical conditions for solving the above Problem.

3. Conditions for MASs consensuability

Before of presenting the MASs consensuability conditions of (3), we have to recast the eigenvalues set $\sigma(A_g)$ of MAS network dynamic matrix A_g.

Lemma 1 Let $\sigma(U)=\{\mu_i\}$ the eigenvalues set of the adjacency matrix U, $\sigma(A_g)$ the eigenvalues set of the MAS dynamical matrix A_g, then results: $\sigma(A_g) = \bigcup_i \sigma(A + \mu_i BC)$ for all $1 \leq i \leq n$.

Proof Let J the Jordan canonical form of U, then it exists a similarity matrix S so that $J = S^{-1}US$. Hence $S \otimes I_n$ is a similarity matrix for the matrices $I_m \otimes A + U \otimes BC$ and $I_m \otimes A + J \otimes BC$. From the Kronecker product (Horn R.A. & Johnson C.R., 1995) results:

$$\begin{gathered}(S \otimes I_n)^{-1}(I_m \otimes A + U \otimes BC)(S \otimes I_n) = \\ (S^{-1} \otimes I_n)(I_m \otimes A + U \otimes BC)(S \otimes I_n) = \\ (I_m \otimes A) + (S^{-1}US \otimes BC) = (I_m \otimes A) + (J \otimes BC)\end{gathered}$$

with J being an upper triangular matrix with $I_m \otimes A + J \otimes BC$ as upper triangular block matrix. Hence the eigenvalues of the matrix $I_m \otimes A + J \otimes BC$ are the union of the eigenvalues of the block matrix on the diagonal.

From the above Lemma 1, the eigenvalues of the MAS dynamic matrix A_g are explicitly function of those of the matrix $A + \mu_i BC$, for all i. So we can decouple the effects of topology structure (by μ_i), the coupling strength BC and node dynamic A on the overall stability of the MAS. This can be used for giving stability MAS condition as function of topology structure, node dynamic and coupling strength as shown by the following Theorem 1:

Theorem 1 Let the MAS composed of n identical MIMO system of order m-*th* and interconnected by the digraph $G = (V, E, U)$ with adjacency matrix U, with eigenvalues $\mu_1 \leq \mu_2 \leq \ldots \mu_n$. If the node dynamic matrix $A = \{a_{ij}\}$ and the coupling matrix $BC = \{c_{ij}\}$ fulfill the conditions:

$$\begin{aligned} a_{ii} + \mu_k c_{ii} &\leq 0 \\ |a_{ii} + \mu_k c_{ii}| &\geq \sum_{j \neq i} |a_{ij} + \mu_k c_{ij}| \end{aligned} \tag{5}$$

$\forall i = 1, 2, ..., m$ and $\forall k = 1, 2, ..., n$, then the MAS (3) is stable.

Proof If the conditions (5) hold, then all eigenvalues of the matrix

$$A + \mu_k BC = \begin{pmatrix} a_{11} + \mu_k c_{11} & a_{12} + \mu_k c_{12} & \cdots & a_{1m} + \mu_k c_{1n} \\ a_{21} + \mu_k c_{21} & a_{22} + \mu_k c_{22} & \cdots & a_{2m} + \mu_k c_{2m} \\ \vdots & \vdots & \vdots & \ddots \\ a_{m1} + \mu_k c_{m1} & a_{m2} + \mu_k c_{m2} & \cdots & a_{mm} + \mu_k c_{mm} \end{pmatrix},$$

$\forall k = 1, 2, \ldots, n$, are located in a convex set in the left complex half plane as result by the application of the Gershgorin's circle theorem (Horn R.A. & Johnson C.R., 1995). Hence, by Lemma 1, the MAS is stable.

The previous Theorem 1 easily yields to the following corollaries.

Corollary 1 Let the MAS composed of n identical MIMO system of order 2 and interconnected by the digraph $G = (V, E, U)$, with adjacency matrix U with eigenvalues $\mu_1 \leq \mu_2 \leq \ldots \mu_n$. If the node dynamic matrix $A = \{a_{ij}\}$ and the coupling matrix $BC = \{c_{ij}\}$ with $c_{ij} \geq 0, i, j = 1, 2$, fulfill the conditions:

$$\begin{aligned} a_{ij} &\geq -c_{ij}\mu_1 \\ a_{ii} &\leq -a_{ij} - (c_{ii} + c_{ij}) \cdot \mu_n \end{aligned} \tag{6}$$

or

$$\begin{aligned} a_{ii} &\leq -c_{ii}\mu_n \\ a_{ij} &\leq -c_{ij}\mu_n \\ a_{ii} &\leq a_{ij} + (c_{ij} - c_{ii}) \cdot \mu_1, \end{aligned} \tag{7}$$

$i, j = 1, 2$, then the MAS (3) is stable.

Because the adjacency matrix U of a graph has both positive and negative eigenvalues, the conditions (6) and (7) implicitly imply the assumption that the single system at the node is stable. In this way, as expected, we derive that it is not possible to stabilize a network of instable systems by acting only on the topological structure. Given a specified node dynamic, coupling strength and bound on the adjacency matrix U, by conditions (6) and (7) we can assess MAS stability. Moreover, the MAS robustness with respect to varying switching topology can be dealt by considering the span of the eigenvalue of the admissible structure topologies. As we will show in the follows, it is possible easily to evaluate the eigenvalues of A_g, given the eigenvalues of U in some simple and representative cases of interest.

Corollary 2 Let the MAS composed of n identical MIMO system of order 1 and interconnected by the digraph $G = (V, E, U)$, with adjacency matrix U with eigenvalues $\mu_1 \leq \mu_2 \leq \ldots \mu_n$. If the node dynamic matrix $A = a$ and the coupling matrix $BC = c$ fulfill the conditions:

$$a \leq -c \cdot \mu_n \quad \text{if} \quad c \geq 0 \tag{8}$$
$$a \leq -c \cdot \mu_1 \quad \text{if} \quad c < 0, \tag{9}$$

then the MAS (3) is stable.

The Corollary 2 reduces the analytical result of Theorem 1 to the case of the consensus of integrator (R.O. Saber & R. Murray, 2004) with coupling gain c. Smaller c, higher is the degree of robustness of the network to the slower node dynamic. In the opposite, higher c reduces the stability margin of the MAS. Finally, for a fixed dynamic at the node, the maximum admissible coupling strength c depends on the maximum and minimum eigenvalues of the adjacency matrix:

$$c \leq -\frac{a}{\mu_n} \quad \text{if} \quad c \geq 0 \tag{10}$$
$$c \geq -\frac{a}{\mu_1} \quad \text{if} \quad c < 0. \tag{11}$$

Corollary 3 Let the MAS of n identical MIMO system of the m-th order, interconnected by the digraph $G = (V, E, U)$, with adjacency matrix U with eigenvalues $\mu_1 \leq \mu_2 \leq \ldots \mu_n$. If the node dynamic matrix $A = \{a_{ij}\}$ and the coupling matrix $BC = \{c_{ij}\}$ are both upper or lower triangular matrix and fulfill the conditions:

$$a_{ii} \leq -c_{ii} \cdot \mu_n \quad \text{if} \quad c_{ii} \geq 0 \tag{12}$$
$$a_{ii} \leq -c_{ii} \cdot \mu_1 \quad \text{if} \quad c_{ii} < 0, \tag{13}$$

then the MAS (3) is stable.

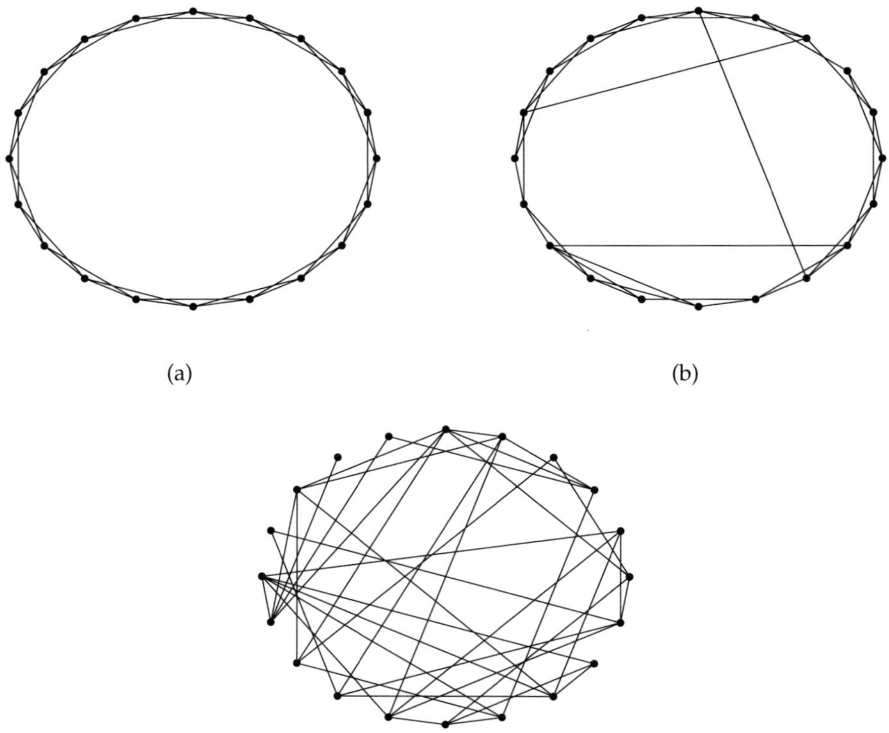

(a) (b) (c)

Fig. 1. Procedure of redirectioning of links in a regular network (a) with increasing probability p. As p increases the network moves from regular (a) to random (c), becoming small world (b) for a critical value of p. n=20, k=4

Notice that if $BC = c \cdot I_n$, (10) and (11) become:

$$c \leq -\frac{\min_i |a_{ii}|}{\mu_n} \qquad \text{If} \quad c_{ii} \geq 0 \tag{14}$$

$$c > -\frac{\min_i |a_{ii}|}{\mu_1} \qquad \text{If} \quad c_{ii} < 0 \tag{15}$$

and hence the stability of MAS is explicitly given as function of the network slowest node dynamic.

Now we would like to point out the case of undirected topology with symmetric adjacency matrix U. If we assume A and BC being symmetric, then A_g is symmetric with real eigenvalue. Moreover from the field value property (Horn R.A. & Johnson C.R., 1995), let $\sigma(A) = \{\alpha_j\}$ and $\sigma(BC) = \{\nu_j\}$ the eigenvalues set of A and BC, then the eigenvalues of $A + \mu_i BC$ are in the interval $[\min_j\{\alpha_j\} + \mu_i \min_j\{\nu_j\}, \max_j\{\alpha_j\} + \mu_i \max_j\{\nu_j\}]$, for every $1 \leq i \leq n$, $1 \leq j \leq m$. In this way, there is a bound need to be satisfied by the topology structure, node dynamic and coupling matrix for MAS stabilization.

In the literature, the MAS consensuability results have been given in terms of Laplacian matrix properties. Here, differently, we have given bounds as function of the adjacency matrix features. Anyway we can use the results on the Laplacian eigenvalue for recasting the bounds given on the adjacency matrix. To this aim, defined the degree d_i of i-*th* node of an undirected graph as $\sum_j u_{ij}$, the Laplacian matrix is defined as $L = D - U$ with D is the diagonal matrix with the degree of node i-*th* in position i-*th*. Clearly L is a zero row sums matrix with non-positive off-diagonal elements. It has at least one zero eigenvalue and all nonzero eigenvalues have nonnegative real parts. So $U = D - L$ and being the minimum and maximum Laplacian eigenvalues respectively bounded by 0 and the highest node degree, we have:

Lemma 2 Let U the adjacency matrix of undirected and connected graph $G = (V, E, U)$, with eigenvalues $\mu_1 \leq \mu_2 \leq \ldots \leq \mu_n$, then results:

$$\mu_1(U) \geq \min_i d_i - \min(\max_{k,j}\{d_k + d_j : (k,j) \in E(G)\}, n) \tag{16}$$

$$\mu_n(U) \leq \max_i d_i \tag{17}$$

Proof Easily follows from the Laplacian eigenvalues bound and the field value property (Horn R.A. & Johnson C.R., 1995).

4. Simulation validation

In the follows we will present a variety of simulations to validate the above theoretical results under different kinds of node dynamic and network topology variations. Specifically the MAS topology variations have been carried out by using the well known Watts-Strogats procedure described in (Watts & S. H. Strogatz, 1998). In particular, starting from the regular network topology ($p = 0$), by increasing the probability p of rewiring the links, it is possible smoothly to change its topology into a random one ($p = 1$), with small world typically occurring at some intermediate value. In so doing neither the number of nodes nor the overall number of edges is changed. In Fig. 1 it shown the results in the case of MAS of 20 nodes with each one having $k = 4$ neighbors.

Among the simulation results we focus our attention on the maximum and minimum eigenvalues of the matrixes U (i.e. μ_n and μ_1) and A_g (i.e. λ_M and λ_m) and their bounds computed by using the results of the previous section. In particular, by Lemma 2, we convey the bounds on U eigenvalues in bounds on A_g eigenvalues suitable for the case of time varying topology structure. We assume in the simulations the matrices A and BC to be symmetric. In this way, if U eigenvalues are in $[v_1, v_2]$, let $\sigma(A) = \{\alpha_i\}$, $\sigma(BC) = \{\nu_i\}$, the eigenvalues of Ag will be in the interval $[\min_i \alpha_i + \min_j\{v_1\nu_j, v_2\nu_j\}, \max_i \alpha_i + \max_j\{v_1\nu_j, v_2\nu_j\}]$ for $i, j = 1, 2, \ldots, n$.

Notice that, known the interval of variation $[v_1, v_2]$ of the eigenvalues set of U under switching topologies, we can recast the conditions (8), (9), (12), (13), (6), (7) and to use it for design purpose. Specifically, given the interval $[v_1, v_2]$ associated to the topology possible variations, we derive conditions on A or BC for MAS consensuability.

We consider a graph of $n = 400$ and $k = 4$. In the evolving network simulations, we started with $k = 4$ and bounded it to the order of $O(log(n))$ for setting a sparse graph. In Tab 1 are drawn the node dynamic and coupling matrices considered in the first set of simulations.

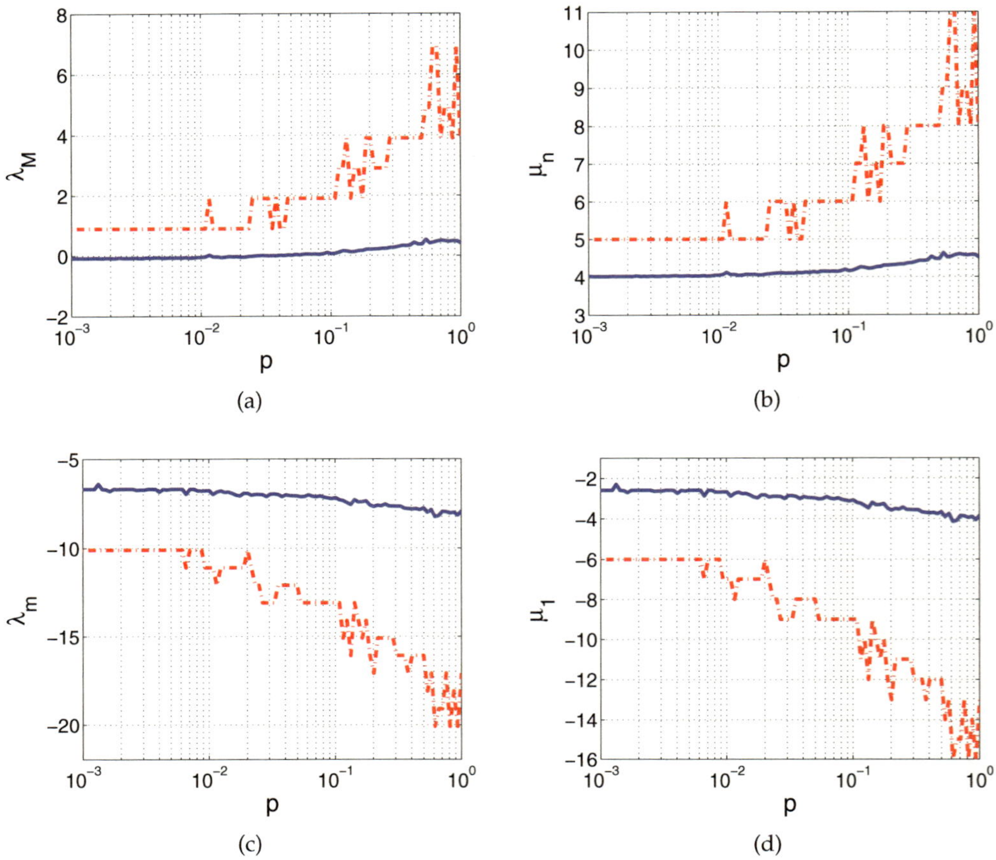

Fig. 2. Case 1. Dashed line: bound on the eigenvalue; continuous line: eigenvalues, (a) Maximum eigenvalue of A_g, (b) Maximum eigenvalue of U, (c) Minimum eigenvalue of A_g, (d) Minimum eigenvalue of U

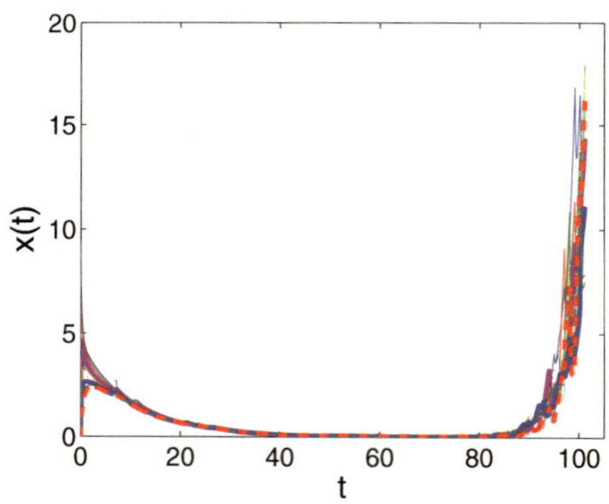

Fig. 3. Case 1: State dynamic evolution in the time

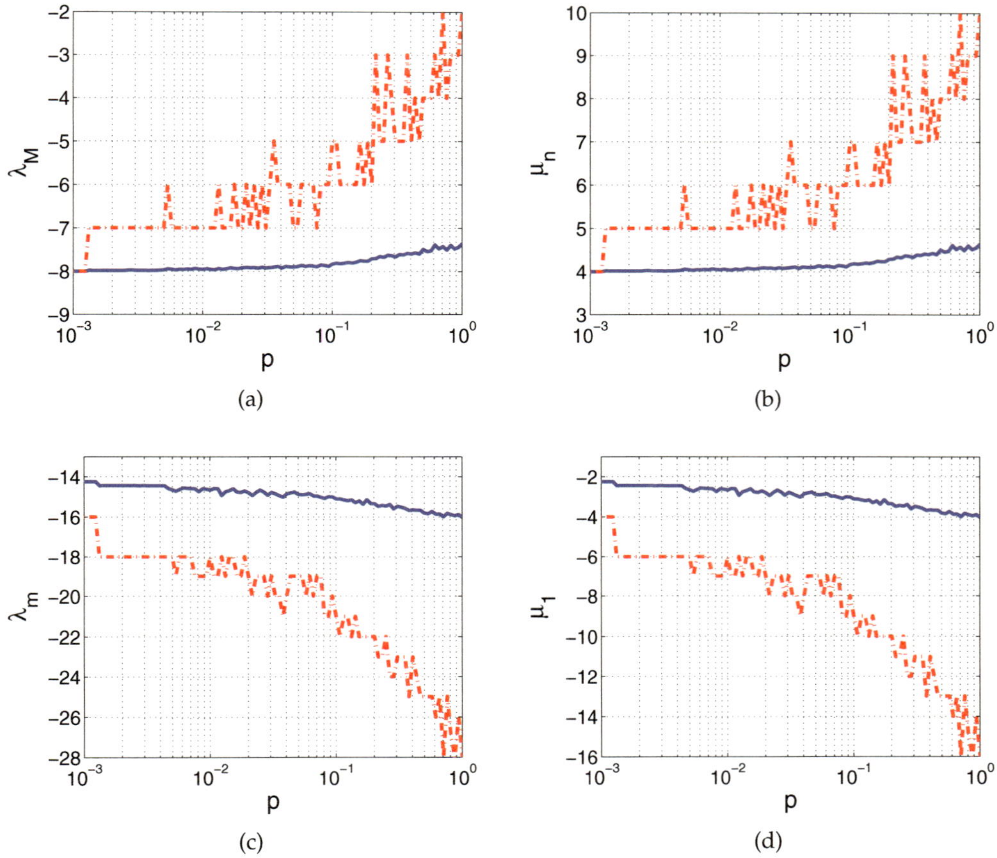

Fig. 4. Case 2. Dashed line: bound on the eigenvalue; continuous line: eigenvalue: (a) Maximum eigenvalue of A_g, (b) Maximum eigenvalue of U, (c) Minimum eigenvalue of A_g, (d) Minimum eigenvalue of U

	A	**B**	**C**
Case 1:	-4.1	1	1
Case 2:	-12	1	1
Case 3:	-6	1	1
Case 4:	-6	2	1

Table 1. Node system matrices (A,B,C)

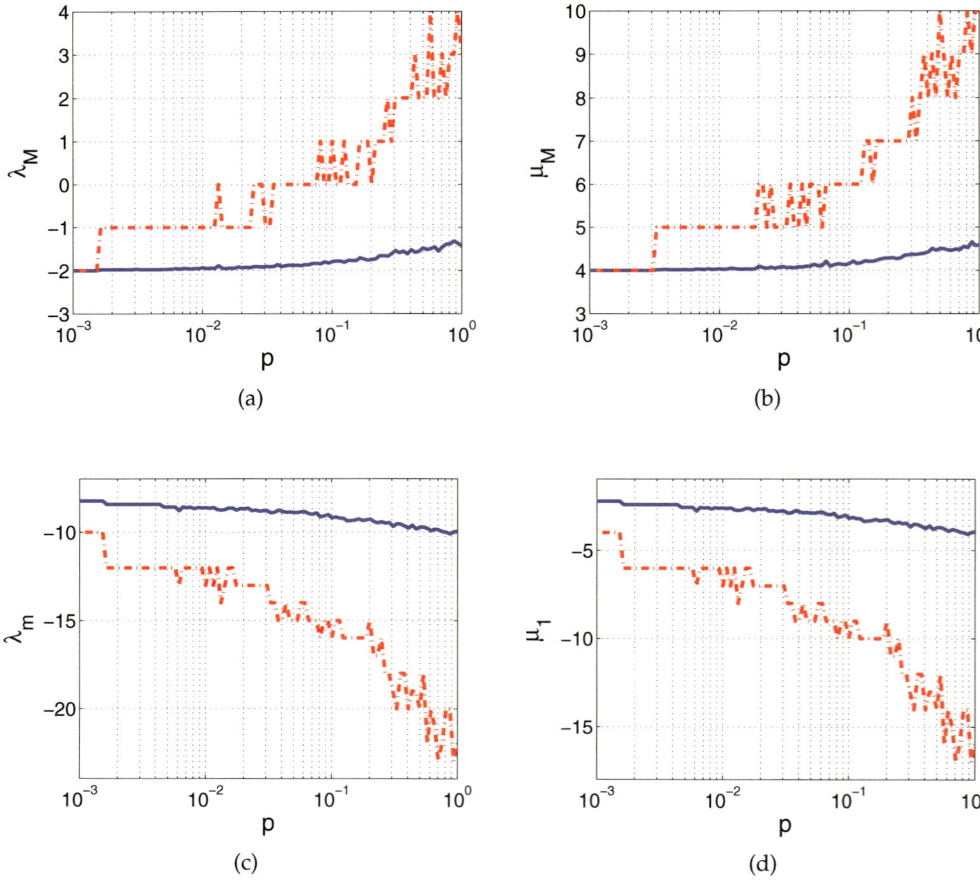

Fig. 5. Case 3. Dashed line: bound on the eigenvalue; continuous line: eigenvalue: (a) Maximum eigenvalue of A_g, (b) Maximum eigenvalue of U, (c) Minimum eigenvalue of A_g, (d) Minimum eigenvalue of U

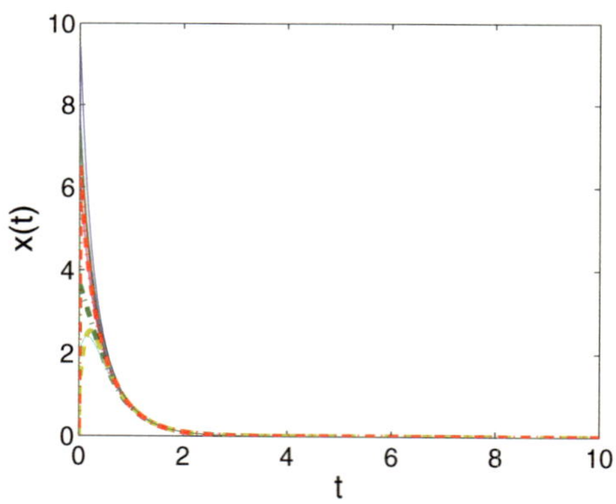

Fig. 6. Case 3: state dynamic evolution in the time

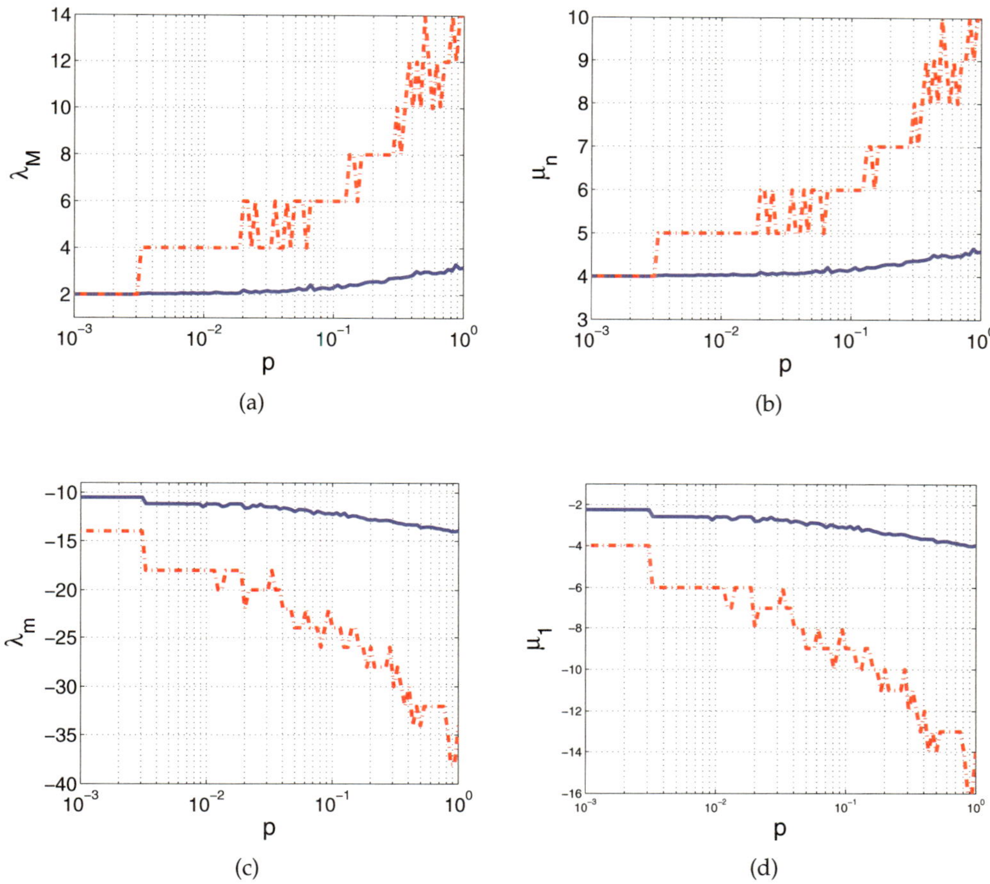

Fig. 7. Case 4. Dashed line: bound on the eigenvalues; continuous line: eigenvalues: (a) Maximum eigenvalue of A_g, (b) Maximum eigenvalue of U, (c) Minimum eigenvalue of A_g, (d) Minimum eigenvalue of U

In the case 1 (Fig 2), we note as although we start from a stable MAS network, the topology variation leads the network instability condition (namely λ_M becomes positive). In Fig. 3 it is shown the time state evolution of the firsts 10 nodes, under the switching frequency of 1 Hz. We note as the MAS converges to the consensus state till it is stable, then goes in instability condition.

In the case 2, we consider a node dynamic faster than the maximum network degree d_M of all evolving network topologies from compete to random graph. Notice that although this assures MAS consensuability as drawn in Fig. 4, it can be much conservative.

In the case 3 (Fig 5), we consider a slower node dynamic than the cases 2. The MAS is robust stable under topology variations. In Fig. 6 the state dynamic evolution is convergent and the settling time is about $4.6/|\lambda_M(A_g)|$.

Then we have varied the value for BC by doubling the B matrix value leaving unchanged the node dynamic matrix. As appears in Fig. 7, the MAS goes in instability condition pointing out

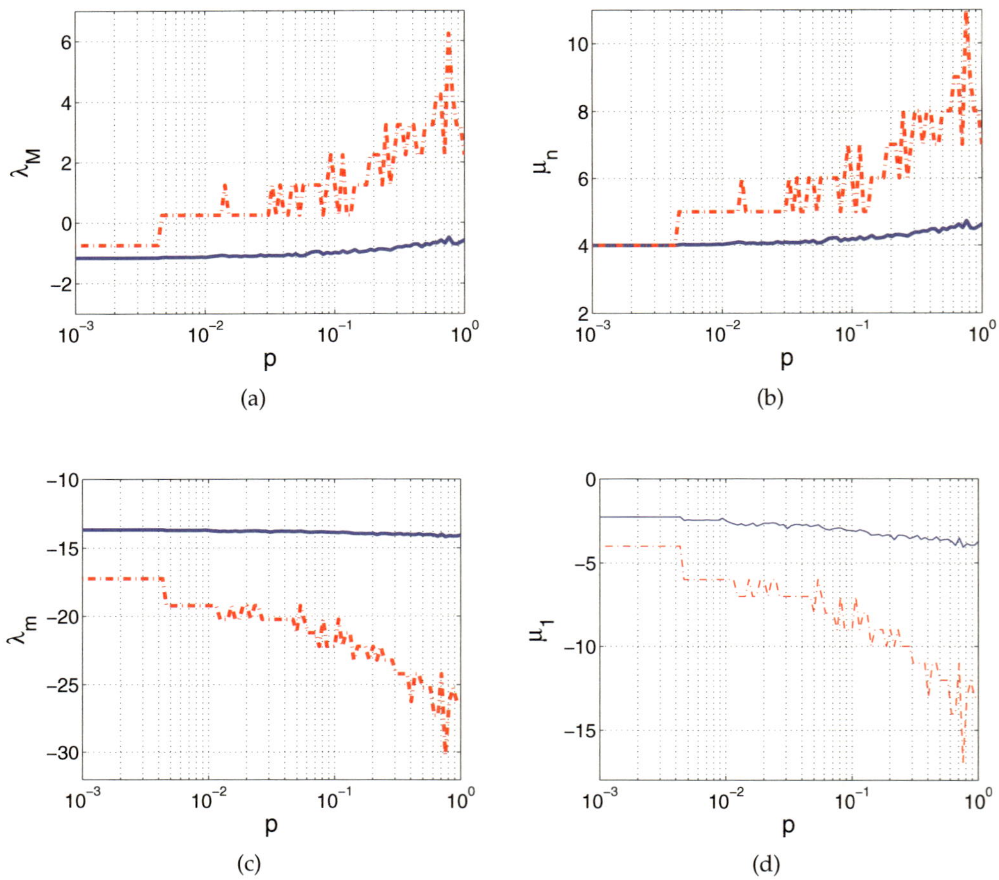

Fig. 8. Case 5. Dashed line: bound on the eigenvalue; continuous line: eigenvalue: (a) Maximum eigenvalue of A_g, (b) Maximum eigenvalue of U, (c) Minimum eigenvalue of A_g, (d) Minimum eigenvalue of U

that also the coupling strength can affect the stability (as stated by the conditions (8), (9)) and that this effect can be amplified by the network topological variations.

	A	B	C
Case 5:	$\begin{bmatrix} -6 & 3 \\ 3 & -12 \end{bmatrix}$	$\begin{bmatrix} 1 \\ 0 \end{bmatrix}$	$\begin{bmatrix} 1 & 0 \end{bmatrix}$
Case 6:	$\begin{bmatrix} -3 & 3 \\ 3 & -6 \end{bmatrix}$	$\begin{bmatrix} 1 \\ 0 \end{bmatrix}$	$\begin{bmatrix} 1 & 0 \end{bmatrix}$
Case 7:	$\begin{bmatrix} -3 & 3 \\ 3 & -6 \end{bmatrix}$	$\begin{bmatrix} 0.25 \\ 0 \end{bmatrix}$	$\begin{bmatrix} 1 & 0 \end{bmatrix}$

Table 2. Node system matrices (A,B,C).

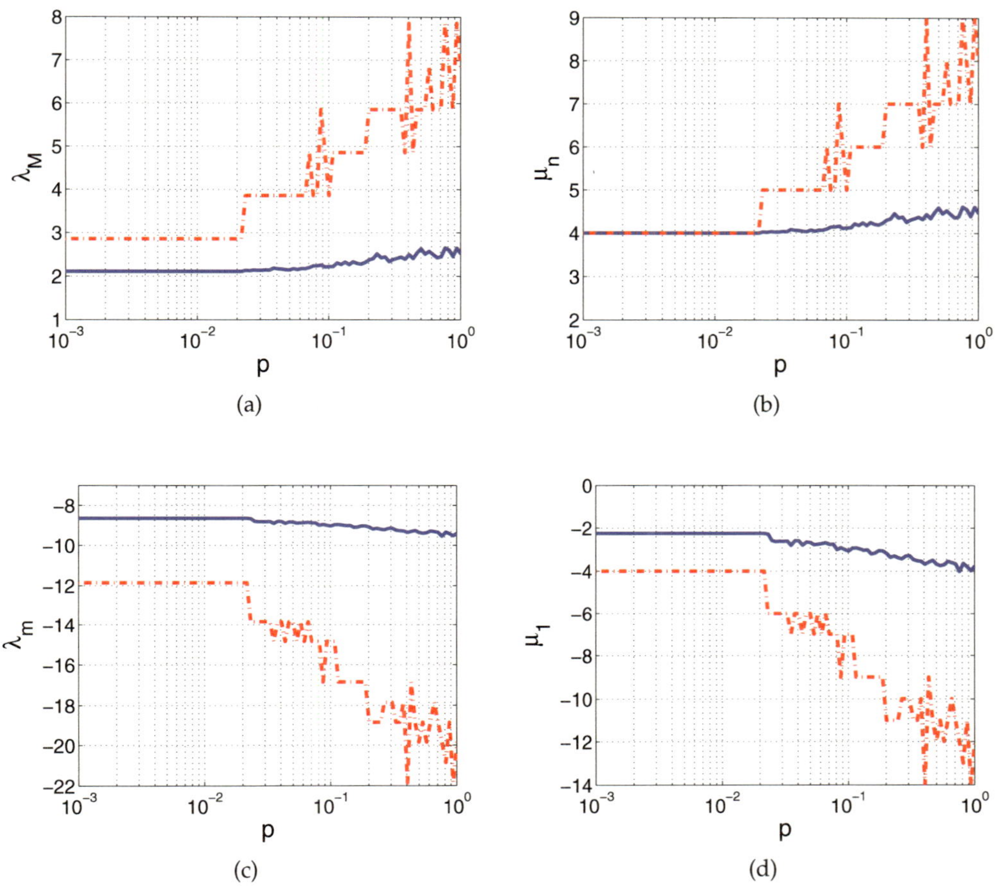

Fig. 9. Case 6. Dashed line: bound on the eigenvalue; continuous line: eigenvalue: (a) Maximum eigenvalue of A_g, (b) Maximum eigenvalue of U, (c) Minimum eigenvalue of A_g, (d) Minimum eigenvalue of U

On the other side, a reduction on BC increases the MAS stability margin. So we can tune the BC value in order to guarantee stability or desired robust stability MAS margin under a specified node dynamic and topology network variations. Indeed if BC has eigenvalues above 1, its effect is to amplify the eigenvalues of U and we need a faster node dynamic for assessing MAS stability. If BC has eigenvalues less of 1, its effect is of attenuation and the node dynamic can be slower without affecting the network stability.
Now we consider SISO system of second order at the node as shown in Tab.2. In this case the matrix BC has one zero eigenvalue being the rows linearly dependent.
In the case 5 the eigenvalues of A are $\alpha_1 = -4.76$ and $\alpha_2 = -13.23$, the eigenvalues of the coupling matrix BC are $\nu_1 = 1$ and $\nu_2 = 0$. In this case the node dynamic is sufficiently fast for guaranteeing MAS consensuability (Fig. 8). In the case 6, we reduce the node dynamic matrix A to $\alpha_1 = -1.15$ e $\alpha_2 = -7.85$. Fig. 9 shows instability condition for the MAS network. We

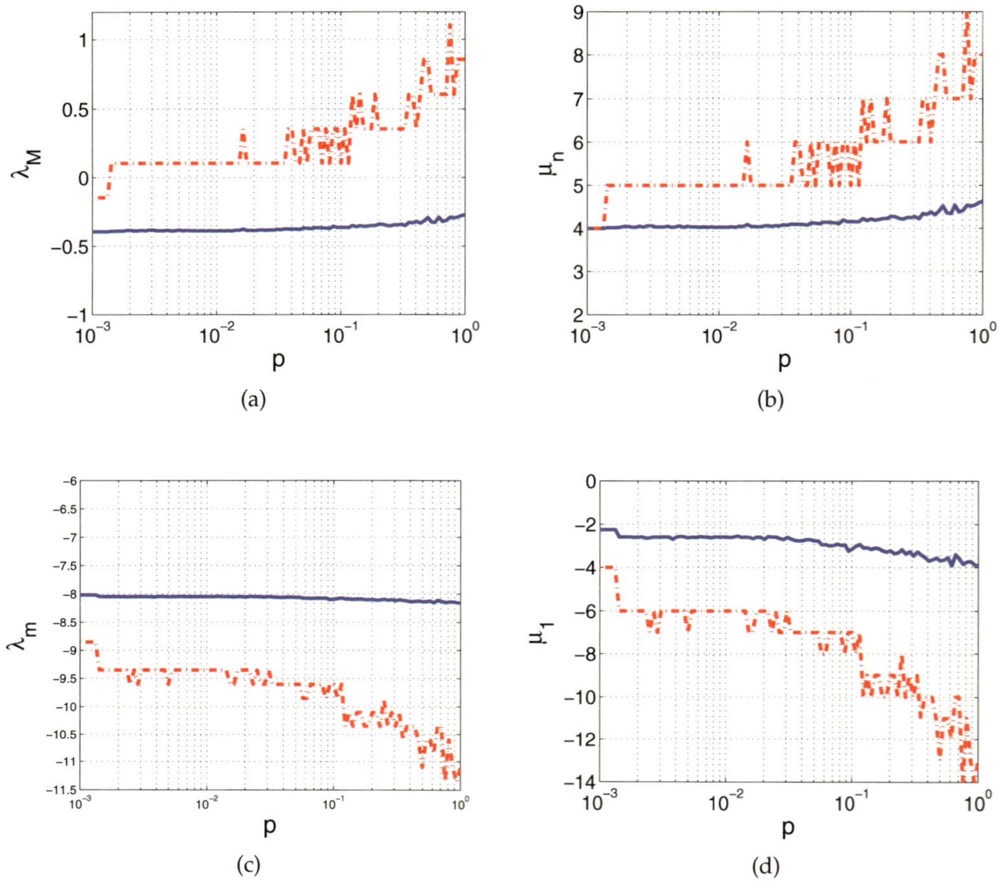

Fig. 10. Case 7. Dashed line: bound on the eigenvalue; continuous line: eigenvalue. (a) Maximum eigenvalue of A_g, (b) Maximum eigenvalue U, (c) Minimum eigenvalue of A_g, (d) Minimum eigenvalue of U

can lead the MAS in stability condition by designing the coupling matrix BC as appear by the case 7 and the associate Fig. 10.

4.1 Robustness to node fault

Now we deal with the case of node fault. We can state the following Theorem.

Theorem 2 Let A and BC symmetric matrix and $G(V, E, U)$ an undirected graph. If the MAS system described by A_g is stable, it is stable also in the presence of node faults. Moreover the MAS dynamic becomes faster after the node fault.

Proof Being the graph undirected and A and BC symmetric then A_g is symmetric. Let $\tilde{A}_g$ the MAS dynamic matrix associated to the network after a node fault. $\tilde{A}_g$ is obtained from A_g by eliminating the rows and columns corresponding to the nodes went down. So $\tilde{A}_g$ is a minor of A_g and for the interlacing theorem (Horn R.A. & Johnson C.R., 1995) it has eigenvalues inside

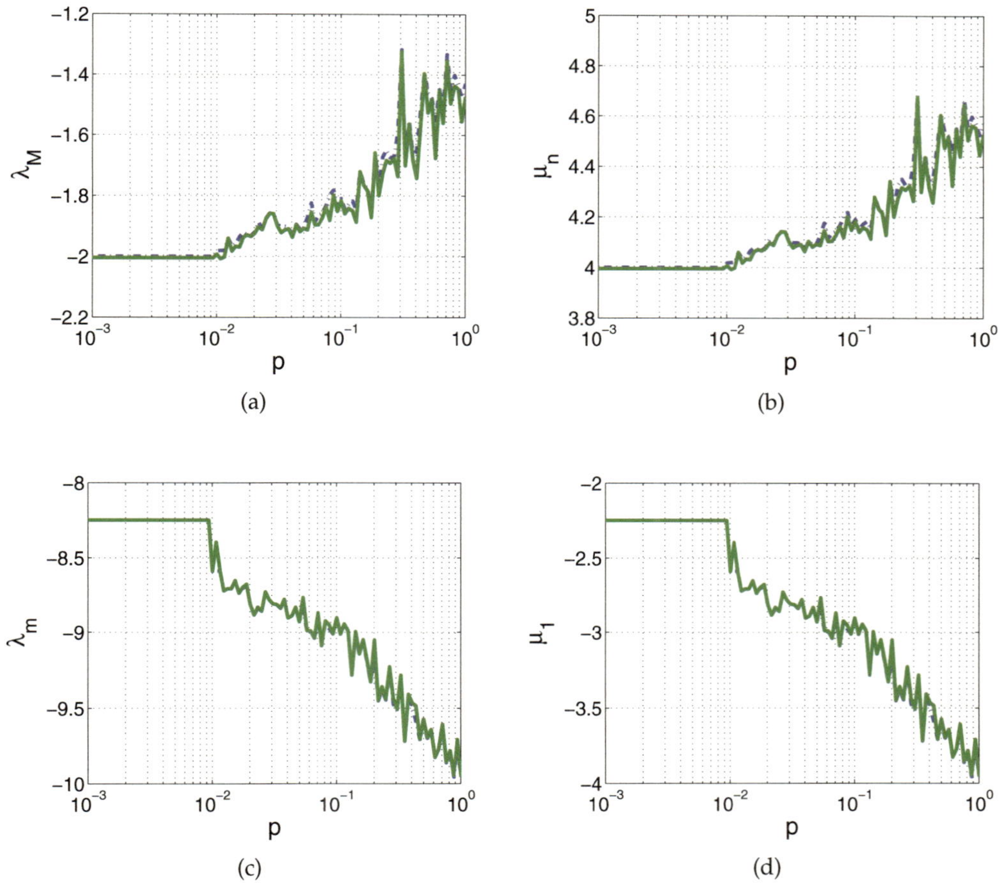

Fig. 11. Eigenvalues in the case $l = 1$. Dashed line: eigenvalue in the case of complete topology with $n = 100$; continuous line: eigenvalue in the case of node fault: (a) Maximum eigenvalue of A_g, (b) Maximum eigenvalue of U, (c) Minimum eigenvalue of A_g, (d) Minimum eigenvalue of U

the real interval with extremes the minimum and maximum A_g eigenvalues. Hence if A_g is stable, $\tilde{A_g}$ is stable too. Moreover, the maximum eigenvalue of $\tilde{A_g}$ is less than one of A_g. So the slowest dynamic of the system $\dot{x}(t) = \tilde{A_g}x(t)$ is faster than the system $\dot{x}(t) = A_g x(t)$.

In the follows we will show the eigenvalues of MAS dynamic in the presence of node fault. We consider MAS network with $n = 100$. We compare for each evolving network topology at each time simulation step, the maximum and minimum eigenvalues of A_g than those ones resulting with the fault of randomly chosen l nodes. Figures 11 and 12 show the eigenvalues of system dynamic for the cases $l = 1$ and $l = 50$.
Notice that as the eigenvalues of U and A_g of fault network are inside the real interval containing the eigenvalues of U and A_g of the complete graph. In Fig. 13 are shown the time evolutions of state of the complete and faulted graphs. Notice that the fault network is faster than the initial network as stated by the analysis of the spectra of A_g and $\tilde{A_g}$.

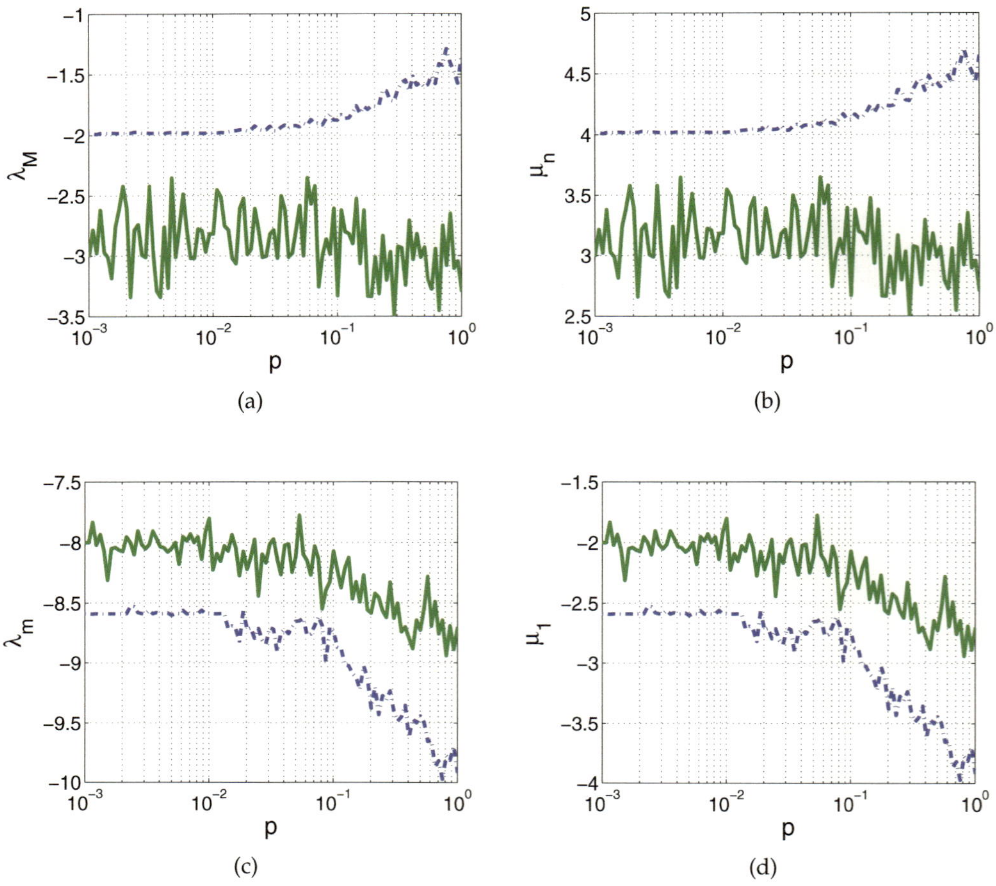

Fig. 12. Eigenvalues in the case of $l = 50$. Dashed line: eigenvalue in the case of complete topology with $n = 100$; continuous line: eigenvalue in the case of node fault: (a) Maximum eigenvalue of A_g, (b) Maximum eigenvalue of U, (c) Minimum eigenvalue of A_g, (d) Minimum eigenvalue of U

5. Conclusions

In this book chapter we have investigated the consensuability of the MASs under both the dynamic agent structure and communication topology variations. Specifically, it has given consensusability conditions of linear MASs as function of the agent dynamic structure, communication topology and coupling strength parameters. The theoretical results are given by transferring the consensusability problem to the stability analysis of LTI-MASs. Moreover, it is shown that the interplay among consensusability, node dynamic and topology must be taken into account for MASs stabilization: consensuability of MASs is assessed for all topologies, dynamic and coupling strength satisfying a pre-specified bound. From the practical point of view the consensuability conditions can be used for both the analysis and planning of MASs protocols to guarantee robust stability for a wide range of possible interconnection topologies, coupling strength and node dynamics. Also, the consensuability

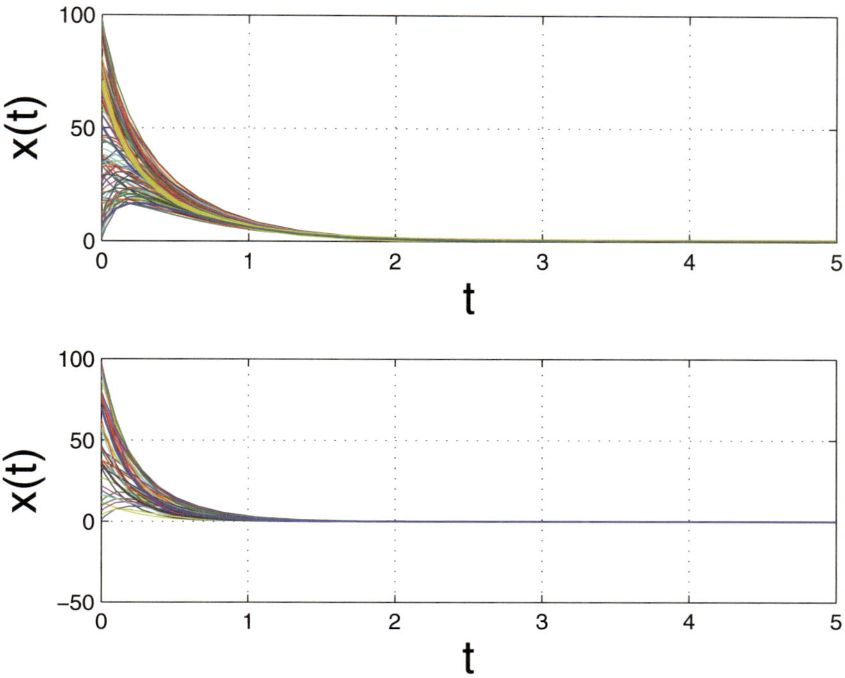

Fig. 13. Time evolution of the state variables for l=50: top Figure: complete graph. Bottom Figure: graph with fault.

of MAS in the presence of node faults has been analyzed. Simulation scenarios are given to validate the theoretical results.

Acknowledgement

The author would like to thank Ms. F. Schioppa for valuable discussion.

6. References

J.K. Hedrick, D.H. McMahon, V.K. Narendran, and D. Swaroop. (1990). *Longitudinal vehical controller design for IVHS systems*. Proceedings of the American Control Conference, pages 3107-3112.

P. Kundur. (1994) *Power System Stability and Control.* McGraw-Hill.

D. Limebeer and Y.S. Hung. (1983). *Robust stability of interconnected systems.* IEEE Trans. Automatic Control, pages 710-716.

A. Michel and R. Miller. (1977). *Qualitative analysis of large scale dynamical systems.* Academic Press.

P.J. Moylan and D.J. Hill. (1978). *Stability criteria for large scale systems.* IEEE Trans. Automatic Control, pages 143-149.

F. Paganini, J. Doyle, and S. Low. (2001). *Scalable laws for stable network congestion control.* Proceedings of the IEEE Conference on Decision and Control, pages 185-190, 2001.

M. Vidyasagar. (1977). *L2 stability of interconnected systems using a reformulation of the passivity theorem.* IEEE Transactions on Circuits and Systems, 24, 637-645.

D. ůSiljak. (1978). *Large-Scale Dynamic Systems.* Elsevier North-Holland.

J.C. Willems. (1976). *Stability of large-scale interconnected systems..*

Saber R.O., Murray R.M. (2004). *Consensus Problems in Networks of Agents with Switching Topology and Time-Delays,* IEEE Transactions on Automatic Control, Vol 49, 9.

Z. Lin, M. Brouke, and B. Francis, (2004). *Local control strategies for groups of mobile autonomous agents.,* Transactions on Automatic Control, 49, vol 4, pages: 622Ű-629.

V. Blondel, J. M. Hendrickx, A. Olshevsky, and J. N. Tsitsiklis, (2005) *Convergence in multiagent coordination, consensus, and flocking,* 44th IEEE Conference on Decision and Control and European Control Conference, pages 2996Ű-3000.

A. V. Savkin, (2004) *Coordinated collective motion of groups of autonomous mobile robots: analysis of VicsekŠs model.,* Transactions on Automatic Control, Vol 49, 6, pages: 981-Ű982.

J. N. Tsitsiklis, D. P. Bertsekas, M. Athans, (1986). *Distributed Asynchronous Deterministic and Stochastic Gradient Optimization Algorithms,* Transactions on Automatic Control, pages. 803-Ű812.

C. C. Cheaha, S. P. Houa, and J. J. E. Slotine, (2009). *Region-based shape control for a swarm of robots.,* Automatica, Vol. 45, 10, pages: 2406-Ű2411.

Ren, W., (2009). *Collective Motion From Consensus With Cartesian Coordinate Coupling .,* IEEE Transactions on Automatic Control. Vol. 54, 6, pages: 1330–1335.

S. E. Tuna, (2008). *LQR-based coupling gain for synchronization of linear systems,* Available at: http://arxiv.org/abs/0801.3390

Luca Scardovi, Rodolphe Sepulchre, (2009) *Synchronization in networks of identical linear systems* Automatica, Volume 45, Issue 11, Pages 2557-2562

W. Ren and R. W. Beard, (2005), *Consensus seeking in multiagent systems under dynamically changing interaction topologies,* IEEE Trans. Automatic Control, vol. 50, no. 5, pp. 655-661.

Ya Zhanga and Yu-Ping Tian, (2009). *Consentability and protocol design of multi-agent systems with stochastic switching topology.,* Automatica, Vol. 45, 5, 2009, Pages 1195–1201.

R. Cogill, S. Lall, (2004). *Topology independent controller design for networked systems,* IEEE Conference on Decision and Control, Atlantis, Paradise Island, Bahamas, Dicembre 2004

D. J. Watts, S. H. Strogatz, (1998). *Collective dynamics of small world networks,* Nature - Macmillan Publishers Ltd, Vol. 393, Giugno 1998.

Horn R.A. and Johnson C.R., (1995). *Topics in Matrix Analysis* Cambridge University Press 1995.

19

On Stabilizability and Detectability of Variational Control Systems

Bogdan Sasu*and Adina Luminiţa Sasu
Department of Mathematics, Faculty of Mathematics and Computer Science, West University of Timişoara, V. Pârvan Blvd. No. 4 300223 Timişoara Romania

1. Introduction

The aim of this chapter is to present several interesting connections between the input-output stability properties and the stabilizability and detectability of variational control systems, proposing a new perspective concerning the interference of the interpolation methods in control theory and extending the applicability area of the input-output methods in the stability theory.

Indeed, let X be a Banach space, let (Θ, d) be a locally compact metric space and let $\mathcal{E} = X \times \Theta$. We denote by $\mathcal{B}(X)$ the Banach algebra of all bounded linear operators on X. If Y, U are two Banach spaces, we denote by $\mathcal{B}(U, Y)$ the space of all bounded linear operators from U into Y and by $\mathcal{C}_s(\Theta, \mathcal{B}(U, Y))$ the space of all continuous bounded mappings $H : \Theta \to \mathcal{B}(U, Y)$. With respect to the norm $|||H||| := \sup_{\theta \in \Theta} ||H(\theta)||$, $\mathcal{C}_s(\Theta, \mathcal{B}(U, Y))$ is a Banach space.

If $H \in \mathcal{C}_s(\Theta, \mathcal{B}(U, Y))$ and $Q \in \mathcal{C}_s(\Theta, \mathcal{B}(Y, Z))$ we denote by QH the mapping $\Theta \ni \theta \mapsto Q(\theta)H(\theta)$. It is obvious that $QH \in \mathcal{C}_s(\Theta, \mathcal{B}(U, Z))$.

Definition 1.1. Let $J \in \{\mathbb{R}_+, \mathbb{R}\}$. A continuous mapping $\sigma : \Theta \times J \to \Theta$ is called *a flow* on Θ if $\sigma(\theta, 0) = \theta$ and $\sigma(\theta, s + t) = \sigma(\sigma(\theta, s), t)$, for all $(\theta, s, t) \in \Theta \times J^2$.

Definition 1.2. A pair $\pi = (\Phi, \sigma)$ is called a *linear skew-product flow* on $\mathcal{E} = X \times \Theta$ if σ is a flow on Θ and $\Phi : \Theta \times \mathbb{R}_+ \to \mathcal{B}(X)$ satisfies the following conditions:
(i) $\Phi(\theta, 0) = I_d$, the identity operator on X, for all $\theta \in \Theta$;
(ii) $\Phi(\theta, t + s) = \Phi(\sigma(\theta, t), s)\Phi(\theta, t)$, for all $(\theta, t, s) \in \Theta \times \mathbb{R}_+^2$ (*the cocycle identity*);
(iii) $(\theta, t) \mapsto \Phi(\theta, t)x$ is continuous, for every $x \in X$;
(iv) there are $M \geq 1$ and $\omega > 0$ such that $||\Phi(\theta, t)|| \leq Me^{\omega t}$, for all $(\theta, t) \in \Theta \times \mathbb{R}_+$.
The mapping Φ is called *the cocycle* associated to the linear skew-product flow $\pi = (\Phi, \sigma)$.

Let $L^1_{loc}(\mathbb{R}_+, X)$ denote the linear space of all locally Bochner integrable functions $u : \mathbb{R}_+ \to X$. Let $\pi = (\Phi, \sigma)$ be a linear skew-product flow on $\mathcal{E} = X \times \Theta$. We consider the variational integral system

$$(S_\pi) \qquad x_\theta(t; x_0, u) = \Phi(\theta, t)x_0 + \int_0^t \Phi(\sigma(\theta, s), t - s)u(s)\, ds, \quad t \geq 0, \theta \in \Theta$$

* The work is supported by The National Research Council CNCSIS-UEFISCSU, PN II Research Grant ID 1081 code 550.

with $u \in L^1_{loc}(\mathbb{R}_+, X)$ and $x_0 \in X$.

Definition 1.3. The system (S_π) is said to be *uniformly exponentially stable* if there are $N, \nu > 0$ such that

$$||x_\theta(t; x_0, 0)|| \leq Ne^{-\nu t}||x_0||, \quad \forall(\theta, t) \in \Theta \times \mathbb{R}_+, \forall x_0 \in X.$$

Remark 1.4. It is easily seen that the system (S_π) is uniformly exponentially stable if and only if there are $N, \nu > 0$ such that $||\Phi(\theta, t)|| \leq Ne^{-\nu t}$, for all $(\theta, t) \in \Theta \times \mathbb{R}_+$.

If $\pi = (\Phi, \sigma)$ is a linear skew-product flow on $\mathcal{E} = X \times \Theta$ and $P \in \mathcal{C}_s(\Theta, \mathcal{B}(X))$, then there exists a unique linear skew-product flow denoted $\pi_P = (\Phi_P, \sigma)$ on $X \times \Theta$ such that this satisfies the variation of constants formula:

$$\Phi_P(\theta, t)x = \Phi(\theta, t)x + \int_0^t \Phi(\sigma(\theta, s), t - s)P(\sigma(\theta, s))\Phi_P(\theta, s)x\, ds \tag{1.1}$$

and respectively

$$\Phi_P(\theta, t)x = \Phi(\theta, t)x + \int_0^t \Phi_P(\sigma(\theta, s), t - s)P(\sigma(\theta, s))\Phi(\theta, s)x\, ds \tag{1.2}$$

for all $(x, \theta, t) \in \mathcal{E} \times \mathbb{R}_+$. Moreover, if M, ω are the exponential growth constants given by Definition 1.2 (iv) for π, then

$$||\Phi_P(\theta, t)|| \leq Me^{(\omega + M\,||P||)t}, \quad \forall(\theta, t) \in \Theta \times \mathbb{R}_+.$$

The perturbed linear skew-product flow $\pi_P = (\Phi_P, \sigma)$ is obtained inductively (see Theorem 2.1 in (Megan et al., 2002)) via the formula

$$\Phi_P(\theta, t) = \sum_{n=0}^{\infty} \Phi_n(\theta, t),$$

where

$$\Phi_0(\theta, t)x = \Phi(\theta, t)x \quad \text{and} \quad \Phi_n(\theta, t)x = \int_0^t \Phi(\sigma(\theta, s), t - s)\, P(\sigma(\theta, s))\, \Phi_{n-1}(\theta, s)x\, ds, n \geq 1$$

for every $(x, \theta) \in \mathcal{E}$ and $t \geq 0$.
Let U, Y be two Banach spaces, let $B \in \mathcal{C}_s(\Theta, \mathcal{B}(U, X))$ and $C \in \mathcal{C}_s(\Theta, \mathcal{B}(X, Y))$. We consider the variational control system (π, B, C) described by the following integral model

$$\begin{cases} x(\theta, t, x_0, u) = \Phi(\theta, t)x_0 + \int_0^t \Phi(\sigma(\theta, s), t - s)B(\sigma(\theta, s))u(s)\, ds \\ y(\theta, t, x_0, u) = C(\sigma(\theta, t))x(\theta, t, x_0, u) \end{cases}$$

where $t \geq 0$, $(x_0, \theta) \in \mathcal{E}$ and $u \in L^1_{loc}(\mathbb{R}_+, U)$.
Two fundamental concepts related to the asymptotic behavior of the associated perturbed systems (see (Clark et al., 2000), (Curtain & Zwart, 1995), (Sasu & Sasu, 2004)) are described by stabilizability and detectability as follows:

Definition 1.5. The system (π, B, C) is said to be:
(i) *stabilizable* if there exists a mapping $F \in \mathcal{C}_s(\Theta, \mathcal{B}(X, U))$ such that the system $(S_{\pi_{BF}})$ is uniformly exponentially stable;
(ii) *detectable* if there exists a mapping $K \in \mathcal{C}_s(\Theta, \mathcal{B}(Y, X))$ such that the system $(S_{\pi_{KC}})$ is uniformly exponentially stable.

Remark 1.6. (i) The system (π, B, C) is stabilizable if and only if there exists a mapping $F \in \mathcal{C}_s(\Theta, \mathcal{B}(X, U))$ and two constants $N, \nu > 0$ such that the perturbed linear skew-product flow $\pi_{BF} = (\Phi_{BF}, \sigma)$ has the property

$$||\Phi_{BF}(\theta, t)|| \leq Ne^{-\nu t}, \quad \forall(\theta, t) \in \Theta \times \mathbb{R}_+;$$

(ii) The system (π, B, C) is detectable if and only if there exists a mapping $K \in \mathcal{C}_s(\Theta, \mathcal{B}(Y, X))$ and two constants $N, \nu > 0$ such that the perturbed linear skew-product flow $\pi_{KC} = (\Phi_{KC}, \sigma)$ has the property

$$||\Phi_{KC}(\theta, t)|| \leq Ne^{-\nu t}, \quad \forall(\theta, t) \in \Theta \times \mathbb{R}_+.$$

In the present work we will investigate the connections between the stabilizability and the detectability of the variational control system (π, B, C) and the asymptotic properties of the variational integral system (S_π). We propose a new method based on input-output techniques and on the behavior of some associated operators between certain function spaces. We will present a distinct approach concerning the stabilizability and detectability problems for variational control systems, compared with those in the existent literature, working with several representative classes of translations invariant function spaces (see Section 2 in (Sasu, 2008) and also (Bennet & Sharpley, 1988)) and thus we extend the applicability area, providing new perspectives concerning this framework.
A special application of our main results will be the study of the connections between the exponential stability and the stabilizability and detectability of nonautonomous control systems in infinite dimensional spaces. The nonautonomous case treated in this chapter will include as consequences many interesting situations among which we mention the results obtained by Clark, Latushkin, Montgomery-Smith and Randolph (see (Clark et al., 2000)) and the authors (see (Sasu & Sasu, 2004)) concerning the connections between stabilizability, detectability and exponential stability.

2. Preliminaries on Banach function spaces and auxiliary results

In what follows we recall several fundamental properties of Banach function spaces and we introduce the main tools of our investigation. Indeed, let $\mathcal{M}(\mathbb{R}_+, \mathbb{R})$ be the linear space of all Lebesgue measurable functions $u : \mathbb{R}_+ \to \mathbb{R}$, identifying the functions equal a.e.

Definition 2.1. A linear subspace B of $\mathcal{M}(\mathbb{R}_+, \mathbb{R})$ is called *a normed function space*, if there is a mapping $|\cdot|_B : B \to \mathbb{R}_+$ such that:
(i) $|u|_B = 0$ if and only if $u = 0$ a.e.;
(ii) $|\alpha u|_B = |\alpha|\ |u|_B$, for all $(\alpha, u) \in \mathbb{R} \times B$;
(iii) $|u + v|_B \leq |u|_B + |v|_B$, for all $u, v \in B$;
(iv) if $|u(t)| \leq |v(t)|$ a.e. $t \in \mathbb{R}_+$ and $v \in B$, then $u \in B$ and $|u|_B \leq |v|_B$.

If $(B, |\cdot|_B)$ is complete, then B is called a *Banach function space.*

Remark 2.2. If $(B, |\cdot|_B)$ is a Banach function space and $u \in B$ then $|u(\cdot)| \in B$.

A remarkable class of Banach function spaces is represented by the translations invariant spaces. These spaces have a special role in the study of the asymptotic properties of the dynamical systems using control type techniques (see Sasu (2008), Sasu & Sasu (2004)).

Definition 2.3. A Banach function space $(B, |\cdot|_B)$ is said to be *invariant to translations* if for every $u : \mathbb{R}_+ \to \mathbb{R}$ and every $t > 0$, $u \in B$ if and only if the function

$$u_t : \mathbb{R}_+ \to \mathbb{R}, \quad u_t(s) = \begin{cases} u(s-t) & , s \geq t \\ 0 & , s \in [0,t) \end{cases}$$

belongs to B and $|u_t|_B = |u|_B$.

Let $C_c(\mathbb{R}_+, \mathbb{R})$ denote the linear space of all continuous functions $v : \mathbb{R}_+ \to \mathbb{R}$ with compact support contained in $\mathbb{R}_+$ and let $L^1_{loc}(\mathbb{R}_+, \mathbb{R})$ denote the linear space of all locally integrable functions $u : \mathbb{R}_+ \to \mathbb{R}$.

We denote by $\mathcal{T}(\mathbb{R}_+)$ the class of all Banach function spaces B which are invariant to translations and satisfy the following properties:
(i) $C_c(\mathbb{R}_+, \mathbb{R}) \subset B \subset L^1_{loc}(\mathbb{R}_+, \mathbb{R})$;
(ii) if $B \setminus L^1(\mathbb{R}_+, \mathbb{R}) \neq \emptyset$ then there is a continuous function $\delta \in B \setminus L^1(\mathbb{R}_+, \mathbb{R})$.

For every $A \subset \mathbb{R}_+$ we denote by χ_A the characteristic function of the set A.

Remark 2.4. (i) If $B \in \mathcal{T}(\mathbb{R}_+)$, then $\chi_{[0,t)} \in B$, for all $t > 0$.
(ii) Let $B \in \mathcal{T}(\mathbb{R}_+)$, $u \in B$ and $t > 0$. Then, the function $\tilde{u}_t : \mathbb{R}_+ \to \mathbb{R}, \tilde{u}_t(s) = u(s+t)$ belongs to B and $|\tilde{u}_t|_B \leq |u|_B$ (see (Sasu, 2008), Lemma 5.4).

Definition 2.5. (i) Let $u, v \in \mathcal{M}(\mathbb{R}_+, \mathbb{R})$. We say that u and v are *equimeasurable* if for every $t > 0$ the sets $\{s \in \mathbb{R}_+ : |u(s)| > t\}$ and $\{s \in \mathbb{R}_+ : |v(s)| > t\}$ have the same measure.

(ii) A Banach function space $(B, |\cdot|_B)$ is *rearrangement invariant* if for every equimeasurable functions $u, v : \mathbb{R}_+ \to \mathbb{R}_+$ with $u \in B$ we have that $v \in B$ and $|u|_B = |v|_B$.

We denote by $\mathcal{R}(\mathbb{R}_+)$ the class of all Banach function spaces $B \in \mathcal{T}(\mathbb{R}_+)$ which are rearrangement invariant.

A remarkable class of rearrangement invariant function spaces is represented by the so-called *Orlicz spaces* which are introduced in the following remark:

Remark 2.6. Let $\varphi : \mathbb{R}_+ \to \mathbb{R}_+$ be a non-decreasing left-continuous function, which is not identically zero on $(0, \infty)$. The *Young function* associated with φ is defined by $Y_\varphi(t) = \int_0^t \varphi(s)\, ds$. For every $u \in \mathcal{M}(\mathbb{R}_+, \mathbb{R})$ let $M_\varphi(u) := \int_0^\infty Y_\varphi(|u(s)|)\, ds$. The set O_φ of all $u \in \mathcal{M}(\mathbb{R}_+, \mathbb{R})$ with the property that there is $k > 0$ such that $M_\varphi(ku) < \infty$, is a linear space. With respect to the norm $|u|_\varphi := \inf\{k > 0 : M_\varphi(u/k) \leq 1\}$, O_φ is a Banach space, called *the Orlicz space* associated with φ.
The Orlicz spaces are rearrangement invariant (see (Bennet & Sharpley, 1988), Theorem 8.9). Moreover, it is well known that, for every $p \in [1, \infty]$, the space $L^p(\mathbb{R}_+, \mathbb{R})$ is a particular case of Orlicz space.

Let now $(X, ||\cdot||)$ be a real or complex Banach space. For every $B \in \mathcal{T}(\mathbb{R}_+)$ we denote by $B(\mathbb{R}_+, X)$, the linear space of all Bochner measurable functions $u : \mathbb{R}_+ \to X$ with the property that the mapping $N_u : \mathbb{R}_+ \to \mathbb{R}_+, N_u(t) = ||u(t)||$ lies in B. Endowed with the norm $||u||_{B(\mathbb{R}_+, X)} := |N_u|_B$, $B(\mathbb{R}_+, X)$ is a Banach space.
Let (Θ, d) be a metric space and let $\mathcal{E} = X \times \Theta$. Let $\pi = (\Phi, \sigma)$ be a linear skew-product flow on $\mathcal{E} = X \times \Theta$. We consider the variational integral system

$$(S_\pi) \qquad x_\theta(t; x_0, u) = \Phi(\theta, t)x_0 + \int_0^t \Phi(\sigma(\theta, s), t-s)u(s)\, ds, \quad t \geq 0, \theta \in \Theta$$

with $u \in L^1_{loc}(\mathbb{R}_+, X)$ and $x_0 \in X$.
An important stability concept related with the asymptotic behavior of dynamical systems is described by the following concept:

Definition 2.7. Let $W \in \mathcal{T}(\mathbb{R}_+)$. The system (S_π) is said to be *completely* $(W(\mathbb{R}_+, X), W(\mathbb{R}_+, X))$*-stable* if the following assertions hold:
(i) for every $u \in W(\mathbb{R}_+, X)$ and every $\theta \in \Theta$ the solution $x_\theta(\cdot\,; 0, u) \in W(\mathbb{R}_+, X)$;
(ii) there is $\lambda > 0$ such that $||x_\theta(\cdot\,; 0, u)||_{W(\mathbb{R}_+, X)} \leq \lambda ||u||_{W(\mathbb{R}_+, X)}$, for all $(u, \theta) \in W(\mathbb{R}_+, X) \times \Theta$.

A characterization of uniform exponential stability of variational systems in terms of the complete stability of a pair of function spaces has been obtained in (Sasu, 2008) (see Corollary 3.19) and this is given by:

Theorem 2.8. *Let $W \in \mathcal{R}(\mathbb{R}_+)$. The system (S_π) is uniformly exponentially stable if and only if (S_π) is completely $(W(\mathbb{R}_+, X), W(\mathbb{R}_+, X))$-stable.*

The problem can be also treated in the setting of the continuous functions. Indeed, let $C_b(\mathbb{R}_+, \mathbb{R})$ be the space of all bounded continuous functions $u : \mathbb{R}_+ \to \mathbb{R}$. Let $C_0(\mathbb{R}_+, \mathbb{R})$ be the space of all continuous functions $u : \mathbb{R}_+ \to \mathbb{R}$ with $\lim\limits_{t\to\infty} u(t) = 0$ and let $C_{00}(\mathbb{R}_+, \mathbb{R}) := \{u \in C_0(\mathbb{R}_+, \mathbb{R}) : u(0) = 0\}$.

Definition 2.9. Let $V \in \{C_b(\mathbb{R}_+, \mathbb{R}), C_0(\mathbb{R}_+, \mathbb{R}), C_{00}(\mathbb{R}_+, \mathbb{R})\}$. The system (S_π) is said to be *completely* $(V(\mathbb{R}_+, X), V(\mathbb{R}_+, X))$*-stable* if the following assertions hold:
(i) for every $u \in V(\mathbb{R}_+, X)$ and every $\theta \in \Theta$ the solution $x_\theta(\cdot\,; 0, u) \in V(\mathbb{R}_+, X)$;
(ii) there is $\lambda > 0$ such that $||x_\theta(\cdot\,; 0, u)||_{V(\mathbb{R}_+, X)} \leq \lambda ||u||_{V(\mathbb{R}_+, X)}$, for all $(u, \theta) \in V(\mathbb{R}_+, X) \times \Theta$.

For the proof of the next result we refer to Corollary 3.24 in (Sasu, 2008) or, alternatively, to Theorem 5.1 in (Megan et al., 2005).

Theorem 2.10. *Let $V \in \{C_b(\mathbb{R}_+, \mathbb{R}), C_0(\mathbb{R}_+, \mathbb{R}), C_{00}(\mathbb{R}_+, \mathbb{R})\}$. The system (S_π) is uniformly exponentially stable if and only if (S_π) is completely $(V(\mathbb{R}_+, X), V(\mathbb{R}_+, X))$-stable.*

Remark 2.11. Let $W \in \mathcal{R}(\mathbb{R}_+) \cup \{C_0(\mathbb{R}_+, X), C_{00}(\mathbb{R}_+, X), C_b(\mathbb{R}_+, X)\}$. If the system (S_π) is uniformly exponentially stable then for every $\theta \in \Theta$ the linear operator

$$P_W^\theta : W(\mathbb{R}_+, X) \to W(\mathbb{R}_+, X), \quad (P_W^\theta u)(t) = \int_0^t \Phi(\sigma(\theta, s), t - s) u(s)\, ds$$

is correctly defined and bounded. Moreover, if $\lambda > 0$ is given by Definition 2.7 or respectively by Definition 2.9, then we have that $\sup_{\theta\in\Theta} ||P_W^\theta|| \leq \lambda$.

These results have several interesting applications in control theory among we mention those concerning the robustness problems (see (Sasu, 2008)) which lead to an inedit estimation of the lower bound of the stability radius, as well as to the study of the connections between stability and stabilizability and detectability of associated control systems, as we will see in what follows. It worth mentioning that these aspects were studied for the very first time for the case of systems associated to evolution operators in (Clark et al., 2000) and were extended for linear skew-product flows in (Megan et al., 2002).

3. Stabilizability and detectability of variational control systems

As stated from the very beginning, in this section our attention will focus on the connections between stabilizability, detectability and the uniform exponential stability. Let X be a Banach space, let (Θ, d) be a metric space and let $\pi = (\Phi, \sigma)$ be a linear skew-product flow on $\mathcal{E} = X \times \Theta$. We consider the variational integral system

$$(S_\pi) \qquad x_\theta(t; x_0, u) = \Phi(\theta, t)x_0 + \int_0^t \Phi(\sigma(\theta, s), t-s)u(s)\, ds, \quad t \geq 0, \theta \in \Theta$$

with $u \in L^1_{loc}(\mathbb{R}_+, X)$ and $x_0 \in X$.

Let U, Y be Banach spaces and let $B \in \mathcal{C}_s(\Theta, \mathcal{B}(U, X))$, $C \in \mathcal{C}_s(\Theta, \mathcal{B}(X, Y))$. We consider the variational control system (π, B, C) described by the following integral model

$$\begin{cases} x(\theta, t, x_0, u) = \Phi(\theta, t)x_0 + \int_0^t \Phi(\sigma(\theta, s), t-s)B(\sigma(\theta, s))u(s)\, ds \\ y(\theta, t, x_0, u) = C(\sigma(\theta, t))x(\theta, t, x_0, u) \end{cases}$$

where $t \geq 0$, $(x_0, \theta) \in \mathcal{E}$ and $u \in L^1_{loc}(\mathbb{R}_+, U)$.
According to Definition 1.5 it is obvious that if the system (S_π) is uniformly exponentially stable, then the control system (π, B, C) is stabilizable (via the trivial feedback $F \equiv 0$) and this is also detectable (via the trivial feedback $K \equiv 0$). The natural question arises whether the converse implication holds.

Example 3.1. Let $X = \mathbb{R}, \Theta = \mathbb{R}$ and let $\sigma(\theta, t) = \theta + t$. Let (S_π) be a variational integral system such that $\Phi(\theta, t) = I_d$ (the identity operator on X), for all $(\theta, t) \in \Theta \times \mathbb{R}_+$. Let $U = Y = X$ and let $B(\theta) = C(\theta) = I_d$, for all $\theta \in \Theta$. Let $\delta > 0$. By considering $F(\theta) = -\delta\, I_d$, for all $\theta \in \Theta$, from relation (1.1), we obtain that

$$\Phi_{BF}(\theta, t)x = x - \delta \int_0^t \Phi_{BF}(\theta, s)x\, ds, \quad \forall t \geq 0$$

for every $(x, \theta) \in \mathcal{E}$. This implies that $\Phi_{BF}(\theta, t)x = e^{-\delta t}x$, for all $t \geq 0$ and all $(x, \theta) \in \mathcal{E}$, so the perturbed system $(S_{\pi_{BF}})$ is uniformly exponentially stable. This shows that the system (π, B, C) is stabilizable.
Similarly, if $\delta > 0$, for $K(\theta) = -\delta\, I_d$, for all $\theta \in \Theta$, we deduce that the variational control system (π, B, C) is also detectable.
In conclusion, the variational control system (π, B, C) is both stabilizable and detectable, but for all that, the variational integral system (S_π) is not uniformly exponentially stable.

It follows that the stabilizability or/and the detectability of the control system (π, B, C) are not sufficient conditions for the uniform exponential stability of the system (S_π). Naturally, additional hypotheses are required. In what follows we shall prove that certain input-output conditions assure a complete resolution to this problem. The answer will be given employing new methods based on function spaces techniques.
Indeed, for every $\theta \in \Theta$, we define

$$P^\theta : L^1_{loc}(\mathbb{R}_+, X) \to L^1_{loc}(\mathbb{R}_+, X), \quad (P^\theta w)(t) = \int_0^t \Phi(\sigma(\theta, s), t-s)w(s)\, ds$$

and respectively

$$B^\theta : L^1_{loc}(\mathbb{R}_+, U) \to L^1_{loc}(\mathbb{R}_+, X), \quad (B^\theta u)(t) = B(\sigma(\theta, t))u(t)$$

$$C^\theta : L^1_{loc}(\mathbb{R}_+, X) \to L^1_{loc}(\mathbb{R}_+, Y), \quad (C^\theta v)(t) = C(\sigma(\theta, t))v(t).$$

We also associate with the control system $S = (\pi, B, C)$ three families of input-output mappings, as follows: *the left input-output operators* $\{L^\theta\}_{\theta\in\Theta}$ defined by

$$L^\theta : L^1_{loc}(\mathbb{R}_+, U) \to L^1_{loc}(\mathbb{R}_+, X), \quad L^\theta := P^\theta B^\theta$$

the right input-output operators $\{R^\theta\}_{\theta\in\Theta}$ given by

$$R^\theta : L^1_{loc}(\mathbb{R}_+, X) \to L^1_{loc}(\mathbb{R}_+, Y), \quad R^\theta := C^\theta P^\theta$$

and respectively *the global input-output operators* $\{G^\theta\}_{\theta\in\Theta}$ defined by

$$G^\theta : L^1_{loc}(\mathbb{R}_+, U) \to L^1_{loc}(\mathbb{R}_+, Y), \quad G^\theta := C^\theta P^\theta B^\theta.$$

A fundamental stability concept for families of linear operators is given by the following:

Definition 3.2. Let Z_1, Z_2 be two Banach spaces and let $W \in \mathcal{T}(\mathbb{R}_+)$ be a Banach function space. A family of linear operators $\{O^\theta : L^1_{loc}(\mathbb{R}_+, Z_1) \to L^1_{loc}(\mathbb{R}_+, Z_2)\}_{\theta\in\Theta}$ is said to be $(W(\mathbb{R}_+, Z_1), W(\mathbb{R}_+, Z_2))$*-stable* if the following conditions are satisfied:
(i) for every $\alpha_1 \in W(\mathbb{R}_+, Z_1)$ and every $\theta \in \Theta$, $O^\theta\alpha_1 \in W(\mathbb{R}_+, Z_2)$;
(ii) there is $m > 0$ such that $||O^\theta\alpha_1||_{W(\mathbb{R}_+,Z_2)} \leq m\, ||\alpha_1||_{W(\mathbb{R}_+,Z_1)}$, for all $\alpha_1 \in W(\mathbb{R}_+, Z_1)$ and all $\theta \in \Theta$.

Thus, we observe that if $W \in \mathcal{R}(\mathbb{R}_+)$, then the variational integral system (S_π) is uniformly exponentially stable if and only if the family $\{P^\theta\}_{\theta\in\Theta}$ is $(W(\mathbb{R}_+, X), W(\mathbb{R}_+, X))$-stable (see also Remark 2.11).

Remark 3.3. Let Z_1, Z_2 be two Banach spaces and let $W \in \mathcal{T}(\mathbb{R}_+)$ be a Banach function space. If $Q \in \mathcal{C}_s(\Theta, \mathcal{B}(Z_1, Z_2))$ then the family $\{Q^\theta\}_{\theta\in\Theta}$ defined by

$$Q^\theta : L^1_{loc}(\mathbb{R}_+, Z_1) \to L^1_{loc}(\mathbb{R}_+, Z_2), \quad (Q^\theta\alpha)(t) = Q(\sigma(\theta, t))\alpha(t)$$

is $(W(\mathbb{R}_+, Z_1), W(\mathbb{R}_+, Z_2))$-stable. Indeed, this follows from Definition 2.1 (iv) by observing that

$$||(Q^\theta\alpha)(t)|| \leq |||Q|||\, ||\alpha(t)||, \quad \forall t \geq 0, \forall \alpha \in W(\mathbb{R}_+, Z_1), \forall \theta \in \Theta.$$

The main result of this section is:

Theorem 3.4. *Let W be a Banach function space such that $W \in \mathcal{R}(\mathbb{R}_+)$. The following assertions are equivalent:*
(i) *the variational integral system (S_π) is uniformly exponentially stable;*
(ii) *the variational control system (π, B, C) is stabilizable and the family of the left input-output operators $\{L^\theta\}_{\theta\in\Theta}$ is $(W(\mathbb{R}_+, U), W(\mathbb{R}_+, X))$-stable;*
(iii) *the variational control system (π, B, C) is detectable and the family of the right input-output operators $\{R^\theta\}_{\theta\in\Theta}$ is $(W(\mathbb{R}_+, X), W(\mathbb{R}_+, Y))$-stable*
(iv) *the variational control system (π, B, C) is stabilizable, detectable and the family of the global input-output operators $\{G^\theta\}_{\theta\in\Theta}$ is $(W(\mathbb{R}_+, U), W(\mathbb{R}_+, Y))$-stable.*

Proof. We will independently prove each equivalence $(i) \iff (ii)$, $(i) \iff (iii)$ and respectively $(i) \iff (iv)$. Indeed, we start with the first one and we prove that $(i) \Longrightarrow (ii)$. Taking into account that (S_π) is uniformly exponentially stable, we have that the family $\{P^\theta\}_{\theta\in\Theta}$ is $(W(\mathbb{R}_+,X),W(\mathbb{R}_+,X))$-stable. In addition, observing that

$$||(L^\theta u)(t)|| \leq \sup_{\theta\in\Theta} ||P^\theta|| \, |||B||| \, ||u(t)||, \quad \forall u \in W(\mathbb{R}_+,U), \forall\theta\in\Theta$$

from Definition 2.1 (iv) we deduce that that the family $\{L^\theta\}_{\theta\in\Theta}$ is $(W(\mathbb{R}_+,U),W(\mathbb{R}_+,X))$-stable.
To prove the implication $(ii) \Longrightarrow (i)$, let $F \in \mathcal{C}_s(\Theta,\mathcal{B}(X,U))$ be such that the system $(S_{\pi_{BF}})$ is uniformly exponentially stable. It follows that the family $\{H^\theta\}_{\theta\in\Theta}$ is $(W(\mathbb{R}_+,X),W(\mathbb{R}_+,X))$-stable, where

$$H^\theta : L^1_{loc}(\mathbb{R}_+,X) \to L^1_{loc}(\mathbb{R}_+,X), \quad (H^\theta u)(t) = \int_0^t \Phi_{BF}(\sigma(\theta,s),t-s)u(s)\,ds, \quad t\geq 0, \theta\in\Theta.$$

For every $\theta\in\Theta$ let

$$F^\theta : L^1_{loc}(\mathbb{R}_+,X) \to L^1_{loc}(\mathbb{R}_+,U), \quad (F^\theta u)(t) = F(\sigma(\theta,t))u(t).$$

Then from Remark 3.3 we have that the family $\{F^\theta\}_{\theta\in\Theta}$ is $(W(\mathbb{R}_+,X),W(\mathbb{R}_+,U))$-stable.
Let $\theta\in\Theta$ and let $u \in L^1_{loc}(\mathbb{R}_+,X)$. Using Fubini's theorem and formula (1.1), we successively deduce that

$$(L^\theta F^\theta H^\theta u)(t) = \int_0^t\int_0^s \Phi(\sigma(\theta,s),t-s)B(\sigma(\theta,s))F(\sigma(\theta,s))\Phi_{BF}(\sigma(\theta,\tau),s-\tau)u(\tau)\,d\tau\,ds =$$

$$= \int_0^t\int_\tau^t \Phi(\sigma(\theta,s),t-s)B(\sigma(\theta,s))F(\sigma(\theta,s))\Phi_{BF}(\sigma(\theta,\tau),s-\tau)u(\tau)\,ds\,d\tau =$$

$$= \int_0^t\int_0^{t-\tau} \Phi(\sigma(\theta,\tau+\xi),t-\tau-\xi)B(\sigma(\theta,\tau+\xi))F(\sigma(\theta,\tau+\xi))\Phi_{BF}(\sigma(\theta,\tau),\xi)u(\tau)\,d\xi\,d\tau =$$

$$= \int_0^t [\Phi_{BF}(\sigma(\theta,\tau),t-\tau)u(\tau) - \Phi(\sigma(\theta,\tau),t-\tau)u(\tau)]\,d\tau =$$

$$= (H^\theta u)(t) - (P^\theta u)(t), \quad \forall t\geq 0.$$

This shows that

$$P^\theta u = H^\theta u - L^\theta F^\theta H^\theta u, \quad \forall u\in L^1_{loc}(\mathbb{R}_+,X), \forall\theta\in\Theta. \tag{3.1}$$

Let m_1 and m_2 be two constants given by Definition 3.2 (ii) for $\{H^\theta\}_{\theta\in\Theta}$ and for $\{L^\theta\}_{\theta\in\Theta}$, respectively. From relation (3.1) we deduce that $P^\theta u \in W(\mathbb{R}_+,X)$, for every $u\in W(\mathbb{R}_+,X)$ and

$$||P^\theta u||_{W(\mathbb{R}_+,X)} \leq m_1(1+m_2|||F|||)\,||u||_{W(\mathbb{R}_+,X)}, \quad \forall u\in W(\mathbb{R}_+,X), \forall\theta\in\Theta.$$

From the above relation we obtain that the family $\{P^\theta\}_{\theta\in\Theta}$ is $(W(\mathbb{R}_+,X),W(\mathbb{R}_+,X))$-stable, so the system (S_π) is uniformly exponentially stable.
The implication $(i) \Longrightarrow (iii)$ follows using similar arguments with those used in the proof of $(i) \Longrightarrow (ii)$. To prove $(iii) \Longrightarrow (i)$, let $K \in \mathcal{C}_s(\Theta,\mathcal{B}(Y,X))$ be such that the system $(S_{\pi_{KC}})$ is uniformly exponentially stable. Then, the family $\{\Gamma^\theta\}_{\theta\in\Theta}$ is $(W(\mathbb{R}_+,X),W(\mathbb{R}_+,X))$-stable, where

$$\Gamma^\theta : L^1_{loc}(\mathbb{R}_+,X) \to L^1_{loc}(\mathbb{R}_+,X), \quad (\Gamma^\theta u)(t) = \int_0^t \Phi_{KC}(\sigma(\theta,s),t-s)u(s)\,ds.$$

For every $\theta \in \Theta$ we define

$$K^\theta : L^1_{loc}(\mathbb{R}_+, Y) \to L^1_{loc}(\mathbb{R}_+, X), \quad (K^\theta u)(t) = K(\sigma(\theta, t))u(t).$$

From Remark 3.3 we have that the family $\{K^\theta\}_{\theta\in\Theta}$ is $(W(\mathbb{R}_+, Y), W(\mathbb{R}_+, X))$-stable.
Using Fubini's theorem and the relation (1.2), by employing similar arguments with those from the proof of the implication $(ii) \Longrightarrow (i)$, we deduce that

$$P^\theta u = \Gamma^\theta u - \Gamma^\theta K^\theta R^\theta u, \quad \forall u \in L^1_{loc}(\mathbb{R}_+, X), \forall \theta \in \Theta. \tag{3.2}$$

Denoting by q_1 and by q_2 some constants given by Definition 3.2 (ii) for $\{\Gamma^\theta\}_{\theta\in\Theta}$ and for $\{R^\theta\}_{\theta\in\Theta}$, respectively, from relation (3.2) we have that $P^\theta u \in W(\mathbb{R}_+, X)$, for every $u \in W(\mathbb{R}_+, X)$ and

$$||P^\theta u||_{W(\mathbb{R}_+,X)} \leq q_1(1 + q_2|||K|||)\, ||u||_{W(\mathbb{R}_+,X)}, \quad \forall u \in W(\mathbb{R}_+, X), \forall \theta \in \Theta.$$

Hence we deduce that the family $\{P^\theta\}_{\theta\in\Theta}$ is $(W(\mathbb{R}_+, X), W(\mathbb{R}_+, X))$-stable, which shows that the system (S_π) is uniformly exponentially stable.
The implication $(i) \Longrightarrow (iv)$ is obvious, taking into account the above items. To prove that $(iv) \Longrightarrow (i)$, let $K \in \mathcal{C}_s(\Theta, \mathcal{B}(Y, X))$ be such that the system $(S_{\pi_{KC}})$ is uniformly exponentially stable and let $\{K^\theta\}_{\theta\in\Theta}$ and $\{\Gamma^\theta\}_{\theta\in\Theta}$ be defined in the same manner like in the previous stage. Then, following the same steps as in the previous implications, we obtain that

$$L^\theta u = \Gamma^\theta B^\theta u - \Gamma^\theta K^\theta G^\theta u, \quad \forall u \in L^1_{loc}(\mathbb{R}_+, X), \forall \theta \in \Theta. \tag{3.3}$$

From relation (3.3) we deduce that the family $\{L^\theta\}_{\theta\in\Theta}$ is $(W(\mathbb{R}_+, U), W(\mathbb{R}_+, X))$-stable. Taking into account that the system (π, B, C) is stabilizable and applying the implication $(ii) \Longrightarrow (i)$, we conclude that the system (S_π) is uniformly exponentially stable. □

Corollary 3.5. *Let $V \in \{C_b(\mathbb{R}_+, \mathbb{R}), C_0(\mathbb{R}_+, \mathbb{R}), C_{00}(\mathbb{R}_+, \mathbb{R})\}$. The following assertions are equivalent:*
(i) the variational integral system (S_π) is uniformly exponentially stable;
(ii) the variational control system (π, B, C) is stabilizable and the family of the left input-output operators $\{L^\theta\}_{\theta\in\Theta}$ is $(V(\mathbb{R}_+, U), V(\mathbb{R}_+, X))$-stable;
(iii) the variational control system (π, B, C) is detectable and the family of the right input-output operators $\{R^\theta\}_{\theta\in\Theta}$ is $(V(\mathbb{R}_+, X), V(\mathbb{R}_+, Y))$-stable
(iv) the variational control system (π, B, C) is stabilizable, detectable and the family of the global input-output operators $\{G^\theta\}_{\theta\in\Theta}$ is $(V(\mathbb{R}_+, U), V(\mathbb{R}_+, Y))$-stable.

Proof. This follows using similar arguments and estimations with those from the proof of Theorem 3.4, by applying Theorem 2.10. □

4. Applications to nonautonomous systems

An interesting application of the main results from the previous section is to deduce necessary and sufficient conditions for uniform exponential stability of nonautonomous systems in terms of stabilizability and detectability. For the first time this topic was considered in (Clark et al., 2000)). We propose in what follows a new method for the resolution of this problem based on the application of the conclusions from the variational case, using arbitrary Banach function spaces.
Let X be a Banach space and let I_d denote the identity operator on X.

Definition 4.1. A family $\mathcal{U} = \{U(t,s)\}_{t\geq s\geq 0} \subset \mathcal{B}(X)$ is called an *evolution family* if the following properties hold:
(*i*) $U(t,t) = I_d$ and $U(t,s)U(s,t_0) = U(t,t_0)$, for all $t \geq s \geq t_0 \geq 0$;
(*ii*) there are $M \geq 1$ and $\omega > 0$ such that $||U(t,s)|| \leq Me^{\omega(t-s)}$, for all $t \geq s \geq t_0 \geq 0$;
(*iii*) for every $x \in X$ the mapping $(t,s) \mapsto U(t,s)x$ is continuous.

Remark 4.2. For every $P \in \mathcal{C}_s(\mathbb{R}_+, \mathcal{B}(X))$ (see e.g. (Curtain & Zwart, 1995)) there is a unique evolution family $\mathcal{U}_P = \{U_P(t,s)\}_{t\geq s\geq 0}$ such that the variation of constants formulas hold:

$$U_P(t,s)x = U(t,s)x + \int_s^t U(t,\tau)P(\tau)U_P(\tau,s)x\,d\tau, \quad \forall t \geq s \geq 0, \forall x \in X$$

and respectively

$$U_P(t,s)x = U(t,s)x + \int_s^t U_P(t,\tau)P(\tau)U(\tau,s)x\,d\tau, \quad \forall t \geq s \geq 0, \forall x \in X.$$

Let $\mathcal{U} = \{U(t,s)\}_{t\geq s\geq 0}$ be an evolution family on X. We consider the nonautonomous integral system

$$(S_{\mathcal{U}}) \qquad x_s(t;x_0,u) = U(t,s)x_0 + \int_s^t U(t,\tau)u(\tau)\,d\tau, \quad t \geq s,\ s \geq 0$$

with $u \in L^1_{loc}(\mathbb{R}_+, X)$ and $x_0 \in X$.

Definition 4.3. The system $(S_{\mathcal{U}})$ is said to be *uniformly exponentially stable* if there are $N, \nu > 0$ such that $||x_s(t;x_0,0)|| \leq Ne^{-\nu(t-s)}||x_0||$, for all $t \geq s \geq 0$ and all $x_0 \in X$.

Remark 4.4. The system $(S_{\mathcal{U}})$ is uniformly exponentially stable if and only if there are $N, \nu > 0$ such that $||U(t,s)|| \leq Ne^{-\nu(t-s)}$, for all $t \geq s \geq 0$.

Definition 4.5. Let $W \in \mathcal{T}(\mathbb{R}_+)$. The system $(S_{\mathcal{U}})$ is said to be *completely* $(W(\mathbb{R}_+, X),$ $W(\mathbb{R}_+, X))$*-stable* if for every $u \in W(\mathbb{R}_+, X)$, the solution $x_0(\cdot;0,u) \in W(\mathbb{R}_+, X)$.

Remark 4.6. If the system $(S_{\mathcal{U}})$ is completely $(W(\mathbb{R}_+, X), W(\mathbb{R}_+, X))$-stable, then it makes sense to consider the linear operator

$$\mathcal{P} : W(\mathbb{R}_+, X) \to W(\mathbb{R}_+, X), \quad \mathcal{P}(u) = x_0(\cdot;0,u).$$

It is easy to see that $\mathcal{P}$ is closed, so it is bounded.

Let now U, Y be Banach spaces, let $B \in \mathcal{C}_s(\mathbb{R}_+, \mathcal{B}(U, X))$ and let $C \in \mathcal{C}_s(\mathbb{R}_+, \mathcal{B}(X, Y))$. We consider the nonautonomous control system $(\mathcal{U}, B, C)$ described by the following integral model

$$\begin{cases} x_s(t;x_0,u) = U(t,s)x_0 + \int_s^t U(t,\tau)B(\tau)u(\tau)\,d\tau, \ t \geq s,\ s \geq 0 \\ y_s(t;x_0,u) = C(t)x_s(t;x_0,u), \qquad t \geq s,\ s \geq 0 \end{cases}$$

with $u \in L^1_{loc}(\mathbb{R}_+, U), x_0 \in X$.

Definition 4.7. The system $(\mathcal{U}, B, C)$ is said to be:
(i) *stabilizable* if there exists $F \in \mathcal{C}_s(\mathbb{R}_+, \mathcal{B}(X, U))$ such that the system $(S_{\mathcal{U}_{BF}})$ is uniformly exponentially stable;
(ii) *detectable* if there exists $G \in \mathcal{C}_s(\mathbb{R}_+, \mathcal{B}(Y, X))$ such that the system $(S_{\mathcal{U}_{GC}})$ is uniformly exponentially stable.

We consider the operators

$$\mathcal{B} : L^1_{loc}(\mathbb{R}_+, U) \to L^1_{loc}(\mathbb{R}_+, X), \quad (\mathcal{B}u)(t) = B(t)u(t)$$

$$\mathcal{C} : L^1_{loc}(\mathbb{R}_+, X) \to L^1_{loc}(\mathbb{R}_+, Y), \quad (\mathcal{C}u)(t) = B(t)u(t)$$

and we associate with the system $(\mathcal{U}, B, C)$ three input-output operators: *the left input-output operator* defined by

$$\mathcal{L} : L^1_{loc}(\mathbb{R}_+, U) \to L^1_{loc}(\mathbb{R}_+, X), \quad \mathcal{L} = \mathcal{P}\mathcal{B}$$

the right input-output operator given by

$$\mathcal{R} : L^1_{loc}(\mathbb{R}_+, X) \to L^1_{loc}(\mathbb{R}_+, Y), \quad \mathcal{R} = \mathcal{C}\mathcal{P}$$

and respectively *the global input-output operator* defined by

$$\mathcal{G} : L^1_{loc}(\mathbb{R}_+, U) \to L^1_{loc}(\mathbb{R}_+, Y), \quad \mathcal{G} = \mathcal{C}\mathcal{P}\mathcal{B}.$$

Definition 4.8. Let Z_1, Z_2 be two Banach spaces and let $W \in \mathcal{T}(\mathbb{R}_+)$ be a Banach function space. An operator $Q : L^1_{loc}(\mathbb{R}_+, Z_1) \to L^1_{loc}(\mathbb{R}_+, Z_2)$ is said to be $(W(\mathbb{R}_+, Z_1), W(\mathbb{R}_+, Z_2))$*-stable* if for every $\lambda \in W(\mathbb{R}_+, Z_1)$ the function $Q\lambda \in W(\mathbb{R}_+, Z_2)$.

The main result of this section is:

Theorem 4.9. *Let W be a Banach function space such that $B \in \mathcal{R}(\mathbb{R}_+)$. The following assertions are equivalent:*
(i) the integral system $(S_{\mathcal{U}})$ is uniformly exponentially stable;
(ii) the control system $(\mathcal{U}, B, C)$ is stabilizable and the left input-output operator $\mathcal{L}$ is $(W(\mathbb{R}_+, U), W(\mathbb{R}_+, X))$-stable;
(iii) the control system $(\mathcal{U}, B, C)$ is detectable and the right input-output operator $\mathcal{R}$ is $(W(\mathbb{R}_+, X), W(\mathbb{R}_+, Y))$-stable;
(iv) the control system $(\mathcal{U}, B, C)$ is stabilizable, detectable and the global input-output operator $\mathcal{G}$ is $(W(\mathbb{R}_+, U), W(\mathbb{R}_+, Y))$-stable.

Proof. We prove the equivalence $(i) \Longleftrightarrow (ii)$, the other equivalences: $(i) \Longleftrightarrow (iii)$ and $(i) \Longleftrightarrow (iv)$ being similar.
Indeed, the implication $(i) \Longrightarrow (ii)$ is immediate. To prove that $(ii) \Longrightarrow (i)$ let $\Theta = \mathbb{R}_+$, $\sigma : \Theta \times \mathbb{R}_+ \to \Theta, \sigma(\theta, t) = \theta + t$ and let $\Phi(\theta, t) = U(t + \theta, \theta)$, for all $(\theta, t) \in \Theta \times \mathbb{R}_+$. Then $\pi = (\Phi, \sigma)$ is a linear skew-product flow and it makes sense to associate with π the following integral system

$$(S_\pi) \qquad x_\theta(t; x_0, u) = \Phi(\theta, t)x_0 + \int_0^t \Phi(\sigma(\theta, s), t - s)u(s)\, ds, \quad t \geq 0, \theta \in \Theta$$

with $u \in L^1_{loc}(\mathbb{R}_+, X)$ and $x_0 \in X$.
We also consider the control system (π, B, C) given by

$$\begin{cases} x(\theta, t, x_0, u) = \Phi(\theta, t)x_0 + \int_0^t \Phi(\sigma(\theta, s), t - s)B(\sigma(\theta, s))u(s)\, ds \\ y(\theta, t, x_0, u) = C(\sigma(\theta, t))x(\theta, t, x_0, u) \end{cases}$$

where $t \geq 0$, $(x_0, \theta) \in \mathcal{E}$ and $u \in L^1_{loc}(\mathbb{R}_+, U)$. For every $\theta \in \Theta$ we associate with the system (π, B, C) the operators P^θ, B^θ and L^θ using their definitions from Section 3.

We prove that the family $\{L^\theta\}_{\theta\in\Theta}$ is $(W(\mathbb{R}_+,U),W(\mathbb{R}_+,X))$-stable. Let $\theta\in\Theta$ and let $\alpha\in W(\mathbb{R}_+,U)$. Since W is invariant to translations the function

$$\alpha_\theta:\mathbb{R}_+\to U,\quad \alpha_\theta(t)=\begin{cases}\alpha(t-\theta), & t\geq\theta\\ 0, & t\in[0,\theta)\end{cases}$$

belongs to $W(\mathbb{R}_+,U)$ and $||\alpha_\theta||_{W(\mathbb{R}_+,U)}=||\alpha||_{W(\mathbb{R}_+,U)}$. Since the operator $\mathcal{L}$ is $(W(\mathbb{R}_+,U),W(\mathbb{R}_+,X))$-stable we obtain that the function

$$\varphi:\mathbb{R}_+\to X,\quad \varphi(t)=(\mathcal{L}\alpha_\theta)(t)$$

belongs to $W(\mathbb{R}_+,X)$. Using Remark 2.4 (ii) we deduce that the function

$$\gamma:\mathbb{R}_+\to X,\quad \gamma(t)=\varphi(t+\theta)$$

belongs to $W(\mathbb{R}_+,X)$ and $||\gamma||_{W(\mathbb{R}_+,X)}\leq||\varphi||_{W(\mathbb{R}_+,X)}$. We observe that

$$(L^\theta\alpha)(t)=\int_0^t U(\theta+t,\theta+s)B(\theta+s)\alpha(s)\,ds=\int_\theta^{\theta+t}U(\theta+t,\tau)B(\tau)\alpha(\tau-\theta)\,d\tau=$$

$$=\int_\theta^{\theta+t}U(\theta+t,\tau)B(\tau)\alpha_\theta(\tau)\,d\tau=(\mathcal{L}\alpha_\theta)(\theta+t)=\gamma(t),\quad \forall t\geq 0.$$

This implies that $L^\theta\alpha$ belongs to $W(\mathbb{R}_+,X)$ and

$$||L^\theta\alpha||_{W(\mathbb{R}_+,X)}=||\gamma||_{W(\mathbb{R}_+,X)}\leq||\varphi||_{W(\mathbb{R}_+,X)}\leq$$

$$\leq||\mathcal{L}||\,||\alpha_\theta||_{W(\mathbb{R}_+,U)}=||\mathcal{L}||\,||\alpha||_{W(\mathbb{R}_+,U)}. \tag{4.1}$$

Since $\theta\in\Theta$ and $\alpha\in W(\mathbb{R}_+,U)$ were arbitrary from (4.1) we deduce that the family $\{L^\theta\}_{\theta\in\Theta}$ is $(W(\mathbb{R}_+,U),W(\mathbb{R}_+,X))$-stable.
According to our hypothesis we have that the system $(\mathcal{U},B,C)$ is stabilizable. Then there is $F\in\mathcal{C}_s(\mathbb{R}_+,\mathcal{B}(X,U))$ such that the (unique) evolution family $\mathcal{U}_{BF}=\{U_{BF}(t,s)\}_{t\geq s\geq 0}$ which satisfies the equation

$$U_{BF}(t,s)x=U(t,s)x+\int_s^t U(t,\tau)B(\tau)F(\tau)U_{BF}(\tau,s)x\,d\tau,\quad \forall t\geq s\geq 0,\forall x\in X \tag{4.2}$$

has the property that there are $N,\nu>0$ such that

$$||U_{BF}(t,s)||\leq Ne^{-\nu(t-s)},\quad \forall t\geq s\geq 0. \tag{4.3}$$

For every $(\theta,t)\in\Theta\times\mathbb{R}_+$, let $\tilde{\Phi}(\theta,t):=U_{BF}(\theta+t,\theta)$. Then, we have that $\tilde{\pi}=(\tilde{\Phi},\sigma)$ is a linear skew-product flow. Moreover, using relation (4.2) we deduce that

$$\int_0^t\Phi(\sigma(\theta,s),t-s)B(\sigma(\theta,s))F(\sigma(\theta,s))\tilde{\Phi}(\theta,s)x\,ds=$$

$$=\int_0^t U(\theta+t,\theta+s)B(\theta+s)F(\theta+s)U_{BF}(\theta+s,\theta)x\,ds=$$

$$=\int_\theta^{\theta+t}U(\theta+t,\tau)B(\tau)F(\tau)U_{BF}(\tau,\theta)x\,d\tau=$$

$$= U_{BF}(\theta+t,\theta)x - U(\theta+t,\theta)x = \tilde{\Phi}(\theta,t)x - \Phi(\theta,t)x \qquad (4.4)$$

for all $(\theta,t) \in \Theta \times \mathbb{R}_+$ and all $x \in X$. According to Theorem 2.1 in (Megan et al., 2002), from relation (4.4) it follows that

$$\tilde{\Phi}(\theta,t) = \Phi_{BF}(\theta,t), \quad \forall(\theta,t) \in \Theta \times \mathbb{R}_+$$

so $\tilde{\pi} = \pi_{BF}$. Hence from relation (4.3) we have that

$$||\Phi_{BF}(\theta,t)|| = ||U_{BF}(\theta+t,\theta)|| \le Ne^{-\nu t}, \quad \forall t \ge 0, \forall\theta \in \Theta$$

which shows that the system $(S_{\pi_{BF}})$ is uniformly exponentially stable. So the system (π, B, C) is stabilizable.

In this way we have proved that the system (S_π) is stabilizable and the associated left input-output family $\{L^\theta\}_{\theta\in\Theta}$ is $(W(\mathbb{R}_+,U), W(\mathbb{R}_+,X))$-stable. By applying Theorem 3.4 we deduce that the system (S_π) is uniformly exponentially stable. Then, there are $\tilde{N}, \delta > 0$ such that

$$||\Phi(\theta,t)|| \le \tilde{N}e^{-\delta t}, \quad \forall t \ge 0, \forall\theta \in \Theta.$$

This implies that

$$||U(t,s)|| = ||\Phi(s,t-s)|| \le \tilde{N}e^{-\delta(t-s)}, \quad \forall t \ge s \ge 0. \qquad (4.5)$$

From inequality (4.5) and Remark 4.4 we obtain that the system $(S_{\mathcal{U}})$ is uniformly exponentially stable. □

Remark 4.10. The version of the above result, for the case when $W = L^p(\mathbb{R}_+,\mathbb{R})$ with $p \in [1,\infty)$, was proved for the first time by Clark, Latushkin, Montgomery-Smith and Randolph in (Clark et al., 2000) employing evolution semigroup techniques.

The method may be also extended for spaces of continuous functions, as the following result shows:

Corollary 4.11. *Let* $V \in \{C_b(\mathbb{R}_+,\mathbb{R}), C_0(\mathbb{R}_+,\mathbb{R}), C_{00}(\mathbb{R}_+,\mathbb{R})\}$. *The following assertions are equivalent:*
(i) the system $(S_{\mathcal{U}})$ *is uniformly exponentially stable;*
(ii) the system $(\mathcal{U}, B, C)$ *is stabilizable and the left input-output operator* $\mathcal{L}$ *is* $(V(\mathbb{R}_+,U), V(\mathbb{R}_+,X))$*-stable;*
(iii) the system $(\mathcal{U}, B, C)$ *is detectable and the right input-output operator* $\mathcal{R}$ *is* $(V(\mathbb{R}_+,X), V(\mathbb{R}_+,Y))$*-stable;*
(iv) the system $(\mathcal{U}, B, C)$ *is stabilizable, detectable and the global input-output operator* $\mathcal{G}$ *is* $(V(\mathbb{R}_+,U), V(\mathbb{R}_+,Y))$*-stable.*

Proof. This follows using Corollary 3.5 and similar arguments with those from the proof of Theorem 4.9. □

5. Conclusions

Stabilizability and detectability of variational/nonautonomous control systems are two properties which are strongly related with the stable behavior of the initial integral system. These two properties (not even together) cannot assure the uniform exponential stability of the initial system, as Example 3.1 shows. But, in association with a stability of certain input-output operators the stabilizability or/and the detectability of the control system (π, B, C) imply the existence of the exponentially stable behavior of the initial system (S_π). Here we have extended the topic from evolution families to variational systems and the obtained results are given in a more general context. As we have shown in Remark 2.6 the spaces involved in the stability properties of the associated input-output operators may be not only L^p-spaces but also general Orlicz function spaces which is an aspect that creates an interesting link between the modern control theory of dynamical systems and the classical interpolation theory.

It worth mentioning that the framework presented in this chapter may be also extended to some slight weak concepts, taking into account the main results concerning the uniform stability concept from Section 3 in (Sasu, 2008) (see Definition 3.3 and Theorem 3.6 in (Sasu, 2008)). More precisely, considering that the system (π, B, C) is weak stabilizable (respectively weak detectable) if there exists a mapping $F \in \mathcal{C}_s(\Theta, \mathcal{B}(X, U))$ (respectively $K \in \mathcal{C}_s(\Theta, \mathcal{B}(Y, X))$) such that the system $(S_{\pi_{BF}})$ (respectively $(S_{\pi_{KC}})$) is uniformly stable, then starting with the result provided by Theorem 3.6 in (Sasu, 2008), the methods from the present chapter may be applied to the study of the uniform stability in terms of weak stabilizability and weak detectability. In authors opinion, the technical trick of the new study will rely on the fact that in this case the families of the associated input-output operators will have to be (L^1, L^∞)-stable.

6. References

Bennett, C. & Sharpley, R. (1988). *Interpolation of Operators*, Pure Appl. Math. 129, ISBN 0-12-088730-4

Clark, S.; Latushkin, Y.; Montgomery-Smith, S. & Randolph, T. (2000). *Stability radius and internal versus external stability in Banach spaces: an evolution semigroup approach*, SIAM J. Control Optim. Vol. 38, 1757–1793, ISSN 0363-0129

Curtain, R. & Zwart, H. J. (1995). *An Introduction to Infinite-Dimensional Linear Control Systems Theory*, Springer-Verlag, New-York, ISBN 0-387-94475-3

Megan, M; Sasu, A. L. & Sasu, B. (2002). *Stabilizability and controllability of systems associated to linear skew-product semiflows*, Rev. Mat. Complutense (Madrid) Vol. 15, 599–618, ISSN 1139-1138

Megan, M.; Sasu, A. L. & Sasu, B. (2005). *Theorems of Perron type for uniform exponential stability of linear skew-product semiflows*, Dynam. Contin. Discrete Impulsive Systems Vol. 12, 23–43, ISSN 1201-3390

Sasu, B. & Sasu, A. L. (2004). *Stability and stabilizability for linear systems of difference equations*, J. Differ. Equations Appl. Vol. 10, 1085–1105, ISSN 1023-6198

Sasu, B. (2008). *Robust stability and stability radius for variational control systems*, Abstract Appl. Analysis Vol. 2008, Article ID 381791, 1–29, ISSN 1085-3375

20

Robust Linear Control of Nonlinear Flat Systems

Hebertt Sira-Ramírez[1], John Cortés-Romero[1,2] and Alberto Luviano-Juárez[1]
[1]*Cinvestav IPN, Av. IPN No. 2508, Departamento de Ingeniería Eléctrica, Sección de Mecatrónica*
[2]*Universidad Nacional de Colombia. Facultad de Ingeniería, Departamento de Ingeniería Eléctrica y Electrónica. Carrera 30 No. 45-03 Bogotá, Colombia*
[1]*México*
[2]*Colombia*

1. Introduction

Asymptotic estimation of external, unstructured, perturbation inputs, with the aim of exactly, or approximately, canceling their influences at the controller stage, has been treated in the existing literature under several headings. The outstanding work of professor C.D. Johnson in this respect, under the name of *Disturbance Accommodation Control* (DAC), dates from the nineteen seventies (see Johnson (1971)). Ever since, the theory and practical aspects of DAC theory have been actively evolving, as evidenced by the survey paper by Johnson Johnson (2008). The theory enjoys an interesting and useful extension to discrete-time systems, as demonstrated in the book chapter Johnson (1982). In a recent article, by Parker and Johnson Parker & Johnson (2009), an application of DAC is made to the problem of decoupling two nonlinearly coupled linear systems. An early application of disturbance accommodation control in the area of Power Systems is exemplified by the work of Mohadjer and Johnson in Mohadjer & Johnson (1983), where the operation of an interconnected power system is approached from the perspective of load frequency control.
A closely related vein to DAC is represented by the sustained efforts of the late Professor Jingqing Han, summarized in the posthumous paper, Han Han (2009), and known as: *Active Disturbance Estimation and Rejection* (ADER). The numerous and original developments of Prof. Han, with many laboratory and industrial applications, have not been translated into English and his seminal contributions remain written in Chinese (see the references in Han (2009)). Although the main idea of observer-based disturbance estimation, and subsequent cancelation via the control law, is similar to that advocated in DAC, the emphasis in ADER lies, mainly, on *nonlinear* observer based disturbance estimation, with necessary developments related to: efficient time derivative computation, practical relative degree computation and nonlinear PID control extensions. The work, and inspiration, of Professor Han has found interesting developments and applications in the work of Professor Z. Gao and his colleagues (see Gao et al. (2001), Gao (2006), also, in the work by Sun and Gao Sun & Gao (2005) and in the article by Sun Sun (2007)). In a recent article, a closely related idea, proposed by Prof. M. Fliess and C. Join in Fliess & Join (2008), is at the core of *Intelligent PID Control*(IPIDC). The mainstream of the IPIDC developments makes use of the Algebraic Method and it implies to resort to first order, or at most second order, *non-phenomenological* plant models. The interesting aspect of this method resides in using suitable algebraic manipulations to

locally deprive the system description of the effects of nonlinear uncertain additive terms and, via further special algebraic manipulations, to efficiently identify time-varying control gains as piece-wise constant control input gains (see Fliess et al. (2008)). An entirely algebraic approach for the control of synchronous generator was presented in Fliess and Sira-Ramírez, Sira-Ramírez & Fliess (2004).
In this chapter, we advocate, within the context of trajectory tracking control for nonlinear flat systems, the use of approximate, yet accurate, state dependent disturbance estimation via linear Generalized Proportional Integral (GPI) observers. GPI observers are the dual counterpart of GPI controllers, developed by M. Fliess *et al.* in Fliess et al. (2002). A high gain GPI observer naturally includes a, self-updating, lumped, time-polynomial model of the nonlinear state-dependent perturbation; it estimates it and delivers the time signal to the controller for on-line cancelation while simultaneously estimating the phase variables related to the measured output. The scheme is, however, approximate since only a small as desired reconstruction error is guaranteed at the expense of high, noise-sensitive, gains. The on-line approximate estimation is suitably combined with linear, estimation-based, output feedback control with the appropriate, on-line, disturbance cancelation. The many similarities and the few differences with the DAC and ADER techniques probably lie in 1) the fact that we do not discriminate between *exogenous* (i.e., external) unstructured perturbation inputs and *endogenous* (i.e., state-dependent) perturbation inputs in the nonlinear input-output model. These perturbations are all lumped into a simplifying time-varying signal that needs to be linearly estimated. Notice that plant nonlinearities generate time functions that are *exogenous* to any observer and, hence, algebraic loops are naturally avoided 2) We emphasize the natural possibilities of *differentially flat systems* in the use of linear disturbance estimation and linear output feedback control with disturbance cancelation (For the concept of flatness see Fliess *et al.* Fliess et al. (1995)) and the book Sira-Ramírez & Agrawal (2004).
This chapter is organized as follows: Section 2 presents an introduction to linear control of nonlinear differentially flat systems via (high-gain) GPI observers and suitable linear controllers feeding back the phase variables related to the output function. The single input-single output synchronous generator model in the form a *swing* equation, is described in Section 3. Here, we formulate the reference trajectory tracking problem under a number of information restrictions about the system. The linear observer-linear controller output feedback control scheme is designed for lowering the deviation angle of the generator. We carry out a robustness test regarding the response to a three phase short circuit. We also carry an evaluation of the performance of the control scheme under significant variations of the two control gain parameters required for an exact cancelation of the gain. Section 4 is devoted to present an experimental illustrative example concerning the non-holonomic car which is also a multivariable nonlinear system with input gain matrix depending on the estimated phase variables associated with the flat outputs.

2. Linear GPI observer-based control of nonlinear systems

Consider the following perturbed nonlinear single-input single input-output, smooth, nonlinear system,

$$y^{(n)} = \psi(t, y, \dot{y}, ..., y^{(n-1)}) + \phi(t, y)u + \zeta(t) \quad (1)$$

The unperturbed system, ($\zeta(t) \equiv 0$) is evidently flat, as all variables in the system are expressible as differential functions of the flat output y.
We assume that the exogenous perturbation $\zeta(t)$ is uniformly absolutely bounded, i.e., it an L_∞ scalar function. Similarly, we assume that for all bounded solutions, $y(t)$, of (1),

obtained by means of suitable control input u, the additive, endogenous, perturbation input, $\psi(t, y(t), \dot{y}(t), ..., y^{(n-1)}(t))$, viewed as a time signal is uniformly absolutely bounded.
We also assume that the nonlinear gain function $\phi(t, y(t))$ is L_∞ and uniformly bounded away from zero, i.e., there exists a strictly positive constant μ such that

$$\inf_t |\phi(t, y(t))| \geq \mu \tag{2}$$

for all smooth, bounded solutions, $y(t)$, of (1) obtained with a suitable control input u. Although the results below can be extended when the input gain function ϕ depends on the time derivatives of y, we let, motivated by the synchronous generator case study to be presented, ϕ to be an explicit function of time and of the measured flat output y. This is equivalent to saying the $\phi(t, y(t))$ is perfectly known.
We have the following formulation of the problem:
Given a desired flat output reference trajectory, $y^*(t)$, devise a linear output feedback controller for system (1) so that regardless of the endogenous perturbation signal $\psi(t, y(t), \dot{y}(t), ..., y^{(n-1)}(t))$ and of the exogenous perturbation input $\zeta(t)$, the flat output y tracks the desired reference signal $y^*(t)$ even if in an approximate fashion. This approximate character specifically means that the tracking error, $e(t) = y - y^*(t)$, and its first, n, time derivatives, globally asymptotically exponentially converge towards a small as desired neighborhood of the origin in the reference trajectory tracking error phase space.
The solution to the problem is achieved in an entirely linear fashion if one conceptually considers the nonlinear model (1) as the following linear perturbed system

$$y^{(n)} = v + \xi(t) \tag{3}$$

where $v = \phi(t, y)u$, and $\xi(t) = \psi(t, y(t), \dot{y}(t), ..., y^{(n-1)}(t)) + \zeta(t)$.
Consider the following preliminary result:

Proposition 1. *The unknown perturbation vector of time signals, $\xi(t)$, in the simplified tracking error dynamics (3), is* observable *in the sense of Diop and Fliess (see Diop & Fliess (1991)).*

Proof The proof of this fact is immediate after writing (3) as

$$\xi(t) = y^{(n)} - v = y^{(n)} - \phi(t, y)u \tag{4}$$

i.e., $\xi(t)$ can be written in terms of the output vector y, a finite number of its time derivatives and the control input u. Hence, $\xi(t)$ is observable.

Remark 2. *This means, in particular, that if $\xi(t)$ is bestowed with an exact linear model; an exact asymptotic estimation of $\xi(t)$ is possible via a linear observer. If, on the other hand, the linear model is only approximately locally valid, then the estimation obtained via a linear observer is asymptotically convergent towards an equally approximately locally valid estimate.*

We assume that the perturbation input $\xi(t)$ may be locally modeled as a $p-1$-th degree time polynomial z_1 plus a residual term, $r(t)$, i.e.,

$$\xi(t) = z_1 + r(t) = a_0 + a_1 t + \cdots + a_{p-1} t^{p-1} + r(t), \text{ for all } t \tag{5}$$

The time polynomial model, z_1, (also called: a Taylor polynomial) is invariant with respect to time shifts and it defines a family of $p-1$ degree Taylor polynomials with arbitrary real

coefficients. We incorporate z_1 as an internal model of the additive perturbation input (see Johnson (1971)).
The perturbation model z_1 will acquire a *self updating* character when incorporated as part of a linear asymptotic observer whose estimation error is forced to converge to a small vicinity of zero. As a consequence of this, we may safely assume that the self-updating residual function, $r(t)$, and its time derivatives, say $r^{(p)}(t)$, are uniformly absolutely bounded. To precisely state this, let us denote by y_j an estimate of $y^{(j-1)}$ for $j = 1, ..., n$.
We have the following general result:

Theorem 3. *The GPI observer-based dynamical feedback controller:*

$$u = \frac{1}{\phi(t,y)}\left[[y^*(t)]^{(n)} - \sum_{j=0}^{n-1}\left(\kappa_j[y_j - (y^*(t))^{(j)}]\right) - \hat{\xi}(t)\right] \tag{6}$$

$$\hat{\xi}(t) = z_1$$

$$\begin{aligned}
\dot{y}_1 &= y_2 + \lambda_{p+n-1}(y - y_1)\\
\dot{y}_2 &= y_3 + \lambda_{p+n-2}(y - y_1)\\
&\vdots\\
\dot{y}_n &= v + z_1 + \lambda_p(y - y_1)\\
\dot{z}_1 &= z_2 + \lambda_{p-1}(y - y_1)\\
&\vdots\\
\dot{z}_{p-1} &= z_p + \lambda_1(y - y_1)\\
\dot{z}_p &= \lambda_0(y - y_1)
\end{aligned} \tag{7}$$

asymptotically exponentially drives the tracking error phase variables, $e_y^{(k)} = y^{(k)} - [y^*(t)]^{(k)}$, $k = 0, 1, .., n-1$ *to an arbitrary small neighborhood of the origin, of the tracking error phase space, which can be made as small as desired from the appropriate choice of the controller gain parameters* $\{\kappa_0, ..., \kappa_{n-1}\}$. *Moreover, the estimation errors:* $\tilde{e}^{(i)} = y^{(i)} - y_i$, $i = 0, ..., n-1$ *and the perturbation estimation error:* $z_m - \xi^{m-1}(t)$, $m = 1, ..., p$ *asymptotically exponentially converge towards a small as desired neighborhood of the origin of the reconstruction error space which can be made as small as desired from the appropriate choice of the controller gain parameters* $\{\lambda_0, ..., \lambda_{p+n-1}\}$.

Proof The proof is based on the fact that the estimation error $\tilde{e}$ satisfies the perturbed linear differential equation

$$\tilde{e}^{(p+n)} + \lambda_{p+n-1}e^{(p+n-1)} + \cdots + \lambda_0\tilde{e} = r^{(p)}(t) \tag{8}$$

Since $r^{(p)}(t)$ is assumed to be uniformly absolutely bounded then there exists coefficients λ_k such that $\tilde{e}$ converges to a small vicinity of zero, provided the roots of the associated characteristic polynomial in the complex variable s:

$$s^{p+n} + \lambda_{p+n-1}s^{p+n-1} + \cdots + \lambda_1 s + \lambda_0 \tag{9}$$

are all located deep into the left half of the complex plane. The further away from the imaginary axis, of the complex plane, are these roots located, the smaller the neighborhood of the origin, in the estimation error phase space, where the estimation error $\tilde{e}$ will remain ultimately bounded (see Kailath Kailath (1979)). Clearly, if $\tilde{e}$ and its time derivatives converge to a neighborhood of the origin, then $z_j - \xi^{(j)}$, $j = 1, 2, ...,$ also converge towards a small vicinity of zero.
The tracking error $e_y = y - y^*(t)$ evolves according to the following linear perturbed dynamics

$$e_y^{(n)} + \kappa_{n-1} e_y^{(n-1)} + \cdots + \kappa_0 e_y = \xi(t) - \hat{\xi}(t) \tag{10}$$

Choosing the controller coefficients $\{\kappa_0, \cdots, \kappa_{n-1}\}$, so that the associated characteristic polynomial

$$s^n + \kappa_{n-1} s^{n-1} + \cdots + \kappa_0 \tag{11}$$

exhibits its roots sufficiently far from the imaginary axis in the left half portion of the complex plane, the tracking error, and its various time derivatives, are guaranteed to converge asymptotically exponentially towards a vicinity of the tracking error phase space. Note that, according to the observer expected performance, the right hand side of (10) is represented by a uniformly absolutely bounded signal already evolving on a small vicinity of the origin. For this reason the roots of (11) may be located closer to the imaginary axis than those of (9). A rather detailed proof of this theorem may be found in the article by Luviano *et al.* Luviano-Juárez et al. (2010)

Remark 4. *The proposed GPI observer (7) is a high gain observer which is prone to exhibiting the "peaking" phenomena at the initial time. We use a suitable "clutch" to smooth out these transient peaking responses in all observer variables that need to be used by the controller. This is accomplished by means of a factor function smoothly interpolating between an initial value of zero and a final value of unity. We denote this clutching function as* $s_f(t) \in [0, 1]$ *and define it in the following (non-unique) way*

$$s_f(t) = \begin{cases} 1 & \text{for } t > \epsilon \\ \sin^q\left(\frac{\pi t}{2\epsilon}\right) & \text{for } t \leq \epsilon \end{cases} \tag{12}$$

where q is a su itably large positive even integer.

2.1 Generalized proportional integral observer with integral injection

Let $\xi(t)$ be a measured signal with an uniformly absolutely bounded iterated integral of order m. The function $\xi(t)$ is a measured signal, whose first few time derivatives are required for some purpose.

Definition 5. *We say that a signal* $\rho_1(t)$ *converges* to a neighborhood *of* $\xi(t)$ *whenever the error signal,* $\xi(t) - \rho_1(t)$*, is ultimately uniformly absolutely bounded inside a small vicinity of the origin.*

The following proposition aims at the design of a GPI observer based estimation of time derivatives of a signal, $\xi(t)$, where $\xi(t)$ is possibly corrupted by a zero mean stochastic process whose statistics are unknown. In order to smooth out the noise effects on the on-line computation of the time derivative, we carry out a double iterated integration of the measured signal, $\xi(t)$, thus assuming the second integral of $\xi(t)$ is uniformly absolutely bounded (i.e., $m = 2$).

Proposition 6. *Consider the following perturbed second order integration system, where the input signal, $\xi(t)$, is a measured (zero-mean) noise corrupted signal satisfying the above assumptions:*

$$\dot{y}_0 = y_1, \;\; \dot{y}_1 = \xi(t) \tag{13}$$

Consider the following integral injection GPI observer for (13) *including an internal time polynomial model of degree r for the signal $\xi(t)$ and expressed as ρ_1,*

$$\begin{aligned} \dot{\hat{y}}_0 &= \hat{y}_1 + \lambda_{r+1}(y_0 - \hat{y}_0) \\ \dot{\hat{y}}_1 &= \rho_1 + \lambda_r(y_0 - \hat{y}_0) \\ \dot{\rho}_1 &= \rho_2 + \lambda_{r-1}(y_0 - \hat{y}_0) \end{aligned} \tag{14}$$

$$\vdots$$

$$\dot{\rho}_r = \lambda_0(y_0 - \hat{y}_0) \tag{15}$$

Then, the observer variables, $\rho_1, \rho_2, \rho_3, ...,$ respectively, asymptotically converge towards a small as desired neighborhood of the disturbance input, $\xi(t)$, and of its time derivatives: $\dot{\xi}(t)$, $\ddot{\xi}(t)$,... provided the observer gains, $\{\lambda_0, ..., \lambda_{r+2}\}$, are chosen so that the roots of the polynomial in the complex variable s.

$$P(s) = s^{r+2} + \lambda_{r+1}s^{r+1} + \cdots + \lambda_1 s + \lambda_0 \tag{16}$$

are located deep into the left half of the complex plane. The further the distance of such roots from the imaginary axis of the complex plane, the smaller the neighborhood of the origin bounding the reconstruction errors.

Proof. Define the twice iterated integral injection error as, $\varepsilon = y_0 - \hat{y}_0$. The injection error dynamics is found to be described by the perturbed linear differential equation

$$\varepsilon^{(r+2)} + \lambda_{r+1}\varepsilon^{(r+1)} + \cdots + \lambda_1\dot{\varepsilon} + \lambda_0\varepsilon = \xi^{(r)}(t) \tag{17}$$

By choosing the observer parameters, $\lambda_0, \lambda_1, \cdots, \lambda_{r+1}$, so that the polynomial (16) is Hurwitz, with roots located deep into the left half of the complex plane, then, according to well known results of solutions of perturbed high gain linear differential equations, the injection error ε and its time derivatives are ultimately uniformly bounded by a small vicinity of the origin of the reconstruction error phase space whose radius of containment fundamentally depends on the smallest real part of all the eigenvalues of the dominantly linear closed loop dynamics (see Luviano *et al.* Luviano-Juárez et al. (2010) and also Fliess and Rudolph Fliess & Rudolph (1997)). □

3. Controlling the single synchronous generator model

In this section, we advocate, within the context of the angular deviation trajectory control for a single synchronous generator model, the use of approximate, yet accurate, state dependent disturbance estimation via linear Generalized Proportional Integral (GPI) observers. GPI observers are the dual counterpart of GPI controllers, developed by M. Fliess *et al.* in Fliess et al. (2002). A high gain GPI observer naturally includes a, self-updating, lumped, time-polynomial model of the nonlinear state-dependent perturbation; it estimates it and delivers the time signal to the controller for on-line cancelation while simultaneously estimating the phase variables related to the measured output. The scheme is, however, approximate since only a small as desired reconstruction error is guaranteed at the expense of high, noise-sensitive, gains. The on-line approximate estimation is suitably combined with linear, estimation-based, output feedback control with the appropriate, on-line, disturbance cancelation. The many similarities and the few differences with the DAC and ADER techniques probably lie in 1) the fact that we do not discriminate between *exogenous* (i.e., external) unstructured perturbation inputs and *endogenous* (i.e., state-dependent) perturbation inputs in the nonlinear input-output model. These perturbations are all lumped into a simplifying time-varying signal that needs to be linearly estimated. Notice that plant nonlinearities generate time functions that are *exogenous* to any observer and, hence, algebraic loops are naturally avoided 2) We emphasize the natural possibilities of *differentially flat systems* in the use of linear disturbance estimation and linear output feedback control with disturbance cancelation (For the concept of flatness see Fliess *et al.* Fliess et al. (1995)) and the book Sira-Ramírez & Agrawal (2004).

3.1 The single synchronous generator model

Consider the swing equation of a synchronous generator, connected to an *infinite bus*, with a series capacitor connected with the help of a thyristor bridge (See Hingorani Hingorani & Gyugyi (2000)),

$$\begin{aligned} \dot{x}_1 &= x_2 \\ \dot{x}_2 &= P_m - b_1 x_2 - b_2 x_3 \sin(x_1) \\ \dot{x}_3 &= b_3(-x_3 + x_3^*(t) + u + \zeta(t)) \end{aligned} \tag{18}$$

x_1 is the load angle, considered to be the measured output. The variable, x_2, is the deviation from nominal, synchronous, speed at the shaft, while x_3 stands for the admittance of the system. The control input, u, is usually interpreted as a quantity related to the fire angle of the switch. $\zeta(t)$ is an unknown, external, perturbation input. The static equilibrium point of the system, which may be parameterized in terms of the equilibrium position for the angular deviation, $\overline{x}_1$, is given by,

$$x_1 = \overline{x}_1,\ \overline{x}_2 = 0,\ \overline{x}_3 = \overline{x}_3^*(t) = \frac{P_m}{b_2 \sin(\overline{x}_1)} \tag{19}$$

We assume that the system parameters, b_2, and, b_3, are known. The constant quantities P_m, b_1 and the time varying quantity, $x_3^*(t)$, are assumed to be completely unknown.

3.2 Problem formulation

It is desired to have the load angular deviation, $y = x_1$, track a given reference trajectory, $y^*(t) = x_1^*(t)$, which remains bounded away from zero, independently of the unknown system parameters and in spite of possible external system disturbances (such as short circuits in the three phase line, setting, momentarily, the mechanical power, P_m, to zero), and other unknown, or un-modeled, perturbation inputs comprised in $\zeta(t)$.

3.3 Main results

The unperturbed system in (18) is *flat*, with flat output given by the load angle deviation $y = x_1$. Indeed, all system variables are differentially parameterizable in terms of the load angle and its time derivatives. We have:

$$
\begin{aligned}
x_1 &= y \\
x_2 &= \dot{y} \\
x_3 &= \frac{P_m - b_1\dot{y} - \ddot{y}}{b_2 \sin(y)} \\
u &= -\frac{b_1\ddot{y} + y^{(3)}}{b_3 b_2 \sin(y)} - \frac{P_m - b_1\dot{y} - \ddot{y}}{b_3 b_2 \sin^2(y)}\dot{y}\cos(y) \\
&\quad + \frac{P_m - b_1\dot{y} - \ddot{y}}{b_2 \sin(y)} - x_3^*(t)
\end{aligned} \tag{20}
$$

The perturbed input-output dynamics, devoid of any zero dynamics, is readily obtained with the help the control input differential parametrization (20). One obtains the following simplified, perturbed, system dynamics, including $\zeta(t)$, as:

$$
y^{(3)} = -\left[b_3 b_2 \sin(y)\right] u + \xi(t) \tag{21}
$$

where $\xi(t)$ is given by

$$
\begin{aligned}
\xi(t) &= -b_1\ddot{y} + b_3\left(P_m - b_1\dot{y} - \ddot{y}\right)\left(1 - \frac{\dot{y}\cos(y)}{b_3 \sin(y)}\right) \\
&\quad - b_3 b_2 \sin(y)\left(x_3^*(t) + \zeta(t)\right)
\end{aligned} \tag{22}
$$

We consider $\xi(t)$ as an unknown but uniformly absolutely bounded disturbance input that needs to be on-line estimated by means of an observer and, subsequently, canceled from the simplified system dynamics via feedback in order to regulate the load angle variable y towards the desired reference trajectory $y^*(t)$. It is assumed that the gain parameters b_2 and b_3 are known.

The problem is then reduced to the trajectory tracking problem defined on the perturbed third order, predominantly, linear system (21) with measurable state dependent input gain and unknown, but uniformly bounded, disturbance input.

We propose the following estimated state feedback controller with a smoothed (i.e., "clutched") disturbance cancelation term, $z_{1s}(t) = s_f(t) z_1(t)$, and smoothed estimated phase variables $y_{js} = s_f(t) y_j(t)$, $j = 1, 2, 3$ with $s_f(t)$ as in equation (12) with a suitable ϵ value.

$$
\begin{aligned}
u &= -\frac{1}{b_3 b_2 \sin(y)}\Big[(y^*(t))^{(3)} - k_2(y_{3s} - \ddot{y}^*(t)) \\
&\quad - k_1(y_{2s} - \dot{y}^*(t)) - k_0(y - y^*(t)) - z_{1s}\Big]
\end{aligned}
$$

The corresponding variables, y_3, y_2 and z_1, are generated by the following linear GPI observer:

$$
\begin{aligned}
\dot{y}_1 &= y_2 + \lambda_5(y - y_1) \\
\dot{y}_2 &= y_3 + \lambda_4(y - y_1) \\
\dot{y}_3 &= -\left(b_3 b_2 \sin(y)\right) u + z_1 + \lambda_3(y - y_1) \\
\dot{z}_1 &= z_2 + \lambda_2(y - y_1) \\
\dot{z}_2 &= z_3 + \lambda_1(y - y_1) \\
\dot{z}_3 &= \lambda_0(y - y_1)
\end{aligned} \tag{23}
$$

where y_1 is the redundant estimate of the output y, y_2 is the shaft velocity estimate and y_3 is the shaft acceleration estimate. The variable z_1 estimates the perturbation input $\xi(t)$ by means of a local, self updating, polynomial model of third order, taken as an internal model of the state dependent additive perturbation affecting the input-output dynamics (21).
The clutched observer variables z_{1s}, y_{2s} and y_{3s} are defined by

$$\theta_s = s_f(t)\theta, \;\; s_f(t) = \begin{cases} \sin^8(\frac{\pi t}{2\epsilon}) & \text{for } t \leq \epsilon \\ 1 & \text{for } t > \epsilon \end{cases} \tag{24}$$

with θ_s being either z_{1s}, y_{2s} or y_{3s}
The reconstruction error system is obtained by subtracting the observer model from the perturbed simplified linear system model. We have, letting $\tilde{e} = e_1 = y - y_1$, $e_2 = \dot{y} - y_2$, etc.

$$\begin{aligned} \dot{e}_1 &= e_2 - \lambda_5 e_1 \\ \dot{e}_2 &= e_3 - \lambda_4 e_1 \\ \dot{e}_3 &= \xi(t) - z_1 - \lambda_3 e_1 \\ \dot{z}_1 &= z_2 + \lambda_2(y - y_1) \\ \dot{z}_2 &= z_3 + \lambda_1(y - y_1) \\ \dot{z}_3 &= \lambda_0(y - y_1) \end{aligned} \tag{25}$$

The reconstruction error, $\tilde{e} = e_1 = y - y_1$, is seen to satisfy the following linear, perturbed, dynamics

$$\tilde{e}^{(6)} + \lambda_5 \tilde{e}^{(5)} + \lambda_4 \tilde{e}^{(4)} + \cdots + \lambda_1 \dot{\tilde{e}} + \lambda_0 \tilde{e} = \xi^{(3)}(t) \tag{26}$$

Choosing the gains $\{\lambda_5, \cdots, \lambda_0\}$ so that the roots of the characteristic polynomial,

$$p_o(s) = s^6 + \lambda_5 s^5 + \lambda_4 s^4 + \cdots + \lambda_1 s + \lambda_0, \tag{27}$$

are located deep into the left half of the complex plane, it follows from the bounded input, bounded output stability theory that the trajectories of the reconstruction error $\tilde{e}$ and those of its time derivatives $\tilde{e}^{(j)}, j = 1, 2, ...$ are uniformly ultimately bounded by a disk, centered at the origin in the reconstruction error phase space, whose radius can be made arbitrarily small as the roots of $p_o(s)$ are pushed further to the left of the complex plane.
The closed loop tracking error dynamics satisfies

$$e_y^{(3)} + \kappa_2 e_y^{(2)} + \kappa_1 \dot{e}_y + \kappa_0 e_y = \xi(t) - z_{1s} \tag{28}$$

The difference, $\xi(t) - z_{1s}$, being arbitrarily small after some time, produces a reference trajectory tracking error, $e_y = y - y^*(t)$, that also asymptotically exponentially converges towards a small vicinity of the origin of the tracking error phase space.
The characteristic polynomial of the predominant linear component of the closed loop system may be set to have poles placed in the left half of the complex plane at moderate locations

$$p_c(s) = s^3 + \kappa_2 s^2 + \kappa_1 s + \kappa_0 \tag{29}$$

3.4 Simulation results

3.4.1 A desired rest-to-rest maneuver

It is desired to smoothly lower the load angle, $y_1 = x_1$, from an equilibrium value of $y = 1$ [rad] towards a smaller value, say, $y = 0.6$ [rad] in a reasonable amount of time, say, $T = 5$ [s], starting at $t = 5$ [s] of an equilibrium operation characterized by (see Bazanella *et al.* Bazanella et al. (1999) and Pai Pai (1989))

$$x_1 = 1,\ x_2 = 0,\ x_3 = 0.8912$$

We used the following parameter values for the system

$$b_1 = 1,\ b_2 = 21.3360,\ b_3 = 20$$

We set the external perturbation input, $\zeta(t)$, as the time signal,

$$\zeta(t) = 0.005e^{(\sin^2(3t)\cos(3t))}\cos(0.3t)$$

The observer parameters were set in accordance with the following desired characteristic polynomial $p_o(s)$ for the, predominantly, linear reconstruction error dynamics. We set $p_o(s) = (s^2 + 2\zeta_o\omega_{no}s + \omega_{no}^2)^3$, with

$$\zeta_o = 1,\ \ \omega_{no} = 20$$

The controller gains $\kappa_2, \kappa_1, \kappa_0$ were set so that the following closed loop characteristic polynomial, $p_c(s)$, was enforced on the tracking error dynamics,

$$p_c(s) = (s^2 + 2\zeta_c\omega_{nc}s + \omega_{nc}^2)(s + p_c)$$

with

$$p_c = 3,\ \omega_{nc} = 3,\ \zeta_c = 1$$

The trajectory for the load angle, $y^*(t)$, was set to be

$$y^*(t) = \overline{x}_{1,\text{initial}} + (\rho(t, t_1, t_2))(\overline{x}_{1,\text{final}} - \overline{x}_{1,\text{initial}})$$

with $\rho(t, t_1, t_2)$ being a smooth Bèzier polynomial achieving a smooth rest-to-rest trajectory for the nominal load angle $y^*(t)$ from the initial equilibrium value $y^*(t_1) = \overline{x}_{1,\text{initial}} = 1$ [rad] towards the final desired equilibrium value $y^*(t_2) = \overline{x}_{1,\text{final}} = 0.6$ [rad]. We set $t_1 = 5.0$ [s], $t_2 = 10.0$ [s]; $\epsilon = 3.0$

The interpolating polynomial $\rho(t, t_1, t_2)$, is of the form:

$$\rho(t) = \tau^8 \Big[r_1 - r_2\tau + r_3\tau^2 - r_4\tau^3 + r_5\tau^4 \\ -r_6\tau^5 + r_7\tau^6 - r_8\tau^7 + r_9\tau^8 \Big]$$

with,

$$\tau = \frac{t - t_1}{t_2 - t_1}$$

The choice,

$$r_1 = 12870,\ r_2 = 91520,\ r_3 = 288288$$
$$r_4 = 524160,\ r_5 = 600600,\ r_6 = 443520$$
$$r_7 = 205920,\ r_8 = 54912,\ r_9 = 6435$$

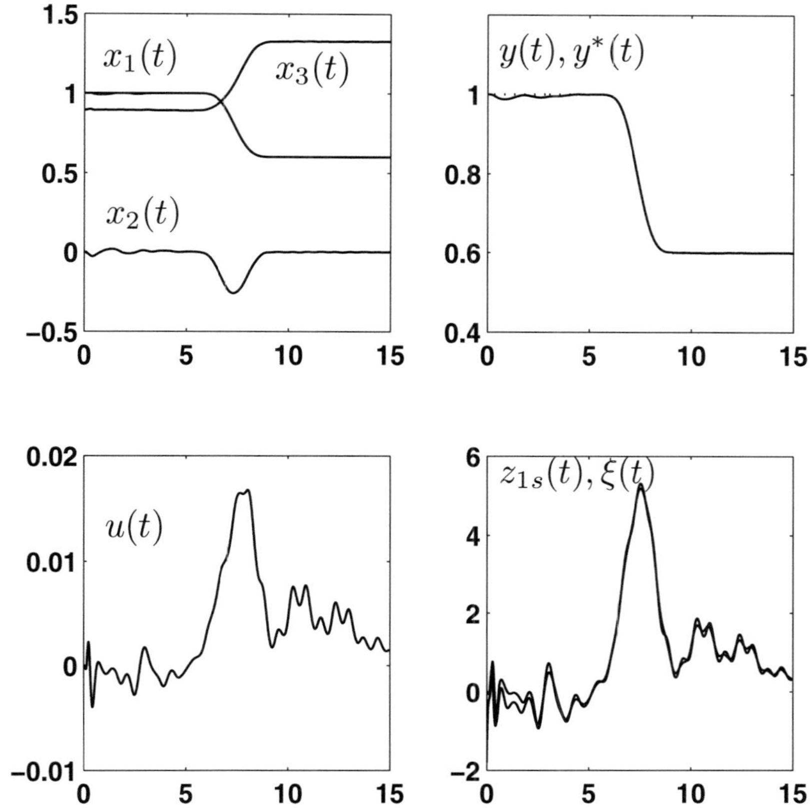

Fig. 1. Performance of GPI observer based linear controller for load angle rest-to-rest trajectory tracking in a perturbed synchronous generator.

renders a time polynomial which is guaranteed to have enough derivatives being zero, both, at the beginning and at the end of the desired rest to rest maneuver.

Figure 1 depicts the closed loop performance of the proposed GPI observer based linear output feedback controller for the forced evolution of the synchronous generator load angle trajectory following a desired rest to rest maneuver.

3.4.2 Robustness with respect to controller gain mismatches

We simulated the behavior of the closed loop system when the gain parameters product, $b_3 b_2$, is not precisely known and the controller is implemented with an estimated (guessed) value of this product, denoted by $\widehat{b_2 b_3}$, and set to be $\widehat{b_2 b_3} = \kappa b_2 b_3$. We determined that κ is a positive factor ranging in the interval $[0.95, \infty]$. However, if we allow independent estimates of the parameters in the form $\hat{b}_2 = \kappa_{b2} b_2$ and $\hat{b}_3 = \kappa_{b3} b_3$, we found that a larger robustness interval of mismatches is allowed by satisfying the empirical relation $\kappa_{b2}\kappa_{b3} \geq 0.95$. The assessment

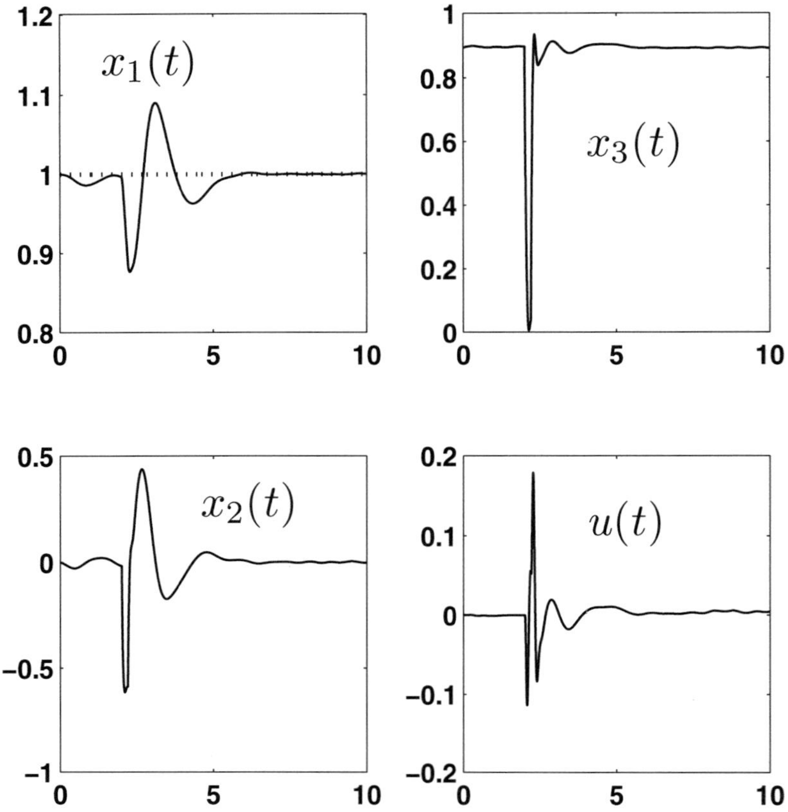

Fig. 2. Performance of GPI observer based controller under a sudden loss of power at t=2 [sec] during 0.2 [sec].

was made in terms of the proposed rest to rest maneuver and possible simulations look about the same.

3.4.3 Robustness with respect to sudden power failures

We simulated an un-modeled sudden three phase short circuit occurring at time $t = 2$ [s]. The power failure lasts for $t = 0.2$ [s]. Figure 3 depicts the performance of the GPI observer based controller in the rapid transient occurring during the recovery of the prevailing equilibrium conditions.

4. Controlling the non-holonomic car

Controlling non-holonomic mobile robots has been an active topic of research during the past three decades due to the wide variety of applications. Several methods have been proposed, and applied, to solve the regulation and trajectory tracking tasks in mobile robots. These methods range from sliding mode techniques Aguilar et al. (1997), Wang et al. (2009),

Yang & Kim (1999), backstepping Hou et al. (2009), neural networks approaches (see Peng et al. (2007) and references therein), linearization techniques Kim & Oh (1999), and classical control approaches (see Sugisaka & Hazry (2007)) among many other possibilities. A classical contribution to this area is given in the work of Canudas de Wit Wit & Sordalen (1992). An excellent book, dealing with some appropriate control techniques for this class of systems, is that of Dixon *et al.* Dixon et al. (2001). A useful approach to control non-holonomic mechanical systems is based on linear time-varying control schemes (see Pomet (1992); Tian & Cao (2007)). In the pioneering work of Samson Samson (1991), smooth feedback controls (depending on an exogeneous time variable) are proposed to stabilize a wheeled cart.

It has been shown that some mobile robotic systems are differentially flat when slippage is not allowed in the model (see Leroquais & d'Andrea Novel (1999)). The differential flatness property allows a complete parametrization of all system variables in terms of the flat outputs an a and a finite number of their time derivatives. Flat outputs constitute a limited set of special, differentially independent, output variables. The reader is referred to the work of Fliess *et al.* Fliess et al. (1995) for the original introduction of the idea in the control systems literature.

From the flatness of the non-holonomic car system, it is possible to reduce the control task to that of a linearizable, extended, multivariable input-output system. The linearization of the flat output dynamics requires the cancelation of the nonlinear input gain matrix, which depends only on the cartesian velocities of the car. To obtain this set of noisy unmeasured state variables, we propose linear Generalized Proportional Integral (GPI) observers consisting of linear, high gain Luenberger-like observers Luenberger (1971) exhibiting an internal polynomial model for the measured signal. These GPI observers, introduced in Sira-Ramírez & Feliu-Battle (2010), can provide accurate, filtered, time derivatives of the injected output signals via an appropriate iterated integral estimation error injection (see also Cortés-Romero et al. (2009)). Since high-gain observers are known to be sensitive to noisy measurements, the iterated integral injection error achieves a desirable low pass filtering effect.

The idealized model of a single axis two wheeled vehicle is depicted in figure 3. The axis is of length L and each wheel of radius R is powered by a direct current motor yielding variable angular speeds ω_1, ω_2 respectively. The position variables are (x_1, x_2) and θ is the orientation angle of the robot. The linear velocities of the points of contact of the wheels respect to the ground are given by $v_1 = \omega_1 R$ and $v_2 = \omega_2 R$. In this case, the only measurable variables are x_1, x_2. This system is subject to non-holonomic restrictions.

The kinematic model of the system is stated as follows

$$\begin{cases} \dot{x}_1 = u_1 \cos\theta, \\ \dot{x}_2 = u_1 \sin\theta, \\ \dot{\theta} = u_2 \end{cases} \tag{30}$$

where:

$$\begin{bmatrix} u_1 \\ u_2 \end{bmatrix} = \begin{bmatrix} R/2 & R/2 \\ -R/L & R/L \end{bmatrix} \begin{bmatrix} \omega_1 \\ \omega_2 \end{bmatrix}$$

The control objective is stated as follows: given a desired trajectory $(x_1^*(t), x_2^*(t))$, devise feedback control laws, u_1, u_2, such that the flat output coordinates, (x_1, x_2), perform an asymptotic tracking while rejecting the un-modeled additive disturbances.

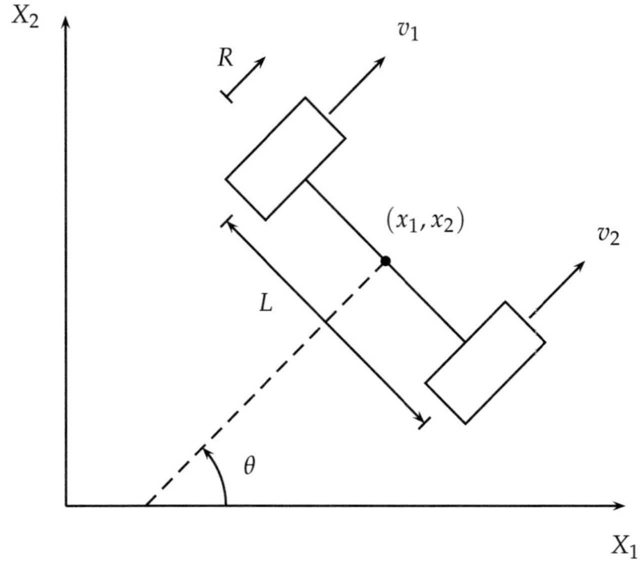

Fig. 3. The one axis car

4.1 Controller design

System (30) is differentially flat, with flat outputs given by the pair of coordinates: (x_1, x_2), which describes the position of the rear axis middle point. Indeed the rest of the system variables, including the inputs are differentially parameterized as follows:

$$\theta = \arctan\left(\frac{\dot{x}_2}{\dot{x}_1}\right),\; u_1 = \sqrt{\dot{x}_1^2 + \dot{x}_2^2},\; u_2 = \frac{\ddot{x}_2\dot{x}_1 - \dot{x}_2\ddot{x}_1}{\dot{x}_1^2 + \dot{x}_2^2}$$

Note that the relation between the inputs and the flat outputs highest derivatives is not invertible due to an ill defined relative degree. To overcome this obstacle to feedback linearization, we introduce, as an extended auxiliary control input, the time derivative of u_1. We have:

$$\dot{u}_1 = \frac{\dot{x}_1\ddot{x}_1 + \dot{x}_2\ddot{x}_2}{\sqrt{\dot{x}_1^2 + \dot{x}_2^2}}$$

This control input extension yields now an invertible control input-to-flat outputs highest derivatives relation, of the form:

$$\begin{bmatrix} \dot{u}_1 \\ u_2 \end{bmatrix} = \begin{bmatrix} \frac{\dot{x}_1}{\sqrt{\dot{x}_1^2+\dot{x}_2^2}} & \frac{\dot{x}_2}{\sqrt{\dot{x}_1^2+\dot{x}_2^2}} \\ \frac{-\dot{x}_2}{\dot{x}_1^2+\dot{x}_2^2} & \frac{\dot{x}_1}{\dot{x}_1^2+\dot{x}_2^2} \end{bmatrix} \begin{bmatrix} \ddot{x}_1 \\ \ddot{x}_2 \end{bmatrix} \tag{31}$$

4.2 Observer-based GPI controller design

Consider the following multivariable feedback controller based on linear GPI controllers and estimated cancelation of the nonlinear input matrix gain:

$$\begin{bmatrix} \dot{u}_1 \\ u_2 \end{bmatrix} = \begin{bmatrix} \dfrac{\hat{\dot{x}}_1}{\sqrt{(\hat{\dot{x}}_1)^2 + (\hat{\dot{x}}_2)^2}} & \dfrac{\hat{\dot{x}}_2}{\sqrt{(\hat{\dot{x}}_1)^2 + (\hat{\dot{x}}_2)^2}} \\ \dfrac{-\hat{\dot{x}}_2}{(\hat{\dot{x}}_1)^2 + (\hat{\dot{x}}_2)^2} & \dfrac{\hat{\dot{x}}_1}{(\hat{\dot{x}}_1)^2 + (\hat{\dot{x}}_2)^2} \end{bmatrix} \begin{bmatrix} v_1 \\ v_2 \end{bmatrix} \tag{32}$$

with the auxiliary control variables, v_1, v_2, given by[1]:

$$\begin{aligned} v_1 &= \ddot{x}_1^*(t) - \left[\frac{k_{12}s^2 + k_{11}s + k_{10}}{s(s+k_{13})}\right](x_1 - x_1^*(t)) \\ v_2 &= \ddot{x}_2^*(t) - \left[\frac{k_{22}s^2 + k_{21}s + k_{20}}{s(s+k_{23})}\right](x_2 - x_2^*(t)) \end{aligned} \tag{33}$$

and where the estimated velocity variables: $\hat{\dot{x}}_1, \hat{\dot{x}}_2$, are generated, respectively, by the variables ρ_{11} and ρ_{12} in the following single iterated integral injection GPI observers (i.e., with $m = 1$),

$$\begin{aligned} \dot{\hat{y}}_{10} &= \hat{y}_1 + \lambda_{13}(y_{10} - \hat{y}_{10}) \\ \dot{\hat{y}}_1 &= \rho_{11} + \lambda_{12}(y_{10} - \hat{y}_{10}) \\ \dot{\rho}_{11} &= \rho_{21} + \lambda_{11}(y_{10} - \hat{y}_{10}) \\ \dot{\rho}_{21} &= \lambda_{10}(y_{10} - \hat{y}_{10}) \\ y_{10} &= \int_0^t x_1(\tau)d\tau \end{aligned} \tag{34}$$

$$\begin{aligned} \dot{\hat{y}}_{20} &= \hat{y}_2 + \lambda_{23}(y_{20} - \hat{y}_{20}) \\ \dot{\hat{y}}_2 &= \rho_{12} + \lambda_{22}(y_{20} - \hat{y}_{20}) \\ \dot{\rho}_{12} &= \rho_{22} + \lambda_{21}(y_{20} - \hat{y}_{20}) \\ \dot{\rho}_{22} &= \lambda_{20}(y_{20} - \hat{y}_{20}) \\ y_{20} &= \int_0^t x_2(\tau)d\tau \end{aligned} \tag{35}$$

Then, the following theorem describes the effect of the proposed integral injection observers, and of the GPI controllers, on the closed loop system:

Theorem 7. Given a set of desired reference trajectories, $(x^*(t), y^*(t))$, for the desired position in the plane of the kinematic model of the car, described by (30); given a set initial conditions, $(x(0), y(0))$, sufficiently close to the initial value of the desired nominal trajectories, $(x^*(0), y^*(0))$, then, the above described GPI observers and the linear multi-variable dynamical feedback controllers, (32)-(35), forces the closed loop controlled system trajectories to asymptotically converge towards a small as desired neighborhood of the desired reference trajectories, $(x_1^*(t), x_2^*(t))$, provided the observer and controller gains

[1] Here we have combined, with an abuse of notation, frequency domain and time domain signals.

are chosen so that the roots of the corresponding characteristic polynomials describing, respectively, the integral injection estimation error dynamics and the closed loop system, are located deep into the left half of the complex plane. Moreover, the greater the distance of these assigned poles to the imaginary axis of the complex plane, the smaller the neighborhood that ultimately bounds the reconstruction errors, the trajectory tracking errors, and their time derivatives.

Proof. Since the system is differentially flat, in accordance with the results in Maggiore & Passino (2005), it is valid to make use of the separation principle, which allows us to propose the above described GPI observers. The characteristic polynomials associated with the perturbed integral injection error dynamics of the above GPI observers, are given by,

$$
\begin{aligned}
P_{\varepsilon 1}(s) &= s^4 + \lambda_{13}s^3 + \lambda_{12}s^2 + \lambda_{11}s + \lambda_{10} \\
P_{\varepsilon 2}(s) &= s^4 + \lambda_{23}s^3 + \lambda_{22}s^2 + \lambda_{21}s + \lambda_{20} \\
s &\in \mathbb{C}
\end{aligned}
$$

thus, the $\lambda_{i,j}$, $i = 1, 2, j = 0, \cdots, 3$, are chosen to identify, term by term, the above estimation error characteristic polynomials with the following desired stable injection error characteristic polynomials,

$$
\begin{aligned}
&P_{\varepsilon 1}(s) = P_{\varepsilon 2}(s) = (s + 2\mu_1\sigma_1 s + \sigma_1^2)(s + 2\mu_2\sigma_2 s + \sigma_2^2) \\
&s \in \mathbb{C}, \mu_1, \mu_2, \sigma_1, \sigma_2 \in \mathbb{R}^+
\end{aligned}
$$

Since the estimated states, $\hat{\dot{x}}_1 = \rho_{11}, \hat{\dot{x}}_2 = \rho_{12}$, asymptotically exponentially converge towards a small as desired vicinity of the actual states: $\dot{x}_1$, $\dot{x}_2$, substituting (32) into (31), transforms the control problem into one of controlling two decoupled double chains of integrators. One obtains the following dominant linear dynamics for the closed loop tracking errors:

$$
e_1^{(4)} + k_{13}e_1^{(3)} + k_{12}\ddot{e}_1 + k_{11}\dot{e}_1 + k_{10}e_1 = 0 \qquad (36)
$$

$$
e_2^{(4)} + k_{23}e_2^{(2)} + k_{22}\ddot{e}_2 + k_{21}\dot{e}_2 + k_{20}e_2 = 0 \qquad (37)
$$

The pole placement for such dynamics has to be such that both corresponding associated characteristic equations guarantee a dominant exponentially asymptotic convergence. Setting the roots of these characteristic polynomials to lie deep into the left half of the complex plane one guarantees an asymptotic convergence of the perturbed dynamics to a small as desired vicinity of the origin of the tracking error phase space. □

4.3 Experimental results

An experimental implementation of the proposed controller design method was carried out to illustrate the performance of the proposed linear control approach. The used experimental prototype was a parallax "Boe-Bot" mobile robot (see figure 5). The robot parameters are the following: The wheels radius is $R = 0.7\ [m]$; its axis length is $L = 0.125\ [m]$. Each wheel radius includes a rubber band to reduce slippage. The motion system is constituted by two servo motors supplied with 6 V dc current. The position acquisition system is achieved by means of a color web cam whose resolution is 352×288 pixels. The image processing was carried out by the MATLAB image acquisition toolbox and the control signal was sent to the robot micro-controller by means of a wireless communication scheme. The main function of the robot micro-controller was to modulate the control signals into a PWM input for the motor. The used micro-controller was a BASIC Stamp 2 with a blue-tooth communication card. Figure 4 shows a block diagram of the experimental framework. The proposed tracking tasks was a six-leaved "rose" defined as follows:

$$x_1^*(t) = \sin(3\omega t + \eta)\sin(2\omega t + \eta)$$
$$x_2^*(t) = \sin(3\omega t + \eta)\cos(2\omega t + \eta)$$

The design parameters for the observers were set to be, $\mu_1 = 1.8$, $\mu_2 = 2.3$, $\sigma_1 = 3$, $\sigma_2 = 4$ and for the corresponding parameters for the controllers, $\zeta_1 = \zeta_3 = 1.2$, $\zeta_2 = \zeta_4 = 1.5$, $\omega_{n1} = \omega_{n3} = 1.8$, $\omega_{n2} = \omega_{n4} = 1.9$. Also, we compared the observer response with that of a GPI observer without the integral injection $(x_{1\_}, x_{2\_})$ Luviano-Juárez et al. (2010). The experimental implementation results of the control law are depicted in figures, 6 and 7, where the control inputs and the tracking task are depicted. Notice that in the case of figure 8, there is a clear difference between the integral injection observer and the usual observer; the filtering effect of the integral observer helped to reduce the high noisy fluctuations of the control input due to measurement noises. On the average, the absolute error for the tracking task, for booth schemes, is less than 1 [cm]. This is quite a reasonable performance considering the height of the camera location and its relatively low resolution.

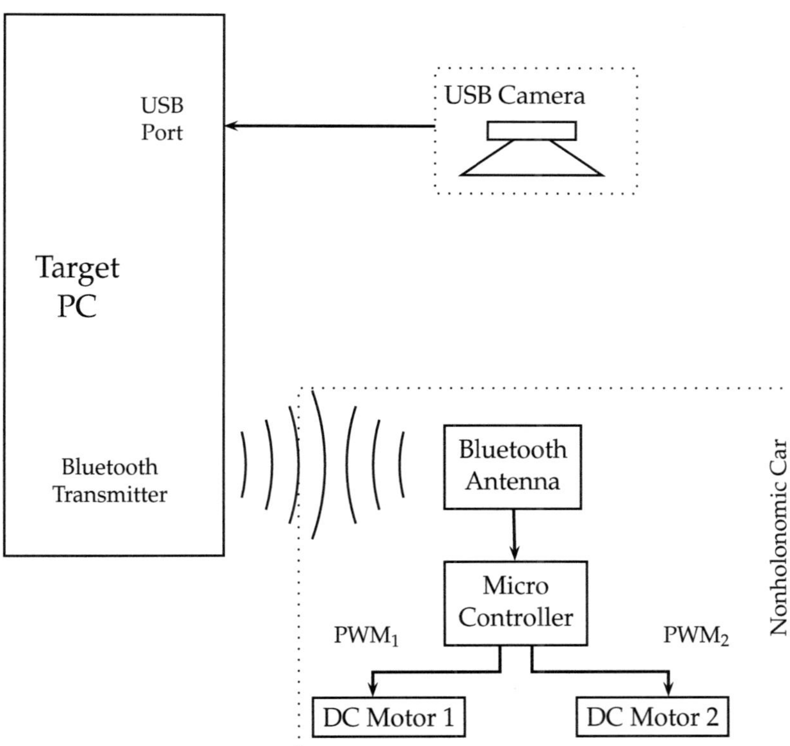

Fig. 4. Experimental control schematics

Fig. 5. Mobile Robot Prototype

5. Conclusions

In this chapter, we have proposed a linear observer-linear controller approach for the robust trajectory tracking task in nonlinear differentially flat systems. The nonlinear inputs-to-flat outputs representation is viewed as a linear perturbed system in which only the orders of integration of the Kronecker subsystems and the control input gain matrix of the system are considered to be crucially relevant for the controller design. The additive nonlinear terms in the input output dynamics can be effectively estimated, in an approximate manner, by means of a linear, high gain, Luenberger observer including finite degree, self updating, polynomial models of the additive state dependent perturbation vector components. This perturbation may also include additional unknown external perturbation inputs of uniformly absolutely bounded nature. A close approximate estimate of the additive nonlinearities is guaranteed to be produced by the linear observers thanks to customary, high gain, pole placement procedure. With this information, the controller simply cancels the disturbance vector and regulates the resulting set of decoupled chain of perturbed integrators after a direct nonlinear input gain matrix cancelation. A convincing simulation example has been presented dealing with a rather complex nonlinear physical system. We have also shown that the method efficiently results in a rather accurate trajectory tracking output feedback controller in a real laboratory implementation. A successful experimental illustration was presented which considered a non-holonomic mobile robotic system prototype, controlled by an overhead camera.

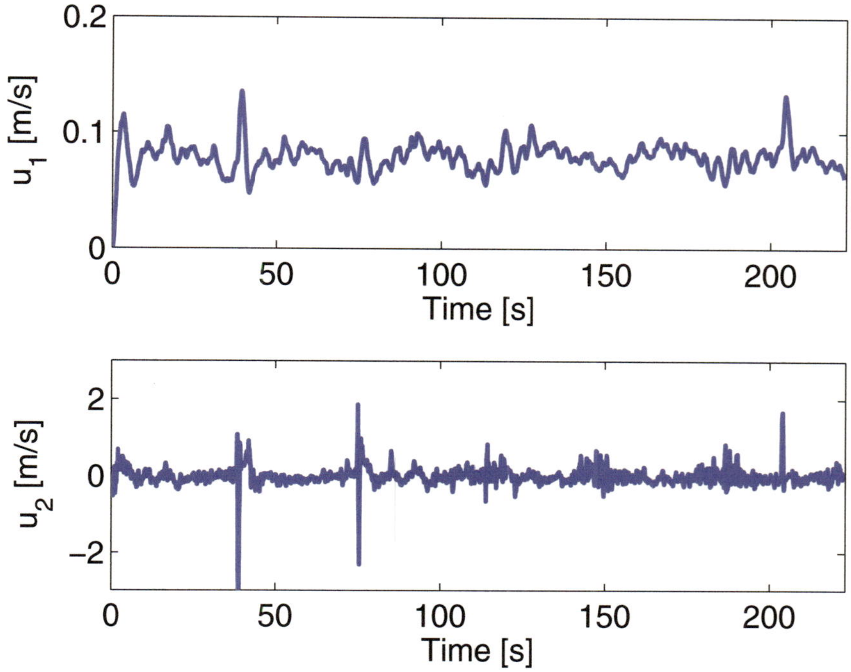

Fig. 6. Experimental applied control inputs

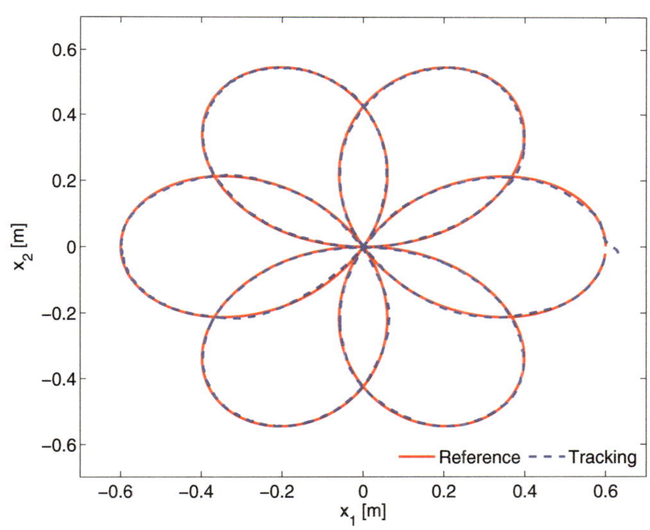

Fig. 7. Experimental performance of GPI observer-based control on trajectory tracking task

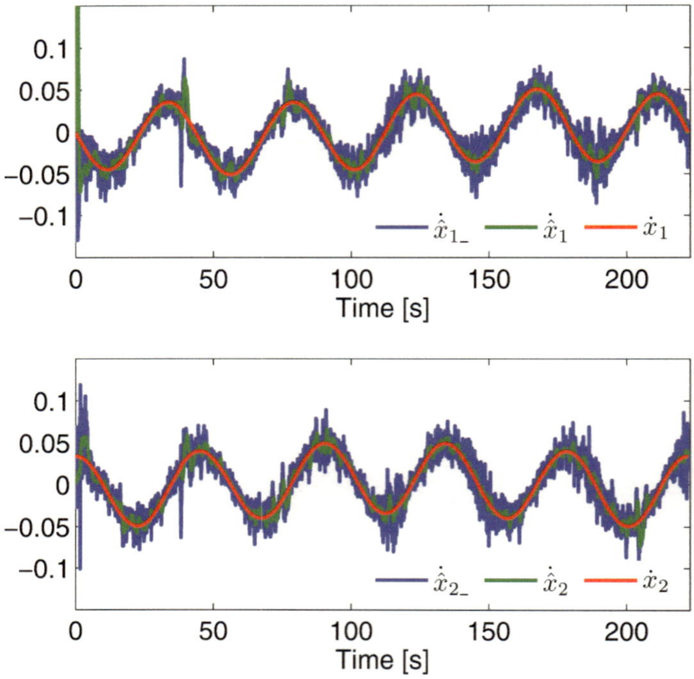

Fig. 8. Noise reduction effect on state estimations via integral error injection GPI observers

6. References

Aguilar, L. E., Hamel, T. & Soueres, P. (1997). Robust path following control for wheeled robots via sliding mode techniques, *International Conference of Intelligent Robotic Systems*, pp. 1389–1395.

Bazanella, A. S., Kokotovic, P. & Silva, A. (1999). A dynamic extension for $l_g v$ controllers, *IEEE Transactions on Automatic Control* 44(3).

Cortés-Romero, J., Luviano-Juárez, A. & Sira-Ramírez, H. (2009). Robust gpi controller for trajectory tracking for induction motors, *IEEE International Conference on Mechatronics*, Málaga, Spain, pp. 1–6.

Diop, S. & Fliess, M. (1991). Nonlinear observability, identifiability and persistent trajectories, *Proceedings of the 36th IEEE Conference on Decision and Control*, Brighton, England, pp. 714 – 719.

Dixon, W., Dawson, D., Zergeroglu, E. & Behal, A. (2001). *Nonlinear Control of Wheeled Mobile Robots*, Vol. 262 of *Lecture Notes in Control and Information Sciences*, Springer, Great Britain.

Fliess, M. & Join, C. (2008). Intelligent PID controllers, *Control and Automation, 2008 16th Mediterranean Conference on*, pp. 326 –331.

Fliess, M., Join, C. & Sira-Ramirez, H. (2008). Non-linear estimation is easy, *International Journal of Modelling, Identification and Control* 4(1): 12–27.

Fliess, M., Lévine, J., Martin, P. & Rouchon, P. (1995). Flatness and defect of non-linear systems: introductory theory and applications, *International Journal of Control* 61: 1327–1361.

Fliess, M., Marquez, R., Delaleau, E. & Sira-Ramírez, H. (2002). Correcteurs proportionnels intègraux généralisés, *ESAIM: Control, Optimisation and Calculus of Variations* 7(2): 23–41.

Fliess, M. & Rudolph, J. (1997). Corps de hardy et observateurs asymptotiques locaux pour systèmes différentiellement plats, *C.R. Academie des Sciences de Paris* 324, Serie II b: 513–519.

Gao, Z. (2006). Active disturbance rejection control: A paradigm shift in feedback control system design, *American Control Conference*, Minneapolis, Minnesota, USA, p. 2399Ű2405.

Gao, Z., Huang, Y. & Han, J. (2001). An alternative paradigm for control system design, *40th IEEE Conference on Decision and Control*, Vol. 5, pp. 4578–4585.

Han, J. (2009). From PID to active disturbance rejection control, *IEEE Transactions on Industrial Electronics* 56(3): 900–906.

Hingorani, N. G. & Gyugyi, L. (2000). *Understanding FACTS*, IEEE Press, Piscataway, N.J.

Hou, Z., Zou, A., Cheng, L. & Tan, M. (2009). Adaptive control of an electrically driven nonholonomic mobile robot via backstepping and fuzzy approach, *IEEE Transactions on Control Systems Technology* 17(4): 803–815.

Johnson, C. D. (1971). Accommodation of external disturbances in linear regulator and servomechanism problems, *IEEE Transactions on Automatic Control* AC-16(6).

Johnson, C. D. (1982). *Control and Dynamic Systems: Advances in Theory and applications*, Vol. 18, Academic Press, NY, chapter Discrete-Time Disturbance-Accommodating Control Theory, pp. 224–315.

Johnson, C. D. (2008). Real-time disturbance-observers; origin and evolution of the idea. part 1: The early years, *40th Southeastern Symposium on System Theory*, New Orleans, LA, USA.

Kailath, T. (1979). *Linear Systems*, Information and System Science Series, Prentice-Hall, Upper Saddle River, N.J.

Kim, D. & Oh, J. (1999). Tracking control of a two-wheeled mobile robot using input-output linearization, *Control Engineering Practice* 7(3): 369–373.

Leroquais, W. & d'Andrea Novel, B. (1999). Modelling and control of wheeled mobile robots not satisfying ideal velocity constraints: The unicycle case, *European Journal of Control* 5(2-4): 312–315.

Luenberger, D. (1971). An introduction to observers, *IEEE Transactions on Automatic Control* 16: 592–602.

Luviano-Juárez, A., Cortés-Romero, J. & Sira-Ramírez, H. (2010). Synchronization of chaotic oscillators by means of generalized proportional integral observers, *International Journal of Bifurcation and Chaos* 20(5): 1509Ű–1517.

Maggiore, M. & Passino, K. (2005). Output feedback tracking: A separation principle approach, *IEEE Transactions on Automatic Control* 50(1): 111–117.

Mohadjer, M. & Johnson, C. D. (1983). Power system control with disturbance-accommodation, *22nd IEEE Conference on Conference on Decision and Control*, San Antonio, Texas.

Pai, M. A. (1989). *Energy Function Analysis for Power System Stability*, Kluwer Academic Publishers.

Parker, G. A. & Johnson, C. D. (2009). Decoupling linear dynamical systems using disturbance accommodation control theory, *41st Southeastern Symposium on System Theory*, Tullahoma, TN, USA.

Peng, J., Wang, Y. & Yu, H. (2007). *Advances in Neural Networks*, Vol. 4491 of *Lecture Notes in Computer Science*, Springer Berlin / Heidelberg, chapter Neural Network-Based Robust Tracking Control for Nonholonomic Mobile Robot, pp. 804–812.

Pomet, J. (1992). Explicit design of time-varying stabilizing control laws for a class of controllable systems without drift., *Systems and Control Letters 1992* 18(2): 147–158.

Samson, C. (1991). *Advanced Robot Control*, Vol. 162 of *Lecture Notes in Control and Information Sciences*, Springer, chapter Velocity and torque feedback control of a nonholonomic cart, pp. 125–151.

Sira-Ramírez, H. & Agrawal, S. (2004). *Differentially flat systems*, Marcel Dekker Inc.

Sira-Ramírez, H. & Feliu-Battle, V. (2010). Robust $\sigma - \delta$ modulation based sliding mode observers for linear systems subject to time polynomial inputs, *International Journal of Systems Science* . To appear.

Sira-Ramírez, H. & Fliess, M. (2004). On the output feedback control of a synchronous generator, *Proceedings of the 43rd. IEEE Conference on Decision and Control*, Bahamas.

Sugisaka, M. & Hazry, D. (2007). Development of a proportional control method for a mobile robot, *Applied Mathematics and Computation* 186: 74–82.

Sun, B. & Gao, Z. (2005). A dsp-based active disturbance rejection control design for a 1-kw h-bridge dcŰdc power converter, *IEEE Transactions on Industrial Electronics* 52(5): 1271Ű1277.

Sun, D. (2007). Comments on active disturbance rejection control, *IEEE Transactions on Industrial Electronics* 54(6): 3428–3429.

Tian, Y. & Cao, K. (2007). Time-varying linear controllers for exponential tracking of non-holonomic systems in chained form, *International Journal of Robust and Nonlinear Control* 17: 631–647.

Wang, Z., Li, S. & Fei, S. (2009). Finite-time tracking control of a nonholonomic mobile robot, *Asian Journal of Control* 11(3): 344–357.

Wit, C. C. D. & Sordalen, O. (1992). Exponential stabilization of mobile robots with nonholonomic constraints, *IEEE Transactions on Automatic Control* 37(11): 1791–1797.

Yang, J. & Kim, J. (1999). Sliding mode control for trajectory tracking of nonholonomic wheeled mobile robots, *IEEE Transactions on Robotics and Automation* 15(3): 578–587.

Part 5

Robust Control Applications

21

Passive Robust Control for Internet-Based Time-Delay Switching Systems

Hao Zhang[1] and Huaicheng Yan[2]

[1]*Department of Control Science and Engineering, Tongji University, Shanghai 200092*

[2]*School of Information Science and Engineering, East China University of Science and Technology,Shanghai 200237*

P R China

1. Introduction

The Internet is playing an important role in information retrieval, exchange, and applications. Internet-based control, a new type of control systems, is characterized as globally remote monitoring and adjustment of plants over the Internet. In recent years, Internet-based control systems have gained considerable attention in science and engineering [1-6], since they provide a new and convenient unified framework for system control and practical applications. Examples include intelligent home environments, windmill and solar power stations, small-scale hydroelectric power stations, and other highly geographically distributed devices, as well as tele-manufacturing, tele-surgery, and tele-control of spacecrafts.

Internet-based control is an interesting and challenging topic. One of the major challenges in Internet-based control systems is how to deal with the Internet transmission delay. The existing approaches of overcoming network transmission delay mainly focus on designing a model based time-delay compensator or a state observer to reduce the effect of the transmission delay. Being distinct from the existing approaches, literatures (7–9) have been investigating the overcoming of the Internet time-delay from the control system architecture angle, including introducing a tolerant time to the fixed sampling interval to potentially maximize the possibility of succeeding the transmission on time. Most recently, a dual-rate control scheme for Internet-based control systems has been proposed in literature (10). A two-level hierarchy was used in the dual-rate control scheme. At the lower level a local controller which is implemented to control the plant at a higher frequency to stabilize the plant and guarantee the plant being under control even the network communication is lost for a long time. At the higher level a remote controller is employed to remotely regulate the desirable reference at a lower frequency to reduce the communication load and increase the possibility of receiving data over the Internet on time. The local and the remote controller are composed of some modes, which mode is enabled due to the time and state of the network. The mode may changes at instant time $k, k \in \{N_+\}$ and at each instant time only one mode of the controller is enabled. A typical dual-rate control scheme is demonstrated in a process control rig (7; 8) and has shown a great potential to over Internet time-delay and bring this new generation of control systems into industries. However, since the time-delay is variable and the uncertainty of the process parameters is unavoidable, a dual-rate Internet-based control system may be unstable for certain control intervals. The interest in the stability of networked control systems have grown in recent years due to its theoretical and practical significance [11-21], but to our knowledge there are very few reports dealing with the robust passive control for such kind of Internet-based control systems. The robust passive control problem for time-delay systems was dealt with in (24; 25). This motivates the present passivity investigation of multi-rate Internet-based switching control systems with time-delay and uncertainties.

In this paper, we study the modelling and robust passive control for Internet-based switching control systems with multi-rate scheme, time-delay, and uncertainties. The controller is switching between some modes due to the time and state of the network, either different time or the state changing may cause the controller changes its mode and the mode may changes at each instant time. Based on remote control and local control strategy, a new class of multi-rate switching control model with time-delay is formulated. Some new robust passive properties of such systems under arbitrary switching are investigated. An example is given to illustrate the effectiveness of the theoretical results.

Notation: Through the paper I denotes identity matrix of appropriate order, and $*$ represents the elements below the main diagonal of a symmetric block matrix. The superscript $\top$ represents the transpose. $L_2[0,\infty)$ refers to the space of square summable infinite vector sequences. The notation $X > 0(\geq,<,\leq 0)$ denotes a symmetric positive definite (positive semi-definite, negative, negative semi-definite) matrix X. Matrices, if not explicitly stated, are assumed to have compatible dimensions. Let $N = \{1,2,\cdots\}$ and $N_+ = \{0,1,2,\cdots\}$ denote the sets of positive integer and nonnegative integer, respectively.

2. Problem formulation

A typical multi-rate control structure with remote controller and local controller can be shown as Fig. 1. The control architrave gives a discrete dynamical system, where plant is in circle with broken line, $x(k) \in R^n$ is the system state, $z(k) \in R^q$ is the output, and $\omega(k) \in R^p$ is the exogenous input, which is assumed to belong to $L_2[0,\infty)$, $r(k)$ is the input and for the passivity analysis one can let $r(k) = 0$, $u_1(k)$ and $u_2(k)$ are the output of remote control and local control, respectively. A_1, B_1, B_2 and C are parameter matrices of the model with appropriate dimensions, K_{2i} and K_{1j} are control gain switching matrices where the switching rules are given by $i(k) = s(x(k),k)$ and $j(k) = \sigma(x(k),k)$, and $i \in \{1,2,\cdots,N_1\}, j \in \{1,2,\cdots,N_2\}, N_1, N_2 \in N$, which imply that the switching controllers have N_1 and N_2 modes, respectively. τ_1 and τ_2 are time-delays caused by communication delay in systems.

For the system given by Fig. 1, it is assumed that, the sampling interval of remote controller is the m multiple of local controller with m being positive integer, and the switching device SW1 closes only at the instant time $k = nm, n \in N_+$, and otherwise, it switches off. Correspondingly, remote controller $u_1(k)$ updates its state at $k = nm, n \in N_+$ only, and otherwise, it keeps invariable. Also, it is assumed that the benchmark of discrete systems is the same as local controller. In this case, the system can be described by the following discrete system with time-delay

$$\begin{cases} x(k+1) = A_1\, x(k) + B_2\, u_2(k) + E\omega(k), \\ u_2(k) \quad = B_1\, u_1(k-\tau_2) - K_{2i}\, x(k), \\ z(k) \quad = C\, x(k) + D\omega(k), \end{cases} \tag{1}$$

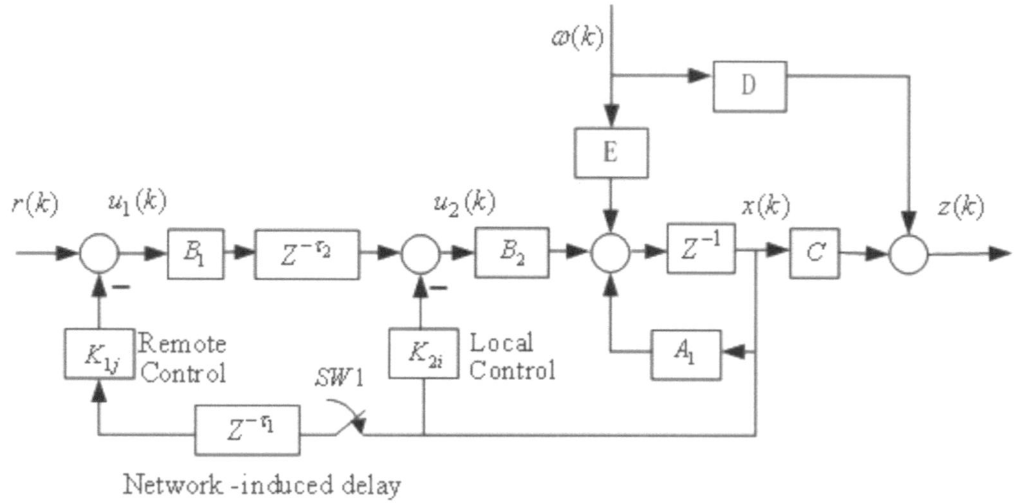

Fig. 1. Multi-rate network control loop with time-delays

where remote controller $u_1(k-\tau_2)$ is given by

$$\begin{cases} u_1(k-\tau_2)=r(k-\tau_2)-K_{1j}\,x(k-\tau_1-\tau_2), k=nm, \\ u_1(k-\tau_2)=r(nm-\tau_2)-K_{1j}\,x(nm-\tau_1-\tau_2), k\in\{nm+1,\cdots,nm+m-1\}, \end{cases} \tag{2}$$

with $i\in\{1,2,\cdots,N_1\}, j\in\{1,2,\cdots,N_2\}, k,n\in N_+$ and $N_1,N_2\in N$. Moreover, it follows from (1) and (2) that, for $k=nm$,

$$\begin{cases} x(k+1)=(A_1-B_2K_{2i})\,x(k)-B_2\,B_1\,K_{1j}\,x(k-\tau_1-\tau_2)+B_2\,B_1\,r(k-\tau_2)+E\omega(k), \\ \quad z(k)\; =C\,x(k)+D\omega(k), \end{cases} \tag{3}$$

and for $k\in\{nm+1,\cdots,nm+m-1\}$,

$$\begin{cases} x(k+1)=(A_1-B_2K_{2i})\,x(k)-B_2\,B_1\,K_{1j}\,x(nm-\tau_1-\tau_2)+B_2\,B_1\,r(nm-\tau_2)+E\omega(k), \\ \quad z(k)\; =C\,x(k)+D\omega(k). \end{cases} \tag{4}$$

For the passivity analysis, one can let $r(k)=0$, and then the system (3) and (4) become

$$\begin{cases} x(k+1)=(A_1-B_2K_{2i})\,x(k)-B_2\,B_1\,K_{1j}\,x(k-\tau)+E\omega(k), k=nm, \\ x(k+1)=(A_1-B_2K_{2i})\,x(k)-B_2\,B_1\,K_{1j}\,x(nm-\tau)+E\omega(k), k\in\{nm+1,\cdots,nm+m-1\}, \\ \quad z(k)\; =C\,x(k)+D\omega(k), \end{cases} \tag{5}$$

where $\tau=\tau_1+\tau_2>0, k\in N_+, n\in N_+, m>0$ is a positive integer.
Obviously, if define $A_i=A_1-B_2K_{2i}$, $B_j=-B_2B_1K_{1j}$, then the controlled system (5) becomes

$$\begin{cases} x(k+1) = A_ix(k)+B_jx(k-\tau)+E\omega(k), k=nm, \\ x(k+1) = A_ix(t)+B_jx(nm-\tau)+E\omega(k), k\in\{nm+1,\cdots,nm+m-1\}, \\ z(k) \quad\; = C\,x(k)+D\omega(k), \end{cases} \tag{6}$$

where A_1, B_1, B_2, C, D, E are matrices with appropriate dimensions, K_{1j} and K_{2i} are mode gain matrices of the remote controller and local controller. At each instant time k, there is only one mode of each controller is enabled. $\tau = \tau_1 + \tau_2 > 0$ and $m > 0$ are integers, $k \in N_+$, $n = 0, 1, 2, \cdots$.
Furthermore, note that, as $k = nm + s$ with $s = 0, 1, \cdots, m-1$, and $nm - \tau = k - (\tau + s)$ then (6) can be rewritten as

$$\begin{cases} x(k+1) = A_i\, x(k) + B_j\, x(k-h) + E\omega(k), \\ z(k) \quad\;\; = C\, x(k) + D\omega(k), \end{cases} \tag{7}$$

with $0 \le \tau \le h \le \tau + m - 1$. Accordingly, for the case of time-varying structured uncertainties (7) becomes

$$\begin{cases} x(k+1) = (A_i + \Delta A(k))\, x(k) + (B_j + \Delta B(k))\, x(k-h) + (E + \Delta E)\omega(k), \\ z(k) \quad\;\; = C\, x(k) + D\omega(k), \end{cases} \tag{8}$$

with $0 \le \tau \le h \le \tau + m - 1$, and $\Delta A(k)$, $\Delta B(k)$ and ΔE being structured uncertainties, and are assumed to have the form of

$$\Delta A(k) = D_1 F(k) E_a, \quad \Delta B(k) = D_1 F(k) E_b, \quad \Delta E(k) = D_1 F(k) E_e, \tag{9}$$

where D_1, E_a, E_b and E_e are known constant real matrices with appropriate dimensions. It is assumed that

$$F^\top(k) F(k) \le I, \quad \forall k. \tag{10}$$

In what follows, the the passive control for the hybrid model (7) and (8) are first studied, and then, an example of systems (8) is investigated.

3. Passivity analysis

On the basis of models (7) and (8), consider the following discrete-time nominal switching system with time-delay:

$$\begin{cases} x(k+1) = A_i x(k) + B_j x(k-h) + E\omega(k), \\ z(k) \quad\;\; = Cx(k) + D\omega(k), \\ x(k) \quad\;\; = \phi(k),\ k \in [-h, 0], \\ i(k) \quad\;\;\, = s(x(k), k), \\ j(k) \quad\;\;\, = \sigma(x(k), k), \end{cases} \tag{11}$$

where s and σ are switching rules, $i \in \{1, \cdots, N_1\}$, $j \in \{1, \cdots, N_2\}$, $N_1, N_2 \in N$, $A_i, B_j \in R^{n\times n}$ are ith and jth switching matrices of system (11), $h \in N$ is the time delay, and $\phi(\cdot)$ is the initial condition.
For the case of structured uncertainties, it can be described by

$$\begin{cases} x(k+1) = A_i(k)x(k) + B_j(k)x(k-h) + E(k)\omega(k), \\ z(k) \quad\;\; = Cx(k) + D\omega(k), \\ x(k) \quad\;\; = \phi(k),\ k \in [-h, 0], \\ i(k) \quad\;\;\, = s(x(k), k), \\ j(k) \quad\;\;\, = \sigma(x(k), k), \end{cases} \tag{12}$$

where $A_i(k) = A_i + \Delta A(k), B_j(k) = B_j + \Delta B(k)$, $E(k) = E + \Delta E(k)$, and it is assumed that (9) and (10) are satisfied. Our problem is to test whether system (11) and (12) are passive with the switching controllers. To this end, we introduce the following fact and related definition of passivity.

Lemma 1 (22). The following inequality holds for any $a \in R^{n_a}, b \in R^{n_b}, N \in R^{n_a \times n_b}, X \in R^{n_a \times n_a}, Y \in R^{n_a \times n_b}$, and $Z \in R^{n_b \times n_b}$:

$$-2a^\top Nb \leq \begin{bmatrix} a \\ b \end{bmatrix}^\top \begin{bmatrix} X & Y-N \\ * & Z \end{bmatrix} \begin{bmatrix} a \\ b \end{bmatrix}, \tag{13}$$

where $\begin{bmatrix} X & Y \\ * & Z \end{bmatrix} \geq 0$.

Lemma 2 (23). Given matrices $Q = Q^\top, H, E$ and $R = R^\top > 0$ of appropriate dimensions,

$$Q + HFE + E^\top F^\top H^\top < 0 \tag{14}$$

holds for all F satisfying $F^\top F \leq R$, if and only if there exists some $\lambda > 0$ such that

$$Q + \lambda HH^\top + \lambda^{-1} E^\top RE < 0. \tag{15}$$

Definition 1 (26) The dynamical system (11) is called passive if there exists a scalar β such that

$$\sum_{k=0}^{k_f} \omega^\top(k) z(k) \geq \beta, \quad \forall \omega \in L_2[0,\infty), \quad \forall k_f \in N,$$

where β is some constant which depends on the initial condition of system.
In the sequel, we provide condition under which a class of discrete-time switching dynamical systems with time-delay and uncertainties can be guaranteed to be passive.
System (11) can be recast as

$$\begin{cases} y(k) = & x(k+1) - x(k), \\ 0 = & (A_i + B_j - I)x(k) - y(k) - B_j \sum\limits_{l=k-h}^{k-1} y(l) + E\omega(k), \\ z(k) = & Cx(k) + D\omega(k), \\ x(k) = & \phi(k),\ k \in [-h, 0] \\ i(k) & = s(x(k), k), \\ j(k) & = \sigma(x(k), k). \end{cases} \tag{16}$$

It is noted that (11) is completely equivalent to (16).

Theorem 1. System (11) is passive under arbitrary switching rules s and σ, if there exist matrices $P_1 > 0, P_2, P_3, W_1, W_2, W_3, M_1, M_2, S_1 > 0, S_2 > 0$ such that the following LMIs hold

$$\Lambda = \begin{bmatrix} Q_1 & Q_2 & P_2^\top B_j - M_1 & P_2^\top E - C \\ * & Q_3 & P_3^\top B_j - M_2 & P_3^\top E \\ * & * & -S_2 & 0 \\ * & * & * & -(D + D^\top) \end{bmatrix} < 0, \tag{17}$$

and

$$\begin{bmatrix} W & M \\ M^\top & S_1 \end{bmatrix} \geq 0, \tag{18}$$

for $i \in \{1, \cdots, N_1\}, j \in \{1, \cdots, N_2\}, N_1, N_2 \in N$, where

$$
\begin{aligned}
Q_1 &= P_2^\top (A_i - I) + (A_i - I)^\top P_2 + hW_1 + M_1 + M_1^\top + S_2,\\
Q_2 &= (A_i - I)^\top P_3 + P_1^\top - P_2^\top + hW_2 + M_2^\top,\\
Q_3 &= -P_3 - P_3^\top + hW_3 + P_1 + hS_1,\\
W &= \begin{bmatrix} W_1 & W_2 \\ * & W_3 \end{bmatrix}, M = \begin{bmatrix} M_1 \\ M_2 \end{bmatrix}.
\end{aligned}
$$

Proof. Construct Lyapunov function as

$$
V(k) = x^\top(k) P_1 x(k) + \sum_{\theta=-h+1}^{0} \sum_{l=k-1+\theta}^{k-1} y^\top(l) S_1 y(l) + \sum_{l=k-h}^{k-1} x^\top(l) S_2 x(l),
$$

then

$$
\begin{aligned}
\Delta V(k) &= V(k+1) - V(k)\\
&= 2x^\top(k) P_1 y(k) + x^\top(k) S_2 x(k) + y^\top(k)(P_1 + hS_1) y(k)\\
&\quad - x^\top(k-h) S_2 x(k-h) - \sum_{l=k-h}^{k-1} y^\top(l) S_1 y(l),
\end{aligned}
\tag{19}
$$

where

$$
2x^\top(k) P_1 y(k) = 2\eta^\top(k) P^\top \{ \begin{bmatrix} y(k) \\ (A_i + B_j - I)x(k) - y(k) + E\omega(k) \end{bmatrix} - \sum_{l=k-h}^{k-1} \begin{bmatrix} 0 \\ B_j \end{bmatrix} y(l) \}, \tag{20}
$$

with $\eta^\top(k) = \begin{bmatrix} x^\top(k) & y^\top(k) \end{bmatrix}, P = \begin{bmatrix} P_1 & 0 \\ P_2 & P_3 \end{bmatrix}$, and

$$
\begin{aligned}
& 2\eta^\top(k) P^\top \begin{bmatrix} y(k) \\ (A_i + B_j - I)x(k) - y(k) + E\omega(k) \end{bmatrix}\\
&= 2\eta^\top(k) P^\top \{ \begin{bmatrix} 0 \\ A_i - I \end{bmatrix} x(k) + \begin{bmatrix} I \\ -I \end{bmatrix} y(k) + \begin{bmatrix} 0 \\ B_j \end{bmatrix} x(k) + \begin{bmatrix} 0 \\ E\omega(k) \end{bmatrix} \}.
\end{aligned}
\tag{21}
$$

According to Lemma 1 we get that

$$
\begin{aligned}
& -2 \sum_{l=k-h}^{k-1} \eta^\top(k) P^\top \begin{bmatrix} 0 \\ B_j \end{bmatrix} y(l)\\
& \le \sum_{l=k-h}^{k-1} \begin{bmatrix} \eta(k) \\ y(l) \end{bmatrix}^\top \begin{bmatrix} W & M - P^\top \begin{bmatrix} 0 \\ B_j \end{bmatrix} \\ * & S_1 \end{bmatrix} \begin{bmatrix} \eta(k) \\ y(l) \end{bmatrix}\\
& = \eta^\top(k) hW \eta(k) + 2\eta^\top(k)(M - P^\top \begin{bmatrix} 0 \\ B_j \end{bmatrix})(x(k) - x(k-h)) + \sum_{l=k-h}^{k-1} y^\top(l) S_1 y(l),
\end{aligned}
\tag{22}
$$

where

$$
\begin{bmatrix} W & M \\ * & S_1 \end{bmatrix} \ge 0.
$$

From (19)-(22) we can get

$$\begin{aligned}\Delta V(k) - 2z^\top(k)\omega(k) = &2\eta^\top(k)P^\top \begin{bmatrix} 0 & I \\ A_i - I & -I \end{bmatrix} \eta(k) + \eta^\top(k)hW\eta(k) \\ &+2\eta^\top(k)Mx(k) + 2\eta^\top(k)(P^\top \begin{bmatrix} 0 \\ B_j \end{bmatrix} - M)x(k-h) + x^\top(k)S_2x(k) \\ &+y^\top(k)(P_1 + hS_1)y(k) - x^\top(k-h)S_2x(k-h) + 2\eta^\top(k)P^\top \begin{bmatrix} 0 \\ E\omega(k) \end{bmatrix} \\ &-2(x^\top(k)C^\top\omega(k) + \omega^\top(k)D^\top\omega(k)).\end{aligned}$$

Let $\xi^\top(k) = [x^\top(k), y^\top(k), x^\top(k-h), \omega^\top(k)]$, then $\Delta V(k) - 2z^\top(k)\omega(k) \leq \xi(k)^\top v\xi(k)$, where

$$v = \begin{bmatrix} \phi & P^\top \begin{bmatrix} 0 \\ B_j \end{bmatrix} - M & \begin{bmatrix} P_2^\top E - C^\top \\ P_3^\top E \end{bmatrix} \\ * & -S_2 & 0 \\ * & * & -(D + D^\top) \end{bmatrix}, \tag{23}$$

and

$$\phi = P^\top \begin{bmatrix} 0 & I \\ A_i - I & -I \end{bmatrix} + \begin{bmatrix} 0 & I \\ A_i - I & -I \end{bmatrix}^\top P + hW + [M\ 0] + \begin{bmatrix} M^\top \\ 0 \end{bmatrix} + \begin{bmatrix} S_2 & 0 \\ 0 & P_1 + hS_1 \end{bmatrix}.$$

If $v < 0$, then $\triangle V(k) - 2z^\top(k)\omega(k) < 0$, which gives

$$\sum_{k=0}^{k_f} \omega^\top(k)z(k) > \frac{1}{2}\sum_{k=0}^{k_f} \triangle V(k) = \frac{1}{2}[V(k_f + 1) - V(0)].$$

Furthermore, since $V(k) = V(x(k)) \geq 0$, it follows that

$$\sum_{k=0}^{k_f} \omega^\top(k)z(k) \geq -\frac{1}{2}V(0) \equiv \beta, \quad \forall\omega \in L_2[0,\infty), \quad \forall k_f \in N,$$

which implies from Definition 1 that the system (11) is passive. Using the Schur complement (23) is equivalent to (17). This complete the proof.

Theorem 2. System (12) is passive under arbitrary switching rules s and σ, if there exist matrices $P_1 > 0, P_2, P_3, W_1, W_2, W_3, M_1, M_2, S_1 > 0, S_2 > 0$ such that the following LMIs holds

$$\begin{bmatrix} Q_1 + E_a^\top E_a & Q_2 & P_2^\top B_j - M_1 + E_a^\top E_b & P_2^\top E - C^\top + E_a^\top E_e & P_2^\top D_1 \\ * & Q_3 & P_3^\top B_j - M_2 & P_3^\top E & P_3^\top D_1 \\ * & * & -S_2 + E_b^\top E_b & E_b^\top E_e & 0 \\ * & * & * & -(D + D^\top) + E_e^\top E_e & 0 \\ * & * & * & * & -I \end{bmatrix} < 0, \tag{24}$$

and

$$\begin{bmatrix} W & M \\ M^\top & S_1 \end{bmatrix} \geq 0, \tag{25}$$

for

$$i \in \{1, \cdots, N_1\}, j \in \{1, \cdots, N_2\}, N_1, N_2 \in N,$$

where Q_1, Q_2, Q_3, W, M are defined in Theorem 1 and E_a, E_b, E_e are given by (9) and (10).

Proof. Replacing A_i, B_j and E in (17) with $A_i + D_1 F(k) E_a$, $B_j + D_1 F(k) E_b$ and $E + D_1 F(k) E_e$, respectively, we find that (17) for (12) is equivalent to the following condition

$$\Lambda + \begin{bmatrix} P_2^\top D_1 \\ P_3^\top D_1 \\ 0 \\ 0 \end{bmatrix} F(k) \begin{bmatrix} E_a & 0 & E_b & E_e \end{bmatrix} + \begin{bmatrix} E_a^\top \\ 0 \\ E_b^\top \\ E_e^\top \end{bmatrix} F^\top(k) \begin{bmatrix} D_1^\top P_2 & D_1^\top P_3 & 0 & 0 \end{bmatrix} < 0.$$

By Lemma 2, a sufficient condition guaranteeing (17) for (12) is that there exists a positive number $\lambda > 0$ such that

$$\lambda\Lambda + \lambda^2 \begin{bmatrix} P_2^\top D_1 \\ P_3^\top D_1 \\ 0 \\ 0 \end{bmatrix} \begin{bmatrix} D_1^\top P_2 & D_1^\top P_3 & 0 & 0 \end{bmatrix} + \begin{bmatrix} E_a^\top \\ 0 \\ E_b^\top \\ E_e^\top \end{bmatrix} \begin{bmatrix} E_a & 0 & E_b & E_e \end{bmatrix} < 0. \qquad (26)$$

Replacing $\lambda P, \lambda S_1, \lambda S_2, \lambda M$ and λW with P, S_1, S_2, M and W respectively, and applying the Schur complement shows that (26) is equivalent to (24). This completes the proof.

4. A numerical example

In this section, we shall present an example to demonstrate the effectiveness and applicability of the proposed method. Consider system (12) with parameters as follows:

$$A_1 = \begin{bmatrix} -6 & -6 \\ 2 & -2 \end{bmatrix}, A_2 = \begin{bmatrix} -4 & -6 \\ 4 & -4 \end{bmatrix}, B_1 = \begin{bmatrix} -1 & -2 \\ 0 & -1 \end{bmatrix}, B_2 = \begin{bmatrix} -2 & 0 \\ -3 & -1 \end{bmatrix},$$
$$B_3 = \begin{bmatrix} -1 & 0 \\ 0 & -1 \end{bmatrix}$$
$$C = \begin{bmatrix} 0.1 & -0.2 \end{bmatrix}, E = \begin{bmatrix} 0.2 \\ 0.1 \end{bmatrix}, E_a = \begin{bmatrix} 0.5 & 0 \\ 0.1 & 0.2 \end{bmatrix}, E_b = \begin{bmatrix} 0.6 & 0 \\ 0 & 0.3 \end{bmatrix}, D_1 = \begin{bmatrix} 0.1 & 0 \\ 0 & 1 \end{bmatrix},$$
$$D = 0.1, h = 5.$$

Applying Theorem 2, with $i \in \{1,2\}, j \in \{1,2,3\}$. It has been found by using software LMIlab that the switching discrete time-delay system (12) is the passive and we obtain the solution as follows:

$$P_1 = 10^{-3} \times \begin{bmatrix} 0.1586 & 0.0154 \\ * & 0.2660 \end{bmatrix}, P_2 = \begin{bmatrix} 0.5577 & 0.3725 \\ -1.6808 & 1.0583 \end{bmatrix}, P_3 = \begin{bmatrix} 0.1689 & -0.0786 \\ -0.0281 & 0.1000 \end{bmatrix},$$
$$S_1 = 10^{-4} \times \begin{bmatrix} 0.4207 & 0.0405 \\ * & 0.6941 \end{bmatrix}, S_2 = \begin{bmatrix} 2.6250 & 0.8397 \\ * & 2.0706 \end{bmatrix}, W_1 = \begin{bmatrix} 0.2173 & -0.0929 \\ * & 0.0988 \end{bmatrix},$$
$$W_2 = \begin{bmatrix} 0.0402 & -0.0173 \\ * & 0.0182 \end{bmatrix}, W_3 = \begin{bmatrix} 0.0075 & -0.0032 \\ * & 0.0034 \end{bmatrix}, M_1 = 10^{-4} \times \begin{bmatrix} -0.0640 & -0.2109 \\ 0.1402 & -0.5777 \end{bmatrix},$$
$$M_2 = 10^{-5} \times \begin{bmatrix} 0.0985 & -0.4304 \\ 0.1231 & -0.9483 \end{bmatrix}.$$

5. Conclusions

In this paper, based on remote control and local control strategy, a class of hybrid multi-rate control models with uncertainties and switching controllers have been formulated and their passive control problems have been investigated. Using the Lyapunov-Krasovskii function approach on an equivalent singular system, some new conditions in form of LMIs have been derived. A numerical example has been shown to verify the effectiveness of the proposed control and passivity methods.

6. Acknowledgements

This work is supported by the Program of the International Science and Technology Cooperation (No.2007DFA10600), the National Natural Science Foundation of China (No.60904015, 61004028), the Chen Guang project supported by Shanghai Municipal Education Commission, Shanghai Education Development Foundation (No.09CG17), the National High Technology Research and Development Program of China (No.2009AA043001), the Shanghai Pujiang Program (No.10PJ1402800), the Fundamental Research Funds for the Central Universities (No.WH1014013) and the Foundation of East China University of Science and Technology (No.YH0142137).

7. References

[1] Huang J, Guan Z H, Wang Z. Stability of networked control systems based on model of discrete-time interval system with uncertain delay. Dynamics of Continuous, Discrete and Impulsive Systems Series B: Applications & Algorithms 2004; 11: 35-44.

[2] Lien C H. Further results on delay-dependent robust stability of uncertain fuzzy systems with time-varying delay. Chaos, Solitons & Fractals 2006; 28(2): 422-427.

[3] Montestruque L A, Antsaklis P J. On the model-based control of networked systems. Automatica 2003; 39: 1837-43.

[4] Montestruque L A , Antsaklis P J.Stability of model-based networked control systems with time-varying transmission times. IEEE Trans. Autom. Control 2005; 49(9): 1562-1573.

[5] Nesic D, Teel A R. Input-to-state stability of networked control systems. Automatica 2004; 40: 2121-28.

[6] Overstreet J. W, Tzes A. An Internet-based real-time control engineering laboratory. IEEE Control Systems Magazine 1999; 9: 320-26.

[7] Yang S H, Chen X , Tan L , Yang L. Time delay and data loss compensation for Internet-based process control systems. Transactions of the Institute of Measurement and Control 2005; 27(2): 103-08.

[8] Yang S H, Chen X ,Alty J L. Design issues and implementation of Internet-based process control systems. Control Eng. Practice 2003; 11: 709-20.

[9] Yang S H, Tan L,Liu G P. Architecture design for Internet-based control systems.Int. J. of Automation and Computing 2005; 1: 1-9.

[10] Yang S H,Dai C.Multi-rate control in Internet based control systems. In Proc. UK Control 2004, Sahinkaya, M.N. and Edge, K.A. (eds), Bath, UK, 2004, ID-053.

[11] Guan Z H, David J H, Shen X. On hybrid impulsive and switching systems and application to nonlinear control.IEEE Trans. Autom. Control 2005; 50(7): 1158-62.

[12] Chen W H, Guan ZH, Lu X M. Delay-dependent exponential stability of uncertian stochastic system with multiple delays: an LMI approach. Systems & Control Letters 2005; 54: 547-55.
[13] Chen W H, Guan ZH, Lu X M. Delay-dependent output feedback guaranteed cost control for uncertain time-delay systems. Automatica 2004; 40: 1263-68.
[14] Huang X, Cao J D, Huang D S. LMI-based approach for delay-dependent exponential stability analysis of BAM neural networks. Chaos, Solitons & Fractals 2005; 24(3): 885-898.
[15] Liu X W, Zhang H B, Zhang F L. Delay-dependent stability of uncertain fuzzy large-scale systems with time delays. Chaos, Solitons & Fractals 2005; 26(1): 147-158.
[16] Li C D, Liao X F, Zhang R. Delay-dependent exponential stability analysis of bi-directional associative memory neural networks with time delay: an LMI approach. Chaos, Solitons & Fractals 2005; 24(4): 1119-1134.
[17] Tu F H, Liao X F, Zhang W. Delay-dependent asymptotic stability of a two-neuron system with different time delays. Chaos, Solitons & Fractals 2006; 28(2): 437-447.
[18] Srinivasagupta D , Joseph B. An Internet-mediated process control laboratory. IEEE Control Systems Magazine 2003; 23: 11-18.
[19] Walsh G C, Ye H, Bushnell L G. Aysmptotic behavior of nonlinear networked control systems.IEEE Trans. Autom. Control 2001; 46: 1093-97.
[20] Zhang L, Shi Y,Chen T, Huang B. A new method for stabilization of networked control systems with random delays. IEEE Trans. Autom. Control 2005; 50(8): 1177-81.
[21] Zhang W, Branicky M S, Phillips S M. Stability of networked control systems. IEEE Control Systems Magazine 2001; 2: 84-99.
[22] Moon Y S, Park P, Koon W H, Lee Y S. Delay-dependent robust stabilization of uncdrtain state-delayed systems. International Journal of Control 2001; 74: 1447-1455.
[23] Xie L. Output feedback H_∞ control of systems with parameter uncertainty. International Journal of Control 1996; 63: 741-750.
[24] Cui B,Hua M, Robust passive control for uncertain discrete-time systems with time-varying delays. Chaos Solitions and Fractals 2006; 29: 331-341.
[25] Mahmoud M,Ismail A. Passiveity analysis and synthesis of discrete-time delay system. Dynam Contin Discrete Impuls Syst Ser A:Math Anal 2004; 11(4): 525-544.
[26] R. Lozano, B. Brogliato, O. Egeland and B. Maschke, *Dissipative Systems Analysis and Control. Theory and Applications*, London, U.K.: CES, Springer, 2000.

22

Robust Control of the Two-mass Drive System Using Model Predictive Control

Krzysztof Szabat, Teresa Orłowska-Kowalska and Piotr Serkies
Wroclaw University of Technology
Poland

1. Introduction

A demand for the miniaturization and reducing the total moment of inertia which allows to shorten the response time of the whole system is evident in modern drives system. However, reducing the size of the mechanical elements may result in disclosure of the finite stiffness of the drive shaft, which can lead to the occurrence of torsional vibrations. This problem is common in rolling-mill drives, belt-conveyors, paper machines, robotic-arm drives including space manipulators, servo-drives and throttle systems (Itoh et al., 2004, Hace et al., 2005 , Ferretti et al. 2005, Sugiura & Hori, Y., 1996, Szabat & Orłowska-Kowalska, 2007, O'Sullivan at al. 2007, Ryvkin et al., 2003 , Wang & Frayman, 2004, Vasak & Peric, 2009, Vukosovic & Stojic, 1998).

To improve performances of the classical control structure with the PI controller, the additional feedback loop from one selected mechanical state variable can be used. The additional feedback allows setting the desired value of the damping coefficient, but the free value of the resonant frequency cannot be achieved simultaneously (Szabat & Orłowska-Kowalska, 2007). According to the literature, the application of the additional feedback from the shaft torque is very common (Szabat & Orłowska-Kowalska, 2007). The design methodology of that system can be divided into two groups. In the first framework the shaft torque is treated as the disturbance. The simplest approach relies on feeding back the estimated shaft torque to the control structure, with the gain less than one. The more advanced methodology, called Resonance Ratio Control (RRC) is presented in (Hori et al., 1999). The system is said to have good damping ability when the ratio of the resonant to antiresonant frequency has a relatively big value (about 2). The second framework consists in the application of the modal theory. Parameters of the control structure are calculated by comparison of the characteristic equation of the whole system to the desired polynomial. To obtain a free design of the control structure parameters, i.e. the resonant frequency and the damping coefficient, the application of two feedbacks from different groups of mechanical state variables is necessary. The design methodology of this type of the systems is presented in (Szabat & Orłowska-Kowalska, 2007).

The control structures presented so far are based on the classical cascade compensation schemes. Since the early 1960s a completely different approach to the analysis of the system dynamics has been developed - the state space methodology (Michels et al., 2006). The application of the state-space controller allows to place the system poles in an arbitrary position so theoretically it is possible to obtain any dynamic response of the system. The suitable location of the closed-loop system poles becomes one of the basic problems of the state space controller application. In (Ji & Sul,, 1995) the selection of the system poles is realized through LQ approach. The authors emphasize the difficulty of the matrices selection in the case of the system parameter variation. The influence of the closed-loop location on the dynamic characteristics of the two-mass system is analyzed in (Qiao et al., 2002), (Suh et al., 2001). In (Suh et al., 2001) it is stated that the location of the system poles in the real axes improve the performance of the drive system and makes it more robust against the parameter changes.

In the case of the system with changeable parameters more advanced control concepts have been developed. In (Gu et al., 2005), (Itoh et al., 2004) the applications of the robust control theory based on the H_∞ and *μ-synthesis* frameworks are presented. The implementation of the genetic algorithm to setting of the control structure parameters is shown in (Itoh et al., 2004). The author reports good performance of the system despite the variation of the inertia of the load machine. The next approach consists in the application of the sliding-mode controller. For example, in paper (Erbatur et al., 1999) this method is applied to controlling the SCARA robot. A design of the control structure is based on the Lyapunov function. The similar approach is used in (Hace et al., 2005) where the conveyer drive is modelled as the two-mass system. The authors claim that the designed structure is robust to the parameter changes of the drive and external disturbances. Other application examples of the sliding-mode control can be found in (Erenturk, 2008). The next two frameworks of control approach relies on the use of the adaptive control structure. In the first framework the controller parameters are adjusted on-line on the basis of the actual measurements. For instance in (Wang & Frayman, 2004) a dynamically generated fuzzy-neural network is used to damp torsional vibrations of the rolling-mill drive. In (Orlowska-Kowalska & Szabat, 2008) two neuro-fuzzy structures working in the MRAS structure are compared. The experimental results show the robustness of the proposed concept against plant parameter variations. In the other framework changeable parameters of the plant are identified and then the controller is retuned in accordance with the currently identified parameters. The Kalman filter is applied in order to identify the changeable value of the inertia of the load machine (Szabat & Orlowska-Kowalska, 2008). This value is used to correct the parameters of the PI controller and two additional feedbacks. A similar approach is presented in (Hirovonen et al., 2006).

The Model Predictive Control (MPC) is one of the few techniques (apart from PI/PID techniques) which are frequently applied to industry (Maciejowski 2002, Cychowski 2009). The MPC algorithm adapts to the current operation point of the process generating an optimal control signal. It is able to directly take into consideration the input and output constraints of the system which is not easy in a control structure using classical structures. Nevertheless, the real time implementations of the MPC are traditionally limited to objects with relatively large time constants (Maciejowski 2002, Cychowski 2009). The application of MPC to industrial processes characterized by fast dynamics, such as those of electrical drives, is complicated by the formidable real-time computational complexity often necessitating the use of high-performance computers and complex software. The state-of-the-art of currently employed predictive control methods in the power electronics and motion control sector is given in (Kennel et al, 2008). Still, there are few works which report the application of the MPC in the control structure of a two-mass system (Cychowski et al. 2009).

The main contribution of this paper is the design and real-time validation of an explicit model predictive controller for a two-mass elastic drive system which is robust to the parameter changes. The explicit version of the MPC algorithm presented here does not involve complex optimization to be performed in a control unit but requires only a piecewise linear function evaluation which can be realized through a simple look-up table approach. This problem is computationally far more attractive than the standard optimization-based MPC and enables the application of complex constrained control algorithms to demanding systems with sampling in the mili/micro second scale. In addition to low complexity, the proposed MPC controller respects the inherent electromagnetic (input) and shaft (output) torque constraints while guaranteeing optimal closed-loop performance. This safety feature is crucial for many two-mass drive applications as violating the shaft ultimate tensile strength may result in damage of the shaft and ultimately in the failure of the entire drive system. Contrary to the previous works of the authors (Cychowski et al. 2009), where the system was working under nominal condition, in the present paper the issues related to the robust control of the drive system with elastic joint are presented.
This paper is divided into seven sections. After an introduction, the mathematical model of the two-mass drive system and utilized control structure are described. In section III the idea of the MPC is presented. Then the whole investigated control structure is described. The simulation results are demonstrated in sections V. After a short description of the laboratory set-up, the experimental results are presented in section VI. Conclusions are presented at the end of the paper.

2. The mathematical model of the two-mass system and the control structure

In technical papers there exist many mathematical models, which can be used for the analysis of the plant with elastic couplings. In many cases the drive system can be modelled as a two-mass system, where the first mass represents the moment of inertia of the drive and the second mass refers to the moment of inertia of the load side. The mechanical coupling is treated as an inertia free. The internal damping of the shaft is sometimes also taken into consideration. Such a system is described by the following state equation (Szabat & Orlowska-Kowalska, 2007) (with non-linear friction neglected):

$$\frac{d}{dt}\begin{bmatrix}\Omega_1(t)\\ \Omega_2(t)\\ M_s(t)\end{bmatrix}=\begin{bmatrix}\frac{-D}{J_1} & \frac{D}{J_1} & \frac{-1}{J_1}\\ \frac{D}{J_2} & \frac{-D}{J_2} & \frac{1}{J_2}\\ K_c & -K_c & 0\end{bmatrix}\begin{bmatrix}\Omega_1(t)\\ \Omega_2(t)\\ M_s(t)\end{bmatrix}+\begin{bmatrix}\frac{1}{J_1}\\ 0\\ 0\end{bmatrix}[M_e]+\begin{bmatrix}0\\ \frac{-1}{J_2}\\ 0\end{bmatrix}[M_L] \qquad (1)$$

where: Ω_1- motor speed, Ω_2- load speed, M_e– motor torque, M_s– shaft (torsional) torque, M_L– load torque, J_1 - inertia of the motor, J_2– inertia of the load machine, K_c– stiffness coefficient, D - internal damping of the shaft.
The schematic diagram of the two-mass system is presented in Fig. 1
The described model is valid for the system in which the moment of inertia of the shaft is much smaller than the moment of the inertia of the motor and the load side. In other cases a more extended model should be used, such as the Rayleigh model of the elastic coupling or even a model with distributed parameters. The suitable choice of the mathematical model is

a compromise between the accuracy and calculation complexity. As can be concluded from the literature, nearly in all cases the simplest shaft-inertia-free model has been used.

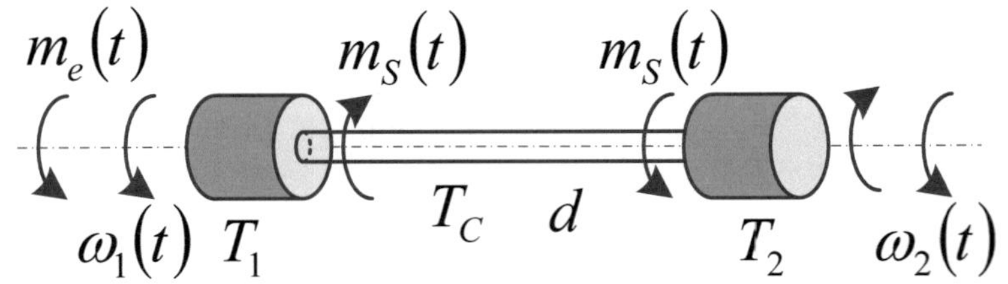

Fig. 1. The schematic diagram of the two-mass system

To simplify the comparison of the dynamical performances of the drive systems of different power, the mathematical model (1) is expressed in per unit system, using the following notation of new state variables:

$$\omega_1 = \frac{\Omega_1}{\Omega_N} \quad \omega_2 = \frac{\Omega_2}{\Omega_N} \quad m_e = \frac{M_e}{M_N} \quad m_s = \frac{M_s}{M_N} \quad m_L = \frac{M_L}{M_N} \tag{2}$$

where: Ω_N - nominal speed of the motor, M_N - nominal torque of the motor, ω_1, ω_2 - motor and load speeds, m_e, m_s, m_L - electromagnetic, shaft and load torques in per unit system.
The mechanical time constant of the motor - T_1 and the load machine - T_2 are thus given as:

$$T_1 = \frac{\Omega_N J_1}{M_N} \quad T_2 = \frac{\Omega_N J_2}{M_N} \tag{3}$$

The stiffness time constant - T_c and internal damping of the shaft - d can be calculated as follows:

$$T_c = \frac{M_N}{K_c \Omega_N} \quad d = \frac{\Omega_N D}{M_N} \tag{4}$$

Taking into account the equations (3)-(5) the state equation of the two-mass system in per-unit value is represented as:

$$\frac{d}{dt}\begin{bmatrix} \omega_1(t) \\ \omega_2(t) \\ m_s(t) \end{bmatrix} = \begin{bmatrix} 0 & 0 & \frac{-1}{T_1} \\ 0 & 0 & \frac{1}{T_2} \\ \frac{1}{T_c} & \frac{-1}{T_c} & 0 \end{bmatrix} \cdot \begin{bmatrix} \omega_1(t) \\ \omega_2(t) \\ m_s(t) \end{bmatrix} + \begin{bmatrix} 1/T_1 \\ 0 \\ 0 \end{bmatrix} \cdot m_e(t) + \begin{bmatrix} 0 \\ -1/T_2 \\ 0 \end{bmatrix} \cdot m_L(t)$$
$$\frac{d}{dt}x_c = A_c x_c + B_c m_e + B_d m_L \tag{5}$$

Usually, due to its small value the internal damping of the shaft d is neglected in the analysis of the two-mass drive system as in eq. (5).

3. Model predictive control and its explicit formulation

In model predictive control, an explicit model of the plant is used to predict the effect of future actions of the manipulated variables on the process output. In the recent literature, the following linear discrete-time state-space model is typically employed [14]

$$\begin{aligned} x(k+1) &= Ax(k) + Bu(k) \\ y(k) &= Cx(k) \end{aligned} \tag{6}$$

where $x(k)$, $u(k)$ and $y(k)$ denote the system state, input and output vectors, respectively. Let y_k represent the value of the output vector at a future time k, given an input sequence u_k, and initial state x_0 of the system. At each time step k, the MPC algorithm solves the following optimization problem:

$$\begin{aligned} &\min_{u_0,\ldots,u_{N_c-1}} \sum_{k=0}^{N_p} y_k^T \mathbf{Q} y_k + \sum_{k=0}^{N_c-1} u_k^T \mathbf{R} u_k \\ &\text{subject to} \\ &u_{\min} \le u_k \le u_{\max}, \quad k = 0,\ldots,N_c - 1 \\ &y_{\min} \le y_k \le y_{\max}, \quad k = 1,\ldots,N_p \\ &x_{k+1} = Ax_k + Bu_k, \quad k \ge 0 \\ &y_k = Cx_k, \qquad k \ge 0 \\ &x_0 = x(0) \end{aligned} \tag{7}$$

where $Q \ge 0$ and $R > 0$ are the weighting matrices, N_p and N_c denote the prediction and control horizon, respectively and u_{min}, u_{max}, y_{min}, and y_{max} are the input and output constraints of the system.

The MPC algorithm based on optimization problem can be implemented in two ways. The traditional approach relies on solving the optimization problem on-line for a given $x(k)$ in a receding-horizon fashion. This means that, at the current time k, only the first element control signal u_k of the optimal input sequence is actually implemented to the plant and the rest of the control moves are discarded. At the next time step, the whole procedure is repeated for the new measured or estimated output $y(k+1)$ (Maciejowski 2002, Cychowski 2009). This strategy can be computationally demanding for systems requiring fast sampling or low-performance computers and hence greatly restricting the scope of applicability to systems with relatively slow dynamics. In the second approach, the problem (7) is first solved off-line for all possible state realizations within some compact set X_f using multi-parametric programming (Maciejowski 2002, Cychowski 2009). Specifically, by treating the state vector $x(k)$ as a parameter vector, it can be shown that the parameter space X_f can be subdivided into characteristic regions, where the optimizer is given as an explicit function of the parameters. This profile is a piecewise affine state feedback law:

$$U(x) = K_r x + g_r, \qquad \forall x \in P_r \tag{8}$$

where P_r are polyhedral sets defined as:

$$P_r = \left\{ x \in \Re^n \mid H_r x \le d_r \right\}, \qquad r = 1,\ldots N_r \tag{9}$$

and N_r denotes the total number of polyhedral regions in the partition. Algorithms for the construction of a polyhedral partition of the state space and computation of a PWA control law are given in (Maciejowski 2002, Cychowski 2009). In its simplest form, the PWA control law (8)–(9) can be evaluated by searching for a region containing current state x in its interior and applying the affine control law associated with this region. More efficient search strategies which offer a logarithmic-type complexity with respect to the total number of regions N_r in the partition have also been developed (Cychowski, 2009, Kvasnica et al. 2004, Tøndel et al. 2003, Spjøtvold et al. 2006). Nonetheless, the implementation of the explicit MPC control law can often be several orders of magnitude more efficient than solving the optimization problem (7) directly. This gain in efficiency is crucial for demanding applications with fast dynamics or high sampling rates in the milli/micro second range, such as the drive system considered in this paper.

4. MPC-based control structure

A typical electrical drive system is composed of a power converter-fed motor coupled to a mechanical system, a microprocessor-based controllers, current, rotor speed and/or position sensors used as feedback signals. Typically, cascade speed control structure containing two major control loops is used, as presented in Fig 2.
The inner control loop performs a motor torque regulation and consists of the power converter, electromagnetic part of the motor, current sensor and respective current or torque controller. As this control loop is designed to provide sufficiently fast torque control, it can be approximated by an equivalent first order term with small time constant. If the control is ensured, the driven machine could be an AC or DC motor, with no difference in the outer speed control loop. The outer loop consists of the mechanical part of the motor, speed sensor, speed controller, and is cascaded to the inner loop. It provides speed control according to the reference value (Szabat & Orlowska-Kowalska, 2007).

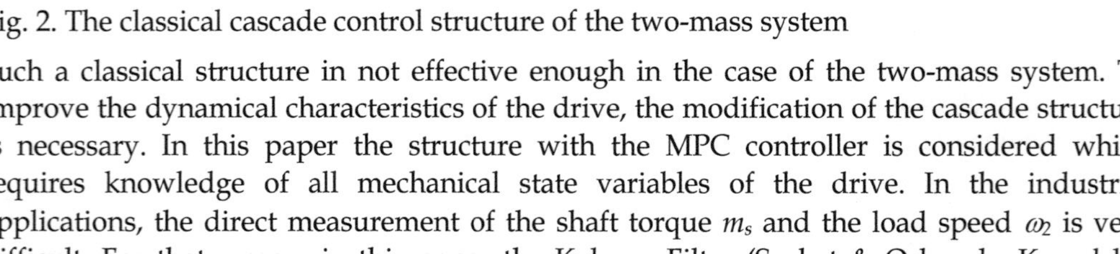

Fig. 2. The classical cascade control structure of the two-mass system

Such a classical structure in not effective enough in the case of the two-mass system. To improve the dynamical characteristics of the drive, the modification of the cascade structure is necessary. In this paper the structure with the MPC controller is considered which requires knowledge of all mechanical state variables of the drive. In the industrial applications, the direct measurement of the shaft torque m_s and the load speed ω_2 is very difficult. For that reason, in this paper the Kalman Filter (Szabat & Orlowska-Kowalska, 2008) is used to provide the information about non-measurable mechanical state variables. Additionally, the load torque m_L is also estimated and used in the MPC based control structure. In Fig. 3 the block diagram of the considered control structure is presented.

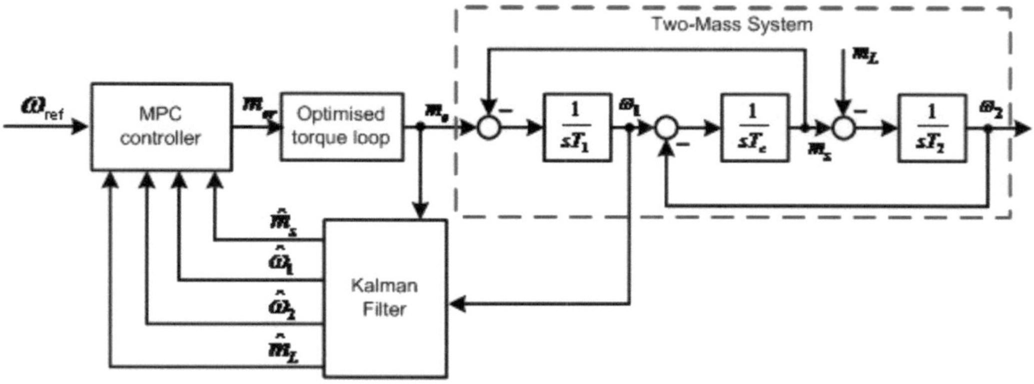

Fig. 3. The block diagram of the MPC-based control structure

5. Simulation study

In this section, the proposed single-loop explicit MPC control strategy for the drive system with an elastic coupling will be evaluated through simulations. A primary design objective for the MPC controller is to ensure that the load speed response follows the set-point with the desired dynamics. This needs to be achieved without generating excessive shaft torque responses and without violating the input and output constraints of the drive. The first two requirements can be addressed by defining the following auxiliary output variables:

$$y_1 = \omega_1 - \omega_{ref} \; y_2 = \omega_2 - \omega_r \; y_3 = m_s - m_L \tag{10}$$

where y_1 and y_2 account for tracking performance and y_3 relates to load-shaft torque imbalance. Due to (8)-(10), the reference speed variable and the disturbance torque need to be directly incorporated into the drive system model.

Decreasing the values of the cost function in the MPC algorithm leads to the minimization of the errors between the reference value of both speeds and reduce the torsional tension in the shaft. In order to calculate the values of the y_1-y_3, the original state vector of the system has to be extended by load torque m_L and the reference speed ω_r. Thus, the new model used in the MPC algorithm is described by the following state equation:

$$\frac{d}{dt}\begin{bmatrix} x_c \\ m_L \\ \omega_{ref} \end{bmatrix} = \begin{bmatrix} A_c & B_d & 0 \\ 0 & 0 & 0 \\ 0 & 0 & A_\omega \end{bmatrix} \cdot \begin{bmatrix} x_c \\ m_L \\ \omega_{ref} \end{bmatrix} + \begin{bmatrix} B_c \\ 0 \\ 0 \end{bmatrix} m_e \tag{11}$$

The dynamics of the reference value is described by the second order term:

$$G = \frac{\omega_o^2}{s^2 + 2\varsigma\omega_o s + \omega_o^2} \tag{12}$$

where ω_o is a reference frequency and the ς is the damping coefficient of the reference model.

The task of the MPC controller is to bring the output variables to zero by manipulating m_{er} while respecting the safety and physical limitations of the drive system, which in the analysed case are set as follows:

$$-3 \le m_{er} \le 3 \quad -1.5 \le m_s \le 1.5 \tag{13}$$

The selection of the prediction and control horizons is a compromise between the drive performance and computational complexity. In practice, $N_c \le N_p$ to avoid large computational burden for the standard MPC and large number of regions for the explicit MPC.

The dynamic of the control system with MPC controller can be adjusted by the changes of the values of the **Q** matrix. In the current work only the elements located in the main diagonal of the matrix have been changed. The form of the matrix **Q** used in the study is presented below:

$$\mathbf{Q} = \begin{bmatrix} q_{11} & 0 & 0 \\ 0 & q_{22} & 0 \\ 0 & 0 & q_{33} \end{bmatrix} \tag{14}$$

Taking into account (10) and (14) the cost function can be presented as follows:

$$J = \sum_{p=1}^{N} \left(q_{11}\left(\omega_1(p) - \omega_{ref}(p)\right)^2 + q_{22}\left(\omega_2(p) - \omega_{ref}(p)\right)^2 + q_{33}\left(m_s(p) - m_L(p)\right)^2 \right) + \sum_{k=0}^{N_u-1} \left(r \cdot m_e(k)^2 \right) \tag{15}$$

The robustness of the MPC algorithm is ensured by the suitable selection of the elements of matrix in (14) with the help of the pattern search algorithm. The cost function used in the optimization algorithm is as follows:

$$F = \min_{q11;\, q22;\, q33} \left[\prod_{i=1}^{3} \left(f_i\left(e_1, e_2, e_3, K_1, K_2\right) \right) \cdot f_4\left(e_4\right) \right] \tag{16}$$

where: e_1- tracking terror of motor speed ω_1, e_2- tracking terror of load speed ω_2, K_1 – penalty coefficient for exceeds of the limit of the shaft torque, K_2 - penalty coefficient for overshoot in the load speed, e_3 – coefficient in the cost function responsible for minimization of the tracking error of the speed for the systems with a different value of the parameter T_2. The terms of the (16) can be represented as in (17):

$$\begin{aligned}
f_1^{T_2=T_{2N}} &= \left\{ \sum_1^n \left(\left| \omega_{ref} - \omega_1 \right| \right) + \sum_1^n \left(\left| \omega_{ref} - \omega_2 \right| \right) + K_1 + K_2 \right\} \\
f_1^{T_2=0.5 \cdot T_{2N}} &= \left\{ \sum_1^n \left(\left| \omega_{ref} - \omega_1 \right| \right) + \sum_1^n \left(\left| \omega_{ref} - \omega_2 \right| \right) + K_1 + K_2 \right\} \\
f_1^{T_2=2 \cdot T_{2N}} &= \left\{ \sum_1^n \left(\left| \omega_{ref} - \omega_1 \right| \right) + \sum_1^n \left(\left| \omega_{ref} - \omega_2 \right| \right) + K_1 + K_2 \right\} \\
f_4 &= \left\{ \sum_1^n \left(\left| \omega_2^{T_2=T_{2N}} - \omega_2^{T_2=0.5T_{2N}} \right| \right) + \sum_1^n \left(\left| \omega_2^{T_2=T_{2N}} - \omega_2^{T_2=2T_{2N}} \right| \right) \right\}
\end{aligned} \tag{17}$$

Penalty coefficient K_1 and K_2 can be expressed as (18):

$$K_1 = \begin{cases} 0 \Rightarrow \; for \; m_s \le m_s^{\max} \\ k_1 \cdot \left(m_s - m_s^{\max} \right) \Rightarrow \; for \; m_s > m_s^{\max} \end{cases}$$
$$K_2 = \begin{cases} 0 \Rightarrow \; for \; \omega_2 \le \omega_2^{ref} \\ k_2 \cdot \left(\omega_2 - \omega_2^{ref} \right) \Rightarrow \; for \; \omega_2 > \omega_2^{ref} \end{cases} \tag{18}$$

The responses of the reference model used under simulation study are shown in Fig. 4.

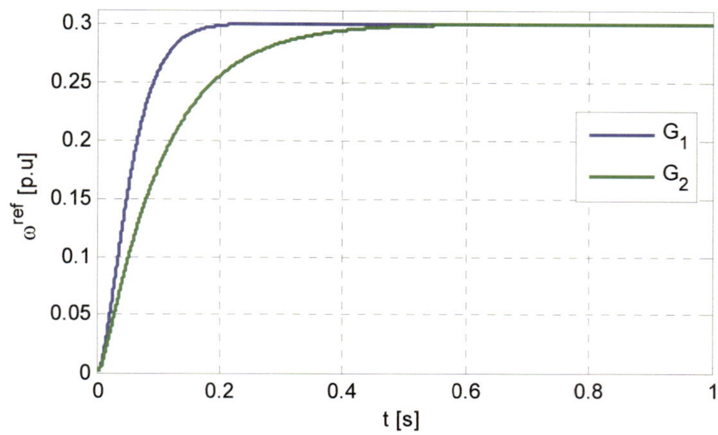

Fig. 4. The responses of the reference models used in the study

As can be concluded from Fig. 4, the settling times for the utilized reference models are 0.2 and 0.4s. These transients determine responses of the system. The pattern search algorithm is looking a one set of values in the matrix **Q** which enables the smallest difference between the speeds and the reference value for different value of the time constant of the load machine. The optimization algorithm has been working with the set value of the reference signal equal to 0.25 of the nominal speed in order to avoid the electromagnetic torque limit.
Transients of the state variables of the system working with the MPC algorithm for slower reference model are presented in Fig.5 . The parameters of the MPC controller are as follows: N=12, N_u=2, numbers of the regions: 121, while values in the matrix are diag(**Q**) = [8.89 0.15 198.2]. The value of cost function in pattern search algorithm is F=1.22e-7.
As can be concluded from the transients presented in Fig. 5, the system is working correctly. The load speed transients for different value of the load side inertia are close to the reference signal. The difference between the motor speed and the reference signal is slightly bigger than - between the reference and the load speed (which comes from the small value of the q_{11}). The application of the load torque causes the speed drop which is eliminated quickly. Those drop is bigger for the system with the smaller value of the load inertia significantly. During this disturbance the electromagnetic torque as well as the shaft torque reach the maximal allowed value for those states. In Fig. 5e the enlarged transients of the load speed errors are presented. It is clearly visible that during the start-up the drive with the biggest inertia has the biggest error.

In the work, a system with increased length of the control horizon has been investigated also. The increase of N_u from 2 to 3 allows to reduce the value of the cost function to 6.78e-8. However, at the same time the number of the controller regions goes up to 381. Due to the large computational complexity (significant in the experiment) the result related to this controller are not presented.

Next the system with faster reference model has been tested. After the optimization procedure the following values of the matrix **Q** were set: diag(**Q**)= [17.22 0.40 398.15]. The transients of the tested system are presented in Fig.6.

The drive systems with different inertia ratio have correct properties. The load speed transients cover the reference value almost perfectly. A much bigger difference exists in the transients of the motor speed. It comes from the small value of q_{11}, as in the previous case. The torsional vibrations are not evident in the system transients. The biggest value of the electromagnetic as well as the shaft torque characterise the system with the biggest inertia. The application of the load torque causes the speed drop but the reaction of the system to the disturbance is very dynamic. The electromagnetic torque reaches its allowed limit (Fig. 6a) in a short while.

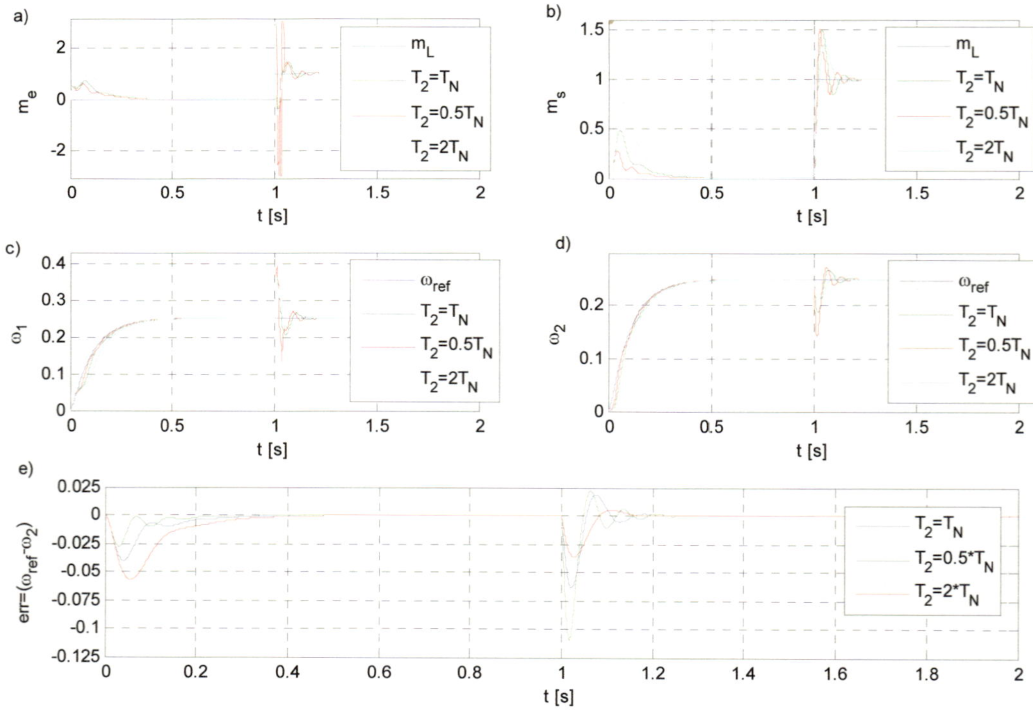

Fig. 5. Transients of the drive system: electromagnetic torque (a), shaft torque (b), motor speed (c), load speed (d), load speed errors (e) for the system with slower reference model and the controller parameters N=12, N_u=2

Finally, the system has been investigated for a bigger value of the reference speed and slower reference model. The values of the parameters of the matrix **Q** remain unchanged

(the same as for the ω_{ref}=0.25). The transients of the state variables of the system are presented in Fig. 7.
As can be seen from Fig. 7, the increase of the value of the reference speed changes the working point of the drive. During the start-up the electromagnetic torque is limited in all cases. Because of this limitation the speeds of the drive cannot follow the reference value. The bigger error appears in the system with T_2=2*T_2 due to the biggest inertia value of the entire drive system. What is evident from the shaft torque transients is that its limitations are in general prevented. Some small exceeds, which come from the applied softening strategy, are visible. However, they are eliminated fast.

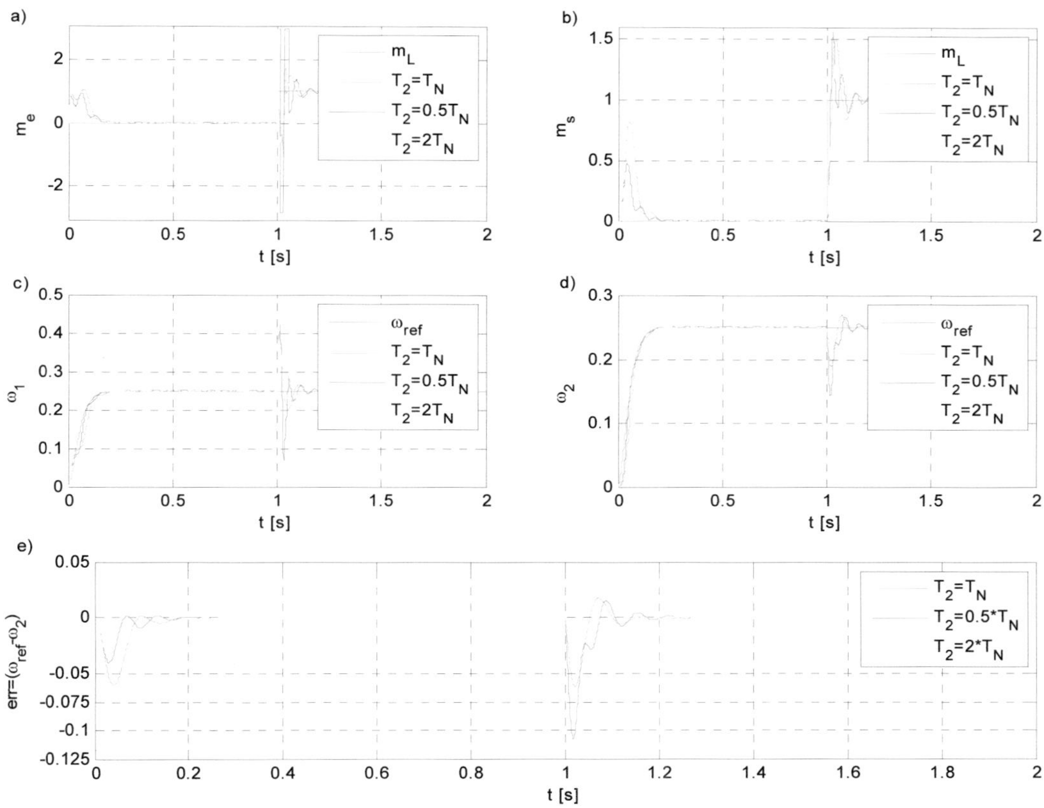

Fig. 6. Transients of the drive system: electromagnetic torque (a), shaft torque (b), motor speed (c), load speed (d), load speed errors (e) for the system with faster reference model and the controller parameters N=12, N_u=2

The pattern search algorithm is not robust against local minimum. In order to eliminate this drawback the starting point of the algorithm has been selected many times. The best three solutions obtained for the three different starting points v_1=[100 100 100]; v_2=[10 10 10]; v_3=[0.1 0.1 0.1] are presented in Tab. 1.
Despite the fact that the value of the cost function is similar for three starting values, the finding points are different. However, the ratios between find values are similar in every

case. It confirms that the find solution is the global solution or very close to it. The parameters of the optimization procedure are presented in Fig. 8.

V	Iteration	F	Finding points		
v_1	57	7.8784e-08	9188	184.0429	206948
v_2	51	7.9072e-08	930	24.1264	21898
v_3	60	8.7297e-08	17.2250	0.4017	398.10

Table 1. Parameters related to the pattern search algorithm

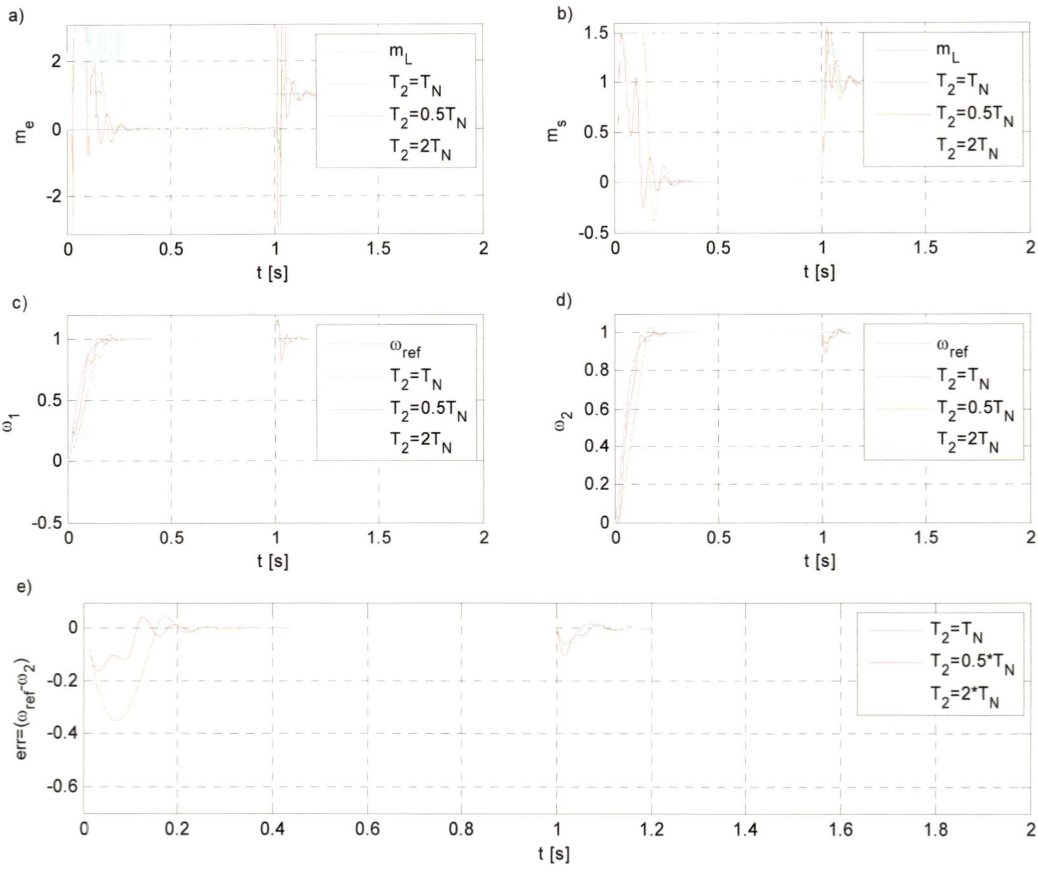

Fig. 7. Transients of the drive system: electromagnetic torque (a), shaft torque (b), motor speed (c), load speed (d), load speed errors (e) for the system with faster reference model and the controller parameters N=12, N_u=2 and nominal value of the reference speed

6. Experimental results

All theoretical considerations have been confirmed experimentally in a laboratory set-up composed of a 0.5kW DC-motor driven by a static converter. The motor is coupled to a load machine by an elastic shaft (a steel shaft of 5mm diameter and 600mm length). The speed

and position of the driven and loading motors have been measured by incremental encoders (36000 pulses per rotation). The mechanical system has a natural frequency of approximately 9.5Hz, while the nominal parameters of the system are T_1=203ms, T_2=203ms, T_c=2.6ms. The picture of the experimental set-up is presented in Fig. 9.
The control structure of the drive is shown in Fig. 3. The sampling time of the electromagnetic torque control as well as the estimator is 100μs in the experimental system. The outer speed control loop has the sampling time equal to 500 μs.

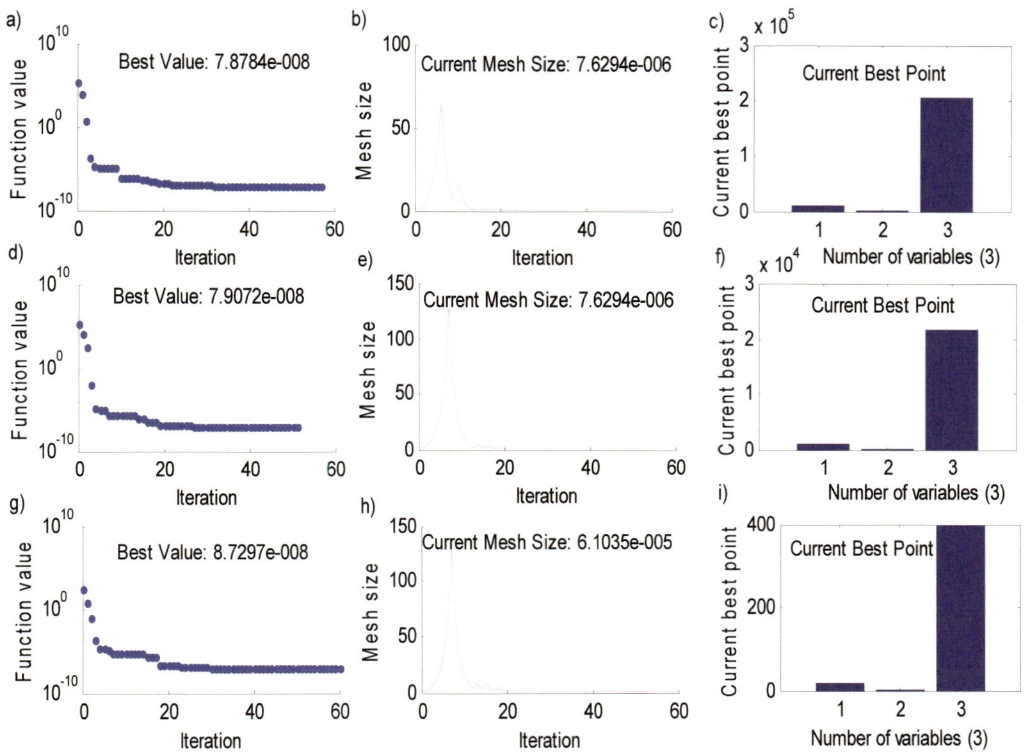

Fig. 8. The parameters of the pattern search algorithm: variation of the cost function (a,d,g), variation of the grid (b,e,h), found values of the parameters (c,f,i) for starting point v1 (a,b,c), v2 (d,e,f), v3 (g,h,i)

In the experimental study the system with a slower reference model has been tested. In Fig. 10 the transients of the motor speeds (a), shaft torques (b), load speed (c) as well as the tracking errors (c) for the drive system with nominal and twice bigger value of the load side inertia are presented. The system was tested for two inertia values of the loading machine.
As can be concluded from the presented transients the drive system works correctly. The shape of the load speeds obtained for different inertia ratio almost perfectly covers the transients of the reference model (Fig. 10). Also the tracking errors between the motor speeds and the reference model are very small. Similarly as in the simulation study, the tracking error during the start-up is bigger for the system with a bigger value of inertia (Fig. 10d). Contrary to this situation, the application of the load torque causes the bigger tracking

error in the system with smaller inertia. The transients of the shaft torque are presented in Fig. 10b. There are no limit exceeds in the shaft torque.

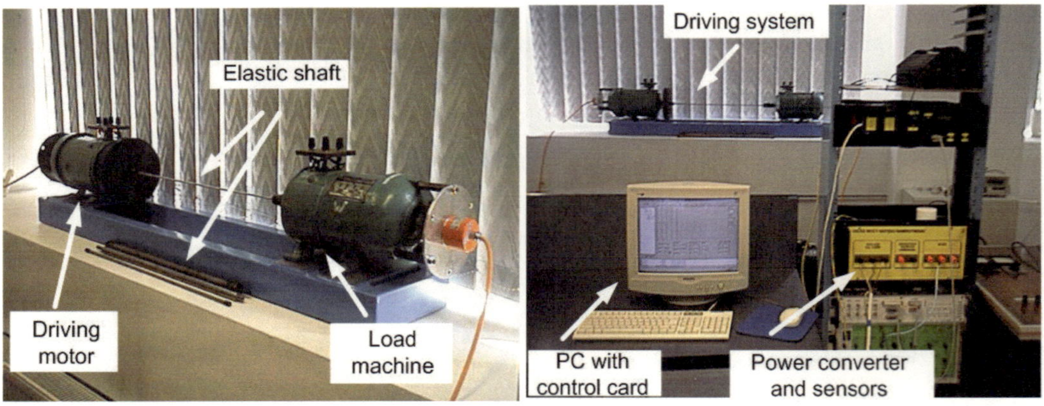

a) b)

Fig. 9. The mechanical part of the laboratory set-up (a) and the general view of the laboratory set-up (b)

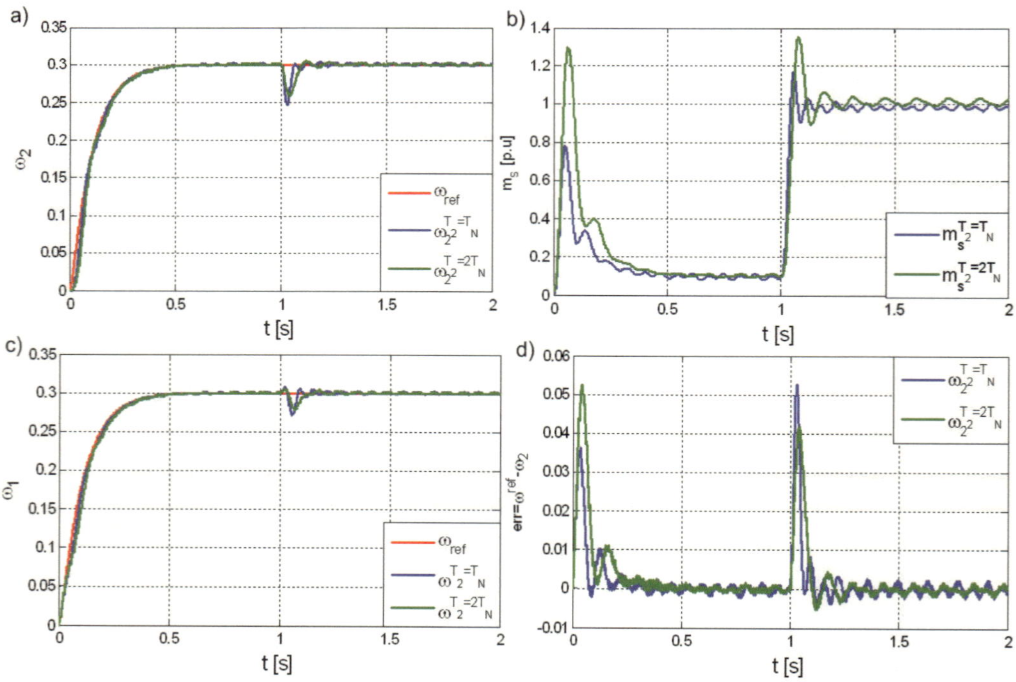

Fig. 10. Transients of the drive system: electromagnetic torque (a), shaft torque (b), motor speed (c), load speed (d), load speed errors (e) for the system with slower reference model and value of the reference speed ω_{ref}=0.25.

After the experiment presented above the system has been examined for nominal value of the reference speed. The drive transients are presented in Fig. 11. As in the previous case, the drive system is working in a stable way. For nominal parameters the motor and the load speeds follow the reference value without noticeable errors (Fig. 11a, c). It steams from the fact that the shaft torque reaches its maximal limits only for a short time (Fig. 11b). During this time the tracking error increases. Then, when the system is below the limit the tracking error goes to zero. The transients of speeds for the system with a bigger value of inertia do not follow the reference value during the start-up because of the limitation of the electromagnetic and shaft torque set in the system. Enlarging the value of these limits will allow to follow the reference system without the error. However, at the same time the mechanical stress could damage the whole drive system.

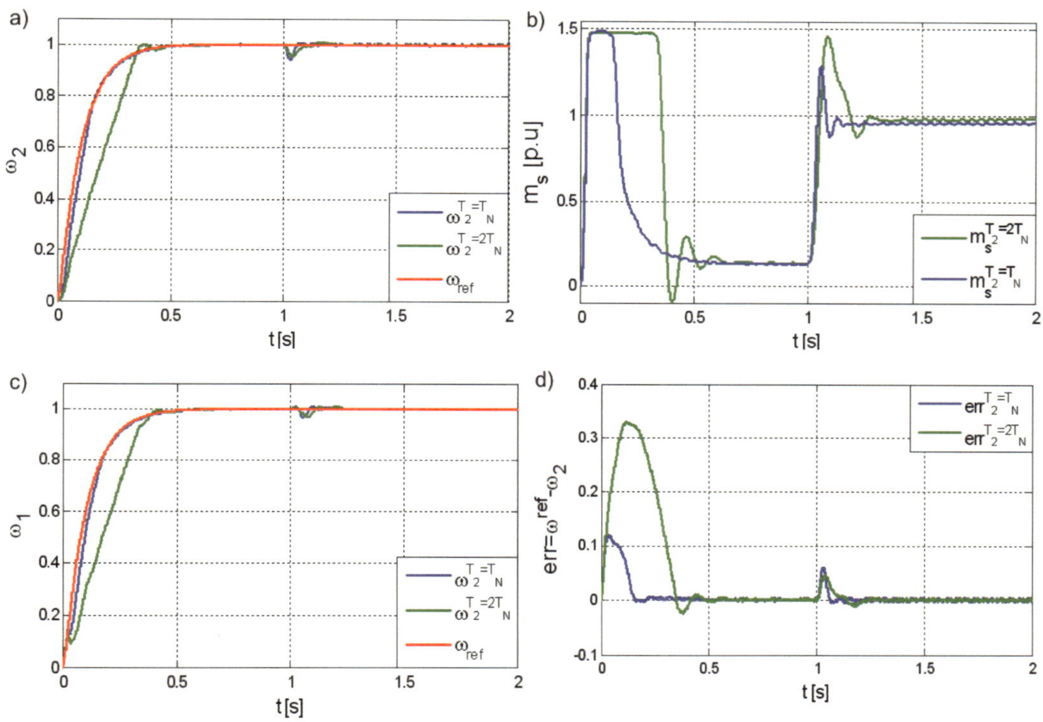

Fig. 11. Transients of the drive system: electromagnetic torque (a), shaft torque (b), motor speed (c), load speed (d), load speed errors (e) for the system with slower reference model and value of the reference speed w_r=0.25.

7. Conclusion

In order to damp the torsional vibrations, which could destroy the mechanical coupling between the driven and loading machine, the control structure with MPC is applied. The coefficients used in MPC are set using the optimization method in order to make the system robust against the changes of the load side inertia. The constraints of the electromagnetic and shaft torques are included during the design of the control algorithm.

As can be concluded from the presented results, the drive system works correctly despite parameter variations. The set control constraints of the shaft torque are not validated. It means that the control structure based on the MPC can ensure safe work in a drive system with uncertain or changeable parameters.
The future work will be devoted to designing of an adaptive MPC control. A part of its work will be the design methodology of a robust Kalman filter used to estimate the mechanical parameters of the drive.

8. Acknowledgements

This research work is supported by the Ministry of Science and Higher Education (Poland) under Grant N N510 352936 (2009-2011)

9. References

Cychowski, M. T. (2009). Robust Model Predictive Control, *VDM Verlag.*

Cychowski, M.T., Szabat, K. & Orlowska-Kowalska T., (2009). Constrained Model Predictive Control of the Drive System with Mechanical Elasticity, *IEEE Trans. Ind. Electronics,* Vol. 56, No. 6, pp 1963-1973.

Erbatur, K., Kaynak, O. & Sabanovic A. (1999). A Study on Robustness Property of Sliding Mode Controllers: A Novel Design and Experimental Investigations, *IEEE Transaction on Industrial Electronics,* Vol. 46, No. 5, pp. 1012-1018.

Erenturk, K. (2008). Nonlinear two-mass system control with sliding-mode and optimised proportional and integral derivative controller combined with a grey estimator, *Control Theory & Applications, IET,* Vol. 2, No. 7, pp. 635 – 642.

Ferretti, G., Magnoni, G. A., Rocco, P., Vigano, L. & Rusconi, A. (2005). On the Use of Torque Sensors in a Space Robotics Application,: *Proc. on the IEEE/RSJ International Conference on Intelligent Robots and Systems IROS 2005,* pp. 1947- 1952, Canada.

Gu D. W., Petkov P. H., Konstantinov M. M. (2005). Robust Control Design with Matlab®, *Springer.*

Hace. A., Jezernik, K. & Sabanovic, A. (2005). Improved Design of VSS Controller for a Linear Belt-Driven Servomechanism, *IEEE/ASME Trans. on Mechatronics,* Vol. 10, No. 4, pp. 385-390.

Hirovonen, M., Pyrhonen, O. & Handroos H. (2006). Adaptive nonlinear velocity controller for a flexible mechanism of a linear motor, Mechatronic, *Elsevier,* Vol. 16, No. 5, pp. 279-290.

Hori, Y., Sawada, H. & Chun, Y. (1999). Slow resonance ratio control for vibration suppression and disturbance rejection in torsional system, *IEEE Trans. on Industrial Electronics,* Vol. 46, No. 1, pp. 162-168.

Itoh D., Iwasaki M., Matsui N. (2004). Optimal Design of Robust Vibration Suppression Controller Using Genetic Algorithms, *IEEE Transaction on Industrial Electronics,* Vol. 51, No. 5, pp. 947-953.

Ji, J. K. & Sul, S. K. (1995). Kalman Filter and LQ Based Speed Controller for Torsional Vibration Suppression in a 2-Mass Motor Drive System, *IEEE Trans. on Industrial Electronics*, Vol. 42, No. 6, pp. 564-571.

Kvasnica, M., Grieder, P., Baotic, M. & Morari, M., (2004) Multi-Parametric Toolbox (MPT), HSCC (Hybrid Systems: Computation and Control), Lecture Notes in Computer Science, Vol. 2993, pp. 448-46.

Kennel, R., Kazmierkowski, M.P., Rodriguez, J. & Cortes, P., (2008). Predictive Control in Power Electronics and Drives, *Tutorial in Int. Symp. on Industrial Electronics*, Cambridge, UK.

Maciejowski J.M. (2002). Predictive Control with Constraints, *Prentice Hall*, UK.

Michels, K., Klawonn, F., Kruse, R. & Nürnberger, A. (2006). Fuzzy Control - Fundamentals, Stability and Design of Fuzzy Controllers, *Springer*.

Orlowska-Kowalska, T. & Szabat, K. (2008). Damping of Torsional Vibrations in Two-Mass System Using Adaptive Sliding Neuro-Fuzzy Approach, *IEEE Transactions on Industrial Informatics*, Vol. 4, No. 1, pp. 47-57.

O'Sullivan, T., Bingham, C. C. & Schofield, N. (2007), Enhanced Servo-Control Performance of Dual-Mass System, *IEEE Trans. on Ind. Electronics*, Vol. 54, No. 3, pp. 1387-1398.

Qiao, R., Zhu, Q. M., Li, S. Y. & Winfield, A. (2002). Torsional Vibration Suppression of a 2-Mass Main Drive System of Rolling Mill with KF Enhanced Pole Placement, *Proc. of the 4th World Congress on Intelligent Control and Automation*, pp. 206-210, China.

Sugiura, K. & Hori, Y. (1996). Vibration Suppression in 2- and 3-Mass System Based on the Feedback of Imperfect Derivative of the Estimated Torsional Torque, *IEEE Trans. on Industrial Electronics,* Vol. 43, No. 2, pp. 56-64.

Suh, G., Hyun, D. S., Park, J. I., Lee, K. D. & Lee, S. G. (2001), Design of a Pole Placement Controller for Reducing Oscillation and Settling Time in a Two-Inertia System, *Proc. of 24th Annual Conference of the IEEE Industrial Electronics Society IECON'01*,pp. 1439-1444, USA.

Szabat, K. & Orłowska-Kowalska, T. (2007). Vibration Suppression in Two-Mass Drive System using PI Speed Controller and Additional Feedbacks - Comparative Study, *IEEE Trans. on Industrial Electronics,* Vol. 54, No. 2, pp.1193-1206.

Szabat, K. & Orlowska-Kowalska, T. (2008). Performance Improvement of Industrial Drives With Mechanical Elasticity Using Nonlinear Adaptive Kalman Filter, *IEEE Transactions on Industrial Electronics*, Vol. 55, No. 3, pp. 1075-1084.

Ryvkin, S., Izosimov, D. & Bayda, S. (2003). Flex mechanical digital control design, *Proceedings of IEEE International Conference on Industrial Technology, IEEE ICIT'03*, Vol.1, pp. 298-302.

Spjøtvold, J., Kerrigan, E.C., Jones, C.N., Tøndel, P., & Johansen, T.A. (2006). On the facet-to-facet property of solutions to convex parametric quadratic programs, *Automatica*, Vol. 42, No. 12, pp. 2209-2214.

Tøndel, P., Johansen, T.A., & Bemporad, A. (2003) An algorithm for multi-parametric quadratic programming and explicit MPC solutions, *Automatica*, Vol. 39, No. 3, pp. 489-497.

Vasak M. & Peric N., (2009), Stability analysis of a patched LQR control system for constrained multi-mass electrical drives, *Przeglad Elektrotechniczny*, Vol. 85, No. 7, pp. 109–114.

Vukosovic, S., N. & Stojic, M. R., (1998). Suppression of Torsional Oscillations in a High-Performance Speed Servo Drive, *IEEE Trans. on Industrial Electronic*, Vol. 45, No. 1, pp. 108-117.

Wang L., Frayman Y. (2002). A Dynamically Generated Fuzzy Neural Network and its Application to Torsional Vibration Control of Tandem Cold Rolling Mill Spindles, *Engineering Applications of Artificial Intelligence*, Vol.15, No. 6, pp. 541-550.

23

Robust Current Controller Considering Position Estimation Error for Position Sensor-less Control of Interior Permanent Magnet Synchronous Motors under High-speed Drives

Masaru Hasegawa and Keiju Matsui
Chubu University University
Japan

1. Introduction

This paper proposes a new current controller for high-speed drives of position sensor-less controlled Interior Permanent Magnet Synchronous Motors (IPMSMs).
Recent demands of motor drive systems are

1. improved efficiency,
2. high torque response,
3. minimum motor size,
4. low cost.

IPMSMs and vector control are widely utilized because of thier efficiency(1) and high torque response(2). In addition, much attention about high speed motors has been attracted from the viewpoint of (3) because the high speed operation makes it possible to achieve smaller motor size for specialized applications such as electric vehicles and home appliances, and so on. High-performance digital control processors cannot be employed to achieve low cost system(4), however, so lower cut-off frequency needs to be achieved because the relatively long control period(– often 500μs to 1ms –) is required. In this situation, the conventional current control system for an IPMSM often degrades and violates stability of the system.
In high-speed drives of AC motors, it has been pointed out that unstable current control tends to occur since coupling terms based on electromotive force impair the characteristics of current control (J.Jung & K.Nam (1999), K.Kondo et al. (1998 (in Japanese)). These papers have proposed a new dynamic decoupling controller, respectively, under the assumption that the controller's coordinate ($\gamma - \delta$) is perfectly aligned with the rotating coordinate fixed to the rotor magnet or rotor flux($d - q$). Hence, it is easily expected that this instability problem tends to be emphasized when position error between these coordinates occurs, which is often visible in the case of position sensor-less control.
In this paper, stability analysis is carried out while considering its application to position sensor-less system (Z.Chen et al. (2003), S.Morimoto et al. (2002), M.Hasegawa & K.Matsui (2008)) , and stable regions are clarified, in which it is especially difficult to control currents on synchronous reference frame at high-speed (K.Tobari et al. (2004 (in Japanese)) . In order

to solve this instability, a simply modified current controller is proposed in this paper. To guarantee both robust stability and current control performance simultaneously, this paper employees two degree of freedom (2DOF) structure fot the current controller, which can enlarge stable region and maintain its performance (Hasegawa et al. (2007)). Finally, some experiments with a disturbance observer for sensor-less control show that the proposed current controller is effective to enlarge high-speed drives for IPMSM sensor-less system.

2. IPMSM model and conventional controller design

IPMSM on the rotational reference coordinate synchronized with the rotor magnet ($d-q$ axis) can be expressed by

$$\begin{bmatrix} v_d \\ v_q \end{bmatrix} = \begin{bmatrix} R+pL_d & -P\omega_{rm}L_q \\ P\omega_{rm}L_d & R+pL_q \end{bmatrix} \begin{bmatrix} i_d \\ i_q \end{bmatrix} + \begin{bmatrix} 0 \\ P\omega_{rm}K_E \end{bmatrix}, \tag{1}$$

in which R means winding resistance, and L_d and $_q$ stand for inductances in d-q axes. ω_{rm} and P express motor speed in mechanical angle and the number of pole pairs, respectively.
In conventional current controller design, the following decoupling controller is usually utilized to independently control d axis current and q axis current:

$$v_d^* = v_d' - P\omega_{rm}L_q i_q , \tag{2}$$

$$v_q^* = v_q' + P\omega_{rm}(L_d i_d + K_E) , \tag{3}$$

where v_d' and v_q' are obtained by amplifying current control errors with proportional - integral controllers to regulate each current to the desired value, as follows:

$$v_d' = \frac{K_{pd}s + K_{id}}{s}(i_d^* - i_d) , \tag{4}$$

$$v_q' = \frac{K_{pq}s + K_{iq}}{s}(i_q^* - i_q) , \tag{5}$$

in which x^* means reference of x. From (1) to (5), feed-back loop for i_d and i_q is constructed, and current controller gains are often selected as follows:

$$K_{pd} = \omega_c L_d , \tag{6}$$

$$K_{id} = \omega_c R , \tag{7}$$

$$K_{pq} = \omega_c L_q , \tag{8}$$

$$K_{iq} = \omega_c R , \tag{9}$$

where ω_c stands for the cut-off frequency for current control. Therefore, the stability of the current control system can be guaranteed, and these PI controllers can play a role in eliminating slow dynamics of current control by cancelling the poles of motor winding $(= -\frac{R}{L_d}, -\frac{R}{L_q})$ by the zero of controllers.
It should be noted, however, that extremely accurate measurement of the rotor position must be assumed to hold this discussion and design because these current controllers are designed and constructed on $d-q$ axis. Hence, the stability of the current control system would easily be violated when the current controller is constructed on $\gamma - \delta$ axis if there exists position error $\Delta\theta_{re}$ (see Fig. 1) due to the delay of position estimation and the parameter mismatches in position sensor-less control system. The following section proves that the instability especially tends to occur in high-speed regions when synchronous motors with large $L_d - L_q$ are employed.

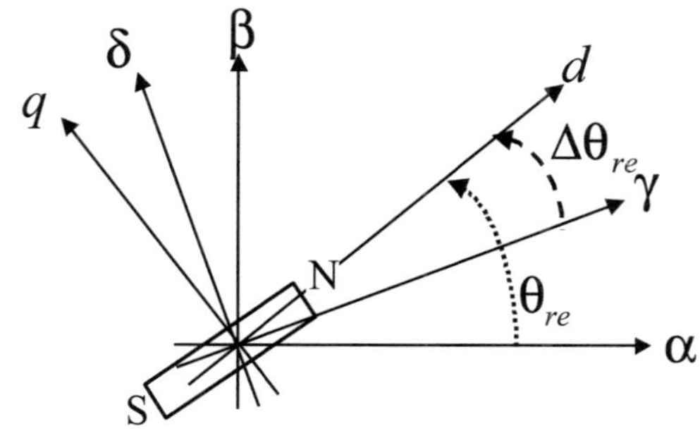

Fig. 1. Coordinates for IPMSMs

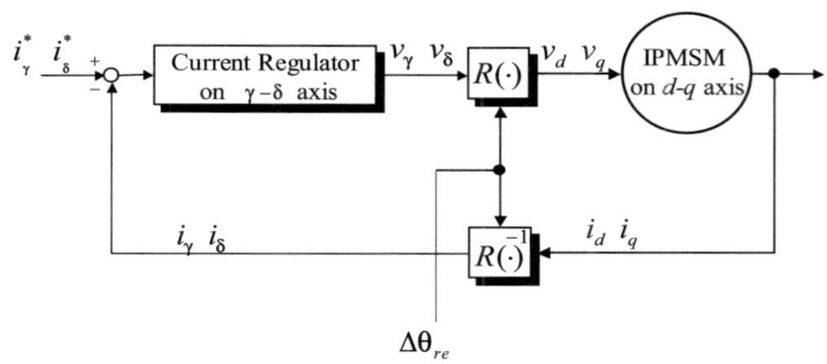

Fig. 2. Control system in consideration of position estimation error

3. Stability analysis of current control system

3.1 Problem Statement

This section analyses stability of current control system while considering its application to position sensor-less system. Let $\gamma - \delta$ axis be defined as a rectangular coordinate away from $d - q$ axis by position error $\Delta\theta_{re}$ shown in Fig.1. This section investigates the stability of the current control loop, which consists of IPMSM and current controller on $\gamma - \delta$ axis as shown in Fig.2.

From (1), IPMSM on $\gamma - \delta$ axis can be rewritten as

$$\begin{bmatrix} v_\gamma \\ v_\delta \end{bmatrix} = \begin{bmatrix} R - P\omega_{rm}L_{\gamma\delta} + L_\gamma p & -P\omega_{rm}L_\delta + L_{\gamma\delta}p \\ P\omega_{rm}L_\gamma + L_{\gamma\delta}p & R + P\omega_{rm}L_{\gamma\delta} + L_\delta p \end{bmatrix} \begin{bmatrix} i_\gamma \\ i_\delta \end{bmatrix} + P\omega_{rm}K_E \begin{bmatrix} -\sin\Delta\theta_{re} \\ \cos\Delta\theta_{re} \end{bmatrix}, \tag{10}$$

in which

$$\begin{aligned} L_\gamma &= L_d - (L_d - L_q)\sin^2\Delta\theta_{re}, \\ L_\delta &= L_q + (L_d - L_q)\sin^2\Delta\theta_{re}, \\ L_{\gamma\delta} &= \frac{L_d - L_q}{2}\sin 2\Delta\theta_{re}. \end{aligned}$$

It should be noted that the equivalent resistances on d axis and q axis are varied as ω_{rm} increases when $L_{\gamma\delta}$ exists, which is caused by $\Delta\theta_{re}$. As a result, $\Delta\theta_{re}$ forces us to modify the current controllers (2) – (5) as follows:

$$v_\gamma^* = v'_\gamma - P\omega_{rm} L_q i_\delta \,, \tag{11}$$

$$v_\delta^* = v'_\delta + P\omega_{rm}(L_d i_\gamma + K_E) \,, \tag{12}$$

$$v'_\gamma = \frac{K_{pd}s + K_{id}}{s}(i_\gamma^* - i_\gamma) \,, \tag{13}$$

$$v'_\delta = \frac{K_{pq}s + K_{iq}}{s}(i_\delta^* - i_\delta) \,. \tag{14}$$

3.2 Closed loop system of current control and stability analysis

This subsection analyses robust stability of the closed loop system of current control. Consider the robust stability of Fig.2 to $\Delta\theta_{re}$. Substituting the decoupling controller (11) and (12) to the model (10) if the PWM inverter to feed the IPMSM can operate perfectly (this means $v_\gamma = v_\gamma^*$, $v_\delta = v_\delta^*$), the following equation can be obtained:

$$\begin{aligned}\begin{bmatrix} v'_\gamma \\ v'_\delta \end{bmatrix} &= \begin{bmatrix} R - P\omega_{rm}L_{\gamma\delta} + L_\gamma p & \Delta Z_{\gamma\delta}(p,\omega_{rm}) \\ \Delta Z_{\delta\gamma}(p,\omega_{rm}) & R + P\omega_{rm}L_{\gamma\delta} + L_\delta p \end{bmatrix} \begin{bmatrix} i_\gamma \\ i_\delta \end{bmatrix} \\ &\quad + P\omega_{rm}K_E \begin{bmatrix} -\sin\Delta\theta_{re} \\ \cos\Delta\theta_{re} - 1 \end{bmatrix} ,\end{aligned} \tag{15}$$

where $\Delta Z_{\gamma\delta}(p,\omega_{rm})$ and $\Delta Z_{\delta\gamma}(p,\omega_{rm})$ are residual terms due to imperfect decoupling control, and are defined as follows:

$$\Delta Z_{\gamma\delta}(p,\omega_{rm}) = -P\omega_{rm}(L_d - L_q)\sin^2\Delta\theta_{re} + L_{\gamma\delta}p \,,$$

$$\Delta Z_{\delta\gamma}(p,\omega_{rm}) = P\omega_{rm}(L_d - L_q)\sin^2\Delta\theta_{re} + L_{\gamma\delta}p \,.$$

It should be noted that the decoupling controller fails to perfectly reject coupled terms because of $\Delta\theta_{re}$. In addition, with current controllers (13) and (14), the closed loop system can be expressed as shown in Fig.3, the transfer function (16) is obtained with the assumption $p\Delta\theta_{re} = 0$, $p\omega_{rm} = 0$ as follows:

$$\begin{bmatrix} i_\gamma \\ i_\delta \end{bmatrix} = \begin{bmatrix} 1 & F_{\gamma\delta}(s) \\ F_{\delta\gamma}(s) & 1 \end{bmatrix}^{-1} \begin{bmatrix} G_\gamma(s)\cdot i_\gamma^* \\ G_\delta(s)\cdot i_\delta^* \end{bmatrix} \tag{16}$$

where

$$F_{\gamma\delta}(s) = \frac{\Delta Z_{\gamma\delta}(s,\omega_{rm})\cdot s}{L_\gamma s^2 + (K_{pd} + R - P\omega_{rm}L_{\gamma\delta})s + K_{id}} ,$$

$$F_{\delta\gamma}(s) = \frac{\Delta Z_{\delta\gamma}(s,\omega_{rm})\cdot s}{L_\delta s^2 + (K_{pq} + R + P\omega_{rm}L_{\gamma\delta})s + K_{iq}} ,$$

$$G_\gamma(s) = \frac{K_{pd}\cdot s + K_{id}}{L_\gamma s^2 + (K_{pd} + R - P\omega_{rm}L_{\gamma\delta})s + K_{id}} ,$$

$$G_\delta(s) = \frac{K_{pq}\cdot s + K_{iq}}{L_\delta s^2 + (K_{pq} + R + P\omega_{rm}L_{\gamma\delta})s + K_{iq}} .$$

Figs.4 and 5 show step responses based on Fig.3 with conventional controller (designed with $\omega_c = 2\pi \times 30$ rad/s) at ω_{rm} =500 min^{-1} and 5000 min^{-1}, respectively. In this simulation, $\Delta\theta_{re}$

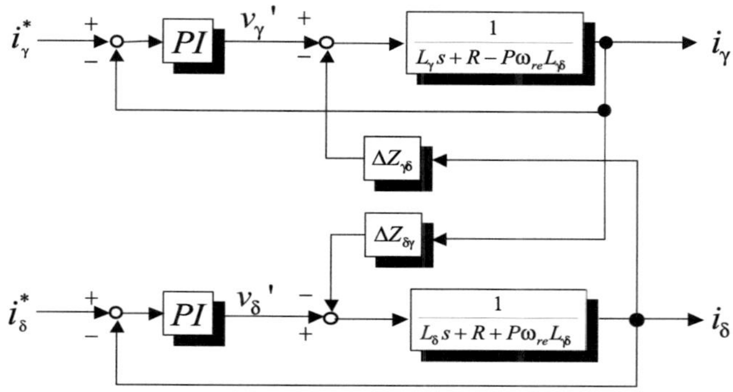

Fig. 3. Closed loop system of current control

Parameters	Value
Rated Power	1.5 kW
Rated Speed	10000 $\min^{-1}$
R	0.061 Ω
L_d	1.44 mH
L_q	2.54mH
K_E	182×10^{-4} V/$\min^{-1}$
P	2 poles

Table 1. Parameters of test IPMSM

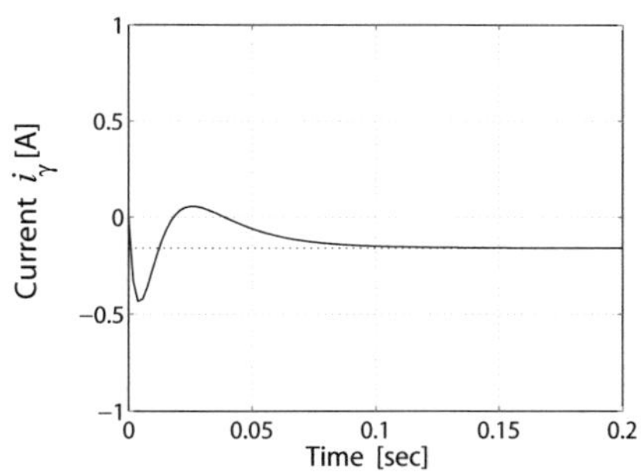

(a) γ axis current response (b) δ axis current response

Fig. 4. Response with the conventional controller ($\omega_{rm} = 500$ $\min^{-1}$)

was intentionally given by $\Delta\theta_{re} = -20°$. i_δ^* was stepwise set to 5 A and i_γ^* was stepwise kept to the value according to maximum torque per current (MTPA) strategy:

$$i_\gamma^* = \frac{K_E}{2(L_q - L_d)} - \sqrt{\frac{K_E^2}{4(L_q - L_d)^2} + \left(i_\delta^*\right)^2}\,. \tag{17}$$

The parameters of IPMSM are shown in Table 1. It can be seen from Fig.4 that each current can be stably regulated to each reference. The results in Fig.5, however, illustrate that each current diverges and fails to be successfully regulated. These results show that the current control system tends to be unstable as the motor speed goes up. In other words, currents diverge and

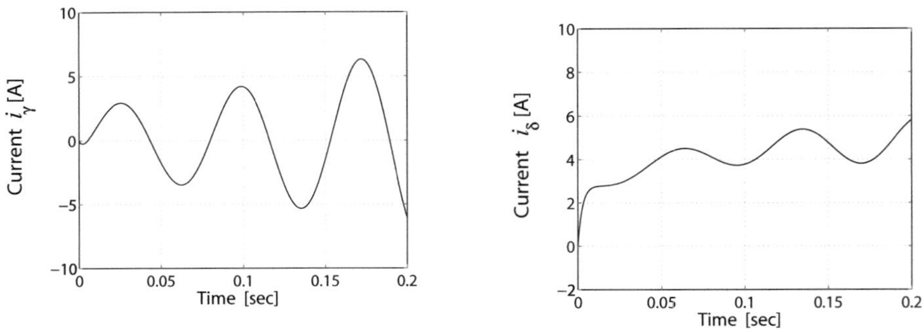

(a) γ axis current response (b) δ axis current response

Fig. 5. Response with conventional controller ($\omega_{rm} = 5000\ \text{min}^{-1}$)

fail to be successfully regulated to each reference in high-speed region because of $\Delta\theta_{re}$, which is often visible in position sensor-less control systems.

Figs.6 and 7 show poles and zero assignment of $G_\gamma(s)$ and $G_\delta(s)$, respectively. It is revealed from Fig.6 that all poles of $G_\gamma(s)$ and $G_\delta(s)$ are in the left half plane, which means the current control loop can be stabilized, and this analysis is consistent with simulation results as previously shown. It should be noted, however, the pole by motor winding is not cancelled by controller's zero, since this pole moves due to $\Delta\theta_{re}$. On the contrary, Fig.7 shows that poles are not in stable region. Hence stability of the current control system is violated, as demonstrated in the aforementioned simulation. This is why one onf the equivalent resistances observed from $\gamma - \delta$ axis tends to become small as speed goes up, as shown in (10), and poles of current closed loop are reassigned by imperfect decoupling control.

It can be seen from $G_\gamma(s)$ and $G_\delta(s)$ that stability criteria are given by

$$K_{pd} + R - P\omega_{rm}L_{\gamma\delta} > 0\,, \tag{18}$$

$$K_{pq} + R + P\omega_{rm}L_{\gamma\delta} > 0\,. \tag{19}$$

Fig.8 shows stable region by conventional current controller, which is plotted according to (18) and (19). The figure shows that stable speed region tends to shrink as motor speed increases, even if position error $\Delta\theta_{re}$ is extremely small. It can also be seen that the stability condition on γ axis (18) is more strict than that on δ axis (19) because of $K_{pd} < K_{pq}$, in which these gains are given by (6) and (8), and $L_d < L_q$ in general. To solve this instability problem, all poles of $G_\gamma(s)$ and $G_\delta(s)$ must be reassigned to stable region (left half plane) even if there exists $\Delta\theta_{re}$. This implies that equivalent resistances in $\gamma - \delta$ axis need to be increased.

4. Proposed current controller with 2DOF structure

4.1 Requirements for stable current control under high-speed region

As described previously, the stability of current control is violated by $\Delta\theta_{re}$. This is because one of the equivalent resistances observed on $\gamma - \delta$ axis tends to become too small, and one of the stability criteria (18) and (19) is not satisfied under high-speed region. To enlarge the stable region, the current controller could, theoretically, be designed with higher performance (larger ω_c). This strategy is, however, not consistent with the aim of achieving lower cost as described in section 1, and thus is not a realistic solution in this case. Therefore, this instability cannot be improved upon by the conventional PI current controller.

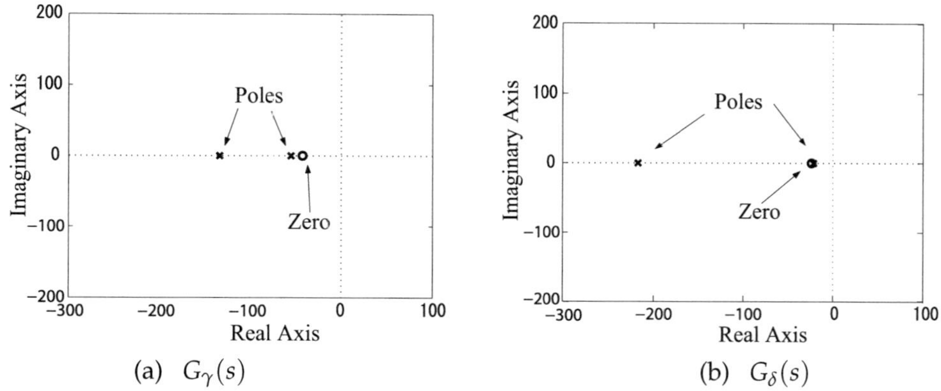

Fig. 6. Poles and zero assignment of $G_\gamma(s)$ and $G_\delta(s)$ at $\omega_{rm} = 500\ \mathrm{min}^{-1}$

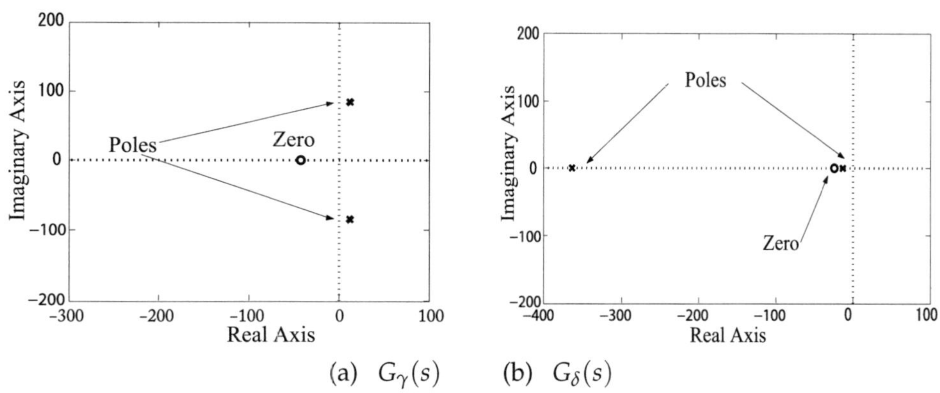

Fig. 7. Poles and zero assignment of $G_\gamma(s)$ and $G_\delta(s)$ at $\omega_{rm} = 5000\ \mathrm{min}^{-1}$

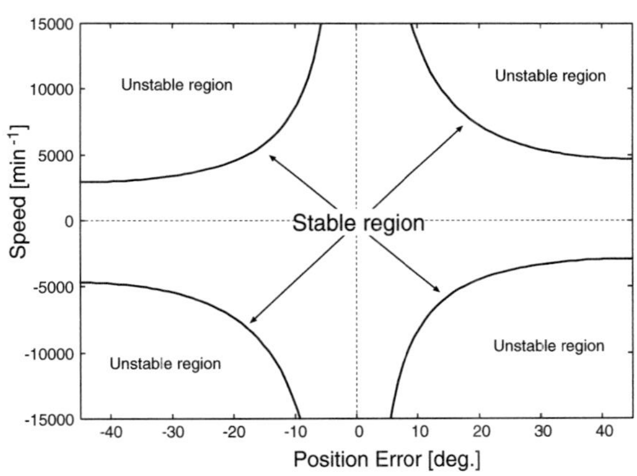

Fig. 8. Stable region by conventional current controller

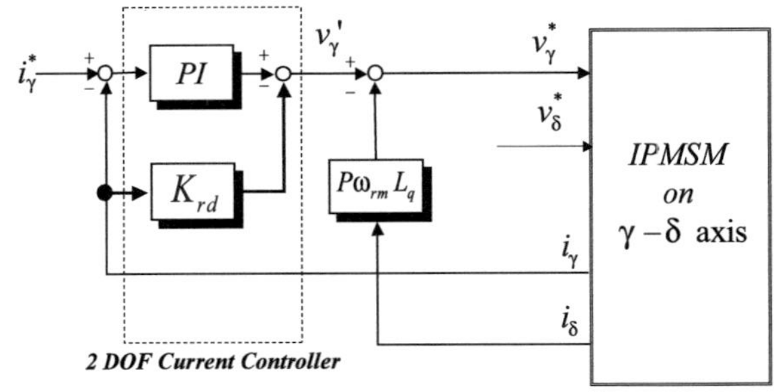

Fig. 9. Proposed current controller with 2DOF structure (only γ axis)

On the other hand, two degree of freedom (2DOF) structure would allow us to simultaneously determine both robust stability and its performance. In this stability improvement problem, robust stability with respect to $\Delta\theta_{re}$ needs to be improved up to high-speed region while maintaining its performance, so that 2DOF structure seems to be consistent with this stability improvement problem of current control for IPMSM drives. From this point of view, this paper employees 2DOF structure in the current controller to enlarge the stability region.

4.2 Proposed current controller

The following equation describes the proposed current controller:

$$v'_\gamma = \frac{K_{pd}s + K_{id}}{s}(i^*_\gamma - i_\gamma) - K_{rd}i_\gamma \,, \tag{20}$$

$$v'_\delta = \frac{K_{pq}s + K_{iq}}{s}(i^*_\delta - i_\delta) - K_{rq}i_\delta \,. \tag{21}$$

Fig. 9 illustrates the block diagram of the proposed current controller with 2DOF structure, where it should be noted that K_{rd} and K_{rq} are just added, compared with the conventional current controller. This current controller consists of conventional decoupling controllers (11) and (12), conventional PI controllers with current control error (13) and (14) and the additional gain on $\gamma - \delta$ axis to enlarge stable region. Hence, this controller seems to be very simple for its implementation.

4.3 Closed loop system using proposed 2DOF controller

Substituting the decoupling controller (11) and (12), and the proposed current controller with 2DOF structure (20) and (21) to the model (10), the following closed loop system can be obtained:

$$\begin{bmatrix} i_\gamma \\ i_\delta \end{bmatrix} = \begin{bmatrix} 1 & F'_{\gamma\delta}(s) \\ F'_{\delta\gamma}(s) & 1 \end{bmatrix}^{-1} \begin{bmatrix} G'_\gamma(s) \cdot i^*_\gamma \\ G'_\delta(s) \cdot i^*_\delta \end{bmatrix}$$

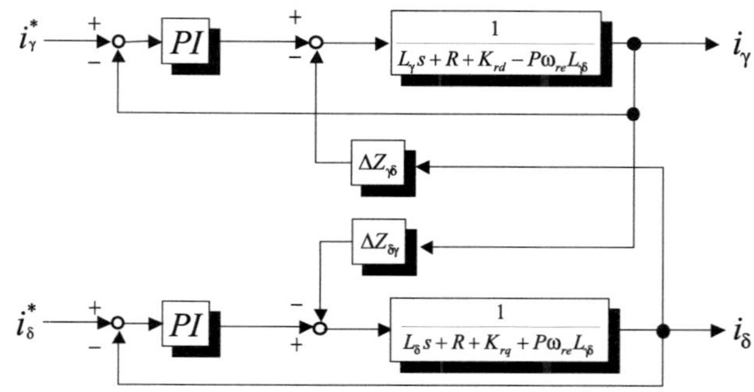

Fig. 10. Current control system with K_{rd} and K_{rq}

where

$$F'_{\gamma\delta}(s) = \frac{\Delta Z_{\gamma\delta}(s,\omega_{rm})\cdot s}{L_\gamma s^2+(K_{pd}+K_{rd}+R-P\omega_{rm}L_{\gamma\delta})s+K_{id}},$$
$$F'_{\delta\gamma}(s) = \frac{\Delta Z_{\delta\gamma}(s,\omega_{rm})\cdot s}{L_\delta s^2+(K_{pq}+L_{rq}+R+P\omega_{rm}L_{\gamma\delta})s+K_{iq}},$$
$$G'_{\gamma}(s) = \frac{K_{pd}\cdot s+K_{id}}{L_\gamma s^2+(K_{pd}+K_{rd}+R-P\omega_{rm}L_{\gamma\delta})s+K_{id}},$$
$$G'_{\delta}(s) = \frac{K_{pq}\cdot s+K_{iq}}{L_\delta s^2+(K_{pq}+K_{rq}+R+P\omega_{rm}L_{\gamma\delta})s+K_{iq}}.$$

From these equations, stability criteria are given by

$$K_{pd}+K_{rd}+R-P\omega_{rm}L_{\gamma\delta} > 0\,, \tag{22}$$
$$K_{pq}+K_{rq}+R+P\omega_{rm}L_{\gamma\delta} > 0\,. \tag{23}$$

The effect of K_{rd} and K_{rq} is described here. It should be noted from stability criteria (22) and (23) that these gains are injected in the same manner as resistance R, so that the current control loop system with K_{rd} and K_{rq} is depicted by Fig.10. This implies that K_{rd} and K_{rq} play a role in virtually increasing the stator resistance of IPMSM. In other words, the poles assigned near imaginary axis ($= -\frac{R}{L_d}, -\frac{R}{L_q}$) are moved to the left ($= -\frac{R+K_{rd}}{L_d}, -\frac{R+K_{rq}}{L_q}$) by proposed current controller, which means that robust current control can be easily realized by designers. In the proposed current controller, PI gains are selected in the same manner as occur in the conventional design:

$$K_{pd} = \omega_c L_d\,, \tag{24}$$
$$K_{id} = \omega_c(R+K_{rd})\,, \tag{25}$$
$$K_{pq} = \omega_c L_q\,, \tag{26}$$
$$K_{iq} = \omega_c(R+K_{rq})\,. \tag{27}$$

This parameter design makes it possible to cancel one of re-assigned poles by zero of PI controller when $\Delta\theta_{re} = 0°$. It should be noted, based this design, that the closed loop dynamics by the proposed controller is identical to that by conventional controller regardless of K_{rd} and K_{rq}:

$$\frac{i_d}{i_d^*} = \frac{i_q}{i_q^*} = \frac{\omega_c}{s+\omega_c}.$$

Therefore, the proposed design can improve robust stability by only proportional gains K_{rd} and K_{rq} while maintaining closed loop dynamics of the current control. This is why the authors have chosen to adopt 2DOF control.

4.4 Design of K_{rd} and K_{rq}, and pole re-assignment results

As previously described, re-assigned poles by proposed controller ($= -\frac{R+K_{rd}}{L_d}, -\frac{R+K_{rq}}{L_q}$) can further be moved to the left in the $s-$plane as larger K_{rd} and K_{rq} are designed. However, employment of lower-performance micro-processor is considered in this paper as described in section 1, and re-assignment of poles by K_{rd} and K_{rq} is restricted to the cut-off frequency of the closed-loop dynamics at most. Hence, K_{rd} and K_{rq} design must satisfy

$$\frac{R+K_{rd}}{L_d} \leq \omega_c, \tag{28}$$

$$\frac{R+K_{rq}}{L_q} \leq \omega_c. \tag{29}$$

As a result, the design of additional gains is proposed as follows:

$$K_{rd} = -R + \omega_c L_d\,, \tag{30}$$

$$K_{rq} = -R + \omega_c L_q\,. \tag{31}$$

Based on this design, characteristics equation of the proposed current closed loop (the denominator of $G'_\gamma(s)$ and $G'_\delta(s)$) is expressed under $\Delta\theta_{re} = 0$ by

$$Ls^2 + 2\omega_c Ls + \omega_c^2 L = 0,$$

where L stands for L_d or L_q. This equation implies that the dual pole assignment at $s = -\omega_c$ is the most desirable solution to improve robust stability with respect to $\Delta\theta_{re}$ under the restriction of ω_c. In other words, this design can guarantee stable poles in the left half plane even if the poles move from the specified assignment due to $\Delta\theta_{re}$.

4.5 Stability analysis using proposed 2DOF controller

Fig.11 shows stable region according to (22) and (23) by proposed current controller designed with $\omega_c = 2\pi \times 30$ rad/s. It should be noted from these results that the stable speed region can successfully be enlarged up to high-speed range compared with conventional current regulator(dashed lines), which is the same in Fig. 8. Point P in this figure stands for operation point at ω_{rm} =5000 min^{-1} and $\Delta\theta_{re} = -20°$. It can be seen from this stability map that operation point P can be stabilized by the proposed current controller with 2DOF structure, despite the fact that the conventional current regulator fails to realize stable control and current diverges, as shown in the previous step response.

Fig.12 demonstrates that stable step response can be realized under ω_{rm} =5000 min^{-1} and $\Delta\theta_{re} = -20°$. These results demonstrate that robust current control can experimentally be realized even if position estimation error $\Delta\theta_{re}$ occurs in position sensor-less control.

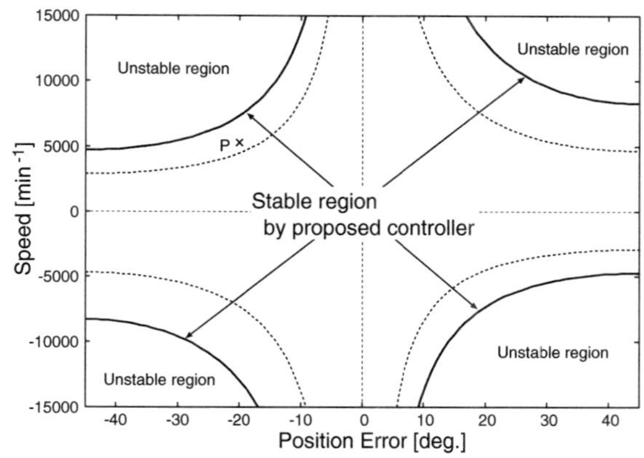

Fig. 11. Stable region by the proposed current regulator with 2DOF structure

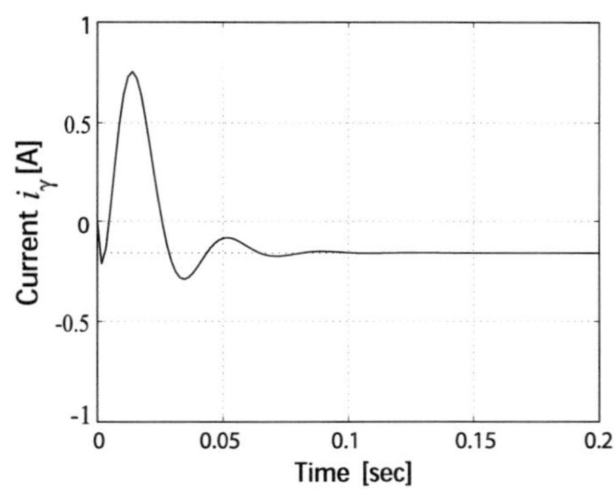

(a) γ axis current response

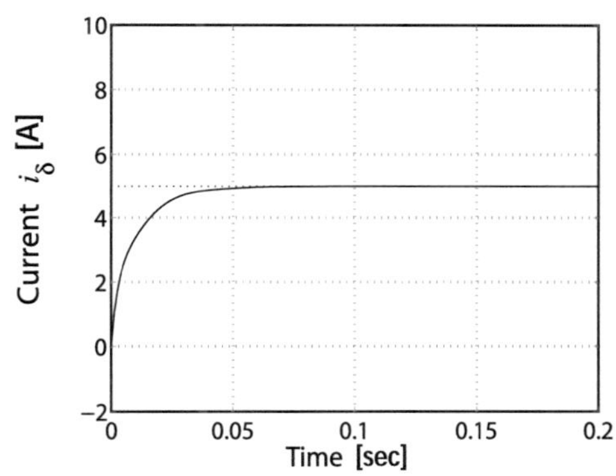

(b) δ axis current response

Fig. 12. Response with proposed controller ($\omega_{rm} = 5000$ min^{-1})

5. Experimental results

5.1 System setup

Experiments were carried out to confirm the effectiveness of the proposed design. The experimental setup shown in Fig.13 consists of a tested IPMSM (1.5 kW) with concentrated winding, a PWM inverter with FPGA and DSP for implementation of vector controller, and position estimator. Also, the induction motor was utilized for load regulation. Parameters of the test IPMSM are shown in Table 1. The speed controller, the current controller, and the coordinate transformer were executed by DSP(TI:TMS320C6701), and the pulse width modulation of the voltage reference was made by FPGA(Altera:EPF10K20TC144-4). The estimation period and the control period were 100 μs, which was set relatively short to experimentally evaluate the analytical results discussed in continuous time domain. The carrier frequency of the PWM inverter was 10 kHz. Also, the motor currents were detected by 14bit ADC. Rotor position was measured by an optical pulse encoder(2048 pulse/rev).

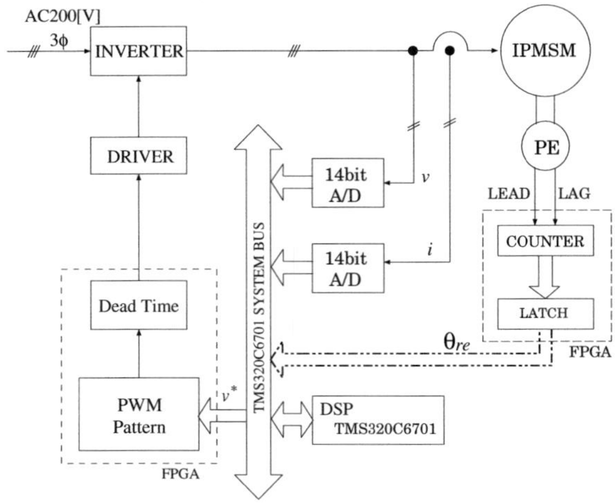

Fig. 13. Configuration of system setup

5.2 Robust stability of current control to rotor position error

The first experiment demonstrates robust stability of the proposed 2DOF controller. In this experiment, the test IPMSM speed was controlled using vector control with position detection in speed regulation mode. The load was kept constant to 75% motoring torque by vector-controlled induction motor. In order to evaluate robustness to rotor position error, $\Delta\theta_{re}$ was intentionally given from 0° to -45° gradually in these experiments.

Figs. 14 and 15 show current control results of the conventional PI controller and the proposed 2DOF controller ($\omega_c = 200\mathrm{rad/s}$) at 4500min^{-1}, respectively. It is obvious from Fig.14 that currents started to be violated at 3.4sec, and they finally were interrupted by PWM inverter due to over-current at 4.2sec. These experimental results showed that $\Delta\theta_{re}$ where currents started to be violated was about -21°, which is consistent with (18) and (19). On the other hand, the proposed 2DOF controller can robustly stabilize current control despite large $\Delta\theta_{re}$ as shown in Fig.15. This result is also consistent with the robust stability analysis discussed in the previous section. Although a current ripple is steadily visible in both experiments, we confirmed that this ripple is primarily the 6th-order component of rotor speed. The tested IPMSM was constructed with concentrated winding, and this 6th-order component cannot be suppressed by lower-performance current controller.

Experimental results at 7000min^{-1} are illustrated in Figs.16 and 17. In the case of conventional controller, current control system became unstable at $\Delta\theta_{re} = -10^\circ$ as shown in Fig.16. Fig.17 shows results of the proposed 2DOF controller, in which currents were also tripped at $\Delta\theta_{re} = -21^\circ$. All $\Delta\theta_{re}$ to show unstable phenomenon is met to (18) and (19), which describes that the robust stability analysis discussed in the previous section is theoretically feasible. This robust stability cannot be improved upon as far as the proposed strategy is applied. In other words, furthermore robust stability improvement necessitates higher cut-off frequency ω_c, which forces us to employ high-performance processor.

5.3 Position sensor-less control

This subsection demonstrates robust stability of current control system when position sensor-less control is applied. As the method for position estimation, the disturbance observer based on the extended electromotive force model (Z.Chen et al. (2003)) was utilized for all experiments. Rotor speed estimation was substituted by differential value of estimated

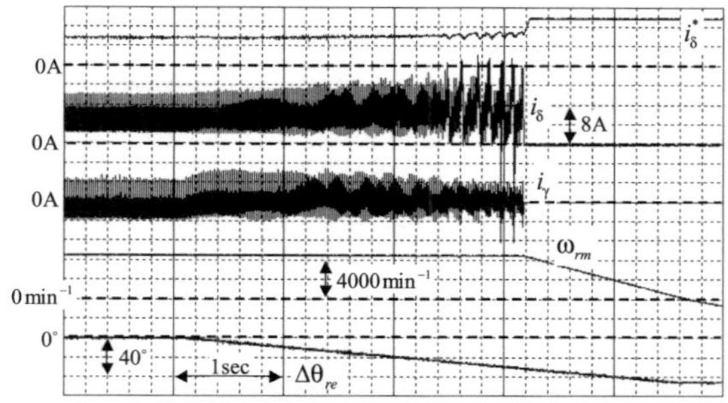

Fig. 14. Current control characteristics by conventional controller at 4500min^{-1}

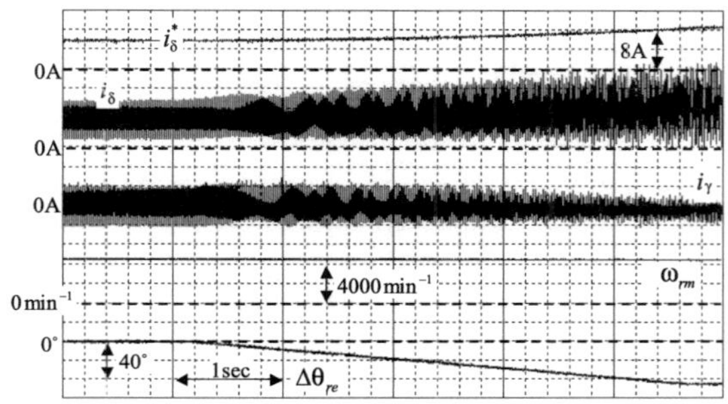

Fig. 15. Current control characteristics by proposed controller at 4500min^{-1}

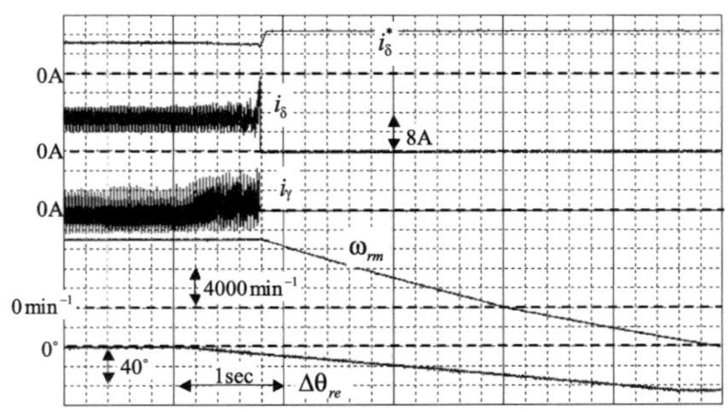

Fig. 16. Current control characteristics by conventional controller at 7000min^{-1}

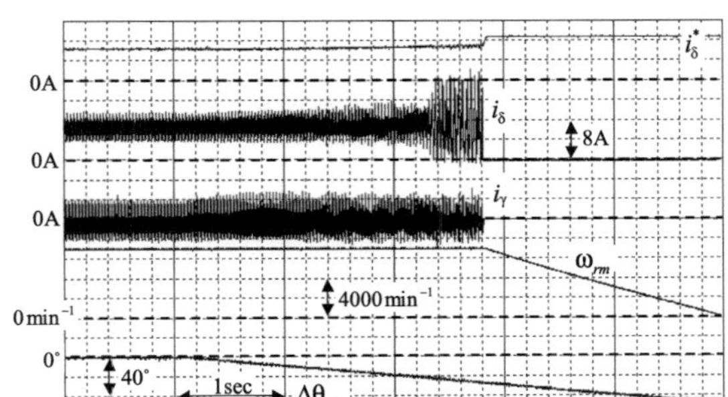

Fig. 17. Current control characteristics by proposed controller at 7000min^{-1}

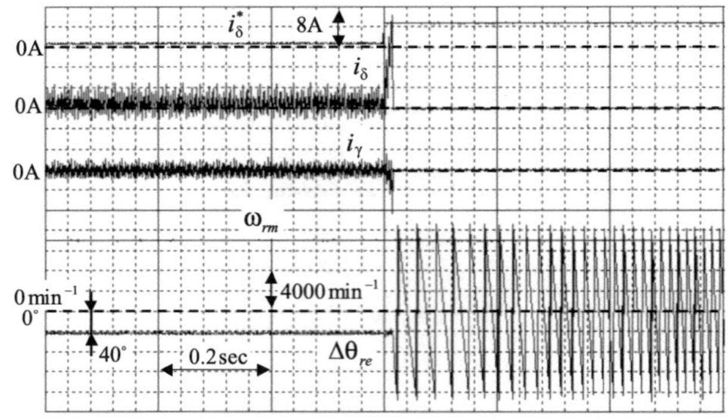

Fig. 18. Current control characteristics by position sensor-less system with conventional controller

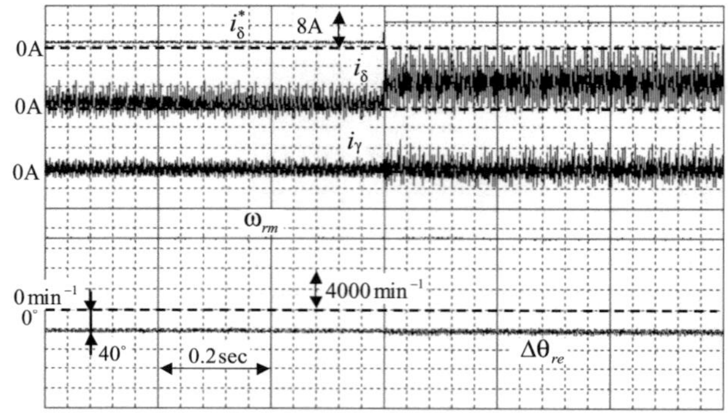

Fig. 19. Current control characteristics by position sensor-less system with proposed controller

rotor position. It should be noted, however, that position estimation delay never fails to occur, especially under high-speed drives, due to the low-pass filter constructed in the disturbance observer. This motivated us to investigate robustness of current control to position estimation delay.

5.3.1 Current step response in position sensor-less control

Figs.18 and 19 show current control results with conventional PI current controller and the proposed controller(designed with $\omega_c = 300\text{rad/s}$), respectively. In these experiments, rotor speed was kept to 7000min^{-1} by the induction motor.

It turns out from Fig.18 that currents showed over-current immediately after current reference i_q^* changed from 1A to 5A, and PWM inverter finally failed to flow the current to the test IPMSM. On the contrary, Fig.19 illustrates that stable current response can be realized even when the current reference is stepwise, which means that the proposed controller is superior to the conventional one in terms of robustness to $\Delta\theta_{re}$.

Also, these figures show that $\Delta\theta_{re}$ of about $-40°$ is steadily caused because of estimation delay in disturbance observer. Needless to say, this error can be compensated since DC component of $\Delta\theta_{re}$ can be obtained in advance according to motor speed and LPF time constant in disturbance observer. $\Delta\theta_{re}$ cannot be compensated, however, at the transient time.

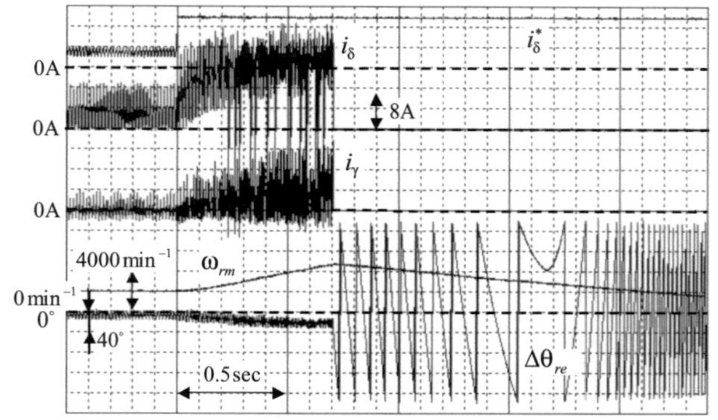

Fig. 20. Speed control characteristics by position sensor-less system with conventional controller

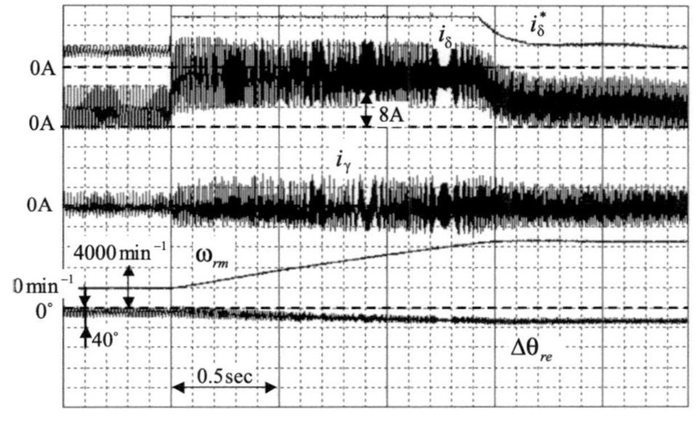

Fig. 21. Speed control characteristics by position sensor-less system with proposed controller

In this study, the authors aimed for robust stability improvement to position estimation error in consideration of transient characteristics such as speed step response and current step response. Hence, $\Delta\theta_{re}$ was not corrected intentionally in these experiments.

5.3.2 Speed step response in position sensor-less control

Figs.20 and 21 show speed step response from $\omega_{rm}^* = 2000\text{min}^{-1}$ to 6500min^{-1} by the conventional PI current controller and proposed controller(designed with $\omega_c = 200\text{rad/s}$), respectively. 20% motoring load was given by the induction motor in these experiments.

It turns out from Fig. 20 that current control begins to oscillate at 0.7sec due to $\Delta\theta_{re}$, and then the amplitude of current oscillation increases as speed goes up. On the other hand, the proposed current controller (Fig. 21) makes it possible to realize stable step response with the assistance of the robust current controller to $\Delta\theta_{re}$.

It should be noted that these experimental results were obtained by the same sensor-less control system except with additional gain and its design of the proposed current controller. Therefore, these sensor-less control results show that robust current controller enables us to improve performances of total control system, and it is important to design robust current controller to $\Delta\theta_{re}$ as well as to realize precise position estimation, which has been surveyed by many researchers over several decades.

6. Conclusions

This paper is summarized as follows:

1. Stability analysis has been carried out while considering its application to position sensor-less system, and operation within stable region by conventional current controller has been analyzed. As a result, this paper has clarified that current control system tends to become unstable as motor speed goes up due to position estimation error.
2. This paper has proposed a new current controller. To guarantee both robust stability and performance of current control simultaneously, two degree of freedom (2DOF) structure has been utilized in the current controller. In addition, a design of proposed controller has also been proposed, that indicated the most robust controller could be realized under the restriction of lower-performance processor, and thus clarifying the limitations of robust performance.
3. Some experiments have shown the feasibility of the proposed current controller with 2DOF structure to realize an enlarged stable region and to maintain its performance.

This paper clarifies that robust current controller enables to improve performances of total control system, and it is important to design robust current controller to $\Delta\theta_{re}$ as well as to realize precise position estimation.

7. References

Hasegawa, M., Y.Mizuno & K.Matsui (2007). Robust current controller for ipmsm high speed sensorless drives, *Proc. of Power Conversion Conference 2007* pp. 1624 –1629.

J.Jung & K.Nam (1999). A Dynamic Decoupling Control Scheme for High-Speed Operation of Induction Motors, *IEEE Trans. on Industrial Electronics* 46(1): 100 – 110.

K.Kondo, K.Matsuoka & Y.Nakazawa (1998 (in Japanese)). A Designing Method in Current Control System of Permanent Magnet Synchronous Motor for Railway Vehicle, *IEEJ Trans. on Industry Applications* 118-D(7/8): 900 – 907.

K.Tobari, T.Endo, Y.Iwaji & Y.Ito (2004 (in Japanese)). Stability Analysis of Cascade Connected Vector Controller for High-Speed PMSM Drives, *Proc. of the 2004 Japan Industry Applications Society Conference* pp. I.171–I.174.

M.Hasegawa & K.Matsui (2008). IPMSM Position Sensorless Drives Using Robust Adaptive Observer on Stationary Reference Frame, *IEEJ Transactions on Electrical and Electronic Engineering* 3(1): 120 – 127.

S.Morimoto, K.Kawamoto, M.Sanada & Y.Takeda (2002). Sensorless Control Strategy for Salient-pole PMSM Based on Extended EMF in Rotating Reference Frame, *IEEE Trans. on on Industry Applications* 38(4): 1054 – 1061.

Z.Chen, M.Tomita, S.Doki & S.Okuma (2003). An Extended Electromotive Force Model for Sensorless Control of Interior Permanent-Magnet Synchronous Motors, *IEEE Trans. on Industrial Electronics* 50(2): 288 – 295.

24

Robust Algorithms Applied for Shunt Power Quality Conditioning Devices

João Marcos Kanieski[1,2,3], Hilton Abílio Gründling[2] and Rafael Cardoso[3]
[1]*Embrasul Electronic Industry*
[2]*Federal University of Santa Maria - UFSM*
[3]*Federal University of Technology - Paraná - UTFPR*
Brazil

1. Introduction

The most common approach to design active power filters and its controllers is to consider the plant to be controlled as the coupling filter of the active power filter. The load dynamics and the line impedances are usually neglected and considered as perturbations in the mathematical model of the plant. Thus, the controller must be able to reject these perturbations and provide an adequate dynamic behavior for the active power filter. However, depending on these perturbations the overall system can present oscillations and even instability. These effects have been reported in literature (Akagi, 1997), (Sangwongwanich & Khositkasame, 1997), (Malesani et al., 1998). The side effects of the oscillations and instability are evident in damages to the bank of capacitors, frequent firing of protections and damage to line isolation, among others (Escobar et al., 2008).

Another problem imposed by the line impedance is the voltage distortion due the circulation of non-sinusoidal current. It degrades the performance of the active power filters due its effects on the control and synchronization systems involved. The synchronization problem under non-sinusoidal voltages can be verified in (Cardoso & Gründling, 2009). The line impedance also interacts with the switch commutations that are responsible for the high frequency voltage ripple at the point of common coupling (PCC) as presented in (Casadei et al., 2000).

Due the effects that line impedance has on the shunt active filters, several authors have been working on its identification or on developing controllers that are able to cope with its side effects. The injection of a small current disturbance is used in (Palethorpe et al., 2000) and (Sumner et al., 2002) to estimate the line impedance. A similar approach, with the aid of Wavelet Tranform is used in (Sumner et al., 2006). Due to line impedance voltage distortion, (George & Agarwal, 2002) proposed a technique based on Lagrange multipliers to optimize the power factor while the harmonic limits are satisfied. A controller designed to reduce the perturbation caused by the mains voltage in the model of the active power filter is introduced in (Valdez et al., 2008). In this case, the line impedances are not identified. The approach is intended to guarantee that the controller is capable to reject the mains perturbation.

Therefore, the line impedances are a concern for the active power filters designers. As shown, some authors choose to measure (estimate or identify) the impedances. Other authors prefer to deal with this problem by using an adequate controller that can cope with this uncertainty or perturbation. In this chapter the authors use the second approach. It is employed a Robust Model Reference Adaptive Controller and a fixed Linear Quadratic Regulator with a new mathematical model which inserts robustness to the system. The new LQR control scheme uses the measurement of the common coupling point voltages to generate all the additional information needed and no disturbance current is used in this technique.

2. Model of the plant

The schematic diagram of the power quality conditioning device, consisting of a DC source of energy and a three-phase/three-legs voltage source PWM inverter, connected in parallel to the utility, is presented in Fig 1.

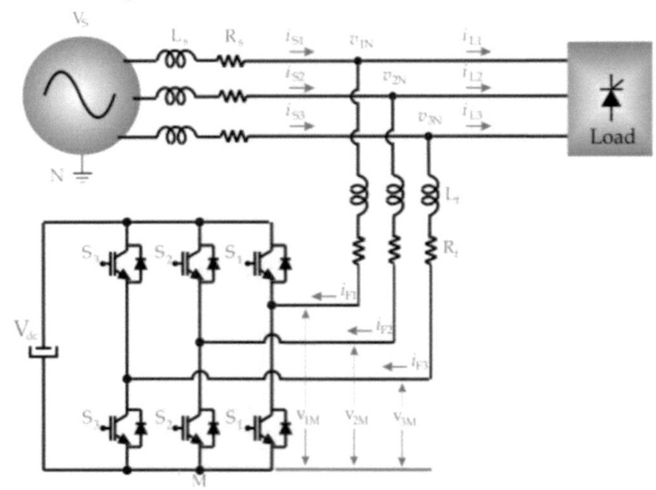

Fig. 1. Schematic diagram of the power quality conditioning device.

The Kirchoff's laws for voltage and current, applied at the PCC, allow us to write the 3 following differential equations in the "123" frame,

$$v_{1N} = L_f \frac{di_{F1}}{dt} + R_f i_{F1} + v_{1M} + v_{MN}, \tag{1}$$

$$v_{2N} = L_f \frac{di_{F2}}{dt} + R_f i_{F2} + v_{2M} + v_{MN}, \tag{2}$$

$$v_{3N} = L_f \frac{di_{F3}}{dt} + R_f i_{F3} + v_{3M} + v_{MN}. \tag{3}$$

The state space variables in the "123" frame have sinusoidal waveforms in steady state. In order to facilitate the control efforts of this system, the model may be transformed to the rotating reference frame "dq". Such frame changing is made by the Park's transformation, given by (4).

$$C_{dqO}^{123} = \frac{2}{3} \begin{bmatrix} \sin(\omega t) & \sin\left(\omega t - \frac{2\pi}{3}\right) & \sin\left(\omega t - \frac{4\pi}{3}\right) \\ \cos(\omega t) & \cos\left(\omega t - \frac{2\pi}{3}\right) & \cos\left(\omega t - \frac{4\pi}{3}\right) \\ \frac{3}{2} & \frac{3}{2} & \frac{3}{2} \end{bmatrix}. \tag{4}$$

The state space variables represented in the 'dq' frame are related to the "123" frame state space variables by equations (5)-(7).

$$\left[v_d\ v_q\ v_O\right]^T = C_{dqO}^{123}\left[v_1\ v_2\ v_3\right]^T, \tag{5}$$

$$\left[i_d\ i_q\ i_O\right]^T = C_{dqO}^{123}\left[i_1\ i_2\ i_3\right]^T, \tag{6}$$

$$\left[d_d\ d_q\ d_O\right]^T = C_{dqO}^{123}\left[d_1\ d_2\ d_3\right]^T. \tag{7}$$

The inverse process is given in equations (8)-(10),

$$\left[v_1\ v_2\ v_3\right]^T = C_{123}^{dqO}\left[v_d\ v_q\ v_O\right]^T, \tag{8}$$

$$\left[i_1\ i_2\ i_3\right]^T = C_{123}^{dqO}\left[i_d\ i_q\ i_O\right]^T, \tag{9}$$

$$\left[d_1\ d_2\ d_3\right]^T = C_{123}^{dqO}\left[d_d\ d_q\ d_O\right]^T, \tag{10}$$

where,

$$C_{123}^{dqO} = {C_{dqO}^{123}}^{-1} = \frac{3}{2}{C_{dqO}^{123}}^{T}, \tag{11}$$

and d is the switching function (Kedjar & Al-Haddad, 2009). As it is a three-phase/three-wire system, the zero component of the rotating frame is always zero, thus the minimum plant model is then given by Eq. (12)

$$\frac{d}{dt}\left[i_{dq}\right] = A\left[i_{dq}\right] + B\left[d_{dq}\right] + E\left[v_{dq}\right], \tag{12}$$

where,

$$A = -\begin{bmatrix} \frac{R_f}{L_f} & -\omega \\ \omega & \frac{R_f}{L_f} \end{bmatrix}, \qquad B = -\begin{bmatrix} \frac{v_{dc}}{L_f} & 0 \\ 0 & \frac{v_{dc}}{L_f} \end{bmatrix},$$

$$E = \begin{bmatrix} \frac{1}{L_f} & 0 \\ 0 & \frac{1}{L_f} \end{bmatrix} \quad and \quad C = \begin{bmatrix} 1 & 0 \\ 0 & 1 \end{bmatrix}.$$

Eq. (12) shows the direct system state variable dependency on the voltages at the PCC, which are presented in the 'dq' frame (v_{dq}). Fig. 2 depicts the plant according to that representation. Based on the block diagram of Fig. 2, it can be seen that the voltages at the PCC have direct influence on the plant output. It suggests that the control designer has also to be careful with those signals, which are frequently disregarded on the project stage.

2.1 Influence of the line impedance on the grid voltages

In power conditioning systems' environment, the line impedance is often an unknown parameter. Moreover, it has a strong impact on the voltages at the PCC, which has its harmonic content more dependent on the load, as the grid impedance increases. Fig. 3 shows the open loop system with a three-phase rectified load connected to the grid through a variable line impedance.

As already mentioned, by increasing the line impedance values, the harmonic content of the voltages at the PCC also increases. Higher harmonic content in the voltages leads to a more

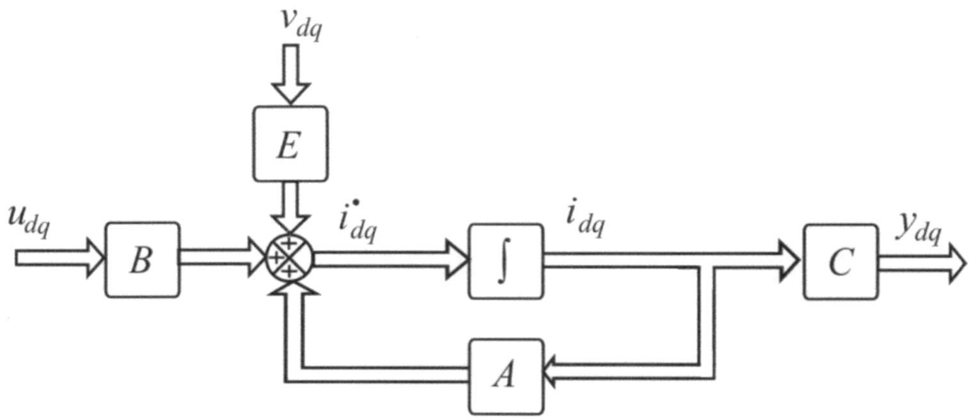

Fig. 2. Block representation of the plant.

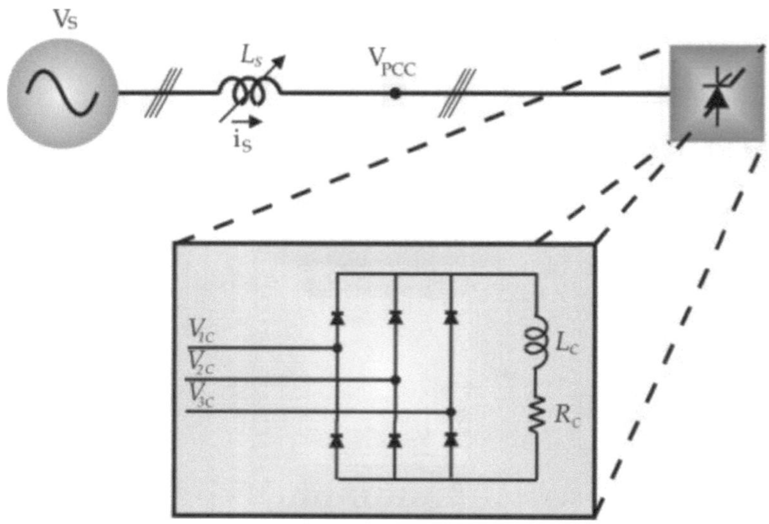

Fig. 3. Open loop system with variable line inductance

distorted waveform. It can be visualized in Fig. 4, that shows the voltage signals v_{123} at the PCC, for a line inductance of $L_S = 2mH$.

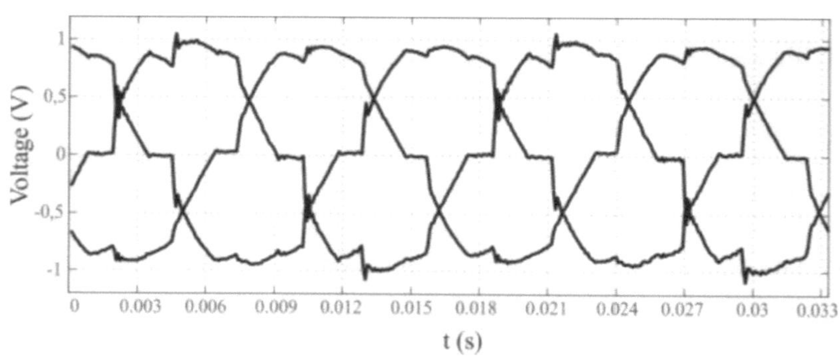

Fig. 4. Open loop voltages at the PCC with line inductance of $L_S = 2mH$.

Fig. 5 shows now an extreme case, with line inductance of $L_S = 5mH$, it is also visually perceptible the significant growth on the voltage harmonic content.

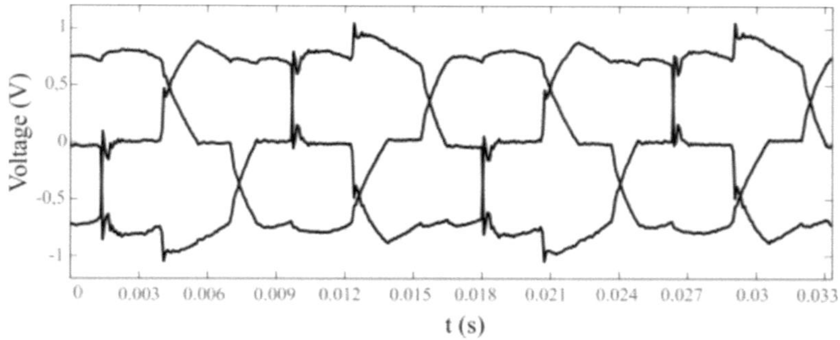

Fig. 5. Open loop voltages at the PCC with line inductance of $L_S = 5mH$.

Concluding, the voltages at the PCC have its dynamic substantially dependent on the line impedance. In other hand, the system dynamic is directly associated with the PCC voltages. Therefore, the control of this kind of system strongly depends on the behavior of the voltages at the PCC. As the output filter of the Voltage Source Inverter (VSI) has generally well-known parameters (they are defined by the designer), which are at most fixed for the linear system operation, one of the greatest control challenges of these plants is associated with the PCC voltages. The text that follows is centered on that point and proposes an adaptive and a fixed robust algorithm in order to control the chosen power conditioner device, even under load unbalance and line with variable or unknown impedance.

3. Robust Model Reference Adaptive Control (RMRAC)

The RMRAC controller has the characteristic of being designed under an incomplete knowledge of the plant. To design such controller it is necessary to obtain a representative mathematical model for the system. The RMRAC considers in its formulation a parametric model with a reduced order modeled part, as well as a multiplicative and an additive term, describing the unmodeled dynamics. The adaptive law is computed for compensating the plant parametric variation and the control strategy is robust to such unmodeled dynamics. In the present application, the uncertainties are due to the variation of the line impedance and load.

3.1 Mathematical model

From the theory presented by Ioannou & Tsakalis (1986) and by Ioannou & Sun (1995), to have an appropriated RMRAC design, the plant should be modeled in the form

$$\begin{aligned} \frac{i_F(s)}{u(s)} &= G(s) = G_0(s)[1 + \mu\Delta_m(s)] + \mu\Delta_a(s) \\ G_0(s) &= k_p \frac{Z_0(s)}{R_0(s)} \end{aligned} \tag{13}$$

where u represents the control input of the system and i_F is the output variable of interest as shown in Fig. 1.

⇒ Assumptions for the Plant

H1.Z_0 is a monic stable polynomial of degree $m(m \leq n-1)$,

H2.R_0 is a monic polynomial of degree n;

H3.The sign of $k_p > 0$ and the values of m, n are known.

For the unmodeled part of the plant it is assumed that:

H4.Δ_m is a stable transfer function;

H5.Δ_a is a stable and strictly proper transfer function;

H6.A lower bound $p_0 > 0$ on the stability margin $p > 0$ for which the poles of $\Delta_m(s-p)$ and $\Delta_a(s-p)$ are stable is known.

3.2 RMRAC strategy

The goal of the model reference adaptive control can be summarized as follows: Given a reference model

$$\frac{y_m}{r} = W_m(s) = k_m \frac{Z_m(s)}{R_m(s)}, \tag{14}$$

it is desired to design an adaptive controller, for $\mu > 0$ and $\mu \in [0, \mu^*)$ where the resultant closed loop system is stable and the plant output tracks, as closer as possible, the model reference output, even under the unmodeled dynamics Δ_m and Δ_a. In (14), r is a uniformly limited signal.

⇒ Assumptions for the model reference:

M1Z_m a monic stable polynomial of degree $m(m \leq n-1)$;

M2R_m is a monic polynomial of degree n.

The plant input is given by

$$u = \frac{\boldsymbol{\theta}^T \boldsymbol{\omega} + c_0 r}{\theta_4} \tag{15}$$

where $\boldsymbol{\theta}^T = \left[\boldsymbol{\theta}_1^T, \boldsymbol{\theta}_2^T, \theta_3\right], \boldsymbol{\omega}^T = \left[\omega_1, \omega_2, y\right] \in \Re^{2n-1}$ and c_0 is the relation between the gain of the open loop system and the gain of the model reference. The input u and the plant output y are used to generate the signals $\omega_1, \omega_2 \in \Re^{n-1}$

$$\omega_1 = \frac{\boldsymbol{\alpha}(s)}{\Lambda(s)} u \quad \text{and} \quad \omega_2 = \frac{\boldsymbol{\alpha}(s)}{\Lambda(s)} y. \tag{16}$$

⇒ Assumptions for the signals ω_1 and ω_2:

R1.The polynomial Λ in (16) is a monic Hurwitz of degree $n-1$, containing stable eigenvalues.

R2.For $n \geq 2$, $\boldsymbol{\alpha} \triangleq \left[s^{n-2}, \ldots, s, 1\right]^T$ and for $n = 1$, $\boldsymbol{\alpha} \triangleq 0$.

For the adaptation of the control action parameters, the following modified gradient algorithm was considered

$$\dot{\boldsymbol{\theta}} = -\sigma \boldsymbol{P} \boldsymbol{\theta} - \frac{\boldsymbol{P} \zeta \varepsilon}{m^2} \tag{17}$$

The σ-modification in 17 is given by

$$\sigma = \begin{cases} 0 & \text{if } \|\boldsymbol{\theta}\| < M_0 \\ \sigma_0 \left(\frac{\|\boldsymbol{\theta}\|}{M_0} - 1 \right) & \text{if } M_0 \leq \|\boldsymbol{\theta}\| < 2M_0 \\ \sigma_0 & \text{if } \|\boldsymbol{\theta}\| > 2M_0 \end{cases} \tag{18}$$

where $\sigma_0 > 0$ is a parameter of design. $\boldsymbol{P} = \boldsymbol{P}^T > 0$, $\varepsilon = y - y_m + \boldsymbol{\theta}^T \zeta - W_m v = \phi^T \zeta + \mu \eta$ and M_0 is an upper limit $\|\boldsymbol{\theta}^*\|$, such that $\|\boldsymbol{\theta}^*\| + \delta_3 \leq M_0$ for a $\delta_3 > 0$. The normalization signal m is given by

$$\dot{m} = -\delta_0 m + \delta_1 \left(|u| + |y| + 1 \right) \tag{19}$$

with $m_0 > \delta_1 / \delta_0$, $\delta_1 \geq 1$ and $\delta_0 > 0$.

The normalization signal m is the parameter which ensures the robustness of the system. Looking to Eq. (15)-(19), it can be seen that when the control action u, the plant output y or both variables are large enough, the $\boldsymbol{\theta}$ parameters decreases and therefore the control action, which depends on the $\boldsymbol{\theta}$ parameters, also has its values reduced, limiting the control action as well as the system output in order to stabilize the system.

3.3 RMRAC applied for the power conditioning device

In the considered power conditioning system, as shows Eq. (12), there is a coupling between the "dq" variables. To facilitate the control strategy, which should consider a multiple input multiple output system (MIMO), it is possible to rewrite Eq. (12) as

$$\begin{aligned} L_f \frac{di_d}{dt} + R_f i_d &= L_f \omega i_q - v_{dc} d_{nd} + v_d \\ L_f \frac{di_q}{dt} + R_f i_q &= -L_f \omega i_d - v_{dc} d_{nq} + v_q \end{aligned} \tag{20}$$

Defining, the equivalent input as in Eq. (21) and (22),

$$u_d = L_f \omega i_q - v_{dc} d_{nd} + v_d \tag{21}$$

and

$$u_q = -L_f \omega i_d - v_{dc} d_{nq} + v_q, \tag{22}$$

the MIMO tracking problem, with coupled dynamics, is transformed in two single input single output (SISO) problems, with decoupled dynamics. Thus, currents i_d and i_q may be controlled independently through the inputs u_d e u_q, respectively. For the presented decoupled plant, the RMRAC controller equations are given by (23) and (24).

$$u_d = \frac{\boldsymbol{\theta}_d^T \boldsymbol{\omega}_d + c_0 r_d}{\theta_{4d}} \tag{23}$$

and

$$u_q = \frac{\boldsymbol{\theta}_q^T \boldsymbol{\omega}_q + c_0 r_q}{\theta_{4q}}. \tag{24}$$

The PWM actions (d_{nd} and d_{nq}), are obtained through Eq. (21) and (22) after computation of (23) and (24).

3.3.1 Design procedure

Before starting the procedure, lets examine the hypothesis ***H1, H2, M1, M2, R1*** and ***R2***. Firstly, as the nominal system, accordingly to Eq. (12), is a first order plant. The degrees n and m are then defined by $n = 1$ and $m = 0$. Therefore, the structure of the model reference and the dynamic of signals ω_1 and ω_2 can be determined. By ***M1*** and ***M2***, the model reference is also a first order transfer function $W_m(s)$, thus

$$W_m(s) = k_m \frac{\omega_m}{s - \omega_m}. \tag{25}$$

Furthermore, from ***R1*** and ***R2***: $\boldsymbol{\alpha} \triangleq 0$; and from Eq. (15), the control law reduces to

$$u_d = \frac{\theta_{3d} i_d + r_d}{\theta_{4d}} \tag{26}$$

and

$$u_q = \frac{\theta_{3q} i_q + r_q}{\theta_{4q}}. \tag{27}$$

From the information of the maximum order harmonic, which has to be compensated by the power conditioning device, it is possible to design the model reference, given in Eq. 25. Choosing, for example, the 35th harmonic, as the last harmonic to be compensated, and $W_m(s)$ with unitary gain, the model reference parameters become $\omega_m = 35 \cdot 2 \cdot \pi \cdot 60 \approx 13195 \frac{rad}{s}$ and $k_m = 1$. Fig. 6 shows the frequency responses of the nominal plant of a power conditioning device, with parameters $L_f = 1mH$ and $R_f = 0.01\Omega$ and of a model reference with the parameters aforementioned.

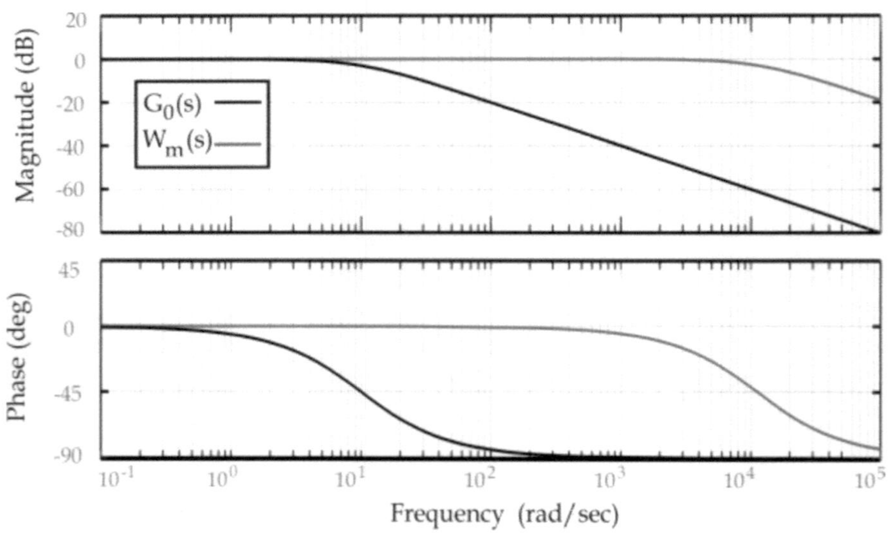

Fig. 6. Bode diagram of $G_0(s)$ and $W_m(s)$.

The vector $\boldsymbol{\theta}$ is obtained by the solution of a Model Reference Controller (MRC) for the modeled part of the plant $G_0(s)$. The design procedure of a MRC is basically to calculate the closed loop system of the nominal plant which has to be equal to the model reference transfer function.

3.3.2 RMRAC results

The RMRAC was applied to the power conditioning device, shown in Fig. 7, to control the compensation currents i_{F123}. Table 1 summarizes the parameters of the system.

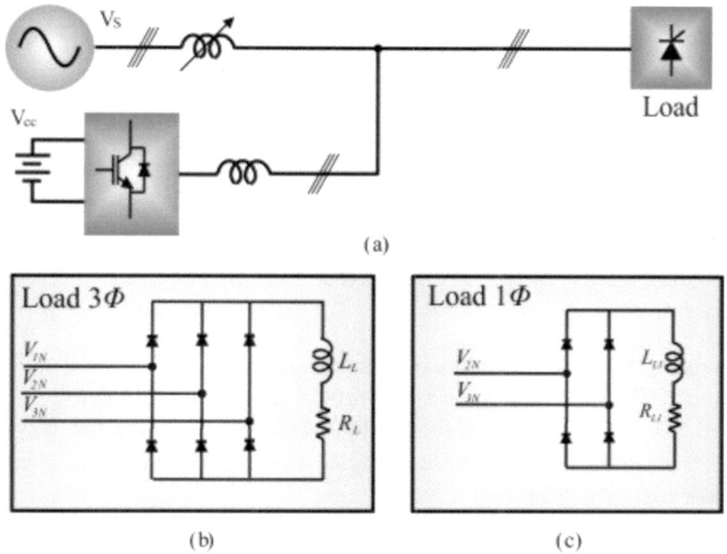

Fig. 7. Block diagram of the system.

Grid Voltage	$380V\ (RMS)$	R_f	0.01Ω
ω	$377rad/s$	L_f	$1mH$
f_s	$12kHz$	R_S	0.01Ω
V_{dc}	$550V$	L_S	$5uH$ / $2mH$ / $5mH$
$\theta_d(0)$	$[-1.02, 0.53]^T$	P	$diag\{0.99, 0.99\}$
$\theta_q(0)$	$[-1.02, 0.53]^T$	k_m	1
c_0	1	ω_m	$13195\frac{rad}{s}$
L_L	$2mH$	L_{L1}	$2mH$
R_L	25Ω	R_{L1}	25Ω

Table 1. Design Parameters

To verify the robustness of the closed loop system, which has to be stable for an appropriated range of line inductance (in the studied case: from $L_S = 5\mu H$ to $L_S = 5mH$), some simulations were carried out considering variations on the line inductance (L_S).

In the first analysis, it was considered a line impedance of $L_S = 5uH$. Fig. 8 (a) shows the load currents as well as the compensated currents, which are provided by the main source. It is also possible to see by Fig. 8 (b) the appropriate reference tracking for the RMRAC controlled system for the case of small line inductance. Fig 8 (b) shows the reference currents in black plotted with the compensation currents in gray.

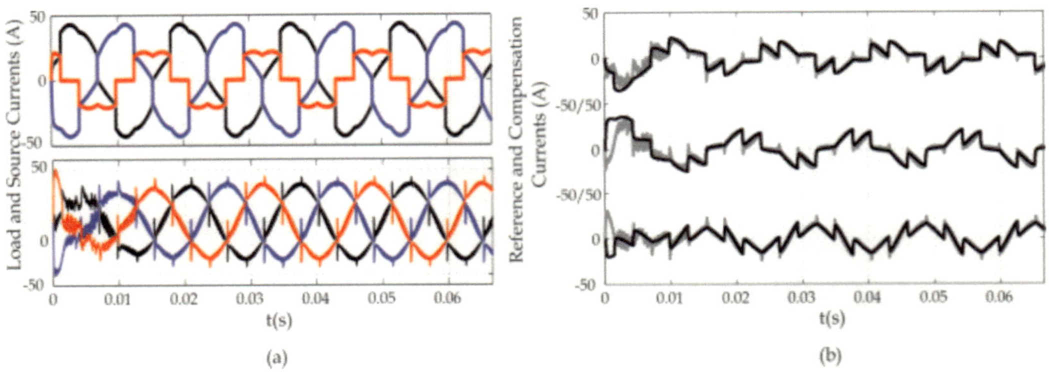

Fig. 8. System currents ($L_S = 5\mu H$): (a) Load and source currents (b) Reference and compensation currents.

In a second analysis, the line impedance was considered $L_S = 2mH$. Fig. 9 (a) shows the load currents and the compensated currents, provided by the main source. Fig. 9 (b) shows the appropriated reference tracking of the RMRAC controlled system for the case of high line inductance. Fig. 9 (b) exhibits the good tracking of the reference currents, shown in black, by the compensation currents, shown in gray.

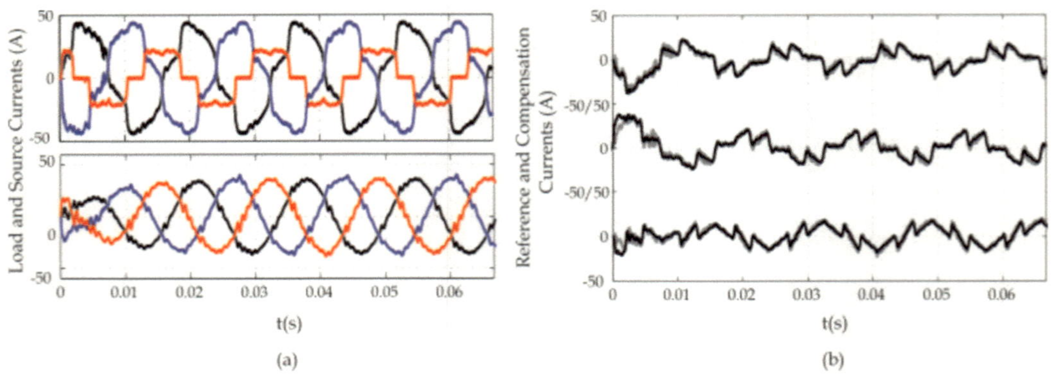

Fig. 9. System currents ($L_S = 2mH$): (a) Load and source currents (b) Reference and compensation currents.

Finally, for an extreme case of line inductance of $L_S = 5mH$, the load currents as well as the compensated currents are shown in Fig. 10 (a). Fig. 10 (b) shows the tracking performance of the RMRAC controlled system under high line impedance circumstances, which remains stable even under high inductance levels. In this figure, the reference is depicted in black and the output current is shown in gray.

Fig. 11 shows the convergence of the $\boldsymbol{\theta_d}$ and $\boldsymbol{\theta_q}$ parameters for each case of line impedance. In black, the parameters θ_{3d}, θ_{3q}, θ_{4d} and θ_{4q} for a line inductance of $L_S = 5\mu H$; in blue for a line inductance of $L_S = 2mH$; and finally, in red, for $L_S = 5mH$. In all cases the controller has an adequate convergence of its parameters.

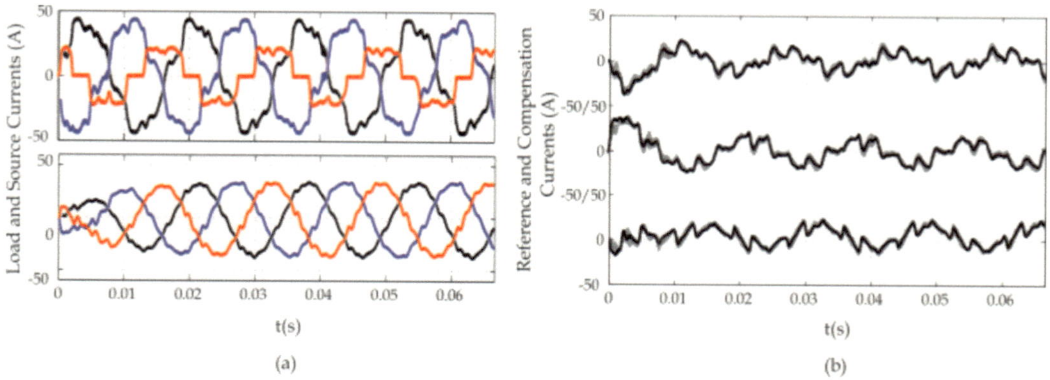

Fig. 10. System currents ($L_S = 5mH$): (a) Load and source currents (b) Reference and compensation currents.

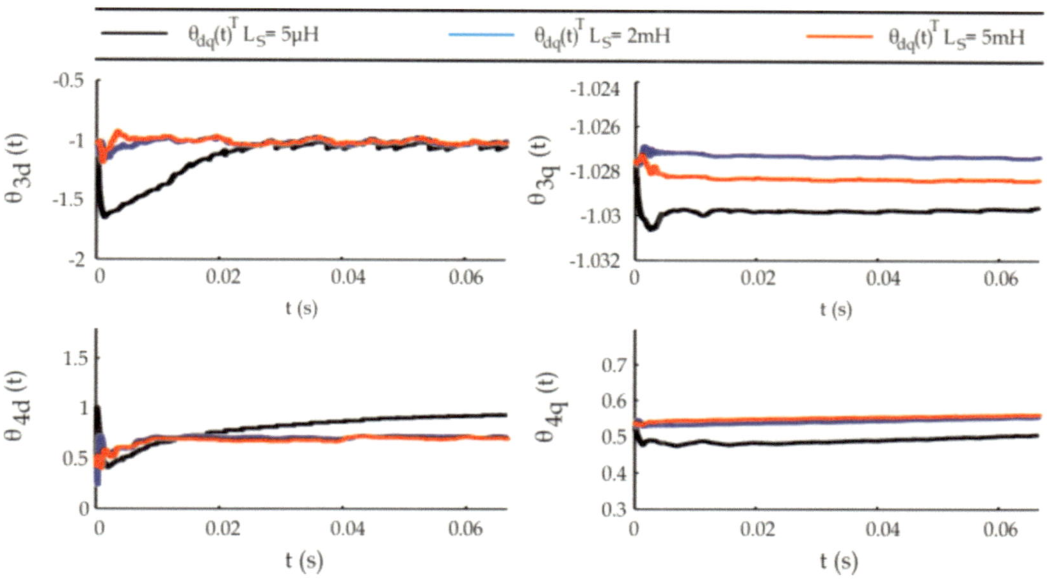

Fig. 11. Control parameters convergence.

4. A fixed robust LQR control

The Linear Quadratic Regulator (LQR) has been widely applied to several applications where optimal control is required. The LQR control strategy implementation uses state feedback where the states weighting can be chosen such the control output is properly designed to satisfy a performance criterion (Phillips & Nagle, 1995). In this control strategy, the states weighting gains are obtained through the solution of an associated algebraic Ricatti equation, which includes a performance index. The advantage of a LQR controller over other controllers found in literature is that it is designed to minimize a performance index, which can reduce the control efforts or keep the energy of some important state variable under control. Moreover, if the plant is accurately modeled, the LQR may be considered a robust controller, minimizing satisfactorily the considered states.

Consider a linear time invariant multivariable controllable system as in Eq. (28)-(29),

$$\dot{x}(t) = Ax(t) + Bu(t) \tag{28}$$

and

$$y(t) = Cx(t). \tag{29}$$

The Zero Order Hold (ZOH) discrete time model of the system with sample period T_s is

$$x_{k+1} = A_d x_k + B_d u_k \tag{30}$$

and

$$y_k = C_d x_k, \tag{31}$$

where,

$$A_d = e^{AT_s}, \quad B_d = A^{-1}(e^{AT_s} - I)B$$

and

$$C_d = C.$$

The LQR control law is given by Eq. (32)

$$u_k = -Kx_k. \tag{32}$$

and the cost function to be minimized is given by Eq. (33)

$$J = \frac{1}{2}\sum_{k=0}^{\infty}\{\, x_k^T\, Q\, x_k + u_k^T R u_k \,\}, \tag{33}$$

where $Q_{m\times m}$ is a positive semidefinite matrix and $R_{n\times n}$ is a positive definite matrix.
The K gains can be obtained solving the algebraic Riccati equation (Phillips & Nagle, 1995),

$$P = {A_d}^T P(A_d - B_d\, K) + Q, \tag{34}$$

$$K = (B_d^T P B_d + R)^{-1} B_d^T P A_d. \tag{35}$$

4.1 Modification on the mathematical model of the system

To achieve an adequate performance, the linear quadratic regulator needs the feedback of all significative states of the system. If there are significative disturbances in the process that can be modeled, it is plausible to include such dynamics in the state space variable set of the plant (Kanieski, Gründling & Cardoso, 2010). Considering the positive sequence of the voltages at the PCC, the equations which represent the behavior of this system can be obtained through the Laplace transform of ϕ radians delayed sines functions, given by Eq. (36):

$$v(s) = V\frac{\omega\cos(\phi) + sen(\phi)\, s}{s^2 + \omega^2}, \tag{36}$$

where V is the magnitude of the waveform considered. The state space representation of that system is given in Eq. (37),

$$\overset{\bullet}{\begin{bmatrix} v(t) \\ \dot{v}(t) \end{bmatrix}} = \begin{bmatrix} 0 & 1 \\ -\omega^2 & 0 \end{bmatrix} \begin{bmatrix} v(t) \\ \dot{v}(t) \end{bmatrix}. \tag{37}$$

Thus, the three-phase sinusoidal waveforms delayed by ϕ radians can be generated by choosing the initial conditions of the voltages and its derivatives, as shown in Eq. (38),

$$\begin{aligned} v(0) &= V sen\,(\phi)\,, \\ \frac{dv(0)}{dt} &= V\omega \cos\,(\phi)\,. \end{aligned} \tag{38}$$

Therefore, the complete model that represents the voltage disturbance at the PCC of the energy storage system in the "123" frame is presented in Eq. (39)

$$\overset{\bullet}{\begin{bmatrix} v_{1N} \\ v_{2N} \\ v_{3N} \\ \dot{v}_{1N} \\ \dot{v}_{2N} \\ \dot{v}_{3N} \end{bmatrix}} = M_{6x6} \begin{bmatrix} \dot{v}_{1N} \\ \dot{v}_{2N} \\ \dot{v}_{3N} \\ v_{1N} \\ v_{2N} \\ v_{3N} \end{bmatrix}, \tag{39}$$

where,

$$M_{6x6} = \begin{bmatrix} 1 & 0 & 0 & 0 & 0 & 0 \\ 0 & 1 & 0 & 0 & 0 & 0 \\ 0 & 0 & 1 & 0 & 0 & 0 \\ 0 & 0 & 0 & -\omega^2 & 0 & 0 \\ 0 & 0 & 0 & 0 & -\omega^2 & 0 \\ 0 & 0 & 0 & 0 & 0 & -\omega^2 \end{bmatrix}.$$

With equations (4), (11) and the matrix differentiation property given by Eq. (40)

$$\frac{d}{dt}\left[C_{123}^{dqO}\left[i_{dqO}\right]\right] = C_{123}^{dqO}\frac{d}{dt}\left[i_{dqO}\right] + \left(\frac{d}{dt}C_{123}^{dqO}\right)\left[i_{dqO}\right], \tag{40}$$

it is possible to derive the following 'dq' frame model for the voltages at the PCC:

$$\overset{\bullet}{\begin{bmatrix} v_d \\ v_q \\ \dot{v}_d \\ \dot{v}_q \end{bmatrix}} = \begin{bmatrix} 0 & 0 & 1 & 0 \\ 0 & 0 & 0 & 1 \\ -\omega^2 & 0 & 0 & \omega \\ 0 & -\omega^2 & -\omega & 0 \end{bmatrix} \begin{bmatrix} v_d \\ v_q \\ \dot{v}_d \\ \dot{v}_q \end{bmatrix}. \tag{41}$$

Thereby, the complete rotating reference frame of the considered system is given by Eq. (42)

$$\overset{\bullet}{\begin{bmatrix} i_d \\ i_q \\ v_d \\ v_q \\ \dot{v}_d \\ \dot{v}_q \end{bmatrix}} = \hat{A} \begin{bmatrix} i_d \\ i_q \\ v_d \\ v_q \\ \dot{v}_d \\ \dot{v}_q \end{bmatrix} + \hat{B} \begin{bmatrix} d_{nd} \\ d_{nq} \end{bmatrix}, \tag{42}$$

where,

$$\hat{A} = \begin{bmatrix} -\frac{R_f}{L_f} & \omega & \frac{1}{L_f} & 0 & 0 & 0 \\ -\omega & -\frac{R_f}{L_f} & 0 & \frac{1}{L_f} & 0 & 0 \\ 0 & 0 & 0 & 0 & 1 & 0 \\ 0 & 0 & 0 & 0 & 0 & 1 \\ 0 & 0 & -\omega^2 & 0 & 0 & \omega \\ 0 & 0 & 0 & -\omega^2 & -\omega & 0 \end{bmatrix}$$

and

$$\hat{B} = -\begin{bmatrix} \frac{v_{dc}}{L_f} & 0 \\ 0 & \frac{v_{dc}}{L_f} \\ 0 & 0 \\ 0 & 0 \\ 0 & 0 \\ 0 & 0 \end{bmatrix}.$$

Hence, the actual model may be seen as depicts Fig. 12 below, where $\hat{C}$ is such that the outputs are the 'dq' frame currents. That is,

$$\hat{C} = \begin{bmatrix} 1 & 0 \\ 0 & 1 \end{bmatrix}.$$

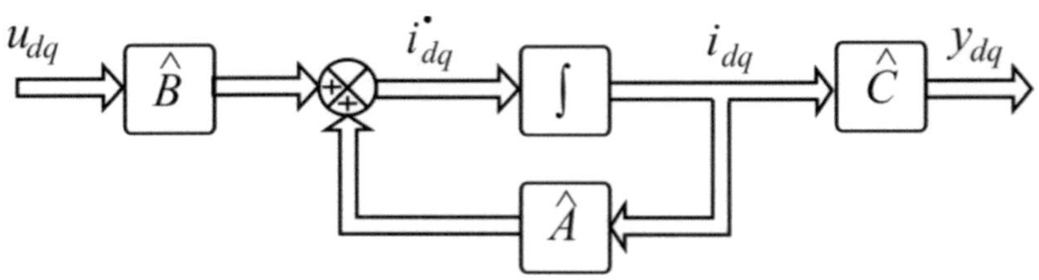

Fig. 12. Model of the system, considering the voltages at PCC as being part of the model.

4.2 LQR tuning

In the LQR tuning case, the cost function to be minimized is given by Eq. (43):

$$J_N = \frac{1}{2}\sum_{k=0}^{\infty}\left(x_{N_k}^T Q_N x_{N_k} + u_{N_k}^T R_N u_{N_k}\right). \tag{43}$$

It is perceptible, on this equation, the presence of the states related to the PCC voltages. Despite of the consideration of the voltage variables at the PCC in the plant model, for the computation of the LQR controller feedback gains, the energy related to those variables are not passible of control. Hence, those variables can not be minimized. Therefore, all the elements of the LQR matrices Q_N and R_N, related to those states, are defined as zero. The other elements are evaluated as in (Kanieski, Carati & Cardoso, 2010).

The dynamic of the states in closed loop is equal to the errors dynamics. Therefore Eq. (44) is applicable,

$$x_{N_k}^T = \left[e_d \ e_q \ v_d^1 \ v_q^1 \ \dot{v}_d^1 \ \dot{v}_q^1\right], \tag{44}$$

which presents, as the feedback vector, the tracking errors of the states of the plant and the states related to the PCC voltages. The vector

$$u_{N_k}^T = \left[d_{nd} \; d_{nq} \right] \tag{45}$$

is the control action of the system in the "dq" frame.
With that, it is possible to obtain the following LQR matrices:

$$Q_N = \begin{bmatrix} q_{N_{11}} & 0 & 0\,0\,0\,0 \\ 0 & q_{N_{22}} & 0\,0\,0\,0 \\ 0 & 0 & 0\,0\,0\,0 \\ 0 & 0 & 0\,0\,0\,0 \\ 0 & 0 & 0\,0\,0\,0 \\ 0 & 0 & 0\,0\,0\,0 \end{bmatrix}, \qquad R_N = \begin{bmatrix} 1\,0 \\ 0\,1 \end{bmatrix}.$$

Fig. 13 shows the structure of the proposed LQR system. In this figure the box "KF" represents the Kalman filter estimator, which gives information about the PCC voltages, needed to compute the LQR K_N gain.

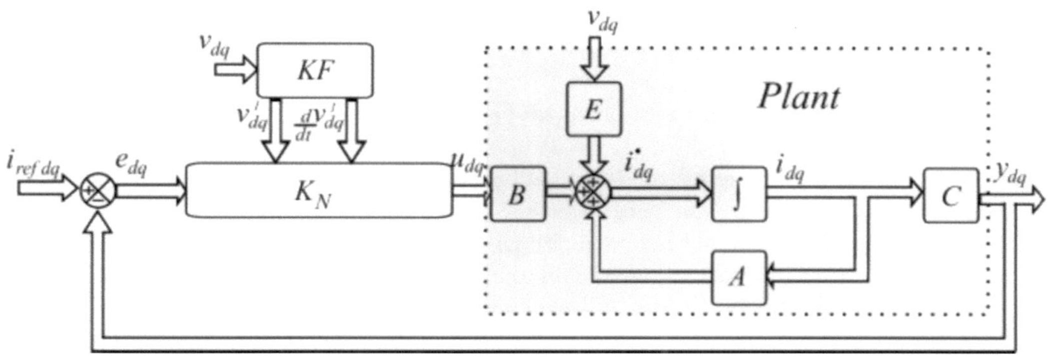

Fig. 13. Block diagram of the proposed LQR system.

5. Optimum extraction of the signal components

Using the same approach as presented in Cardoso (2008), the harmonic components of a distorted signal can be optimally extracted using a Kalman filter with an appropriate mathematical model describing the evolution of such a signal.

5.1 Modeling a signal with harmonics

The use of Kalman filter implies a model that describes the evolution of the process to be filtered. As presented by Cardoso et al. (2007) and Cardoso et al. (2008), a linear signal S_k with n harmonic components, that is,

$$S_k = \sum_{i=1}^{n} A_{i_k} \sin(i\omega_k t_k + \theta_{i_k}) \tag{46}$$

where A_{i_k}, $i\omega_k$ and θ_{i_k} are the amplitude, angular frequency and phase of each harmonic component i at the time instant t_k, has the following state-variable representation

$$\begin{bmatrix} x_1 \\ x_2 \\ \vdots \\ x_{2n-1} \\ x_{2n} \end{bmatrix}_{k+1} = \begin{bmatrix} M_1 & \cdots & 0 \\ \vdots & \ddots & \vdots \\ 0 & \cdots & M_n \end{bmatrix}_k \begin{bmatrix} x_1 \\ x_2 \\ \vdots \\ x_{2n-1} \\ x_{2n} \end{bmatrix}_k + \begin{bmatrix} \gamma_1 \\ \gamma_2 \\ \vdots \\ \gamma_{2n-1} \\ \gamma_{2n} \end{bmatrix}_k, \tag{47}$$

$$y_k = \begin{bmatrix} 1 \; 0 \cdots 1 \; 0 \end{bmatrix} \begin{bmatrix} x_1 \\ x_2 \\ \vdots \\ x_{2n-1} \\ x_{2n} \end{bmatrix}_k + \nu_k, \tag{48}$$

where

$$M_i = \begin{bmatrix} \cos(i\omega_k T_s) & \sin(i\omega_k T_s) \\ -\sin(i\omega_k T_s) & \cos(i\omega_k T_s) \end{bmatrix}, \tag{49}$$

$$x_{(2i-1)_k} = A_{i_k} \sin(i\omega_k t_k + \theta_{i_k}) \tag{50}$$

and

$$x_{2i_k} = A_{i_k} \cos(i\omega_k t_k + \theta_{i_k}). \tag{51}$$

In Eq. (47) it is considered a perturbation vector $[\gamma_1 \; \gamma_2 \; \cdots \; \gamma_{2n-1} \; \gamma_{2n}]_k^T$ that models amplitude or phase changes in the signal. In Eq. (48) ν_k represents the measurement noise. At the same time that the mathematical model given by equations (47)-(49) describes a signal with harmonics, it has the appropriate form necessary for the use in the Kalman filter.

The Kalman filter algorithm mentioned above will be used to generate the current references by measuring the load currents and extracting the references from it. Moreover, by measuring the voltages at the PCC, it is also possible to filter its harmonic components and, as suggests Eq. (51), the quadrature component as well. Therefore, considering that the voltage at the PCC is predominantly at the fundamental frequency, the voltage modeled in Eq. (39) is directly obtained by Eq. (50) and the derivative component is obtained by Eq. (51) multiplied by the angular frequency ω.

Fig. 14 shows the response of a Kalman filter estimator, when extracting a voltage signal and its derivative. V_{Grid} and $\dot{V}_{Grid}$, in gray, are the normalized grid voltage and its derivative. In black, $\hat{V}_{Grid}$ and $\hat{\dot{V}}_{Grid}$ are the normalized Kalman filter estimated voltage and its derivative, which, as it can be seen, presents a good performance.

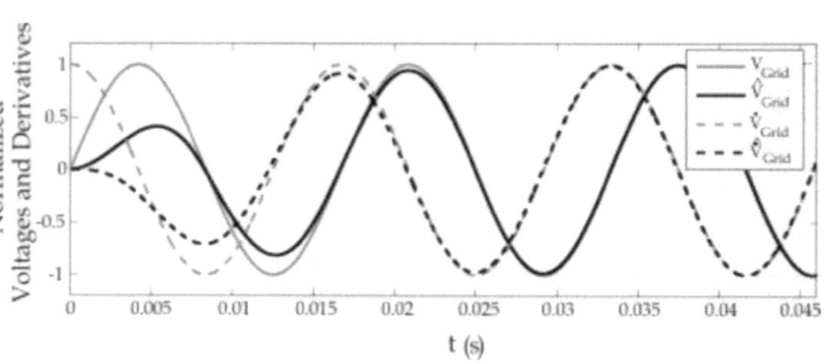

Fig. 14. Grid voltage and its derivative with their estimates provided by the Kalman filter.

5.1.1 LQR Results

The LQR control of the compensation currents i_{F123} was implemented in the same platform as presented in the RMRAC results section (section 3.3.2). Table 1 also shows the parameters used in the analysis made on this section. The Q_N and R_N matrices of the LQR controller are given by:

$$Q_N = \begin{bmatrix} 1500 & 0 & 0\,0\,0\,0 \\ 0 & 2000 & 0\,0\,0\,0 \\ 0 & 0 & 0\,0\,0\,0 \\ 0 & 0 & 0\,0\,0\,0 \\ 0 & 0 & 0\,0\,0\,0 \\ 0 & 0 & 0\,0\,0\,0 \end{bmatrix}, \qquad R_N = \begin{bmatrix} 1\,0 \\ 0\,1 \end{bmatrix}.$$

The same procedure that was used for the verification of the RMRAC controller was used to analyze the LQR controller performance. At first, with a small line inductance of $L_S = 5uH$, Fig. 15 (a) shows the load currents and the compensated currents, provided by the main source. The appropriated reference tracking is verified in Fig. 15 (b) for the case of small line inductance. In Fig. 15 (b), it is shown the reference currents in black, while the compensation currents are presented in gray.

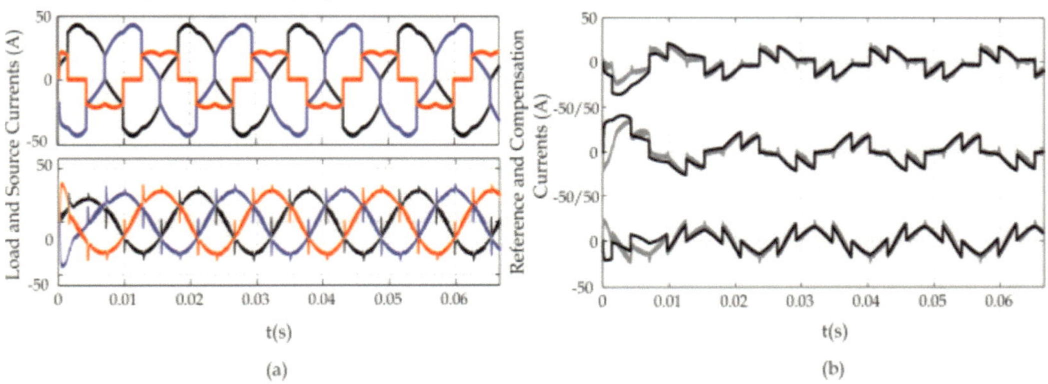

Fig. 15. System currents ($L_S = 5\mu H$): (a) Load and source currents (b) Reference and compensation currents.

In a second analysis, an inductor of $L_S = 2mH$ represents the line impedance. Fig. 16 depicts the results for this case.
Fig. 16 (a) shows the load currents and the compensated currents, provided by the main source. Fig. 16 (b) depicts good reference tracking for the system controlled by the proposed LQR scheme for the case of high line inductance.
Finally, for the case of line inductance of $L_S = 5mH$, the load currents and the compensated currents are shown in Fig. 17 (a). Fig. 17 (b) shows the tracking performance of the LQR controller, under high line impedance circumstances, which remains stable and with a good tracking performance, even under high inductance levels.

6. Conclusion

In this chapter, it was presented two robust algorithms frequently discussed in the literature, which are chosen to control the power quality conditioner due to its well known features of

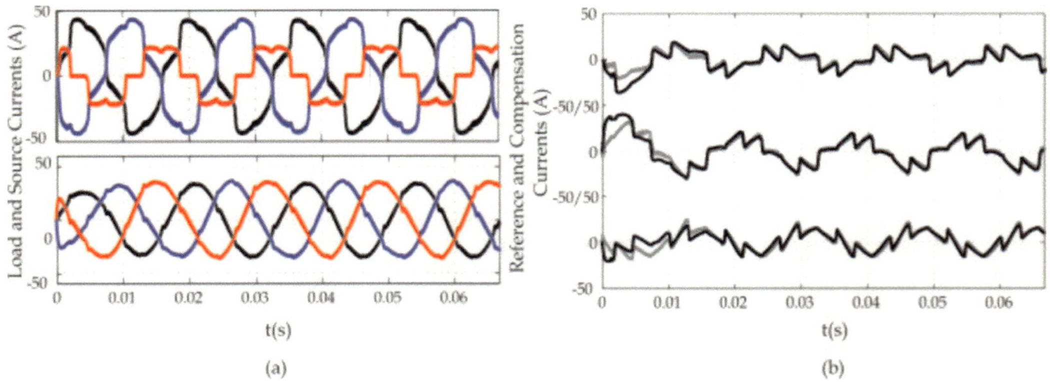

Fig. 16. System currents ($L_S = 2mH$): (a) Load and source currents (b) Reference and compensation currents.

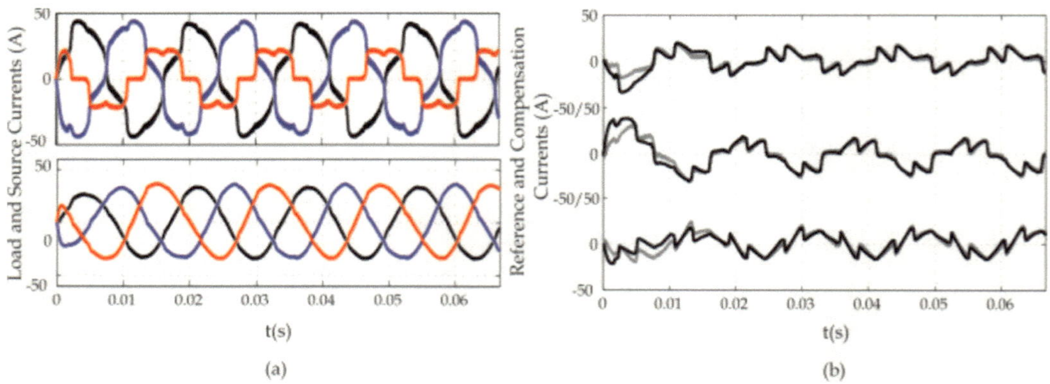

Fig. 17. System currents ($L_S = 5mH$): (a) Load and source currents (b) Reference and compensation currents.

performance and robustness: The Robust Model Reference Adaptive Controller and the fixed Linear Quadratic Regulator.
The RMRAC controller guarantees the robustness of the closed loop system by acting on the θ parameter values. The evolution of the direct axis θ parameters were presented as well as the tracking performance of the controller, for each case of line impedance ($L_S = 5\mu H$, $L_S = 2mH$ and $L_S = 5mH$), showing adequate convergence of the closed loop dynamics into the designed model reference.
For the Linear Quadratic Regulator a novel modeling approach was presented and applied to the power quality conditioning system. The scheme aimed, compared to other techniques found in literature, to have a more realistic representation of the system, by using a resonant model of the PCC voltages instead of considering it as a disturbance. The great advantage of this approach lies in having a fixed controller capable to deal with voltages disturbances, resulted from high harmonic content of load, in conditions of high line impedance. The presented mathematical model decouples the whole system, formed by the conditioner device, the main source and the load, providing an easy manner of considering just the power quality device output filter, the voltages at the PCC and its derivatives on the controller

project. The tracking performance of this fixed controller was presented, for the same three cases of line impedance tested in the RMRAC system.
The results obtained for the two cases, presented during the text, illustrated that the developed control structures exhibit good performance and robustness regarding to line inductance variation, for a huge class of loads, including those with unbalance.

7. References

Akagi, H. (1997). Control strategy and site selection of a shunt active filter for damping of harmonic propagation in power distribution systems, *IEEE Transactions on Power Delivery* 12(1): 354–363.

Cardoso, R. (2008). *Synchronization Algorithms, Power Quality Analysis And Reference Generation For Active Power Filters: A Stochastic Approach*, PhD in Electrical Engineering, Federal University of Santa Maria (UFSM).

Cardoso, R., Figueiredo, R. C., Pinheiro, H. & Gründling, H. A. (2008). Kalman filter based synchronization methods, *IET Generation, Transmission & Distribution* 2(4): 542–555.

Cardoso, R. & Gründling, H. A. (2009). Grid synchronization and voltage analysis based on the Kalman filter, *in* V. M. Moreno & A. Pigazo (eds), *Kalman Filter: Recent Advances and Applications*, I-Tech, Viena, Austria, pp. 439–460.

Cardoso, R., Kanieski, J. M., Pinheiro, H. & Gründling, H. A. (2007). Reference generation for shunt active power filters based on optimum filtering theory, *Conference Record of the 2007 IEEE Industry Applications Conference. 42nd IAS Annual Meeting*, IEEE, New Orleans, USA.

Casadei, D., Grandi, G. & Rossi, C. (2000). Effects of supply voltage non-idealities on the behavior of an active power conditioner for cogeneration systems, *Proceedings of the 31st IEEE Power Electronics Specialists Conference*, IEEE, s.l., pp. 1312–1317.

Escobar, G., Valdez, A. A. & Ortega, R. (2008). An adaptive controller for a shunt active filter considering load and line impedances, *Proceedings of the 11th IEEE International Power Electronics Congress*, IEEE, s.l., pp. 69–74.

George, S. & Agarwal, V. (2002). A novel technique for optimising the harmonics and reactive power under non-sinusoidal voltage conditions, *Proceedings of the 28th IEEE Annual Conference of the Industrial Electronics Society*, IEEE, s.l., pp. 858–863.

Ioannou, P. A. & Sun, J. (1995). *Robust Adaptive Control*, Prentice-Hall.

Ioannou, P. A. & Tsakalis, K. S. (1986). A robust direct adaptive controller, *IEEE Transactions on Automatic Control* AC-31(11): 1033–1043.

Kanieski, J. M., Carati, E. G. & Cardoso, R. (2010). An energy based lqr tuning approach applied for uninterruptible power supplies, *Proceedings of IEEE Latin American Symposium on Circuits and Systems*, IEEE, Foz do Iguaçu, Brazil.

Kanieski, J. M., Gründling, H. A. & Cardoso, R. (2010). A new lqr modeling approach for power quality conditioning devices, *Proceedings of the 36th Annual Conference of the IEEE Industrial Electronics Society*, IEEE, Phoenix, USA.

Kedjar, B. & Al-Haddad, K. (2009). Dsp-based implementation of an lqr with integral action for a three-phase three-wire shunt active power filter, *IEEE Transactions on Industrial Electronics* 56(8): 2821–2828.

Malesani, L., Mattavelli, P. & Buso, S. (1998). On the applications of active filters to generic load, *Proceedings of the International Conference on Harmonic and Quality of Power ICHQP8*, IEEE, s.l., pp. 310–319.

Palethorpe, B., Sumner, M. & Thomas, D. W. P. (2000). System impedance measurement for use with active filter control, *Proceedings of the Power Electronics and Variable Speed Drives Conference*, IEE, s.l., pp. 24–28.

Phillips, C. L. & Nagle, H. T. (1995). *Digital Control System Analysis and Design*, Prentice-Hall.

Sangwongwanich, S. & Khositkasame, S. (1997). Design of harmonic current detector and stability analysis of a hybrid parallel active filter, *Proceedings of the Power Conversion Conference*, IEEE, Nagaoka, Japan, pp. 181–186.

Sumner, M., Palethorpe, B., Zanchetta, P. & Thomas, D. W. P. (2002). Experimental evaluation of active filter control incorporating on-line impedance measurement, *Proceedings of the 10th International Harmonics and Quality of Power*, IEEE, s.l., pp. 501–506.

Sumner, M., Thomas, D. W. P. & nd P. Zanchetta, A. A. (2006). Power system impedance estimation for improved active filter control, using continuous wavelet transforms, *Proceedings of the IEEE Transmission and Distribution Conference and Exhibition*, IEEE, s.l., pp. 653–658.

Valdez, A. A., Escobar, G. & Ortega, R. (2008). A controller for the active filter considering load and line impedances, *Proceedings of the 47th IEEE Conference on Decision and Control*, IEEE, Cancun, Mexico, pp. 3749–3754.

25

Robust Bilateral Control for Teleoperation System with Communication Time Delay - Application to DSD Robotic Forceps for Minimally Invasive Surgery -

Chiharu Ishii
Hosei University
Japan

1. Introduction

Minimally invasive surgery (MIS) has excellent characteristics that can reduce the burden on patients. However, surgeons experience great difficulties in operation due to limitations in dexterity imposed by the surgical instruments and the small work space. Therefore, the development of surgical assistance devices with the application of robotic and mechatronic technology is in high demand (Taylor & Stoianovici, 2003).

Recently, robotic surgical support systems such as `da VINCI' are in clinical use (Guthart & Salisbury, 2000). In particular, the development of multi-DOF robotic forceps manipulators capable of reproducing complex human finger movements in laparoscopic surgery is one of the most important issues in the field of robotic surgical systems.

A large number of conventional multi-DOF robotic forceps manipulators currently available for MIS are of the wire actuation type (Ikuta et al., 2003). However, the rigidity and the durability of wires are poor. Furthermore, cleaning and sterilization of the wire are problematic.

In order to improve the rigidity and the sterilization capability of the manipulator, multi-DOF robotic forceps manipulators which use methods different from wire actuation for bending motion have been developed. These are roughly divided into two types. The first type is where two-DOF bending is achieved by combining independent joints which perform yaw and pitch motions, respectively. The second type is where omnidirectional two-DOF bending is achieved by inclination of the entire bending part of the forceps. Many manipulators of the first type are linkage-driven forceps manipulators. In (Yamashita et al., 2005), an endoscopic forceps manipulator using a multi-slider linkage mechanism is developed without using wires for bending motion. However, a wire is used for gripping motion. In (Arata et al., 2005), a linkage-driven forceps manipulator which does not use wires for either bending or gripping motions is developed.

On the other hand, as one of the omnidirectional driven-type forceps manipulators, an active forceps manipulator in the form of a tripodal platform is developed in (Kobayashi et al., 2002). Although it has high rigidity, its bending range is 40 to 50 degrees, and it is difficult to expand the bending range due to constraints inherent in the mechanism.

We have developed a multi-DOF robotic forceps manipulator for minimally invasive surgery incorporating a novel omnidirectional bending technique with a screw drive mechanism, termed Double-Screw-Drive (DSD) mechanism, so far (Ishii et al., 2010). A robotic forceps manipulator incorporating the DSD mechanism (DSD robotic forceps) can bend without using wires. Without wires, it has high rigidity, and it can bend at 90 degrees in any arbitrary direction. In addition, the gripper of the DSD robotic forceps can perform rotational motion. Opening and closing motions of the gripper are attained by wire actuation.

In order to improve the operability of the robotic surgical support systems and to help surgeon's dexterity, development of haptic forceps teleoperation systems is required. Most recently, haptic forceps manipulator for minimally invasive surgery has been proposed in (Seibold et al., 2005) and (Zemiti et al., 2007), in which operation force is measured by sensor and force feedback is provided. In addition, the motion scaling, which can adequately reduce or enlarge the movements and tactile senses of the operator and the robot, is necessary to assure safety of the surgery.

On the other hand, communication time delay is inevitable in teleoperation systems, which may causes instability of the teleoperation systems. Therefore, stability of the system must be guaranteed in the presence of the communication time delay between master device and slave device. For bilateral teleoperation systems with constant time delay, stabilization method based on scattering transformation is proposed in (Anderson & Spong, 1989). (Chopra & Spong, 2005) proposed a passivity based control scheme which guarantees delay dependent exponential stability of the position and velocity tracking error. However, coupling torques are given as a function of position and velocity, and is not a function of force. Hence, motion scaling in force tracking cannot be achieved.

In this chapter, improving the control scheme proposed in (Chopra & Spong, 2005), such a passivity based bilateral control scheme that enables motion scaling in both position tracking and force tracking, and guarantees the stability of the teleoperation system in the presence of constant time delay, is proposed. This can be achieved by adding force tracking error terms to the coupling torques.

Then, the proposed bilateral control scheme is applied to a haptic control of bending motion of the DSD robotic forceps teleoperation system with constant time delay. However, the proposed bilateral control law is applicable only to the one-DOF bending motion of the DSD robotic forceps. Therefore, using the change of coordinates, the proposed bilateral control scheme is extended so that it may become applicable to the omnidirectional bending motion of the DSD robotic forceps.

Experimental works were carried out using the proposed bilateral control scheme, and experimental results showed the effectiveness of the proposed control scheme.

2. DSD robotic forceps

In this section, details of the DSD robotic forceps are explained. Overview of the developed DSD robotic forceps manipulator is shown in Fig. 1, and the configuration of its bending part is shown in Fig. 2.

2.1 Specifications

The total length of the DSD robotic forceps manipulator is 635 mm, and its gross weight is 1050 g. The main specifications of the DSD robotic forceps are given as follows.

1. In order to insert a forceps into a trocar, the diameter of the rod of the forceps must be 10 mm or less since the diameter of the trocar is 12 mm.
2. The bending force, defined as the lifting force at the tip of the forceps, must be larger than 4 N, which would allow the forceps to lift 1/3 of an average human liver. This ability is required during operations of internal organs under the liver.
3. The bending range must be 180 (-90 to +90) degrees or more in both horizontal and vertical direction. This ability is required in order to obtain a sufficient degree of freedom in limited work space.
4. The gripper must be able to perform opening and closing motions smoothly. This operational requirement is necessary for the proper holding and releasing of medical needles.
5. In order to perform suturing in a small work space, such as the opposite or the far side of internal organs, the gripper of the forceps must be able to rotate.

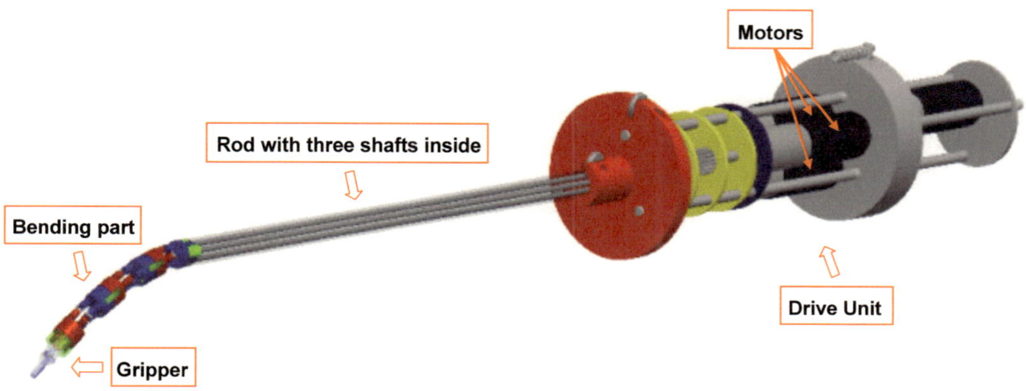

Fig. 1. Overview of DSD robotic forceps manipulator

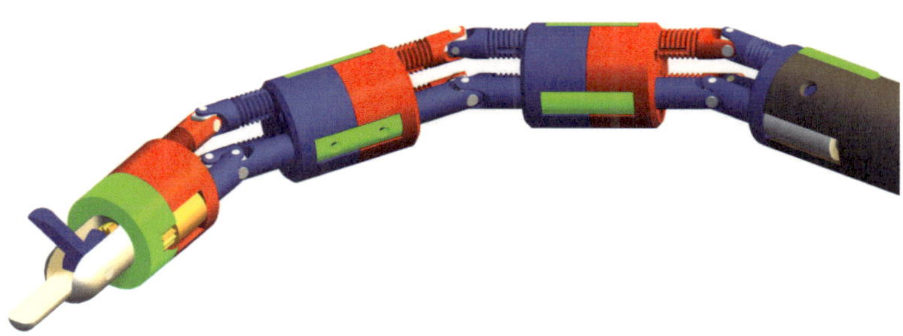

Fig. 2. Bending part of DSD robotic forceps manipulator

2.2 Bending mechanism

One module of the bending mechanism is shown in Fig.3.

The DSD mechanism has three linkages, and when examined in cross-sectional view, each linkage is 120 degrees apart from the other linkages and 6 mm from the center of the cross-section. Let us denote the group consisting of part ⑤ and part ⑥ as a "bending linkage" and the group consisting of part ① and part ② as a "grasping linkage". Bending motion is achieved by rotating the two bending linkages, and grasping linkage is used for actuating the gripper. The key point of this mechanism is that one side of part ⑤ is a left-handed screw and the other side is a right-handed screw. When a DSD module is connected to another module, a joint is formed. The principle of the bending motion for such a joint is illustrated in Fig. 4.

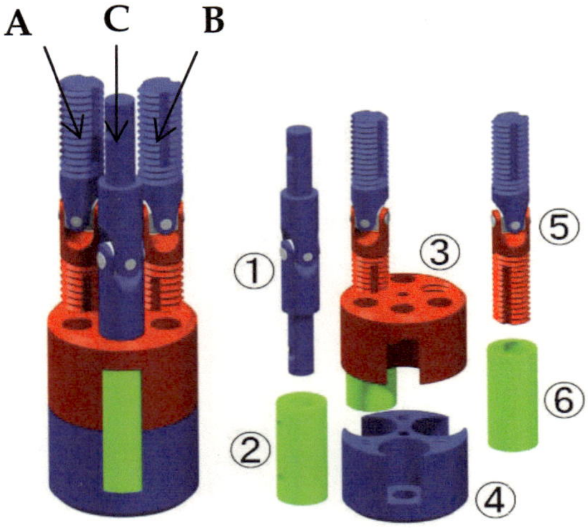

A and B: Bending linkage, C: Grasping linkage

① Universal joint shaft
② Coupling
③ Plate with left-handed threaded hole
④ Plate with right-handed threaded hole
⑤ Universal joint of the screw drive
⑥ Spline nut

Fig. 3. One module of DSD mechanism

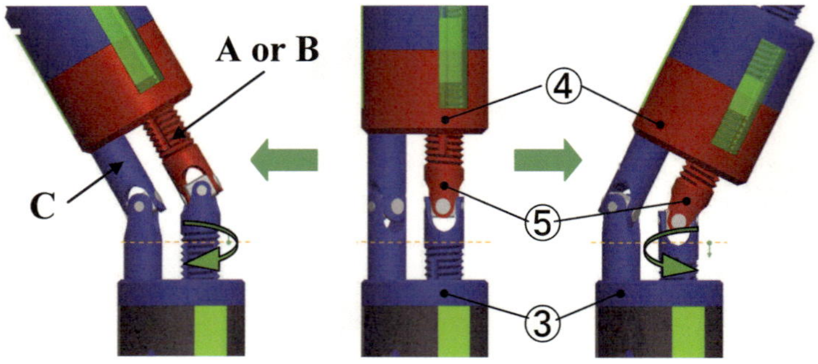

Fig. 4. Principle of bending motion

The left-handed screw of part ⑤ connects to part ③, and the right-handed screw of part ⑤ connects to part ④ of another module. The rotation of the linkage changes the connecting length of the screw and the plate at both ends of part ⑤. As a result, an angle is formed between part ③ and part ④. For example, when the linkage rotates clockwise, part ③ and part ④ approach each other, and when the linkage rotates counterclockwise, they move away from each other. Thus, bending motion is achieved. The maximum bending angle of one joint is between -30 and +30 degrees since this is the allowable bending angle of the universal joint. One bending linkage allows for one-DOF bending motion, and by using two bending linkages and controlling their rotation angles, arbitrary omnidirectional bending motion can be attained. The total length of the bending part is 59 mm excluding a gripper.

2.3 Attachment and rotary gripper

The gripper is exchangeable as an end effector and can be replaced with tools such as scalpels or surgical knives. Fig. 5 shows the attachment of the end effecter and mechanism of the rotary gripper. Gear 1 is on the tip of the grasping linkage and gear 2 is at the root of the jaw mesh. The gripper is turned by rotation of the grasping linkage. Although the rotary gripper can rotate arbitrary degrees, it should be rotated within 360 degrees to avoid winding of the wire which drives the jaw.

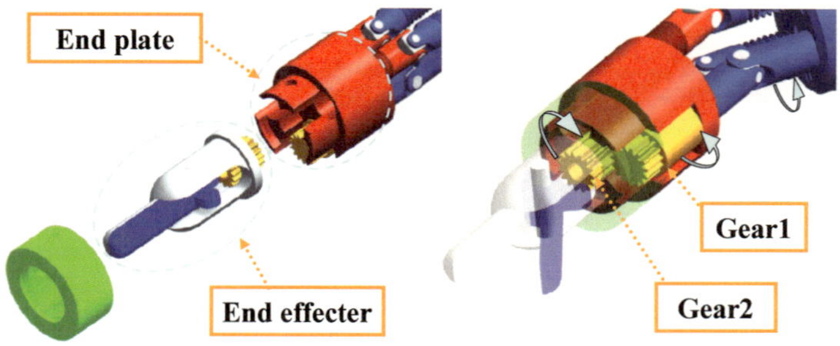

Fig. 5. Attachment and rotation of gripper

2.4 Open and close of jaws

The opening and closing motions of the gripper are achieved by wire actuation. Only one side of the jaws can move, and the other side is fixed. The wire for actuation connects to the drive unit through the inside of the DSD mechanism and the rod, and is pulled by the motor. The open and closed states of the gripper are shown in Fig.6.

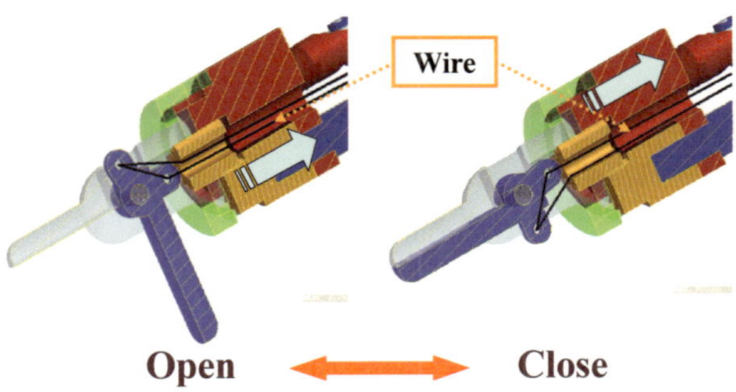

Fig. 6. Grasping of gripper

2.5 Drive unit

The feature of a drive unit for the DSD robotic forceps manipulator is shown in Fig.7. The total length of the drive unit is 274 mm, its maximum diameter is 50 mm, and its weight is 935 g. Driving forces from motors are transmitted to the linkages through the gears. There are four motors in the drive unit. Three motors are mounted at the center of the drive unit. Two of them are used for inducing bending motion and the third one is used for inducing rotary motion of the gripper. The fourth motor, which is mounted in the tail, is for the opening and closing motions of the gripper actuated by wire. The wire capstan is attached to the motor shaft of the forth motor and acts as a reel for the wire. The spring is used for maintaining the tension of the wire. DC micromotors 1727U024C (2.25W) produced by FAULHABER Co. were selected for the bending motion and the rotary motion of the gripper. For the opening and closing motions of the gripper, a DC micro motor 1727U012C (2.25W) produced by FAULHABER Corp. was selected. A reduction gear and a rotary encoder are installed in the motor.

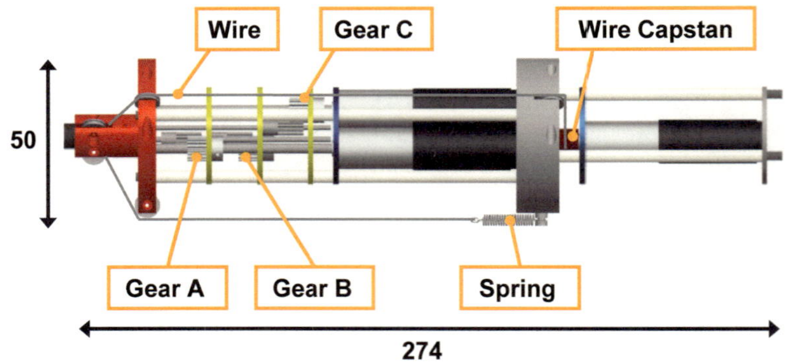

Fig. 7. Drive unit

The inside part of the rod, as shown in Fig. 1, consists of three shafts, each 2 mm in diameter and 300 mm long. Each motor in the drive unit and each linkage in the DSD mechanism are connected to each other through a shaft. Therefore, the rotation of each motor is transmitted to each respective linkage through a shaft.

2.6 Built DSD robotic forceps manipulator

The proposed DSD robotic forceps manipulator was built from stainless steel SUS303 and SUS304 to satisfy bio-compatibility requirements. The miniature universal joints produced by Miyoshi Co., LTD. were selected. The universal joints have a diameter of 3 mm and are of the MDDS type. The screws on both sides of the yokes were fabricated by special order.
The built DSD robotic forceps manipulator is shown in Fig. 8. Its maximum diameter from the top of the bending part to the root of the rod is 10 mm. The total length of the bending part, including the gripper, is 85 mm.

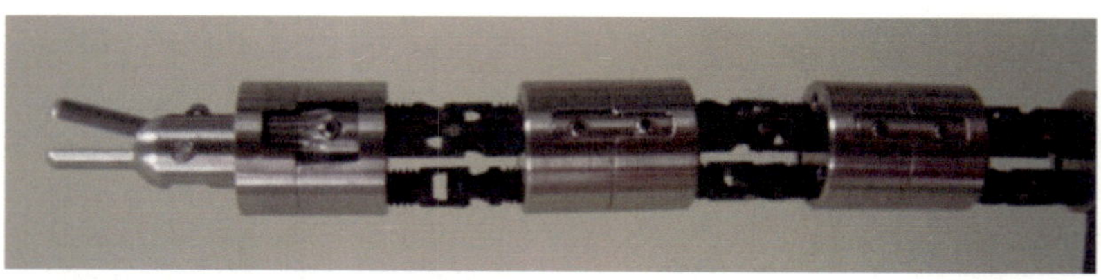

Fig. 8. Built DSD robotic forceps manipulator

A transition chart of the rotary gripper is shown in Fig.9.

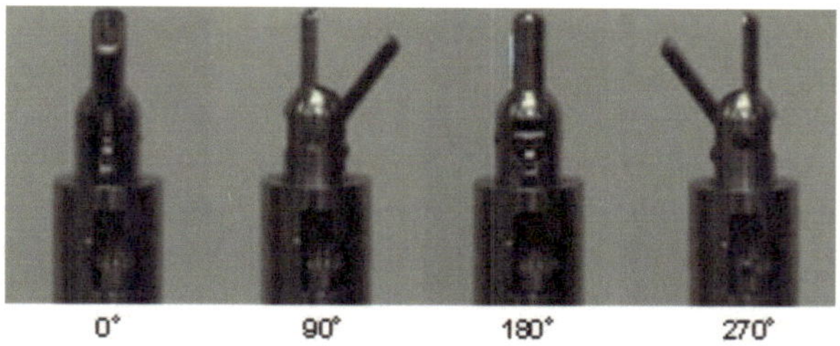

Fig. 9. Transition chart of the rotary gripper

2.7 Master manipulator for teleoperation

In a laparoscopic surgery, multi-DOF robotic forceps manipulators are operated by remote control. In order to control the DSD robotic forceps as a teleoperation system, the joy-stick type master manipulator for teleoperation was designed and built in (Ishii et al., 2010) by reconstruction of a ready-made joy-stick combined with the conventional forceps, which enables to control bending, grasping and rotary motions of the DSD robotic forceps manipulator. In addition, the built joy-stick type master manipulator was modified so that the operator can feel reaction force generated by the electric motors. The teleoperation system and the force feedback mechanisms for the bending force are illustrated in Fig.10. The operation force is detected by the strain gauges, and variation of the position is measured by the encoders mounted in the electric motors.

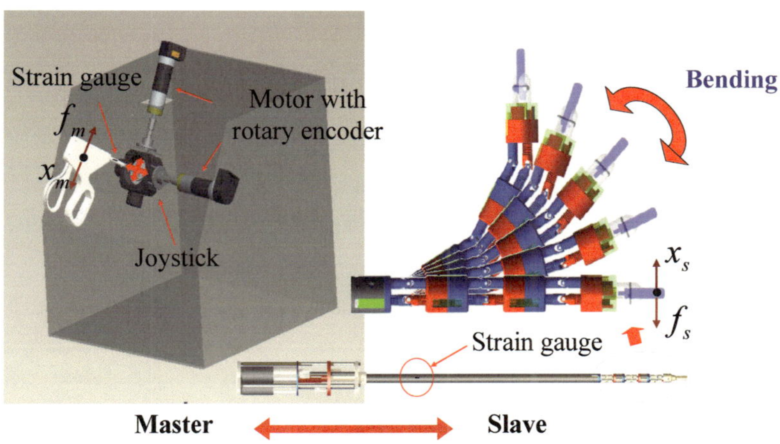

Fig. 10. DSD robotic forceps teleoperation system

3. Bilateral control for one-DOF bending

In this section, bilateral control law for one-DOF bending of the DSD robotic forceps teleoperation system with communication time delay is derived.

3.1 Derivation of Control Law

Let the dynamics of the one-DOF master-slave teleoperation system be given by

$$m_m \ddot{x}_m + b_m \dot{x}_m + c_m x_m = \tau_m + f_m , \tag{1}$$

$$m_s\ddot{x}_s + b_s\dot{x}_s + c_s x_s = \tau_s - f_s, \tag{2}$$

where subscripts m and s denote master and slave respectively. x_m and x_s represent the displacements, m_m and m_s the masses, b_m and b_s the viscous coefficients, and c_m and c_s the spring coefficients of the master and slave devices. f_m stands for the force applied to the master device by human operator, f_s the force of the slave device due to the mechanical interaction between slave device and handling object, and τ_m and τ_s are input motor toques.
As shown in Fig.11, there exists constant time delay T in the network between the master and the slave systems.

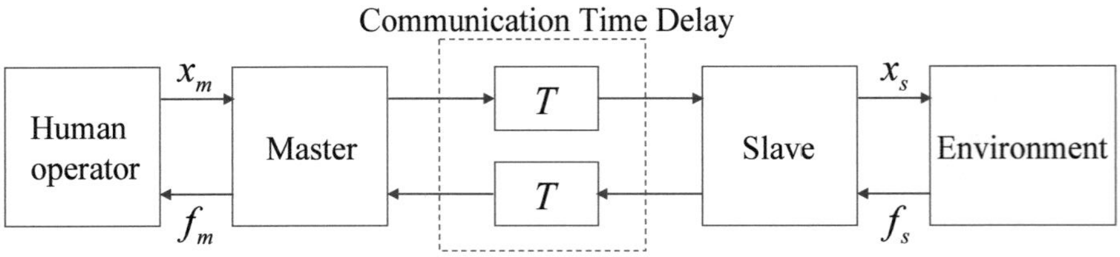

Fig. 11. Communication time delay in teleoperation systems

Define motor torques as

$$\tau_m = \bar{\tau}_m - m_m\lambda\dot{x}_m - b_m\lambda x_m + c_m x_m, \tag{3}$$

$$\tau_s = \bar{\tau}_s - m_s\lambda\dot{x}_s - b_s\lambda x_s + c_s x_s, \tag{4}$$

where λ is a positive constant, and $\bar{\tau}_m$ and $\bar{\tau}_s$ are coupling torques. Then, the dynamics are rewritten as follows.

$$m_m\dot{r}_m + b_m r_m = \bar{\tau}_m + f_m, \tag{5}$$

$$m_s\dot{r}_s + b_s r_s = \bar{\tau}_s - f_s, \tag{6}$$

where r_m and r_s are new variables defined as

$$r_m = \dot{x}_m + \lambda x_m, \tag{7}$$

$$r_s = \dot{x}_s + \lambda x_s. \tag{8}$$

Control objective is described as follows.

[Design Problem] Find a bilateral control law which satisfies the following two specifications.

Specification 1: In both position tracking and force tracking, the motion scaling, which can adequately reduce or enlarge the movements and tactile senses of the master device and the slave device, is achievable.

Specification 2: The stability of the teleoperation system in the presence of the constant communication time delay between master device and slave device, is guaranteed.

Assume the following condition.

Assumption: The human operator and the remote environment are passive.

In the presence of the communication time delay between master device and slave device, the following fact is shown in (Chopra et al., 2003).

Fact: In the case where the communication time delay T is constant, the teleoperation system is passive.

From Assumption and Fact, the following inequalities hold.

$$-\int_0^t r_m f_m d\tau \geq 0, \quad \int_0^t r_s f_s d\tau \geq 0, \tag{9}$$

$$-\int_0^t r_s f_m(\tau - T) d\tau \geq 0, \quad \int_0^t r_m f_s(\tau - T) d\tau \geq 0. \tag{10}$$

Using inequalities (9) and (10), define a positive definite function V as follows.

$$\begin{aligned} V = {} & m_m r_m^2 + G_p m_s r_s^2 + K_1 \int_{t-T}^t \left(r_m^2 + G_p^2 r_s^2\right) d\tau \\ & -2(K_m + 1) \int_0^t r_m f_m d\tau + 2G_p\left(G_f K_s + 1\right) \int_0^t r_s f_s d\tau \\ & -2G_p K_s \int_0^t r_s f_m(\tau - T) d\tau + 2G_f K_m \int_0^t r_m f_s(\tau - T) d\tau \end{aligned}, \tag{11}$$

where K_1, K_m and K_s are feedback gains, and $G_p \geq 1$ and $G_f \geq 1$ are scaling gains for position tracking and force tracking, respectively.

The derivative of V along the trajectories of the systems (5) and (6) is given by

$$\begin{aligned} \dot{V} = {} & 2m_m r_m \dot{r}_m + 2G_p m_s r_s \dot{r}_s + K_1\left(r_m^2 + G_p^2 r_s^2\right) \\ & - K_1\left(r_m^2(t-T) + G_p^2 r_s^2(t-T)\right) \\ & - 2(K_m + 1) r_m f_m + 2G_p\left(G_f K_s + 1\right) r_s f_s \\ & - 2G_p K_s r_s f_m(t-T) + 2G_f K_m r_m f_s(t-T) \\ = {} & 2r_m\left(-b_m r_m + \overline{\tau}_m + f_m\right) + 2G_p r_s\left(-b_s r_s + \overline{\tau}_s - f_s\right) \\ & - K_1\left(G_p r_s(t-T) - r_m\right)\left(G_p r_s(t-T) + r_m\right) \\ & - K_1\left(r_m(t-T) - G_p r_s\right)\left(r_m(t-T) + G_p r_s\right) \\ & - 2(K_m + 1) r_m f_m + 2G_p\left(G_f K_s + 1\right) r_s f_s \\ & - 2G_p K_s r_s f_m(t-T) + 2G_f K_m r_m f_s(t-T). \end{aligned} \tag{12}$$

Let the coupling torques be given as follows.

$$\overline{\tau}_m = K_1\left(G_p r_s(t-T) - r_m\right) - K_m\left(G_f f_s(t-T) - f_m\right), \tag{13}$$

$$\overline{\tau}_s = K_1\left(r_m(t-T) - G_p r_s\right) + K_s\left(f_m(t-T) - G_f f_s\right). \tag{14}$$

Using (13) and (14), (12) is rewritten as follows.

$$\begin{aligned}
\dot{V} &= -2b_m r_m^{\,2} + 2r_m\overline{\tau}_m + 2r_m f_m - 2G_p b_s r_s^{\,2} + 2G_p r_s \overline{\tau}_s - 2G_p r_s f_s \\
&\quad + \left\{\overline{\tau}_m + K_m\left(G_f f_s(t-T) - f_m\right)\right\}\left[2r_m + K_1^{-1}\left\{\overline{\tau}_m + K_m\left(G_f f_s(t-T) - f_m\right)\right\}\right] \\
&\quad - \left\{\overline{\tau}_s - K_s\left(f_m(t-T) - G_f f_s\right)\right\}\left[2G_p r_s + K_1^{-1}\left\{\overline{\tau}_s - K_s\left(f_m(t-T) - G_f f_s\right)\right\}\right] \\
&\quad - 2(K_m+1) r_m f_m + 2G_p\left(G_f K_s + 1\right) r_s f \\
&\quad - 2G_p K_s r_s f_m(t-T) + 2G_f K_m r_m f_s(t-T) \\
&= -2b_m r_m^{\,2} - 2G_p b_s r_s^{\,2} \\
&\quad - K_1^{-1}\left\{\overline{\tau}_m + K_m\left(G_f f_s(t-T) - f_m\right)\right\}^2 - K_1^{-1}\left\{\overline{\tau}_s - K_s\left(f_m(t-T) - G_f f_s\right)\right\}^2 \\
&\le -K_1\left(G_p r_s(t-T) - r_m\right)^2 - K_1\left(r_m(t-T) - G_p r_s\right)^2.
\end{aligned} \tag{15}$$

Thus, stability of the teleoperation system is assured in spite of the presence of the constant communication time delay, and delay independent exponential convergence of the tracking errors of position to the origin is guaranteed.
Finally, motor torques (3) and (4) are given as follows.

$$\begin{aligned}
\tau_m &= K_1 G_p \dot{x}_s(t-T) + \lambda K_1 G_p x_s(t-T) - K_m G_f f_s(t-T) \\
&\quad -(K_1 + \lambda m_m)\dot{x}_m + (c_m - \lambda(K_1 + b_m))x_m + K_m f_m
\end{aligned}, \tag{16}$$

$$\begin{aligned}
\tau_s &= K_1 \dot{x}_m(t-T) + \lambda K_1 x_m(t-T) + K_s f_m(t-T) \\
&\quad -\left(K_1 G_p + \lambda m_s\right)\dot{x}_s + \left(c_s - \lambda(K_1 G_p + b_s)\right)x_s - K_s f_s
\end{aligned}. \tag{17}$$

3.2 Experiments

In order to verify an effectiveness of the proposed control law, experimental works were carried out for the developed DSD robotic forceps teleoperation system. Here, only vertical direction of the bending motion is considered. Namely, bending motion of the DSD robotic forceps is restricted to one degree of freedom. Then, the dynamics of the master-slave teleoperation system are given by equations (1) and (2), since only one bending linkage is used. Parameter values of the system are given as m_m = 0.07 kg, m_s = 0.025 kg, b_m = 0.25 Nm/s, b_s = 2.5 Nm/s, c_m = 9 N/s and c_s = 9 N/s. The control system is constructed under the MATLAB/Simulink software environment.
In the experiments, 200g weights pet bottle filled with water was hung up on the tip of the forceps, and lift and down were repeated in vertical direction. Appearance of the experiment is shown in Fig. 12.
First, in order to see the effect of the motion scaling, experimental works with the following conditions were carried out.

a. Verification of the effect of the motion scaling.
   i) $G_p = G_f = 1$ and $T = 0$
   ii) $G_p = 2$, $G_f = 3$ and $T = 0$

Second, in order to see the effect to the time delay, comparison of the proposed bilateral control scheme and conventional bilateral control method was performed.

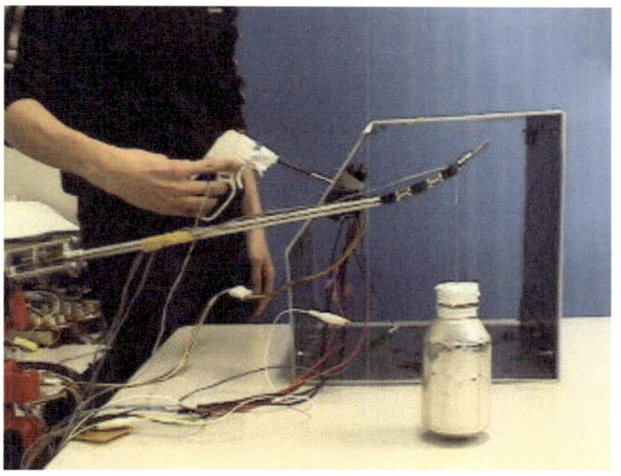

Fig. 12. Appearance of experiment

b. Verification of the effect to the time delay.
   i) $G_p = G_f = 1$ and $T = 0.125$
   ii) Force reflecting servo type bilateral control law with constant time delay $T = 0.125$

In b-ii), the force reflecting servo type bilateral control law is given as follows.

$$\tau_m = K_f\left(f_m - f_s(t-T)\right), \tag{18}$$

$$\tau_s = K_p\left(x_m(t-T) - x_s\right), \tag{19}$$

where K_f and K_p are feedback gains of force and position. The time delay T = 0.125 is intentionally generated in the control system, whose value was referred from (Arata et al., 2007) as the time delay of the control signal between Japan and Thailand: approximately 124.7 ms.

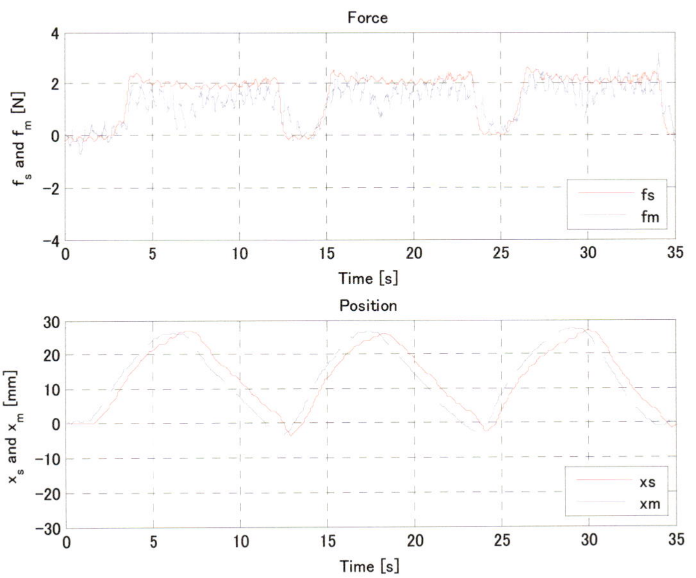

Fig. 13. Experimental result for a-i)

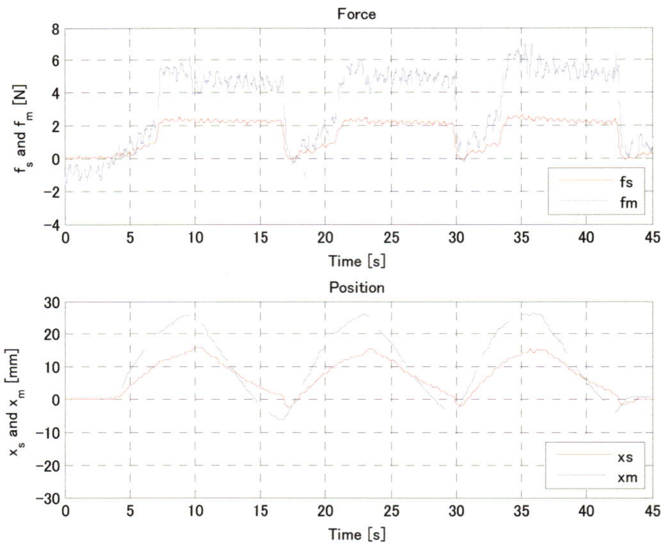

Fig. 14. Experimental result for a-ii)

Note that the proposed bilateral control scheme guarantees stability of the teleoperation system in the presence of constant time delay, however, stability is not guaranteed in use of the force reflecting servo type bilateral control law in the presence of constant time delay.
Feedback gains were adjusted by trial and error through repetition of experiments, which were determined as $\lambda = 3.8$, $K_1 = 30$, $K_m = 400$, $K_s = 400$, $K_p = 60$ and $K_f = 650$. Experimental results for condition a) are shown in Fig. 13 and Fig. 14.
As shown in Fig. 13 and Fig. 14, it is verified that the motion of slave tracks the motion of master with specified scale in both position tracking and force tracking.
Experimental results for condition b) are shown in Fig. 15 and Fig. 16.

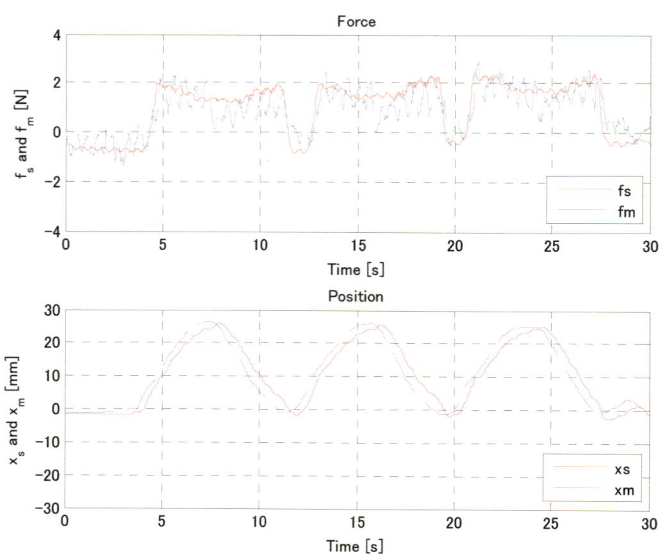

Fig. 15. Experimental result for b-i)

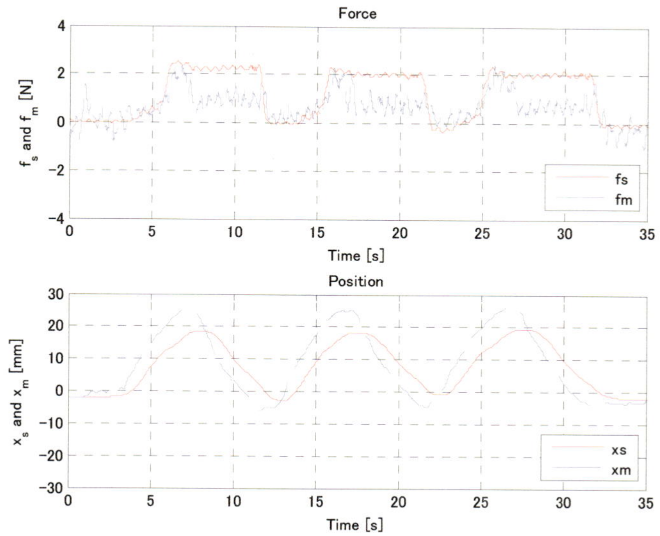

Fig. 16. Experimental result for b-ii)

As shown in Fig. 15 and Fig. 16, tracking errors of both position and force in Fig. 15 are smaller than those of Fig. 16. From the above observations, the effectiveness of the proposed control law for one-DOF bending motion of the DSD robotic forceps was verified.

4. Bilateral control for omnidirectional bending

In this section, the bilateral control scheme described in the former session is extended to omnidirectional bending of the DSD robotic forceps teleoperation system with constant time delay.

4.1 Extension to omnidirectional bending

As shown in Fig.10, master device is modified joy-stick type manipulator. Namely, this is different structured master-slave system. The cross-section views of shaft of the joy-stick and the DSD robotic forceps are shown in Fig.17.

Due to the placement of strain gauges and motors with encoder of the master device, the dynamics of the master device are given in x-y coordinates as follows.

$$m_m\ddot{x}_m + b_m\dot{x}_m + c_m x_m = \tau_{xm} + f_{xm}, \tag{20}$$

$$m_m\ddot{y}_m + b_m\dot{y}_m + c_m y_m = \tau_{ym} + f_{ym}. \tag{21}$$

When only motor A drives, bending direction of the DSD robotic forceps is along A-axis, and when only motor B drives, bending direction of the DSD robotic forceps is along B-axis. Thus, due to the arrangement of the bending linkages, the dynamics of the slave device are given in A-B coordinates as follows.

$$m_s\ddot{A}_s + b_s\dot{A}_s + c_s A_s = \tau_{As} - f_{As}, \tag{22}$$

$$m_s\ddot{B}_s + b_s\dot{B}_s + c_s B_s = \tau_{Bs} - f_{Bs}. \tag{23}$$

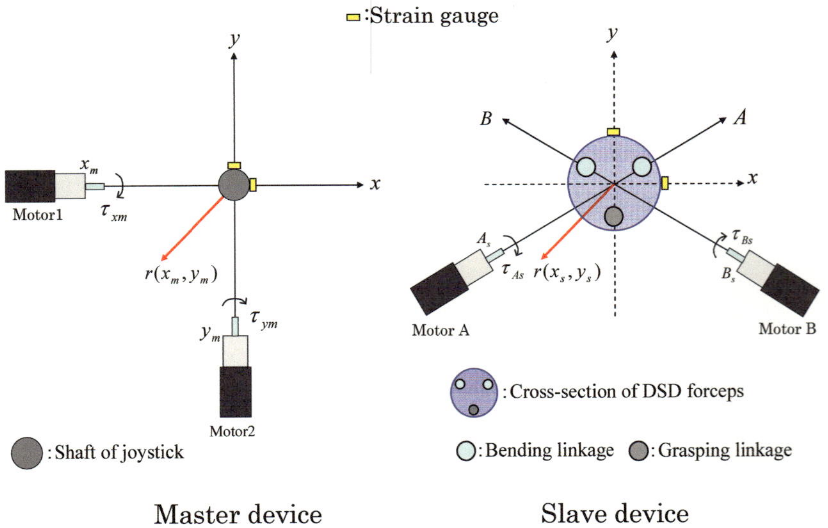

Fig. 17. Coordinates of master device and slave device

In order to extend the proposed bilateral control law to the omnidirectional bending motion of the DSD robotic forceps, the coordinates must be unified.

As shown in Fig. 17, x_m and y_m are measured by encoders. f_{xm}, f_{ym}, f_{xs}, and f_{ys} are measured by strain gauges. τ_{xm}, τ_{ym}, τ_{xs} and τ_{ys} are calculated from the bilateral control laws. These values are obtained in x-y coordinates. Therefore, consider to unify the coordinates in x-y coordinates. While, displacement of the slave A_s and B_s are measured by encoder, which are obtained in A-B coordinates. These values must be changed into x-y coordinates.

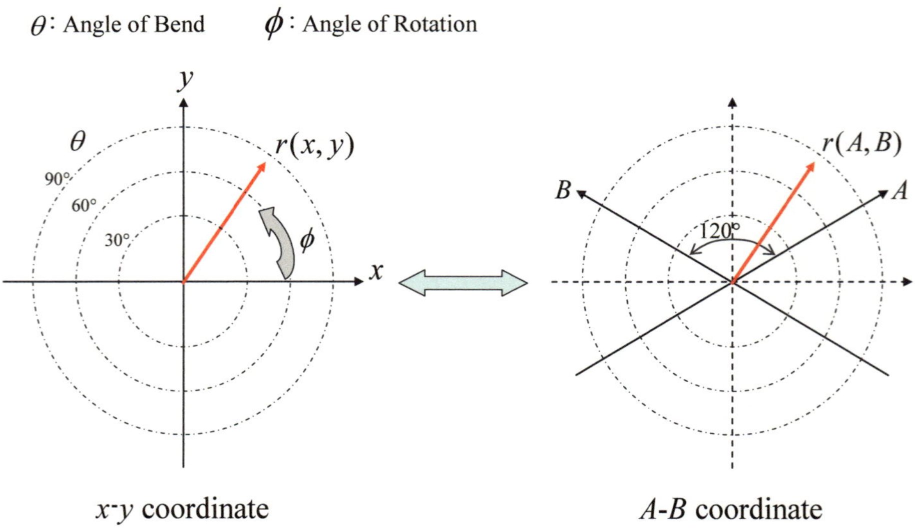

Fig. 18. Change of coordinates

The change of coordinates for position $r(A,B)$ given in A-B coordinates to $r(x,y)$ given in x-y coordinates (Fig. 18) is given as follows.

$$\begin{bmatrix} x \\ y \end{bmatrix} = \frac{1}{2}\begin{bmatrix} \sqrt{3} & -\sqrt{3} \\ 1 & 1 \end{bmatrix}\begin{bmatrix} A \\ B \end{bmatrix}. \tag{24}$$

Thus, the dynamics of the slave device given in A-B coordinates are converted into x-y coordinates. Finally, the dynamics of the two-DOF DSD robotic forceps teleoperation system in horizontal direction and vertical direction are described as follows.

$$\begin{cases} m_m\ddot{x}_m + b_m\dot{x}_m + c_m x_m = \tau_{xm} + f_{xm} \\ m_s\ddot{x}_s + b_s\dot{x}_s + c_s x_s = \tau_{xs} - f_{xs} \end{cases} \tag{25}$$

$$\begin{cases} m_m\ddot{y}_m + b_m\dot{y}_m + c_m y_m = \tau_{ym} + f_{ym} \\ m_s\ddot{y}_s + b_s\dot{y}_s + c_s y_s = \tau_{ys} - f_{ys} \end{cases} \tag{26}$$

For each direction, the bilateral control law derived in the former session, which is developed for one-DOF bending of the DSD robotic forceps, is applied.
However, as shown in Fig. 17, the actual torque inputs to the motors in the slave device are τ_{As} and τ_{Bs}. Therefore, input torque of the slave must be given in A-B coordinates. τ_{As} and τ_{Bs} can be obtained from τ_{xs} and τ_{ys} through an inverse transformation of (24), which is given by

$$\begin{bmatrix} \tau_{As} \\ \tau_{Bs} \end{bmatrix} = \begin{bmatrix} 1/\sqrt{3} & 1 \\ -1/\sqrt{3} & 1 \end{bmatrix}\begin{bmatrix} \tau_{xs} \\ \tau_{ys} \end{bmatrix}. \tag{27}$$

Thus, bilateral control for the omnidirectional bending motion of the DSD robotic forceps is realized.

4.2 Experiments

Experimental works were carried out using the proposed bilateral control laws. The parameter values of the system are given as same value as described in subsection 3.2.
In the experiments, 100g weight pet bottle filled with water was hung up on the tip of the forceps, and the pet bottle was lifted by vertical bending motion of the forceps. Then, the forceps was controlled so that the tip of the forceps draws a quarter circular orbit counterclockwise, and the PET bottle was landed on the floor.
Experimental works were carried out under the communication time delay T = 0.125. The control gains were determined by trial and error through the repetition of experiments, which are given as λ = 5.0, K_1 = 40, K_m = 80, and K_s = 80. Scaling gains were chosen as G_p= G_f = 1. Experimental results are shown in Fig. 19.
In Fig. 19, the top two figures show force and position in x coordinates, and the bottom two figures show force and position in y coordinates. In the experiment, the PET bottle was lifted at around 4 seconds, and landed on the floor at around 20 seconds. The counterclockwise rotation at the tip of the forceps has begun from around 12 seconds.
Although small tracking errors can be seen, the reaction forces which acted on the slave device in x-y directions were reproducible to the master manipulator as tactile sense. In terms of above observations, it can be said that the effectiveness of the proposed control scheme was verified.

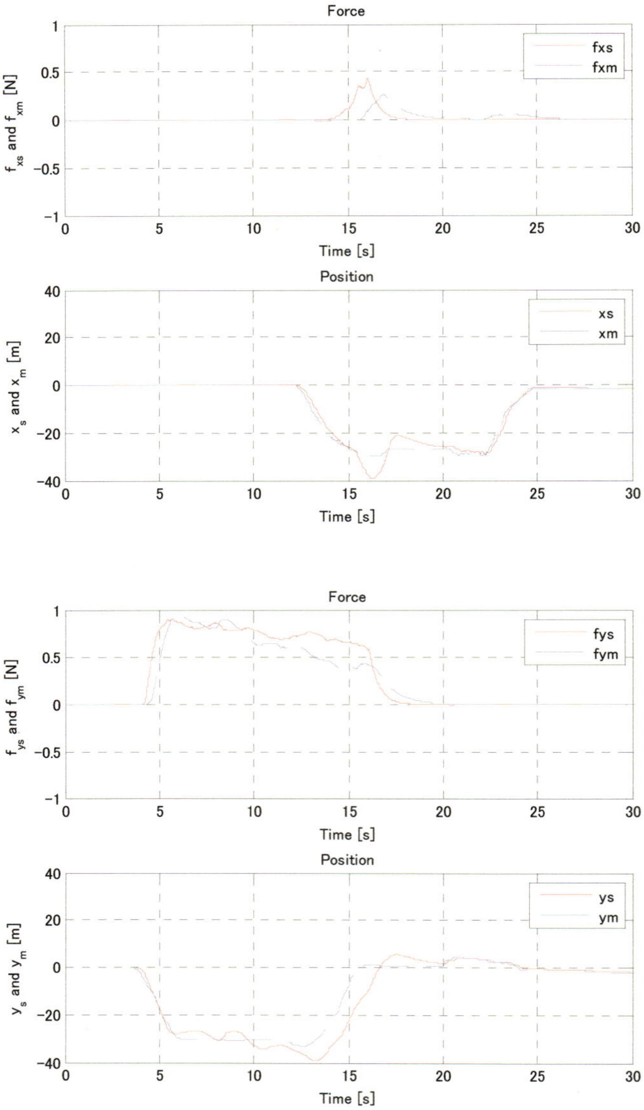

Fig. 19. Experimental results for omnidirectional bending of DSD robotic forceps

5. Conclusion

In this chapter, robust bilateral control for teleoperation systems in the presence of communication time delay was discussed. The Lyapunov function based bilateral control law that enables the motion scaling in both position tracking and force tracking, and guarantees stability of the system in the presence of the constant communication time delay, was proposed under the passivity assumption.

The proposed control law was applied to the haptic control of one-DOF bending motion of the DSD robotic forceps teleoperation system with constant time delay, and experimental works were executed.

In addition, the proposed bilateral control scheme was extended so that it may become applicable to the omnidirectional bending motion of the DSD robotic forceps. Experimental works for the haptic control of omnidirectional bending motion of the DSD robotic forceps teleoperation system with constant time delay were carried out. From the experimental results, the effectiveness of the proposed control scheme was verified.

6. Acknowledgement

The part of this work was supported by Grant-in-Aid for Scientific Research(C) (20500183). The author thanks H. Mikami for his assistance in experimental works.

7. References

Anderson, R. & Spong, M. W. (1989). Bilateral Control of Teleoperators with Time Delay, *IEEE Transactions on Automatic Control*, Vol.34, No. 5, pp.494-501

Arata, J., Mitsuishi, M., Warisawa, S. & Hashizume, M. (2005). Development of a Dexterous Minimally-Invasive Surgical System with Augumented Force Feedback Capability, *Proceedings of 2005 IEEE/RSJ International Conference on Intelligent Robots and Systems*, pp.3207-3212

Arata, J., Takahashi, H., Pitakwatchara, P., Warisawa, S., Tanoue, K., Konishi, K., Ieiri, S., Shimizu, S., Nakashima, N., Okamura, K., Fujino, Y., Ueda, Y., Chotiwan, P., Mitsuishi, M. & Hashizume, M. (2007). A Remote Surgery Experiment Between Japan and Thailand Over Internet Using a Low Latency CODEC System, *Proceedings of IEEE International Conference on Robotics and Automation*, pp.953-959

Chopra, N., Spong, M. W., Hirche, S. & Buss, M. (2003). Bilateral Teleoperation over the Internet: the Time Varying Delay Problem, *Proceedings of the American Control Conference*, pp.155-160

Chopra, N. & Spong, M.W. (2005). On Synchronization of Networked Passive Systems with Time Delays and Application to Bilateral Teleoperation, *Proceedings of SICE Annual Conference* 2005

Guthart, G. & Salisbury, J. (2000). The Intuitive Telesurgery System: Overview and Application, *Proceedings of 2000 IEEE International Conference on Robotics and Automation*, San Francisco, CA, pp.618-621

Ikuta, K., Yamamoto, K. & Sasaki, K. (2003). Development of Remote Microsurgery Robot and New Surgical Procedure for Deep and Narrow Space, *Proceedings of 2003 IEEE International Conference on Robotics & Automation*, Taipei, Taiwan, pp.1103-1108

Ishii, C.; Kobayashi, K.; Kamei, Y. & Nishitani, Y. (2010). Robotic Forceps Manipulator with a Novel Bending Mechanism, *IEEE/ASME Transactions on Mechatronics*, TMECH.2009.2031641, Vol.15, No.5, pp.671-684

Kobayashi, Y., Chiyoda, S., Watabe, K., Okada, M. & Nakamura, Y. (2002). Small Occupancy Robotic Mechanisms for Endoscopic Surgery, *Proceedings of International Conference on Medical Computing and Computer Assisted Intervention*, pp.75-82

Seibold, U., Kubler, B. & Hirzinger, G. (2005). Prototype of Instrument for Minimally Invasive Surgery with 6-Axis Force Sensing Capability, *Proceedings of 2005 IEEE International Conference on Robotics and Automation*, pp.496-501

Taylor, R. & Stoianovici, D. (2003). Medical Robotics in Computer-Integrated Surgery, *IEEE Transactions on Robotics and Automation*, Vol.19, No.5, pp.765-781

Yamashita, H., Iimura, A., Aoki, E., Suzuki, T., Nakazawa, T., Kobayashi, E., Hashizume, M., Sakuma, I. & Dohi, T. (2005). Development of Endoscopic Forceps Manipulator Using Multi-Slider Linkage Mechanisms, *Proceedings of 1st Asian Symposium on Computer Aided Surgery - Robotic and Image guided Surgery -*

Zemiti, N., Morel, G., Ortmaier, T. & Bonnet, N. (2007). Mechatronic Design of a New Robot for Force Control in Minimally Invasive Surgery, *IEEE/ASME Transactions on Mechatronics*, Vol.12, No.2, pp.143-153

26

Robust Vehicle Stability Control Based on Sideslip Angle Estimation

Haiping Du[1] and Nong Zhang[2]
[1]School of Electrical, Computer and Telecommunications Engineering
University of Wollongong, Wollongong, NSW 2522
[2]Mechatronics and Intelligent Systems, Faculty of Engineering
University of Technology, Sydney, P.O. Box 123, Broadway, NSW 2007
Australia

1. Introduction

Vehicle stability control is very important to vehicle active safety, in particular, during severe driving manoeuvres. The yaw moment control has been regarded as one of the most promising means of vehicle stability control, which could considerably enhance vehicle handling and stability (Abe, 1999; Mirzaei, 2010). Up to the date, different strategies on yaw moment control, such as optimal control (Esmailzadeh et al., 2003; Mirzaei et al., 2008), fuzzy logic control (Boada et al, 2005; Li & Yu 2010), internal model control (IMC) (Canale et al., 2007), flatness-based control (Antonov et al, 2008), and coordinated control (Yang et al, 2009), etc., have been proposed in the literature.

It is noticed that most existing yaw moment control strategies rely on the measurement of both sideslip angle and yaw rate. However, the measurement of sideslip angle is hard to be done in practice because the current available sensors for sideslip angle measurement are all too expensive to be acceptable by customers. To implement yaw moment controller without increasing too much cost on a vehicle, the estimation of sideslip angle based on measurement available signals, such as yaw rate and lateral acceleration, etc., is becoming necessary. And, the measurement noise should also be considered so that the estimation based controller is more robust. On the other hand, most of the existing studies use a linear lateral dynamics model with nominal cornering stiffness for the yaw moment controller design. Since the yaw moment control obviously relies on the tyre lateral force and the tyre force strongly depends on tyre vertical load and road conditions which are very sensitive to the vehicle motion and the environmental conditions, the tyre cornering stiffness must have uncertainties. Taking cornering stiffness uncertainties into account will make the controller being more robust to the variation of road conditions. In addition, actuator saturation limitations resulting from some physical constraints and tyre-road conditions must be considered so that the implementation of the controller can be more practical.

In this chapter, a nonlinear observer based robust yaw moment controller is designed to improve vehicle handling and stability with considerations on cornering stiffness uncertainties, actuator saturation limitation, and measurement noise. The yaw moment controller uses the measurement of yaw rate and the estimation of sideslip angle as feedback signals, where the sideslip angle is estimated by a Takagi-Sugeno (T-S) fuzzy model-based observer. The design objective of this observer based controller is to achieve optimal performance on sideslip angle and estimation error subject to the cornering stiffness uncertainties, actuator saturation limitation, and measurement noise. The design of such an observer based controller is implemented in a two-step procedure where linear matrix inequalities (LMIs) are built and solved by using available software Matlab LMI Toolbox. Numerical simulations on a vehicle model with nonlinear tyre model are used to validate the control performance of the designed controller. The results show that the designed controller can achieve good performance on sideslip angle responses for a given actuator saturation limitation with measurement noise under different road conditions and manoeuvres.

This chapter is organised as follows. In Section 2, the vehicle lateral dynamics model is introduced. The robust observer-based yaw moment controller design is introduced in Section 3. In Section 4, the simulation results on a nonlinear vehicle model are discussed. Finally, conclusions are presented in Section 5.

The notation used throughout the paper is fairly standard. For a real symmetric matrix M the notation of M>0 (M<0) is used to denote its positive- (negative-) definiteness. $\|.\|$ refers to either the Euclidean vector norm or the induced matrix 2-norm. I is used to denote the identity matrix of appropriate dimensions. To simplify notation, * is used to represent a block matrix which is readily inferred by symmetry.

2. Vehicle dynamics model

In spite of its simplicity, a bicycle model of vehicle lateral dynamics, as shown in Fig. 1, can well represent vehicle lateral dynamics with constant forward velocity and is often used for controller design and evaluation.

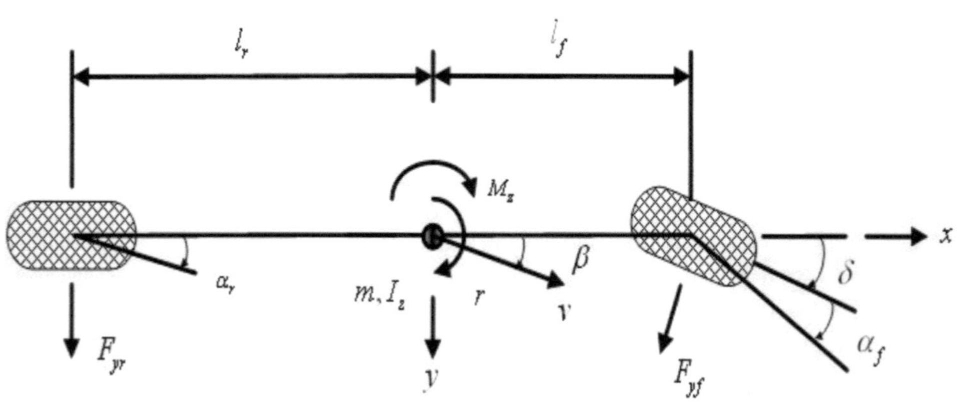

Fig. 1. Vehicle lateral dynamics model

In this model, the vehicle has mass m and moment of inertia I_z about yaw axis through its center of gravity (CG). The front and rear axles are located at distances l_f and l_r, respectively, from the vehicle CG. The front and rear lateral tyre forces F_{yf} and F_{yr} depend on slip angles α_f and α_r, respectively, and the steering angle δ changes the heading of the front tyres.

When lateral acceleration is lower, the tyres operate in the linear region and the lateral forces at the front and rear can be related to slip angles by the cornering stiffnesses of the front and rear tyres as

$$F_{yf}=-C_{\alpha f}\alpha_f, \qquad F_{yr}=-C_{\alpha r}\alpha_r \tag{1}$$

where $C_{\alpha f}$ and $C_{\alpha r}$ are cornering stiffnesses of the front and rear tyres, respectively. With using Newton law and the following relationships

$$\alpha_f=\beta+\frac{l_f r}{v}-\delta, \qquad \alpha_r=\beta-\frac{l_r r}{v} \tag{2}$$

vehicle lateral dynamics model can be written in state space equation as

$$\begin{bmatrix}\dot{\beta}\\ \dot{r}\end{bmatrix}=\begin{bmatrix}-\frac{C_{\alpha f}+C_{\alpha r}}{mv} & -1-\frac{l_f C_{\alpha f}-l_r C_{\alpha r}}{mv^2}\\ -\frac{l_f C_{\alpha f}-l_r C_{\alpha r}}{I_z} & -\frac{l_f^2 C_{\alpha f}-l_r^2 C_{\alpha r}}{I_z v}\end{bmatrix}\begin{bmatrix}\beta\\ r\end{bmatrix}+\begin{bmatrix}\frac{C_{\alpha f}}{mv}\\ \frac{l_f C_{\alpha f}}{I_z}\end{bmatrix}\delta+\begin{bmatrix}0\\ \frac{1}{I_z}\end{bmatrix}M_z \tag{3}$$

where β is vehicle sideslip angle, r is yaw rate, M_z is yaw moment, v is forward velocity. Equation (3) can be further written as

$$\dot{x}=Ax+B_1 w+B_2\bar{u} \tag{4}$$

where

$$A=\begin{bmatrix}-\frac{C_{\alpha f}+C_{\alpha r}}{mv} & -1-\frac{l_f C_{\alpha f}-l_r C_{\alpha r}}{mv^2}\\ -\frac{l_f C_{\alpha f}-l_r C_{\alpha r}}{I_z} & -\frac{l_f^2 C_{\alpha f}-l_r^2 C_{\alpha r}}{I_z v}\end{bmatrix},\quad B_1=\begin{bmatrix}\frac{C_{\alpha f}}{mv}\\ \frac{l_f C_{\alpha f}}{I_z}\end{bmatrix}$$
$$B_2=\begin{bmatrix}0\\ \frac{1}{I_z}\end{bmatrix},\quad x=\begin{bmatrix}\beta\\ r\end{bmatrix},\quad w=\delta,\ u=M_z \tag{5}$$

and

$$\bar{u}=\text{sat}(u)=\begin{cases}-u_{lim} & \text{if } u<-u_{lim}\\ u & \text{if } -u_{lim}\le u\le u_{lim}\\ u_{lim} & \text{if } u>u_{lim}\end{cases} \tag{6}$$

which is used to define the saturation state of control input and u_{lim} is the limitation of available yaw moment in practice.

It is noticed that the linear relationship between tyre lateral force and slip angle in equation (1) can only exist when lateral acceleration is lower (less than about 0.4 g). When lateral acceleration increases, the relationship goes into nonlinear region as shown in Fig. 2 where change of lateral tyre force to sideslip angle generated from Dugoff tyre model is depicted.

Therefore, cornering stiffnesses are no longer constant values but time-varying variables, and relationship between tyre lateral force and slip angle is a nonlinear function of sideslip angle. To describe this nonlinear relationship, cornering stiffnesses need to be measured or estimated. However, either way is difficult to be implemented due to cost or accuracy consideration although some approaches have been proposed for the estimation of cornering stiffnesses.

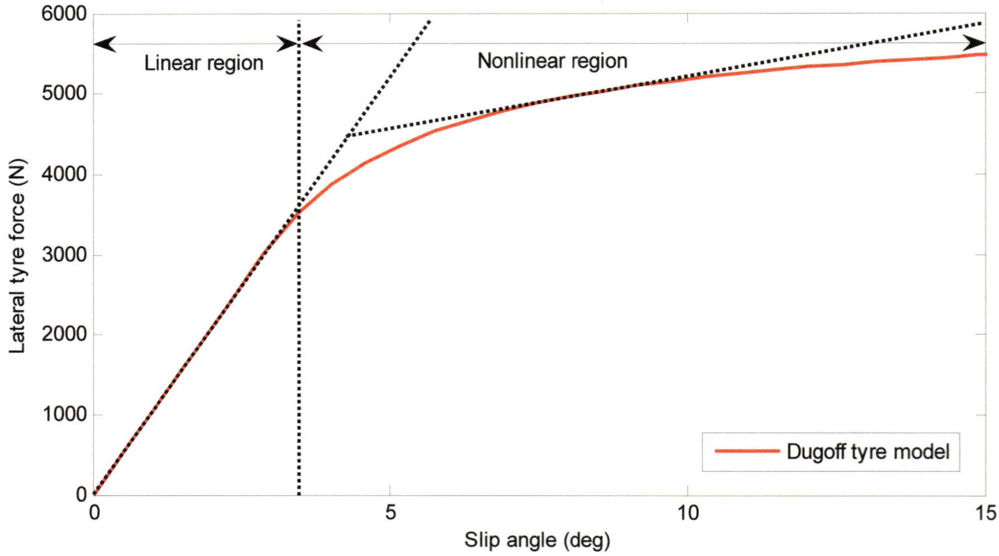

Fig. 2. Tyre lateral force characteristics.

Since Takagi-Sugeno (T-S) fuzzy model has been effectively applied to approximate nonlinear functions in many different applications (Tanaka & Wang, 2001), instead of estimating cornering stiffness, we use T-S fuzzy model to describe the nonlinear relationship between tyre lateral force and sideslip angle in the vehicle lateral dynamics model. The plant rules for the T-S fuzzy lateral dynamics model are built as

IF Δr is N_1 THEN

$$\dot{x}=A_1x+B_{11}w+B_2\bar{u} \tag{7}$$

IF Δr is N_2 THEN

$$\dot{x}=A_2x+B_{12}w+B_2\bar{u} \tag{8}$$

where N_1 and N_2 are fuzzy sets, Δr is premise variable which is defined by deviation of yaw rate as

$$\Delta r=\left[1+\left(\frac{v}{v_c}\right)^2\right]\frac{(r_{ref}-r)l}{v}\frac{|r|}{r} \tag{9}$$

where v_c is characteristic velocity, $l=l_f+l_r$, and the reference yaw rate r_{ref} is defined as

$$r_{ref}=\frac{v}{l}\frac{\delta}{1+\left(\frac{v}{v_c}\right)^2} \tag{10}$$

The deviation of yaw rate is used as a premise variable in this T-S fuzzy model because it can approximately show the degree of nonlinear state and can be used to judge whether the vehicle is in linear or nonlinear region (Fukada, 1999).
By fuzzy blending, the final output of the T-S fuzzy model is inferred as follows

$$\dot{x}=\sum_{i=1}^{2}h_i(\Delta r)(A_i x+B_{1i}w+B_2\bar{u}) \tag{11}$$

where $h_i(\Delta r)=\mu_i(\Delta r)/\sum_{i=1}^{2}\mu_i(\Delta r)$, $\mu_i(\Delta r)$ is the degree of the membership of Δr in N_i. In general, triangular membership function can be used for fuzzy set N_i, and we have $h_i(\Delta r)\geq 0$ and $\sum_{i=1}^{2}h_i(\Delta r)=1$. A_i and B_{1i} are sub-matrices which are obtained by substituting cornering stiffness values for linear and nonlinear regions, respectively.

3. Observer based robust controller design

It was pointed in many previous research works that both sideslip angle and yaw arte are useful information for effective vehicle handling and stability control. However, sensors for measuring sideslip angle are really expensive and cannot be used in stability control for commercial automotives. Therefore, estimation of slip angle is a cost-effective way to solve this problem. On the contrary, measurement of yaw rate is relatively easy and cheap, and gyroscopic sensor can be used to do it. Base on the measurable yaw rate signal, sideslip angle can be estimated and then used for full state feedback control signal.
In a real application, the state measurements can not be perfect. Thus, the measured state variables should be corrupted by measurement noises as

$$y=Cx+n \tag{12}$$

where y is the measured output, n denotes the measurement noise, C is a constant matrix (if all the state variables are measured, C is an identity matrix). To estimate the state variables from noisy measurements, we construct a T-S fuzzy observer as

$$\begin{aligned}\dot{\hat{x}}&=\sum_{i=1}^{2}h_i(\Delta r)[A_i\hat{x}+B_2\bar{u}+L_i(y-\hat{y})]\\ \hat{y}&=C\hat{x}\end{aligned} \tag{13}$$

where $\hat{x}$ is observer state vector, L_i is observer gain matrix to be designed, $\hat{y}$ is observer output.
By defining the estimation error

$$e=x-\hat{x} \tag{14}$$

we obtain

$$\dot{e}=\dot{x}-\dot{\hat{x}}=\sum_{i=1}^{2}h_i(\Delta r)[(A_i-L_iC)e+B_{1i}w-L_in] \tag{15}$$

To making the estimation error as small as possible, we define one control output as

$$z_o=C_ee \tag{16}$$

where C_e is constant matrix. The objective of observer design is to find L_i such that the H_∞ norm of $\|T_{ow}\|$, which denotes the closed-loop transfer function from the steering input w to the control output z_o (estimation error e) and is defined as

$$\|T_{ow}\|_\infty=\sup_{\|w\|_2\neq 0}\frac{\|z_o\|_2}{\|w\|_2} \tag{17}$$

where $\|z_o\|_2^2=\int_0^\infty z_o^T(t)z_o(t)dt$ and $\|w\|_2^2=\int_0^\infty w^T(t)w(t)dt$, is minimised.

On the other hand, to realise good handling and stability, the sideslip angle and the yaw rate need to be controlled to the desired values. Generally, the desired sideslip angle is given as zero and the desired yaw rate is defined in terms of vehicle speed and steering input angle (Zheng, 2006). For simplicity, we only consider to control sideslip angle as small as possible, which in most cases can also lead to satisfied yaw rate. Thus, we define another control output as

$$z_\beta=C_\beta x \tag{18}$$

where C_β =[1 0], and the objective is to design a robust T-S fuzzy controller based on the estimated state variables as

$$u=\sum_{i=1}^{2}h_i(\Delta r)K_i\hat{x} \tag{19}$$

where K_i is control gain matrix to be designed, such as the H_∞ norm of $\|T_{\beta w}\|$, which denotes the closed-loop transfer function from the steering input w to the control output z_β, is minimised. Together with control output (16), the control output for both observer and controller design is defined as

$$\tilde{z}=\tilde{C}_z\tilde{x}=\begin{bmatrix}C_\beta & C_\beta\\ 0 & C_e\end{bmatrix}\begin{bmatrix}\hat{x}\\ e\end{bmatrix} \tag{20}$$

where $\tilde{x}=[\hat{x}^T\ e^T]^T$ is the augmented system state vector. It can be seen from (20) that C_e can be used to make the compromise between z_β and z_o in the control objective.

To derive the conditions for obtaining K_i and L_i, we now define a Lyapunov function as

$$V=\hat{x}^TP\hat{x}+e^TQe \tag{21}$$

where $P = P^T > 0$, $Q = Q^T > 0$. Taking the time derivative of V along (13) and (15) yields

$$
\begin{aligned}
\dot{V}&=\dot{\hat{x}}^T P\hat{x}+\hat{x}^T P\dot{\hat{x}}+\dot{e}^T Qe+e^T Q\dot{e}\\
&=\sum_{i=1}^{2} h_i\left\{\begin{array}{l}2\left[A_i x+B_2\dfrac{1+\varepsilon}{2}u+B_2\left(\bar{u}-\dfrac{1+\varepsilon}{2}u\right)+L_iCe+L_in\right]^T P\hat{x}\\+2\left[(A_i-L_iC)e+B_{1i}w-L_in\right]^T Qe\end{array}\right\}\\
&\le\sum_{i=1}^{2} h_i\left\{\begin{array}{l}2\left[A_i x+B_2\dfrac{1+\varepsilon}{2}u+L_iCe+L_in\right]^T P\hat{x}+2\left[(A_i-L_iC)e+B_{1i}w-L_in\right]^T Qe\\+\kappa\left(u-\dfrac{1+\varepsilon}{2}u\right)^T\left(u-\dfrac{1+\varepsilon}{2}u\right)+\kappa^{-1}\hat{x}^T PB_2B_2^T P\hat{x}\end{array}\right\}\\
&\le\sum_{i=1}^{2} h_i\left\{\begin{array}{l}2\left[A_i x+B_2\dfrac{1+\varepsilon}{2}u+L_iCe+L_in\right]^T P\hat{x}+2\left[(A_i-L_iC)e+B_{1i}w-L_in\right]^T Qe\\+\kappa\left(\dfrac{1-\varepsilon}{2}\right)^2 u^Tu+\kappa^{-1}\hat{x}^T PB_2B_2^T P\hat{x}\end{array}\right\}\\
&=\sum_{i=1}^{2} h_i\left\{\begin{array}{l}\hat{x}^T\left[A_i^TP+PA_i+\left(\dfrac{1+\varepsilon}{2}B_2K_i\right)^T P+\dfrac{1+\varepsilon}{2}PB_2K_i+\kappa\left(\dfrac{1-\varepsilon}{2}\right)^2K_i^TK_i+\kappa^{-1}PB_2B_2^TP\right]\hat{x}(t)\\+e^T\left[(A_i-L_iC)^TQ+Q(A_i-L_iC)\right]e+\hat{x}^TPL_iCe+e^TC^TL_i^TP\hat{x}\\+w^TB_{1i}^TQe+e^TQB_{1i}w+\hat{x}^TPL_in+n^TL_i^TP\hat{x}-e^TQL_in-n^TL_i^TQe\end{array}\right\}\\
&=\sum_{i=1}^{2} h_i\left[\tilde{x}^T\Phi_i\tilde{x}+\tilde{x}^T\Gamma_i\tilde{w}+\tilde{w}^T\Gamma_i^T\tilde{x}\right]
\end{aligned}
\tag{22}
$$

where definition (19) and inequalities $X^TY+Y^TX\le\kappa X^TX+\kappa^{-1}Y^TY$ for any matrices X and Y and positive scalar κ (Du et al, 2005) and $\left(\bar{u}-\frac{1+\varepsilon}{2}u\right)^T\left(\bar{u}-\frac{1+\varepsilon}{2}u\right)\le\left(\frac{1-\varepsilon}{2}\right)^2u^Tu$ for any $0<\varepsilon<1$ (Kim & Jabbari, 2002) are applied, and $\tilde{w}=[w^T\ n^T]^T$,

$$
\Phi_i=\begin{bmatrix}A_i^TP+PA_i+\left(\dfrac{1+\varepsilon}{2}B_2K_i\right)^TP+\dfrac{1+\varepsilon}{2}PB_2K_i\\+\kappa\left(\dfrac{1-\varepsilon}{2}\right)^2K_i^TK_i+\kappa^{-1}PB_2B_2^TP & PL_iC\\ * & (A_i-L_iC)^TQ+Q(A_i-L_iC)\end{bmatrix}
\tag{23}
$$

$$
\Gamma_i=\begin{bmatrix}0 & PL_i\\ QB_{1i} & -QL_i\end{bmatrix}
\tag{24}
$$

By adding $\tilde{z}^T\tilde{z}-\gamma^2w^Tw$ to two sides of (22) yields

$$
\begin{aligned}
&\dot{V}+\tilde{z}^T\tilde{z}-\gamma^2\tilde{w}^T\tilde{w} \\
&\le \sum_{i=1}^{2} h_i\left[\tilde{x}^T\Phi_i\tilde{x}+\tilde{x}^T\Gamma_i\tilde{w}+\tilde{w}^T\Gamma_i^T\tilde{x}+\tilde{x}^T\tilde{C}_z^T\tilde{C}_z\tilde{x}-\gamma^2\tilde{w}^T\tilde{w}\right] \\
&=\sum_{i=1}^{2} h_i\bar{x}^T\Theta_i\bar{x}
\end{aligned}
\tag{25}
$$

where $\bar{x}=[\tilde{x}^T \ \tilde{w}^T]^T$, and

$$
\Theta_i=\begin{bmatrix}
\begin{array}{l}A_i^TP+PA_i+\left(\frac{1+\varepsilon}{2}B_2K_i\right)^TP+\frac{1+\varepsilon}{2}PB_2K_i \\ +\kappa\left(\frac{1-\varepsilon}{2}\right)^2K_i^TK_i+\kappa^{-1}PB_2B_2^TP+C_\beta^TC_\beta\end{array} & PL_iC+C_\beta^TC_\beta & 0 & PL \\
* & \begin{array}{c}(A_i-L_iC)^TQ+Q(A_i-L_iC) \\ +C_e^TC_e+C_\beta^TC_\beta\end{array} & QB_{1i} & -QL \\
* & * & -\gamma^2 & 0 \\
* & * & * & -\gamma^2
\end{bmatrix}
\tag{26}
$$

It can inferred from (25) that if $\Theta_i < 0$, then $\dot{V}+\tilde{z}^T\tilde{z}-\gamma^2\tilde{w}^T\tilde{w}<0$. Thus, the closed-loop system augmented by (13) and (15) is stable when the disturbance $\tilde{w}=0$ and the H_∞ performance on $\|T_{\tilde{z}w}\|$ is satisfied when $\tilde{x}(0)=0$.
By the Schur complement, $\Theta_i < 0$ is equivalent to

$$
\begin{bmatrix}
\begin{array}{l}A_i^TP+PA_i+\left(\frac{1+\varepsilon}{2}B_2K_i\right)^TP+\frac{1+\varepsilon}{2}PB_2K_i \\ +\kappa\left(\frac{1-\varepsilon}{2}\right)^2K_i^TK_i+\kappa^{-1}PB_2B_2^TP+C_\beta^TC_\beta\end{array} & PL_iC+C_\beta^TC_\beta \\
* & \begin{array}{c}(A_i-L_iC)^TQ+Q(A_i-L_iC) \\ +C_e^TC_e+C_\beta^TC_\beta\end{array}
\end{bmatrix}
+\gamma^{-2}\begin{bmatrix} 0 & PL_i \\ QB_{1i} & -QL_i\end{bmatrix}\begin{bmatrix} 0 & PL_i \\ QB_{1i} & -QL_i\end{bmatrix}^T<0
\tag{27}
$$

which can be written as

$$
\begin{bmatrix}\Omega_{11} & \Omega_{12} \\ * & \Omega_{22}\end{bmatrix}<0
\tag{28}
$$

where

$$
\begin{aligned}
\Omega_{11}&=A_i^TP+PA_i+\left(\frac{1+\varepsilon}{2}B_2K_i\right)^TP+\frac{1+\varepsilon}{2}PB_2K_i+\kappa\left(\frac{1-\varepsilon}{2}\right)^2K_i^TK_i+\kappa^{-1}PB_2B_2^TP \\
&\quad +C_\beta^TC_\beta+\gamma^{-2}PL_iL_i^TP \\
\Omega_{12}&=PL_iC+C_\beta^TC_\beta-\gamma^{-2}PL_iL_i^TQ \\
\Omega_{22}&=(A_i-L_iC)^TQ+Q(A_i-L_iC)+C_e^TC_e+C_\beta^TC_\beta+\gamma^{-2}Q(B_{1i}B_{1i}^T+L_iL_i^T)Q
\end{aligned}
\tag{29}
$$

It is noted from (28) that P, Q, K_i, L_i, and κ are unknown parameters in the inequality that need to be determined. Because they are coupled together, no effective algorithms for solving them simultaneously can be found by now. Therefore, a two-step procedure is applied. Note that (28) means that $\Omega_{22} < 0$. So, in the first step, we solve $\Omega_{22} < 0$. By defining $X_i = QL_i$ and using the Schur complement, from $\Omega_{22} < 0$, we obtain

$$\begin{bmatrix} A_i^T Q^T - C^T X_i^T + QA_i - X_i C + C_\beta^T C_\beta & C_e^T & QB_{1i} & X_i \\ * & -I & 0 & 0 \\ * & * & -\gamma^2 I & 0 \\ * & * & * & -\gamma^2 I \end{bmatrix} < 0 \tag{30}$$

which are LMIs and can be solved by means of the Matlab LMI Toolbox software. Then, we can obtain L_i by using $L_i = Q^{-1}X_i$ for a given γ.

In the second step, by defining $W = P^{-1}$, pre- and post-multiplying (28) by $\text{diag}(W\ I)^T$ and its transpose, respectively, we obtain

$$\begin{bmatrix} \begin{matrix} WA_i^T + A_i W + W\left(\frac{1+\varepsilon}{2}B_2 K_i\right)^T + \frac{1+\varepsilon}{2}B_2 K_i W \\ +\gamma^{-2}L_i L_i^T + \kappa\left(\frac{1-\varepsilon}{2}\right)^2 WK_i^T K_i W + \kappa^{-1}B_2 B_2^T + WC_\beta^T C_\beta W \end{matrix} & L_i C + WC_\beta^T C_\beta - \gamma^{-2}L_i L_i^T Q \\ & \Omega_{22} \end{bmatrix} < 0 \tag{31}$$

After defining $Y_i = K_i W$ and using the Schur complement, we obtain

$$\begin{bmatrix} \begin{matrix} WA_i^T + A_i W + \frac{1+\varepsilon}{2}B_2 Y_i + \frac{1+\varepsilon}{2}Y_i B_2 \\ +\kappa\left(\frac{1-\varepsilon}{2}\right)^2 Y_i^T Y_i + \kappa^{-1}B_2 B_2^T + \gamma^{-2}L_i L_i^T \end{matrix} & WC_\beta^T & \begin{matrix} L_i C + WC_\beta^T C_\beta \\ -\gamma^{-2}L_i L_i^T Q \end{matrix} \\ * & -I & 0 \\ * & * & \Omega_{22} \end{bmatrix} < 0 \tag{32}$$

which are LMIs and can be solved by means of the Matlab LMI Toolbox software to obtain $K_i = Y_i W^{-1}$ for a given γ.

On the other hand, from (19), the constraint

$$\left|\sum_{i=1}^{2} h_i(\Delta r) K_i \hat{x}\right| \le \frac{u_{lim}}{\varepsilon} \tag{33}$$

can be expressed as $|K_i \hat{x}| \le \frac{u_{lim}}{\varepsilon}$.

Let $\Psi(K_i)=\left\{\hat{x}\left|\left|\hat{x}K_i^TK_i\hat{x}\right|\le\left(\frac{u_{lim}}{\varepsilon}\right)^2\right.\right\}$, the equivalent condition for an ellipsoid $\Psi(P,\rho)=\left\{\hat{x}\left|\left|\hat{x}^TPx\le\rho\right|\right.\right\}$ being a subset of $\Psi(K_i)$, i.e., $\Psi(P,\rho)\subset\Psi(K_i)$, is given as (Cao & Lin, 2003)

$$K_i\left(\frac{P}{\rho}\right)^{-1}K_i^T\le\left(\frac{u_{lim}}{\varepsilon}\right)^2 \tag{34}$$

By the Schur complement, inequality (34) can be written as

$$\begin{bmatrix}\left(\frac{u_{lim}}{\varepsilon}\right)^2 I & K_i\left(\frac{P}{\rho}\right)^{-1} \\ * & \left(\frac{P}{\rho}\right)^{-1}\end{bmatrix}\ge 0 \tag{35}$$

Using the definitions $W = P^{-1}$ and $Y_i = K_iW$, inequality (35) is equivalent to

$$\begin{bmatrix}\left(\frac{u_{lim}}{\varepsilon}\right)^2 I & Y_i \\ * & \rho^{-1}W\end{bmatrix}\ge 0 \tag{36}$$

In summary, the procedure for the observer based robust controller design is given as: (1) give initial value for γ; (2) solve LMIs (30), (32), and (36) to obtain L_i and K_i; (3) decrease γ and repeat the previous two steps until no feasible solutions can be found; (4) construct the observer and controller in terms of L_i and K_i.

4. Numerical simulations

To evaluate the effectiveness of the proposed observer based controller design approach, numerical simulations on a yaw-plane 2DOF vehicle dynamics model with nonlinear Dugoff tyre model will be done in this section. The parameters used for the vehicle model are given as m=1298.9 kg, I_z=1627 kg.m^2, l_f=1.0 m, l_r=1.454 m. The robust observer based controller is designed using the above introduced approach, where C_f= C_r =60000 N.rad^{-1} is used when tyre sideslip angle is in linear region and C_f= C_r =6000 N.rad^{-1} is used when tyre sideslip angle is in nonlinear region, and the saturation limit is assumed as 3000 Nm, i.e., u_{lim}=3000 Nm. By choosing ε =0.024, ρ =9.8, we obtain the controller matrices as $K_1=10^4[2.2258\ -2.6083]$ and $K_2=10^4[-1.1797\ -1.6864]$, and observer gain matrices as L_1=[8.1763 165.4576] and L_2=[8.4599 162.6120].

To testify the vehicle lateral dynamics performance, a J-turn manoeuvre, which is produced from the ramp steering input (the maximum degree is 6 deg), is used. To validate the effectiveness of the designed observer based controller, we first assume the vehicle is driving on a snow surface road (road friction is assumed as 0.5) with forward velocity 20 m/s, and only yaw rate is measurable without measurement noise. To see the observer performance clearly, we define different initial values for the vehicle model and observer. Fig. 3 shows sideslip angle responses under J-turn manoeuvre for the uncontrolled system (without any controller), the controlled system (with the designed controller), and the sideslip angle observer.

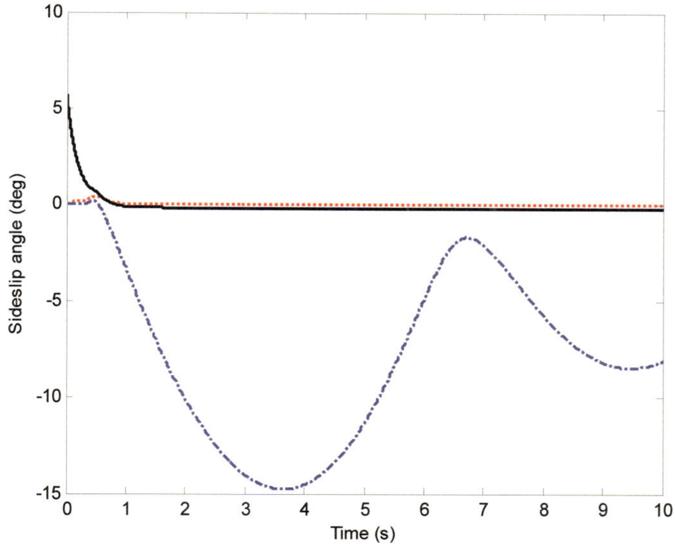

Fig. 3. Sideslip angle responses under J-turn manoeuvre on a snow road without measurement noise. Dashed-dotted line is sideslip angle for uncontrolled system. Dotted line is sideslip angle for controlled system with the designed controller, and solid line is sideslip angle estimated from observer.

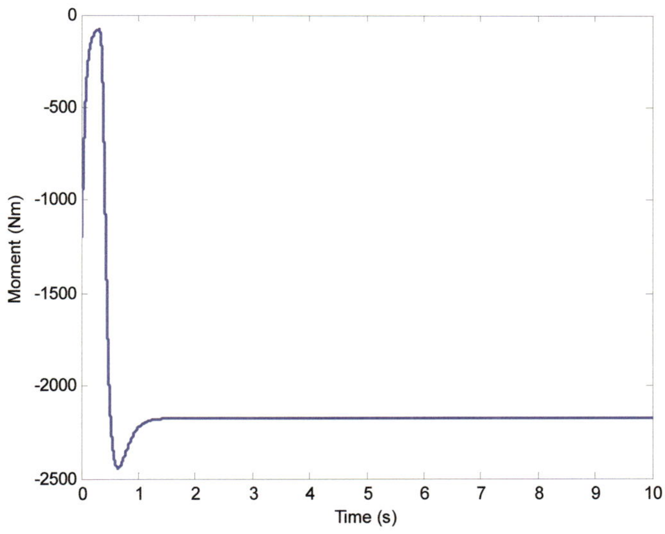

Fig. 4. Yaw moment under J-turn manoeuvre on a snow road without measurement noise.

It can be seen from Fig. 3 that the sideslip angle of the controlled system converges to the desired sideslip value, zero degree. On the contrary, the sideslip angle of the uncontrolled system is big which may cause vehicle unstable motion (Mirzaei, 2010). It is also observed from Fig. 3 that the estimated sideslip angle is converging to the real sideslip angle quickly even though there is big difference on the initial values of observer and vehicle model. Fig. 4 shows the required yaw moment, which is within the defined saturation limit. As demonstrated by the simulation results, the designed observer based controller effectively improves the vehicle handling and stability performance with using yaw rate measurement.

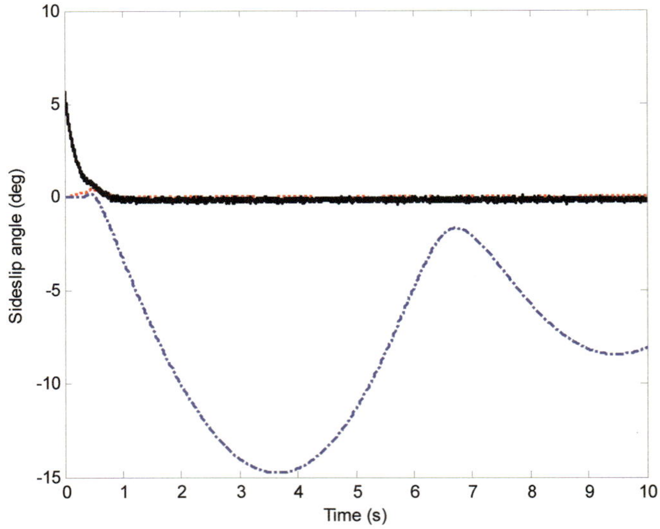

Fig. 5. Sideslip angle responses under J-turn manoeuvre on a snow road with measurement noise. Dashed-dotted line is sideslip angle for uncontrolled system. Dotted line is sideslip angle for controlled system with the designed controller, and solid line is sideslip angle estimated from observer.

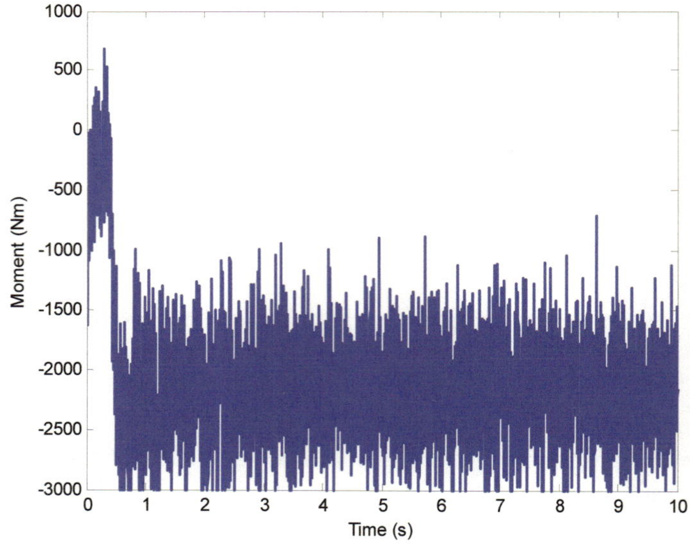

Fig. 6. Yaw moment under J-turn manoeuvre on a snow road with measurement noise.

To validate the robustness of the designed controller, we now add measurement noise on yaw rate. The sideslip angle responses under J-turn manoeuvre for uncontrolled system, controlled system, and observer are shown in Fig. 5. It can been seen from Fig. 5 that the sideslip angle of the controlled system is still approaching to the desired sideslip angle in spite of small effect caused by the measurement noise. The required yaw moment is shown in Fig. 6, where big variations on the yaw moment caused by measurement noise can be observed.

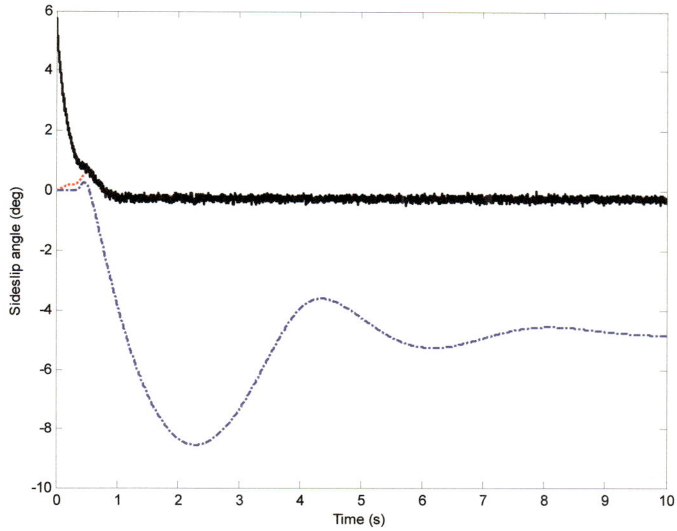

Fig. 7. Sideslip angle responses under J-turn manoeuvre on a dry surface road with measurement noise. Dashed-dotted line is sideslip angle for uncontrolled system. Dotted line is sideslip angle for controlled system with the designed controller, and solid line is sideslip angle estimated from observer.

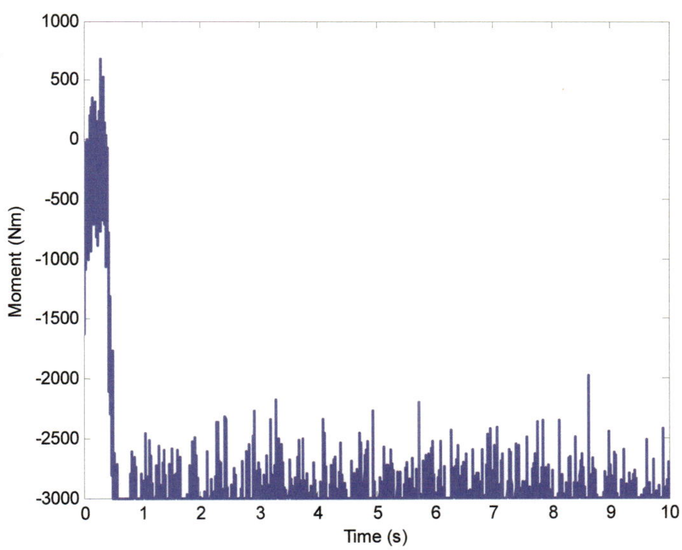

Fig. 8. Yaw moment under J-turn manoeuvre on a dry road with measurement noise.

To further validate the robustness of the designed controller, we now assume the vehicle is driving on a dry surface road (road friction is assumed as 0.9) with forward velocity 20 m/s. The sideslip angle responses and yaw moment are shown in Figs. 7 and 8, respectively. It can be seen the designed controller can achieve good performance with the limited yaw moment no matter the change of road condition.

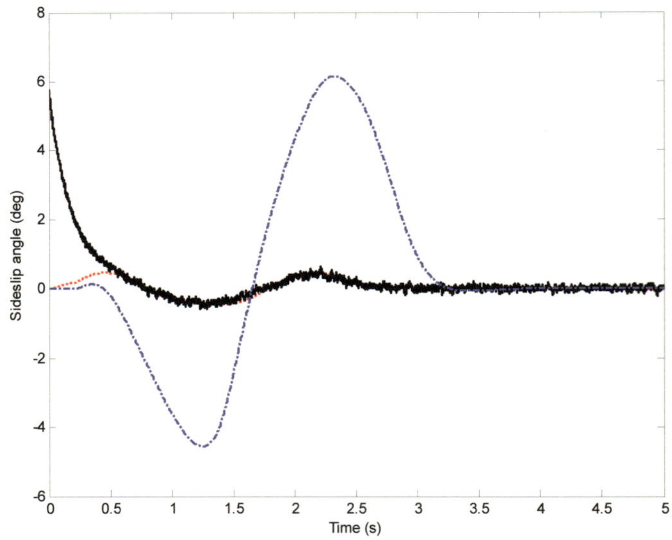

Fig. 9. Sideslip angle responses under lane change manoeuvre on a dry road with measurement noise. Dashed-dotted line is sideslip angle for uncontrolled system. Dotted line is sideslip angle for controlled system with the designed controller, and solid line is sideslip angle estimated from observer.

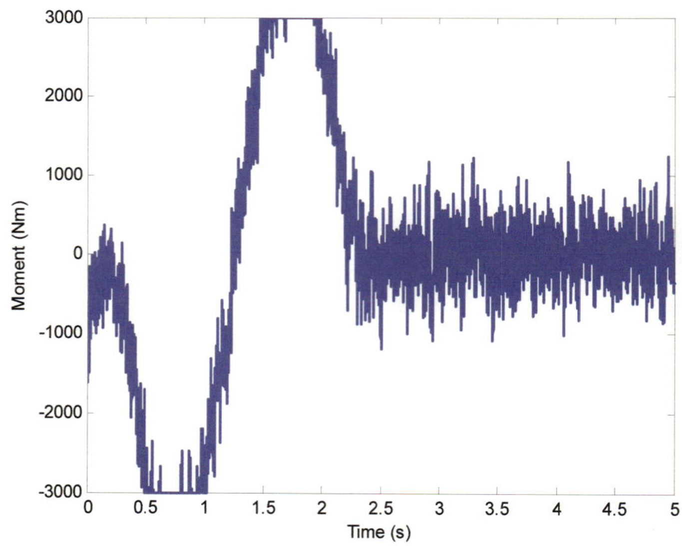

Fig. 10. Yaw moment under lane change manoeuvre on a dry road with measurement noise.

Finally, a lane change manoeuvre is applied to validate the effectiveness of the designed controller. As shown in Fig. 9 on sideslip angle and Fig. 10 on yaw moment, similar conclusion can be obtained on the performance achieved by the designed controller.

5. Conclusion

In this chapter, the practical design of a robust direct yaw moment controller for vehicle to improve lateral dynamics stability and handling with considering tyre cornering stiffness uncertainties, actuator saturation, measurement noise, and estimation of sideslip angle is studied. A two-step procedure is used to solve the observer and controller design problem, which can further be expressed as LMIs and can be solved very efficiently using currently available software like Matlab LMI Toolbox. Numerical simulations are applied to check the performance of the designed controller. The results show that the designed controller can improve vehicle handling and stability regardless of the measurement noise, changes of road conditions and manoeuvres. Further study on this topic will consider parameter uncertainties such as mass, moment of inertia, and forward velocity, etc., and study the reflection of road friction on vehicle model with choosing the most appropriate premise variables and defining the optimal membership functions.

6. References

Abe, M. (1999). Vehicle dynamics and control for improving handling and active safety: from four-wheel steering to direct yaw moment control, *Proc. Instn. Mech. Engrs. Part K: J. Multi-body Dynamics* Vol. 213, 87-101

Antonov, S., Fehn, A. and Kugi, A. (2008). A new flatness-based control of lateral vehicle dynamics, *Vehicle System Dynamics* Vol. 46, No. 9, 789-801

Boada, B. L., Boada, M. J. L. and Diaz, V. (2005). Fuzzy-logic applied to yaw moment control for vehicle stability, *Vehicle System Dynamics* Vol. 43, No. 10, 753-770

Canale, M., Fagiano, L., Milanese, M. and Borodani, P. (2007). Robust vehicle yaw control using an active differential and IMC techniques, *Control Engineering Practice* Vol. 15, 923-941

Cao, Y.-Y. and Lin, Z. (2003). Robust stability analysis and fuzzy-scheduling control for nonlinear systems subject to actuator saturation, *IEEE Transactions on Fuzzy Systems* 11(1): 57-67

Du, H., Lam, J. and Sze, K. Y. (2005). $H\infty$ disturbance attenuation for uncertain mechanical systems with input delay, *Transactions of the Institute of Measurement and Control* 27(1): 37-52

Esmailzadeh, E., Goodarzi, G. R. and Vossoughi, G. R. (2003). Optimal yaw moment control law for improved vehicle handling, *Mechatronics* Vol. 13, 659-675

Fukada, Y. (1999). Slip-angle estimation for vehicle stability control, *Vehicle System Dynamics*, Vol. 32, 375-388

Kim, J. H. and Jabbari, F. (2002). Actuator saturation and control design for buildings under seismic excitation, *Journal of Engineering Mechanics* Vol. 128, No. 4, 403-412.

Li, B. and Yu, F. (2010). Design of a vehicle lateral stability control system via a fuzzy logic control approach, *Proc. Instn. Mech. Engrs. Part D: J. Automobile Engineering* Vol. 224, 313-326.

Mirzaei, M., Alizadeh, G., Eslamian, M. and Azadi, S. (2008). An optimal approach to nonlinear control of vehicle yaw dynamics, *Proc. Instn. Mech. Engrs. Part I: J. Systems and Control Engineering* Vol. 222, 217-229.

Mirzaei, M. (2010). A new strategy for minimum usage of external yaw moment in vehicle dynamic control system, *Transportation Research Part C: Emerging Technologies* Vol. 18, No. 2, 213–224.

Tanaka, K. andWang, H. O. (2001). *Fuzzy control systems design and analysis: A linear matrix inequality approach*, John Wiley & Sons, Inc., New York.

Yang, X., Wang, Z. and Peng, W. (2009). Coordinated control of AFS and DYC for vehicle handling and stability based on optimal guaranteed cost theory, *Vehicle System Dynamics*, Vol. 47, No. 1, 57-79

Zheng, S.; Tang, H., Han, Z. and Zhang, Y. (2006). Controller design for vehicle stability enhancement, *Control Engineering Practice*, Vol. 14, 1413-1421

27

QFT Robust Control of Wastewater Treatment Processes

Marian Barbu and Sergiu Caraman
"Dunarea de Jos" University of Galati
Romania

1. Introduction

Wastewater treatment issues are extremely important for humanity. Their consideration becomes more than a necessity, a responsibility and every producer must improve their treatment processes. The efficiency increasing of these processes has been done in two ways:

1. by technological way - various types of treatment were developed during the past years and this domain has almost no technological secrets;
2. by using control methods – which currently represent a real challenge for researchers.

Wastewater treatment processes consist of a series of physical, chemical or biological processes that allow the separation between some particles (solid or dissolved, organic compounds, minerals etc.) and water, aiming to obtain a "clean" water able to meet certain standards for discharge or domestic/industrial consumption. In Europe, the water purity standards are established by the Directive no. 2000/60/EC. In the same time, the standards that are currently in use, defined by *water law* from February 3th, 1992, modified by the ordinance from February 2nd, 1998, are added to this directive. These rules define the maximum concentrations for each harmful compound from the wastewater. Generally the admissible concentrations are functions of the daily effluent flow.

Currently, new rules are applied regularly to the wastewater treatment. Global indicators for treatment efficiency, such as COD (*Chemical Oxygen Demand*), BOD (*Biochemical Oxygen Demand*), TOC (*Total Organic Carbon*) and for nutrients removal (phosphorus, ammonia nitrogen, total nitrogen etc.), whose normative are increasingly stringent, are taken into account. New compounds such as pigments, heavy metals, organic compounds, chlorinated solvents etc. are also considered for removal. For the waters coming from different industries and to be discharged into nature, the treatment rules are not the same. They depend on the receiving water and the type of the industry from which the wastewater results. For example, in metallurgical industry the wastewater containing heavy metals dominates, unlike the food industry, where the water containing organic compounds prevails.

Biological treatment processes are characterized by a number of specific features that make these processes real challenges for the specialists in control (Olsson & Newell, 1999):

- the daily volume of wastewater treated can be huge;
- the disturbances in the influent are enormous compared to most industries;
- the influent must be accepted and treated, there is no returning it to the supplier;
- the concentrations of nutrients (pollutants) are very small, even challenging sensors;

- the process depends on microorganisms, which have a definite mind of their own;
- wastewater treatment processes are very complex, strongly non-linear and characterized by uncertainties regarding its parameters (Goodman & Englande, 1974).

In the literature there are many models that try to capture as closely as possible the evolution of the wastewater treatment processes with activated sludge (Henze et al., 1987, 1995, 2000). The modelling of these processes is made globally, considering the nonlinear dynamics, but trying in the same time to simplify the models for their use in control (Barbu, 2009). One can state that the problem of wastewater treatment process control is difficult due to the factors mentioned before. The low repeatability rate, slow responses and the lack or high cost of the measuring instruments for the state variables of bioprocesses (biomass concentration, COD concentration etc.) also contribute to the difficulty of wastewater treatment process control. Therefore advanced and robust control algorithms that usually include in their structure state and parameter observers are currently used to control these processes.

Accordingly to (Larsson & Skogestad, 2000) two approaches in choosing the process control structure are taken into consideration: the approach oriented to the process and the one based on mathematical model. The first approach assumes the separated control of the main interest variables: dissolved oxygen concentration, nitrate and phosphate. One of the major and oldest problems encountered in wastewater treatment processes with direct impact on performance requirements is the dissolved oxygen concentration control. One can state that a satisfactory level of the dissolved oxygen concentration allows the developing of the microorganism' populations (the sludge) used in the process (Olsson, 1985), (Ingildsen, 2002). Taking into account the importance of this problem, there are many approaches regarding the dissolved oxygen control in the literature: PI and PID-control, fuzzy logic, robust control, model based control etc. (Garcia-Sanz et al., 2008), (Olsson & Newell, 1999). Recently the control problem of nitrate and phosphate level also became a priority. The control based on mathematical model of the wastewater treatment process has known many developing, depending on the type of the mathematical model used in the control algorithm design, as in the case of state estimators. So, the model described in (Olsson & Chapman, 1985) allowed the use of classic and modern techniques. It can be mentioned the classic structures of PI and PID type (Katebi et al., 1999) where the non-linear model linearized around an operating point is used for controller design, up to exact linearizing control, multivariable or in an adaptive version together with a state and parameter estimator (Nejjari et al., 1999). The use of this model leads to the design of an indirect control structure of the process. It can be concluded that the control of the dissolved oxygen concentration in the aerated tank practically assures a satisfactory level for the organic substrate. This problem - the control of the dissolved oxygen concentration - has been approached with good results in the control of a non-linear organic substrate removal process using multi-model techniques (Barbu et al., 2004).

The use of ASM1 model (*Activated Sludge Model 1*) determined by a work group belonging to IAWQ (*International Association of Water Quality*) makes the control problem more difficult and the results are less numerous. Based on ASM1 model in (Brdys et al., 2001a) a non-linear predictive control technique for the indirect control of organic substrate through the control of dissolved oxygen concentration has been used. For the same model (Brdys et al., 2001b) proposes a hierarchic control structure. This structure contains three levels: a higher level where a stable trajectory for the process on a time horizon is calculated, a mean level where the optimization of the trajectories for dissolved oxygen concentration, the recycled activated sludge flow and the recycled nitrate flow takes place and the lower level where the control of dissolved oxygen concentration based on the setpoint imposed by the mean level is done. Another approach that now is very appropriated is artificial intelligence based control. It uses the knowledge and the expertise of the specialists about the process management. Expert systems, fuzzy and neuro-fuzzy systems have been used for the wastewater treatment processes control (Manesis et al., 1998), (Yagi et al., 2002).

In the present chapter the authors propose the use of a robust control method (QFT - *Quantitative Feedback Theory*) for wastewater treatment processes control. Generally, wastewater treatment processes, as well as biotechnological processes, are characterized by parametric uncertainties that are determined by the operating conditions and the biomass growth. QFT method is a linear method frequently used for the processes described by variable parameter models. In this case, the transfer function with variable parameters will include both modifications caused by changing the operating point and parametric uncertainties that affect the process.

The chapter is structured as follows: the second section presents a few aspects regarding wastewater treatment process modelling (subsection 2.1 describes the wastewater treatment pilot plant with which some experiments were carried out in different operating conditions: different types of wastewaters, different concentrations of the influent and biomass etc. aiming to control the dissolved oxygen concentration in the aerated tank despite the variability of the operating conditions; in subsection 2.2 the simplified version of ASM1 model for ammonium removal is presented); the third section deals with the theoretical aspects regarding QFT robust control method; the fourth section shows the results obtained in the case of two control applications: the first is the control of dissolved oxygen concentration (experimentally validated) and the second is the control of ammonium concentration in the wastewater (validated through numerical simulations). In both control applications the robust control method QFT was used. The last section is dedicated to the conclusions.

2. A few aspects regarding wastewater treatment process modelling

This section deals with the wastewater treatment pilot plant used for carrying out the experiments for the design of dissolved oxygen robust control loop (subsection 2.1) and with simplified version of ASM1 model used for the ammonium removal (subsection 2.2).

2.1 Wastewater treatment pilot plant

A wastewater treatment pilot plant which is completely controlled by the computer (Figure 1) was conceived for studying and implementing various control algorithms in a national research project managed by "Dunarea de Jos" University of Galati.

The objective of the pilot plant was the efficiency improvement of the biological treatment processes of various types of wastewaters in aerobic conditions using control methods. This concept leads to a flexible design which allows us to interchange easily the treatment profiles (Barbu et al., 2010).

The feeding tank [1] has the capacity of 100 L and the ability to maintain the wastewater inside at almost constant characteristics due to its refrigeration equipment (1 - 6°C). The feeding flow can be strictly controlled through a peristaltic pump with a 12 Lph maximum flow. Before being pumped into the tanks the wastewater can be heated in a small expansion tank. The aeration tank [2] is the heart of the biological treatment process. Here the wastewater is mixed with the activated sludge and to fulfil the process it is also mixed with air. The air is bubbled into the aeration tank through a set of air ejectors which have also a mixing role. To be able to control the medium homogeneity the aeration tank is also equipped with a mechanical paddle mixer with three working regimes: 60rpm, 180rpm and 300rpm. The aeration tank working volume is 35L. The treatment temperature can be on-line monitored and controlled through a temperature probe and an electric heating resistance both mounted inside the tank. The pH can also be on-line monitored and controlled through a pH electrode connected to a pH controller and two peristaltic pumps, one for acid and the other for base (acid tank [3] and base tank [4]). The turbidity can be on-line monitored with a dedicated optical electrode. The evolution of biomass can be indirectly estimated through the turbidity values; the correlation between the two variables is usually made off-line by measuring the sludge dry matter. The aeration tank is also provided with an ORP (oxide-reduction potential) transducer. ORP potential can be correlated, in some cases, with the COD of the wastewater. The anoxic tank [5] can be used in the advanced nitrification - denitrification processes or it can be used in a sludge stabilization stage. In our experiments this tank remained unused. The sludge flocks formed in the aeration tank are allowed to settle in the clarifier [6]. This tank is provided with an ultrasonic level transducer which gives the flexibility to work at different retention times according to the chosen treatment scheme. From the bottom of the clarifier the settled sludge is recycled with a peristaltic pump back into the aeration tank.

Fig. 1. Wastewater treatment pilot plant

One of the most important variables in an aerobic treatment process is the DO (dissolved oxygen) concentration which is controlled by a cascade control structure. The cascade control system contains an inner loop (air flow control loop) that has a fast dynamics and an outer loop (the DO control loop) that has a slower dynamics. The air flow is on-line measured with a flow meter and it is controlled with an electric continuous valve. The DO concentration is on-line measured with an electrochemical electrode mounted in the aeration tank and it is controlled using the aeration rate as a control variable. The transducer signals are captured by a PCI data acquisition board. A HMI (*Human-Machine Interface*) facilitates the process control and monitoring. The data can be stored in a data base and processed thereafter.

2.2 Mathematical model of the wastewater treatment processes that include the nitrogen removal

The most popular model in literature of the wastewater treatment processes that includes the carbon and nitrogen removal is ASM1, proposed in 1987 (Henze et al., 1987). The model is extremely complex, it captures eight phenomena occurring in the anoxic and aerated reactors:

P_1	Aerobic growth of heterotrophic biomass - the process converts readily biodegradable substrate, dissolved oxygen and ammonium in the heterotrophic biomass;
P_2	Anoxic growth of heterotrophic biomass – the process converts readily biodegradable substrate, nitrate and ammonium in heterotrophic biomass;
P_3	Aerobic growth of autotrophic biomass – the process converts the dissolved oxygen, and ammonium in autotrophic biomass;
P_4	Heterotrophic decomposition - heterotrophic biomass is decomposed into slowly biodegradable substrate and other particles;
P_5	Autotrophic decomposition - autotrophic biomass is decomposed into slowly biodegradable substrate and other particles;
P_6	Ammonification - biodegradable organic nitrogen is converted to ammonium;
P_7	Hydrolysis of the organic matter - slowly biodegradable substrate is converted into readily biodegradable substrate;
P_8	Hydrolysis of organic nitrogen - solid biodegradable organic nitrogen is converted into soluble biodegradable organic nitrogen.

Table 1. The eight phenomena occurring in the anoxic and aerated reactor

The main deficiency of the model ASM1 is its complexity, making it virtually useless in control issues. A simplified version of the model ASM1 is proposed in (Jeppsson, 1996). Thus, in this version, only the significant variables for an average time scale (several hours to several days) are considered. Therefore, variables with a slow variation in time are considered constant, while those with a fast variation will be neglected. Based on these considerations, the processes of autotrophic and heterotrophic growth could be seen as slow events, so the processes denoted by P_4 and P_5 can be neglected within the model. The ammonification and hydrolysis processes (P_6, P_7 and P_8) will also be neglected, because under normal operating conditions these processes have a constant evolution.

The model ASM1 contains 13 state variables, as follows:

S_I	Soluble inert organic matter;
S_S	Readily biodegradable soluble substrate;
X_I	Various independent particles of inert organic matter and other particles;
X_S	Readily biodegradable soluble substrate;
$X_{B,H}$	Activated heterotrophic biomass;
$X_{B,A}$	Activated autotrophic biomass;
X_P	Different particles resulting from the biomass decomposition;
S_O	Dissolved oxygen concentration
S_{NO}	Soluble nitrate;
S_{NH}	Soluble ammonium;
S_{ND}	Soluble biodegradable organic nitrogen;
X_{ND}	Various particulate of biodegradable organic nitrogen;
S_{ALK}	Alkalinity

Table 2. State variables of ASM1 model

As a consequence, from the eight processes initially modelled by ASM1, only three of them will be used in the simplified model. The treatment process will be modelled as a system with two tanks, an anoxic one and an aerated one. The assumption that the amount of dissolved oxygen concentration in the anoxic tank is equal to zero is done: $S_O(1)=0$. In these circumstances, the simplified ASM1 model is described by the following equations:

$$\frac{dS_{NH}(1)}{dt}=\frac{Q}{V_1}S_{NH,in}-\frac{Q+Q_i}{V_1}S_{NH}(1)+\frac{Q_i}{V_1}S_{NH}(2)-i_{XB}P_2(1) \tag{1}$$

$$\frac{dS_{NH}(2)}{dt}=\frac{Q+Q_i}{V_2}S_{NH}(1)-\frac{Q+Q_i}{V_2}S_{NH}(2)-i_{XB}P_1(2)-\left(i_{XB}+\frac{1}{Y_A}\right)P_3(2) \tag{2}$$

$$\frac{dS_{NO}(1)}{dt}=-\frac{Q+Q_i}{V_1}S_{NO}(1)+\frac{Q_i}{V_1}S_{NO}(2)-\frac{1-Y_H}{2.86Y_H}P_2(1) \tag{3}$$

$$\frac{dS_{NO}(2)}{dt}=\frac{Q+Q_i}{V_2}S_{NO}(1)-\frac{Q+Q_i}{V_2}S_{NO}(2)+\frac{1}{Y_A}P_3(2) \tag{4}$$

$$\frac{dS_S(1)}{dt}=\frac{Q}{V_1}S_{S,in}-\frac{Q+Q_i}{V_1}S_S(1)+\frac{Q_i}{V_1}S_S(2)-\frac{1}{Y_H}P_2(1) \tag{5}$$

$$\frac{dS_S(2)}{dt}=\frac{Q+Q_i}{V_2}S_S(1)-\frac{Q+Q_i}{V_2}S_S(2)-\frac{1}{Y_H}P_1(2) \tag{6}$$

$$P_1(1)=\mu_H\frac{S_S(1)}{K_S+S_S(1)}\frac{S_O(1)}{K_{O,H}+S_O(1)}X_{B,H} \tag{7}$$

$$P_1(2) = \mu_H \frac{S_S(2)}{K_S + S_S(2)} \frac{S_O(2)}{K_{O,H} + S_O(2)} X_{B,H} \quad (8)$$

$$P_2(1) = \mu_H \frac{S_S(1)}{K_S + S_S(1)} \frac{K_{O,H}}{K_{O,H} + S_O(1)} \frac{S_{NO}(1)}{K_{NO} + S_{NO}(1)} \eta_g X_{B,H} \quad (9)$$

$$P_2(2) = \mu_H \frac{S_S(2)}{K_S + S_S(2)} \frac{K_{O,H}}{K_{O,H} + S_O(2)} \frac{S_{NO}(2)}{K_{NO} + S_{NO}(2)} \eta_g X_{B,H} \quad (10)$$

$$P_3(1) = \mu_A \frac{S_{NH}(1)}{K_{NH} + S_{NH}(1)} \frac{S_O(1)}{K_{O,A} + S_O(1)} X_{B,A} \quad (11)$$

$$P_3(2) = \mu_A \frac{S_{NH}(2)}{K_{NH} + S_{NH}(2)} \frac{S_O(2)}{K_{O,A} + S_O(2)} X_{B,A} \quad (12)$$

Observation: index 1 refers to the anoxic tank and index 2 - to the aerated tank.
Further on the input and output process variables are presented:

- input variables: internal recirculating flow, Qi, dissolved oxygen concentration in the aerated tank, $S_O(2)$, and external carbon dosage, $S_{Sdosage}$.
- output variables (measurable variables): ammonium concentration at the output, $S_{NH}(2)$, (equal to ammonium concentration from the aerated tank) and nitrate concentration at the output, $S_{NO}(2)$, (equal to nitrate concentration from the aerated tank).

The two process output variables are quality variables too. Thus the purpose of the control structure will be the obtaining of an effluent having an output ammonium concentration less than 1 gN/m^3 and an output nitrate concentration less than 6 gN/m^3.
For the model described by equations (1) - (12) the following parameters were taken into consideration:
V_1=2000 m^3, V_2=3999 m^3, Q=18446 m^3/day, $S_{NH,in}$=30 gN/m^3, η_g=0.8, i_{XB}=0.08, $S_{S,in}$=115+ $S_{Sdosage}$ gCOD/m^3, K_{NH}=1 gNH$_3$-N/m^3, K_{NO}=0.5 gNO$_3$-N/m^3, Y_A=0.24, Y_H=0.67, $K_{O,H}$=0.2 gO$_2$/m^3, $K_{O,A}$=0.4 O$_2$/m^3, K_S=10 gCOD/m^3, μ_A=0.6 day^{-1}, μ_H=5 day^{-1}, $X_{B,A}$=110 gCOD/m^3, $X_{B,H}$=2200 gCOD/m^3.

Figure 2 presents the simulation results regarding the free dynamics of the simplified ASM1 model. The simulation was done considering the following initial conditions: $S_{NH}(1)(0)$=10 gN/m^3, $S_{NH}(2)(0)$=9.7 gN/m^3, $S_{NO}(1)(0)$=0.9 gN/m^3, $S_{NO}(2)(0)$=2.15 gN/m^3, $S_S(1)(0)$=2.8 gCOD/m^3, $S_S(2)(0)$=0.9 gCOD/m^3.

The following values of the input variables were also taken into consideration: $S_O(2)$=1.5 mg/l, Q_i=40000 m^3/day, $S_{Sdosage}$=40 gCOD/ m^3.

3. Robust control of monovariable processes using QFT method

QFT is a robust control method proposed by Horowitz in 1973 and it was designed for the control of the processes described by linear models with variable parameters (Horowitz, 1973). QFT is a technique that uses Nichols frequency characteristics aiming to ensure a robust design over a specified uncertainty area of the process. The method can be also applied for nonlinear processes through their linearization around several operating points. It results a linear model with variable parameters describing the nonlinear process behaviour in every point of the operating area. The limits of variation of the linear model parameters obtained through linearization can be extended to incorporate the effect of the parametric uncertainties that affect the nonlinear process. For this linear model a robust controller using QFT method is then designed.

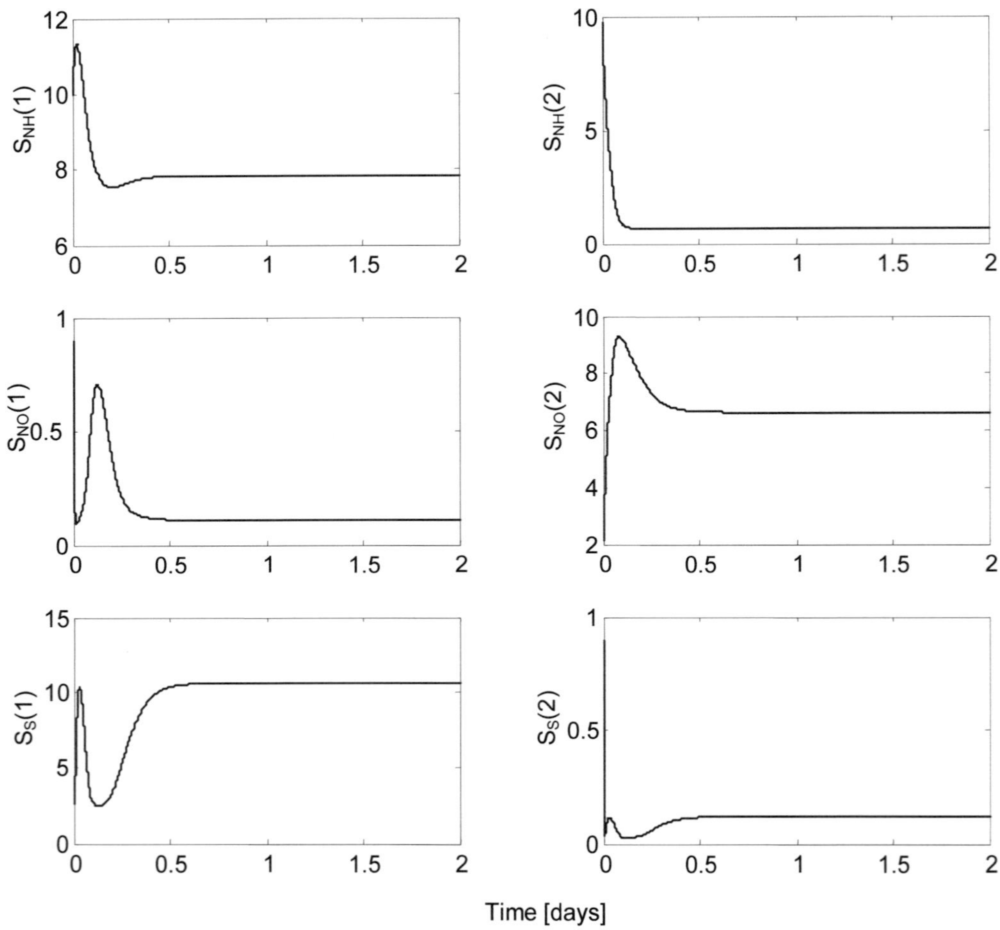

Fig. 2. Simulation results of the simplified ASM1 model

In most control cases, the evolution of the output variable, $y(t)$, of the closed-loop system must be bounded by an upper and a lower limit, as presented in Figure 3, where both limits of the response to a step signal were shown. QFT method ensures the operation of a linear system with variable parameters within the imposed domain of evolution.
It is considered a process described by a variable parameter transfer function of the following type:

$$P(s) = \frac{Ka}{s(s+a)} \tag{13}$$

where parameters K and a varies due to the operating conditions, so $K \in [K_{\min}, K_{\max}]$ and $a \in [a_{\min}, a_{\max}]$.

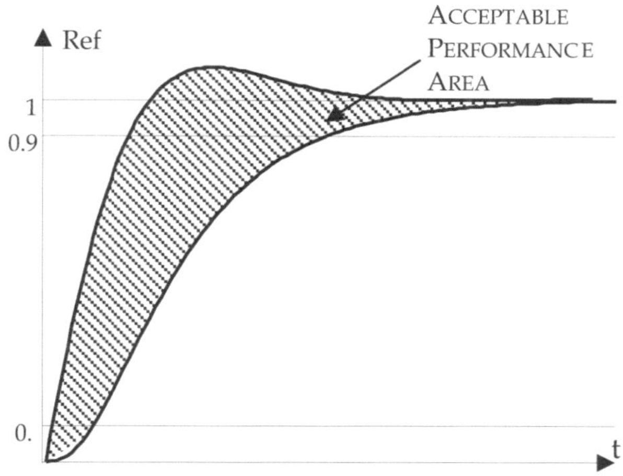

Fig. 3. Upper and lower bounds of the system output

QFT method consists in the synthesis of a compensator G(s) and a prefilter F(s) so that the behaviour of the closed-loop system is between the bounds imposed to the system. Figure 4 presents the control structure:

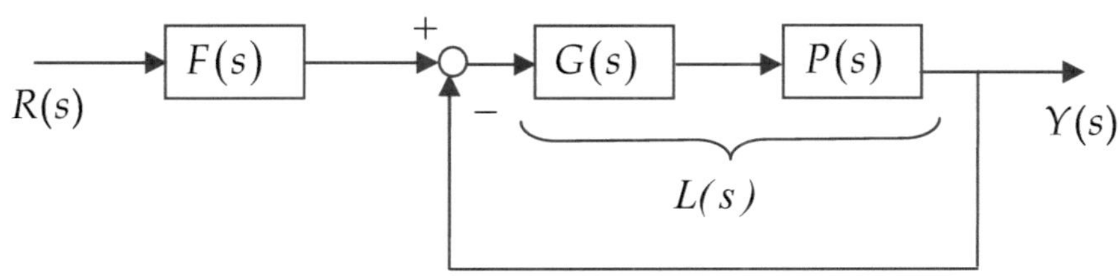

Fig. 4. Compensated linear system

The steps of robust design using QFT method for a tracking problem are the following (Houpis & Rasmussen, 1999):

Step 1. The synthesis of the desired tracking model.

The synthesis of the tracking model consists in defining the performance specifications through two invariant linear transfer functions, which set upper and lower design limits. In this way a series of closed-loop system performances which will result from the design are imposed. The considered performances are the rising time, the response time and the overshoot. The tracking specifications are referring to the tracking system which, in closed-loop, has the following transfer function:

$$H_u(s) = \frac{F(s)G(s)P(s)}{1+G(s)P(s)} = \frac{F(s)L(s)}{1+L(s)} \tag{14}$$

Since the linear model parameters change depending on the operating regime, the closed-loop system characteristics will have some variations. One imposes that these changes be within certain limits defined by an „upper" and „lower" gain characteristic:

$$\left|H_{ri}(j\omega)\right| \le \left|H_{u}(j\omega)\right| \le \left|H_{rs}(j\omega)\right| \tag{15}$$

in which, usually, the upper tracking model corresponds to the response of a second order system with overshoot, while the lower tracking model corresponds to a first order step response. Thus $H_{ri}(s)$ and $H_{rs}(s)$ have the expressions (Houpis & Rasmussen, 1999):

$$H_{rs}(s) = \frac{\omega_n^2}{s^2 + 2\varsigma\omega_n s + \omega_n^2} \tag{16}$$

$$H_{ri}(s) = \frac{a_1 a_2}{(s+a_1)(s+a_2)} \tag{17}$$

In (16) and (17) it has to take into account the constraint regarding the steady transfer coefficient, that always must be equal to 1. Thus, at each frequency ω_i a bandwidth $\delta_u(j\omega_i)$ is provided, as shown in Figure 5.

In the transfer function of the upper limit a zero close to the origin could be introduced, with an effect as low as possible on the response time. This zero produces the increasing of the bandwidth $\delta_u(j\omega_i)$ at high frequencies. The bandwidth can be increased further by adding a pole near the origin. This pole does not significantly modify the response time of the lower limit transfer function. By introducing these additional elements one seeks for an easier fitting of the parametric uncertainties into the higher frequencies domain and thus the problem of prefilter synthesis $F(s)$ is simplified.

Step 2. Description of the linearized process through a set of N invariant linear models, which define the parametric uncertainty of the model.

The linearized process is described through a set of N invariant linear models which define the parametric uncertainty of the model. The parametric uncertainties of the linear model are determined by the range of operating and parametric uncertainties of the nonlinear model.

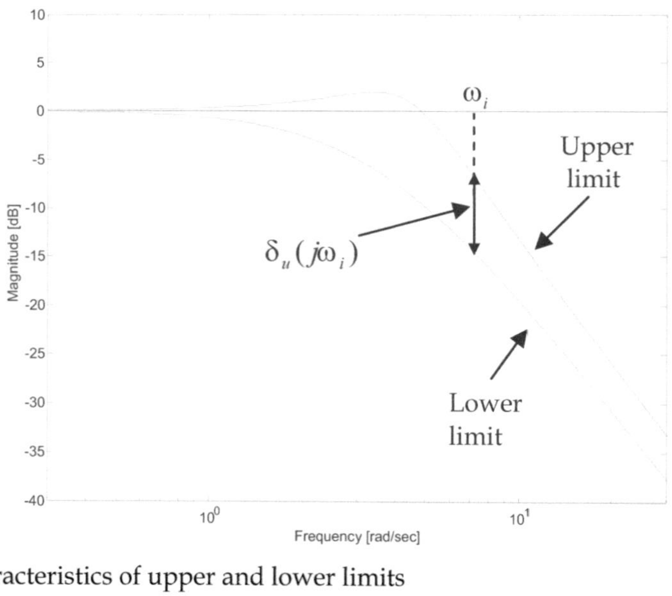

Fig. 5. Bode characteristics of upper and lower limits

Step 3. The obtaining of the templates at specified frequencies which graphically describe the parametric uncertainty area of the process on Nichols characteristic.

The N characteristics (gain and phase) of the considered models are represented on Nichols diagram for every frequency value. These N points define a closed contour, named template, which limits the variation range of parametric uncertainty.

Step 4. Selection of the nominal process, $P_0(s)$.

Although any process can be chosen, in practice the process whose point on the Nichols characteristic represents the bottom left corner of the templates for all frequencies used in the design procedure is chosen.

Step 5. Determination of the stability contour - the contour U - on Nichols characteristic.

The performance specifications referring to stability and robust tracking define the limits within which the transfer function of the tracking system can vary, when the linear model varies in the uncertainty area. The stability of the feedback loop, regardless of how the model parameters vary in the uncertainty region is ensured by the stability specifications. The transfer function of the closed-loop system is:

$$H_0(s) = \frac{G(s)P(s)}{1+G(s)P(s)} = \frac{L(s)}{1+L(s)} \tag{18}$$

One imposes that in the considered bandwidth, the gain characteristics associated to the closed-loop transfer function to not exceed a value of the upper limit (Horowitz, 1991):

$$|H_0| = \left|\frac{GP}{1+GP}\right| \le M_L \tag{19}$$

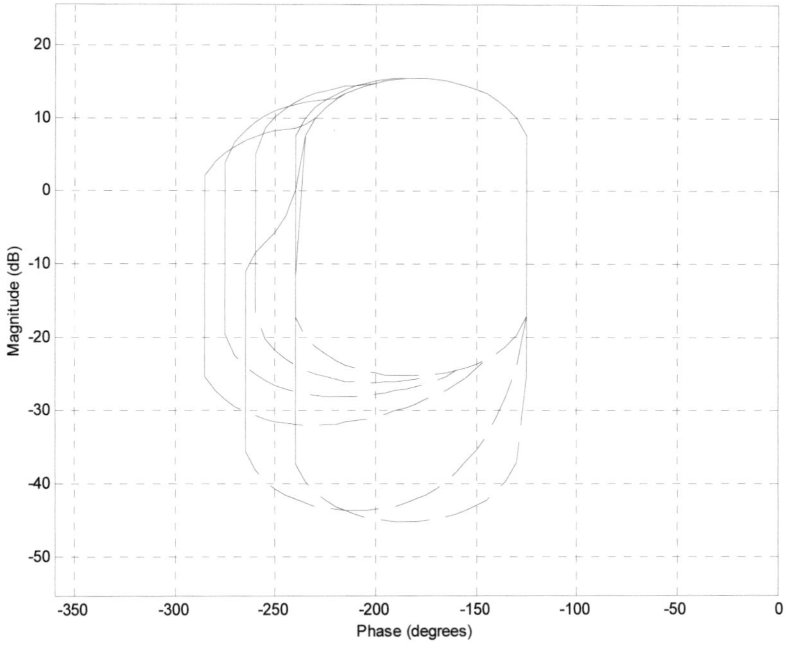

Fig. 6. Stability contours corresponding to the model given by equation (13)

In these conditions, a region that cannot be penetrated by the templates and the transmission functions $L(j\omega)$ for all frequencies ω is established on Nichols characteristic. This region is bounded by the contour M_L. The stability margins are determined using a frequency vector covering the area of interest. These margins differ from one frequency to another. Figure 6 presents the stability margins of the linear model given by equation (13).

Step 6. Determination of the robust tracking margins on Nichols characteristic.

The robust tracking margins must be chosen such that the placing of the loop transmission on this margin or above it ensures the robust tracking condition imposed by equation (15) to be met at every chosen frequency. This practically means that for each frequency the difference between the gain of the extreme points from the process template must be less than or equal to the maximum bandwidth $\delta_u(j\omega_i)$. Figure 7 illustrates the robust tracking margins of the linear model given by equation (13) with the tracking models (16) and (17).

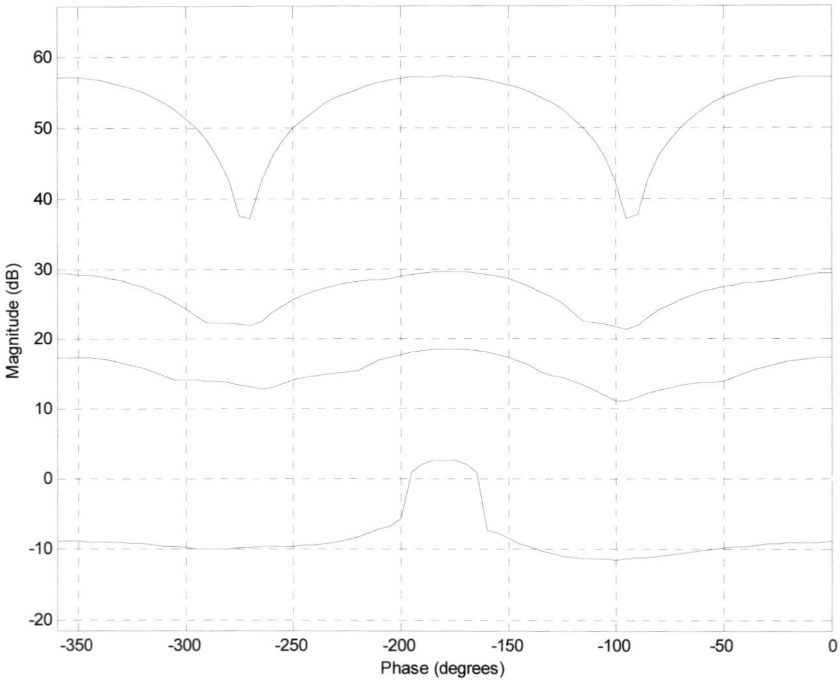

Fig. 7. Robust tracking margins corresponding to the model given by equation (13)

Step 7. Determination of the optimal margins on Nichols characteristics.

The optimal tracking margins are obtained from the intersection between the stability contours and the robust tracking margins for the frequencies considered of interest, taking into account the constraints that are imposed to the loop transmission. Thus the stability contour resulted at a certain frequency cannot be violated, so only the domains from the tracking margin that are not within the stability boundaries (18) will be taken into consideration. Figure 8 illustrates the optimal margins of the linear model given by equation (13).

Step 8. Synthesis of the nominal loop transmission, $L_0(s) = G(s)P_0(s)$, that satisfies the stability contour and the tracking margins.

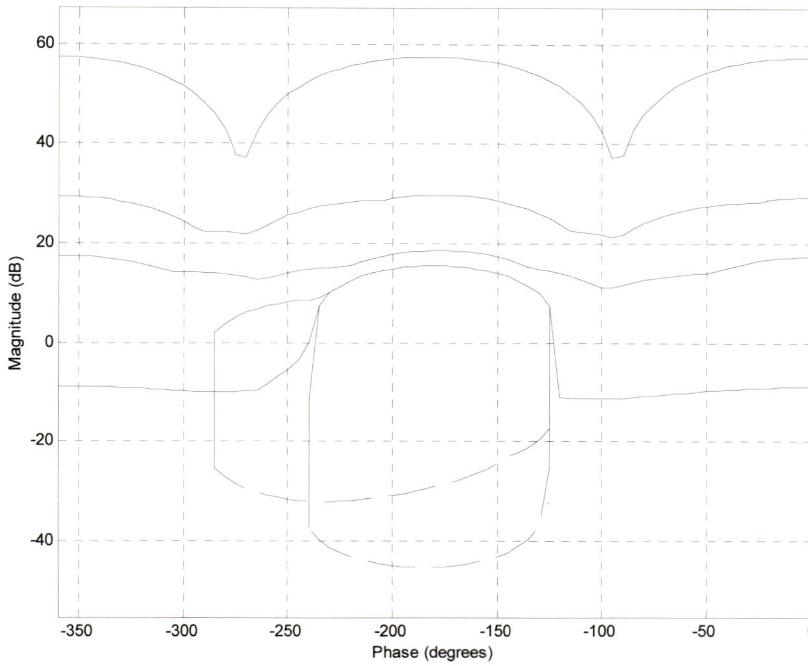

Fig. 8. Optimal tracking margins

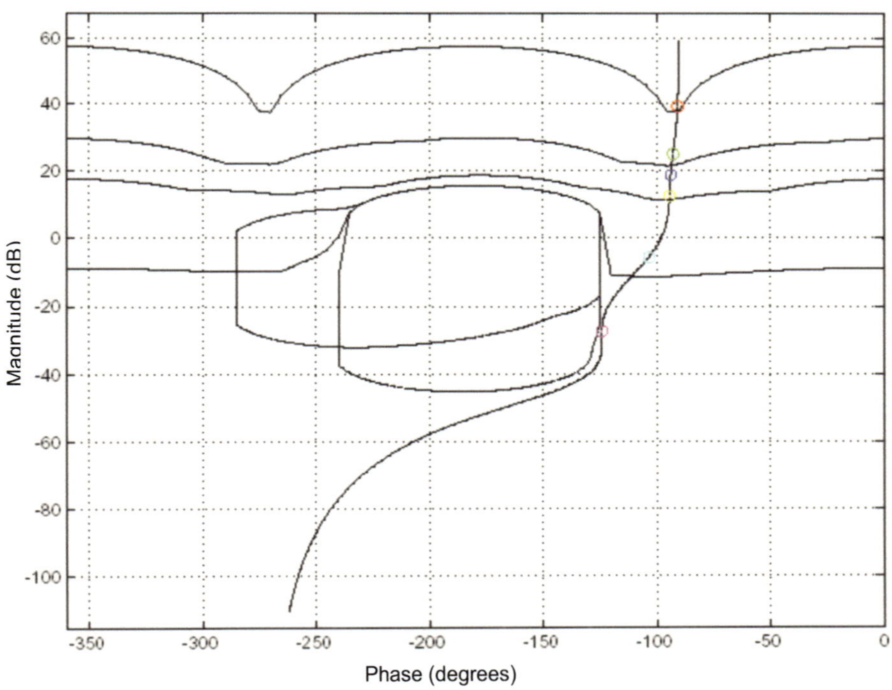

Fig. 9. Synthesis of the controller $G(s)$

Starting from the optimal tracking margins, the transmission of the nominal loop is also represented on Nichols diagram, corresponding to the nominal model, $P_0(s)$, considering initial expression of the controller $G(s)$. The transmission loop is designed such as not to penetrate the stability contours and the gain values must be kept on or above the robust tracking margins corresponding to the considered frequency. Figure 9 presents the optimal margins and the transmission on the nominal loop which has been obtained in its final form. It can be noticed that the transmission values within the loop, for the six considered frequencies, are distinctly marked, with respect to the condition that the first four values must be placed above the corresponding tracking margins.

Step 9. Synthesis of the prefilter *F*(*s*).

Figure 10 presents Bode characteristic of the closed-loop system without filter. It can be noticed that the band defined by the tracking limits of the closed-loop system (solid lines) is smaller than the band defined by performance specification limits (dotted lines) but Bode characteristic also evolves outside limits imposed by the performance specifications. In order to bring the system within the envelope defined by the performance specification limits, the filter *F*(*s*) is used. Figure 11 presents Bode characteristic of the closed-loop system with compensator and prefilter. It can be seen that the system respects the performance specifications of robust tracking (the envelope defined by solid lines is inside the envelope defined by dotted lines). Thus the robust closed-loop system respects the stability and robust specifications in range of variation of the model parametric uncertainties.

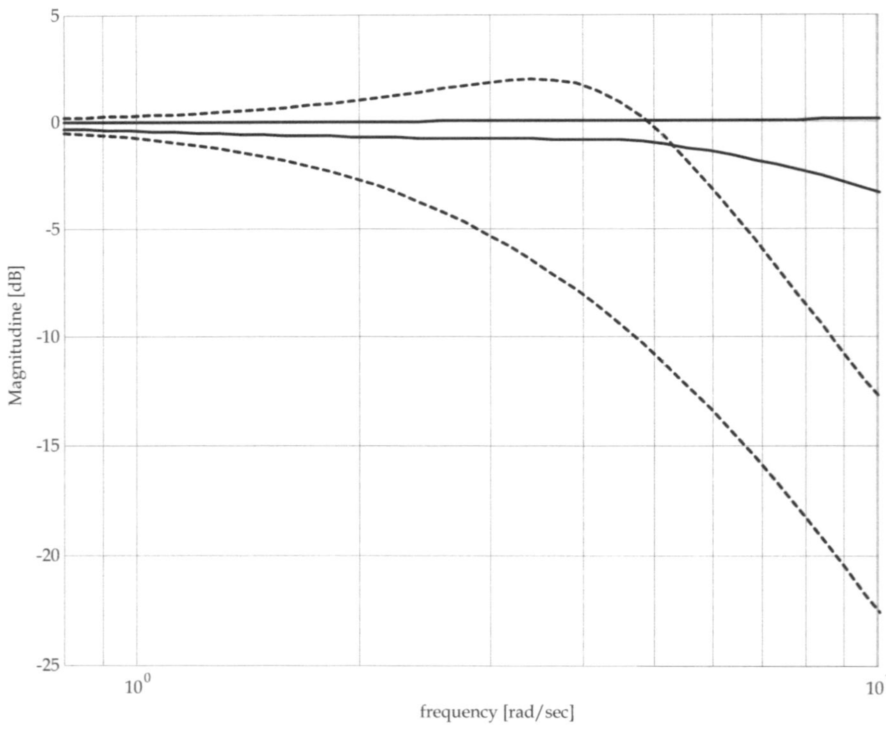

Fig. 10. Closed-loop system response with compensator

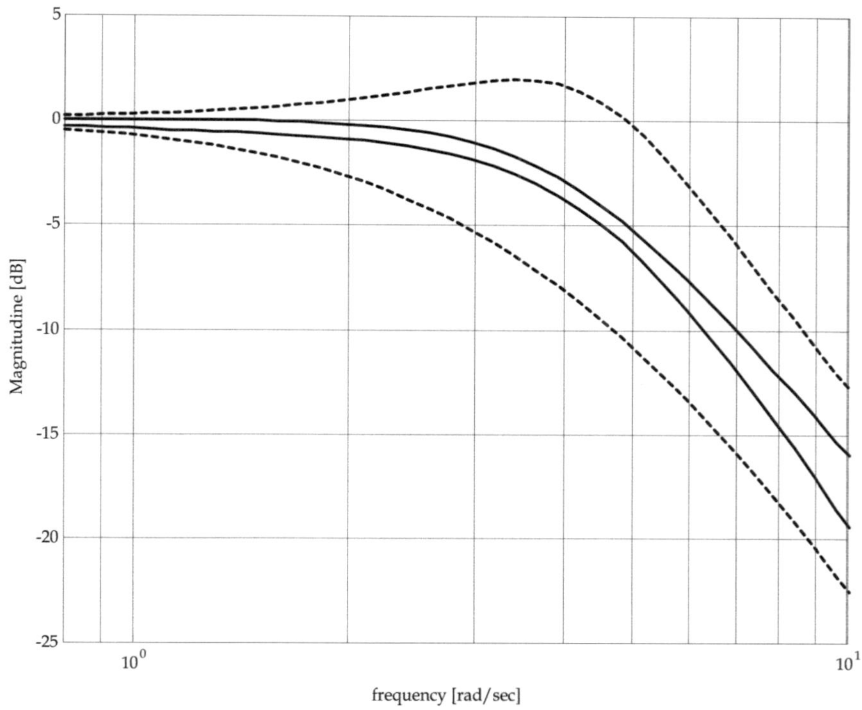

Fig. 11. Closed-loop system response with compensator and prefilter

4. Robust control of the wastewater treatment processes using QFT method

The control structure of a wastewater treatment process contains a first level with local control loops (temperature, pH, dissolved oxygen concentration etc.), which is intended to establish the nominal operating point, over which is superposed a second control level (global) for the removal of various pollutants such as organic substances, ammonium etc. For this reason the models used for developing control structures range from the simplest models for local control loops, up to very complicated models such as ASM models, as it is mentioned in section 1. Thus, subsection 4.1 will present the identification of dissolved oxygen concentration control loop and subsection 4.2 will present the control of ammonium concentration using the simplified version of ASM1 model. All the design steps of QFT algorithm were implemented using QFT Matlab® toolbox.

4.1 Dissolved oxygen concentration control in a wastewater treatment plant with activated sludge

To identify the dissolved oxygen concentration control loop a sequence of steps of various amplitudes was applied to the control variable that is the aeration rate. Figure 12 presents the sequence of steps applied to the dissolved oxygen concentration control system, while Figure 13 shows the evolution of the dissolved oxygen concentration. Analyzing the results presented in Figure 13 it can be concluded that the evolution of the dissolved oxygen concentration corresponds to the evolution of a first order system. At the same time, it can be seen in the same figure that the evolution of the dissolved oxygen concentration is strongly influenced by biomass and organic substrate evolutions. Thus, depending on

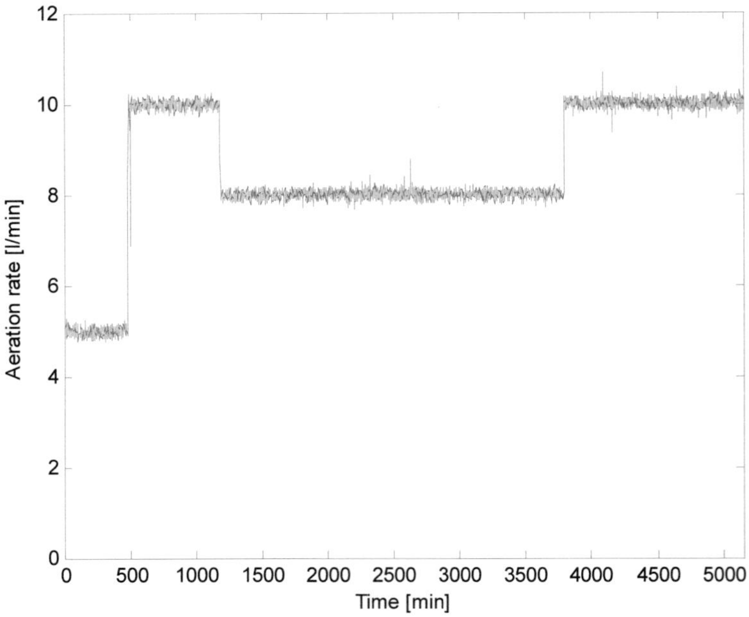

Fig. 12. Step sequence of the control variable: aeration rate

the oxygen consumption of microorganisms, the dissolved oxygen concentration from the aerated tank has different dynamics, each corresponding to different parameters of a first-order system.

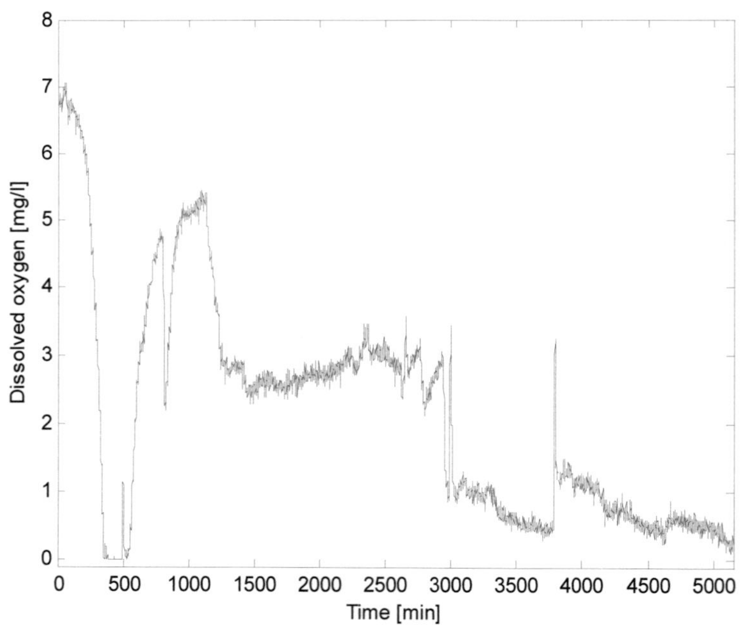

Fig. 13. Evolution of the dissolved oxygen concentration in the case when the aeration rate evolves according to Figure 12

In addition, considering that the microbial activity from the wastewater treatment process is influenced by the environmental conditions under which the process unfolds (temperature, pH etc.) and the type of substrate used in the process (in the pilot plant will be used organic substrates derived from milk and beer industries, substrates having different biochemical composition) it results that more transfer functions are necessary, aiming to model the evolution of the dissolved oxygen concentration in the aerated tank depending on the aeration rate. One possibility to model the dissolved oxygen concentration depending on the aeration rate is to take into consideration a first order transfer function with variable parameters (Barbu et al., 2010):

$$H(s) = \frac{K}{Ts + 1} \tag{20}$$

where, as a result of the identification experiments performed on data collected from different experiments carried out with the pilot plant, it was taken into consideration that the gain factor K varies in the range $K \in [0.8 \;\; 1.4]$ and the time constant of the first-order element varies in the range $T \in [1700 \;\; 2500]$.

The closed-loop system should have a behaviour between the two imposed limits, that give the accepted performance area. Taking into account the variation limits of the linear model parameters considered before, the two tracking models (the lower and upper bounds) were established:

$$H_{rs}(s) = \frac{10(s + 0.1)}{(s + 0.007 \pm j \cdot 0.007)} \tag{21}$$

$$H_{ri}(s) = \frac{1}{(300s + 1)(310s + 1)(30s + 1)} \tag{22}$$

Based on the linear model with variable parameters, given by equation (20), and on the tracking models, given by equations (21) and (22), all the steps provided in the design methodology using QFT robust method for a setpoint tracking problem has been completed. The transfer functions of the controller and prefilter are:

$$G(s) = \frac{0.22143\ (s + 0.00039)}{s\ (s + 0.01217)} \tag{23}$$

$$F(s) = \frac{0.0068}{(s + 0.0068)} \tag{24}$$

Analyzing the controller transfer function $G(s)$, given by equation (23), it can be noticed that it also includes an integral component. Since the control variable is limited to a higher value given by the air generator used to provide the aeration - in the case of this pilot plant: 25 l/min - and the controller includes an integral component, it was necessary to introduce an antiwind-up structure. This structure prevents the saturation of the control variable (the achievement of some unacceptable values for the integrator), helping to improve the dynamic regime of the controller.

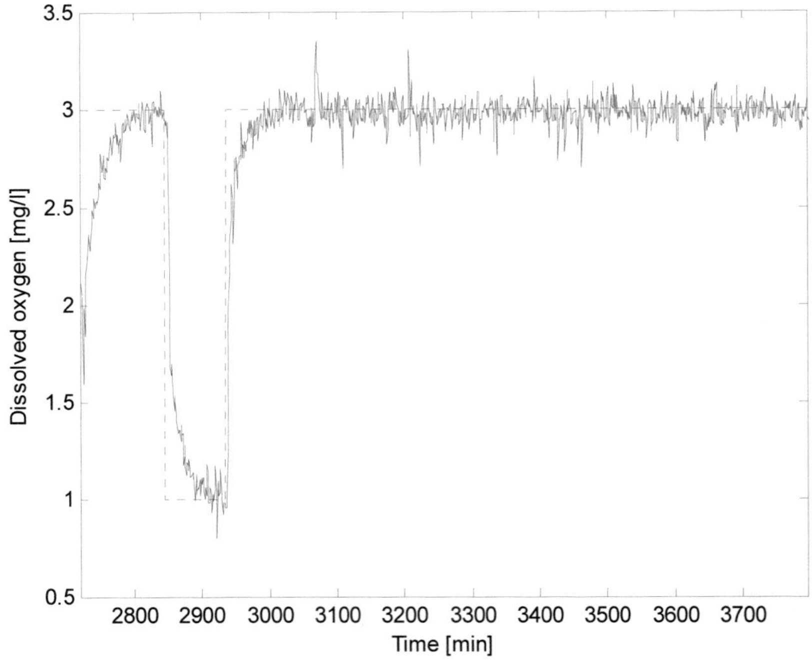

Fig. 14. Evolution of the dissolved oxygen concentration: solid line – pilot plant, dotted line – setpoint

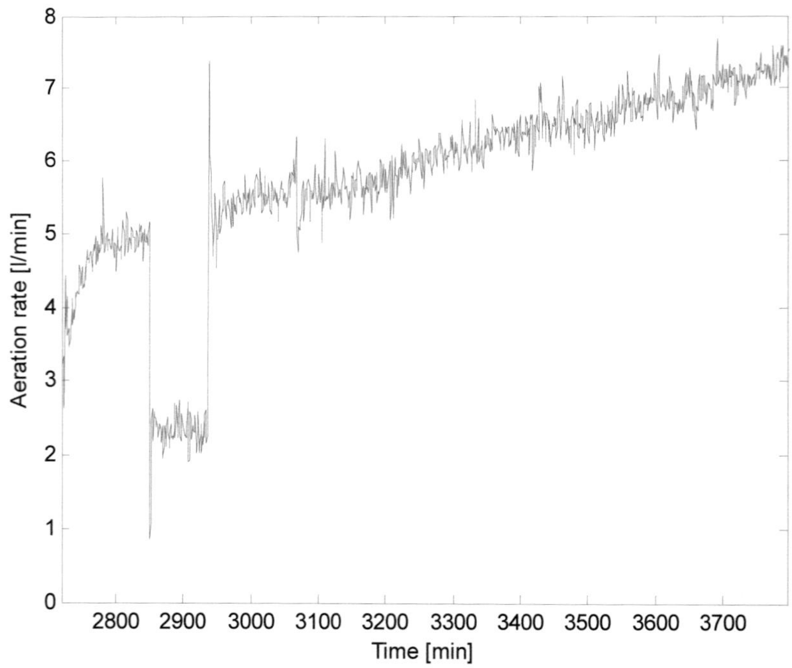

Fig. 15. Evolution of the control variable

The QFT proposed control structure was tested in the case of two experiments. The purpose was to observe the behaviour of the QFT controller in the case of two types of different wastewaters and when the process is in different stages of evolution from the biomass developing point of view. The first experiment was made considering the wastewater from the milk industry. Within this experiment, values of the dissolved oxygen setpoint ranging between 1mg/l and 3mg/l were taken into consideration. Figure 14 presents the evolution of the output variable (the DO concentration) and Figure 15 presents the evolution of the control variable (the air flow). The second experiment was made considering wastewater from the beer industry and in this experiment the biomass concentration developed in the aerated tank was monitored too. The results obtained in this experiment are shown in Figures 16, 17 and 18.

As a conclusion, the results obtained in the present chapter are very good in both cases, the QFT robust control structure succeeding to keep the dissolved oxygen setpoint imposed in the case of both types of wastewater considered in the experiments, from beer and milk industry, without being affected by the modification of the microorganism's concentration developed in the aerated tank during the experiments. This justifies the choice to use a robust controller as is the one designed by QFT method. At the same time, from the analysis of the evolution diagrams of the aeration rate and the dissolved oxygen concentration, it can be noticed that for maintaining a constant setpoint of the dissolved oxygen concentration in the aerated tank, the aeration rate will be directly influenced by the concentration of microorganisms that consume oxygen in the aerated tank.

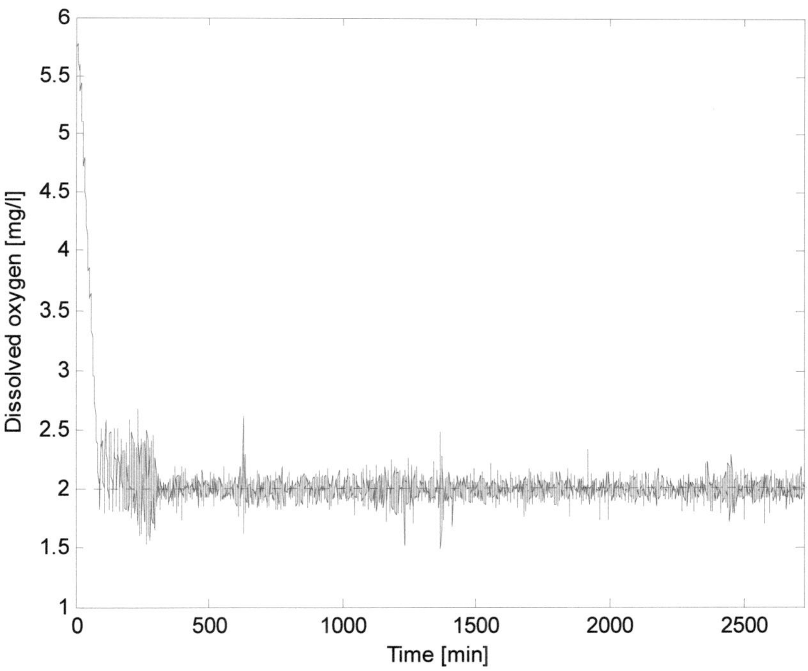

Fig. 16. Evolution of the dissolved oxygen concentration: solid line - pilot plant, dotted line - setpoint

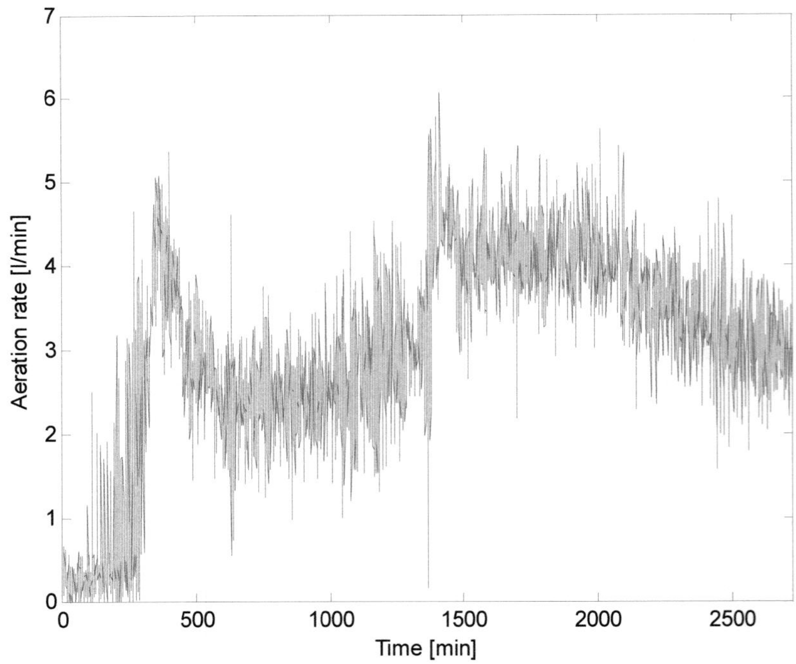

Fig. 17. Evolution of the control variable

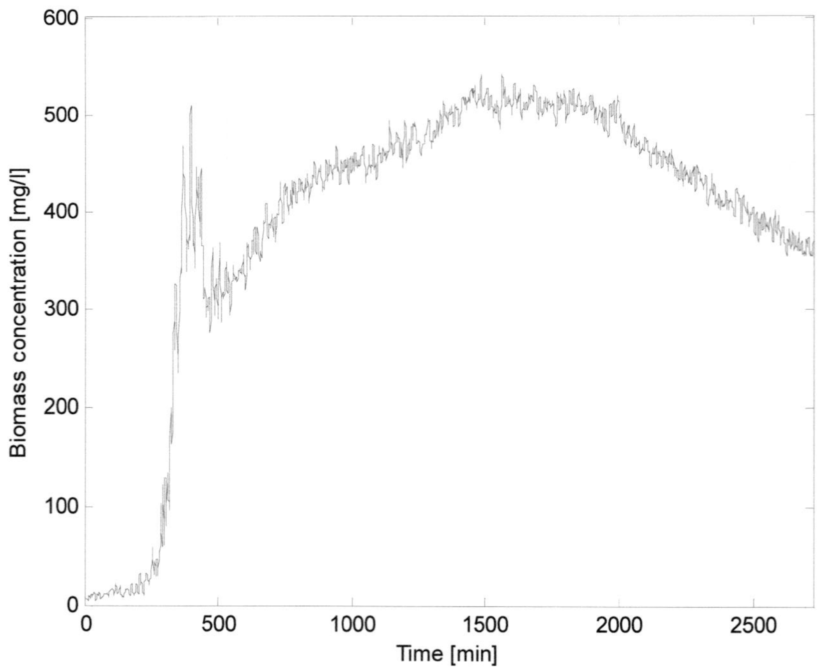

Fig. 18. Evolution of the biomass concentration

4.2 QFT multivariable control of a biological wastewater treatment process using ASM1 model

Within this section the robust linear control method QFT is used for the control of a nonlinear wastewater treatment process with activated sludge. The considered model for the wastewater treatment process is a simplified version of ASM1 model which has been presented in subsection 2.2. For this purpose the non-linear model was linearized in different operating points, resulting a linear model with variable parameters that approximates the behaviour of the non-linear process in all its operating points. The control variables of the multivariable process are: internal recycled flow, Q_i, and dissolved oxygen concentration from the aerated tank, $S_O(2)$. The output variables are the following: ammonium concentration at the output, $S_{NH}(2)$, equal to ammonium concentration from the aerated tank and nitrate concentration at the output, $S_{NO}(2)$, equal to nitrate concentration from the aerated tank. The purpose of the control structure is to obtain an effluent having an ammonium concentration at the output under 1 gN/m^3 and a nitrate concentration at the output under 6 gN/m.^{3}

In (Barbu & Caraman, 2007) an analysis of the channel interaction, using RGA (Relative Gain Array) method was performed. This analysis indicates the fact that a control structure based on decentralized loops, considering as main channels - the control channels and as secondary channels - the disturbance channels, could be adopted. From the same analysis it results the following control channels: the dissolved oxygen concentration from the aerated tank – the ammonium concentration at the output ($S_O(2)-NH(2)$) and the recycle rate – the nitrate concentration at the output (Q_i - $NO(2)$). The secondary channels with a very weak interaction between them are: the recycle rate - the nitrate concentration at the output (Q_i - $NH(2)$) and the dissolved oxygen concentration from the aerated tank - the nitrate concentration at the output ($S_O(2)-NO(2)$).

The non-linear wastewater treatment process can be linearized taking into consideration three main functioning points (Barbu & Caraman, 2007):

1. rain - $S_{NH,in}$ =25 gN/m^3, $S_O(2)$ =1.5 mg/l, Q_i =30000 m^3/day;
2. normal - $S_{NH,in}$ =30 gN/m^3, $S_O(2)$ =1.5 mg/l, Q_i =40000 m^3/day;
3. drought - $S_{NH,in}$ =35 gN/m^3, $S_O(2)$ =2 mg/l, Q_i =50000 m^3/day.

The transfer functions obtained in the case of the three operating regimes were simplified through a frequency analysis and they have the following expressions:

1. Rain:

$$P_{S_O(2)-NH(2)}(s)=\frac{23.664}{s+115} \tag{25}$$

$$P_{Q_i-NO(2)}(s)=\frac{0.00149(s^2+36.03s+2596)}{(s+115)(s+11.82)(s+22.56)} \tag{26}$$

2. Normal:

$$P_{S_O(2)-NH(2)}(s)=\frac{15.036}{s+115.4} \tag{27}$$

$$P_{Q_i-NO(2)}(s)=\frac{0.00156(s^2+41.81s+2924)}{(s+115.4)(s+13.75)(s+26.92)} \tag{28}$$

3. Drought:

$$P_{S_O(2)-NH(2)}(s)=\frac{11.32}{s+109.6} \tag{29}$$

$$P_{Q_i-NO(2)}(s)=\frac{0.00149(s^2+42.66s+2584)}{(s+109.6)(s+13.68)(s+29.13)} \tag{30}$$

Taking into account the transfer functions obtained for the three operating regimes, it can be seen that the main channel, the dissolved oxygen concentration from the aerated tank – the ammonium concentration at the output ($S_O(2)-NH(2)$) can be described by the following transfer function with variable parameters:

$$P_{S_O(2)-NH(2)}(s)=\frac{K_1}{s+a_1} \tag{31}$$

where: $K_1 \in [10\ \ 20], a_1 \in [109\ \ 116]$.
The tracking models imposed for this control channel are given by the following transfer functions:

$$H_{rs}=\frac{20\cdot(s+100)}{(s+20\pm j\cdot 40)} \tag{32}$$

$$H_{ri}=\frac{73500}{(s+30)(s+35)(s+70)} \tag{33}$$

As a result of applying the QFT algorithm, the following robust controller results:

$$G_{S_O(2)-NH(2)}(s)=\frac{270.2621}{s+0.063} \tag{34}$$

and the prefilter:

$$F_{S_O(2)-NH(2)}(s)=\frac{45.205}{s+45.205} \tag{35}$$

The control channel, the recycled rate – the nitrate concentration at the output (Q_i - $NO(2)$), is described by the following transfer function with variable parameters:

$$P_{Q_i-NO(2)}(s)=\frac{K_2(s^2+a_2s+b_2)}{(s+c_2)(s+d_2)(s+e_2)} \tag{36}$$

with:

$$K_2 \in [0.0014\ 0.0016], a_2 \in [36\ 43], b_2 \in [2580\ 930], c_2 \in [109\ 116], d_2 \in [11.5\ 14], e_2 \in [22\ 29.5]$$

The tracking models imposed for this control channel are given by the following transfer functions:

$$H_{rs} = \frac{10 \cdot (s+50)}{(s+10 \pm j \cdot 20)} \tag{37}$$

$$H_{ri} = \frac{12000}{(s+15)(s+20)(s+40)} \tag{38}$$

As a result of applying the QFT algorithm, the following robust controller results:

$$G_{Q_i-NO(2)}(s) = \frac{10000(0.965s+13.255)}{s+0.0067} \tag{39}$$

and the prefilter:

$$F_{Q_i-NO(2)}(s) = \frac{1.07s+25.194}{s+25.194} \tag{40}$$

The robust control structure proposed in this chapter has been tested trought numerical simulation in the case of each of the three operating regimes. In Figures 19 and 20 the simulation results for the two extreme operating regimes (rain and drought) are presented. It was also tested an operating sequence when the three operating regimes alternate, as is presented in Figure 21. All this figures show that the robust multivariable control structure is able to track the setpoints imposed for the output variables and the biodegradable substrate is efficiently treated. This is achieved despite the fact that the multivariable nonlinear process modifies its operating point, both in terms of the inflow and the organic matter load.

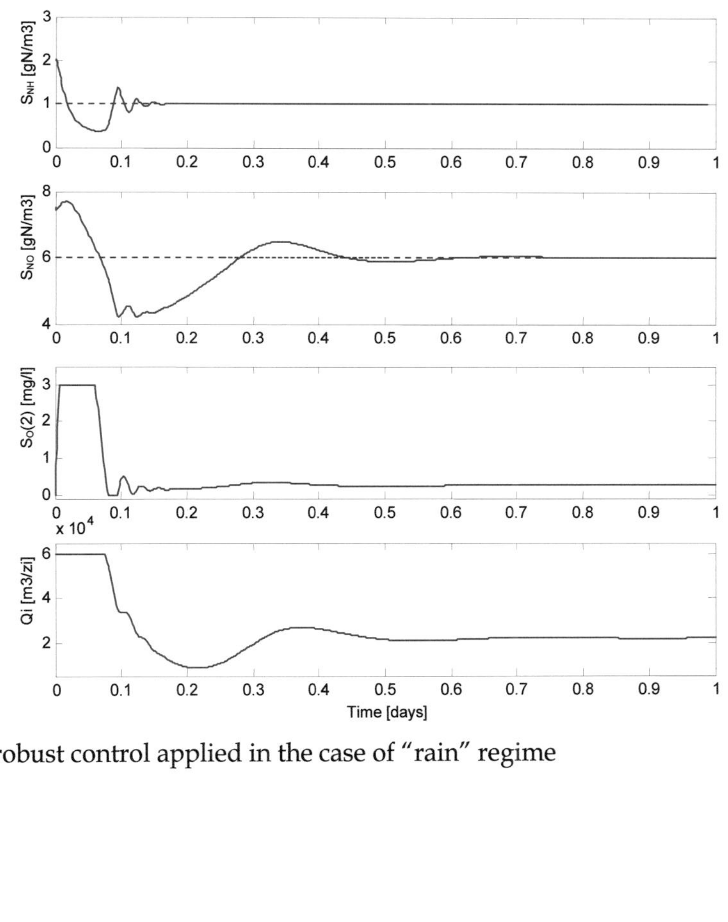

Fig. 19. QFT robust control applied in the case of "rain" regime

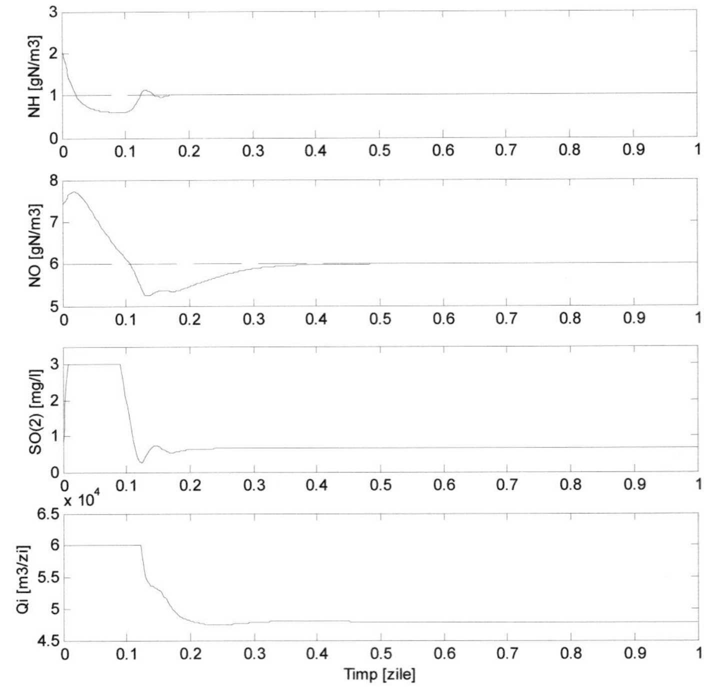

Fig. 20. QFT robust control applied in the case of "drought" regime

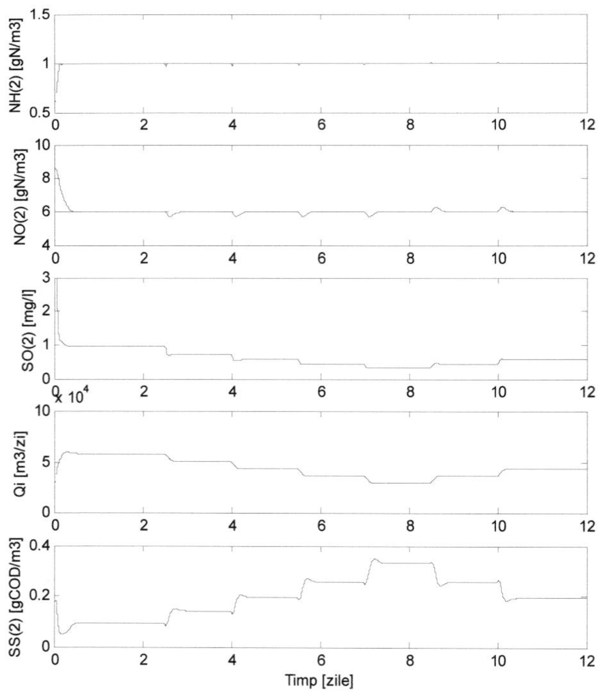

Fig. 21. QFT robust control of the wastewater treatment process

5. Conclusions

The present chapter deals with the robust control of wastewater treatment processes with activated sludge using QFT method aiming to increase their efficiency. The paper shows that QFT method is suitable for the control of these processes, taking into account the complexity, nonlinearity and the high degree of uncertainty that characterize biological wastewater treatment processes. QFT robust control method proved its effectiveness to be applied with good results both in local control loops, such as dissolved oxygen concentration control loop, as well as in the overall biological treatment algorithm, such as the control of ammonium concentration from the wastewater.

In order to design the QFT control law in the case of dissolved oxygen concentration control an analysis of the control loop dynamics was performed. It was concluded that the process can be approximated by linear models in different operating points. The testing of QFT control structure was done on a pilot plant for biological wastewater treatment, also presented in the paper.

In order to design the QFT control law in the case of the control of ammonium concentration in the effluent a simplified version of the ASM1 model was used. This model was linearized in three relevant operating points (rain, drought and normal). For each linear model the corresponding control structure has been designed. The results were validated through numerical simulation.

In both applications developed in this work it can be seen that QFT control structures offers good results, that is the output variables are tracking the imposed setpoints despite the fact that the nonlinear process modifies its operating point, both in terms of the inflow and the organic matter load.

6. Acknowledgement

Marian Barbu acknowledges the support of CNCSIS-UEFISCSU, Project 9 PN II-RU 79/2010.

7. References

Barbu, M., Barbu, G. & Ceanga, E. (2004). The Multi-model Control of the Wastewater Treatement Process with Activated Sludge, *Proceedings of 12th Mediterranean Conference on Control and Automation - MED'04*, Kusadasi, Turkey, CD-ROM.

Barbu, M. and Caraman S. (2007). QFT Multivariabil Control of a Biotechnological Wastewater Treatment Process Using ASM1 Model, *Proceedings of 10th IFAC Symposium on Computer Applications in Biotechnology*, Cancun.

Barbu, M., (2009). *Automatic Control of Biotechnological Processes*, Galati University Press, Romania.

Barbu, M., Ifrim, G., Caraman, S. & Bahrim G. (2010), QFT Control of Dissolved Oxygen Concentration in a Wastewater Treatment Pilot Plant, *Proceedings of 11th IFAC Symposium on Computer Applications in Biotechnology*, Leuven, Belgium, Pp. 58.

Brdys, M.A. & Zhang, Y. (2001a). Robust Hierarchical Optimising Control of Municipal Wastewater Treatment Plants, *Preprints of the 9th IFAC/IFORS/IMACS/IFIP Symposium Large Scale Systems: Theory & Applications – LSS'2001*, Bucharest, Romania, Pp. 540-547.

Brdys, M.A. & Konarczak, K. (2001b) Dissolved Oxygen Control for Activated Sludge Processes, *Preprints of the 9th IFAC/IFORS/IMACS/IFIP Symposium Large Scale Systems: Theory & Applications – LSS'2001*, Bucharest, Romania, Pp. 548-553.

Garcia-Sanz, M., Eguinoa, I., Gil, M., Irizar. I & Ayesa, E. (2008). MIMO Quantitative Robust Control of a Wastewater Treatment Plant for Biological Removal of Nitrogen and Phosphorus, *16th Mediterranean Conference on Control and Automation*, Corcega.

Goodman, B.L. & Englande, A.J. (1974). A Unified Model of the Activated Sludge Process, *Journal of Water Pollution Control Fed.*, Vol. 46, Pp. 312-332.

Henze, M., et al. (1987). *Activated Sludge Model No. 1*, IAWQ Scientific and Technical Report No. 1, IAWQ, Great Britain.

Henze, M., et al. (1995). *Activated Sludge Model No. 2*, IAWQ Scientific and Technical Report No. 3, IAWQ, Great Britain.

Henze, M., et al. (2000). Henze, M., et al., *Activated Sludge Models ASM1, ASM2, ASM2d and ASM3*, IWA Publishing, London, Great Britain.

Horowitz, I.M. (1973). Optimum Loop Transfer Function in Single-Loop Minimum Phase Feedback Systems, *International Journal of Control*, Vol. 22, Pp. 97-113.

Horowitz, I.M. (1991). Survey of Quantitative Feedback Theory (QFT), *International Journal of Control*, Vol. 53, No. 2, Pp. 255-291.

Houpis, C.H. & Rasmussen, S.J. (1999). *Quantitative Feedback Theory Fundamentals and Applications*, Marcel Dekker, Inc., New York.

Ingildsen, P. (2002). *Realising Full-Scale Control in Wastewater Treatment Systems Using In Situ Nutrient Sensors*, Ph.D. Thesis, Department of Industrial Electrical Engineering and Automation, Lund University, Sweden.

Jeppsson, U. (1996). *Modelling aspects of wastewater treatment processes*, Ph.D. thesis, Dept. of Industrial Electrical Eng. and Automation, Lund University, Sweden.

Katebi, M.R., Johnson, M.A. & Wilke J. (1999). *Control and Instrumentation for Wastewater Treatment Plant*, Springer-Verlag, London.

Larsson, T. & Skogestad, S. (2000). Plantwide control - A review and a new design procedure, *Modelling, Identification and Control*, Vol. 21, No. 4, Pp. 209-240.

Manesis, S.A., Sapidis, D.J. & King, R. E. (1998). Intelligent Control of Wastewater Treatment Plants, *Artificial Intelligence in Engineering*, Vol. 12, No. 3, pp. 275-281.

Nejjari, F., et al. (1999). Non-linear multivariable adaptive control of an activated sludge wastewater treatment process, *International Journal of Adaptive Control and Signal Processing*, Vol. 13, Issue 5, Pp. 347-365.

Olsson, G. (1985). Control strategies for the activated sludge process, *Comprehensive Biotechnology*, Editor M. Moo-Young, Pergamon Press, Pp. 1107-1119.

Olsson, G. & Chapman, D. (1985). Modelling the dynamics of clarifier behaviour in activated sludge systems, *Advances in Water Pollution Control*, IAWPRC, Pergamon Press.

Olsson, G. & Newell, B. (1999). *Wastewater treatment systems – modelling, diagnosis and control*, IWA Publishing, Great Britain.

Yagi, S. et al. (2002). Fuzzy Control of a Wastewater Treatment Plant for Nutrients Removal, *Proceedings of the International Conference on Artificial Intelligence in Engineering & Technology*, Sabah, Malaysia.

28

Control of a Simple Constrained MIMO System with Steady-state Optimization

František Dušek and Daniel Honc
University of Pardubice
Czech Republic

1. Introduction

This chapter covers two issues (along many others) relating to complex systems control. The main theme is connected with control of Multi-Input Multi-Output (MIMO) systems. If the controlled system has more inputs than outputs (further labelled as MI+MO) there exist many combinations of the inputs for one combination of the outputs. We are able to reach desired system outputs (the main control aim) with many input combinations. This situation is very interesting from practical point of view. Usually optimal inputs combination exists - from some point of view. This combination leads for example to minimum energy consumption, maximum production efficiency or minimum machinery load etc. In practice the set of possible inputs combination is reduced because of constrains and the best feasible combination lies very often on constrain. It would be suitable to extend the controller design to include supplementary demand simultaneously with the fulfilment of the main control aim - to ensure best feasible input combination, too. The common advanced controller like LQ controller has no problem with MIMO system which has different number of inputs and outputs in contrary to standard controllers designed as decentralized control. However the constrains respecting within the standard LQ controller design is not possible. Another advanced controller - Model Predictive Controller (MPC) allows constrains handling (Camacho & Bordons, 1999), (Maciejowski, 2002), (Rossiter, 2003) but the standard controller design doesn't solve which combination of inputs will occur in the steady-state in case of system with more inputs than outputs.
One possibility how to achieve optimal inputs combination is to formulate one term of the cost function connected with inputs penalization as a deviation from ideal inputs combination. This approach is used e.g. in (Novák, 2009) but according to the opinion of authors this approach isn't as universal as following proposal. We have suggested adding another term into the cost function of predictive controller - terminal state in a form of the deviation from desired terminal state. The desired terminal state is chosen that it corresponds to feasible optimal inputs combination and a value of the set-point at the end of the control horizon. Authors call this technique *Steady-state optimization* because the influence of the terminal state deviation comes to light namely in steady-state when the main control aim (desired output combination) is or has been already fulfilled - see also (Dušek & Honc, 2009). The controller ensures both main and supplementary control aims - achievement of desired outputs and inputs moving to an optimal combination. An incorporation of the terminal state into the cost function has also another advantage. The

addition of the terminal state into the cost function is one of the possibilities how to ensure closed-loop stability (Mayne et al., 2000). In (Bitmead et al., 1990) it is proposed using the quadratic form $\mathbf{x}_t^T\mathbf{P}\mathbf{x}_t$, as a terminal cost function where vector $\mathbf{x}_t$ is state at the end of control horizon (terminal state) and the matrix $\mathbf{P}$ is terminal value of the Riccati difference equation. We propose incorporate the terminal cost function in the form of desired ($\mathbf{x}_w$) and terminated state ($\mathbf{x}_t$) deviation - $(\mathbf{x}_w - \mathbf{x}_t)^T\mathbf{Q}_x(\mathbf{x}_w - \mathbf{x}_t)$ (see Chapter 4.2). The determination of a desired state based on the controlled system steady-state gain matrix is shown for the case of general MIMO system in the article (Dušek & Honc, 2008b) and in detail in (Dušek & Honc, 2008a in Czech). The computation of desired state for the case of MI+MO systems is described in Chapter 4.3.

Suggested technique is applied on the thermostatic bath control. The idealized thermostatic bath (see sketch in Fig. 1) is an example of one of the simplest real constrained system with more inputs than outputs. On this example it is possible to demonstrate another problem we can meet by the control of complex systems - in some cases an advanced controller improves control quality only slightly in comparison with very simple controller. The problem usually arises when one property of the controlled system is dominant. In that case a simple controller respecting the dominant feature can provide satisfactory control. But situation can change dramatically if some specific (or additional) information about the system is available or additional control demands are formulated. Manipulated variables asymmetric constraints are dominant feature of controlled system in our case. Very simple on-off controller based on knowledge of constrains provides similar control quality (from performance measures and control costs points of view) as the sophisticated MPC controller even based on full knowledge of MIMO system dynamic. This holds for the case that we do not know nor do not use information about future reference signal course in MPC controller design. Simple controller do not allow to use such information on the contrary to advanced controller - predictive controller respecting constrains and using future reference course knowledge.

2. Problem formulation, solution fundamentals

In the following text we will show two different control designs for an ideal thermostatic bath. It is possible to describe controlled system behaviour by continues dynamical fourth order mathematical model with four inputs (three are manipulated and constrained) and one controlled output. The model derivation is based on first principle approach (energy conservation law) and a few simplified assumptions. Model parameters are chosen so that the model behaviour is realistic for needs of simulated control experiments. The continuous-time model is numerically transformed into discrete-time state space form for chosen sample time.

The aim of the control is to follow as good as possible a reference signal with respecting the manipulated variable constrains with minimum control cost - energy consumption. Two very different controllers have been designed. The first one is a couple of very simple discrete-time on-off controllers based on system specific feature - two asymmetrically constrained manipulated inputs. The second one is an advanced discrete-time predictive controller with quadratic cost function, finite horizon and banded constrains based on a discrete-time linear state space MI+MO model (TISO - system with two manipulated inputs and one controlled output). The controller cost function is supplemented by a quadratic terminal cost function of the desired and actual terminal state deviation - ensuring steady-state optimization. An addition of a desired terminal state into controller cost function allows including the demand on minimal energy cost. The minimisation of cost function is made by quadratic programming. The behaviour of both controllers is demonstrated on simulated discrete-time control experiments with continuous-time model of ideal thermostatic bath. Results of simulated controls by on-off controller and proposed predictive controller are discussed. Control responses of the predictive controller without knowledge of future course of the reference signal (only an actual set-point is known) and when the future course is known are compared too. All the computations, results evaluation and visualisation have been made in MATLAB environment.

3. Controlled system

The controlled system is the ideal thermostatic bath which principal sketch is drawn in Fig. 1. Similar real devices are used for controlled heating or cooling of some element. This device is one of the simplest real systems with more inputs then outputs. It is represented by a partially isolated vessel filled with water (denoted C) and placed element D - its temperature T_D is controlled. It is possible to increase the water temperature T_C with electric heating (denoted A). Heating power E is controlled continuously. Cooling helix (denoted B) is used to decrease water temperature - water flows with flow-rate Q through a pipe. Inlet temperature T_{B0} must be lower than a placed element desired temperature T_D. Temperature T_C is affected also by ambient temperature T_o (heating exchange with surroundings because of imperfect isolation). Ambient temperature T_o can cool down the bath if $T_o < T_C$ or heat it if $T_o > T_C$.

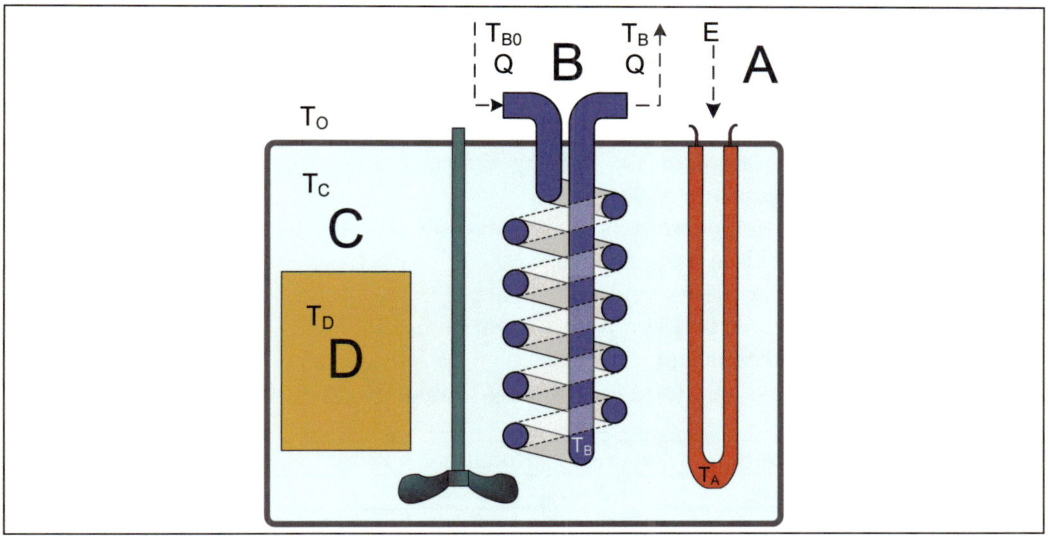

Fig. 1. Thermostatic bath scheme

An ambient temperature T_o is supposed to be constant. A cooling water flow-rate must be within the range $0 \le Q \le Q_{max}$, cooling water input temperature $T_{B0min} \le T_{B0} \le T_{B0max}$ and heating power $0 \le E \le E_{max}$. These asymmetrical constrain lead to special actuating of inputs - it is possible to increase or decrease the state variables only with the particular input.

3.1 Derivation of mathematical model of the plant

The thermostatic bath can be divided into four parts (thermal capacities) according to the scheme in Fig. 1. The state of every part is approximated by "characteristic or average" temperature. The introduction of characteristic temperatures leads to the essential simplification of a process description and hence partial differential equations using is not necessary. Based on the energy balance of every part the whole system can be described under another simplified assumptions (ideal mixing, constant heat transfer coefficients etc.) with a four ordinary differential equations - mathematical model of the plant. The model has eight time depending variables - four input variables (cooling water flow-rate Q with input temperature T_{B0}, heating power E, ambient temperature T_0) and four state variables (characteristic temperature of the heating element T_A, cooling water characteristic temperature T_B, water characteristic temperature T_C and placed element characteristic temperature T_D).

If we put together thermal balances mentioned above and introduce simplified assumptions we get relatively simple dynamic mathematical model of the thermostatic bath as a set of four ordinary differential equations written as

$$E = \alpha_A S_{AC}(T_A - T_C) + m_A c_A \mathrm{d}T_A/\mathrm{d}t \tag{1a}$$

$$Q c_B T_{B0} + \alpha_B S_{BC}(T_C - T_B) = Q c_B T_B + m_B c_B \mathrm{d}T_B/\mathrm{d}t \tag{1b}$$

$$\alpha_A S_{AC}(T_A - T_C) = \alpha_B S_{BC}(T_C - T_B) + \alpha_C S_{C0}(T_C - T_o) + \alpha_D S_{DC}(T_C - T_D) + m_C c_C \mathrm{d}T_C/\mathrm{d}t \tag{1c}$$

$$\alpha_D S_{DC}(T_C - T_D) = m_D c_D \mathrm{d}T_D/\mathrm{d}t \tag{1d}$$

where

T_o is ambient temperature,

$E(t)$ is heating power in the range $0 \leq E \leq E_{max}$ (increases temperature of A),

$Q(t)$ is flow-rate of the cooling water in the range $0 \leq Q \leq Q_{max}$ (decreases temperature of B),

$T_{B0}(t)$ is input temperature of the cooling water in the range $T_{B0min} \leq T_{B0} \leq T_{B0max}$ (decreases temperature of B),

$T_x(t)$ is characteristic temperature (state variables $T_A \dots T_D$),

m_x is mass of individual part,

c_x is specific heat capacity of individual part,

S_{xy} is heat transfer area between two adjacent parts and

α_x is heat transfer coefficient.

Parameters given in Table 1 are used in following simulation experiments.

	units	heating A	cooling B	water C	element D
m_x	kg	0.3	0.1567	4.0	8.93
c_x	$\mathrm{J \cdot kg^{-1} \cdot K^{-1}}$	452	4180	4180	383
S_{xy}	$\mathrm{m^2}$	0.0095	0.065	0.24	0.06
α_x	$\mathrm{J\, m^{-2} \cdot s^{-1} \cdot K^{-1}}$	750	500	5	500

Table 1. Process model parameters

The graphs in Fig. 2 demonstrate the basic dynamic behaviour of the system with parameters according to Table 1. In this figure it is depicted the temperature response of placed element T_D (upper graph) to 10 minutes wide pulse of maximal heating power E (middle graph) and 10 minutes wide pulse of minimal cooling water temperature T_{B0} (lower graph). The experiment starts from a system steady-state when all temperatures are the same and equal to the ambient temperature T_0. This steady-state corresponds to heating power equal zero and cooling water temperature equal to ambient temperature. The constant cooling water flow-rate is 0.5 litres per minute. From graphs it is evident that maximal heating is more powerful than maximal cooling.

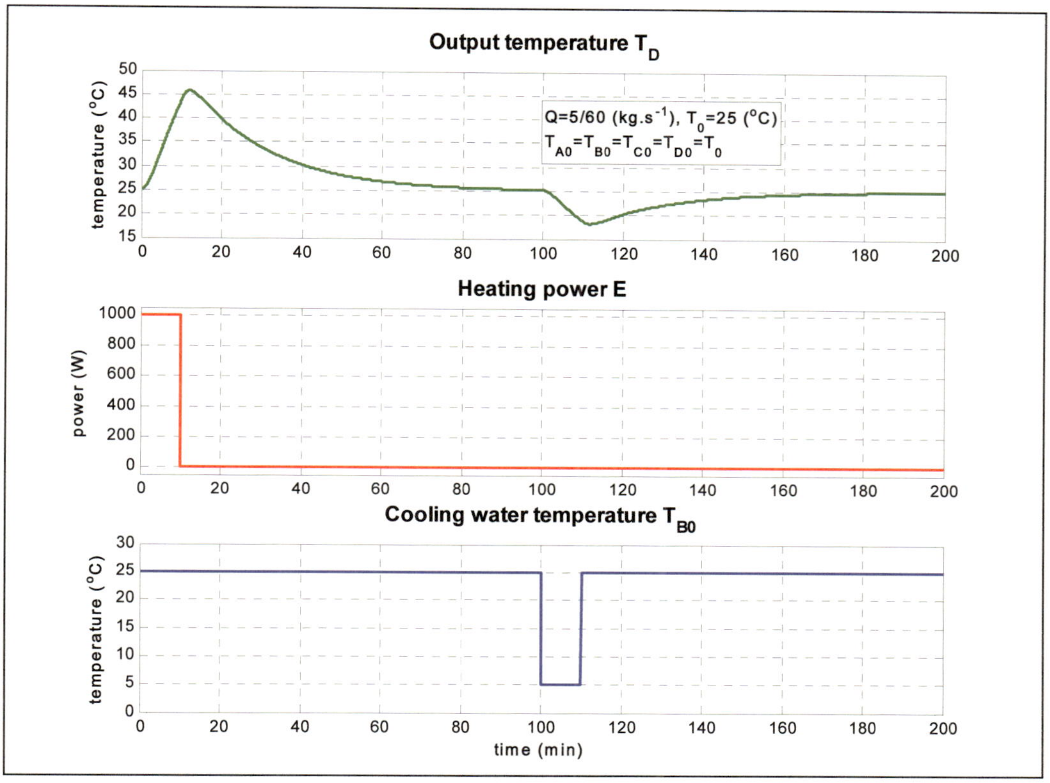

Fig. 2. Output variable response to inputs changes

3.2 Continuous-time mathematical model in a standard form

From the control point of view the system has three manipulated variables (Q, T_{B0}, E), one measured disturbance (T_0), four state variables (T_A, T_B, T_C and T_D) and one controlled variable (T_D). To get linear system suitable for the control design we choose only the input temperature T_{B0} and heating power E as manipulated variables. The dynamic of input temperature T_{B0} refrigerating is neglected to simplify the thermostatic bath description. The cooling water flow-rate Q is supposed to be constant. This "non practical" choice is made due to simplification of predictive controller design - to avoid problem with nonlinear system control design. For needs of this text it isn't important whether the manipulated variable is cooling water flow-rate or temperature.

The equations (1a) - (1d) can be rewritten in a matrix form of standard continuous-time state space model as

$$dx/dt = \mathbf{A}_c\mathbf{x} + \mathbf{B}_c\mathbf{u} \tag{2a}$$

$$y = \mathbf{C}_c\mathbf{x} \tag{2b}$$

Integral part of the process description is information about the manipulated variables constrains.

$$0 \le E \le E_{max} \tag{2c}$$

$$T_{B0min} \le T_{B0} \le T_{B0max} \tag{2d}$$

where

$\mathbf{x}(t)$ is vector of state variables T_A, T_B, T_C and T_D,

$$\mathbf{x}(t) = [T_A(t), T_B(t), T_C(t), T_D(t)]^T$$

$\mathbf{u}(t)$ is vector of inputs E, T_{B0} and T_0,

$$\mathbf{u}(t) = [E(t), T_{B0}(t), T_0(t)]^T$$

$y(t)$ is output variable T_D and

$\mathbf{A}_c$, $\mathbf{B}_c$, $\mathbf{C}_c$ are matrices of continuous-time state space model parameters (see Eq. 2e)

$$\mathbf{A}_c = \begin{bmatrix} -\frac{\alpha_A S_{AC}}{m_A c_A} & 0 & \frac{\alpha_A S_{AC}}{m_A c_A} & 0 \\ 0 & -\frac{\alpha_B S_{BC}}{m_B c_B} - \frac{Q}{m_B} & \frac{\alpha_B S_{BC}}{m_B c_B} & 0 \\ \frac{\alpha_A S_{AC}}{m_C c_C} & \frac{\alpha_B S_{BC}}{m_C c_C} & -\frac{\alpha_A S_{AC} + \alpha_B S_{BC} + \alpha_C S_{C0} + \alpha_D S_{DC}}{m_C c_C} & \frac{\alpha_D S_{DC}}{m_C c_C} \\ 0 & 0 & \frac{\alpha_D S_{DC}}{m_D c_D} & -\frac{\alpha_D S_{DC}}{m_D c_D} \end{bmatrix}$$
$$\mathbf{B}_c = \begin{bmatrix} \frac{1}{m_A c_A} & 0 & 0 \\ 0 & \frac{Q}{m_B} & 0 \\ 0 & 0 & \frac{\alpha_C S_{C0}}{m_C c_C} \\ 0 & 0 & 0 \end{bmatrix} \qquad \mathbf{C}_c = \begin{bmatrix} 0 & 0 & 0 & 1 \end{bmatrix} \tag{2e}$$

The continuous-time mathematical model (2) with parameters given by Table 1 was used in simulation control experiments as a plant (process) model.

3.3 Discrete-time mathematical model for MPC control design

A standard predictive controller design is based on a discrete-time linear time invariant (LTI) process model. If we suppose constant cooling water flow-rate Q than the matrices $\mathbf{A}_c$ and $\mathbf{B}_c$ in (2e) are constant (time invariant) for given values of thermostatic bath parameters. Now we can transform the linear continues-time model (2) into equivalent linear discrete-time state space model (3) or an input-output model under the "zero order hold" assumption - that the value of inputs between two equidistant sample times are constant. We get the values of discrete-time state space model matrices **A**, **B** and **C** for given sample time T numerically (in MATLAB with function c2d)

$$\mathbf{x}(k+1) = \mathbf{A}\mathbf{x}(k) + \mathbf{B}\mathbf{u}(k) \tag{3a}$$

$$y(k) = \mathbf{C}\mathbf{x}(k) \tag{3b}$$

$$\mathbf{u}_{\min} \leq \mathbf{u}(\mathrm{k}) \leq \mathbf{u}_{max} \tag{3c}$$

where

$\mathbf{x}(k)$ is vector of sampled state variables T_A, T_B, T_C and T_D,

$$\mathbf{x}(k) = [T_A(k), T_B(k), T_C(k), T_D(k)]^T$$

$\mathbf{u}(k)$ is vector of discrete-time inputs E, T_{B0} and T_0,

$$\mathbf{u}(k) = [E(k), T_{B0}(k), T_0(k)]^T$$

$$\mathbf{u}_{\min} = [0, T_{B0\min}, T_0]^T$$

$$\mathbf{u}_{\max} = [E_{\max}, T_{B0\max}, T_0]^T$$

$y(k)$ is sampled output variable T_D and

$\mathbf{A}, \mathbf{B}, \mathbf{C}$ are discrete-time model parameters (matrices).

3.4 Prediction equations in matrix form

If we use cost function in a general matrix form then it is suitable to formulate future process output directly in a matrix form and not in the original iterative form (3a). Because we will also need a state prediction at the end of the prediction horizon we will formulate the state prediction equation for N sample step ahead, too. Based on knowledge of the actual state $\mathbf{x}(k)$ and a vector of future inputs $\mathbf{u}_N$ we can write these two prediction matrix equations as

$$\mathbf{y}_N = \mathbf{S}_{yx}\mathbf{x}(k) + \mathbf{S}_{yu}\mathbf{u}_N \tag{4a}$$

$$\mathbf{x}(k+N) = \mathbf{S}_{xx}\mathbf{x}(k) + \mathbf{S}_{xu}\mathbf{u}_N \tag{4b}$$

where

$\mathbf{y}_N$ is vector of future output T_D at time k, k+1, …, k+N-1,

$$\mathbf{y}_N = [T_D(k+1), T_D(k+2), \ldots, T_D(k+N)]^T$$

$\mathbf{x}(\mathrm{k})$ is vector of state variables T_A, T_B, T_C and T_D at time k,

$\mathbf{u}_N$ is vector of future inputs E, T_{B0} and T_0 at time k, k+1, …, k+N-1,

$$\mathbf{u}_N = [\mathbf{u}^T(k), \mathbf{u}^T(k+1), \ldots, \mathbf{u}^T(k+N-1)]^T \text{ and}$$

$\mathbf{S}_{xx}$, $\mathbf{S}_{xu}$, $\mathbf{S}_{yx}$, $\mathbf{S}_{xu}$ are constant matrices depending on the process matrices $\mathbf{A}$, $\mathbf{B}$ and $\mathbf{C}$

$$\mathbf{S}_{xx} = \mathbf{A}^N \qquad \mathbf{S}_{xu} = \begin{bmatrix} \mathbf{A}^{N-1}\mathbf{B} & \mathbf{A}^{N-2}\mathbf{B} & \ldots & \mathbf{AB} & \mathbf{B} \end{bmatrix}$$

$$\mathbf{S}_{yx} = \begin{bmatrix} \mathbf{CA} \\ \mathbf{CA}^2 \\ \vdots \\ \mathbf{CA}^{N-2} \\ \mathbf{CA}^{N-1} \end{bmatrix} \qquad \mathbf{S}_{xu} = \begin{bmatrix} \mathbf{CB} & 0 & \ldots & 0 & 0 \\ \mathbf{CAB} & \mathbf{CB} & \ldots & 0 & 0 \\ \vdots & \vdots & \ddots & \vdots & \vdots \\ \mathbf{CA}^{N-3}\mathbf{B} & \mathbf{CA}^{N-4}\mathbf{B} & \ldots & \mathbf{CB} & 0 \\ \mathbf{CA}^{N-2}\mathbf{B} & \mathbf{CA}^{N-3}\mathbf{B} & \ldots & \mathbf{CAB} & \mathbf{CB} \end{bmatrix} \tag{4c}$$

4. Control design

The main control objectives are to follow the reference signal, to respect manipulated variables constrains and at the same time to minimize energy costs for heating and refrigerating of the cooling water (we do not consider dynamics of the cooling water input temperature). It is evident that the set-point of steady-state output temperature is reachable with many combinations of heating power and cooling water temperature. The system has different overall energy consumption for each combination. From energy consumption point of view the ideal combination in steady-state is when heating power equals to zero and temperature of cooling water equals to surrounding temperature. This ideal combination with zero energy cost is feasible only in the situation that desired temperature is equal to surrounding temperature. Because of the imperfect thermostat insulation it is necessary either permanently to heat or to cool in all other cases. Hence if it is necessary to heat then refrigeration must be off and vice versa. This idea is the principle of the simplest on-off controller without any tuneable parameters described in Chapter 4.1. Based on this idea it is also possible to design many other simple controllers with some solution for the system with more inputs than outputs. Well known is for example the technique called split range in which the output of a controller is split into two or more manipulated variables.
But these solutions are made ad hoc. More systematic and general way is to use MIMO controller. Such a controller based on principles of model predictive control is described in Chapter 4.2.

4.1 On-off controller

It is possible to control the thermostatic bath with objectives and conditions mentioned above by an on-off controller (to switch between minimal and maximal cooling water input temperature and heating power according to the sign of the control error). This approach uses only one dominant characteristic - asymmetrical manipulated variable actuating. The control error performance measure is comparable with a sophisticated predictive controller without using any information about the future set-point. This strategy uses the only information about manipulated variables constraints (E_{max}, T_{B0min}, $T_{B0,max}$), actual value of output variable T_D and actual value of set point w at discrete time k and there are no tuneable controller parameters. The resulting very simple thermostatic bath on-of controller is given by (5)

$$e = w(k) - T_D(k) \tag{5a}$$

$$\text{if } e<0 \text{ then } \{E=E_{max}\ T_{B0}=T_{B0,max}\} \text{ else } \{E=0\ T_{B0}=T_{B0,min}\} \tag{5b}$$

where

$w(k)$ is an actual set-point - desired value of output T_D at time k,
e is an actual control error,
E is actual heating power and
T_{B0} is actual cooling water input temperature.

Such a very primitive strategy has interesting features. It is a feedback control with a huge feedback gain and closed loop stability is ensured by respecting the constraints. It is simple variant of adaptive control approach called in literature as Self-Oscillating Adaptive Systems (Åström & Wittenmark, 1995). Big feedback gain causes controller insensitivity to changing of process properties (Wellstead & Zarrop, 1991) and we can suppose operation without problems even in case when the cooling water flow-rate is used as a manipulated variable - in case of nonlinear system. The disadvantage is permanent control variables switching between minimal and maximal values and thereby permanently alternating of the controlled variable in the steady-state. The control quality and control costs are worse than in case of controller with continuous output - see Fig. 3.

4.2 Predictive controller

Predictive controller design is open methodology and it allows incorporating many of control demands and other information. The control objective is formulated as a minimization of a discrete-time cost function that is constrained. It means that the dependencies given by the process model have to be respected. From math point of view it is a task of finding constrained extreme. If the cost function is quadratic with finite horizons, process model is linear and variables are unconstrained then the analytic solution of the cost function minimization exists in a form of matrix equations. If inputs, outputs or states are linearly constrained then it is possible to solve arising task numerically with quadratic programming techniques.

We formulate the discrete-time quadratic cost function on finite horizon of length N steps (both predictive and control horizon) in matrix form (6a) and inputs constrains (6b) as

$$\begin{aligned} & J(\Delta\mathbf{u}_N,\mathbf{w}_N,\mathbf{x}_w;N) = \mathbf{e}_N^T\mathbf{Q}\mathbf{e}_N + \Delta\mathbf{u}_N^T\mathbf{R}\Delta\mathbf{u}_N + \Delta\mathbf{x}^T(N)\mathbf{Q}_x\Delta\mathbf{x}(N) \\ & \mathbf{e}_N = \mathbf{w}_N - \mathbf{y}_N \qquad \mathbf{u}_N = \mathbf{u}_{N,0} + \Delta\mathbf{u}_N \qquad \Delta\mathbf{x}(N) = \mathbf{x}_w - \mathbf{x}(k+N) \end{aligned} \tag{6a}$$

$$\mathbf{u}_{N,\min} \le \mathbf{u}_N \le \mathbf{u}_{N,\max} \tag{6b}$$

where

$\mathbf{y}_N$ is vector of predicted process outputs (see Eq. 4a),
$\mathbf{w}_N$ is vector of future reference signal,
$\mathbf{u}_N$ is vector of future process inputs,
$\mathbf{u}_{N,0}$ is vector of supposed (known) future process inputs,
$\Delta\mathbf{u}_N$ is vector of computed deviations from supposed process inputs (this vector contains only the manipulated inputs),
$\mathbf{x}_w$ is desired terminal state (see Chapter 4.3),
$\mathbf{x}(k+N)$ is predicted terminal state (see Eq. 4b),
N is length of control and prediction horizon (number of samples),
$\mathbf{Q}$, $\mathbf{Q}_x$, $\mathbf{R}$ are square weighting matrices and
$\mathbf{u}_{N,\min}$, $\mathbf{u}_{N,\max}$ are vectors of input constrains.

The cost function (6a) is composed of three parts. All parts are quadratic function of adequate deviations. The first two parts are functions of the all points over the whole horizon and the last part is a function of the last point of horizon only. The first part is a function of control error (the deviation between output and reference signal). It ensures a satisfaction of the main control aim - following the reference signal as close as possible.

The second part is a function of manipulated variables and ensures that the main control aim isn't fulfilled at any cost - infinite or very large values of manipulated variables. The disadvantage of standard form (without deviations) is arising of a steady-state control error. We use this term in a form of deviations of inputs from supposed future inputs. The deviation decreases a steady-state control error and incorporating of supposed course of

inputs $\mathbf{u}_{N,0}$ allows involving known unmanipulated inputs or disturbances (ambient temperature in our case). Supposed inputs can be also used for optimization of inputs values combination in case of MI+MO systems – similar way as in (Novák, 2009).
The third part is the deviation of a desired and actual state at the end of horizon (terminal state). Adding a terminal cost function ensures closed-loop stability (Mayne et al., 2000) but also leads to arising of a steady-state control error. Proposal of a terminal cost function in the quadratic form was made by (Bitmead et al., 1990). We propose the quadratic terminal cost function of desired and terminal state deviation. The deviation decreases a steady-state control error and desired terminal state allows taking into account additional control requirement. We use the desired state to steady-state optimization in case of MI+MO systems. The desired state computation for MI+MO system is described in following chapter 4.3.
If we use prediction equations (4a), (4b) to eliminate process output $\mathbf{y}_N$ and terminal state $\mathbf{x}(k+N)$ then the cost function (6a) and constraints (6b) can be rewritten into form

$$J(\Delta\mathbf{u}_N) = \Delta\mathbf{u}_N^T\mathbf{M}\Delta\mathbf{u}_N + \Delta\mathbf{u}_N^T\mathbf{m}_k + \mathbf{m}_k^T\Delta\mathbf{u}_N + c = \Delta\mathbf{u}_N^T\mathbf{M}\Delta\mathbf{u}_N + 2\mathbf{m}_k^T\Delta\mathbf{u}_N + c$$
$$\mathbf{M} = \mathbf{R} + \mathbf{S}_{yu}^T\mathbf{Q}\mathbf{S}_{yu} + \mathbf{S}_{xu}^T\mathbf{Q}_x\mathbf{S}_{xu} \tag{7a}$$
$$\mathbf{m} = -\left\{\mathbf{S}_{yu}^T\mathbf{Q}\left[\mathbf{w}_N - \mathbf{S}_{yx}\mathbf{x}(k) - \mathbf{S}_{yu}\mathbf{u}_{N,0}\right] + \mathbf{S}_{xu}^T\mathbf{Q}_x\left[\mathbf{x}_w - \mathbf{S}_{xx}\mathbf{x}(k) - \mathbf{S}_{xu}\mathbf{u}_{N,0}\right]\right\}$$

$$\mathbf{u}_{N,\min} - \mathbf{u}_{N,0} \le \Delta\mathbf{u}_N \le \mathbf{u}_{N,\max} - \mathbf{u}_{N,0} \tag{7b}$$

The minimization of (7a) regarding to $\Delta\mathbf{u}_N$ without constrains (7b) is possible in explicit form (7c) on condition that the matrix **M** is positive definite and symmetric.

$$\Delta\mathbf{u}_N = -\mathbf{M}^{-1}\mathbf{m} \tag{7c}$$

The minimization of (7a) with constrains (7b) is a task of quadratic programming. In both cases we get a vector of future manipulated inputs deviation $\Delta\mathbf{u}_N$ that in combination of supposed inputs $\mathbf{u}_{N,0}$ gives vector of optimal process inputs $\mathbf{u}_N$. The calculated value of optimal inputs depends on actual state $\mathbf{x}$(k), future course of reference signal $\mathbf{w}_N$, desired terminal state $\mathbf{x}_w$ and supposed future inputs $\mathbf{u}_{N,0}$. If the actual state isn't measured, then a state estimator based on state space model (3a), (3b) can be used. If the future course of reference signal isn't known then the actual set point is used as a future course of constant reference signal. The calculation of desired terminal state is described in next chapter. The vector of supposed inputs $\mathbf{u}_{N,0}$ can be constructed from actual values of inputs which are supposed to be constant in the future or if we know the future course of some inputs (as known future disturbances) then we can add this information in corresponding part of $\mathbf{u}_{N,0}$.
We apply only the control actions of the first member $\mathbf{u}$(k) from the optimal vector $\mathbf{u}_N$ and the minimization is repeated in the next sample time.

4.3 Calculation of desired state

The calculation of the desired state $\mathbf{x}_w$ is a fundamental part of steady-state optimization. It is based on a non square steady-state gain matrix **Z** of a MI+MO system model (3)

$$y_\infty = \mathbf{Z}\mathbf{u} \tag{8a}$$

$$\mathbf{Z} = \mathbf{C}(\mathbf{I}-\mathbf{A})^{-1}\mathbf{B} \tag{8b}$$

The calculation is described by equations (9a - 9d) and the solution is valid for steady state. The minimization of quadratic cost function (9a) represents the requirement of a minimal quadratic distance between ideal inputs values $\mathbf{u}_{ideal}$ and accessible inputs $\tilde{\mathbf{u}}$. At the same time the equation (9b) has to be respected. This equation arising from (3) represents a requirement that accessible inputs $\tilde{\mathbf{u}}$ lead the system (in steady-state) to set-point at the end of horizon $\mathbf{w}(k+N)$. These two equations formulate a standard task of constrained extreme minimization. The solution of this task is a value of $\tilde{\mathbf{u}}$ which can be recalculated to desired state $\mathbf{x}_w$ using the equation (9d).

$$\min_{\tilde{\mathbf{u}}}\left[\left(\mathbf{u}_{ideal}-\tilde{\mathbf{u}}\right)^T \mathbf{I}\left(\mathbf{u}_{ideal}-\tilde{\mathbf{u}}\right)\right] \tag{9a}$$

$$\mathbf{Z}\tilde{\mathbf{u}} = \mathbf{w}(k+N) \tag{9b}$$

$$\mathbf{u}_{\min} \le \tilde{\mathbf{u}} \le \mathbf{u}_{\max} \tag{9c}$$

$$\mathbf{x}_w = \left(\mathbf{I}-\mathbf{A}\right)^{-1}\mathbf{B}\tilde{\mathbf{u}} \tag{9d}$$

We can get the solution of unconstrained extreme task - by considering only equations (9a) and (9b) - in explicit form by Lagrange's multipliers. If we rewrite (9a), (9b) into form of (10a) then we get the searched input $\tilde{\mathbf{u}}$ as part of the solution of matrix expression (10b)

$$\left(\mathbf{u}_{ideal}-\tilde{\mathbf{u}}\right)^T \mathbf{I}\left(\mathbf{u}_{ideal}-\tilde{\mathbf{u}}\right)+\left[\mathbf{Z}\tilde{\mathbf{u}}-\mathbf{w}(k+N)\right]^T \boldsymbol{\lambda} = 0 \tag{10a}$$

$$\begin{bmatrix}\tilde{\mathbf{u}}\\ \boldsymbol{\lambda}\end{bmatrix} = \begin{bmatrix}2\mathbf{I} & \mathbf{Z}^T\\ \mathbf{Z} & 0\end{bmatrix}^{-1}\begin{bmatrix}\mathbf{u}_{ideal}\\ \mathbf{w}(k+N)\end{bmatrix} \tag{10b}$$

In case of constrained inputs - by considering also equation (9c) - the problem is formulated as quadratic programming task and the searched input $\tilde{\mathbf{u}}$ has to be computed numerically.

5. Simulated control experiments

The simulated experiments demonstrate the discrete-time control of continuous-time MI+MO system (thermostatic bath) with simple on-off controller (5) and predictive controller without (7) and with steady-state optimization (8). Placed element temperature T_D is controlled - responses of simulated reference signal tracking are depicted in Figures 3-5. The control by predictive controller is shown for two situations - without and with future reference signal course knowledge. All experiments are made under identical conditions and the control performance is evaluated by two measures. First measure $C_{quality}$ (quality) represents a value of quadratic control error area and the second measure C_{cost} (cost) is price of total energy consumption for the heating E_{heat} and cooling E_{cool}. The energy consumption for refrigerating E_{cool} (cooling water temperature decreasing at chosen constant flow-rate) is supposed (for needs of following simulations) to be equivalent to energy consumption for cooling water heating about same temperature difference with efficiency e_f=0.5 (50%). These two measures can be written as

$$C_{quality} = T\sum_{k=0}^{N_s}\left[w(k) - T_D(k)\right]^2 \tag{11a}$$

$$C_{cost} = k_{price}\left(E_{heat} + E_{cool}\right)$$

$$E_{heat} = T\sum_{k=0}^{N_s} E(k) \qquad E_{cool} = T\frac{Qc_B}{e_f}\sum_{k=0}^{N_s-1}\left(T_o - T_{BO}(k)\right) \tag{11b}$$

where

k_{price} is electric energy price per unit,

T is a controller sample time,

e_f is efficiency of cooling and

N_s is number of samples during experiment.

The ideal thermostatic bath is simulated as a continuous-time system (Eq. 1) with parameters given in Table 1. All control experiments start from steady-state ($\mathbf{x}_0$, $\mathbf{u}_0$) and respect inputs variables ranges. Conditions and constrains are listed in Table 2. The input values $\mathbf{u}_0$ leading to the steady-state $\mathbf{x}_0$ are no optimal from energy consumption point of view. These values were chosen to show the influence of the steady-state optimization for case of predictive controller.

	Input variables ranges and initial inputs $\mathbf{u}_0$				Steady-state $\mathbf{x}_0$, $\mathbf{u}_0$	
	E [W]	Q [kg s^{-1}]	T_{B0} [°C]	T_o [°C]	T_A [°C]	64.63
$\mathbf{u}_{max}$	1000	0.5/60	25	25	T_B [°C]	22.02
$\mathbf{u}_0$	250	0.5/60	15	25	T_C [°C]	29.54
$\mathbf{u}_{min}$	0	0.5/60	5	25	T_D [°C] = y_0	29.54

Table 2. Input variables and steady-state

The on-off controller is realized as discrete-time system (with zero-order hold terms on the outputs with sample time T = 10 s). Its response is depicted in Fig. 3. The achieved values of control quality and costs measures in this experiment are used as a standard and marked as 100%. The control quality is apparently bad (the output oscillates) but the computed control quality value is comparable with predictive controller without knowledge of future reference signal course. The on-off controller responds immediately to changes in reference signal with maximal values of heating or cooling and hence the output response is as quick as possible. In spite of the fact that the on-off controller ensures that the heating and cooling doesn't actuate concurrently the energy consumption is high because of heating and cooling switching to their maximal values.

Predictive controller with steady-state optimization and inputs constrains (6) is realized as a discrete-time system (with zero-order hold terms on the outputs with sample time T = 10 s). The control horizon is N = 60 samples (that is N×T = 600 s = 10 min). Control response of the predictive controller without future reference signal knowledge is depicted in Fig. 4. It means that the controller has information about actual value of set-point and the future reference signal is assumed to be constant and equal to set-point at current time. If the actual set-point changes then the constant future reference signal over the whole control horizon changes too. The values of control measures are relative to corresponding values achieved in control with on-off controller and expressed as percentages.

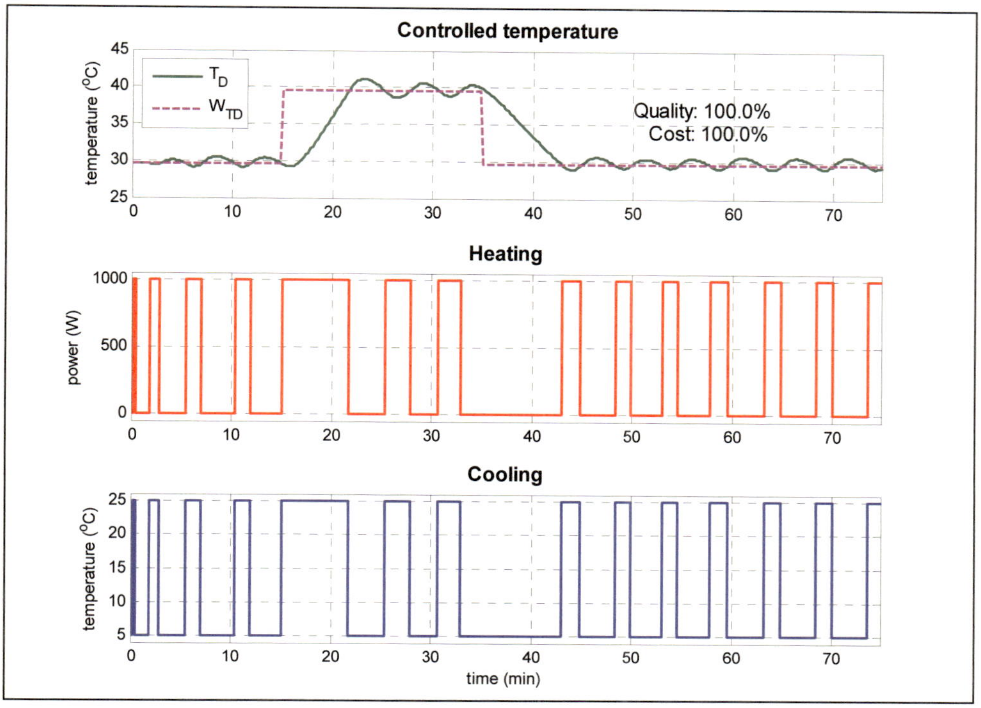

Fig. 3. On-off controller

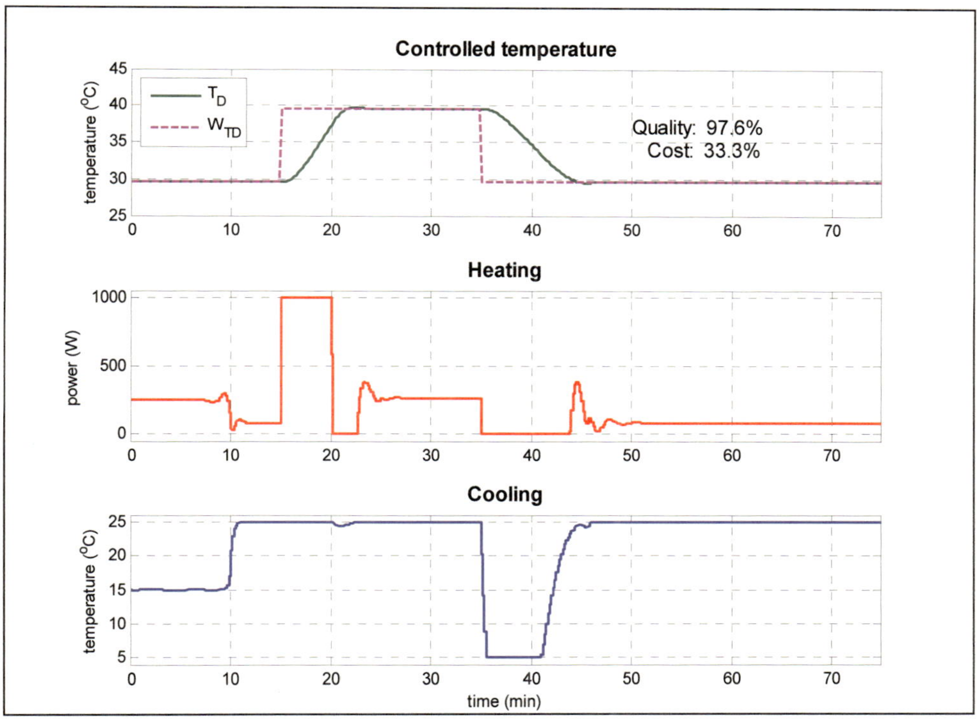

Fig. 4. Predictive controller - without future reference signal knowledge

The first 15 minutes of experiment depicted in the Fig. 4 demonstrates the steady-state optimization - the controller manipulates inputs without output change. The inputs achieve their optimal values after time corresponding to length of horizon (10 minutes). Because of the absent of a future reference signal knowledge the controller react only to actual set-point. The control quality (97.6%) is comparable with on-off controller but the energy consumption is significantly better (33.3%).
A control response of the predictive controller with steady-state optimization and with knowledge of the future reference signal is depicted in the Fig. 5. This experiment demonstrates best control approach from the point of control quality and energy consumption. The controller uses maximum of accessible information. Due to prediction horizon and future reference signal knowledge the controller can act before the actual set-point change. The time of advance controller reaction depends on both system dynamic and constrains. Hence it can be different when the set-point changes up and down. On this experiment we can also see that the control quality is preferred before control cost. There are parts of control where heating and cooling act simultaneously. We can see this in transient state only. This behaviour also depends on the choice of weighting matrices in the cost function (6a).

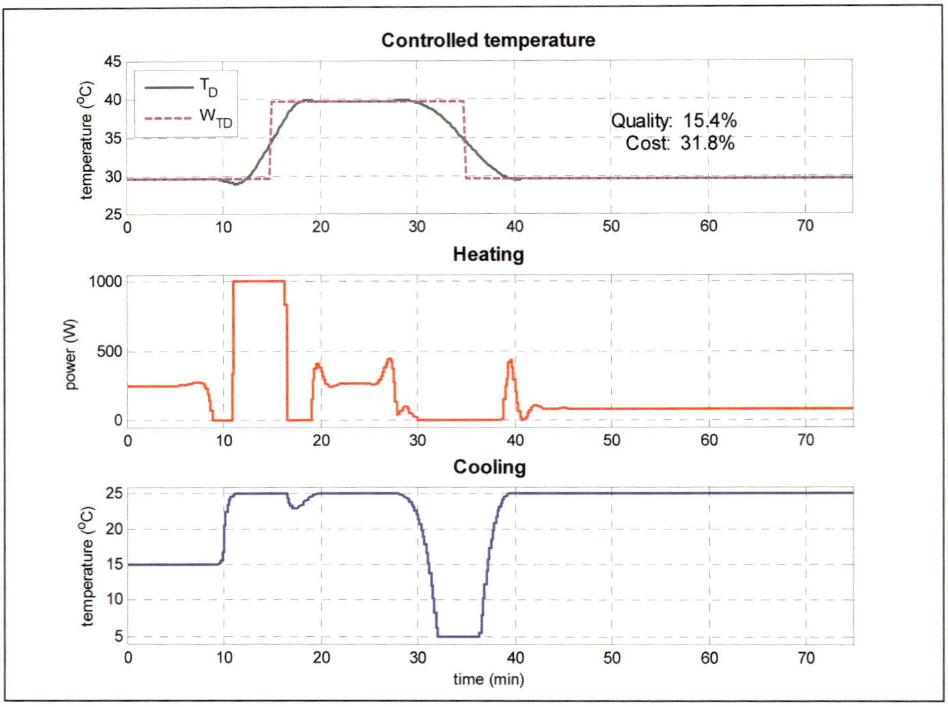

Fig. 5. Predictive controller – with future reference signal knowledge

6. Conclusion

Control design is often "made-to-measure problem" especially if one feature of the controlled process is dominant and therefore affecting control possibilities. Even quite sophisticated generally designed controller does not improve control quality compared to simple solution respecting the dominant properties. This was illustrated on an example of a MI+MO system thermostatic bath - a system with two constrained inputs and one controlled output. If the set-point changes significantly the controller can not do anything else than to set both control variables on their appropriate limits because of constrains on heating and cooling. Control response of quite complicated predictive controller will be improved if additional information and requirements are implemented within the control design. Process dynamics knowledge including cross couplings, was fully used only if we considered known future reference signal.

Another problem connected with systems with more inputs then outputs was illustrated on the mentioned example of MI+MO system. To solve the problem of indeterminate inputs combinations in case of MI+MO processes control we propose to add the "steady-state optimization" to controller design. Under the steady-state optimization we understand that we need to find such an inputs combination that is as close as possible to ideal process inputs and at the same time reaching the set-point in steady-state. We can observe the effect of the steady-state optimization during the first 15 minutes of the control response in Fig 4. The "ideal" desired input variables combination for the steady-state in our case is zero heating power and maximal cooling water input temperature - that is a combination with lowest energy cost. Future control error and terminal state error is minimized in every time instant as a result of the cost function form with respect to manipulated variables constrains. The effect of steady-state optimization is nice to see in steady-state but it takes effect continuously.

To add the steady-state optimization to a predictive controller design we use the terminal cost function. The quadratic terminal cost function was originally introduced to ensure controller stability. We modified the criterion so that the deviation of a desired and the predicted terminal state is used instead the terminal state only. The computation of the desired terminal state is based on a desired input variables combination, value of set-point at the end of control horizon and no square steady-state gain matrix. The solution is formulated as a standard constrained extreme finding task where the inputs constraints can be included, too.

7. Acknowledgements

The research has been supported in the program of Ministry of Education of Czech Republic MSM 0021627505 in part "Řízení, optimalizace a diagnostika složitých systémů". This support is gratefully acknowledged.

8. References

Åström, K.J. & Wittenmark, B. (1995). *Adaptive Control*. 2nd ed. Addison-Wesley, ISBN 0-201-55866-1, Boston, USA

Bitmead, R. R. ; Gevers, M. & Wertz, M. (1990). *Adaptive optimal control – The thinking man's GPC*, Englewood Cliffs, Prentice-Hall, New Jersey USA

Camacho, E. F. & Bordons, C. (1999). *Model Predictive Control*, Springer Verlag, ISBN 3-540-76241-8, London, UK

Dušek, F. & Honc, D. (2008a). Transformace soustav s různým počtem vstupů a výstupů pro decentralizované řízení (in Czech). *Automatizace*, 51, 7-8, 2008, 458-462, ISSN 0005-125X

Dušek, F. & Honc, D. (2008b). Static compensator for non-square MIMO systems. *Proceedings of 8th International Scientific-Technical Conference Process Control 2008*, pp. 13 (full text on CD), ISBN 978-80-7395-077-4, Kouty nad Desnou, June 9 - 12 2008. University of Pardubice, Czech Republic

Dušek, F. & Honc, D. (2009). Terminal State in a Predictive Controller Cost Function. In: *17th International Conference on Process Control '09*, pp. 409-413, ISBN 978-80-227-3081-5, Štrbské pleso, June 9 – 12 2009, Slovak University of Technology in Bratislava, Slovak Republic

Goodwin, G.C. ; Graebe, S.F. & Salgado, M.E. (2001). *Control System Design*, Prentice-Hall, ISBN 0272-1708, New Jersey, USA

Maciejowski, J.M. (2002). *Predictive Control with Constrains*, Pearson Education Ltd., ISBN 0-201-39823-0, Harlow, UK.

Mayne, D. Q. ; Rawlings, J. B. ; Rao, C. V. & Scokaert, P. O. M. (2000). Constrained model predictive control: Stability and optimality. *Automatica*, 36, 6, 789-814, ISSN 0005-1098

Novák, J. ; Chalupa, P. & Bobál V. (2009). Local Model Networks for Modelling and Predictive Control of Nonlinear Systems, *Proceedings of 23rd European Conference on Modelling and Simulation*, pp. 557-562, Universidad Rey Juan Carlos Madrid, June 9-12 2009, Madrid, Spain

Rossiter, J. A. (2003). *Model-Based Predictive Control, a Practical Approach*, CRC Press LLC, ISBN 0-8493-1291-4, Florida, USA

Wellstead, P. E. & Zarrop, M. B. (1991). *Self-tuning Systems: Control and Signal Processing.* John Wiley&Sons, ISBN 0-471-92883-6, Chichester, England

29

Robust Inverse Filter Design Based on Energy Density Control

Junho Lee and Young-Cheol Park
Yonsei University
Korea

1. Introduction

3-D audio systems, which provide a listener with 3-D sound illusion at arbitrary locations, are an important part of immersive interfaces. 3-D audio systems can use headphones or loudspeakers to present the 3-D sound. A main limitation of producing 3-D sound through loudspeakers is distortion caused by room acoustics. Various methods of designing inverse filters that can equalize the room response have been suggested. One of the common problems of the previous methods is the restriction on the listener to sit in a relatively narrow equalization zone. The problems were mainly caused by the fact that equalization was conducted by controlling sound pressure at discrete points, and the points were not ideal locations to obtain global control of the pressure field.

Most inverse filtering approaches are based on the cost function defined by using acoustic pressure. These inverse filtering systems typically minimize the squared acoustic pressure at a control point using the least square (LS) optimization Nelson et al. (1992)Nelson et al. (1995)Kirkeby et al. (1998). These systems, however, often produce distortions such as boosting at certain frequencies in the vicinity of the control point, because the room transfer function (RTF), being defined in an acoustic pressure field, changes drastically with variation in source and receiver positions inside a space. Thus, a listener's slight movement easily harms the inverse filtering performance. To overcome the problem of distortion, an alternative equalization method called joint LS Abe et al. (1997)Ward (2000), was presented, in which a sum of the squared pressures at several control points is minimized. A major disadvantage of this approach is that global control over control points can be partly effective, sometimes; it is not guaranteed. Recently, the filters for crosstalk cancellation were designed using the minimax optimization Rao et al. (2007). The minimax approach is known to provide better channel separation in low-frequency than LS approach with marginal improvement Rao et al. (2007). But it still inherits the distortion problems mentioned previously, since it is also based on acoustic pressure at the control point.

Another approach to the problem of room transfer function (RTF) variations with source and receiver position is equalization via a vector quantization (VQ) method Mourjopoulos (1994), in which an inverse filter is updated during operation using an RTF codebook generated by using the VQ method. Although the inverse filtering using VQ methods can resolve the problems of previous methods by making them effective for all possible source and receiver positions inside the enclosure, the method needs large sets of off-line measurements of RTFs to make the RTF codebook and additional tracking modules to find receiver positions.

In this chapter, we present an alternative approach to the problem of room equalization. This approach utilizes a new performance function based on energy density. The idea of energy density control has been developed in the field of active noise control for the global attenuation of broadband noise fields Sommerfeldt & Nashif (1994). It was proven that the energy density control system outperforms the squared pressure system since the former system is capable of observing more modes of the pressure field in an enclosure than the latter. More specifically, if the magnitude of the potential energy in the form of pressure density associated with a particular mode goes to zero at a control point, the kinetic energy in the form of particle velocity will approach a maximum. Thus the algorithm is useful in widening the control zones. We will begin with reviewing the previous approaches to provide robust inverse filtering and their problems. Later, details of the energy density control will be described in application, such as in room equalization.

2. Inverse filtering for multichannel sound reproduction system

When a sound source generates a sound field in a room, a large number of echoes build up and then slowly decay as the sound is absorbed by the walls and the air, creating reverberation. It is a desirable property of auditoriums to the extent that it helps to overcome the inverse square law dropoff of sound intensity in the enclosure. However, if it is excessive, it makes the sounds run together with loss of articulation - the sound becomes muddy. In addition, they are also undesirable when reproducing a desired sound field in a room. A digital equalization filter can be used to compensate for deficiencies in a loudspeaker-room frequency response for sound reproduction systems. In order to design a sound reproduction system of this kind, one essentially has to invert the transfer function of the reproduction environment.

2.1 Previous approaches

2.1.1 Equalization based on joint LS optimization

Fig. 1 shows the general form of the inverse filtering network for controlling L points where $\mathbf{h}_{p,ml}(n), m = 1, \cdots, M, l = 1, \cdots, L, n = 0, \cdots, N_h - 1$, represents $N_h \times 1$ the acoustic impulse response vector of the path from the mth loudspeaker to the lth control point. To measure the acoustic impulse responses, in general, microphones are located at the control points and a test signal is radiated through loudspeakers. Thus, the impulse responses are obtained in the acoustic pressure field. Given the impulse responses, the objective of designing an equalization system is to find FIR filters $\mathbf{w}_m(n), n = 0, \cdots, N_w - 1$, such that the recorded signals are reproduced perfectly at the control points by making the equalized response as close as possible to the desired one.

The equalized response between the desired and actual impulse responses at the lth control point from M sources can be expressed as

$$\hat{d}_{p,l}(n) = \sum_{m=1}^{M} \sum_{k=0}^{N_h+N_w-2} w_m(k) h_{p,ml}(n-k), \ n = 0, \cdots, N_h + N_w - 1. \tag{1}$$

This can be written in a matrix form as

$$\hat{\mathbf{d}}_{p,l} = \begin{bmatrix} \mathbf{H}_{p,1l} \ \mathbf{H}_{p,2l} \cdots \mathbf{H}_{p,Ml} \end{bmatrix} \begin{bmatrix} \mathbf{w}_1 \\ \mathbf{w}_2 \\ \vdots \\ \mathbf{w}_M \end{bmatrix}, \tag{2}$$

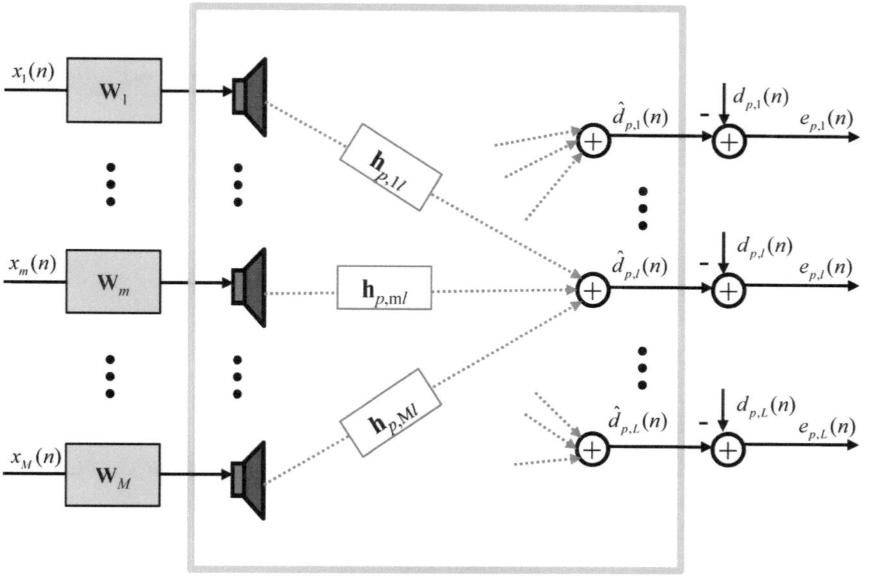

Fig. 1. Block diagram of the inverse filtering network.

where $\hat{\mathbf{d}}_{p,l} = [\hat{d}_{p,l}(0)\ \hat{d}_{p,l}(1)\ \cdots\ \hat{d}_{p,l}(N_h + N_w - 2)]^T$ is an $(N_h + N_w - 1) \times 1$ equalized response vector, $\mathbf{w}_m = [w_m(0)\ w_m(1)\ \cdots\ w_m(N_w - 1)]^T$ is an $N_w \times 1$ weight vector of the equalization filter, and the matrix $\mathbf{H}_{p,ml}$ is an $(N_h + N_w - 1) \times N_w$ convolution matrix defined using the acoustic impulse response as follows:

$$\mathbf{H}_{p,ml} = \begin{bmatrix} h_{p,ml}(0) & & 0 \\ \vdots & \ddots & \vdots \\ h_{p,ml}(N_h - 1) & \ddots & h_{p,ml}(0) \\ \vdots & \ddots & \vdots \\ 0 & & h_{p,ml}(N_h - 1) \end{bmatrix}. \tag{3}$$

Thus, the equalized response for M sound channels and L control points can be stacked as

$$\begin{bmatrix} \hat{\mathbf{d}}_{p,1} \\ \hat{\mathbf{d}}_{p,2} \\ \vdots \\ \hat{\mathbf{d}}_{p,L} \end{bmatrix} = \begin{bmatrix} \mathbf{H}_{p,11} & \mathbf{H}_{p,21} & \cdots & \mathbf{H}_{p,M1} \\ \mathbf{H}_{p,12} & \mathbf{H}_{p,22} & \cdots & \mathbf{H}_{p,M2} \\ \vdots & \vdots & \ddots & \vdots \\ \mathbf{H}_{p,1L} & \mathbf{H}_{p,2L} & \cdots & \mathbf{H}_{p,ML} \end{bmatrix} \begin{bmatrix} \mathbf{w}_1 \\ \mathbf{w}_2 \\ \vdots \\ \mathbf{w}_M \end{bmatrix} \tag{4}$$

or, more compactly as

$$\hat{\mathbf{d}}_p = \mathbf{H}_p \mathbf{w}. \tag{5}$$

The vector of error between the desired and actual impulse responses at the L control points can now be represented as

$$\mathbf{e}_p = \mathbf{d}_p - \mathbf{H}_p \mathbf{w} \tag{6}$$

where $\mathbf{d}_p = [\mathbf{d}_{p,1}^T \ \mathbf{d}_{p,2}^T \ \cdots \ \mathbf{d}_{p,L}^T]^T$ represent the desired impulse responses. Finally, an optimal weight vector can be obtained by minimizing the error between the desired and actual impulse responses. In an LS sense, the equalization filters are designed using a cost function:

$$J_{SP}(\mathbf{w}) = \left\|\mathbf{d}_p - \mathbf{H}_p\mathbf{w}\right\|_2 \tag{7}$$

where $\|\cdot\|_2$ denotes the vector 2-norm. The optimum set of coefficients in this case is given by

$$\mathbf{w}_{SP,o} = \mathbf{H}_p^+\mathbf{d}_p \tag{8}$$

where $^+$ denotes pseudo inverse, so that $\mathbf{H}_p^+ = \left(\mathbf{H}_p^T\mathbf{H}_p\right)^{-1}\mathbf{H}_p^T$. Because the equalization filters in this case jointly minimize the sum of squared errors at the multiple control points, it is referred to as the joint LS method Ward (2000)Abe et al. (1997). Note that, for $L = 1$, the design method reduces to LS method Nelson et al. (1992)Nelson et al. (1995)Kirkeby et al. (1998).

The optimization method based on squared pressure is widely used because it guarantees to have a unique global minimum. However, it is found that the equalized response away from the error sensor position can be worse than the unequalized response in such a design method Elliott & Nelson (1989).

2.1.2 Equalization based on minimax optimization

An alternative approach to the design of equalization filters is to use minimax optimization techniques. Now, the cost function becomes

$$J_{PM}(\mathbf{w}) = \left\|\mathbf{d}_p - \mathbf{H}_p\mathbf{w}\right\|_\infty \tag{9}$$

where $\|\cdot\|_\infty$ denotes the L_∞ norm. It was originally proposed to design crosstalk cancellation filters Sturm (1999) but it can be applicable for designing equalization filters. The second-order cone programming (SOCP) approach can be used to design equalization filters in the minimax sense. The SOCP provides the optimization problem can be solved using interior point solvers such as the Self-Dual-Minimization (SeDuMi) toolbox of MATLAB Sturm (1999).

According to the results in Sturm (1999), the minimax approach gives better channel separation at low frequencies than the LS method. The same can be expected when it is used for the design of equalization filters, but it is also easily harmed by the movement of the listener's changes in location since this method inherits the robustness problem of equalization in the pressure field.

2.1.3 Equalization by vector quantization (VQ)

All-pole modeling of room responses can achieve reduction in the room transfer function (RTF) and the resulting equalizer order Mourjopoulos & Paraskevas (1991). According to this method, the all-pole model of the RTF with coefficients, $a_k, k = 1, 2, \cdots, K$, is defined as

$$H_{ap}(z) = \frac{G}{1 + \sum\limits_{k=1}^{K} a_k z^{-k}} \tag{10}$$

where G is an arbitrary gain constant, K is the model order. And the following equation is the all-pole RTF equalizer's impulse response when $G = 1$:

$$w(n) = Z^{-1}\left[\frac{1}{H_{ap}(z)}\right]. \tag{11}$$

It is clear that an all-pole RTF equalizer will not achieve perfect equalization, because the all-pole model succeeds in representing poles of RTF, which correspond to resonances of RTF. This method, however, can achieve a required reduction in the equalizer order at some expense of its performance Toole & Olive (1988) and be used as the first stage of RTF processing at a second stage VQ . The use of vector quantization (VQ) can optimally classify such responses, obtained at different source and receiver positions Mourjopoulos (1994). Fig. 2 shows a block diagram for application of a VQ equalizer. By using the VQ method, the extremely large set of possible RTFs inside the enclosure will be classified into a smaller number of groups, so that a three-dimensional codebook of RTFs can be established, which can be used for equalizer design. During equalizer operation, the coefficients for equalization will be downloaded into the equalizer when it is detected that the listener is moving into a location. The combination of all-pole RTF modeling and the VQ method can solve the problems of the previous methods by making them effective for all possible source and receiver positions inside the enclosure.

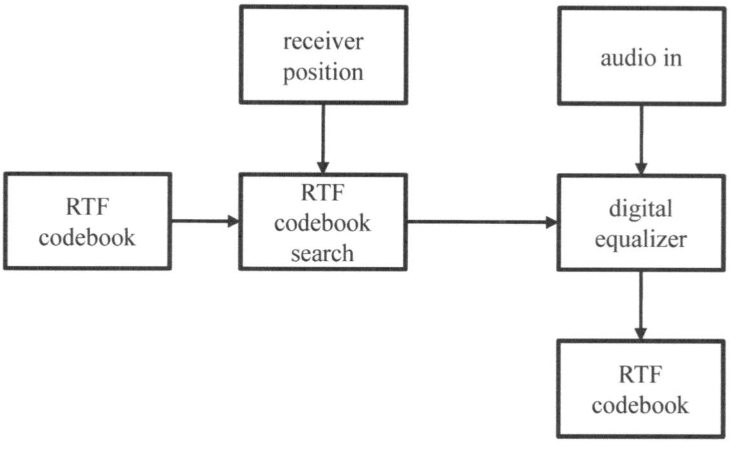

Fig. 2. Block diagram for application of VQ equalizer Mourjopoulos (1994).

2.2 Practical problems

As stated previously, most room equalization research is based on the cost function defined by using acoustic pressure and these equalization systems typically minimize the squared acoustic pressure at a control point using LS optimization Nelson et al. (1992)Nelson et al. (1995)Kirkeby et al. (1998)Abe et al. (1997)Ward (2000)Mourjopoulos (1994)Elliott & Nelson (1989). However, by controlling the acoustic pressure, the observability problem that leads to performance degradation happens. This is due to the magnitude of potential energy in the form of pressure associated with a particular mode goes to zero at the control point. Previous studies have shown that the geometry of the loudspeakers have a significant effect on the robustness of the inverse filtering Ward & Elko (1999). At certain frequencies, the sound signal arriving from the contralateral loudspeaker is delayed by approximately a half-period when compared with the signal coming from the ipsilateral loudspeaker. In a typical stereo setup with loudspeaker angle of 30° relative to the listener, the difference of the propagation path lengths between one loudspeaker and two listening points corresponding to the ear positions is $80 - 100mm$. Thus, in such a setup, one of the frequencies being involved in the signal cancellation is 1700Hz which corresponds to $190mm$ wavelength. At such frequencies, the inverse filtering based on pressure control is associated with numerical problems that seriously impair the robustness of the control system. In Ward & Elko (1999), the effect of loudspeaker position on the robustness of crosstalk cancellers was analyzed and a simple expression for determining the optimum loudspeaker positions was derived.

On the other hand, the VQ method requires previous large sets of off-line measurements of RTFs in order to design the enclosure's codebook and an additional tracking module is necessary to deal with the listener's movement. In Gardner (1997), Gardner employed a head tracking module using a camera in order to solve the performance degradation of the binaural synthesizer and the crosstalk canceller being caused by head movement. Fig. 3 shows a head-tracked 3-D loudspeaker audio system. The binaural synthesis block is to synthesize the ear signals corresponding to the target scene by appropriately encoding directional cues, and the crosstalk cancellation network delivers these signals to the listener without distortions by inverting the acoustic impulse response of the path from loudspeakers to the listener. When the listener moves away from the listening point, the crosstalk canceller and the binaural synthesis module are steered to the location of the tracked listener with the help of the head tracker module. In such a way, the 3-D audio system can preserve the 3-D illusion over a large listening area.

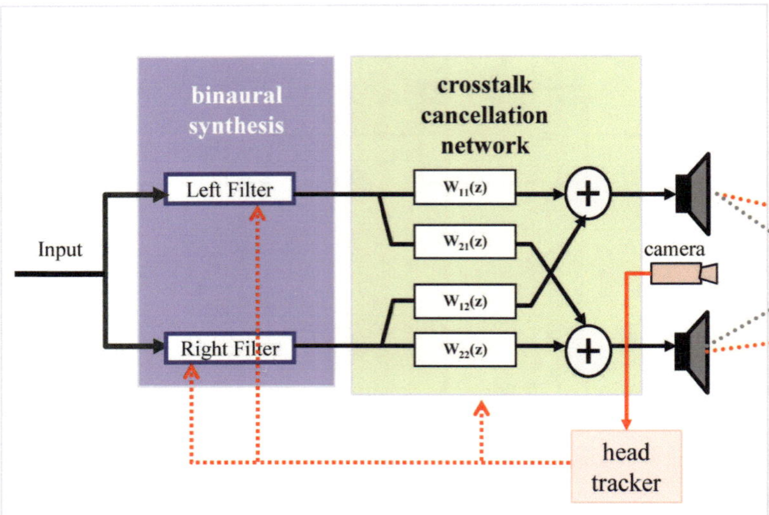

Fig. 3. Block diagram of head-tracked 3-D loudspeaker audio system.

3. Robust inverse filtering for multichannel sound reproduction system

Acoustic energy density function is defined using acoustic pressure and particle velocity. By controlling acoustic energy density, the observability problems that often limit the performance when controlling the pressure field are effectively overcome. To control acoustic energy density, however, a velocity sensor or equivalent estimation method is required.

3.1 Acoustic energy density

The time-averaged acoustic energy density at a point in space, $\mathbf{x} = (x, y, z)$, is defined as

$$\xi(\mathbf{x}) = \frac{1}{4\rho c^2} |p(\mathbf{x})|^2 + \frac{\rho}{4} |\vec{v}(\mathbf{x})|^2 \tag{12}$$

where ρ is the ambient fluid density, c is the acoustic wave speed, $p(\mathbf{x})$ is the acoustic pressure, and $\vec{v}(\mathbf{x}) = (v_x(\mathbf{x}), v_y(\mathbf{x}), v_z(\mathbf{x}))$ is the acoustic velocity vector. Note that acoustic energy density consists of potential energy density in the form of pressure and kinetic energy density in the form of particle velocity. Thus, it can be said that systems based on the squared pressures use only half of the acoustic information. Minimizing the sum of the squared pressure, which is part of the potential energy, at discrete points in space may significantly increase both the kinetic energy at those points as well as the total energy in the enclosure. The squared pressure system therefore often yields only local control. On the other hand, minimizing the sum of the total energy density at discrete points can yield improved equalization over a wide area covered by the control points since the energy has been definitely reduced at least at the specified points in space Parkins et al. (2000).

As previously mentioned, the acoustic pressure-based control inherently suffers from the observability problem that limits performance. One way of overcoming this problem is to control the acoustic energy density that is expected to provide robust equalization due to fairly uniform distribution of acoustic energy density.

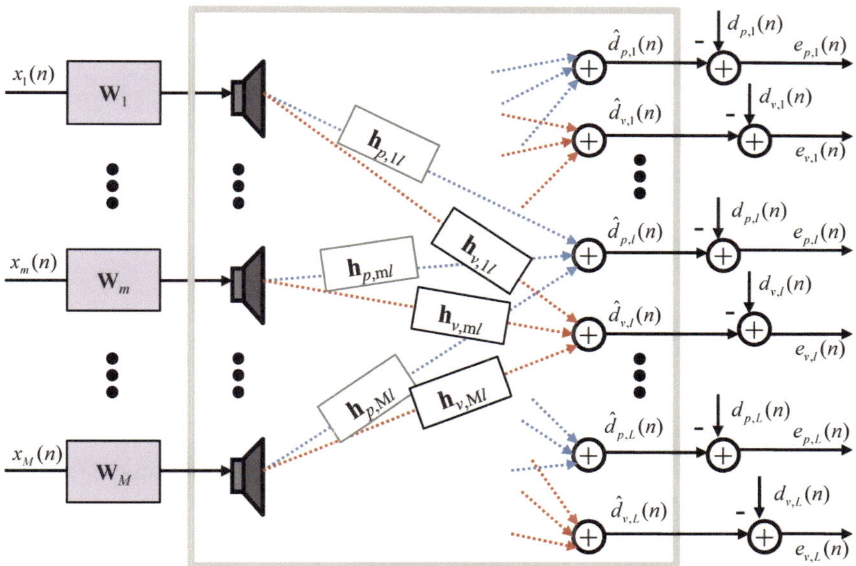

Fig. 4. Block diagram of the inverse filtering based on energy density control.

3.2 Equalization based on energy density control

Fig. 4 shows a block diagram of the inverse filtering system with L control points. The acoustic impulse responses in the velocity field at the lth control point due to M sources are described as

$$\hat{d}_{v_x,l}(n) = \sum_{m=1}^{M} \sum_{k=0}^{N_h+N_w-2} w_m(k) h_{v_x,ml}(n-k), \tag{13}$$

$$\hat{d}_{v_y,l}(n) = \sum_{m=1}^{M} \sum_{k=0}^{N_h+N_w-2} w_m(k) h_{v_y,ml}(n-k), \tag{14}$$

$$\hat{d}_{v_z,l}(n) = \sum_{m=1}^{M} \sum_{k=0}^{N_h+N_w-2} w_m(k) h_{v_z,ml}(n-k), \tag{15}$$

where the subscript v_x, v_y, and v_z refer to the x, y, and z-directional components of velocity, respectively. Let $\hat{\mathbf{d}}_{v,l}^T$ and $\mathbf{H}_{v,ml}$, respectively, denote the 3×1 velocity vector and $3(N_h + N_w - 1) \times N_w$ convolution matrix as given by

$$\hat{\mathbf{d}}_{v,l}^T = \left[\hat{d}_{v_x,l}(n)\ \hat{d}_{v_y,l}(n)\ \hat{d}_{v_z,l}(n) \right]^T, \tag{16}$$

$$\mathbf{H}_{v,ml} = \begin{bmatrix} \mathbf{H}_{v_x,ml} \\ \mathbf{H}_{v_y,ml} \\ \mathbf{H}_{v_z,ml} \end{bmatrix}. \tag{17}$$

The elements $\mathbf{H}_{v_x,ml}$, $\mathbf{H}_{v_y,ml}$, and $\mathbf{H}_{v_z,ml}$ are matrices defined similarly to Eq. (3). Now, the equalized velocity response at the control point l can be written as

$$\hat{\mathbf{d}}_{v,l}^T = \left[\mathbf{H}_{v,1l}\ \mathbf{H}_{v,2l} \cdots \mathbf{H}_{v,Ml} \right] \begin{bmatrix} \mathbf{w}_1 \\ \mathbf{w}_2 \\ \vdots \\ \mathbf{w}_M \end{bmatrix}. \tag{18}$$

The equalized velocity responses can be stacked in a matrix as

$$\begin{bmatrix} \hat{\mathbf{d}}_{v,1} \\ \hat{\mathbf{d}}_{v,2} \\ \vdots \\ \hat{\mathbf{d}}_{v,L} \end{bmatrix} = \begin{bmatrix} \mathbf{H}_{v,11} & \mathbf{H}_{v,21} & \cdots & \mathbf{H}_{v,M1} \\ \mathbf{H}_{v,12} & \mathbf{H}_{v,22} & \cdots & \mathbf{H}_{v,M2} \\ \vdots & \vdots & \ddots & \vdots \\ \mathbf{H}_{v,1L} & \mathbf{H}_{v,2L} & \cdots & \mathbf{H}_{v,ML} \end{bmatrix} \begin{bmatrix} \mathbf{w}_1 \\ \mathbf{w}_2 \\ \vdots \\ \mathbf{w}_M \end{bmatrix} \tag{19}$$

or

$$\hat{\mathbf{d}}_v = \mathbf{H}_v \mathbf{w}. \tag{20}$$

Now, the vector of error between the desired and equalized responses in the velocity field at the L control points is given as

$$\mathbf{e}_v = \mathbf{d}_v - \mathbf{H}_v \mathbf{w} \tag{21}$$

where $\mathbf{d}_v = [\mathbf{d}_{v,1}^T\ \mathbf{d}_{v,2}^T\ \cdots\ \mathbf{d}_{v,L}^T]^T$ represent desired impulse responses in velocity fields. Using Eqs. (6) and (21), the acoustic energy density at the control points is expressed as

$$\xi = \frac{1}{2\rho c^2} \left[\mathbf{e}_p^T \mathbf{e}_p + (\rho c)^2 \mathbf{e}_v^T \mathbf{e}_v \right]. \tag{22}$$

For controlling the energy density, the optimal weight vector is determined by the following cost function:

$$J_{ED}(\mathbf{w}) = \left\| \begin{bmatrix} \mathbf{d}_p \\ (\rho c)\, \mathbf{d}_v \end{bmatrix} - \begin{bmatrix} \mathbf{H}_p \mathbf{w} \\ (\rho c)\, \mathbf{H}_v \mathbf{w} \end{bmatrix} \right\|_2. \tag{23}$$

Note that the modified energy density, i.e. $(2\rho c^2)\xi$, is chosen as the cost function. The optimum filter coefficients are then

$$w_{ED,o} = \begin{bmatrix} \mathbf{H}_p \\ (\rho c)^2 \mathbf{H}_v \end{bmatrix}^+ \begin{bmatrix} \mathbf{d}_p \\ (\rho c)^2 \mathbf{d}_v \end{bmatrix}. \tag{24}$$

3.3 Velocity components estimation

As shown in Eq. (19), we need the x, y, and z-components of the acoustic velocity to implement the energy density control method. To this end, we can use a particle velocity sensor such as a laser vibrometer or velocity microphone. But a more convenient method of doing the same is approximated estimation using two pressure sensors (microphones). In this method, it is assumed that two microphones are highly phase-matched.

Euler's equation in one dimension relates the gradient of the acoustic pressure to the time-derivative of the acoustic velocity at a point as

$$\rho \frac{\partial v_x(x,t)}{\partial t} = -\frac{\partial p(x,t)}{\partial x}. \tag{25}$$

Thus, the acoustic velocity is obtained using

$$\hat{v}_x(x,t) = -\frac{1}{\rho} \int_{-\infty}^{t} \frac{\partial p(x,t)}{\partial x} dt. \tag{26}$$

By approximating the pressure gradient as the pressure difference in a small distance, Eq. (35) can be approximated as

$$\hat{v}_x(x,t) \approx -\frac{1}{\rho} \int_{-\infty}^{t} \frac{p_2(t) - p_1(t)}{\Delta x} dt \tag{27}$$

where $p_1(t)$ and $p_2(t)$ are the pressures measured by two closely spaced microphones with a distance Δx. Integration can be performed using a digital integrator Hodges et al. (1990) expressed in a simple recursive form:

$$\hat{v}_x(n) = \hat{v}_x(n-1) - \frac{1}{\rho \Delta x f_s} \left[p_2(n) - p_1(n) \right] e^{-1/f_s} \tag{28}$$

where f_s denotes the sampling frequency.

3.4 Robustness analysis

For ease of analysis, we define the transfer function (TF) between the mth loudspeaker and the lth control point as Ward & Elko (1999)

$$\mathcal{H}_{p,ml}(\omega) = e^{j2\pi\lambda^{-1}\Delta_{ml}}, \tag{29}$$

where λ is the wavelength and Δ_{ml} is the distance between the loudspeaker and the control point. It should be noted that this model disregards both propagation attenuation and the head shadow effect. Assuming a transaural system, the transfer functions between the two loudspeakers and the two microphones are collectively expressed as

$$\mathbb{H}_p(\omega) = \begin{bmatrix} \mathcal{H}_{p,11}(\omega) & \mathcal{H}_{p,12}(\omega) \\ \mathcal{H}_{p,21}(\omega) & \mathcal{H}_{p,22}(\omega) \end{bmatrix}. \tag{30}$$

Then the robustness of the equalization system is reflected by the condition number of the matrix $\mathbb{H}_p(\omega)$ Ward & Elko (1999) defined as

$$cond\left\{\mathbb{H}_p(\omega)\right\} = \frac{\sigma_{\max}\left(\mathbb{H}_p^H(\omega)\mathbb{H}_p(\omega)\right)}{\sigma_{\min}\left(\mathbb{H}_p^H(\omega)\mathbb{H}_p(\omega)\right)}, \tag{31}$$

where $\sigma_{min}(\cdot)$ and $\sigma_{max}(\cdot)$ denote the smallest and largest singular values, respectively. Suppose that the TF matrix is acoustically symmetric so that $\mathcal{H}_{p,11}(\omega) = \mathcal{H}_{p,22}(\omega)$ and $\mathcal{H}_{p,21}(\omega) = \mathcal{H}_{p,12}(\omega)$. We now have

$$\mathbb{H}_p^H(\omega)\mathbb{H}_p(\omega) = 2\left|\mathcal{H}_{p,11}(\omega)\right|^2 \begin{bmatrix} 1 & \cos(2\pi\lambda^{-1}\Delta) \\ \cos(2\pi\lambda^{-1}\Delta) & 1 \end{bmatrix}, \tag{32}$$

where Δ denotes the interaural path difference given by $\Delta_{11} - \Delta_{12}$. Singular values can be found from the following characteristic equation:

$$(1-k)^2 - \cos^2(2\pi\lambda^{-1}\Delta) = 0. \tag{33}$$

By the definition of robustness, the equalization system will be the most robust when $\cos(2\pi\lambda^{-1}\Delta) = 0$ ($\mathbb{H}_p(\omega)$ is minimized) and the least robust when $\cos(2\pi\lambda^{-1}\Delta) = \pm 1$ ($\mathbb{H}_p(\omega)$ is maximized) Ward & Elko (1999).
A similar analysis can be applied to acoustic energy density control. The composite transfer function between the two loudspeakers and the two microphones in the pressure and velocity fields becomes

$$\mathbb{H}_{ed}(\omega) = \begin{bmatrix} \mathcal{H}_{p,11}(\omega) & \mathcal{H}_{p,21}(\omega) \\ (\rho c)\mathbb{H}_{v,11}(\omega) & (\rho c)\mathbb{H}_{v,21}(\omega) \\ \mathcal{H}_{p,12}(\omega) & \mathcal{H}_{p,22}(\omega) \\ (\rho c)\mathbb{H}_{v,12}(\omega) & (\rho c)\mathbb{H}_{v,22}(\omega) \end{bmatrix}, \tag{34}$$

where $\mathbb{H}_{v,ml}(\omega)$ is the frequency-domain matrix corresponding to $\mathbf{H}_{v,ml}$. Note that the pressure and velocity at a point in space $\mathbf{x} = (x, y, z)$, $\vec{v}(\mathbf{x})$, and $p(\mathbf{x})$ are related via

$$j\omega\rho\vec{v}(\mathbf{x}) = -\nabla p(\mathbf{x}), \tag{35}$$

where ∇ represents a gradient. Using this relation, the velocity component for the x direction can be written as

$$\mathcal{H}_{v_x,ml}(\omega) = \frac{1}{\rho c} \cdot \frac{\Delta x_{ml}}{d}\mathcal{H}_{p,ml}(\omega), \tag{36}$$

where d and Δx_{ml} denote the distance and the x component of the displacement vector between the mth loudspeaker and the lth control point, respectively. Note that the velocity component for the y and z directions can be expressed similarly. Now we have

$$\mathbb{H}_{ed}^H(\omega)\mathbb{H}_{ed}(\omega) = 2\begin{bmatrix} 2 & Q\cos(2\pi\lambda^{-1}\Delta) \\ Q\cos(2\pi\lambda^{-1}\Delta) & 2 \end{bmatrix}, \tag{37}$$

where

$$Q = 1 + \frac{\Delta x_{11}\Delta x_{12} + \Delta y_{11}\Delta y_{12} + \Delta z_{11}\Delta z_{12}}{d_{11}(d_{11}+\Delta)}. \tag{38}$$

Singular values can be obtained from the following characteristic equation:

$$(2-k)^2 - \left(Q\cos\left(2\pi\lambda^{-1}\Delta\right)\right)^2 = 0. \tag{39}$$

From Eqs. (33) and (39), it can be noted that the maximum condition number of $\mathbb{H}_p(\omega)$ equals to infinity, while that of $\mathbb{H}_{ed}(\omega)$ is $(2+Q)/(2-Q)$, when $\cos(2\pi\lambda^{-1}\Delta) = \pm 1$. Eq. (38) also shows that the maximum condition number of the energy density field becomes smaller as Δ increases because Q approaches to 1. Now, by comparing the maximum condition numbers,

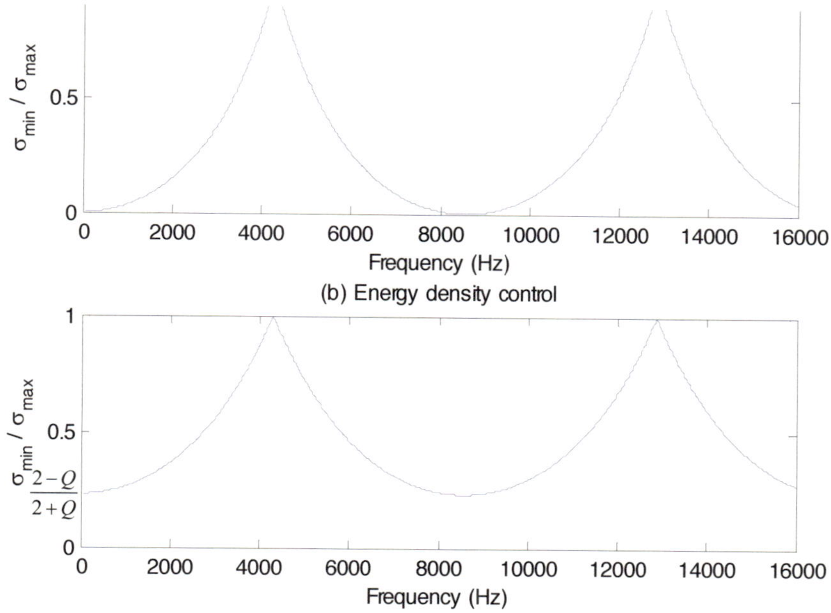

Fig. 5. The reciprocal of the condition number.

the robustness of the control system can be inferred. Fig. 5 shows the reciprocal condition number for the case where the loudspeaker is symmetrically placed at a 1 m and 30° relative to the head center. The reciprocal condition number of the pressure control approaches to zero, but the energy density control has the reciprocal condition numbers that are relatively significant for entire frequencies. Thus, it can be said that the equalization in the energy density field is more robust than the equalization in the pressure field.

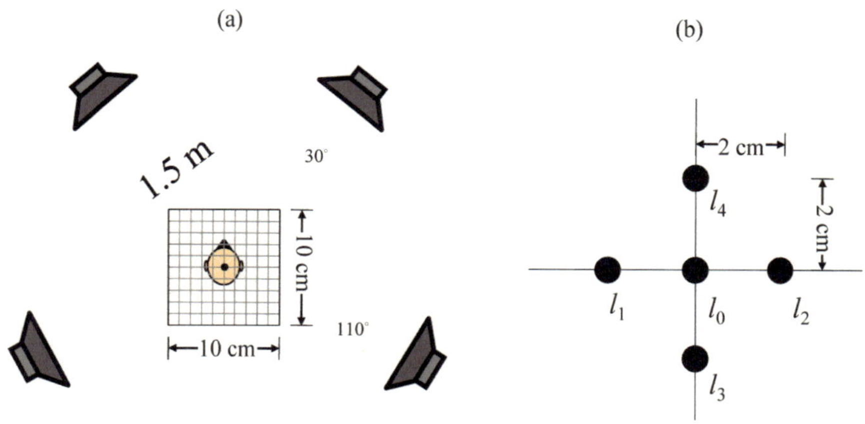

Fig. 6. Simulation environments. (a) Configuration for the simulation of a multichannel sound reproduction system. (b) Control points in the simulations. l_0 corresponds to the center of the listener's head.

4. Performance Evaluation

We present simulation results to validate energy density control. First, the robustness of an inverse filtering for multichannel sound reproduction system is evaluated by simulating the acoustic responses around the control points corresponding to the listener's ears. The performance of the robustness is objectively described in terms of the spatial extent of the equalization zone.

4.1 Simulation result

In this simulation, we assumed a multichannel sound reproduction system consisting of four sound sources ($M = 4$) as shown in Fig. 6(a). Details of the control points are depicted in Fig. 6(b). We assumed a free field radiation and the sampling frequency was 48 kHz. Impulse responses from the loudspeakers to the control points were modeled using 256-tap FIR filters ($N_h = 256$), and equalization filters were designed using 256-tap FIR filters ($N_w = 256$). The conventional LS method was tried by jointly equalizing the acoustic pressure at l_1, l_2, l_3, and l_4 points, and the energy density control was optimized only for the l_0 point. The delayed Dirac delta function was used for the desired response, i.e., $d_{p,l_0}(n) = \cdots = d_{p,l_4}(n) = \delta(n - n_0)$.

Center frequency	The control point (cm)				
	(0, 0)	(0, 5)	(2.5, 2.5)	(5, 0)	(5, 5)
500 Hz	0.06	-0.28	-0.13	-0.42	-0.28
1 kHz	0.30	-1.39	-0.60	-1.91	-3.55
2 kHz	1.26	-7.61	-2.76	-14.53	-10.25

Table 1. The error in dB for the pressure control system based on joint LS optimization at each center frequency.

Center frequency	The control point (cm)				
	(0, 0)	(0, 5)	(2.5, 2.5)	(5, 0)	(5, 5)
500 Hz	0.00	0.25	0.09	-0.21	0.03
1 kHz	0.00	0.25	0.06	-0.95	-0.76
2 kHz	0.00	0.25	-0.69	-4.50	-4.58

Table 2. The error in dB for the energy density control system at each center frequency.

We scanned the equalized responses in a 10 cm square region around the l_0 position, and results are shown in Fig. 7. Note that only the upper right square region was evaluated due to the symmetry. For the energy density control, velocity x and y were used. Velocity z was not used. As evident in Fig. 7, the energy density control shows a lower error level than the joint LS-based squared pressure control over the entire region of interest except at the points corresponding to l_2 $(2\ cm, 0\ cm)$ and l_4 $(0\ cm, 2\ cm)$, where the control microphones for the joint LS control were located.

Next, an equalization error was measured as the difference between the desired and actual responses defined by

$$C(dB) = 10 \log \left\{ \frac{\sum_{\omega=\omega_{min}}^{\omega_{max}} \left| \mathcal{D}(\omega) - \hat{\mathcal{D}}(\omega) \right|^2}{\sum_{\omega=\omega_{min}}^{\omega_{max}} \left| \mathcal{D}(\omega) \right|^2} \right\}, \tag{40}$$

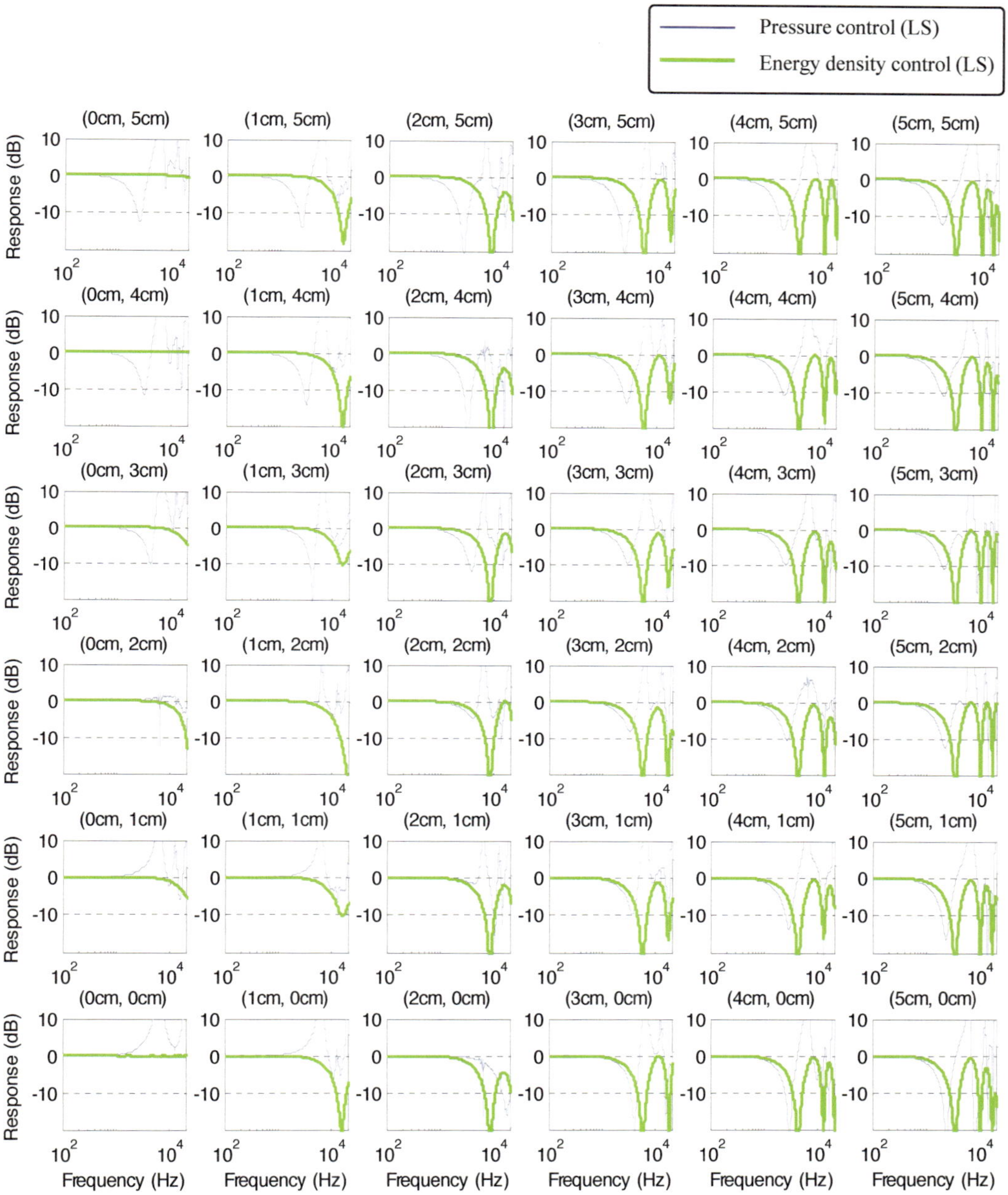

Fig. 7. The spatial extent of equalization by controlling pressure based joint LS optimization and energy density.

where ω_{min} and ω_{max} denote the minimum and maximum frequency indices of interest, respectively. In order to compare the robustness of equalization, we evaluated the pressure level in the vicinity of the control points. The equalization errors are summarized in Tables 1 and 2. Results show that the energy density control has a significantly lower equalization error than the joint LS-based squared pressure control, especially at 2 kHz where there are $7 \sim 10$ dB differences.

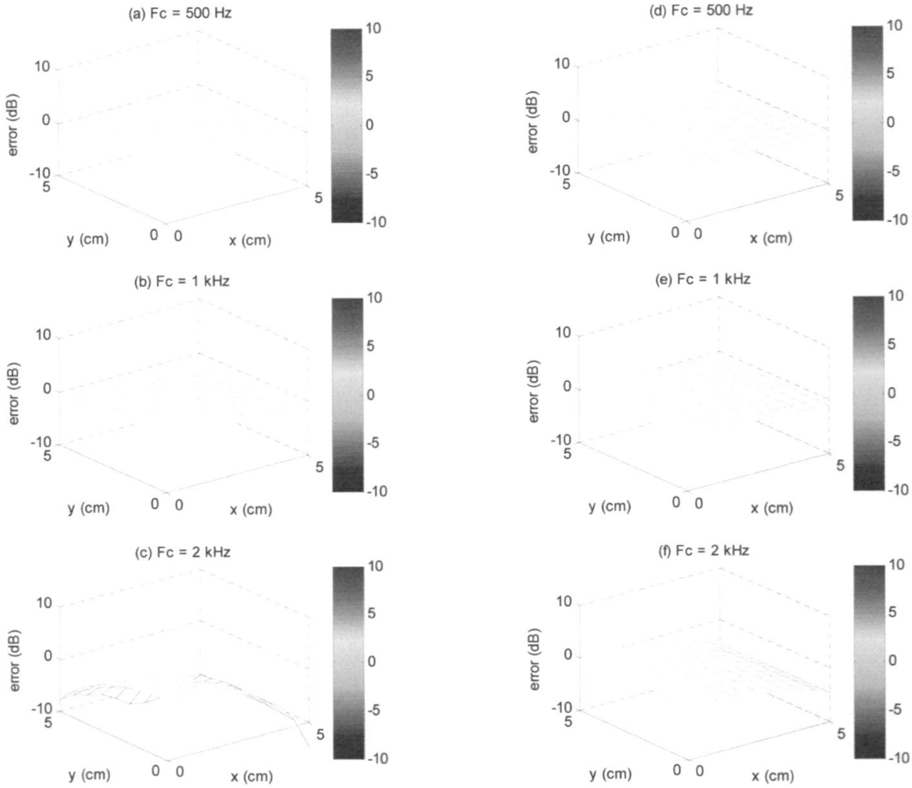

Fig. 8. A three-dimensional plot of the error surface for the pressure control (left column) and the energy density control (right column) at different center frequencies.

Finally, three-dimensional contour plots of the equalization errors are presented in Fig. 8. Fig. 8(a) and (d) show both methods have similar equalization performance at 1 kHz due to the relatively long wavelength. However, Figs. 8 (a), (b), and (c) indicate that the error of the pressure control rapidly increases as the frequency increased. On the other hand, the energy density control provides a more stable equalization zone, which implies that the energy density control can overcome the observability problem to some extent. Thus, it can be concluded that the energy density control system can provide a wider zone of equalization than the pressure control system.

4.2 Implementation consideration

It should be mentioned that it is necessary to have the acoustic velocity components to implement the energy density control system. It has been demonstrated that the two-microphone approach yields performance which is comparable to that of ideal energy density control in the field of the active noise control system Park & Sommerfeldt (1997). Thus, it is expected that the energy density control being implemented using the two-microphone approximation maintains the robustness of room equalization observed in the previous simulations.

To examine this, we applied two microphone techniques, which were described in section 3.3 to determine the acoustic velocity along an axis. By using Eq. (28), simulations were conducted for the case of $\Delta x = 2cm$ to evaluate the performance of the two-sensor implementation. Here, l_0 and l_2 are used for estimating the velocity component for x direction and l_0 and l_4 are used for estimating the velocity component for y direction; the velocity component for z direction was not applied. The results obtained by using the ideal velocity signal and two microphone technique are shown in Fig. 9. It can be concluded that the energy density system employing the two microphone technique provides comparable performance to the control system employing the ideal velocity sensor.

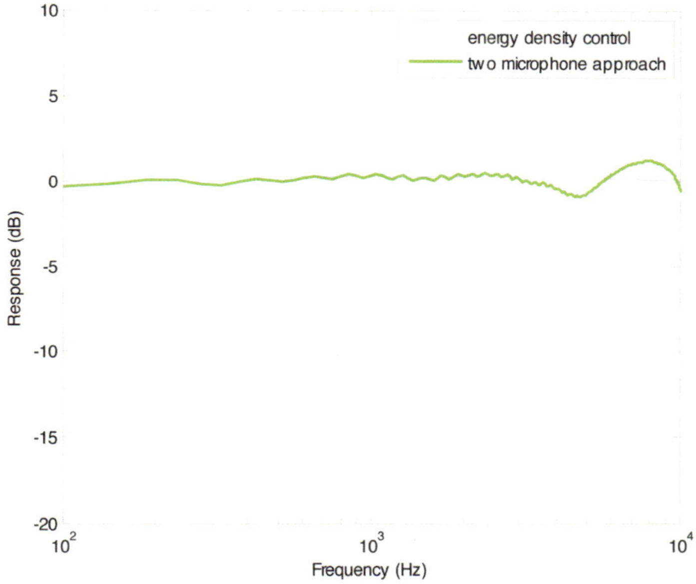

Fig. 9. The performance of the energy density control algorithm being implemented using the two microphone technique.

5. Conclusion

In this chapter, a method of designing equalization filters based on acoustic energy density was presented. In the proposed algorithm, the equalization filters are designed by minimizing the difference between the desired and produced energy densities at the control points. For the effective frequency range for the equalization, the energy density-based method provides more robust performance than the conventional squared pressure-based method. Theoretical analysis proves the robustness of the algorithm and simulation results showed that the proposed energy density-based method provides more robust performance than the conventional squared pressure-based method in terms of the spatial extent of the equalization zone.

6. References

Abe, K., Asano, F., Suzuki, Y. & Sone, T. (1997). Sound field reproduction by controlling the transfer functions from the source to multiple points in close proximity, *IEICE Trans. Fundamentals* E80-A(3): 574–581.

Elliott, S. J. & Nelson, P. A. (1989). Multiple-point equalization in a room using adaptive digital filters, *J. Audio Eng. Soc.* 37(11): 899–907.

Gardner, W. G. (1997). Head-tracked 3-d audio using loudspeakers, *Proc. IEEE Workshop on Applications of Signal Processing to Audio and Acoustics*, New Paltz, NY, USA.

Hodges, T., Nelson, P. A. & Elliot, S. J. (1990). The design of a precision digital integrator for use in an active vibration control system, *Mech. Syst. Sign. Process.* 4(4): 345–353.

Kirkeby, O., Nelson, P. A., Hamada, H. & Orduna-Bustamante, F. (1998). Fast deconvolution of multichannel systems using regularization, *IEEE Trans. on Speech and Audio Process.* 6(2): 189–195.

Mourjopoulos, J. (1994). Digital equalization of room acoustics, *J. Audio Eng. Soc.* 42(11): 884–900.

Mourjopoulos, J. & Paraskevas, M. (1991). Pole-zero modelling of room transfer functions, *J. Sound and Vib.* 146: 281–302.

Nelson, P. A., Bustamante, F. O. & Hamada, H. (1995). Inverse filter design and equalization zones in multichannel sound reproduction, *IEEE Trans. on Speech and Audio Process.* 3(3): 185–192.

Nelson, P. A., Hamada, H. & Elliott, S. J. (1992). Adaptive inverse filters for stereophonic sound reproduction, *IEEE Trans. on Signal Process.* 40(7): 1621–1632.

Park, Y. C. & Sommerfeldt, S. D. (1997). Global control of broadband noise fields using energy density control, *J. Acoust. Soc. Am.* 101: 350–359.

Parkins, J. W., Sommerfeldt, S. D. & Tichy, J. (2000). Narrowband and broadband active control in an enclosure using the acoustic energy density, *J. Acoust. Soc. Am.* 108(1): 192–203.

Rao, H. I. K., Mathews, V. J. & Park, Y.-C. (2007). A minimax approach for the joint design of acoustic crosstalk cancellation filters, *IEEE Trans. on Audio, Speech and Language Process.* 15(8): 2287–2298.

Sommerfeldt, S. D. & Nashif, P. J. (1994). An adaptive filtered-x algorithm for energy-based active control, *J. Acoust. Soc. Am.* 96(1): 300–306.

Sturm, J. F. (1999). Using sedumi 1.02, a matlab toolbox for optimization over symmetric cones, *Optim. Meth. Softw.* 11-12: 625–653.

Toole, F. E. & Olive, S. E. (1988). The modification of timbre by resonances: Perception and measurement, *J. Audio Eng. Soc.* 36: 122–141.

Ward, D. B. (2000). Joint least squares optimization for robust acoustic crosstalk cancellation, *IEEE Trans. on Speech and Audio Process.* 8(2): 211–215.

Ward, D. B. & Elko, G. W. (1999). Effect of loudspeaker position on the robustness of acoustic crosstalk cancellation, *IEEE Signal Process. Lett.* 6(5): 106–108.

Robust Control Approach for Combating the Bullwhip Effect in Periodic-Review Inventory Systems with Variable Lead-Time

Przemysław Ignaciuk and Andrzej Bartoszewicz
Institute of Automatic Control, Technical University of Łódź
Poland

1. Introduction

It is well known that cost-efficient management of production and goods distribution systems in varying market conditions requires implementation of an appropriate inventory control policy (Zipkin, 2000). Since the traditional approaches to inventory control, focused mainly on the statistical analysis of long-term variables and (static) optimization performed on averaged values of various cost components, are no longer sufficient in modern production-inventory systems, new solutions are being proposed. In particular, due to the resemblance of inventory management systems to engineering processes, the methods of control theory are perceived as a viable alternative to the traditional approaches. A summary of the initial control-theoretic proposals can be found in (Axsäter, 1985), whereas more recent results are discussed in (Ortega & Lin, 2004) and (Sarimveis et al., 2008). However, despite a considerable research effort, one of the utmost important, yet still unresolved (Geary et al., 2006) problems observed in supply chain is the bullwhip effect, which manifests itself as an amplification of demand variations in order quantities.

We consider an inventory setting in which the stock at a distribution center is used to fulfill an unknown, time-varying demand imposed by customers and retailers. The stock is replenished from a supplier which delivers goods with delay according to the orders received from the distribution center. The design goal is to generate ordering decisions such that the entire demand can be satisfied from the stock stored at the distribution center, despite the latency in order procurement, referred to as lead-time delay. The latency may be subject to significant fluctuations according to the goods availability at the supplier and transportation time uncertainty. When demand is entirely fulfilled any cost associated with backorders, lost sales, and unsatisfied customers is eliminated. Although a number of researchers have recognized the need to explicitly consider the delay in the controller design and stability analysis of inventory management systems, e.g. Hoberg et al. (2007), robustness issues related to simultaneous delay and demand fluctuations remain to a large extent unexplored (Dolgui & Prodhon, 2007). A few examples constitute the work of Boukas et al. (2000), where an H_∞-norm-based controller has been designed for a production-inventory system with uncertain processing time and input delay, and Blanchini et al. (2003), who concentrated on the stability analysis of a production system with uncertain demand and process setup. Both papers are devoted to the control of manufacturing systems, rather than optimization of goods flow in supply chain, and do not consider rate smoothening as an explicit design goal. On the contrary, in this work, we focus on the supply chain dynamics and provide formal methods for obtaining a smooth, non-oscillatory ordering signal, what is imperative for reducing the bullwhip effect (Dejonckheere et al., 2003).

From the control system perspective we may identify three decisive factors responsible for poor dynamical performance of supply chains and the bullwhip effect: 1) abrupt order changes in response to demand fluctuations, typical for the traditional order-up-to inventory policies, as discussed in (Dejonckheere et al., 2003); 2) inherent delay between placing of an order and shipment arrival at the distribution center which may span several review periods; and finally, 3) unpredictable variations of lead-time delay. Therefore, to avoid (or combat) the bullwhip effect, the designed policy should smoothly react to the changes in market conditions, and generate order quantities which will not fluctuate excessively in subsequent review intervals even though demand exhibits large and unpredictable variations. This is achieved in this work by solving a *dynamical optimization* problem with quadratic performance index (Anderson & Moore, 1989). Next, in order to eliminate the negative influence of delay variations, a compensation technique is incorporated into the basic algorithm operation together with a saturation block to explicitly account for the supplier capacity limitations. It is shown that in the inventory system governed by the proposed policy the stock level never exceeds the assigned warehouse capacity, which means that the potential necessity for an expensive emergency storage outside the company premises is eliminated. At the same time the stock is never depleted, which implies the 100% service level. The controller demonstrates robustness to model uncertainties and bounded external disturbance. The applied compensation mechanism effectively throttles undesirable quantity fluctuations caused by lead-time changes and information distortion thus counteracting the bullwhip effect.

2. Problem formulation

We consider an inventory system faced by an unknown, bounded, time-varying demand, in which the stock is replenished with delay from a supply source. Such setting, illustrated in Fig. 1, is frequently encountered in production-inventory systems where a common point (distribution center), linked to a factory or external, strategic supplier, is used to provide goods for another production stage or a distribution network. The task is to design a control strategy which, on one hand, will minimize lost service opportunities (occurring when there is insufficient stock at the distribution center to satisfy the current demand), and, on the other hand, will ensure smooth flow of goods despite model uncertainties and external disturbances. The principal obstacle in providing such control is the inherent delay between placing of an order at the supplier and goods arrival at the center that may be subject to significant fluctuations during the control process. Another factor which aggravates the situation is a possible inconsistency of the received shipments with regard to the sequence of orders. Indeed, it is not uncommon in practical situations to obtain the goods from an earlier order after the shipment arrival from a more recent one. In addition, we may encounter other types of disturbances affecting the replenishment process related to organizational issues and quality of information (Zomerdijk & de Vries, 2003) (e.g. when a shipment arrives on time but is registered in another review period, or when an incorrect order is issued from the distribution center). The time-varying latency of fulfilling of an order will be further referred to as lead-time or lead-time delay.

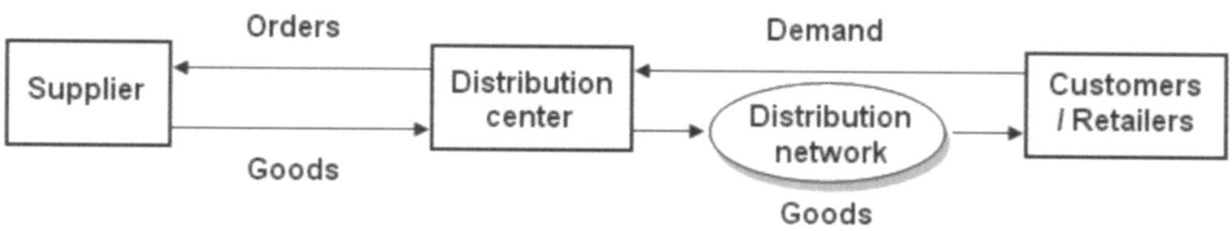

Fig. 1. Inventory system with a strategic supplier

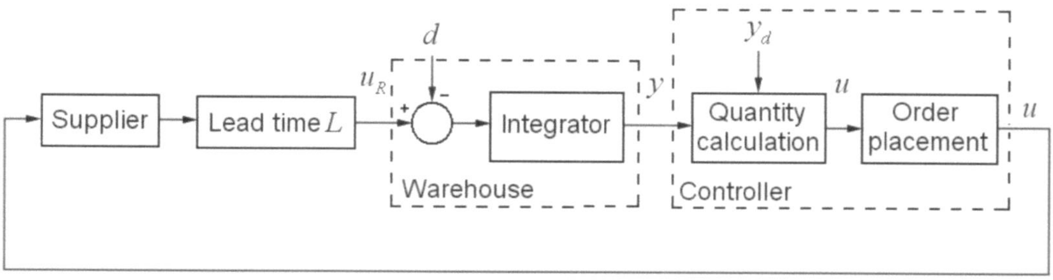

Fig. 2. System model

The schematic diagram of the analyzed periodic-review inventory system is depicted in Fig. 2. The stock replenishment orders u are issued at regular time instants kT, where T is the review period and $k = 0, 1, 2,...,$ on the basis of the on-hand stock (the current stock level in the warehouse at the distribution center) $y(kT)$, the target stock level y_d, and the history of previous orders. Each non-zero order placed at the supplier is realized with lead-time delay $L(k)$, assumed to be a multiple of the review period, i.e. $L(k) = n(k)T$, where $n(k)$ and its nominal value $\bar{n}$ are positive integers satisfying

$$(1-\delta)\bar{n} \leq n(k) \leq (1+\delta)\bar{n} \tag{1}$$

and $0 \leq \delta < 1$. Notice that (1) is the only constraint imposed on delay variations, which means that within the indicated interval the actual delay of a shipment may accept any statistical distribution. This implies that consecutive shipments sent by the supplier may arrive out of order at the distribution center and concurrently with other shipments which were sent earlier or afterwards. Since the presented model does not require stating the cause of lead-time variations, neither specification of a particular function $n(k)$ or its distribution, it allows for conducting the robustness study in a broad spectrum of practical situations with uncertain latency in delivering orders.

The imposed demand (the number of items requested from inventory in period k) is modeled as an *a priori* unknown, bounded function of time $d(kT)$,

$$0 \leq d(kT) \leq d_{\max}. \tag{2}$$

Notice that this definition of demand is quite general and it accounts for any standard distribution typically analyzed in the considered problem. If there is a sufficient number of items in the warehouse to satisfy the imposed demand, then the actually met demand $h(kT)$

(the number of items sold to customers or sent to retailers in the distribution network) will be equal to the requested one. Otherwise, the imposed demand is satisfied only from the arriving shipments, and additional demand is lost (we assume that the sales are not backordered, and the excessive demand is equivalent to a missed business opportunity). Thus, we may write

$$0 \le h(kT) \le d(kT) \le d_{\max}. \tag{3}$$

The dynamics of the on-hand stock y depends on the amount of arriving shipments $u_R(kT)$ and on the satisfied demand h. Assuming that the warehouse is initially empty, i.e. $y(kT) = 0$ for $k < 0$, and the first order is placed at $kT = 0$, then for any $kT \ge 0$ the stock level at the distribution center may be calculated from the following equation

$$y(kT) = \sum_{j=0}^{k-1} u_R(jT) - \sum_{j=0}^{k-1} h(jT) = \sum_{j=0}^{k-1} u[jT - L(j)] - \sum_{j=0}^{k-1} h(jT). \tag{4}$$

Let us introduce a function $\xi(kT) = \xi_+(kT) - \xi_-(kT)$, where

- $\xi_+(kT)$ represents the sum of these surplus items which arrive at the distribution center by the time kT earlier than expected since their delay experienced in the neighborhood of kT is smaller than the nominal one, and
- $\xi_-(kT)$ denotes the sum of items which should have arrived by the time kT, but which cannot reach the center due to the (instantaneous) delay greater than the nominal one.

Assuming that the order quantity is bounded by some positive value $u_{\max}$ (e.g. the maximum number of items the supplier can accumulate and send in one review period), which is commonly encountered in practical systems, then on the basis of (1),

$$\forall_{k\ge 0} \ |\xi(kT)| \le \xi_{\max} = u_{\max}\delta\bar{L}, \tag{5}$$

where $\bar{L} = \bar{n}T$ is the nominal lead-time. With this notation we can rewrite (4) in the following way

$$y(kT) = \sum_{j=0}^{k-1} u[(j - \bar{n})T] + \xi(kT) - \sum_{j=0}^{k-1} h(jT). \tag{6}$$

It is important to realize that because lead-time is bounded, it suffices to consider the effects caused by its variations (represented by function $\xi(\cdot)$ in the model) only in the neighborhood of kT implied by (1). Since the summing operation is commutative, all the previous shipments, i.e. those arriving before $(k - \delta\bar{n})T$, can be added as if they had actually reached the distribution center on time and this will not change the overall quantity of the received items. In other words, delay variations of shipments acquired in the far past do not inflict perturbation on the current stock.

The discussed model of inventory management system can also be presented in the state space. The state-space realization facilitates adaptation of formal design techniques, and is selected as a basis for the control law derivation described in detail in Section 3.

State-space representation

In order to proceed with a formal controller design we describe the discrete-time model of the considered inventory system in the state space:

$$\begin{aligned} &\mathbf{x}\left[(k+1)T\right]=\mathbf{A}\mathbf{x}(kT)+\mathbf{b}u(kT)+\mathbf{v}_1 h(kT)+\mathbf{v}_2\xi(kT),\\ &y(kT)=\mathbf{q}^T\mathbf{x}(kT), \end{aligned} \tag{7}$$

where $\mathbf{x}(kT) = [x_1(kT)\ x_2(kT)\ x_3(kT) \dots\ x_n(kT)]^T$ is the state vector with $x_1(kT) = y(kT)$ representing the stock level in period k and the remaining state variables $x_j(kT) = u[(k - n + j - 1)T]$ for any $j = 2, \dots, n$ equal to the delayed input signal u. $\mathbf{A}$ is $n \times n$ state matrix, $\mathbf{b}$, $\mathbf{v}_1$, $\mathbf{v}_2$, and $\mathbf{q}$ are $n \times 1$ vectors

$$\mathbf{A}=\begin{bmatrix}1&1&0&\dots&0\\0&0&1&\dots&0\\\vdots&\vdots&\vdots&\ddots&\vdots\\0&0&0&\dots&1\\0&0&0&\dots&0\end{bmatrix},\quad \mathbf{b}=\begin{bmatrix}0\\0\\\vdots\\0\\1\end{bmatrix},\quad \mathbf{v}_1=\begin{bmatrix}-1\\0\\\vdots\\0\\0\end{bmatrix},\quad \mathrm{v}_2=\begin{bmatrix}1\\0\\\vdots\\0\\0\end{bmatrix},\quad \mathbf{q}=\begin{bmatrix}1\\0\\\vdots\\0\\0\end{bmatrix}, \tag{8}$$

and the system order $n = \bar{n} + 1$. For convenience of the further analysis, we can rewrite the model in the alternative form

$$\begin{cases} x_1\left[(k+1)T\right]=x_1(kT)+x_2(kT)-h(kT)+\xi(kT),\\ x_2\left[(k+1)T\right]=x_3(kT),\\ \quad\vdots\\ x_{n-1}\left[(k+1)T\right]=x_n(kT),\\ x_n\left[(k+1)T\right]=u(kT). \end{cases} \tag{9}$$

Relation (9) shows how the effects of delay are accounted for in the model by a special choice of the state space in which the state variables contain the information about the most recent order history. The desired system state is defined as

$$\mathbf{x_d}=\begin{bmatrix}x_{d1}\\x_{d2}\\\vdots\\x_{dn-1}\\x_{dn}\end{bmatrix}=\begin{bmatrix}x_{d1}\\0\\\vdots\\0\\0\end{bmatrix}, \tag{10}$$

where $x_{d1} = y_d$ denotes the demand value of the first state variable, i.e. the target stock level. By choosing the desired state vector as

$$\mathbf{x_d} = [y_d\ 0\ 0\ \dots\ 0]^T$$

we want the first state variable (on-hand stock) to reach the level y_d, and to be kept at this level in the steady-state. For this to take place all the state variables $x_2 \dots x_n$ should be zero once $x_1(kT)$ becomes equal to y_d, precisely as dictated by (10).

In the next section, equations (7)–(10) describing the system behavior and interactions among the principal system variables (ordering signal, on-hand stock level and imposed demand) will be used to develop a discrete control strategy goverining the flow of goods between the supplier and the distribution center.

3. Proposed inventory policy

In this section, we formulate a new inventory management policy and discuss its properties related to handling the flow of goods. First, the nominal system is considered, and the controller parameters are selected by solving a linear-quadratic (LQ) optimization problem. Afterwards, the influence of perturbation is analyzed and an enhanced, nonlinear control law is formulated which demonstrates robustness to delay and demand variations. The key element in the improved controller structure is the compensator which reduces the effects caused by delay fluctuations and information distortion.

3.1 Optimization problem

From the point of view of optimizing the system dynamics, we may state the aim of the control action as bringing the currently available stock to the target level without excessive control effort. Therefore, we seek for a control $u_{opt}(kT)$, which minimizes the following cost functional

$$J(u) = \sum_{k=0}^{\infty} \left\{ u^2(kT) + w\left[y_d - y(kT)\right]^2 \right\}, \tag{11}$$

where w is a positive constant applied to adjust the influence of the controller command and the output variable on the cost functional value. Small w reduces excessive order quantities, but lowers the controller dynamics. High w, in turn, implies fast tracking of the reference stock level at the expense of large input signals. In the extreme case, when $w \to \infty$, the term $y_d - y(kT)$ prevails and the developed controller becomes a dead-beat scheme. From the managerial point of view the application of a quadratic cost structure in the considered problem of inventory control has similar effects as discussed in (Holt et al., 1960) in the context of production planning. It allows for a satisfactory tradeoff between fast reaction to the changes in market conditions (reflected in demand variations) and smoothness of ordering decisions. As a result, the controller will track the target inventory level y_d with good dynamics, yet, at the same time, it will prevent rapid demand fluctuations from propagating in supply chain. A huge advantage of our approach based on dynamical optimization over the results proposed in the past is that the smoothness of ordering decisions is ensured by the controller structure itself. This allows us to avoid signal filtering and demand averaging, typically applied to decrease the degree of ordering variations in supply chain, and thus to avoid errors and inaccuracies inherently implied by these techniques.

Applying the standard framework proposed in (Zabczyk, 1974), to system (7)-(8), the control $u_{opt}(kT)$ minimizing criterion (11) can be presented as

$$u_{opt}(kT) = -\mathbf{g}\mathbf{x}(kT) + r, \tag{12}$$

where

$$\begin{aligned} \mathbf{g} &= \mathbf{b}^T\mathbf{K}\left(\mathbf{I}_n + \mathbf{b}\mathbf{b}^T\mathbf{K}\right)^{-1}\mathbf{A}, \\ r &= \mathbf{b}^T\left[\mathbf{K}\left(\mathbf{I}_n + \mathbf{b}\mathbf{b}^T\mathbf{K}\right)^{-1}\mathbf{b}\mathbf{b}^T - \mathbf{I}_n\right]\mathbf{k}, \\ \mathbf{k} &= -\mathbf{A}^T\left[\mathbf{K}\left(\mathbf{I}_n + \mathbf{b}\mathbf{b}^T\mathbf{K}\right)^{-1}\mathbf{b}\mathbf{b}^T - \mathbf{I}_n\right]\mathbf{k} - w\mathbf{q}y_d, \end{aligned} \tag{13}$$

and semipositive, symmetric matrix $\mathbf{K}_{n\times n}$, $\mathbf{K}^T = \mathbf{K} \geq 0$, is determined according to the following Riccati equation

$$\mathbf{K} = \mathbf{A}^T\mathbf{K}\left(\mathbf{I}_n + \mathbf{b}\mathbf{b}^T\mathbf{K}\right)^{-1}\mathbf{A} + w\mathbf{q}\mathbf{q}^T. \tag{14}$$

Finding the parameters of the LQ optimal controller for the considered system with delay is a challenging task, as it involves solving an nth order matrix Riccati equation. Nevertheless, by applying the approach presented in (Ignaciuk & Bartoszewicz, 2010) we are able to solve the problem analytically and obtain the control law in a closed form. Below we summarize major steps of the derivation.

3.2 Solution to the optimization problem

We begin with the most general form of matrix $\mathbf{K}$ which can be presented as

$$\mathbf{K}_0 = \begin{bmatrix} k_{11} & k_{12} & k_{13} & \dots & k_{1n} \\ k_{12} & k_{22} & k_{23} & \dots & k_{2n} \\ k_{13} & k_{23} & k_{33} & \dots & k_{3n} \\ \vdots & \vdots & \vdots & \ddots & \vdots \\ k_{1n} & k_{2n} & k_{3n} & \dots & k_{nn} \end{bmatrix}. \tag{15}$$

In the first iteration, we place $\mathbf{K}_0$ directly in (14), and after substituting matrix $\mathbf{A}$ and vector $\mathbf{b}$ as defined by (8), we seek for similarities between the elements k_{ij} on either side of the equality sign in (14). In this way we find the relations among the first four elements in the upper left corner of $\mathbf{K}$: $k_{12} = k_{22} = k_{11} - w$ (note that $k_{21} = k_{12}$ since $\mathbf{K}$ is symmetric). Consequently, after the first analytical iteration, we obtain the following form of $\mathbf{K}$

$$\mathbf{K}_1 = \begin{bmatrix} k_{11} & k_{11}-w & k_{13} & \dots & k_{1n} \\ k_{11}-w & k_{11}-w & k_{23} & \dots & k_{2n} \\ k_{13} & k_{23} & k_{33} & \dots & k_{3n} \\ \vdots & \vdots & \vdots & \ddots & \vdots \\ k_{1n} & k_{2n} & k_{3n} & \dots & k_{nn} \end{bmatrix}. \tag{16}$$

Now we substitute $\mathbf{K}_1$ given by (16) into the expression on the right hand side of (14) and compare with its left hand side. This allows us to represent the elements k_{i3} (i = 1, 2, 3) in terms of k_{11} as $k_{13} = k_{23} = k_{33} = k_{11} - 2w$. Concisely in matrix form we have

$$\mathbf{K}_2 = \begin{bmatrix} k_{11} & k_{11}-w & k_{11}-2w & k_{14} & \dots & k_{1n} \\ k_{11}-w & k_{11}-w & k_{11}-2w & k_{24} & \dots & k_{2n} \\ k_{11}-2w & k_{11}-2w & k_{11}-2w & k_{34} & \dots & k_{3n} \\ k_{14} & k_{24} & k_{34} & k_{44} & \dots & k_{4n} \\ \vdots & \vdots & \vdots & \vdots & \ddots & \vdots \\ k_{1n} & k_{2n} & k_{3n} & k_{4n} & \dots & k_{nn} \end{bmatrix}. \tag{17}$$

We proceed with the substitutions until a general pattern is determined, i.e. until all the elements of $\mathbf{K}$ can be expressed as functions of k_{11} and the system order n. We get $k_{ij} = k_{11} -$

$(j - 1)w$ for $j \geq i$ (the upper part of **K**) and $k_{ij} = k_{11} - (i - 1)w$ for $j < i$ (the lower part of **K**). In matrix form

$$\mathbf{K} = \begin{bmatrix} k_{11} & k_{11} - w & k_{11} - 2w & \cdots & k_{11} - (n-1)w \\ k_{11} - w & k_{11} - w & k_{11} - 2w & \cdots & k_{11} - (n-1)w \\ k_{11} - 2w & k_{11} - 2w & k_{11} - 2w & \cdots & k_{11} - (n-1)w \\ \vdots & \vdots & \vdots & \ddots & \vdots \\ k_{11} - (n-1)w & k_{11} - (n-1)w & k_{11} - (n-1)w & \cdots & k_{11} - (n-1)w \end{bmatrix}. \tag{18}$$

If we substitute (18) into the right hand side of equation (14) and compare the first element in the upper left corner of the matrices on either side of the equality sign, we get the expression from which we can determine k_{11}:

$$k_{11} = nw + 1 - \left[k_{11} - (n-1)w + 1\right]^{-1}. \tag{19}$$

Equation (19) has two roots

$$k_{11}^{'} = \sqrt{w}\left[(2n-1)\sqrt{w} - \sqrt{w+4}\right]/2 \text{ and } k_{11}^{''} = \sqrt{w}\left[(2n-1)\sqrt{w} + \sqrt{w+4}\right]/2. \tag{20}$$

Since $\det(\mathbf{K}) = w^{n-1}[k_{11} - (n - 1)w]$, only $k_{11}^{''} = \sqrt{w}\left[(2n-1)\sqrt{w} + \sqrt{w+4}\right]/2 \geq (n-1)w$ guarantees that **K** is semipositive definite. Consequently, we get matrix **K** (18) with $k_{11} = k_{11}^{''}$. This concludes the solution of the Riccati equation.

Having found **K**, we evaluate **g**,

$$\mathbf{g} = [1 \;\; 1 \;\; 1 \;\; \cdots \;\; 1]\left\{1 - \left[k_{11} - (n-1)w + 1\right]^{-1}\right\}. \tag{21}$$

Vector **k** is determined by substituting matrix **K** given by (18) into the last equation in set (13). We obtain

$$\mathbf{k} = \left[k_1 \;\; k_1 + wy_d \;\; k_1 + 2wy_d \;\; \cdots \;\; k_1 + (n-1)wy_d\right]^T, \tag{22}$$

where

$$k_1 = -wy_d\left\{n + \left[k_{11} - (n-1)w\right]^{-1}\right\}. \tag{23}$$

Then, using the second equation in set (13), and substituting (23), we calculate r,

$$r = -\frac{k_1 + (n-1)wy_d}{k_{11} - (n-1)w + 1} = \frac{wy_d}{k_{11} - (n-1)w}. \tag{24}$$

Finally, using (21) and (24), the optimal control $u_{opt}(kT)$ can be presented in the following way:

$$u_{opt}(kT) = -\mathbf{g}\mathbf{x}(kT) + r = -\left(1 - \frac{1}{k_{11} - (n-1)w + 1}\right)\sum_{j=1}^{n} x_j(kT) + \frac{wy_d}{k_{11} - (n-1)w}. \tag{25}$$

Substituting $k_{11} = \sqrt{w}\left[(2n-1)\sqrt{w} + \sqrt{w+4}\right]/2$, we arrive at

$$u_{opt}(kT) = \alpha\left[y_d - x_1(kT) - \sum_{j=2}^{n} x_j(kT)\right], \tag{26}$$

where the gain $\alpha = (\sqrt{w(w+4)} - w)/2$. From (9) the state variables x_j (j = 2, 3,..., n) may be expressed in terms of the control signal generated at the previous n - 1 samples as

$$x_j(kT) = u\left[(k-n+j-1)T\right]. \tag{27}$$

Recall that we introduced the notation $x_1(kT) = y(kT)$. Then, substituting (27) into (26), we obtain

$$u_{opt}(kT) = \alpha\left[y_d - y(kT) - \sum_{j=k-\bar{n}}^{k-1} u(jT)\right], \tag{28}$$

which completes the design of the inventory policy for the nominal system. The policy can be interpreted in the following way: the quantity to be ordered in each period is proportional to the difference between the target and the current stock level ($y_d - y(kT)$), decreased by the amount of open orders (the quantity already ordered at the supplier, but which has not yet arrived at the warehouse due to lead-time delay). It is tuned in a straightforward way by the choice of a single parameter a, i.e. smaller a implies more dampening of demand variations (for a detailed discussion on the selection of a refer to (Ignaciuk & Bartoszewicz, 2010)).

3.3 Stability analysis of the nominal system

The nominal discrete-time system is asymptotically stable if all the roots of the characteristic polynomial of the closed-loop state matrix $\mathbf{A}_c = [\mathbf{I}_n - \mathbf{b}(\mathbf{c}^T\mathbf{b})^{-1}\mathbf{c}^T]\mathbf{A}$ are located within the unit circle on the z-plane. The roots of the polynomial

$$\det(z\mathbf{I}_n - \mathbf{A}_c) = z^n + (\alpha-1)z^{n-1} = z^{n-1}\left[z-(1-\alpha)\right], \tag{29}$$

are located inside the unit circle, if $0 < a < 2$. Since for every n and for every w the gain satisfies the condition $0 < a \leq 1$, the system is asymptotically stable. Moreover, since irrespective of the value of the tuning coefficient w the roots of (29) remain on the nonnegative real axis, no oscillations appear at the output. By changing w from 0 to ∞, the nonzero pole moves towards the origin of the z-plane, which results in faster convergence to the demand state. In the limit case when $w = \infty$, all the closed-loop poles are at the origin ensuring the fastest achievable response in a discrete-time system offered by a dead-beat scheme.

3.4 Robustness issues

The order calculation performed according to (28) is based on the nominal delay which constitutes an estimate of the true (variable) lead-time set according to the contracting agreement with the supplier. The controller designed for the nominal system is robust with respect to demand fluctuations, yet may generate negative orders in the presence of lead-time variations. In order to eliminate this deficiency and at the same time account for the supplier capacity limitations, we introduce the following modification into the basic algorithm

$$u(kT)=\begin{cases}0, & \text{if } \varphi(kT)<0,\\ \varphi(kT), & \text{if } 0\le\varphi(kT)\le u_{\max},\\ u_{\max}, & \text{if } \varphi(kT)>u_{\max},\end{cases} \tag{30}$$

where $u_{\max} > d_{\max}$ is a constant denoting the maximum order quantity that can be provided by the supplier in a single review period. Function $\varphi(\cdot)$ is defined as

$$\varphi(kT)=\alpha\left\{y_d-y(kT)-\sum_{j=k-\bar{n}}^{k-1}u(jT)+\varepsilon\sum_{j=0}^{k-1}\left[u_R(jT)-u\left(jT-\bar{L}\right)\right]\right\}. \tag{31}$$

It consists of two elements:

- LQ optimal controller as given by (28), and
- delay variability compensator tuned by the coefficient $\varepsilon \in [0, 1]$, which accumulates the information about the differences between the number of items which actually arrived at the distribution center and those which were expected to arrive.

3.5 Properties of the robust policy

The properties of the designed nonlinear policy (30)-(31) will be formulated as two theorems and analyzed with respect to the most adverse conditions (the extreme fluctuations of demand and delay). The first proposition shows how to adjust the warehouse storage space to always accommodate the entire stock and in this way eliminate the risk of (expensive) emergency storage outside the company premises. The second theorem states that with an appropriately chosen target stock level there will be always goods in the warehouse to meet the entire demand.

Theorem 1. If policy (30)-(31) is applied to system (7)-(8), then the stock level at the distribution center is always upper-bounded, i.e.

$$\underset{k\ge 0}{\forall}\; y(kT)\le y_{\max}=y_d+u_{\max}+(1+\varepsilon)\xi_{\max}. \tag{32}$$

Proof. Based on (4), (5), and the definition of function $\xi(\cdot)$, the term compensating the effects of delay variations in (31) satisfies the following relation

$$\sum_{j=0}^{k-1}\left[u_R(jT)-u\left(jT-\bar{L}\right)\right]=\sum_{j=0}^{k-1}\left\{u\left[jT-L(j)\right]-u\left(jT-\bar{L}\right)\right\}=\sum_{j=0}^{k-1}\xi(jT)=\xi(kT). \tag{33}$$

Therefore, we may rewrite function $\varphi(\cdot)$ as

$$\varphi(kT)=\alpha\left[y_d-y(kT)-\sum_{j=k-\bar{n}}^{k-1}u(jT)+\varepsilon\xi(kT)\right]. \tag{34}$$

It follows from the algorithm definition and the system initial conditions that the warehouse at the distribution center is empty for any $k \le (1-\delta)\bar{n}$. Consequently, it is sufficient to show that the proposition holds for all $k > (1-\delta)\bar{n}$. Let us consider some integer $l > (1-\delta)\bar{n}$ and the value of $\varphi(\cdot)$ at instant lT. Two cases ought to be analyzed: the situation when $\varphi(lT) \ge 0$, and the circumstances when $\varphi(lT) < 0$.

Case 1. We investigate the situation when $\varphi(lT) \ge 0$. Directly from (34), we get

$$y(lT) \le y_d + \varepsilon\xi(lT) - \sum_{j=l-\bar{n}}^{l-1} u(jT). \tag{35}$$

Since u is always nonnegative, we have

$$y(lT) \le y_d + \varepsilon\xi(lT). \tag{36}$$

Moreover, since $\xi(lT) \le \xi_{\max}$, we obtain

$$y(lT) \le y_d + \varepsilon\xi_{\max} \le y_{\max}, \tag{37}$$

which ends the first part of the proof.

Case 2. In the second part of the proof we analyze the situation when $\varphi(lT) < 0$. First, we find the last instant $l_1T < lT$ when $\varphi(\cdot)$ was nonnegative. According to (34), $\varphi(0) = ay_d > 0$, so the moment l_1T indeed exists, and the value of $y(l_1T)$ satisfies the inequality similar to (35), i.e.

$$y(l_1T) \le y_d + \varepsilon\xi(l_1T) - \sum_{j=l_1-\bar{n}}^{l_1-1} u(jT). \tag{38}$$

The stock level at instant lT can be expressed as

$$y(lT) = y(l_1T) + \sum_{j=l_1-\bar{n}}^{l-\bar{n}-1} u(jT) + \xi(lT) - \sum_{j=l_1}^{l-1} h(jT), \tag{39}$$

which after applying (38) leads to

$$\begin{aligned} y(lT) &\le y_d + \varepsilon\xi(l_1T) - \sum_{j=l_1-\bar{n}}^{l_1-1} u(jT) + \sum_{j=l_1-\bar{n}}^{l-\bar{n}-1} u(jT) + \xi(lT) - \sum_{j=l_1}^{l-1} h(jT) \\ &\le y_d + \varepsilon\xi(l_1T) + \xi(lT) + \sum_{j=l_1}^{l-\bar{n}-1} u(jT) - \sum_{j=l_1}^{l-1} h(jT). \end{aligned} \tag{40}$$

The algorithm generated a nonzero quantity for the last time before lT at l_1T, and this value can be as large as $u_{\max}$. Consequently, the sum $\sum_{j=l_1}^{l-\bar{n}-1} u(jT) = u(l_1T) \le u_{\max}$. From inequalities (3) and the condition $\xi(lT) \le \xi_{\max}$ we obtain the following stock estimate

$$\begin{aligned} y(lT) &\le y_d + \varepsilon\xi(l_1T) + \xi(lT) + u(l_1T) \\ &\le y_d + \varepsilon\xi_{\max} + \xi_{\max} + u_{\max} = y_{\max}, \end{aligned} \tag{41}$$

which concludes the second part of the reasoning and completes the proof of Theorem 1. □
Theorem 1 states that the warehouse storage space is finite and never exceeds the level of y_{max}. This means that irrespective of the demand and delay variations the system output $y(\cdot)$ is bounded, and the risk of costly emergency storage is eliminated. The second theorem, formulated below, shows that with the appropriately selected target stock y_d we can make the on-hand stock positive, which guarantees the maximum service level in the considered system with uncertain, variable delay.

Theorem 2. If policy (30)–(31) is applied to system (7)–(8), and the target stock level satisfies

$$y_d > u_{max}(\bar{n} + 1/\alpha + 1) + (1+\varepsilon)\xi_{max}, \tag{42}$$

then for any $k \geq (1+\delta)\bar{n} + T_{max}/T$, where $T_{max} = Ty_{max}/(u_{max} - d_{max})$, the stock level is strictly positive.

Proof. The theorem assumption implies that we deal with time instants $kT \geq (1+\delta)\bar{n}T + T_{max}$. Considering some $l \geq (1+\delta)\bar{n} + T_{max}/T$ and the value of signal $\varphi(lT)$, we may distinguish two cases: the situation when $\varphi(lT) < u_{max}$, and the circumstances when $\varphi(lT) \geq u_{max}$.

Case 1. First, we consider the situation when $\varphi(lT) < u_{max}$. We obtain from (34)

$$y(lT) > y_d - \frac{u_{max}}{\alpha} - \sum_{j=l-\bar{n}}^{l-1} u(jT) + \varepsilon\xi(lT). \tag{43}$$

The order quantity is always bounded by u_{max}, which implies

$$y(lT) > y_d - u_{max}/\alpha - u_{max}\bar{n} + \varepsilon\xi(lT). \tag{44}$$

Since $\xi(\cdot) \geq -\xi_{max}$, we get

$$y(lT) > y_d - u_{max}/\alpha - u_{max}\bar{n} - \varepsilon\xi_{max}. \tag{45}$$

Using assumption (42), we get $y(lT) > 0$, which concludes the first part of the proof.

Case 2. In the second part of the proof we investigate the situation when $\varphi(lT) \geq u_{max}$. First, we find the last period $l_1 < l$ when function $\varphi(\cdot)$ was smaller than u_{max}. It comes from Theorem 1 that the stock level never exceeds the value of y_{max}. Furthermore, the demand is limited by d_{max}. Thus, the maximum interval T_{max} during which the controller may continuously generate the maximum order quantity u_{max} is determined as $T_{max} = Ty_{max} / (u_{max} - d_{max})$, and instant l_1T does exist. Moreover, from the theorem assumption we get $l_1T \geq (1+\delta)\bar{n}T$, which means that by the time l_1T the first shipment from the supplier has already reached the distribution center, no matter the delay variation.
The value of $\varphi(l_1T) < u_{max}$. Thus, following similar reasoning as presented in (43)–(45), we arrive at $y(l_1T) > 0$ and

$$\begin{aligned} y(lT) &> y_d - \frac{u_{max}}{\alpha} - \sum_{j=l_1-\bar{n}}^{l_1-1} u(jT) + \varepsilon\xi(l_1T) + \sum_{j=l_1-\bar{n}}^{l-\bar{n}-1} u(jT) + \xi(lT) - \sum_{j=l_1}^{l-1} h(jT) \\ &= y_d - \frac{u_{max}}{\alpha} + \varepsilon\xi(l_1T) + u(l_1T) + \sum_{j=l_1+1}^{l-1} u(jT) - \sum_{j=l-\bar{n}}^{l-1} u(jT) + \xi(lT) - \sum_{j=l_1}^{l-1} h(jT). \end{aligned} \tag{46}$$

Recall that l_1T was the last instant before lT when the controller calculated a quantity smaller than u_{max}. This quantity, $u(l_1T)$, could be as low as zero. Afterwards, the algorithm generates the maximum order and the first sum in (46) reduces to $u_{max}(l - 1 - l_1)$. Moreover, since for any k, $u(kT) \le u_{max}$, the second sum is upper-bounded by $u_{max}\overline{n}$, which implies

$$y(lT) > y_d - u_{max} / \alpha + \varepsilon\xi(l_1T) + 0 + u_{max}(l - 1 - l_1) - u_{max}\overline{n} + \xi(lT) - \sum_{j=l_1}^{l-1} h(jT). \quad (47)$$

According to (3), the realized demand satisfies $0 \le h(\cdot) \le d_{max}$, hence

$$y(lT) > y_d - u_{max} / \alpha + u_{max}(l - 1 - l_1) - u_{max}\overline{n} + \varepsilon\xi(l_1T) + \xi(lT) - d_{max}(l - l_1). \quad (48)$$

Since $\xi(lT) \ge - \xi_{max}$, we get

$$y(lT) > y_d - u_{max} / \alpha + u_{max}(l - 1 - l_1) - u_{max}\overline{n} - \varepsilon\xi_{max} - \xi_{max} - d_{max}(l - l_1). \quad (49)$$

Finally, using the theorem assumption (42), we may estimate the stock level at instant lT in the following way

$$y(lT) > (u_{max} - d_{max})(l - l_1). \quad (50)$$

Since $l > l_1$, and by assumption $u_{max} > d_{max}$, we get $y(lT) > 0$. This completes the proof of Theorem 2. □

Remark. Theorem 2 defines the warehouse storage space which needs to be provided to ensure the maximum service level. The required warehouse capacity is specified following the worst-case uncertainty analysis (for an instructive insight how this methodology relates to production-distribution systems see e.g. (Blanchini et. al., 2003) and (Sarimveis et al., 2008)). Notice, however, that the value given in (42) scales linearly with the maximum order quantity related to demand by the inequality $u_{max} > d_{max}$. Therefore, in the situation when the mean demand differs significantly from the maximum one, it may be convenient to substitute u_{max} with some positive $d_L < d_{max} < u_{max}$. In such a case the 100% service level is no longer ensured, yet the average stock level, and as a consequence the holding costs, will be reduced.

4. Numerical example

We verify the properties of the nonlinear inventory policy (30)–(31) proposed in this work in a series of simulation tests. The system parameters are chosen in the following way: review period $T = 1$ day, nominal lead-time $\overline{L} = \overline{n}T = 8$ days, tolerance of delay variation $\delta = 0.25$, the maximum daily demand at the distribution center d_{max} = 50 items, and the maximum order quantity u_{max} = 55 items. In order to provide fast response yet with a smooth ordering signal, the controller gain should not exceed 0.618, which corresponds to the balanced optimziation case with $w = 1$. Since, additionally, we should account for ordering oscillations caused by delay changes, in the tests the gain is adjusted to $a(w) = a(0.5) = 0.5$. We consider two scenarios reflecting the most common market situations.

Scenario 1. In the first series of simulations we test the controller performance in response to the demand pattern illustrated in Fig. 3, which shows a trend in the demand with abrupt seasonal changes. It is assumed that lead-time fluctuates according to

$$L(k) = \lfloor [1 + \delta \sin(2\pi kT / \bar{n})] \bar{n}T \rfloor = \lfloor [1 + 0.25 \sin(\pi k / 4)] 8 \rfloor, \tag{51}$$

where $\lfloor f \rfloor$ denotes the integer part of f. The actual delay in procurring orders is illustrated in Fig. 4.

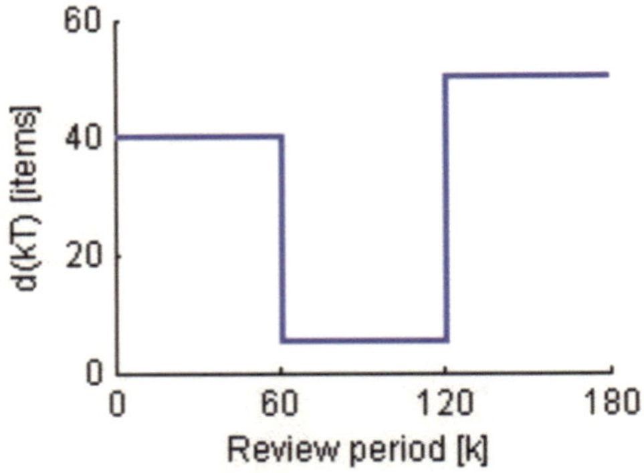

Fig. 3. Market demand - seasonal trend

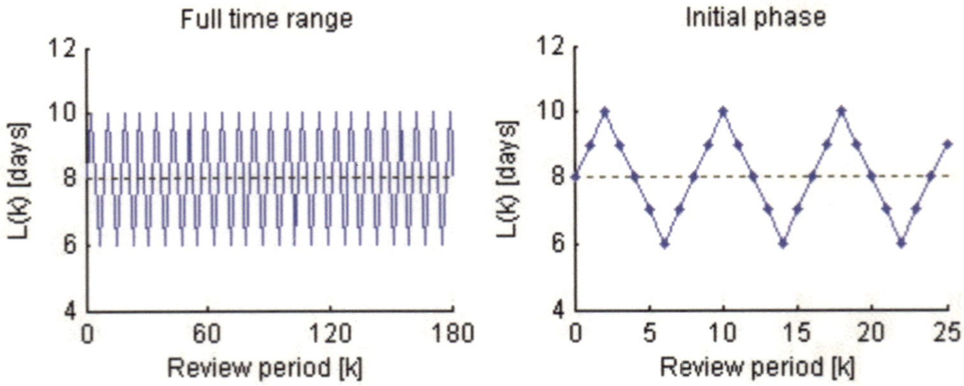

Fig. 4. Lead-time delay

In order to elaborate on the adverse effects of delay variations, and assess the quality of the proposed compensation mechanism, we run two tests. In the first one (curve (a) in the graphs), we show the controller performance with compensation turned off, i.e. with $\varepsilon = 0$, and in the seond test, we consider the case of a full compensation in action with ε set equal to 1 (curve (b) in the graphs). The target stock level y_d is adjusted according to the guideliness provided by Theorem 2 so that the maximum service level is obtained, and the storage space y_{max} is reserved according to the condition stipulated in Theorem 1. The actual values used in the simulations are summarized in Table 1.

The test results are shown in Figs. 5–7: the ordering signal generated by the controller in Fig. 5, the received orders in Fig. 6, and the resultant on-hand stock in Fig. 7. It is clear from the graphs that the proposed controller quickly responds to the sudden changes in the demand trend. Moreover, the stock does not increase beyond the warehouse capacity, and it never drops to zero after the initial phase which implies the 100% service level. If we compare the curves representing the case of a full compensation (b) and the case of the compensation turned-off (a) in Figs. 5 and 7, we can notice that the proposed compensation mechanism eliminates the oscillations of the control signal originating from delay variations. This allows for smooth reaction to the changes in market trend, and an ordering signal which is easy to follow by the supplier. We can learn from Fig. 7 that the obtained smooth ordering signal also permits reducing the on-hand stock while keeping it positive. This means that the maximum service level is achieved, but with decreased holding costs.

Compensation {on/off}	**Target stock** y_d [items]	**Storage space** y_{max} [items]
off: $\varepsilon = 0$	720 > 715	885
on: $\varepsilon = 1$	830 > 825	1105

Table 1. Controller parameters in Scenario 1

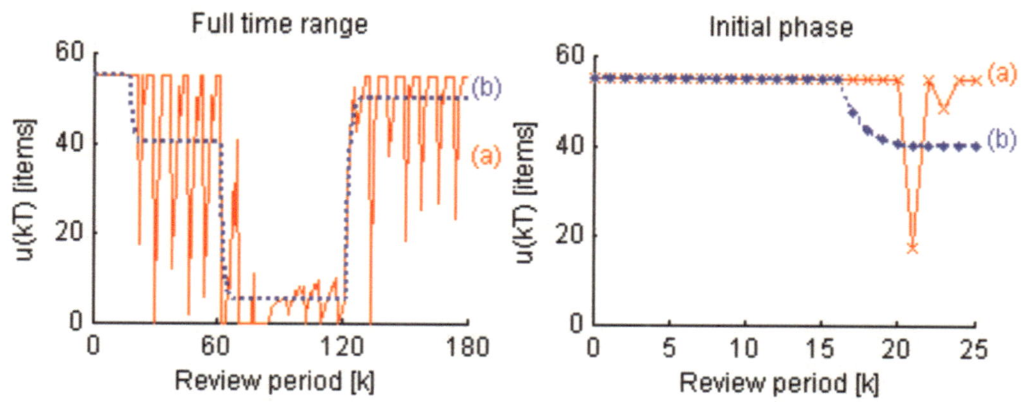

Fig. 5. Generated orders

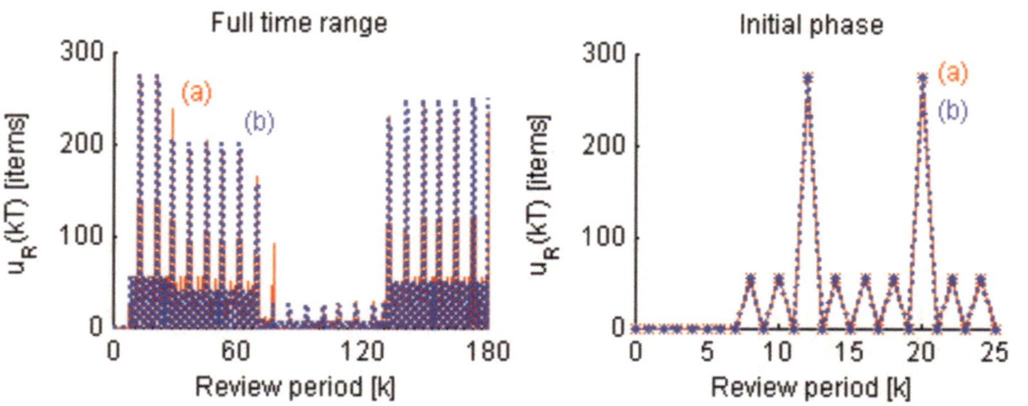

Fig. 6. Received shipments

Scenario 2. In the second scenario, we investigate the controller behavior in the presence of highly variable stochastic demand. Function $d(\cdot)$ following the normal distribution with mean d_μ = 25 items and standard deviation d_δ = 25 items, $D_{norm}(25, 25)$, is illustrated in Fig. 8.

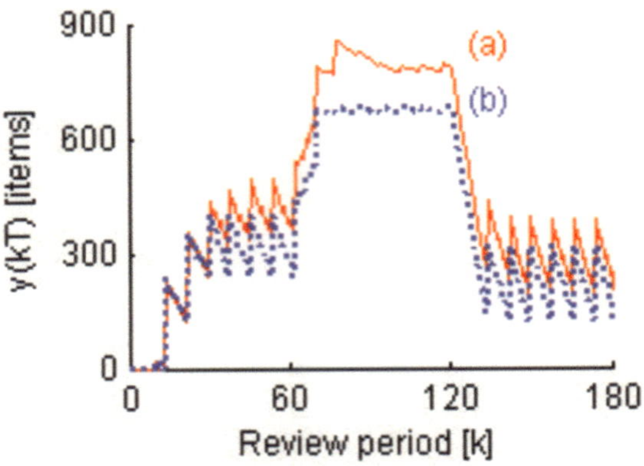

Fig. 7. On-hand stock

Since the mean demand in the stochastic pattern significantly differs from the maximum value, we adjust the target stock according to (42) with $u_{max} > d_{max}$ replaced by $d_{\mu} = 25$ items. This results in $y_d = 375$ items (with $\varepsilon = 1$). Although it is no longer guaranteed to satisfy all of the customer demand (the service level decreases to 98%), the holding costs are nearly halved. For the purpose of comparison we also run the tests for a classical order-up-to (OUT) policy (order up to a target value y_{OUT} if the total stock - equal to the on-hand stock plus open orders - drops below y_{OUT}). In order to compare the controllers in a fair way, we apply the same compensation mechanism for the OUT policy as is used for our, LQ-based scheme. We also reduce the value of the target stock level for the OUT policy y_{OUT} setting $\alpha = 1$ in (42). The controller parameters actually used in the test are grouped in Table 2. Lead-time is assumed to follow the normal distribution D_{norm}(8 days, 2 days). The actual delay in procurring orders is illustrated in Fig. 9.

Policy	**Target stock** y_d | y_{OUT} [items]	**Storage space** y_{max} [items]
LQ-based	375	500
OUT	350	475

Table 2. Controller parameters in Scenario 2

The orders generated by both policies are shown in Fig. 10, the received shipments in Fig. 11, and the on-hand stock in Fig. 12. It is evident from the plots that in contrast to the OUT policy (a), our scheme (b) successfully dampens demand fluctuations at the very first stage of supply chain, and it results in a smaller on-hand stock. Performing statistical analysis we obtain 261 items^2 order variance for the OUT policy and 99 items^2 for our controller. Consequently, according to the most popular (Miragliota, 2006) measure of the bullwhip effect proposed by Chen et al. (2000), which is the ratio of variances of orders and demand, we obtain for our scheme 0.44, which corresponds to 2.27 attenuation of demand variations. The ratio of variances for the OUT policy equals 1.16 > 1 which implies amplified variations and the bullwhip effect. This clearly shows the benefits of application of formal control concepts, in particular dynamical optimization and disturbance compensation, in alleviating the adverse effects of uncertainties in supply chain.

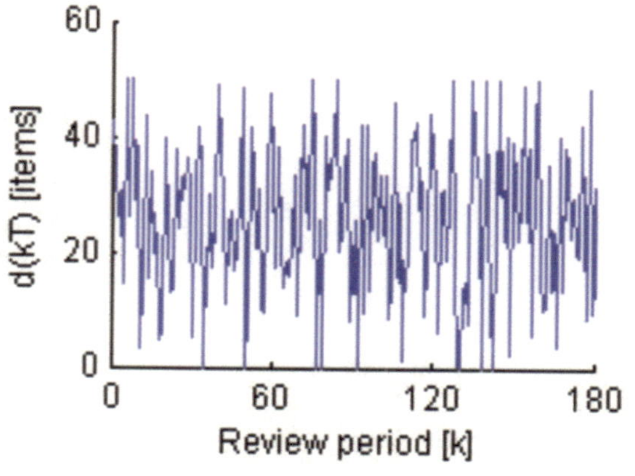

Fig. 8. Market demand following the normal distribution with mean and standard deviation equal to 25 items

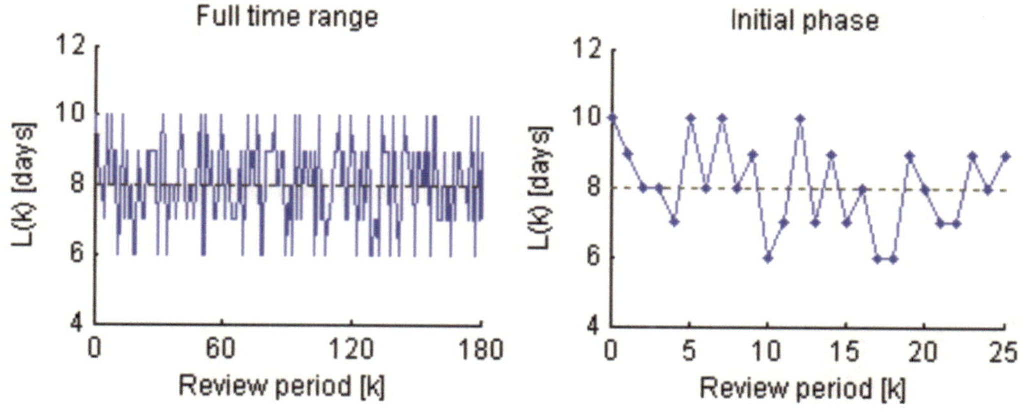

Fig. 9. Lead-time delay following the normal distribution with mean 8 days and standard deviation 2 days

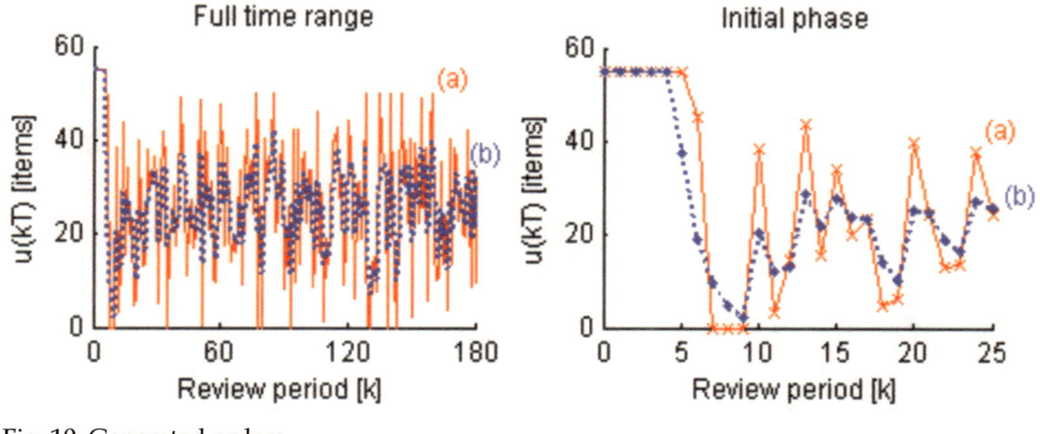

Fig. 10. Generated orders

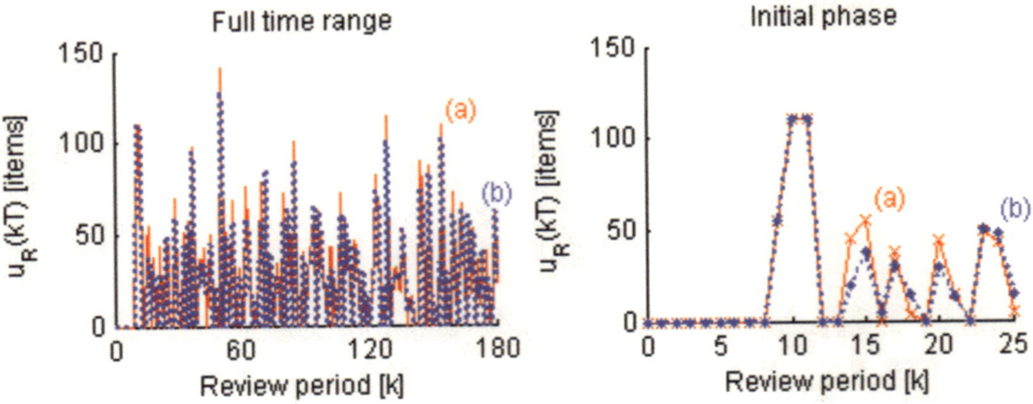

Fig. 11. Received shipments

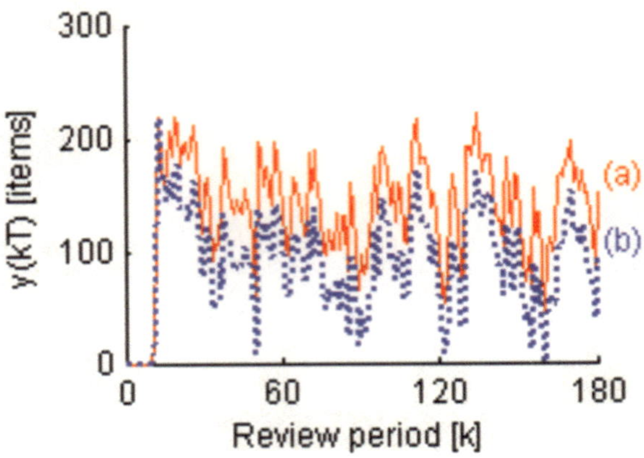

Fig. 12. On-hand stock

5. Conclusion

In this chapter, we presented a robust supply policy for periodic-review inventory systems. The policy is designed based on sound control-theoretic foundations with the aim of reducing the bullwhip effect. The proposed policy successfully counteracts the increase of order oscillations in the presence of highly variable demand, lead-time fluctuations, and supplier capacity constraints. It guarantees that the incoming shipments will not cause warehouse overflow, implying that emergency storage is never required. Moreover, the presented policy ensures that all of the demand is satisfied from the on-hand stock, thus eliminating the risk of missed service opportunities and necessity for backorders.

6. Acknowledgement

This work was financed by the Polish State budget in the years 2010–2012 as a research project N N514 108638 "Application of regulation theory methods to the control of logistic processes". P. Ignaciuk gratefully acknowledges financial support provided by the

Foundation for Polish Science (FNP). He is also a scholarship holder of the project entitled "Innovative Education without Limits - Integrated Progress of the Technical University of Łódz" supported by the European Social Fund.

7. References

Anderson, B. D. O. & Moore, J. B. (1989). *Optimal Control: Linear Quadratic Methods*, Prentice-Hall, ISBN: 0-13-638651-2, Englewood Cliffs, New Jersey

Axsäter, S. (1985). Control theory concepts in production and inventory control. *International Journal of Systems Science*, Vol. 16, No. 2, 161–169, ISSN: 0020-7721

Blanchini, F.; Miani, S.; Pesenti, R. & Rinaldi F. (2003). Stabilization of multi-inventory systems with uncertain demand and setups. *IEEE Transactions on Robotics and Automation*, Vol. 19, No. 1, 103–116, ISSN: 1042-296X

Boukas, E. K.; Shi, P. & Agarwal, R. K. (2000). An application of robust control technique to manufacturing systems with uncertain processing time. *Optimal Control Applications and Methods*, Vol. 21, No. 6, 257–268, ISSN: 0143-2087

Chen, C.; Drezner, Z.; Ryan, J. K. & Simchi-Levi D. (2000). Quantifying the bullwhip effect in a simple supply chain: the impact of forecasting, lead times, and information, *Management Science*, Vol. 46, No. 3, 436–443, ISSN: 0025-1909

Dejonckheere, J.; Disney, S. M.; Lambrecht, M. R. & Towill, D. R. (2003). Measuring and avoiding the bullwhip effect: A control theoretic approach. *European Journal of Operational Research*, Vol. 147, No. 3, 567–590, ISSN: 0377-2217

Dolgui, A. & Prodhon, C. (2007). Supply planning under uncertainties in MRP environments: A state of the art. *Annual Reviews in Control*, Vol. 31, No. 2, 269–279, ISSN: 1367-5788

Geary, S. ; Disney, S. M. & Towill, D. R. (2006). On bullwhip in supply chains - historical review, present practice and expected future impact. *International Journal of Production Economics*, Vol. 101, No. 1, 2–18, ISSN: 0925-5273

Hoberg, K.; Bradley, J. R. & Thonemann, U. W. (2007). Analyzing the effect of the inventory policy on order and inventory variability with linear control theory, *European Journal of Operational Research*, Vol. 176, No. 3, 1620–1642, ISSN: 0377-2217

Holt, C. C.; Modigliani, F.; Muth, J. F. & Simon, H. A. (1960). *Planning Production, Inventories, and Work Force*, Prentice-Hall, Englewood Cliffs, New Jersey

Ignaciuk, P. & Bartoszewicz, A. (2010). LQ optimal sliding mode supply policy for periodic review inventory systems. *IEEE Transactions on Automatic Control*, Vol. 55, No. 1, 269–274, ISSN: 0018-9286

Miragliotta, G. (2006). Layers and mechanisms: A new taxonomy for the bullwhip effect. *International Journal of Production Economics*, Vol. 104, No. 2, 365–381, ISSN: 0377-2217

Ortega, M. & Lin, L. (2004). Control theory applications to the production-inventory problem: A review. *International Journal of Production Research*, Vol. 42, No. 11, 2303–2322, ISSN: 0020-7543

Sarimveis, H.; Patrinos, P.; Tarantilis, C. D. & Kiranoudis, C. T. (2008). Dynamic modeling and control of supply chain systems: A review. *Computers & Operations Research*, Vol. 35, No. 11, 3530–3561, ISSN: 0305-0548

Zabczyk, J. (1974). Remarks on the control of discrete-time distributed parameter systems. *SIAM Journal on Control*, Vol. 12, No. 4, 721–735, ISSN: 0036-1402.

Zipkin, P. H. (2000). *Foundations of Inventory Management*, McGraw-Hill, ISBN: 0-256-11379-3, New York

Zomerdijk, L. G. & de Vries, J. (2003). An organizational perspective on inventory control: Theory and a case study. *International Journal of Production Economics*, Vol. 81–82, No. 1, 173–183, ISSN: 0377-2217

31

Robust Control Approaches for Synchronization of Biochemical Oscillators

Hector Puebla[1], Rogelio Hernandez Suarez[2], Eliseo Hernandez Martinez[3] and Margarita M. Gonzalez-Brambila[4]

[1,3,4]*Universidad Autónoma Metropolitana*

[2]*Instituto Mexicano del Petróleo.*

México

1. Introduction

Synchronization is an important property in fundamental biological processes. Synchronization of biochemical oscillations confers positive functional advantages to the organism, including temporal organization, spatial organization, and efficiency for communication between cells (Berridge et al., 1998; Fall et al., 2002; Goldbeter, 2002; Keener & Sneyd, 1998). Indeed, the relevance of synchronization has been stressed frequently. For instance synchronized circadian rhythms may influence the pharmacology and the tolerability of anticancer drugs and/or their antitumor efficacy (Petty, 2004; Fu & Lee, 2003). In the heart, the impulses coming through the vagus nerve trigger the contraction of the heart only if they are properly synchronized (Keener & Sneyd, 1998; diBernardo et al., 1998). Synchronized behavior of calcium oscillators is believed that enables communication from one side of a cell to another, or between cells, and can serve to synchronize a global, multicellular, response to a local stimulus (Berridge et al., 1998; Perc & Marhl, 2004). Moreover, there are some evidences which support that coherent oscillations play an important role in sensory processing (Izhikevich, 2007).

Understanding both the processes that influence the synchronization of individual biochemical oscillators and how the behaviors of living cells arise out of the properties of coupled populations of biological oscillators are important goals in the study of biological systems, and a field of research with enormous practical application. For instance, elucidating how and why local biochemical oscillators separated by different distances fluctuate in synchrony and the study of conditions under which spatiotemporal patterns of biochemical oscillators can be generated and suppressed (Mikhailov & Hess, 1995; Wolkenhauer et al., 2003; Walleczek, 2003). Indeed, clarifying the mechanisms behind spatial synchrony represents a challenge for biologist and also could ultimately provide critical information to exploit the synchronized behavior in living organisms. For instance, the application of the knowledge of dynamical systems in biology and medicine is giving rise to new therapeutic approaches, such as the treatment of Parkinson's disease by means of neuronal desynchronization (Tass, 2002), or the indications for the development of new drugs based on the collective dynamical instabilities in living cells (Petty, 2004).

Different approaches have been used to synchronize individual biochemical oscillators (Afraimovich et al., 1997; Boccaletti et al., 2002; Canavier et al., 1999; Collins & Stewart, 1994; Goldbeter, 1996; Mirollo & Strogatz, 1990; Morgul & Solak, 1996; Nijmeijer & Mareels, 1997; Pikovsky et al., 1996; Zhou et al., 2008). Classical synchronization approaches includes different coupling approaches and the periodic modulation of an external forcing (periodical or noisy). Despite that synchronization of nonlinear oscillators has been addressed from control theory community, few papers have been addressed the control and synchronization problem of biochemical oscillators. In particular, from control theory perspective, there are basically two ways that are used for synchronization of nonlinear systems. The first is related with observer based synchronization which is applied for coupling identical systems (*i.e.*, same structure and order) and different initial conditions (Alvarez-Ramirez et al., 2002; Nijmeijer & Mareels, 1997; Morgul & Solak, 1996). In these cases, identical synchronization is reached which implies the coincidence of the states of the coupled systems. The second approach from control theory is the application of control laws allows to achieve the synchronization between nonlinear oscillators, with different structure and order, where the variable states of the slave system are forced to follow the trajectories of the master system, such that this approach can be seen as a tracking problem (Fradkov & Pogromsky, 1998; Alvarez-Ramirez et al., 2001). For control designs the presence of disturbances, dynamic uncertainties, and nonlinearities in biochemical models pose great challenges. In particular, biochemical systems have a high degree of uncertainties.

Relevant contributions using control and system theory approaches are the following. Sontag (2004) has been establishes global asymptotic stability results using small gain theorems for a class of biochemical systems. Kimura and Nishigaki (2005) have been established an analogy of circadian rhythm with the PLL framework. Iglesias (2003) has been addressed the feedback mechanism in chemotaxis using control theory concepts. Steeling et al. (2004) have been introduced a robustness analysis and a model predictive control approach for circadian oscillations. Takeuchi *et al.* (2006) have been also addressed the generation and suppression of circadian oscillations with control theory tools. We have previously showed that both modeling error compensation approach and high-order sliding mode control approach can be used to robust synchronize intracellular calcium oscillators and excitable media (Puebla, 2005, Puebla et al., 2009; Puebla et al., 2010; Aguilar-Lopez et al., 2010).

In this chapter we extend the application of robust controllers for the synchronization of three benchmark models of biochemical oscillators: (i) Goodwin model of genetic oscillations (Goodwin, 1965), (ii) FitzHugh-Nagumo model of neurons (FitzHugh, 1961); and (iii) a model of circadian rhythms in Drosphila (Goldbeter, 1996). We introduce three robust control approaches for the synchronization of biochemical oscillators: (i) A modeling error compensation approach (Alvarez-Ramirez, 1999), (ii) integral sliding mode control (Levant, 2001), and (iii) geometric linearizing control (Alvarez-Ramirez, 1999; Hangos et al., 2004). The proposed controllers have two nice features for biological applications: (*i*) robustness against model uncertainties, and (*ii*) simplicity in the resulting controller. We show how a certain class of cellular processes can be dynamically synchronized by appropriate input signals.

This chapter is organized as follows: In Section 2, for the sake of clarity in presentation, we briefly provide some issues on the phenomenology, modeling and nonlinear dynamics in cellular systems. In Section 3 we review classical synchronization approaches of biochemical oscillations that have been reported in the literature. In Section 4 we present the synchronization problem addressed in this chapter and the robust control approaches for the synchronization of biochemical oscillations. Three numerical benchmark examples in Section

5 shows the implementation of the proposed feedback control approaches. Finally, in Section 6 we close this chapter with some concluding remarks.

2. Modeling of biochemical oscillators

In this section we define the class of biochemical oscillators that we are studying. First, we briefly discuss the phenomenology of biological mechanism underlying in biochemical systems. Next we present some ideas of the modeling of biochemical systems. Finally, we introduce the class of biochemical systems under consideration in this chapter.

2.1 Biological mechanisms

The processes that underlie cellular behavior are organized in complexly coupled biochemical reaction networks, where feedforward and feedback information flows provide the links between the different levels in the hierarchy of cell biochemical network organization (Arkin & Ross, 1994; Goldbeter, 1996; Glass & Mackey, 1988). Theoretical models of biochemical reaction networks have been proposed that simulate, for example, cellular dynamics of Ca oscillations, interactions between different cell signaling pathways, genetic regulatory circuits, cellular control networks for DNA replication and cellular division (Segel, 1980; Goldbeter, 1996; Keener & Sneyd, 1998; Smolen et al., 2003).

Cells are equipped with exquisite sensing systems which allow them to be continuously aware of the conditions in their environment and react appropriately to these conditions. The basic elements of a cellular signaling system are a sensor protein, made of a receptor domain and a transmitter domain, and a response regulator, consisting of a receiver domain and a regulator domain (Keener & Sneyd, 1998; Blumenfeld & Tikhonov, 1994). Stimulation of the sensor (normally bound to the cell membrane) leads to activation of the transmitter, which produces an intracellular signal. This signal is processed by a cascade of molecules and finally arrives at the receiver, which in turn activates the regulator. Regulators produce a response by modulating gene expression or enzyme activities. The key components in this transfer of information are proteins, which form networks and are able to perform computational tasks (Goldbeter, 1996; Glass & Mackey, 1998; Fall et al., 2002). Proteins can change their state by interaction with other proteins or by biochemical modifications (such as phosphorylations) catalyzed by other proteins. Another common mechanism is the release of small molecules called second messengers, which diffuse in the cell and activate other proteins (Berridge, 1998; Keener & Sneyd, 1998).

2.2 Modeling of cellular processes

In contrast to the high complexity of the cell, simple mathematical models have been developed, mostly based on experimental observations describing phenomena like limitation, activation, inhibition, saturation, multiple substrate uptake, bottlenecks and multiplicity of metabolic steady states (De Jong, 2002; Fall et al., 2002; Goldbeter, 2002; Segel, 1980). Mathematical models of the intracellular complexity of cellular systems are often based on systems of nonlinear ordinary differential equations (ODEs). These models are usually valid for a limited, but often sufficiently large range of operating conditions. Of course, the level of complexity of the mathematical description depends on the application. When the problem is taken with all its complexity, for instance, if we require that the model accounts for spatial inhomogeneity, diffusion processes and transport delay, then we deal with partial differential

equations and time delay (De Jong, 2002; Smolen et al., 2003; Asthagiri & Lauffenburger, 2001). In this chapter, we restrict ourselves to the simpler case of ODEs.
In cells, most biochemical reactions of interest are catalyzed by enzymes, and a variety of mathematical descriptions have been developed for these reactions. Many enzymatic reactions have complex kinetic mechanisms, and specialized equations are needed to describe their rates in detail. Two typical rate models are the Michelis-Menten kinetics and the allosteric Hill function (Keener & Sneyd, 1998; Segel, 1980).

1. Michaelis-Menten model: This kinetic model is relevant to situations where there is no intermediate or product inhibition, and there is no allostericity or cooperativity. The kinetic model is defined by,

$$\mu_{\max}\frac{S}{k_s + S} \tag{1}$$

$\mu_{\max}$ is the maximal growth rate and k_s the half-saturation constant.

2. Allosteric interactions: Binding of small molecules can alter an enzyme's conformation and alter the rate of the reaction catalyzed by the enzyme. Allosteric interactions can therefore mediate feedback and feedforward interactions within a biochemical pathway, as well as crosstalk between pathways. In models of enzyme regulation, allosteric interactions are commonly represented by Hill functions. These are saturable functions of the concentration of the effector molecule. With the concentration of effector denoted by L, if L activates an enzyme, the enzyme activity is taken as proportional to the following increasing function of the n-th power of L:

$$\frac{L^n}{L^n + K_H^n} \tag{2}$$

The parameter n is called the Hill coefficient. Greater values of n correspond to steeper sigmoids, that is, to a narrowing of the range of L over which the enzyme activity is significantly above 0 and also significantly below 1. If L inhibits an enzyme, the enzyme activity is taken as proportional to a decreasing function of L:

$$\frac{K_H^n}{L^n + K_H^n} \tag{3}$$

2.3 Nonlinear dynamics in cellular systems

Nonlinear phenomena including multiple steady states, periodic or chaotic temporal evolution and self-organization can be supported by the dynamical cellular system since functional kinetics are nonlinear in the descriptive variables and the system is maintained far from equilibrium. The variety of functional dynamics is a consequence of the nonlinearities inherent in multiple modes of biochemical regulation, such as cooperativity and kinetics at the levels of gene expression, protein synthesis, enzyme activity, receptor function, and transport processes (Keener & Sneyd, 1998; Blumenfeld & Tikhonov, 1994; Goldbeter, 2002; Glass, 2001).

1. Simple oscillations: Oscillations occur at every level of a biological organization, with periods ranging from milliseconds (neurons) to seconds (cardiac cells), minutes (oscillatory enzymes), hours (pulsatile hormone secretion), days (circadian rhythms), weeks (ovarian cycle) and even to years (predator-prey interactions in ecology). Oscillatory behavior

often originates at the cellular level from regulatory feedback loops which involve many parameters and interacting variables. More generally, oscillations in reaction rates and concentrations commonly rely, on negative feedback to sustain oscillations. Oscillations have been observed in the metabolic flux through glycolysis and also in the rates of secretion of hormones such as insulin (Goldbeter, 2002; Glass, 2001).

2. Bursting and chaos: Bursting represents one type of complex oscillations that is particularly common in neurobiology. An active phase of spike generation is followed by a quiescent phase, after which a new active phase begins. Chaos is a common mode of complex oscillatory behavior that has been studied intensively in physical, chemical and biological systems. It has been discussed the existence of two main routes to complex oscillatory phenomena. The first relies on forcing a system that displays simple periodic oscillations by a periodic input. In an appropriate range of input frequency and amplitude, one can often observe the transition from simple to complex oscillatory behavior such as bursting and chaos. For other frequencies and amplitudes of the forcing, entrainment or quasi-periodic oscillations occur. In the second route complex oscillatory phenomena may arise through the interplay between several instability-generating mechanisms, each of which is capable of producing sustained oscillations (Goldbeter, 2002; Glass, 2001).

Oscillatory dynamic is not the only possible outcome of nonlinear equations. Indeed, nonlinear systems are in general classified within three categories: bistable, excitable, and oscillatory. Bistable systems are characterized by the existence of two different stable states. Excitable systems posses a unique stable fixed point; however, if they are affected by a perturbation which overcomes a certain threshold amplitude, they are able to perform an excursion in the phase space before returning to the stable fixed point. That is, they do not relax immediately to the stationary state, but keep the excitation for a finite time (Ferrel, 2002; Fall et al., 2002; Mikhailov & Hess, 1995).

2.4 The class of biochemical oscillators

As the basic single biochemical oscillator we consider single-input nonlinear systems in the form,

$$\begin{aligned} \frac{dy}{dt} &= f_1(y,z) + g(y,z)u \\ \frac{dz}{dt} &= f_2(y,z) \end{aligned} \tag{4}$$

where $f_1(y,z) \in \mathbb{R}$, $f_2(y,z) \in \mathbb{R}^{n-1}$, and $g(y,z) \in \mathbb{R}$, are smooth functions of their arguments, $y \in \mathbb{R}$, is the measured output of the system, $z \in \mathbb{R}^{n-1}$, is the internal state, and u can be manipulated for synchronization purposes.

Suppose that there are N subsystems in a lattice y_i, $i = 1, ..., N$, and, in the absence of coupling, the dynamics of y_i is given by the biochemical oscillator (4). That is, the dynamics of y_i satisfies,

$$\begin{aligned} \frac{dy_i}{dt} &= f_{1,i}(y_i,z_i) + g_i(y_i,z_i)u_i, \; i = 1, ..., N \\ \frac{dz_i}{dt} &= f_{2,i}(y_i,z_i) \end{aligned} \tag{5}$$

Its not hard to see that several published models of biochemical oscillators can be described by model (5) (Keener & Sneyd, 1998; Goldbeter, 2002; Tyson et al., 2003; De Jong, 2002).

3. Synchronization of biochemical oscillations

Classical theory of synchronization distinguishes between forced synchronization by an external periodic driving force and synchronization via the coupling between oscillators. In both cases manifestations of synchronization are the same. In this section we briefly review both external forcing and coupling based synchronization approaches proposed in the literature for biochemical oscillations.

3.1 Synchronization of biochemical oscillations via coupling

Consider that the N subsystems are coupled,

$$\frac{dy_i}{dt} = C_i(y) + f_{1,i}(y_i, z_i) + g_i(y_i, z_i)u_i, \; i = 1, ..., N \tag{6}$$
$$\frac{dz_i}{dt} = f_{2,i}(y_i, z_i)$$

where $y = [y_1, ..., y_N]^T$ and $C(y)$ is a coupling function.

3.1.1 Diffusive coupling

Consider that the coupling function $C(y)$ is described via a local diffusive (nearest neighborhood) coupling, such that,

$$C(y) = \sigma(y_{i-1} - 2y_i + y_{i+1}) \tag{7}$$

where σ is the coupling strength. This case is quite interesting since it can be seen as a lattice approximation to reaction-diffusion systems,

$$\frac{\partial^2 y}{\partial t^2} = \sigma \frac{\partial^2 y}{\partial \xi^2} + f(y, z) + g(y, z)u \tag{8}$$

where $u = [u_1, ..., u_N]^T \in \mathbb{R}^N$ and ξ is the spatial coordinate.

Local coupling provides the system with the notion of vicinity and distance. This is, each element directly interacts only with its neighbors, which then transmit the interaction to their own neighbors. Thus, a localized perturbation spreads through the system affecting first its close proximity and later reaching the farther parts of the system. This is a crucial property of reaction-diffusion systems.

Diffusive coupling via gap junctions is considered as the natural form of coupling in many cellular processes (di Bernando et al., 1998; Fall et al., 2002; Glass, 2001; Mirollo & Strogatz, 1990). Gap junctions are composed of arrays of small channels that permit small molecules to shuttle from one cell to another and thus directly link the interior of adjacent cells. Importantly, gap junctions allow electrical and metabolic coupling among cells because signals initiated in one cell can readily propagate to neighboring cells (Keener & Sneyd, 1998; Izhikevich, 2007). Thus, gap junctions between cells and electrical coupling can be considered as a particular form of diffusive coupling.

In the domain of biological systems, nonlocal coupling can be present as well. Coupling is nonlocal if diffusion is such that the substance released by one cell can reach and affect not only its neighbors, but even cells which are located far away from it.

3.1.2 Random coupling

In random coupling the coupling function is described as follows,

$$\mathcal{C}(y) = \sigma A y \tag{9}$$

where the elements A_{kl} of the matrix A are either 0 or 1 and are assigned in a random way. This is,

$$A_{kl} = \begin{cases} 0 \text{ if } r_{kl} < r_{\min} \\ 1 \text{ if } r_{kl} \geq r_{\min} \end{cases} \tag{10}$$

where $r_{kl} \in [0,1]$ is a uniformly distributed random number and the threshold $r_{\min} \in (0,1)$. This coupling structure resembles that of neural networks (Izhikevich, 2007).

3.1.3 Kuramoto coupling

A successful approach to the problem of synchronization consists of modeling each member of the population as a phase oscillator. Kuramoto analyzed a model of phase oscillators running at arbitrary intrinsic frequencies, and coupled through the sine of their phase differences (Kuramoto, 1984). The Kuramoto model is simple enough to be mathematically tractable, yet sufficiently complex to be non-trivial. The model is rich enough to display a large variety of synchronization patterns and sufficiently flexible to be adapted to many different contexts.
The Kuramoto model consists of a population of N coupled phase oscillators, $\theta_i(t)$, having natural frequencies ω_i distributed with a given probability density $g(\omega)$, and whose dynamics is governed by,

$$\frac{d\theta_i}{dt} = \omega_i + \sum_{j=1}^{N} K_{ij} \sin(\theta_j - \theta_i), \quad i = 1, ..., N \tag{11}$$

where K_{ij} is the coupling matrix. When the coupling is sufficiently weak, the oscillators run incoherently whereas beyond a certain threshold collective synchronization emerges spontaneously. Many different models for the coupling matrix K_{ij} have been considered such as nearest-neighbor coupling, hierarchical coupling, random long-range coupling, or even state dependent interactions (Kuramoto, 1984).

3.2 Applications

Classical synchronization approaches have been applied successfully for the synchronization of biochemical oscillators. Winfree (2002) has suggested that such critical perturbations applied at the appropriate phase of a limit cycle should stop the clock, at least transiently, if the perturbation brings the oscillator back into the vicinity of the steady state. Ueda *et al* (2002) studied a model for circadian rhythms in Drosophila. As a single cell oscillator, they used a more detailed model incorporating 10 variables. They then apply a local coupling through each possible variable, and show that for some of them, synchronization occurs. Interestingly, they assessed the effect of fluctuations in parameter values and show that the coupled system is relatively robust to noise. Another theoretical model of coupled circadian oscillators through local coupling has been proposed by Kunz and Achermann (2003). Using

the van der Pol model, they described possible spatial effects, including wave propagation and pattern formation. Gonze *et al* (2005) proved that a mean field approach can be an effective way to couple a population of circadian oscillators, where the global coupling drives oscillators, which would be damped under a constant forcing.
Gap junctions are tacitly postulated as a sufficient means of intercellular communication for synchronizing Ca^{2+} transients (Berridge, 1998; Perc & Marhl, 2004). Ca^{2+} ions may pass through gap junction channels to the neighboring cell by passive diffusion. Recently, it has been shown that individual hepatocytes can have very different intrinsic oscillation frequencies but become phase-locked when coupled by gap junctions (Hofer, 2003; Tang & Othmer, 1995). It is shown that junctional calcium fluxes are effective in synchronizing calcium oscillations in coupled hepatocytes. Many neuronal and non-neuronal systems exhibit synchronized oscillatory behavior in networks of electrically coupled cells (Fall et al., 2002). Experimental findings have revealed that in some of these systems electrical coupling is essential for the generation of oscillations and not only for their modulation (FitzHugh, 1961; Winfree, 2001; Izhikevich, 2007).

3.3 Synchronization of biochemical oscillations via an external forcing

The intrinsic nonlinearity of living systems is of great significance to scientists who study the response of cells, tissues and whole organisms to natural or artificial stimuli. External or artificial stimuli of biological systems by time variation of appropriate control parameters is of great importance from a general point of view. Forced or tuned oscillators are not only considered to be important in cellular rhythms, but also in technical applications involving biochemical reaction systems external control may be of great benefit for improving performance criteria of bioengineering processes (Greenman et al., 2004).
External modulated forcing has been applied for synchronization purposes in some contributions. For example a population of chaotic amoebae was subjected to a small-amplitude periodic forcing, which appeared to be sufficient to transform chaotic behavior into periodic (Goldbeter, 1996). In many organisms, the source of external forcing has been identified to be a variation of the light due to night and day cycles. Indeed, the molecular basis of the effect of light on different circadian biochemical networks has been unraveled (Gonze & Goldbeter, 2000; Jewett et al., 1991). The question on whether such external forcing is enough to induce the synchronization between circadian cells usually observed in experiments, or if coupling between the cells is needed, is still open.

4. Robust control approaches for synchronization of biochemical oscillators

In this section the synchronization problem framed as a tracking feedback control problem is presented. Three robust control approaches are then briefly described: (i) the modeling error compensation, (ii) the sliding mode control, and (iii) geometric linearizing control.

4.1 Synchronization problem

The synchronization problem consists of making two or more systems oscillate in a synchronized way. This synchronization problem is cast as a control problem where the control objective is tracking with respect to a desired single synchronization signal $y_{ref}(t)$ via manipulation of an external input u.
The synchronization problem description is completed by the following assumptions:

A1 The measurement of the variable to be synchronized y, is available for synchronization design purposes.

A2 Nonlinear functions $f_{1,j}(x_i)$ and $g_{1,j}(x_i)$ are uncertain, and can be available rough estimates of these terms.

The following comments are in order:

- $A1$ is a reasonable assumption. For instance, in neurons the measurement of the membrane potential is standard. Free intracellular calcium (Ca^{2+}) can be also measured using florescence techniques. Even in the absence of such measurements, a state estimator can be designed. On the other hand, cell must have some internal mechanism to knows perfectly its behavior. Indeed, it has been reported elsewhere that Ca^{2+} acts as an intracellular messenger, relaying information within cells to regulate their activity, such that should be exist some internal mechanism in the cells to knows its behavior (Berridge, 1998).
- $A2$ considers that functions $f_{1,j}(x_i)$ and $g_{1,j}(x_i)$ can contain uncertain parameters, or in the worst case the whole terms are unknown. Indeed, parameters in biochemical systems have some degree of uncertainties, as these parameter values commonly are estimated from experimental data, which contain errors due to both the estimation procedure adopted to fit data and the experimental errors of the data themselves (De Jong, 2002; Keener & Sneyd, 1998). From a practical viewpoint, the assumption of model uncertainties in our control methodology allows to design a controller that uses only the minimum system information in order to control the calcium nonlinear dynamics and the resulting control can be easily interpreted from a biological viewpoint and implemented.
- The use of an external input as the manipulable variable is realistic. Indeed, several experimental studies have shown that the synchronization of individual biochemical oscillators depends on external stimulus properties (FitzHugh, 1961; Glass, 2001; Gonze & Goldbeter, 2000; Jewett et al., 1991; Marhl & Schuster, 2003; Izhikevich, 2007). An external electrical stimuli can be modeled including an applied current in the current balance equation. Chemical stimuli can be modeled either by varying concentrations of relevant agents or by varying parameters which are believed to be correlated to the stimulating chemical.

The proposed feedback and synchronization arrangements are shown in Fig. 1. A sensor measures a time-varying output from the cell, $y(t)$, and feeds it to a controller. The controller produces a signal, $u(t)$, which drives an actuator to produce a time-varying input to individual biochemical oscillators to get the desired synchronized dynamic behavior. On the other hand, a reference or master oscillator provides the desired reference to individual or slave oscillators, that will be driven by individual external inputs to follow the desired reference behavior.

4.2 Modeling error compensation approach

Sun et al. (1994) proposed a robust controller design method for single-input/single-output (SISO) minimum-phase linear systems. The design approach consists of a modeling error compensator (MEC). The central idea is to compensate the error due to uncertainty by determining the modeling error via plant input and output signals and use this information in the design. In addition to a nominal feedback, another feedback loop is introduced using the modeling error and this feedback action is explicitly proportional to the parametric error which is the source of uncertainty. The MEC approach was extended for a class of

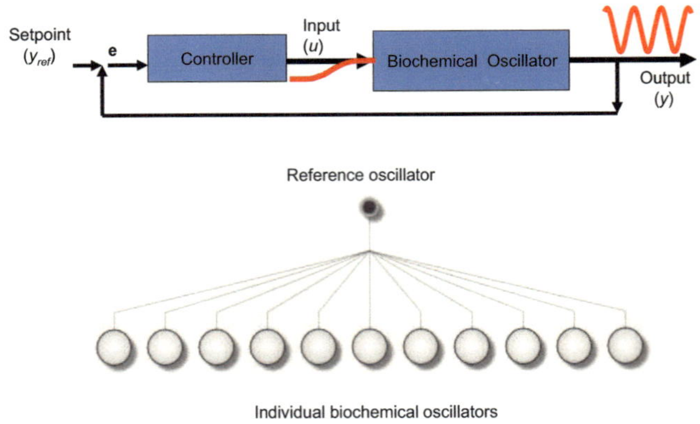

Fig. 1. Feedback and synchronization system.

linear time-varying and nonlinear linearizable lumped parameter systems with uncertain and unknown terms by Alvarez-Ramirez (1999), where instead of designing a robust state feedback to dominate the uncertain term, the uncertain term is viewed as an extra state that is estimated using a high-gain observer. The estimation of the uncertain term gives the control system some degree of adaptability. The extension of the MEC approach to distributed parameter systems has been applied by Puebla (2005) and Puebla et al. (2009, 2010) for a class of biological distributed parameter systems. The underlying idea behind MEC control designs is to lump the input-output uncertainties into a term, which is estimated and compensated via a suitable algorithm.
Consider the class of biochemical oscillators described in Section 2:

$$\frac{dy_i}{dt} = f_{1,i}(y_i, z_i) + g_i(y_i, z_i)u_i, \; i = 1, ..., N \tag{12}$$
$$\frac{dz_i}{dt} = f_{2,i}(y_i, z_i)$$

Let $e_i = y_i - y_{ref}$ be the tracking error, and define the modeling error function η_i as,

$$\eta_i = (\widetilde{f}_{1,i} - f_{1,i}) + (\widetilde{g}_i - g_i)u_i \tag{13}$$

where $\widetilde{f}_{1,i}$ and $\widetilde{g}_i$ are rough estimates of uncertain functions $f_{1,i}$ and g_i respectively. System (12) can be written as,

$$\dot{e}_i = \eta_i - \widetilde{f}_{1,i} - \widetilde{g}_i u - \dot{y}_{ref} \tag{14}$$

where $\dot{y}_{ref}$is the first derivative of y_{ref}. Consider the inverse dynamics control law,

$$u_i = \widetilde{g}_i^{-1}(\eta_i - \widetilde{f}_{1,i} - \dot{y}_{ref} + \tau_c^{-1} e_i) \tag{15}$$

where $\tau_c > 0$ is a closed-loop time constant. Under the inverse-dynamics control law (15), the closed-loop system dynamics is $de_i/dt = -\tau_c^{-1} e_i$, so that the error dynamic behavior is given as $e(t) = e(0)\exp(-t/\tau_c)$. In this way, the asymptotic convergence $e(t) \to 0$, and so $y \to y_{ref}$, is guaranteed.

In order to implement the control input an estimate of the real uncertain term is computed using a high-gain reduced-order observer,

$$\dot{\tilde{\eta}}_i = \tau_e^{-1}(\eta_i - \tilde{\eta}_i) \tag{16}$$

where $\tau_e > 0$ is the estimation time constant. After some direct algebraic manipulations the reduced order observer (16) can be written as,

$$\begin{aligned} \dot{w}_i &= \tilde{f}_{1,i} + \tilde{g}_i u_i + \dot{y}_{ref} - \tilde{\eta}_i \\ \tilde{\eta}_i &= \tau_e^{-1}(w_i + e_i) \end{aligned} \tag{17}$$

The final form of the controller is given by the feedback function (18) and the modeling error estimator (17),

$$u_i = \tilde{g}_i^{-1}(\tilde{\eta}_i - \tilde{f}_{1,i} - \dot{y}_{ref} + \tau_c^{-1} e_i) \tag{18}$$

the resulting feedback controlled depends only on the measure y and the estimated values of uncertain terms $\tilde{f}_{1,i}$ and $\tilde{g}_i$. Notice that in a worst-case design, one can choose $\tilde{f}_{1,i} = 0$.
The above model-based control approach has only two control design parameters, *i.e.*, τ_c and τ_e. The closed-loop parameter τ_c can be chosen as the inverse of the dominant frequency of the open-loop dynamics. On the other hand, the estimation parameter $\tau_e > 0$, which determines the smoothness of the modeling error can be chosen as $\tau_e < \frac{1}{2}\tau_c$. On the other hand, system (12) is of relative grade one. However, straight extensions of the MEC control design to both autonomous third and second order systems can be found in Puebla *et al.* (2003), and Alvarez-Ramirez, respectively.

4.3 Sliding mode control approaches

Sliding mode control techniques have long been recognized as a powerful robust control method (Hangos et al., 2004; Levant, 2001; Sira-Ramirez, 2002). Sliding-mode control schemes, have shown several advantages like allowing the presence of matched model uncertainties and convergence speed over others existing techniques as Lyapunov-based techniques, feedback linearization and extended linearization, however standard sliding-mode controllers have some drawbacks: the closed-loop trajectory of the designed solution is not robust even with respect to the matched disturbances on a time interval preceding the sliding motion, the classical sliding-mode controllers are robust in the case of matched disturbances only, the designed controller ensures the optimality only after the entrance point into the sliding mode. To try to avoid the above a relatively new kind of sliding-mode structures have been proposed as the named high-order sliding-mode technique, these techniques consider a fractional power of the absolute value of the tracking error coupled with the sign function, this structure provides several advantages as simplification of the control law, higher accuracy and chattering prevention (Hangos et al., 2004; Levant, 2001; Sira-Ramirez, 2002). In this section we present some ideas of the integral high order sliding mode control (IHOSMC).
Sliding mode control design consists of two phases. In the first phase the sliding surface is to be reached (reaching mode), while in the second the system is controlled to move along the sliding surface (sliding mode). In fact, these two phases can be designed independently from each other. Reaching the sliding surface can be realized by appropriate switching elements (Hangos et al., 2004).

Defining

$$\sigma(e) = e_i = y_i - y_{ref}$$

as the sliding surface, we have that the continuous part of the sliding mode controller is given by,

$$u_{eq,i} = -g_i^{-1}(f_{1,i} - \dot{y}_{ref})$$

such that,

$$\dot{\sigma}(e) = 0$$

where $\dot{y}_{ref}$is the time-derivative of the desired trajectory signal. Once on the surface, the dynamic response of the system is governed by $de_i/dt = 0$. To force the system trajectory to converge to the sliding surface in the presence of both model uncertainties and disturbances, with chattering minimization and finite-time convergence, the sliding trajectory is proposed as (Levant, 2001; Aguilar-Lopez et al., 2010),

$$u_{eq,i} = -g_i^{-1}[\delta_1 e_i + \delta_2 \int_0^t \left\{ sign(e_i) \left| e_i \right|^{1/p} \right\} d\tau] \tag{19}$$

where δ_1 and δ_2 are control design parameters. The final IHOSMC is given by,

$$u_i = -g_i^{-1}(f_{1,i} - \dot{y}_{ref} + \delta_1 e_i + \delta_2 \int_0^t \left\{ sign(e_i) \left| e_i \right|^{1/p} \right\} d\tau) \tag{20}$$

The synthesis of the above control law requires accurate knowledge of both $f_{1,i}$ and dy_{ref}/dt to be realizable. To enhance the robust performance of the above control laws, the uncertain terms is lumped in single terms and compensated with a reduced-order observer. However, by exploiting the properties of the sliding part of the sliding-mode type controllers to compensates uncertain nonlinear terms, the knowledge of the nonlinear term $f_{1,i}$ can be avoided.

Summarizing, the IHOSMC is composed by a proportional action, which has stabilizing effects on the control performance, and a high order sliding surface, which compensates the uncertain nonlinear terms to provide robustness to the closed-loop system. This behavior is exhibited because, once on the sliding surface, system trajectories remain on that surface, so the sliding condition is taken and make the surface and invariant set. This implies that some disturbances or dynamic uncertainties can be compensated while still keeping the surface an invariant set.

4.4 Robust geometric linearizing control

Differential geometry is an essential tool for the study of the structural properties of nonlinear control systems. Differential geometric techniques of nonlinear control include static and dynamic feedback linearization, input-output linearization, nonlinear state observers and disturbance decoupling (Hangos et al., 2004). In exact linearization the main idea is to apply a suitable nonlinear coordinate transformation to a nonlinear system in order to obtain a linear one in the new co-ordinates and between the original output and the newly introduced transformed input. The coordinates transformation must be supplemented by a static nonlinear feedback to achieve linearization. After linearization any controller design method can be used to stabilize the system or modify its dynamic properties.

Exact linearization via state feedback is a limited technique for control of nonlinear systems because it is only applicable for systems satisfying a relative degree condition. Indeed, the relative degree of the system needs to be equal to the number of state variables, i.e. $r = n$. Therefore the exact linearization may not be applicable or may not be feasible in practical cases. Input–output linearization is an alternative way of achieving linear behavior of a system by nonlinear coordinate transformation (Hangos et al., 2004).
A main drawback in the use of differential geometric control techniques is that depends on the exact cancelation of the nonlinear dynamics in order to obtain an input-output linear dynamic behavior. As a consequence, the perfect knowledge of the system is required. Robustness of geometric differential approaches has received attention in the literature. In this section we describe a robust geometric input-output linearizing control, where the presence of modeling errors, unmeasured disturbances and parametric uncertainties are considered in the controller design.
Consider the class of biochemical oscillators described in Section 2 with $y_i = h_i(y_i, z_i)$. An input-output linearizing controller u_i is given by,

$$\begin{aligned} u_i &= \frac{1}{\mathcal{L}_g \mathcal{L}_f^{r-1} h_i(y_i, z_i)} \left(-\mathcal{L}_f^r h_i(y_i, z_i) + v_i\right) \\ &= \alpha_i(y_i, z_i) + \beta_i(y_i, z_i) v_i \\ \alpha_i(y_i, z_i) &= \frac{-\mathcal{L}_f^r h_i(y_i, z_i)}{\mathcal{L}_g \mathcal{L}_f^{r-1} h_i(y_i, z_i)} \\ \beta_i(y_i, z_i) &= \frac{1}{\mathcal{L}_g \mathcal{L}_f^{r-1} h_i(y_i, z_i)} \end{aligned} \tag{21}$$

where $\mathcal{L}_g$ and $\mathcal{L}_f$ are the lie derivatives of g_i and $f_{1,i}$ respectively, and v_i is a new external input.
Under the input-output linearizing controller we have,

$$\frac{dy_i}{dt} = f_{1,i}(y_i, z_i) + g_i(y_i, z_i)\alpha_i(y_i, z_i) + \beta_i(y_i, z_i)v_i, \ i = 1, ..., N \tag{22}$$

$$\frac{dz_i}{dt} = f_{2,i}(y_i, z_i) \tag{23}$$

The linearizing input-output controller decomposes the system into two parts: (i) a linear subsystem of order r which is influenced by the chosen input u_i (22), and (ii) a nonlinear subsystem described by the zero dynamics. Thus the main applicability condition of input–output linearization is to have a stable zero dynamics in a wide domain of the state-space, or even better, a globally stable zero dynamics (Hangos et al., 2004).
If exact cancelation of nonlinear terms is achieved then the closed-loop system is given by,

$$\begin{aligned} \frac{dy_i}{dt} &= v_i \\ \frac{dz_i}{dt} &= f_{2,i}(y_i, z_i) \end{aligned} \tag{24}$$

It is relatively simple to device a feedback control law for v_i, which stabilizes the output of the system, y_i, to the desired reference, y_{ref}. A valid choice of the new control input is a simple linear input, $v_i = -\tau_c^{-1}(y_i - y_{ref}) + \overset{\bullet}{y}_{ref}$, that guarantees the stability of the overall system provided that the zero dynamics is stable, i.e.,

$$\frac{de_i}{dt} = -e_i \tag{25}$$

where τ_c is controller design parameter, $e_i = y_i - y_{ref}$, is the tracking error.
The linearizing input-output controller needs accurate knowledge of the nonlinear dynamics of the system, hence, turns to be inapplicable if the model for the process includes uncertainties. This fact is behind the motivation to provide robustness properties of the above linearizing input-output controller. In order to provide robustness against inexact model cancelations of nonlinear terms, unmodeled dynamics, and external perturbation we proceed as in the approach of modeling error compensation approach (Alvarez-Ramirez, 1999).
Consider system (22) subject to model uncertainties η_i,

$$\frac{dy_i}{dt} = \tilde{f}_{1,i}(y_i, z_i) + \tilde{g}_i(y_i, z_i)\alpha_i(y_i, z_i) + \beta_i(y_i, z_i)v_i + \eta_i,\ i = 1, ..., N \tag{26}$$

where η_i is defined as,

$$\eta_i = (f_{1,i}(y_i, z_i) - \tilde{f}_{1,i}(y_i, z_i)) + (g_i(y_i, z_i) - \tilde{g}_i(y_i, z_i))\alpha_i(y_i, z_i) \tag{27}$$

where $\tilde{f}_{1,i}$ and $\tilde{g}_i$ are rough estimates of terms $f_{1,i}$ and g_i all the uncertain terms associated to the biochemical system model are lumped. The uncertain function η_i can be estimated using a state observer (Alvarez-Ramirez, 1999). We introduce a reduced order observer given by (16) to this end. After some direct algebraic manipulations we get the robust linearizing input-output controller as,

$$\begin{aligned}
\frac{dw_i}{dt} &= -\tilde{f}_{1,i}(y_i, z_i) - \tilde{g}_i(y_i, z_i)\alpha_i(y_i, z_i) - \beta_i(y_i, z_i)v_i - \overline{\eta}_i,\ i = 1, ..., N \\
\overline{\eta}_i &= \tau_e^{-1}(w_i + y_i) \\
v_i &= -\beta_i(y_i, z_i)^{-1}[\overline{\eta}_i - \tau_c^{-1}e_i] \\
u_i &= -\alpha_i(y_i, z_i) + \beta_i(y_i, z_i)v_i
\end{aligned} \tag{28}$$

Comparing the above robust linearizing input-output controllers with the controller derived via a MEC approach we can exploit the tunning guidelines of the MEC approach to provide some guidelines for the tunning of controller parameters τ_c and τ_e (Alvarez-Ramirez, 1999).

5. Applications

In this section we consider three examples of the implementation of the proposed synchronization approach with the robust feedback control approaches presented in the above section. The examples are: (i) the Goodwin model, (ii) a Fitz-Hugh-Nagumo neuron model, and (iii) circadian rhythms in Drosphila.

5.1 Goodwin model for genetic oscillators

Synchronization of coupled genetic oscillators has important biological implications and potential engineering applications from both theoretical and experimental viewpoints, and it is also essential for the understanding of the rhythmic phenomena of living organisms at both molecular and cellular levels. The Goodwin model (Goodwin, 1965) is a benchmark model of genetic oscillations that contains three simple biochemical components (nuclear messenger, cytoplasmic messenger, and repressor). In the original model, a clock gene mRNA produces a clock protein, which activates a transcriptional inhibitor, which inhibits the transcription of the clock gene, thus forming a negative feedback loop.

Using the notation previously introduced, we consider the following external forcing modification of the Goodwin model that consists of the following set of three ordinary differential equations (Goodwin, 1965; Keener & Sneyd, 1998),

$$\frac{dy}{dt} = \frac{c_1}{1+z_2^p} - c_2 y + u \tag{29}$$

$$\frac{dz_1}{dt} = c_3 y - c_4 z_1$$

$$\frac{dz_2}{dt} = c_5 z_1 - c_6 z_2$$

where y, z_1 and z_2 represent respectively the concentrations of the mRNA, the enzyme and the product of the reaction of the enzyme and a substrate, assumed to be available at a constant level. All c_i are constant positive parameters. The creation of y is inhibited by the product z_2 and is degraded according to first-order kinetics, while z_1 and z_2 are created and degraded by first-order kinetics. We also assumed that u is a plausible manipulated variable.

The synchronization objective is to synchronize an ensemble of two independent genetic oscillators, to the dynamics generated by a reference Goodwin genetic oscillator, via an external forcing u to the mRNA concentration y. Figure 2 shows the synchronization performance for the three proposed robust control approaches: MEC control, IHOSMC, and the GLC, in the upper, middle and bottom parts of Figure 2 respectively. It can be seen from Figure 2 that the synchronization objective is achieved for all robust control approaches. MEC approach uses less control effort than IHOSMC and GLC. The control input for the IHOSMC displays a switching type behavior typical of SMC approaches. The modulation of external inputs depends on the measured state such that a feedback mechanism is established and modifies the natural dynamic behavior of the controlled biochemical oscillators.

5.2 FitzHugh-Nagumo model of neurons

The central nervous system can display a wide spectrum of spatially synchronized, rhythmic oscillatory patterns of activity with frequencies in the range from 0.5Hz (rhythm), 20Hz (rhythm), to 30-80 Hz (rhythm) and even higher up to 200Hz (Izhikevich, 2007). In the past decade it has been shown that synchronized activity and temporal correlation are fundamental tools for encoding and exchanging information for neuronal information processing in the brain (Izhikevich, 2007). In particular, it has been suggested that clusters of cells organize spontaneously into flexible groups of neurons with similar firing rates, but with a different temporal correlation structure.

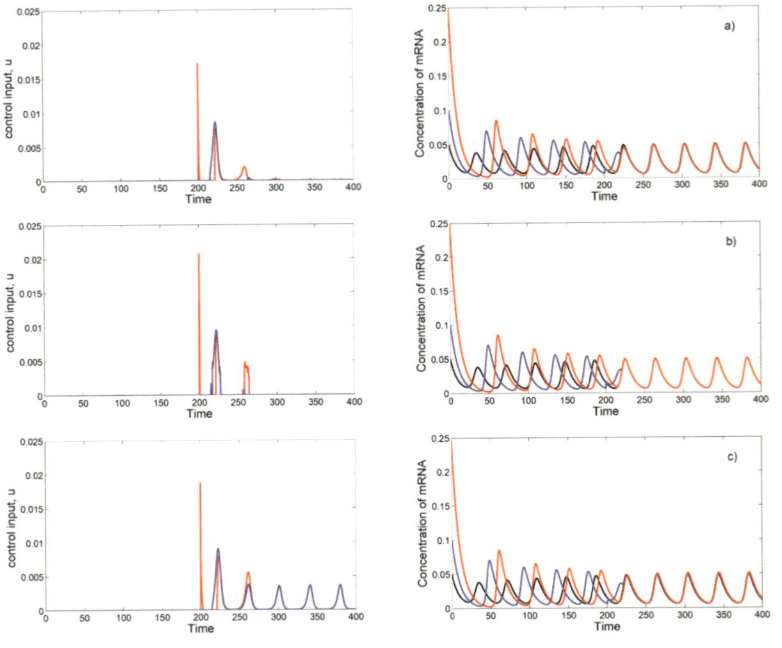

Fig. 2. Synchronization of Goodwin model for genetic oscillators via (a) MEC, (b) IHOSMC and (c) GLC

A benchmark model of neural activity was proposed by FitzHugh and Nagumo (FHN) as a mathematical representation of the firing behavior of neuron (FitzHugh, 1961). The neural FHN model is an excitable media (Keener & Sneyd, 1998). Excitable media are systems that sit at a steady state and are stable to small disturbances. If, however, they receive a disturbance (such as a sudden increase in the concentration of the feedback species) above some critical or threshold value, then they respond with an excitation event (which corresponds to the reaction front). The FHN model and its modifications served well as simple but reasonable models of excitation propagation in nerve, heart muscle and other biological excitable media (Izhikevich, 2007).

The FHN neuron model with external current u studied in this paper is described by the following set of two ordinary differential equations,

$$\frac{dy}{dt} = -y(y-c_1)(y-1) - z + I_0 + I\cos(c_4 t) + u \qquad (30)$$

$$\frac{dz}{dt} = \beta(c_5 y - z)$$

where y is the potential difference across the membrane, z is a recovery variable which measures the state of excitability of the cell. Parameters c_i are positive constants, I_0 stands for the ionic current inside the cell, I is the amplitude of the external current.

We apply a control approach by injecting the external signal at each individual oscillator in order to track the desired synchronized signal. In this case, the desired synchronized signal is a periodic signal. Figures 3 and 4 shows the synchronization performance for the MEC approach

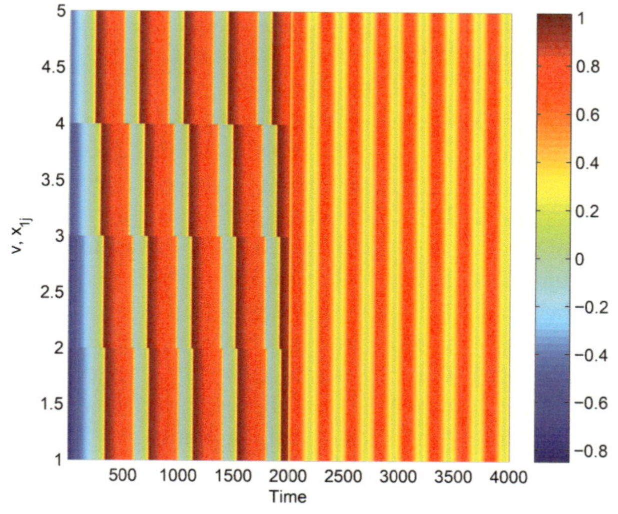

Fig. 3. Synchronization of 5 individual oscillators for FHN model of neurons.

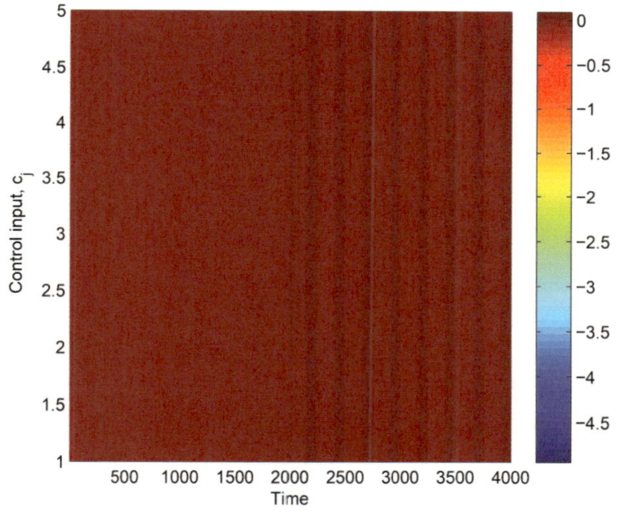

Fig. 4. Corresponding control input for Figure 3.

for an ensemble of 5 individual oscillators. It can be seen that, after a short transient, the array of FHN neurons synchronizes about the desired periodical dynamical behavior. Figure 4 shows that by using periodic applied current we can force the periodicity of the synchronized neurons. The applied input depends on the current state of the neuron which receives the external impulse.

5.3 Circadian rhythms in Drosphila

The biological functions of most living organisms are organized along an approximate 24-h time cycle or circadian rhythm (Goldbeter, 1996). Circadian rhythms, are endogenous because they can occur in constant environmental conditions, e.g. constant darkness. Circadian rhythm can also be entrained by external forcing of modified light-darkness cycles or phase-shifted when exposed to light pulses (Goldbeter, 1996; Fu & Lee, 2003; Jewett et al., 1991).

Circadian rhythms are centrally regulated by the suprachiasmatic nucleus (SCN) of the hypothalamus. Most neurons in the SCN become active during the day and are believed to comprise the biological clock. Dispersed SCN cells exhibit sustained circadian oscillations with periods ranging from 20 to 28 hours, but on the tissue level, SCN neurons display a significant degree of synchrony. Over time, the development of a circadian rhythm might impart larger benefits to the organism. In cyanobacteria, for example, matching of the free-running period to the light-dark cycle time provides a selective advantage, which is presumably the basis for its evolution (Ouyang et al., 1998). In Arabidopsis, matching between the circadian period and the light-dark cycle results in plants that fix carbon at a higher rate and grow and survive better than those that lack such a match (Dodd et al., 2005).

Concerning the modeling of this phenomenon, it has to be stressed that the mechanism can be considerably different for the different living beings in which it has been studied, ranging from unicellular organisms to mammalians, going through fungi and flies. Some of the most recent models have a high degree of complexity and involve up to 16 differential equations. However, it seems to be accepted that the central mechanism causing oscillations is represented by a negative feedback exerted by a protein on the expression of its corresponding gene.

We consider as the single biochemical oscillator a simple five-variable model proposed for circadian rhythms for the central clock of fruit fly Drosophila (Gonze & Goldbeter, 2000),

$$
\begin{aligned}
\frac{dy}{dt} &= u\frac{K_I^n}{K_I^n + z_4^n} - v_m\frac{y}{K_m + y} \\
\frac{dz_1}{dt} &= k_s y - V_1\frac{z_1}{K_1 + z_1} + V_2\frac{z_2}{K_2 + z_2} \\
\frac{dz_2}{dt} &= V_1\frac{z_1}{K_1 + z_1} - V_2\frac{z_2}{K_2 + z_2} - V_3\frac{z_2}{K_3 + z_2} + V_4\frac{z_3}{K_4 + z_4} \\
\frac{dz_3}{dt} &= V_3\frac{z_2}{K_3 + z_2} - V_4\frac{z_3}{K_4 + z_4} - k_1 z_3 + k_2 z_4 - v_d\frac{z_3}{K_d + z_3} \\
\frac{dz_4}{dt} &= k_1 z_3 - k_2 z_4
\end{aligned}
$$

where y, z_1, z_2, z_3 and z_4 denote, respectively, the concentrations of mRNA, PER protein, mono- and di-phosphorylated forms of PER protein, and the amount of phosphorylated protein located in the cells. Once in the nucleus, PER protein down-regulates mRNA translation, leading to the observed oscillating behavior. The manipulated variable u denotes the maximal speed of transcription of y. It seems that progresses in gene manipulation techniques make it reasonable to think of modifying of this parameter. Definition of other parameters can be found in Goldbeter, (1996). Kinetic parameters can differ from one oscillator to the other and thus holds variability in individual circadian oscillators.

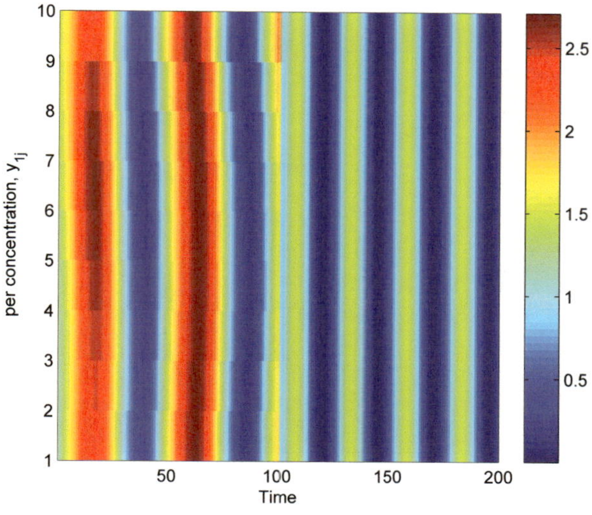

Fig. 5. Synchronization of the circadian rhythms in Drosphila using a periodic modulation of the external input.

The synchronization objective is fix a nominal 24-h period of the circadian oscillations for an ensemble of individual circadian oscillators. In this case we have implemented the GLC. Figures 5 and 6 shows that by using a periodic modulation of the external input, we can force the circadian periodicity. As was stated above, synchronization of circadian rhythms has been achieved via the periodic modulation of a light sensitive parameter. In this case, the parameter modulation requires the periodic manipulation of the maxima speed of transcription of mRNA, which should be addressed using gene manipulation techniques, and is beyond of the scope of this contribution.

6. Conclusions and perspectives

In this chapter we have discussed the synchronization problem of biochemical oscillators and we have addressed this problem via three robust feedback control approaches. In this section we provide some concluding remarks and a perspective on the synchronization of biochemical systems.

6.1 Concluding remarks

One interesting phenomenon in biological systems is the collective rhythm of all dynamic cells. Synchronization occurs in many populations of biological oscillators. From the general synchronization point of view, synchronization approaches can be classified into two general groups: (i) natural coupling (self-synchronization), and (ii) artificial coupling forced via periodic modulation or explicit feedback control approaches. Classical methods are determined by an interplay of time scales by phase locking or, respectively, natural frequency entrainment or due to suppression of inherent frequencies. Artificial coupling where an external input can be manipulated can be looked as control synthesis issue and studied within the control theory framework developed in this work.

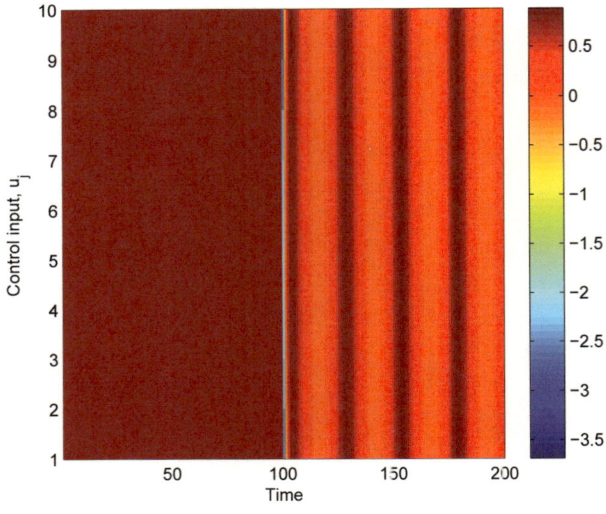

Fig. 6. Corresponding control input for Figure 5.

In this chapter we have shown that external stimulation with robust feedback control can effectively synchronize populations of individual oscillators. We have introduced three robust feedback control approaches: (i) the MEC approach, that leads to simple practical control design with good robustness and performance capabilities, (ii) sliding mode control approach that leads to a simple design with the feature of switching type action that can be appropriate for biochemical systems, and (iii) a robust geometric linearizing input-output control, that can be useful to establish a relation between neural processing behavior in cells and the mathematical formalism of geometric differential methods. Numerical simulations results indicate good tracking performance of the proposed robust control approaches. The three robust control schemes are based on a minimum information from the cell model (output), not on the precise details of the model (*e.g.*, kinetic parameters). Thus, our control scheme is likely to be effective in the more complicated models of cell dynamics.
From a general point of view external forcing of cellular processes is important in many application areas ranging from bioengineering to biomedicine. At the level of biology the problem is to supply an input to the cell such that the biochemical processes of the cell achieve specified control objectives. At the level of control theory the biological problem amounts to the construction of a control law such the control objectives are achieved. In this way, the results in this work must be seen as a first approach to addressing the systematic design of control systems in cellular processes.

6.2 Perspectives

Feedback control and synchronization for cells is in its infancy, with numerous challenges and opportunities ahead. For instance, an implicit assumption of the control frameworks discussed in this article is that the control law is implemented without regard the actuator and sensor constraints for cells. Besides, we have considered cellular systems described by ordinary differential systems, without delays. Delays are however known to be involved in biological systems, because for example mRNA synthesis and transport (in eukaryotic cells) are certainly not instantaneous. Systems with delays are however most difficult to analyze and control, because they are differential systems of infinite dimensions, to which mathematical tools are more involved.

Feedback control theory in combination with biological knowledge can lead to a better understand of the complex dynamics of cellular processes. Indeed, the design of closed-loop system in biological systems is a first step to gain insights of the suppression and generation of oscillatory behavior, and the closed-loop response can resembles the features of the behavior of biological processes. Current work is in progress in order to study various synchronization mechanisms by investigating the effects of various biologically plausible couplings and several kinds of noise from the viewpoint of feedback control theory.

7. References

Afraimovich, V. S.; Chow, S. N. & Hale, J. K. (1997). Synchronization in lattices of coupled oscillators. *Physica D*, 103, 442.

Aguilar-Lopez, R.; Martinez-Guerra, R.; Puebla, H. & Hernadez-Suarez, R. (2010). High order sliding-mode dynamic control for chaotic intracellular calcium oscillations. *Nonlinear Analysis -B: Real World Applications*, 11, 217-231.

Alvarez-Ramirez, J. (1999). Adaptive control of feedback linearizable systems: a modelling error compensation approach. *Int. J. Robust Nonlinear Control*, 9, 361.

Alvarez-Ramirez, J.; Puebla, H. & Espinosa, G. (2001). Linear control in a lattice of coupled second order oscillators. *Int. J. Bifurcation Chaos*, 11, Vol. 1, 185-195.

Alvarez-Ramirez, J.; Puebla, H. & Cervantes, I. (2002). Stability of observed-based chaotic communications for a class of Lur'e systems. *Int. J. Bifurcation Chaos*, 12, Vol. 7, 1605-1618.

Arkin, A.P. & Ross, J. (1994). Computational functions in biochemical reaction networks. *Biophys. J.*, 67, 560.

Asthagiri, R. & Lauffenburger, D.A. (2001). A computational study of feedback effects on signal dynamics in a mitogen-activated protein kinase (MAPK) pathway model. *Biotechnol. Prog.*, 17, 227.

Berridge, M.J.; Bootman M.D. & Lipp D. (1998). Calcium a life and death signal. *Nature*, 395, 645.

Blumenfeld, L.A. & Tikhonov, A.N. (1994). Biophysical Thermodynamics of Intracellular Processes, *Springer Verlag*, New York.

Boccaletti, S.; Kurths, J.; Osipov, G.; Valladares, D.L. & Zhou, C.S. (2002). The synchronization of chaotic systems, *Phys. Rep.*, 36, 1.

Canavier, C.C.; Baxter, D.A.; Clark, J.W. & Byrne, J.H. (1999). Control of multistability in ring circuits of oscillators. *Biol.Cybern.*, 80, 87.

Collins, J.J. & Stewart, I. N. (1994). A group-theoretic approach to rings of coupled biological oscillators. *Biol. Cybern.*, 71, 95.

di Bernardo, D.; Signorini, M. R. & Cerutti, S. (1998). A model of two nonlinear coupled oscillators for the study of heartbeat dynamics. *Int. J. Bifurcation and Chaos*, 8.

De Jong, H. (2002). Modeling and simulation of genetic regulatory systems: a literature review. *J. Comp. Biol.*, 9, 67.

Dodd, A.N.; Salathia, N., Hall, A.; Kevei, E.; Toth, R.; Nagy, F.; Hibberd, J.M., Millar, A.J. & Webb, A.A.R. (2005). Plant circadian clocks increase photosynthesis, growth, survival, and competitive advantage. *Science*, 309, 630-633.

Ferrell, J.E. (2002). Self-perpetuating states in signal transduction: positive feedback, double-negative feedback and bistability. *Curr. Opin. Cell. Biol.*, 14, 140.

Fall, C.P.; Marland, E.S.; Wagner, J.M. & Tyson, J.J. (2002). Computational Cell Biology. *Springer-Verlag*, New York.

FitzHugh, R. (1961). Impulses and physiological states in theoretical models of nerve membrane. *Biophys. J.*, 1, 445-466.

Fradkov, A.L. & Pogromsky A.Y. (1998). Introduction to Control of Oscillations and Chaos. *World Scientific Publishing*, USA.

Fu, L.N. & Lee, C.C. (2003). The circadian clock: pacemaker and tumor suppressor. *Nature Rev.*, 3, 350-361.

Goodwin BC. (1965) Oscillatory behaviour in enzymatic control processes. *Adv Enzyme Regul.*, 3, 425-438.

Glass, L. & Mackey, M. C. (1988). From clocks to chaos. *Princeton University Press*, Princeton, New Jersey.

Glass, L. (2001). Synchronization and rhythmic processes in physiology. *Nature*, 410, No. 8, 277-284.

Goldbeter, A. (2002). Computational approaches to cellular rhythms. *Nature*, 420, 238.

Goldbeter, A. (1996). Biochemical Oscillations and Cellular Rhythms. *Cambridge University Press*, Cambridge, UK.

Goldbeter, A. (1991). A minimal cascade model for the miotic oscillator involving cyclin and cdc2 kinase. *Proc. Natl. Acad. Sci.* USA 88, 9107.

Greenman, J.; Kamo & Boots, M.. (2004). External forcing of ecological and epidemiological systems: a resonanace approach. *Phys. D*, 190, 136.

Hangos, K.M.; Bokor, J. & Szederkényi, G. (2004). Analysis and Control of Nonlinear Process Systems. *Springer-Verlag*, London.

Hofer, T. (2003). Model of intracellular calcium oscillations in hepatocytes: synchronization of heterogeneous cells. *Biophysical J.*, 77, 1244-1256.

Iglesias, P. (2003). Feedback control in intracellular signaling pathways: regulating chemostaxis in Disctyostelium discoideum. *European J. Control*, 9, 227.

Izhikevich E.M. (2007). Dynamical Systems in Neuroscience: The Geometry of Excitability and Bursting. *The MIT press*, USA.

Jewett, M.E.; Kronauer, R.E. & Czeisler, C.A. (1991). Light-induced suppression of endogenous circadian amplitude in humans. *Nature*, 350, 59-62.

Kunz, H. & Achermann, P. (2003). Simulation of circadian rhythm generation in the suprachiasmatic nucleus with locally coupled self sustained oscillators. *J. Theor. Biol.*, 224, 63-78.

Kuramoto, Y. (1984). Chemical oscillations, waves and turbulence *Springer Verlag*, New York.

Keener, J. & Sneyd. J. (1998). Mathematical Physiology. *Springer Verlag*, Berlin.

Kimura, H. & Nishigaki, Y. (2005). Circadian rhythm as a phase locked loop. In *Proc. IFAC World Congress.*

Levant, A. (2001). Universal single-input-single-output sliding mode controllers with finite-time convergence. *IEEE Autom. Control*, 46, 1447-1451.

Marhl, M. & Schuster, S. (2003). Under what conditions signal transduction pathways are highly flexible in response to external forcing? a case study on calcium oscillations. *J. Theor. Biol.*, 224, 491.

Mirollo, R. E. & Strogatz, S. H. (1990). Synchronization of pulse-coupled biological oscillators. *SIAM J. Appl. Math.*, 50, 1645.

Mikhailov, A.S. & Hess B. (1995). Fluctuations in living cells and intracellular traffic. *J. Theor. Biol.*, 176, 185-192.

Morgul, O. & Solak, E. (1996). Observed based synchronization of chaotic systems. *Phys. Rev. E*, 54, 4803-4811.

Nijmeijer, H. & Mareels, M. Y. (1997). An observer looks at synchronization, *IEEE Trans. Circuits Syst.-I*, 44, 882.

Ouyang, Y.; Andersson, C.R., Kondo, T., Golden, S.S. & Johnson, C.H. (1998). Resonating circadian clocks enhance fitness in cyanobacteria. *Proc. Natl. Acad. Sci.*, 95, 8660-8664.

Petty, H.R. (2004). Dynamic chemical instabilities in living cells may provide a novel route in drug development, *Chem. Bio. Chem.*, 5, 1359–1364.

Perc, M. & Marhl M. (2004). Synchronization of regular and chaotic oscillations: the role of local divergence and the slow passage effect. *Int. J. Bif. Chaos*, 14, 2735-2751.

Pikovsky, A.S; Kurths, J. & Rosenblum, M.G. (1996). Synchronization in a population of globally coupled chaotic oscillators, *Europhys. Lett.*, 34, 165.

Puebla, H. (2005). Controlling intracellular calcium oscillations and waves. *J. Biol. Sys.*, 13, 173.

Puebla, H.; Aguilar-Lopez, R.; Ramirez-Castelan, E.; Hernandez-Martinez, E. & Alvarez-Ramirez, J. (2010). Control and synchronization of Hodgkin-Huxley neurons. In *BIOMAT 2009 International Symposium on Mathematical and Computational Biology*, World Scientific Publishing Co. 125-135.

Puebla, H.; Alvarez-Ramirez, J. & Cervantes, I. (2003). A simple tracking control for Chua's circuit. *IEEE Trans. Circ. Sys.*, 50, 280.

Puebla, H.; Martin, R.; Alvarez-Ramirez, J. & Aguilar-Lopez, R. (2009). Controlling nonlinear waves in excitable media. *Chaos. Solitons Fractals*, 39, 971-980.

Sira-Ramirez, H. (2002). Dynamic second order sliding-mode control of the Hovercraft vessel. *IEEE Trans. Control Syst. Tech.*,10, 860-865.

Segel, L.A. (1980). Mathematical Models in Molecular and cellular Biology. *Cambridge University Press*, Cambridge, U.K.

Smolen, P.D.; Baxter, D.A. & Byrne, J.H. (2003). Mathematical modeling and analysis of intracellular signaling pathways, *From Molecules to Networks: An Introduction to Cellular and Molecular Neuroscience*, USA.

Sontag, E.D. (2004). Some new directions in control theory inspired by systems biology. *Syst. Biol.*, 1, 9.

Stelling, J.; Gilles, E.D. & Doyle, F.J. III. (2004). Robustness properties of circadian clock architectures. *Proc. Natl. Acad. Sci.*, 101, 13210-13215.

Sun, J.; Olbrot, A.W. & Polis. M.P. (1994). Robust stabilization and robust performance using model reference control and modeling error compensation. *IEEE Transactions Automatic Control*, 39, 630-635.

Takeuchi, T.; Hinohara, T.; Uchida, K. & Shibata, S. (2006). Control theoretic views on circadian rhythms. In *Proc. IEEE CCA*, 2006, 1740-1745.

Tang, Y. & Othmer, H.G. (1995). Frequency encoding in excitable systems with applications to calcium oscillations. *Proc. Natl. Acad. Sci. USA* 92, 7869-7873.

Tass, P.A. (2002). Effective desynchronization with a stimulation technique based on soft phase resetting, *Europhys. Lett.*, 57, 164-170.

Tyson, J.J.; Chen, K.C. & Novak, B. (2003). Sniffers, buzzers, toggles and blinkers: dynamics of regulatory and signaling pathways in the cell. *Curr. Op. Cell Biol.*, 15, 221.

Ueda, H.R.; Hirose, K. & Iino, M. (2002). Intercellular coupling mechanism for synchronized and noise-resistant circadian oscillators. *J. Theor. Biol.* 216, 501-512.

Walleczek, J. (2003). The frontiers and challenges of biodynamics research, *Self-Organized biological dynamics and nonlinear control*, Cambridge, U.K.

Winfree, A.T. (2001). The Geometry of Biological Time. *Springer Velag*, 2nd edn., New York.

Wolkenhauer, O.; Kitano, H. & Cho, K.H. (2003). Systems biology. *IEEE Control Sys. Mag.*, 38.

Yi, T.M.; Huang, Y.; Simon, M. I. & Doyle, J. (2000). Robust perfect adaptation in bacterial chemotaxis through integral feedback control. *Proc. Natl. Acad. Sci.*, 97, 4649.

Zhou, T.; Zhang, J.; Yuan, Z. & Chen, L. (2008). Synchronization of genetic oscillators, *Chaos*, 18, 037126-890.